Lecture Notes in Computer Science   10100

Commenced Publication in 1973
Founding and Former Series Editors:
Gerhard Goos, Juris Hartmanis, and Jan van Leeuwen

More information about this series at http://www.springer.com/series/7409

Peter Brusilovsky · Daqing He (Eds.)

Social Information Access

Systems and Technologies

 Springer

Editors
Peter Brusilovsky
University of Pittsburgh
Pittsburgh, PA
USA

Daqing He
University of Pittsburgh
Pittsburgh, PA
USA

ISSN 0302-9743   ISSN 1611-3349 (electronic)
Lecture Notes in Computer Science
ISBN 978-3-319-90091-9   ISBN 978-3-319-90092-6 (eBook)
https://doi.org/10.1007/978-3-319-90092-6

Library of Congress Control Number: 2018940144

LNCS Sublibrary: SL3 – Information Systems and Applications, incl. Internet/Web, and HCI

Printed on acid-free paper

This Springer imprint is published by the registered company Springer International Publishing AG
part of Springer Nature
The registered company address is: Gewerbestrasse 11, 6330 Cham, Switzerland

Preface

Social information access (SIA) is a stream of research that explores methods for organizing the past interactions of users in a community in order to provide future users with better access to information. SIA covers a wide range of different systems and technologies that operate on a different scale, which can range from a small closed corpus site to the whole Web. Although the technologies located on the different parts of this stream may not even recognize each other as being a part of the same whole, the whole stream is driven by the same goal: to use the power of a user community to improve information access.

As a type of information access that can offer multiple benefits while being relatively easy to organize and maintain, SIA has been attracting more and more attention from researchers and practitioners. The overarching goal of this book is to provide a comprehensive hands-on overview of modern social information access technologies and systems. The book is designed with two audiences in mind. On the one hand, it can help students and young researchers who are interested in learning about this new field; and on the other hand, it is able to assist more experienced researchers and practitioners in the development of new social information access technologies and applications. To meet this goal, each chapter has a dual nature. To support novices, it provides a review of a specific group of techniques and technologies, and to support practitioners, it explains critical algorithms and systems or reviews a set of case studies.

Starting with an overview chapter, the book offers an extensive coverage of social information access techniques for the three main types of information access: search, browsing, and recommendation. Four of the book's chapters focus on social search, two focus on social navigation and browsing, and seven focus on recommendation. Within each group, the chapters are organized by sources of social information that are used to enhance the information access. The book also presents chapters on privacy issues in social information access and social Q&A.

April 2018

Peter Brusilovsky
Daqing He

Contents

1

Introduction to Social Information Access

Peter Brusilovsky⬤ and Daqing He$^{(\boxtimes)}$⬤

School of Computing and Information, University of Pittsburgh,
Pittsburgh, PA 15213, USA
{peterb,dah44}@pitt.edu

Abstract. This chapter offers an introduction to the emerging field of
social information access. Social information access focuses on technologies that organize users past interaction with information in order to
provide future users with better access to information. These technologies have become increasingly more popular in all areas of information
access, including search, browsing, and recommendation. Starting with a
definition of the new field and a brief history of social information access,
this chapter introduces a multi-aspect classification of social information
access technologies. The two important factors for our classification are
the types of information access involved and the source of the social information that has been leveraged to support information access. These two
factors are the angles we use in this chapter to create a map of the field, as
well as to introduce the book structure and the role of the remaining book
chapters in covering social information access topics and technologies.

1 Social Information Access

The social Web (or Web 2.0), through various platforms, such as Wikis, blogs,
Twitter, or Facebook, have changed the role of Web users from simply information consumers to their emergence as key information producers for content
sharing and community building. Recognizing users as information producers
has also attracted attention to novel information access technologies supported
by "collective wisdom," distilled from actions of those who worked with this
information earlier. The ideas of so-called "social information access" has been
explored by a number of research groups worldwide as part of the effort for
developing techniques to help users obtain the right information for satisfying
their information needs [9,15,33,54,56,59].

In this book, we define *social information access* as a stream of research that
explores methods for organizing the past interactions of users in a community
in order to provide future users with better access to information. Social information access covers a wide range of different systems and technologies that
operate on a different scale, which can range from a small closed corpus site to
the whole Web. Although the technologies located on the different sides of this
stream may not even recognize each other as being a part of the same whole, the

© Springer International Publishing AG, part of Springer Nature 2018
P. Brusilovsky and D. He (Eds.): Social Information Access, LNCS 10100, pp. 1–18, 2018.
https://doi.org/10.1007/978-3-319-90092-6_1

whole stream is driven by the same goal: to use the power of a user community to improve information access.

Self-organization is a vital feature of all social information access systems. These systems are able to work properly despite little or no involvement of human indexers, organizers, or other kinds of experts. They are truly powered by a community of users. Due to this feature, social information access technologies are frequently considered as an alternative to the traditional (content-oriented) information access technologies. In most cases, social information access can run in parallel with traditional types of information access, and may help users to find resources that would be hard to find in a traditional way. In other cases where traditional information access is hard to organize (for example, in a collection of non-indexed images), social mechanisms (such as tagging) can serve as a handy replacement. However, it has been more and more frequently demonstrated that most of the benefits could be obtained by integrating social and traditional technologies. For example, hybrid recommender systems can integrate collaborative and content-based recommender mechanisms.

As a type of information access that can offer multiple benefits while being relatively easy to organize and maintain, social information access has been attracting more and more attention from researchers and practitioners. The overarching goal of this book is to provide a comprehensive hands-on overview of modern social information access technologies. The book is designed with two audiences in mind. On the one hand, it could help students and young researchers who are interested in learning about this new field; and on the other hand, it can assist more experienced researchers and practitioners in the development of new social information access technologies and applications. To support this goal, each chapter carries a dual nature. To support novices, it provides a review of a specific group of techniques, and to support practitioners, it explains critical algorithms or reviews a set of case studies.

The role of this chapter is to introduce the field of social information access to the readers and to explain the layout of the book itself. We start our introduction with a brief history of social information access and follow with a multi-aspect classification to define the space of social information access. The classification provides the readers with the necessary knowledge for comprehending and distinguishing different kinds of social information access techniques. Based on the same classification, we also introduce the content and the structure of the book to provide the readers with a "big picture" of how different chapters cover various dimensions of the classification.

2   The Emergence of Social Information Access

The ideas that underpin social information access can be traced back to several visionary projects. Vannevar Bush's seminal paper on *Memex* introduced the idea of "trails" through information space, which Memex users could create and share with others [13]. The *Superbook* project [53] demonstrated the benefits of directly engaging end-users into document indexing. The *Edit Wear and Read Wear* project [32] introduced the concept of a history-rich information space and demonstrated that social information access could be based on implicit, rather than explicit, actions of past users.

Besides these influential pioneering projects, it was the opportunities and the needs produced by the rapidly expanding World Wide Web (WWW) that led to the emergence of social information access as a research area. On the one hand, the World Wide Web, with its increased volume of users, has enabled the collection of social information at scale. On the other hand, it has opened access to information to many inexperienced users who could benefit from additional support. The "Web push" led to a rapid expansion of social information access research between 1994 and 2000. This period brought many innovations as research teams investigated new approaches to help users in the rapidly expanding information space. In the context of this book, we will examine four main streams of research that established the field of social information access at the turn of the centuries: collaborative filtering, social navigation, social search, and social bookmarking.

Collaborative filtering [24,39,54] attempted to propagate information items between users with similar interests. The emergence of collaborative filtering is typically traced back to the *Information Tapestry* project [24], which coined the term collaborative filtering. Information Tapestry introduced an approach that was later called pull-active collaborative filtering, where users had to actively query the community feedback left by earlier users in order to receive social guidance. Later, several projects expanded the scope of collaborative filtering. For example, *Lotus Notes recommender* [45] proposed push-active collaborative filtering, where users are encouraged to send interesting documents directly to their colleagues. However, the majority of pioneering work in this area has focused on indirect approaches to collaborative filtering based on *automatic* matching users with similar interests and cross-recommending positively-rated items. This stream of work includes such pioneer systems as *GroupLens* [54], *Ringo* [56], and *Video Recommender* [31].

In its early form, *social navigation* [15,18,20] attempted to visualize the aggregate or individual actions of community users. It was motivated by observing users' navigation in real space where they frequently follow the footprints of others. Proponents of social navigation in information space argued that "digital footprints" could also help future users to navigate through information space [16,62]. Inspired by the "footprint" examples provided in *Read Wear and Edit Wear* system [32] and the concept of social navigation in information space introduced by Dourish and Chalmers [18], early pioneers of social navigation developed and evaluated several well-cited systems, such as *Juggler* [16], *Footprints* [62], and *EFOL* [60].

The work on *social search* also expanded rapidly between 1994 and 2000. At that point, the accumulation of social data embedded in search engine logs and Web structure led to an explosion of creative approaches on how to use this data to help Web searchers [4,8,14,21,37,47,52]. The most influential one was *PageRank* [8], a novel ranking approach based on social data encapsulated in the global structure of Web links. Promoted by the success of the Google search engine, this technology inspired a large number of works on social search [12].

Social bookmarking was another important early avenue of research that happened between 1995 and 2000. It was motivated by the need to support the organization of personal information space as well as to share valuable online resources with others (i.e., the same motivation that encouraged early work on active collaborative filtering [24,45]). Pioneer systems that focused on social bookmarking, such as *Siteseer* [55], *WebTagger* [38], *WDB* [61], and *PowerBookmarks* [41] explored different ways of organizing and sharing bookmarks. Among these, collaborative tagging, which was originally explored in WebTagger [38], emerged as the most efficient way to help new users locate useful information that has already been discovered and classified by others. In fewer than 10 years, social bookmarking and tagging systems, popularized by systems like del.icio.us and flickr.com, grew into a new major Internet technology [25,28].

There were several attempts to bring together researchers working on different types of social information access during the first decade of research in the field. Several workshops that gathered like-minded researchers, as well as the publications that resulted from these workshops [15,33,44,48] have clearly contributed to the expansion of social information access ideas and the conceptualization of social feedback as a source of knowledge in assisting users. However, these integration attempts were based on a limited volume of work and failed to include work on social search and social bookmarking. Our book represents another attempt to bring together a diverse set of research on social information access. While our main goal is to provide an overview of the current state of each major stream of research on social information access, we base the structure of the book on an integrative multi-dimensional classification of social information access techniques. This classification, which is introduced in Sect. 3, highlights both the similarities between different groups of social information access techniques and the opportunities to support users across multiple kinds of information access.

3 Classifying Social Information Access Technologies

The term "Social information access" contains two parts: "information access" and "social"; as a result, the most natural way to classify social information access technologies is by answering two questions: "What kind of information access is considered?" and "How this information access is made social?" In this section, we expand this idea into a multi-dimensional classification framework. Our goal of designing this framework is to make it compatible with older classification attempts, and at the same time, to make it rich and expressive enough to classify a large variety of modern social information access techniques. In the next section, the suggested framework is immediately applied to introduce and classify the social information techniques that are presented in the remaining chapters of this book.

3.1 Types of Information Access

Following earlier classification attempts introduced in [9,10], the first dimension of our framework for classifying social information access techniques is the type of information access. "Access" is a reasonably studied concept that refers to an interactive process, which starts with a user noticing their needs and ends with the user obtaining the necessary information. It is an iterative process with multiple stages and possible back loops. However, there are four different ways to obtain the information, which results in the four core types of information access [9,10]: *ad-hoc information retrieval, information filtering (recommendation), hypertext browsing (navigation)*, and *information visualization.*

In ad-hoc information retrieval (IR), users achieve access to relevant information by issuing a query to an IR system and then analyzing/accessing a ranked list of returned information items (for example, book records). An information filtering (IF) or recommender system also returns a ranked list of information items in response, not to an ad-hoc query, but to a user profile that has usually been accumulated over a longer period of time. Traditional IF systems match a user-provided profile against a flow of incoming documents (for example, news articles) to select the most relevant items for the user. In contrast, modern recommender systems construct and maintain dynamic user profiles by observing user's interactions to produce new recommendations, even in stable document collections. In hypertext browsing, a user attempts to find relevant documents by browsing links that connect documents in a collection. In information visualization, a set of documents is presented to the user using a certain visualization metaphor in either two or three dimensions; the user observes or (in the case of interactive visualization) interacts with the visualized set to find the most relevant information items.

Since a review [9] suggested to distinguish these four types of information access in 2008, social media have introduced many new ways of accessing information. These new ways could be called *human-driven*, because they focus on automating traditional human ways of information exchange. To differentiate from these human-driven information access, the original four ways could be called *system-driven*. In most cases, these new ways of information access do not introduce new *kinds* of information access, but rather introduce new *sources* of information. For example, a Twitter feed could be searched, browsed, or accessed through a standard information filtering interface. However, there is at least one exception: information access through questions and answers powered by modern Q&A systems. We suggest that this should be considered as the fifth basic type of social information access.

From the point of view of classification, it is important to recognize that the types of information access have strongly influenced the development of certain social information access technologies. For example, browsing-based access encourages research on navigation support systems that can help users to select a link to follow among many links on the current page. The natural approach to using community wisdom in this context is to show "where did the people go" [16,62] by augmenting links with digital "wear" indicators. The natural approach to collect this knowledge was to track user page visits [11] or link traversals [62]. Consequently, social navigation technologies (history-enriched environments) have been developed for supporting browsing-based access in social context.

Another example is social search technologies that were developed to support traditional IR information access. In this context, users expect to see a ranked list of relevant resources. The natural approach to using community wisdom in this area is by re-ranking results using community wisdom [8,35,63] or by inserting community-relevant links into the list or results [59] so that the returned documents reflect not only query relevance, but also the degree of their appreciation by the community. A reliable approach to collecting this wisdom is to track connections between queries and items selected or rated by the community members in the context of these queries [35,63].

3.2 Making Information Access Social

The "social" aspect of information access stresses its ability to transfer information that comes from one group of the user community ("providers") to another group ("recipients"). This information transfer is frequently called "collaboration"; although, in most cases, no real collaboration takes place.

By its nature, this information transfer or collaboration could be classified along two important dimensions: *intent* (direct–indirect) and *concurrency* (synchronous–asynchronous). This classification was originally introduced by Dieberger et al. [15] in a social navigation context, which at that time covered both browsing and recommendation. Eight years later, these two dimensions were also introduced by Golovchinsky et al. [26] as a part of a taxonomy for collaborative search. While the latter work used an explicit–implicit dichotomy instead of a direct–indirect dichotomy for classifying intent, it expressed the same meaning (i.e., intent) as the earlier direct–indirect dichotomy in [15]. In the following, we suggest the use of the direct–indirect dimension in its original form to classify intent and reserve the explicit–implicit dichotomy to categorize the types of information traces (see Sect. 3.3).

Direct vs. indirect information transfer determines whether the transfer of information (or a collaboration) is intentional.

In the case of a *direct transfer*, the "providers" directly communicate information to the recipients (or guide the recipients to the appropriate information) with the goal of assisting others. In many cases, to initiate this transfer, a recipient with an information need is expected to also directly solicit information from "providers". However, indirect approaches could be used to determine the correct "provider" to ask.

In case of *indirect transfer*, "the providers" do not directly provide information to "recipients". In fact, their work with information only aims to satisfy their own needs. It is the traces of their own work with information that could be processed and used to help the recipients in finding the most relevant information. In other words, information is indirectly collected from the community to help other users. In social information access, indirect transfer is much more common. It also can leverage a larger diversity of "social wisdom" than a direct transfer.

Synchronous vs. asynchronous transfer determines whether providers and recipients coexist in time.

With *synchronous information transfer*, providers and recipients work on their information access tasks at the same time, and information directly or indirectly generated by the providers is immediately used to help the recipients. In synchronous context, the same user frequently works as both a provider and a recipient: they use the information produced by others, and they also generate information to help others.

With *asynchronous information transfer*, the recipients are supported in their information access tasks by "social wisdom" produced by those providers who worked with information earlier in the process. Among the two alternatives, this is the more commonly found case in social information access. Unless a collection of social information has just been started, the fraction of users who work with information at any given time and the volume of social wisdom provided by these users are many times smaller than the volume of all past users and the information that they directly or indirectly contributed.

While this two-dimensional scheme has been useful to the research community for many years, it creates an imbalanced classification, because the vast majority of social information access techniques fall into the *indirect-asynchronous corner*. For example, among the technologies represented in this book (see Table 1), Social Q&A (Chap. 3 [50]) is an example of direct-asynchronous access, collaborative search (Chap. 4 [64]) offers examples of direct and indirect synchronous access, and social navigation (Chap. 5 [20]) provides examples of indirect-synchronous, direct-asynchronous, and indirect-asynchronous groups. The rest of the chapters all focus solely on indirect asynchronous technologies. Consequently, there should be another classification that focuses on the "social" aspect of information access that can specifically help to distinguish various indirect-asynchronous technologies from one another. We offer such a classification in Sect. 3.3.

Table 1. Classification of social information access techniques by intent and concurrency

		Concurrency	
		Synchronous	Asynchronous
Intent	Direct	Collaborative search [64]	Social Q&A [50] Social navigation [20] Recommendation [39]
	Indirect	Collaborative search [64] Social navigation [20]	Social search [12, 29, 49] Social navigation [20] Tag-based navigation [17] Recommendation [6, 7, 27, 34, 39, 40, 51]

3.3   Types and Sources of Social Information

3.3.1   Explicit and Implicit Traces of User Activity

As mentioned in the previous section, it is important to classify social information access techniques by using the concepts of intent and concurrency in the process of collecting social information traces and passing them to new users; but only these two concepts are not sufficient to differentiate a wide variety of modern social information access techniques. To overcome this, Brusilovsky [10] suggested to further classify users' past actions leveraged by social information access. These actions are called *users' feedback* in the field of personalized and social systems. User feedback can be explicit, in which the users explicitly express some opinions about an information item. An example of explicit feedback is a user's rating. Although user ratings are still popular sources of information in some social information access systems (e.g., collaborative recommendation systems), it has long been recognized that user ratings form a comparatively small fraction of user interactions with information. Consequently, recent work has focused on implicit feedback [34], where various users' actions are collected and analyzed to infer their attitudes. The most popular source of implicit feedback is search or browsing logs with a sequence of clicks and dwell time (also known as a clickstream). Although there is a risk that clickstreams and other implicit sources of evidence might be less reliable, they are more readily available in various contexts.

Explicit ratings and clickstreams are the two extreme ends of the implicit-explicit continuum of social information. Nowadays, the gap between these two extremes has been filled by a whole range of user actions collected in social information access systems. For example, at the explicit part of the spectrum, users can take actions such as annotation, commenting, and tagging. Yet, unlike ratings, these actions usually do not quantify the degree of the match between the user's need and the annotated item. On the implicit side, it is possible to have actions, such as purchasing a product online, listening to a digital music track, or eating at a local restaurant. All these types of implicit feedbacks can provide more reliable evidence about a user's interests than a clickstream, because each action is associated with a larger commitment of time and/or money.

Due to the rapid increase of the variety of information traces collected by modern social systems, it has become harder and harder to offer an extensive classification. Therefore, this book attempts to separately discuss and classify explicit and implicit information sources in three broad contexts: browsing, search, and recommendation, which correspond to Chap. 5 [20], Chap. 7 [12], and Chap. 14 [34], respectively. We urge the reader to examine these chapters for detailed discussion of each issue.

In the remainder of this section, we will briefly examine the connections between the main *types* of social information and the main groups of *sources* (i.e., type of systems), and we will use Table 2 to highlight our analysis. The list of types and the classification of sources do not pretend to be exhaustive because it is evident that social information access systems will continue exploring new sources and new kinds of social information. Our goal here is to make this list helpful in distinguishing and classifying the majority of existing techniques.

Table 2. Sources of social information in web and social systems

Social information	Search engines	Hypertext / Web	Communication	Annotation systems	Curation systems	Social networking	Social bookmarking	Recommender systems	Consumption systems	Location based
Queries & SERP clicks	✓						✓	✓		
Browsing trails		✓	✓	✓	✓	✓	✓	✓	✓	✓
References (links)		✓	✓		✓	✓	✓			
Annotations, comments				✓	✓	✓	✓		✓	✓
Tags, categorizations				✓		✓	✓	✓	✓	✓
Social links			✓			✓	✓	✓	✓	✓
Ratings				✓				✓	✓	✓
Consumption actions									✓	
Real world trails										✓

3.3.2   Search Engines

Search engines receive users' search queries and generate search engine result pages (SERPs). By tracking individual users through their search sessions, a search engine can archive successful sequences of *queries and SERP clicks*; namely, those search results that users decided to explore further. Both queries and SERP clicks are useful social information, and mining their accumulation can be used to generate social wisdom. Currently, this is one of the most powerful approaches for improving search.

3.3.3   The Linked Web

The open Web and many specialized Web-based systems (i.e. Wikipedia) allow users to create information pages and link them to one another. They offer activity traces of two kinds of users - page authors and Web surfers. Page authors extensively use *references* (i.e., Web links) to other pages when creating their pages. These links were one of the earliest sources of social information that was used to improve information access through the better ranking of search results [8]. Web surfers leave *browsing trails* as they navigate, where each click offers a small evidence to indicate that the selected link is the most attractive for the given user on the traversed page. Within a single Web site or Web system, clicked links are easily accumulated in Web logs. Across sites and systems, traces of user Web browsing behavior can be aggregated by using browsing agents [42], intermediaries [3], browser plugins [58], and other approaches (see [23] for a review). An advanced user tracking approach could augment browsing trails with *in-page behavior* (such as scrolling or mousing, among others) - a valuable source of social wisdom that could be used for both distinguishing the most useful pages and guiding users to the most relevant parts of a page [30].

3.3.4   Communication Systems

Various communication systems, such as bulletin boards, discussion forums, e-mail, chat, blogs, and microblogs accumulate large volumes of social wisdom. Open discussion sites, such as bulletin boards, forums, and blogs can be easily crawled and mined. Private e-mail and chat traces are generally harder to use, but given that many Web mail systems are maintained by companies that also operate search engines, the e-mails also emerge as a valuable source for search improvement. Communication systems are also used as sources for useful *references* and implicit *social links*. A reference to a Web link in any kind of message or discussion is a good evidence of the importance of the link. These links could be simply extracted or associated with a discussion topic and surrounding text. Replying to, commenting, or forwarding actions all offer the evidence of a social link between users. Modern blogs and microblogs could also serve as sources of *explicit social links*: users can establish social links in the form of "watching" other users' updates. In these systems, posts could be also extended with social tags.

3.3.5   Annotation Systems

The Web was originally envisioned as having the ability to provide comments and annotations for every Web page, but this infrastructure has never been fully implemented. Instead, the task of Web page annotation has been taken over by various Web annotation systems including the original Annotea project from WWW Consortium [36]. These systems allowed every Web user to add comments for a Web page or its fragment, or simply to mark up the most valuable fragments. Motivated by the research on Web page annotations, page annotation functionality was implemented in a number of Web systems, including Web-based books, textbooks, and digital libraries [19, 43, 46]. In modern social systems, commenting and annotation functionality have been applied to a broad range of items beyond Web pages. With these systems, users can add comments, annotations, and reviews to hotels, movies, books, and many other items. Web annotation systems and "item-focused" social systems could offer three types of social information. Firstly, an annotation can be treated as a sign of the item's importance, which could be used to attract attention to it on SERPs [2] or on a Web site. Moreover, a within-page annotation system can collect user in-page behaviors, which are used in guiding future users to the most valuable part of the page [19]. Secondly, the content of page annotations or comments describes a page or an item from the prospect of the annotation author. These comments could be used for search and recommendation [51]. Thirdly, many modern "item-focused" systems collect not only item annotations, but also *item ratings*.

3.3.6   Social Networking Systems

Facebook and LinkedIn started as platforms to connect people, but they have gradually included elements from microblogs and social bookmarking systems. Modern social linking systems serve as the primary source of *social links* while also contributing *item links* and *comments*.

3.3.7   Curation Systems

Since the early days of the Web, there have been multiple attempts to engage Web users into adding additional levels of organization to the Web. The most remarkable among these projects are those that intend to build a hierarchically organized directory of Web pages, pioneered by Yahoo.com and expanded by the Open Directory Project (dmoz.org). Another important group of Web organization systems are various guided path systems [22]. Guided path systems allowed their users to build and publish Web paths, which are annotated sequenced of Web pages. Pages connected by the path are usually conceptually similar to each other and deliver a common narration. In addition to path systems, there are other simpler social systems that allow users to contribute social wisdom by grouping together similar pages without the need to provide comments or sequences [1,5,57]. All these Web organization systems offer a good source of *references*: the very fact of page or item sharing is a usual sign of its value. In addition, curation systems could provide other social information: page *comments*, page *categorization*, and user-judged *similarity* between pages, where the similarity criteria can be pages that were contributed under the same category, group, or path.

3.3.8   Social Bookmarking Systems

Social bookmarking systems could be considered to be one of the most successful curation systems. Integrating ideas from several earlier streams of research, including Web annotations, bookmark lists, classification systems, and recommender systems, social bookmarking systems have introduced a new way to organize and navigate socially contributed Web information [25,28]. Social bookmarking systems allow their users to openly share information items (Web pages, photographs, research papers) while providing text *comments* and annotating these items with a set of free *tags*. These tags offer a nice balance for resource organization between unstructured comments and formal hierarchical classification systems, like the Open Directory Project. Modern social bookmarking systems, such as CiteULike, Flickr, or Pinterest also support one or more types of social connections; usually an ability to watch other users and form groups or communities. As a result, these systems have become a valuable source of various *social links*.

3.3.9   Consumption Systems and Recommender Systems

Consumption systems and recommender systems are two related groups of online systems. Consumption systems refer to all systems where users can access and "consume" content. Some examples include online shopping systems where users can purchase goods, online journals sites/digital libraries where users can download content to read, and online music and video services that allow users to stream selected content. These systems differ from other Web systems, since obtaining an item in a consumption system requires a higher level of commitment from the user than a simple click (i.e., purchasing or downloading). These *consumption* actions left by past users allow the systems to accumulate more reliable social evidence of the value of the item. To help future users with higher-commitment decisions, consumption systems usually encourage post-consumption items *ratings* and *comments*. The last two aspects make consumer systems similar to classic recommender systems. The difference between these two types of systems is small: ratings and comments in consumption systems are directly used by end users, whereas such information in a recommender system is used by the recommender engine to proactively suggest relevant items for users to explore. In addition, classic recommender systems rarely offer the immediate ability to consume (purchase, play) recommended items. Nowadays, this difference has nearly disappeared, with most recommender systems being integrated into consumption systems and most consumption systems offering some form of recommendation. In addition, modern recommender and consumption systems frequently support certain forms of *social links*, which allow users to watch each other or form *groups*.

3.3.10   Location-Based Systems

The newest group of social systems are location-based systems, where users leave various feedback about objects located in a real space, such as restaurants, stores, cafes, or other physical objects. The feedback may range from simple check-ins to extensive reviews. Location-based offer several traditional types of social wisdom explored by other social systems, such as establishing *social links*, using *tags*, and others. However, they also add a unique new source of social information – real-world user traces. This information could be used to generate a whole new type of social recommendations [7].

4   The Book Structure

When assembling this book, our goal was to provide a broad overview of modern research on social information access. To ensure good coverage, we followed the classification of social information access techniques introduced in the previous section. In other words, the book represents an attempt to provide examples for every aspect of the introduced classification, and often covers the most important combinations of the aspects as well.

One particularly important goal for us is to provide sufficient coverage of social approaches for the three main types of information access: search, browsing, and recommendation. As shown in Table 3, four of the book's chapters focus on social search, two focus on social navigation and browsing, and seven focus on recommendation. While we are not able to offer a chapter dedicated to social visualization, we ensured that social browsing chapters address some visualization techniques. We also provided a chapter that focuses on social Q&A.

Table 3. Book chapters organized by the type of information access

Access type	Groups of technologies
Search	Chapters 4, 7–9 [12, 29, 49, 64]
Browsing	Chapters 5, 6 [17, 20]
Recommendation	Chapters 10–16 [6, 7, 27, 34, 39, 40, 51]
Visualization	Chapters 5, 6 [17, 20]
Q&A	Chapter 3 [50]

While the majority of modern social information access techniques could be classified as indirect-asynchronous, we also want to ensure that the book provides examples of direct and synchronous social information access (Table 1). For a browsing type of access, Chap. 5 [20] specifically discusses examples of direct-asynchronous, indirect-synchronous, and indirect-asynchronous social navigation. For search-based access, Chap. 4 [64] focuses on synchronous approaches in social search (more commonly known as collaborative search) and covers both direct and indirect collaboration. Chapters 7 [12], 8 [29], and 9 [49] focus on indirect-asynchronous techniques. There is no chapter focused on direct-asynchronous social search, but this area is covered by the chapter on Social Q&A [50] and some brief discussions in Chap. 4 [64]. For modern recommender technologies, which are asynchronous by their nature, Chap. 10 [39] focuses on classic rating-based recommendation, whose coverage is at the direct-asynchronous corner. Traditional classifications consider ratings as directly provided social feedback, however, it is less obvious nowadays since ratings in modern recommendations are frequently provided to get better recommendations rather than to recommend items for other users. Chapter 14 [34] offers a good discussion on this issue. The remaining recommendation chapters all focus on indirect-asynchronous approaches.

Our last goal in respect to the coverage is to ensure that the chapters cover the major types of social traces, both explicit and implicit. Including this aspect helps to uncover deep similarities among approaches from different groups when the comparison of different social information access is based on the same type of social traces. Table 4 explains how the book chapters cover most of the popular types of social traces.

For the search and browsing types of information access, Chaps. 5 [20] and 7 [12] provide coverage of most types of explicit and implicit sources. Both chapters offer a useful discussion and classification on covered sources (see Table 4 for more details). Similarly, on the recommendation side, Chap. 15 covers a range of information sources for *people recommendation*. Among the explicit types of social information, this book pays special attention to tags and links, due to their historical and practical importance. Three separate chapters focus on using tags for search (Chap. 9 [49]), navigation (Chap. 6 [17]), item recommendation (Chap. 12 [6]), and people recommendation (Chap. 15 [27]). The book also offers a dedicated chapter on using information and social links for search (Chap. 8 [29]), as well as using social links for item recommendation (Chap. 11 [40]) and people recommendation (Chap. 15 [27]). The recommendation side of the book also provides dedicated chapters for two other types of explicit traces: ratings (Chap. 10 [39]) and text-based feedback (Chap. 13 [51]). The former represents the classic stream of research on collaborative filtering, while the latter focuses on a source that is rapidly increasing in both volume and practical value.

The use of all implicit information sources for recommendation is covered in Chap. 14 [34]. At the same time, there is a dedicated chapter for recommendations based on "real world trails" (Chap. 16 [7]). This chapter plays a special role in the book, as it also serves as the closing chapter. As mentioned in Sect. 2, user navigation in the real world served as a motivation for the pioneers of social navigation, who wanted to visualize traces and the presence of other users in an information space, just as they are visible in real space. It could be considered a sign of the field's maturity that social information access techniques developed to help people navigate in information spaces have now been brought back to real spaces and are able to help guide users to the most relevant places.

Table 4. Main sources of social information (from implicit to explicit) and their coverage in the book

Source of social information	Search	Browsing	Recommendation
Search engine logs	Chapter 7 [12]	Chapter 5 [20]	
Browsing trails	Chapter 7 [12]	Chapter 5 [20]	Chapter 14 [34]
Real world trails			Chapter 16 [7]
Information links	Chapters 7, 8 [12, 29]	Chapter 5 [20]	
Annotations and comments	Chapter 7 [12]	Chapter 5 [20]	Chapter 13 [51]
Tags	Chapters 7, 9 [12, 49]	Chapter 6 [17]	Chapters 12,15 [6, 27]
Social Links	Chapters 7, 8 [12, 29]		Chapters 11, 15 [27, 40]
Ratings			Chapter 10 [39]

As previously stated, the book chapters align nicely with our classification of social information access techniques, which provides a sound guide for reading and finding information in the book. In addition, to cover each topic, we invited top experts in the field with extensive knowledge on specific types of social information access techniques. In the process of preparing this book, each chapter went through several cycles of review and feedback among the editors, the authors of other chapters, and a team of PhD students who served as "pilot readers". We hope that this book will serve as a good reference to the literature of social information access, as well as a handbook that can help readers in developing their own social information access approaches.

References

1. Abel, F., Frank, M., Henze, N., Krause, D., Plappert, D., Siehndel, P.: GroupMe! - where semantic web meets web 2.0. In: Aberer, K., et al. (eds.) ASWC/ISWC-2007. LNCS, vol. 4825, pp. 871–878. Springer, Heidelberg (2007). https://doi.org/10.1007/978-3-540-76298-0_63
2. Ahn, J., Farzan, R., Brusilovsky, P.: Social search in the context of social navigation. J. Korean Soc. Inf. Manag. **23**(2), 147–165 (2006)
3. Barrett, R., Maglio, P.P.: Intermediaries: an approach to manipulating information streams. IBM Syst. J. **38**(4), 629–641 (1999)
4. Beeferman, D., Berger, A.: Agglomerative clustering of a search engine query log. In: Sixth ACM SIGKDD International Conference on Knowledge Discovery and Data Mining, pp. 407–416 (2000)
5. Bernstein, M.: Web research: the Eastgate Web Squirrel. SIGWEB Newsl. **5**(1), 6 (1996)
6. Bogers, T.: Tag-based recommendation. In: Brusilovsky, P., He, D. (eds.) Social Information Access. LNCS, vol. 10100, pp. 441–479. Springer, Cham (2017)
7. Bothorel, C., Lathia, N., Picot-Clemente, R., Noulas, A.: Location recommendation with social media data. In: Brusilovsky, P., He, D. (eds.) Social Information Access. LNCS, vol. 10100, pp. 624–653. Springer, Cham (2017)
8. Brin, S., Page, L.: The anatomy of a large-scale hypertextual (web) search engine. In: Ashman, H., Thistewaite, P. (eds.) Seventh International World Wide Web Conference, vol. 30, pp. 107–117. Elsevier Science B.V. (1998)
9. Brusilovsky, P.: Social information access: the other side of the social web. In: Geffert, V., Karhumäki, J., Bertoni, A., Preneel, B., Návrat, P., Bieliková, M. (eds.) SOFSEM 2008. LNCS, vol. 4910, pp. 5–22. Springer, Heidelberg (2008). https://doi.org/10.1007/978-3-540-77566-9_2
10. Brusilovsky, P.: The other side of the social web: a taxonomy for social information access. In: The 18th Brazilian Symposium on Multimedia and the Web, pp. 1–4. ACM Press (2012). http://dl.acm.org/citation.cfm?id=2382638
11. Brusilovsky, P., Chavan, G., Farzan, R.: Social adaptive navigation support for open corpus electronic textbooks. In: De Bra, P.M.E., Nejdl, W. (eds.) AH 2004. LNCS, vol. 3137, pp. 24–33. Springer, Heidelberg (2004). https://doi.org/10.1007/978-3-540-27780-4_6
12. Brusilovsky, P., Smyth, B., Shapira, B.: Social search. In: Brusilovsky, P., He, D. (eds.) Social Information Access. LNCS, vol. 10100, pp. 213–276. Springer, Cham (2017)
13. Bush, V., et al.: As we may think. The Atl. Mon. **176**(1), 101–108 (1945)
14. Chakrabarti, S., Dom, B., Raghavan, P., Rajagopalan, S., Gibson, D., Kleinberg, J.: Automatic resource compilation by analyzing hyperlink structure and associated text. In: Proceedings of the Seventh International Conference on World Wide Web 7, vol. 30, pp. 65–74. Elsevier Science Publishers B.V. (1998)
15. Dieberger, A., Dourish, P., Höök, K., Resnick, P., Wexelblat, A.: Social navigation: techniques for building more usable systems. Interactions **7**(6), 36–45 (2000)
16. Dieberger, A.: Supporting social navigation on the World Wide Web. Int. J. Hum Comput Stud. **46**(6), 805–825 (1997)
17. Dimitrov, D., Helic, D., Strohmaier, M.: Tag-based navigation and visualization. In: Brusilovsky, P., He, D. (eds.) Social Information Access. LNCS, vol. 10100, pp. 181–212. Springer, Cham (2017)

18. Dourish, P., Chalmers, M.: Running out of space: models of information navigation. In: SIGCHI Conference on Human Factors in Computing Systems, pp. 23–26 (1994)
19. Farzan, R., Brusilovsky, P.: Annotated: a social navigation and annotation service for web-based educational resources. New Rev. Hypermedia Multimed. **14**(1), 3–32 (2008)
20. Farzan, R., Brusilovsky, P.: Social navigation. In: Brusilovsky, P., He, D. (eds.) Social Information Access. LNCS, vol. 10100, pp. 142–180. Springer, Cham (2017)
21. Fitzpatrick, L., Dent, M.: Automatic feedback using past queries: social searching? In: Proceedings of the 20th Annual International ACM SIGIR Conference on Research and Development in Information Retrieval, pp. 306–313 (1997)
22. Furuta, R., Shipman III, F.M., Marshall, C.C., Brenner, D., Hsieh, H.: Hypertext paths and the World-Wide Web: experience with Walden's paths. In: Bernstein, M., Carr, L., sterbye, K. (eds.) Eight ACM International Hypertext Conference (Hypertext 1997), pp. 167–176. ACM (1997)
23. Gauch, S., Speretta, M., Chandramouli, A., Micarelli, A.: User profiles for personalized information access. In: Brusilovsky, P., Kobsa, A., Nejdl, W. (eds.) The Adaptive Web. LNCS, vol. 4321, pp. 54–89. Springer, Heidelberg (2007). https://doi.org/10.1007/978-3-540-72079-9_2
24. Goldberg, D., Nichols, D., Oki, B.M., Terry, D.: Using collaborative filtering to weave an information tapestry. Commun. ACM **35**(2), 61–70 (1992)
25. Golder, S.A., Huberman, B.A.: Usage patterns of collaborative tagging systems. J. Inf. Sci. **32**(2), 198–208 (2006)
26. Golovchinsky, G., Pickens, J., Back, M.: A taxonomy of collaboration in online information seeking. In: 1st International Workshop on Collaborative Information Seeking (2008). https://arxiv.org/abs/0908.0704
27. Guy, I.: People recommendation on social media. In: Brusilovsky, P., He, D. (eds.) Social Information Access. LNCS, vol. 10100, pp. 570–623. Springer, Cham (2017)
28. Hammond, T., Hannay, T., Lund, B., Scott, J.: Social bookmarking tools (I): a general review. D-Lib Mag. **11**(4) (2005). https://urldefense.proofpoint. com/v2/url?u=http-3A__www.dlib.org_dlib_april05_hammond_04hammond. html&d=DwIBaQ&c=vh6FgFnduejNhPPD0fl_yRaSfZy8CWbWnIf4XJhSqx8& r=UyK1_569d50MjVlUSODJYRW2epEY0RveVNq0YCmePcDz4DQHW- CkWcttrwneZ0md&m=8USpmwWQapLg5kc22S3fbLNcKx4X7l5lz5l3P9_xg1A& s=TNPKGVSVp4iidBHHcJ7b-Qh6slsuJGc9nB83u6Uh1_c&e=
29. Han, S., He, D.: Network-based social search. In: Brusilovsky, P., He, D. (eds.) Social Information Access. LNCS, vol. 10100, pp. 277–309. Springer, Cham (2017)
30. Han, S., He, D., Yue, Z., Brusilovsky, P.: Supporting cross-device web search with social navigation-based mobile touch interactions. In: Ricci, F., Bontcheva, K., Conlan, O., Lawless, S. (eds.) UMAP 2015. LNCS, vol. 9146, pp. 143–155. Springer, Cham (2015). https://doi.org/10.1007/978-3-319-20267-9_12
31. Hill, W., Stead, L., Rosenstein, M., Furnas, G.: Recommending and evaluating choices in a virtual community of use. In: SIGCHI Conference on Human Factors in Computing Systems, CHI 1995, pp. 194–201. ACM (1995). http://portal.acm. org/citation.cfm?id=223904.223929
32. Hill, W.C., Hollan, J.D., Wroblewski, D., McCandless, T.: Edit wear and read wear. In: Proceedings of the SIGCHI Conference on Human Factors in Computing Systems, CHI 1992, pp. 3–9. ACM Press (1992)
33. Höök, K., Benyon, D., Munro, A.J.: Designing Information Spaces: The Social Navigation Approach. Springer, Berlin (2003). https://doi.org/10.1007/978-1-4471-0035-5

34. Jannach, D., Lerche, L., Zanker, M.: Recommending based on implicit feedback. In: Brusilovsky, P., He, D. (eds.) Social Information Access. LNCS, vol. 10100, pp. 510–569. Springer, Cham (2017)
35. Joachims, T.: Optimizing search engines using clickthrough data. In: Proceedings of the Eighth ACM SIGKDD International Conference on Knowledge Discovery and Data Mining (KDD 2002), pp. 133–142. ACM (2002)
36. Kahan, J., Koivunen, M.R., Prud'Hommeaux, E., Swick, R.R.: Annotea: an open RDF infrastructure for shared web annotations. Comput. Netw. **39**(5), 589–608 (2002)
37. Kantor, P.B., Boros, E., Melamed, B., Mekov, V., Shapira, B., Neu, D.J.: Capturing human intelligence in the net. Commun. ACM **43**(8), 112–116 (2000)
38. Keller, R.M., Wolfe, S.R., Chen, J.R., Rabinowitz, J.L., Mathe, N.: A bookmarking service for organizing and sharing URLs. In: Sixth International World Wide Web Conference, pp. 1103–1114 (1997). Computer Networks and ISDN Systems **29**(8–13) (1997). http://www.ra.ethz.ch/CDstore/www6/Technical/Paper189/Paper189.html
39. Kluver, D., Ekstrand, M., Konstan, J.: Rating-based collaborative filtering: algorithms and evaluation. In: Brusilovsky, P., He, D. (eds.) Social Information Access. LNCS, vol. 10100, pp. 344–390. Springer, Cham (2017)
40. Lee, D., Brusilovsky, P.: Recommendations based on social links. In: Brusilovsky, P., He, D. (eds.) Social Information Access. LNCS, vol. 10100, pp. 391–440. Springer, Cham (2017)
41. Li, W.S., Vu, Q., Agrawal, D., Hara, Y., Takano, H.: PowerBookmarks: a system for perzonalizable web information organization, sharing, and management. In: 8th International World Wide Web Conference, pp. 297–311. Elsevier (1999). http://www8.org/w8-papers/3b-web-doc/power/power.pdf
42. Lieberman, H.: Letizia: an agent that assists web browsing. In: The Fourteenth International Joint Conference on Artificial Intelligence, pp. 924–929 (1995)
43. Liesaputra, V., Witten, I.H.: Seeking information in realistic books: a user study. In: Joint Conference on Digital Libraries, JCDL 2008, pp. 29–38 (2008)
44. Lueg, C., Fisher, D.: From Usenet to CoWebs: Interacting with Social Information Spaces. Springer, Berlin (2003). https://doi.org/10.1007/978-1-4471-0057-7
45. Maltz, D., Ehrlich, K.: Pointing the way: active collaborative filtering. In: Proceedings of the SIGCHI Conference on Human Factors in Computing Systems, CHI 1995, pp. 202–209. ACM (1995)
46. Marshall, C.C., Bly, S.: Sharing encountered information: digital libraries get a social life. In: The 4th ACM/IEEE-CS Joint Conference on Digital Libraries (JCDL 2004), pp. 218–227 (2004)
47. McBryan, O.A.: GENVL and WWWW: tools for taming the web. In: the 1st International World Wide Web Conference, p. 7990 (1994)
48. Munro, A.J., Höök, K., Benyon, D.: Social Navigation of Information Space. Springer, Berlin (1999). https://doi.org/10.1007/978-1-4471-0837-5
49. Navarro Bullock, B., Hotho, A., Stumme, G.: Accessing information with tags: search and ranking. In: Brusilovsky, P., He, D. (eds.) Social Information Access. LNCS, vol. 10100, pp. 310–343. Springer, Cham (2017)
50. Oh, S.: Social Q&A. In: Brusilovsky, P., He, D. (eds.) Social Information Access. LNCS, vol. 10100, pp. 75–107. Springer, Cham (2017)
51. O'Mahoney, M., Smyth, B.: From opinions to recommendations. In: Brusilovsky, P., He, D. (eds.) Social Information Access. LNCS, vol. 10100, pp. 480–509. Springer, Cham (2017)

52. Raghavan, V.V., Sever, H.: On the reuse of past optimal queries. In: 18th Annual International ACM SIGIR Conference on Research and Development in Information Retrieval, pp. 344–350. ACM (1995)
53. Remde, J.R., Gomez, L.M., Landauer, T.K.: SuperBook: an automatic tool for information exploration hypertext? In: The ACM Conference on Hypertext, Hypertext 1987, pp. 175–188 (1987)
54. Resnick, P., Iacovou, N., Suchak, M., Bergstrom, P., Riedl, J.: GroupLens: an open architecture for collaborative filtering of netnews. In: ACM 1994 Conference on Computer Supported Cooperative Work, pp. 175–186. ACM Press (1994). https://doi.org/10.1145/192844.192905
55. Rucker, J., Polano, M.J.: Siteseer: personalized navigation for the web. Commun. ACM **40**(3), 73–75 (1997)
56. Shardanand, U., Maes, P.: Social information filtering: algorithms for automating "word of mouth". In: Katz, I., Mack, R., Marks, L. (eds.) The SIGCHI Conference on Human Factors in Computing Systems, CHI 1995, pp. 210–217. ACM (1995)
57. Shipman, F.M., Marshall, C.C.: Spatial hypertext: an alternative to navigational and semantic links. ACM Comput. Surv. **31**(4es), 1–5 (1999)
58. Smyth, B., Coyle, M., Briggs, P.: HeyStaks: a real-world deployment of social search. In: Sixth ACM Conference on Recommender Systems, RecSys 2012, Dublin, Ireland, 9–13 September 2012, pp. 289–292 (2012)
59. Smyth, B., Freyne, J., Coyle, M., Briggs, P., Balfe, E.: I-SPY — anonymous, community-based personalization by collaborative meta-search. In: Coenen, F., Preece, A., Macintosh, A. (eds.) Research and Development in Intelligent Systems XX, pp. 367–380. Springer, London (2004). https://doi.org/10.1007/978-0-85729-412-8_27
60. Svensson, M., Höök, K., Laaksolahti, J., Waern, A.: Social navigation of food recipes. In: Proceedings of the SIGCHI Conference on Human Factors in Computing Systems, CHI 2001, pp. 341–348. ACM (2001)
61. Takano, H., Winograd, T.: Dynamic bookmarks for the www: Managing personal navigation space by analysis of link structure and user behavior. In: Grnbk, K., Mylonas, E., Shipman III, F.M. (eds.) Ninth ACM International Hypertext Conference (Hypertext 1998), pp. 297–298. ACM Press (1998)
62. Wexelblat, A., Maes, P.: Footprints: history-rich tools for information foraging. In: Proceedings of the SIGCHI Conference on Human Factors in Computing Systems, CHI 1999, pp. 270–277. ACM (1999)
63. Xue, G., Zeng, H., Chen, Z., Yu, Y., Ma, W., Xi, W., Fan, W.: Optimizing web search using web click-through data. In: Proceedings of the Thirteenth ACM International Conference on Information and Knowledge Management, pp. 118–126. ACM (2004)
64. Yue, Z., He, D.: Collaborative information search. In: Brusilovsky, P., He, D. (eds.) Social Information Access. LNCS, vol. 10100, pp. 108–141. Springer, Cham (2017)

2
Privacy in Social Information Access

Bart P. Knijnenburg(✉)(iD)

Clemson University, Clemson, SC, USA
bartk@clemson.edu

Abstract. Social information access (SIA) systems crucially depend on user-provided information, and must therefore provide extensive privacy provisions to encourage users to share their personal data. Even though the information SIA systems use is usually considered public, they often use this information in novel ways, and the outcomes of this process may at times lead to unintended consequences for their users' privacy. Indeed, even if a SIA system is deemed generally beneficial, privacy concerns can play a limiting role in its adoption. This chapter analyzes the privacy implications of several types of SIA systems (aggregators, public content systems, and social network-based systems) from various angles, and discusses a wide range of solutions (both technical and decision-support solutions) to potential privacy threats. Acknowledging that SIA systems are not just a threat to users' privacy, the chapter concludes with a discussion of the use of social information access as a solution to privacy threats, i.e. by using it to provide social justifications, or by means of adaptive privacy decision support.

1 Introduction

How can we have meaningful interactions with the world while maintaining a certain level of personal privacy? This is an age-old problem, the implications of which have skyrocketed in the modern information age [262]. The Internet has not only vastly improved our ability to interact; it has also been the harbinger of many new privacy problems (as well as a catalyst for many existing ones). Privacy is a particularly important aspect of social information access (SIA) systems, which help users get to the right information using the actions, preferences and/or contributions of other users [36][1]. In the process of providing the improved access functionality, these actions, preferences, and contributions (which may be privacy-sensitive) may be disclosed to the system, selected other users (cf. "contacts"), or even the general public. These disclosures raise privacy concerns, and researchers have demonstrated that such privacy concerns can play a limiting role in users' adoption of SIA systems and services, even if those services are potentially beneficial [16,147,248,274]. To put it simply: If users think that their privacy is being violated, they will stop contributing [16,147,248], or

---

[1] See Chap. 1 of this book for a more detailed definition [37].

© Springer International Publishing AG, part of Springer Nature 2018
P. Brusilovsky and D. He (Eds.): Social Information Access, LNCS 10100, pp. 19–74, 2018.
https://doi.org/10.1007/978-3-319-90092-6_2

even leave the system altogether [274]. As such, SIA systems that do not take privacy into account are very likely to fail. In this chapter we will analyze the privacy implications of SIA systems, and look at possible solutions to potential privacy threats that these systems pose.

Many definitions of privacy exist [113,258], but for the purpose of this Chap. I will define it as a state of limited access to one's personal information[2]. In this sense, a "privacy decision" is a decision to regulate (i.e., restrict or allow) the flow of personal information to—or the use of this information by–one or more recipients. Within the SIA domain, this definition spans a wide variety of privacy-related situations. Let us first organize these situations by mapping out the dimensions along which they vary. Figure 1 shows an overview of these dimensions.

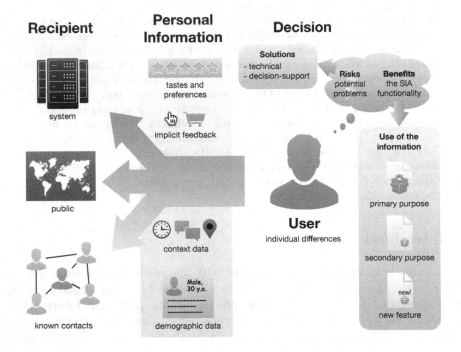

Fig. 1. An overview of the privacy dimensions of SIA systems.

1.1   The Recipient

Users' privacy decisions depend crucially on the *recipient(s)* of their personal information [141,167,208,288]. As almost every SIA system is in some sense a mediator of social online interactions, the recipients may include the system

---

[2] "Knowledge of one's actions" can also be considered information, and this is included in the definition as well.

itself, the public, or a network of known contacts. The relative importance of these three types of recipients depends on the type of SIA systems.

The system. Users of "aggregators" such as social navigation (Chap. 5 of this book [81]), social search engines (Chaps. 7–9 of this book [38,68,97]) and recommender systems (Chaps. 10–15 of this book [30,95,116,134,168,209]) typically disclose some personal information (e.g., ratings, clicks, search queries, social connections) to the system, but the system only reveal this information to other users in aggregated (e.g. stats) or derivative (e.g. recommendations) form. In this case, the systems themselves are the only recipients of personal identifiable information.

The public. Users of "public content systems" such as (micro)blogs (often used as a content source for SIA systems), social Q&A systems (Chap. 3 of this book [206]), and certain social tagging systems (Chap. 6 of this book [68]) create a publicly accessible stream of information (consisting of e.g., tweets, questions and answers, tags). This adds a second type of recipient: the public.

Known contacts. Users of "network-based systems" such as social networks (another type of system that is often used as a content source for SIA systems), collaborative systems (Chap. 4 of this book [323]), and certain location-sharing systems (Chap. 16 of this book [31]), can choose to share (part of) their information (e.g. status updates, "likes", location updates) within a closed network of contacts. This adds a third type of recipient: known contacts.

Our motivations for sharing information with these different types of recipients, as well as the associated privacy concerns, are fundamentally different [113, 202,214]. They will therefore be discussed in separate sections of this chapter.

1.2   The Personal Information

Another important dimension of users' privacy management activity is the type of information they (are asked to) disclose. Many existing privacy studies treat each piece of personal information as an independent decision [6,124], or as a summated composite score that essentially represents a unidimensional "disclosure propensity" [119,125], that may have different degrees of sensitivity [186–188]. However, recent work has shown that people's information disclosure decisions differ not only in *degree* but also in *kind*. In other words, people have fundamentally different preferences regarding different types of personal information [142].

Tastes and preferences. Explicit tastes or preferences are generally considered the least private aspects of one's personal information [1]. However, the aggregation of preferences may result in inferences about personality or lifestyle that the user is unconformable disclosing (e.g., it is possible to predict sexual orientation based on 5–10 Facebook likes [154]). Users are intuitively aware of this threat of aggregation, and indeed seem to get increasingly wary as disclosures accumulate [21,139,157].

Implicit feedback. Implicit feedback also concerns tastes and preferences, only in this case they are automatically collected from users' behavioral traces (clicks, purchases). The collection of this type of information is relatively unobtrusive, but also opaque: compared to manually provided information, it is much harder for a user to understand the potential consequences of a system tracking their behavior, let alone to control this [21,50,281,325]. Finally, with regard to public or network-based systems, automatically gathered preferences may reveal embarrassing information about the user [266].

Context data. Context data is also automatically collected, but concerns behavior that is not directly representative of users' tastes and preferences, such as a user's interaction with other users, location, calendar events, etc. The field of context-aware recommender systems has shown that this type of information can be used to improve the accuracy of predictions about users' tastes and preferences [8]. Note though, that users may worry that the system may make incorrect inferences based on the data [139]. Similar to implicit feedback, the automatic collection and display of context data may result in embarrassing situations in public or network-based systems [213].

Demographic data. SIA systems mainly leverage user preferences and/or behaviors, and therefore often do not use demographic data directly. Note though, that demographics (e.g., gender, marital status, and income) can be used to find nearest neighbors or to estimate preference vectors, and as such they are increasingly used in prediction algorithms (e.g., to overcome cold-start problems) [169,170,205,326]. Demographics information is sometimes usurped in one go from the user's Facebook or Google Plus account [73,298]; at other times the system straight up asks demographic questions [139,149,325].

Demographic data ranges from very innocuous (e.g., one's age and gender are usually disclosed without hesitation) to very private (users are much less willing to disclose information about sexual, medical, or financial topics) [139,141,149]. Most demographics, though, fall in between preferences [1] and context data [139] in terms of overall sensitivity. Note though, that demographic data privacy concerns are *multi-dimensional*: different users have different concerns regarding different types of information [142]. This makes dealing with demographic data privacy particularly challenging.

1.3   The Use of the Information

SIA systems use social information in novel ways to enable interesting and useful new experiences. The way information is used is however also an important privacy dimension. Already in 1993, Culnan [61] pointed out that "strategic uses of information technology based on personal information may raise privacy concerns among consumers if these applications do not reflect a common set of values" (p. 341). This means that SIA systems need to carefully consider whether users agree to accept the privacy implications of the provided functionality; something that is more straightforward when the SIA functionality is the primary purpose of the system, but less so when it is a secondary purpose or when it is introduced to a system at a later stage.

SIA as the primary purpose. Legally speaking users' agreement with the privacy implications of a SIA system is implied when users ostensibly provide the data in return for the SIA functionality. For example: when a recommender system collects rating feedback, users know at the time of disclosure that this is the primary purpose for which the disclosed information will be used. Note that this exchange of personal information can be seen as an "implied social contract", cf. [62, 193, 194, 228], which does not necessarily require that users fully understand the privacy implications of using the system (i.e., informed consent). Indeed, as users often do not fully understand the privacy implications (because the algorithmic nature of most SIA functionality can result in unpredictable and sometimes unwanted outcomes [139]) this scenario can still result in privacy violations [7]. However, the very act of collecting and using the data for the intended SIA purpose is not a violation in itself.

SIA as a secondary purpose. In many cases, the SIA functionality is not the primary purpose of the system [115]. In this case, the SIA feature uses personal information that is disclosed not for the purpose of the SIA feature, but for the main functionality of the system. For example: users enter search queries into a search engine to get search results, or use the tagging feature in a digital library to organize their book collection; the subsequent use for personalization is secondary. Even rating data can fall under secondary use when its primary purpose is not personalization (e.g. seller feedback on an e-commerce site). Privacy experts argue that such secondary use of the information should be explicitly communicated to the users, otherwise they may be surprised to find out about it, and feel that their privacy is violated [280]. They also suggest giving users the opportunity to opt out of the SIA functionality, lest they feel that the privacy threats of the SIA functionality outweigh the benefits of the main functionality [61, 115, 280].

SIA as a new feature. In some cases, the SIA functionality was originally not part of the system; it was implemented after the fact as a new feature. Facebook is notorious for such new features, e.g. News Feed, Beacon, and See Friendship [239, 250, 312]. Its latest invention in this realm is a feature where a user's product "likes" are used to automatically advertise the products to the user's friends [104, 236, 292]. Since this feature was implemented ad hoc, users could not have known about this at the time they pressed the "like" button (especially since the feature essentially uses a "like" like a "share"). In such a case, the developers of the system are advised not to activate the feature by default, but to instead invite the user to opt in[3] [61]. Otherwise, this unauthorized secondary use of the user's data (cf. [259]) is invariably seen as a strong privacy violation [100, 115, 192].

---

[3] This is a trade-off; having users opt in to a feature might be an impediment to the adoption of the feature (cf. [282]).

1.4   The Decision

Once the type of recipient (*to whom?*), information (*what?*), and use (*how?*) are known, the next question is how users come to a decision about whether to use the system (i.e., disclose their information) or not. Laufer and Wolfe [163,164] were the first to argue that users engage in a "calculus of behavior" in which they explicitly trade off the *benefits* and *risks* of disclosure—a process which researchers now call the "privacy calculus" [100,171,194,318]. Researchers have demonstrated empirically that benefits and risks are important [171,228]—if not the only [82]—antecedents of information disclosure intentions, and in turn, actual disclosure behavior [133,222].

Li [172] argues that the privacy calculus can be seen as a privacy-specific instance of decision-making theories like the utility maximization or expectancy-value theory [16,240]. The expectancy-value theory states that people gather information about various aspects of each choice option, and assign a value to each of these aspects [85]. Utility maximization, in turn, states that people will trade off the different aspects and then choose the option that maximizes their utility [27]. The aspect of *risk* is defined as users' perception of the potential loss of control over the requested information, such as when the information may be used without permission [82], while the aspect of *benefit* is defined as users' perception of the relevance of information requests in the context in which they are made [269]. While there is surprisingly no agreed-upon formal definition of the privacy calculus, the term itself has become a well-established concept in privacy research [172,223,258].

When arguing about the privacy calculus, many researchers note that users can only make an effective tradeoff if they are given comprehensive control over their information disclosure, and adequate information about the implications of their decisions. They make these claims based on survey data [40,278,319], behavior logs [47,241], as well as empirical tests [41,166]. Indeed, one may argue that some level of control is necessary to engage in a risk/benefit tradeoff, and that people can only make an informed tradeoff if they are given adequate information.

Is providing transparency and control sufficient to warrant accurate privacy decisions? Transparency and control require people to be rational decision makers who will use the provided information and controls to their best advantage. Unfortunately, though, like many decisions, privacy decisions are often not very rational [4,5], and fall prey to all sorts of decision fallacies, such as "herding effects" [6], the "default and framing effects" [119,120,158], and "context effects" [143]. Because of this, transparency and control sometimes have counter-intuitive effects on users' privacy decisions [33,139]. Moreover, while users claim to want transparency and control, they often avoid the hassle of actually exploiting it (see [23,56,290] for an overview, and [24,92,118] for experimental evidence).

In sum, it seems that users' privacy decision-making practices can range from purely heuristic to predominantly deliberate, and only users who feel motivated and capable of making deliberate decisions are more likely to do so [13,149,

176, 321, 329]. In supporting users' privacy decisions or overcoming their privacy concerns via technical means, it is therefore important to understand what the consequences of such interventions are for both deliberate and heuristic privacy decision-makers [149].

1.5   The User

In making privacy-related risk/benefit tradeoffs, users also rely on their personal preferences. One of the most widely accepted findings in privacy research (and an ongoing theme in recurrent privacy surveys [101, 304, 305]) is that there are **individual differences between users** regarding their tendency to share personal information [57, 246, 279]. Westin famously described three types of individuals: the unconcerned, privacy fundamentalists, and the pragmatic majority [305]. More recent work show that users' sharing tendencies do not just vary in extent, but also in kind: users can be clustered (cf. [208]) into distinct disclosure profiles (cf. [142]), or even into fundamentally distinct privacy management strategies (cf. [310, 313]).

These individual differences have important consequences for the privacy support that systems can provide: transparency and control may get overly complicated when users differ extensively in their main privacy management strategy [307], and any automated decision support would need to take users' personal preferences into account [136]. That said, the existence of a small but comprehensive set of privacy profiles suggests that privacy support systems can be modeled as Social Information Access systems themselves. We explore this idea in Sect. 5.

1.6   The Problems

Finally, we can create a taxonomy of the kinds of privacy problems may occur in the context of SIA systems. We limit ourselves to problems that are inherent to the functionality of SIA systems; more general problems such as security breaches (e.g. hacking) are outside the scope of this chapter.

Incorrect predictions and inferences. While users value the personalized services provided by most SIA systems, they get annoyed when the system makes an incorrect prediction or inference about their goals and preferences [259]. In effect, researchers suggest that users should have the opportunity to scrutinize [131] and correct [83] potential mistakes in the system's predictions.

Unwanted or creepy correct predictions. Possibly worse than incorrect predictions are unwanted or creepy correct predictions [251, 282]. Accordingly, users seem to actively engage in some kind of "reputation management" when using personalized systems. For example, in interviews regarding the data collected by a mobile app recommender, Knijnenburg and Kobsa [139] found that users would occasionally decide not to disclose a certain piece of information because "it doesn't accurately represent me as a person". The problem of incorrect and unwanted correct predictions is most prominent when the system uses context data, because this data is a step removed from users' actual demographics and preferences [139].

Unintended disclosures. In systems where users' activities are open to the public or known contacts, unintended disclosures can be a privacy problem. In some cases, the users themselves make disclosures they subsequently regret [255,302]. In worse cases, the system may unintentionally reveal things about the user through its suggestions or recommendations [71]. The latter is more likely to happen when the SIA system uses private data to display something publicly. For example, a social navigation feature on a blogging website might publicly display related blogs based on the blogger's browsing of other blogs. This might result in a list of "related blogs" that publicly disclose the blogger's private interests.

Unwanted interactions. In systems that connect people (e.g. "people recommendation", Chap. 15 of this book [95]), it is important to avoid unwanted interactions. For example, it is awkward if a system recommends someone to become "friends" with their ex-spouse's new partner. In a more subtle variety, Page et al. [213] found that some users may interpret the access provided by a SIA system as an invitation to interact, which may result in awkward unwanted interactions.

Re-identification. Some systems allow users to use a pseudonym as their username. This allows users create an online identity that is separate from their real-world identity, which may lead to more extensive and frank interactions with the system [151]. Note, though, that it is sometimes possible to "de-anonymize" (i.e., re-identify) a user. This can happen when an adversary knows a (possibly imprecise) subset of the user's data, (e.g. items, ratings, or times of interaction) [198] that can be correlated with the pseudonymous data.

1.7   Outline of the Chapter

The remainder of this chapter will discuss the privacy implications of the different types of SIA systems discussed in this book. We group these systems by recipient type as "aggregators" (disclosure to the system only) in Sect. 2, "public content systems" (disclosure to the system and the public) in Sect. 3, and "network-based systems" (disclosure to the system and known contacts) in Sect. 4. For each recipient type I will discuss:

- the **motivations** for disclosure or non-disclosure;
- **examples** of privacy concerns or violations originating from primary, secondary, or unintended data use;
- factors that cause **individual differences between users** in terms of their privacy concerns and their privacy-related behavior;
- **technical solutions** to potential privacy threats;
- **decision-support solutions** that help users make better privacy decisions.

While the technical solutions differ per system type, the decision-support solutions can almost invariably be applied to all types of systems. That said, most decision-support solutions have been most prominently researched in the context of a single type of system, so I will discuss them under the header of that system type. Similarly, some examples fit under multiple system types, because commercial systems sometimes transcend the boundaries of our categorization (e.g. Twitter can be classified as either a public content system or a network-based system; Facebook, a network-based system, is increasingly turning into an aggregator). Cross-references between sections are therefore inevitable; for the convenience of the reader, Table 1 provides an overview.

Table 1. Overview of different system types

Type of system	SIA functionality	Potential recipients	Technical solutions
Aggregators (Sect. 2)	Social navigation (Chap. 5)	Only the system itself	Pseudonyms
	Social search (Chaps. 7–9)		Differential privacy
	Recommendation (Chaps. 10–15)		Client-side computation
Public content systems (Sect. 3)	Micro-blogging (e.g. Twitter)	The system + public	Right to be forgotten
	Social Q& A (Chap. 3)		
	Social tagging (Chap. 6)		
Network-based systems (Sect. 4)	Social networking (e.g. Facebook)	The system + known contacts (+ public)	Contact categorization
	Collaboration (Chap. 4)		Plausible deniability
			Deniable plausibility

After discussing the privacy aspects of SIA for the three main recipient types, Sect. 5 will discuss the opposite angle: using SIA to improve privacy decision-making.

2   Disclosure to a System: Aggregators

"Aggregators" are SIA systems that collect personal information, but only reveal it to other users in aggregated or derivative form. Since the system shields the user data from other users and the public, the system itself is the only recipient that the user will need to trust. Aggregators include social navigation (Chap. 5 of this book [81]), social search engines (Chaps. 7–9 of this book [38,97,199]) and recommender systems (Chaps. 10–15 of this book [30,95,116,134,168,209]).

2.1 Motivations

In line with the privacy calculus, research has shown that users are willing to disclose some of their personal information if the system provides sufficient benefits [100,207,228], such as content relevance, time savings, enjoyment and novelty [105,111,147]. Many systems ask users to disclose information *before* they get to enjoy any benefits, though. In these situations, users may assess the *anticipated* (rather than observed) benefits [139,149].

Factors that influence the "risk"-side of the privacy calculus take the form of perceived privacy threat (i.e., system-specific privacy concerns), perceived privacy protection, and trust in the developer of the SIA system [139,147,149, 274]. Perceived privacy threat and trust in turn depend on the understandability of and control over the personalization process [136]. Perceived privacy threat can also be an *outcome* of disclosure, with users perceiving more threat as they disclose more information, especially when the disclosure happened outside the user's conscious awareness [136].

The inherent tradeoff of risks and benefits that governs users' disclosure to personalized SIA systems has been dubbed the "privacy-personalization paradox" [16,84]. Importantly, despite the benefits of social information access, users may not agree with the data-collection required to make the system work if the perceived risks are too high. Hence, in deciding whether to collect and use certain private information, the developers of the system are advised to make sure that both benefits and privacy meet a certain threshold [287], or that they are at least in balance [49,317,318]. This is contrary to the Big Data "collect everything mentality" [268], which permeates the current online landscape. The "paradox" suggests that this mentality is not sustainable, and indeed, the rise of SIA information systems coincides with a growing concern over unauthorized secondary use [103].

2.2 Examples

2.2.1 Filter Bubbles

One important privacy problem with aggregators is unwanted stereotyping. The most famous example of this problem was published in the 2002 Wall Street Journal article "If TiVo Thinks You Are Gay, Here's How to Set It Straight" [324]. The article describes how TiVo—a digital video recorder with a built-in recommender system—sometimes persistently records TV shows with gay themes after the user watches particular items; its algorithm apparently "overfitting" a previously encountered social information pattern. Such incorrectly stereotyped recommendations can lead to embarrassing situations when other people (e.g.

family members or visitors) get to observe these recommendations. Note, though, that even correct stereotypes—the primary function of a personalized system—can be harmful when applied without discretion. Researchers have argued that heavily filtered content, such as the output of Google's pervasive social search functionality, may isolate us from a diversity of viewpoints, content, and experiences, and thus make us less likely to discover and learn new things (a phenomenon known as the "Filter Bubble" [217]). The Filter Bubble can be thought of as a privacy threat because it intrudes upon our ability to experience the world from an unbiased perspective [264]. At its worst, stereotyped recommendations can create a "positive feedback loop" [161], where users unknowingly try to fit the stereotype. This leads to a very worrying concern that recommender algorithms may gradually replace human creativity and understanding; a scenario reminiscent of the seminal privacy novel *1984* [144,210].

2.2.2   Unwanted Predictions

More prominent privacy violations can occur when users are unknowingly subject to personalization. Customer loyalty programs are a good example of this. Companies usually track the shopping patterns of loyalty card carrying customers and use this data to analyze and optimize marketing efforts, inventory management and in-store product placement (cf. [25,51]). In addition, they send these customers direct mail advertisements. In recent years, though, companies have started to send customers *personalized* advertisements that are based on the collected shopping patterns. For example, Target, the second-largest discount retailer in the United States, has hired data scientists to analyze the online and in-store shopping patterns of its customers to make highly detailed predictions about their shopping needs, mainly for advertisement purposes. These data scientists have, for example, uncovered fifteen products that allow them to assign each female loyalty card carrying shopper a "pregnancy prediction score" [71]. Shoppers who pass a certain threshold receive advertisements for baby products in the mail. While this targeted advertisement practice is described in their loyalty program privacy policy, most Target customers are unaware of this personalization. It is therefore no surprise that the father of a 14-year-old girl found out via Target that his daughter was pregnant: Target had predicted the pregnancy and sent the girl an ad leaflet with baby products; her dad caught this leaflet in the mail, and thereby found out that the girl was pregnant before she had found the courage to come clean about it [71]. Such unintentional privacy violations occasionally occur when users are unaware of the secondary use of their information for personalization purposes. More pervasive privacy violations exist in the use of personalization to create price discrimination [190,191].

2.2.3   De-anonymization

Due to the current "collect everything mentality" [268] in Big Data, there are numerous situations where a new personalization feature is introduced *after* the system has collected a veritable amount of user data. This is where the most serious privacy violations occur, because users are not prepared for their data being used for personalization—and often, neither is the system or the company providing it. An example is the ill-fated second installment of the Netflix Prize, in which Netflix released anonymized user data as part of a \$1 million contest to improve its recommendation algorithm. Within two weeks of releasing the data, researchers were able to "de-anonymize" the data by cross-referencing ratings with (public) IMDb profiles [198]. In response, a closeted lesbian mother sued Netflix, alleging that the de-anonymization procedure could "out" her based on her viewing behavior. Netflix paid a settlement and prematurely ended the contest [254]. It is clear that both the plaintiff and Netflix would likely have acted differently had they anticipated the novel applications for which their data could be used.

2.3   Individual Differences Between Users

Regarding "aggregator"-style SIA systems, users seem to have separate disclosure tendencies for demographic and contextual information [139]. Context data is directly relevant for social navigation, social search engines, and recommender systems, as it represents the behavior to which aggregation mechanisms are applied. Demographic information, on the other hand, is used by some SIA systems to supplement behavioral data, e.g. in cold start situations.

Users differ in their disclosure of context and demographic data. While demographic data disclosure is based primarily on users' system-specific privacy concerns, context data disclosure depends more on users' (anticipated) satisfaction with the SIA system [149]. Arguably, this is because contextual data can be more easily misinterpreted by a system, hence users only trust the most capable systems with their contextual data. Moreover, Knijnenburg et al. [142] cluster users on their context and demographic data disclosure tendencies, and find—aside from groups with low, medium, and high disclosure tendencies for both types of data—a group of users with a high tendency to disclose demographic data, but a low tendency to disclose contextual data. SIA system developers are advised to take extra care of this subset of users who are particularly concerned about context data disclosure.

Interestingly, in research where context data was collected for a mobile app recommender [142], users who are less likely to use their phone to browse the Internet or check their e-mail are *more* concerned about context data disclosure, despite the fact that they generate less context data overall. This arguably means that a lack of familiarity with the platform that collects their data causes users to be more sceptical towards context data disclosure. Given this lack of familiarity, technical solutions such as client-side personalization (see Sect. 2.4) may not work very well, so decision-support solutions (e.g. justifications, request ordering, or control-inducing design interventions; see Sect. 2.5) may work better.

2.4   Technical Solutions

A possible mitigation of privacy concern with SIA systems is to allow users to remain anonymous. Fully anonymous interaction with aggregating SIA systems is difficult though, since the aggregation functionality crucially depends on the systems' ability to recognize the user across interactions [244]. More realistically, users can be allowed to interact with the system under a pseudonym [15,151]. Note, though, that the high dimensionality and sparsity of the data typically collected by SIA systems can be exploited to re-identify users [198]. Moreover, existing tools for anonymity have to deal with their own unique set of usability issues [204].

De-anonymization can be reduced by not giving others access to any of the user data. Note, though, that even without such access, it may be possible for a third party to make inferences based on the output of the system. For example, Calandrino et al. [42] show that given some background knowledge on the behavior of a target user, adversaries can create fake accounts that are similar to the target user, which an aggregator SIA system may identify as neighbors of that user. This allows the adversary to isolate the target user's data from the recommendations provided to the fake accounts. A means to overcome this problem is differential privacy, a privacy model that inserts carefully calibrated noise into the user profile computation. The noise masks the influence that any difference in a particular record could have on the outcome of the computation [178,184,235,330].

A final technical solution that has recently become popular abandons the assumption that personal data collected on users' local devices must be sent to a remote site for the aggregation function to take place. In this "client-side" solution, all necessary calculations take place on the user's own device [46,126,197]. For example, client-side recommender systems usually employ content-based recommendation strategies, but distributed and hybrid versions of collaborative filtering algorithms do exist [44,162,252,293]. From a conceptual and technical point of view, preventing anyone from accessing personal data enhances the privacy of the user [263]. However, the inference methods that can be used are limited (e.g. if-then rules, simple classification), since the users' personal data never leaves the client. The used rules and classification profiles could stem from prior market or user research, or be based on data of users who did not opt to keep their data client-side only. Research in recommender systems shows that users indeed prefer client-side methods as a means to alleviate privacy concerns [149,274]. Client-side solutions could have a similar effect for social navigation and search, but client-side solutions in these types of systems are an under-explored area of research.

2.5 Decision-Support Solutions

Since users' privacy decisions are not always very rational (cf. [4,5]), considerable research effort has gone into helping users decide whether to give SIA systems access to their information. One of such decision-support solutions is to give users *explanations* regarding how their data will be used by the system. This is particularly important for social navigation, social search engines, and recommender systems, which often have a rather opaque operating mechanism [88,90,284]. For example, research in the area of recommender systems has found that explanations increase users' understanding of the recommendation process [90,295], which may in turn increase their trust in the system [299]. Note, though, that explanations may inadvertently focus the user on potential privacy issues, and thus lead to overall lower levels of disclosure, trust and satisfaction [139]. Other "justification" mechanisms (such as appealing to social norms [6,26,221], see Fig. 2, or the benefits of disclosure [152]) have shown to backfire as well [139]. As different users are susceptible to different justification methods, a better approach seems to be to *tailor* the justification method to the user [138] (see Sect. 5).

Fig. 2. Justifications like this lead to overall lower levels of disclosure, trust and satisfaction [139].

In situations where a system specifically asks the user to provide access to certain pieces of information (or types of information, such as when a mobile app asks for access to different types of phone data), then the *order* of such requests can significantly influence users' disclosure behavior. For example, Acquisti et al. [6] show that users disclose more information when such requests are made in a decreasing (as opposed to increasing) order of intrusiveness. Users in that situation may perceive a higher level of privacy threat, though [136]. Another option is to request the most useful items first (given that the system can determine which items or data types are most useful to its operation) [182,195,231]. A trade-off between usefulness and sensitivity is also possible. Finally, it is possible to adapt this trade-off to the user's disclosure tendency; this is something I explore in Sect. 5.

Finally, certain design interventions can cause users to think more carefully about their information disclosure. For example, users of "Web form auto-completion"—a feature available in modern browsers that automatically fills out forms—are usually not very careful in their disclosure decision-making [229]. Auto-completion makes it too easy to submit forms without carefully weighing the risks and benefits of disclosing each piece of potentially private information that the Web form requests. An alternative design that adds buttons to the end of each field that allow the user to remove the information from that specific field has been found to make users more considerate of these specific risks and benefits, which makes their information disclosure more purpose-specific [141] (see Fig. 3).

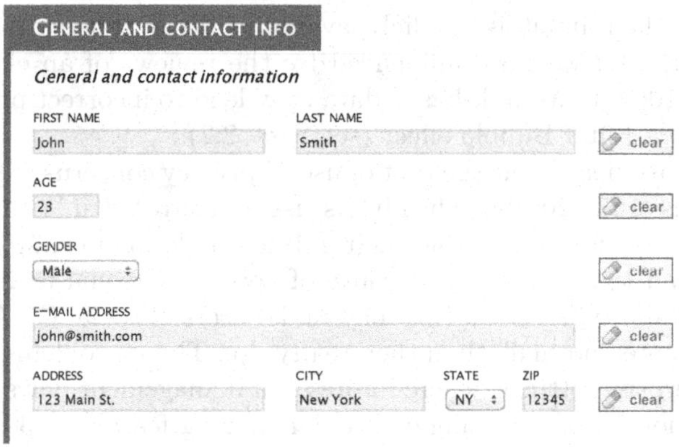

Fig. 3. A Web form auto-completion tool with buttons that allow the user to remove the information from each specific field.

3   Disclosure to Public: Public Content Systems

"Public content systems" are SIA systems that are public by default—in fact, the public availability of their content (e.g. answers to questions, "tweets", "pins", or social bookmarks) is often crucial to their operation. In disclosing to these systems, users have to consider two types of recipients: the system, and other people. Public content systems include social Q&A systems (Chap. 3 of this book [206]), and certain social tagging systems (Chap. 6 of this book [68]). Microblogs (e.g. Twitter) are also public content systems, but the addition of a following/follower mechanism also makes them similar to social network-based systems (see Sect. 4).

3.1   Motivations

Aside from the motivations discussed under Aggregators (Sect. 2.1), an important motivation for disclosure in public content systems is self-presentation [165, 245]. For example, people who answer questions in social Q&A systems want to show off their expertise, and systems can capitalize on this desire using a reputation system that rewards them for giving high-quality answers [175, 234]. In social tagging systems, self-presentation presents itself in users' choice of tagged items: for example, on Pinterest many users tend to tag items for *aspirational* rather than *practical* purposes [327]. This means that SIA system developers need to be careful when they want to make real-world recommendations (e.g., targeted advertisements) based on users' tags: just because someone pins Ferrari cars does not mean they can afford one. A better recommendation would be a cheaper car that makes the owner *feel like* they are driving a Ferrari.

Another problem with self-presentation is that users' actions and opinions may be biased by the crowd, the system, or certain highly reputable individuals. A "herding" or "groupthink" effect (where users follow the popular opinion) is not uncommon in recommender systems [53, 59, 108], and it may be even stronger when their input is publicly available [314]. Indeed, a recent study shows that Yelp reviewers are influenced by the reviews of an elite and active friend in Yelp [10]. Groupthink-based data may lead to incorrect predictions (see Sect. 1.6), or to a Filter Bubble effect (see Sect. 2.2).

Self-presentation is also at the root of users' privacy concerns regarding public content systems [155, 320]. Specifically, as users try to "fit in", they may selectively hide certain information from their public profiles. For example, Huberman et al. [110] found that people's disclosure of certain personal traits depends on the trait's desirability. As such, this may make users of social Q&A and tagging systems seem more "normal" than they really are. These problems may be exacerbated in enterprise settings, where impression management plays an important role in users' motivation for using e.g. social tagging features [230].

Finally, it is interesting to note that people sometimes share more personal information in a public content system than they are willing to share with their friends or family [102, 181, 213]. This seemingly paradoxical behavior can be explained by the concept of "imagined audiences" [173, 181]: people do not imagine their friends or family as a potential audience for their disclosure to the public content system. Indeed, it makes a big difference whether a piece of information is passively available on a user's "Wall", versus actively pushed towards a friend's "News Feed" (see the second example below).

3.2 Examples

3.2.1 Disclosure Regret

An important privacy problem with public content systems is "disclosure regret" (cf. [220, 255, 302]). This happens when a user makes a public disclosure that, on second thought, can be misinterpreted or taken out of context. SIA functionality—especially on microblogs, which often restrict users' ability to express themselves to a limited number of characters—can sometimes strip information from its context, and can thus act as a catalyst for disclosure regret. A painful example of this phenomenon happened on December 20, 2013, when the top PR person for a New York media conglomerate tweeted "Going to Africa. Hope I don't get AIDS. Just kidding. I'm white!" [277]. Meant as an acerbic joke to her 170 followers, the tweet was retweeted thousands of times, and by the time her flight landed, her boss had already responded publicly to calls for her resignation. The problem is that the "retweet" functionality is designed so that the audience sees the original tweet outside of its original its context: in this situation, they viewed the tweet as an ignorant statement from a high-ranked PR person, rather than part of a string of distasteful private jokes [238]. While the retweet functionality is a powerful SIA feature that provides a social way to filter and discover information, the inherent lack of contextual information that often accompanies such functionality can be the cause of many misunderstandings and other regretful situations in online communication [237,297].

3.2.2   Altered Contexts

The above example describes a situation where a tweet that was meant as a private joke ended up having a public audience. The opposite can also be a privacy problem: a publicly available piece of information can easily be misinterpreted when it is actively pushed towards a friend. The Facebook "News Feed", for example, is a SIA functionality that uses the social graph to selectively present content generated by one's friends. From the friend's perspective, though, this SIA functionality *highlights* (rather than filters) the information, and presents it *out of the context* of the their Facebook Wall [237,297]. When users are not yet familiar this functionality, a generic comment can be misinterpreted as a personal message. It is no wonder, then that the News Feed functionality was initially despised by many Facebook users. Over time, though, Facebook users have learned to adapt their disclosure to the presence of this functionality [239]. SIA system developers should be careful about the transition period when a new functionality is introduced that changes the presentation context of disclosed information.

3.2.3   Context-Aware Spam

A final privacy problem with public data is that it can be used freely by anyone, with no restrictions on the application. This is a very problematic situation with regard to identity theft [34]. Even if the typical information used to steal someone's identity—social security number, answers to security questions, banking information—are often not disclosed online, the information that *is* disclosed can be used in targeted phishing attacks (i.e. spear phishing [34,106]) by generating "context-aware spam" that is tailored to certain attributes of the user (e.g. their school, hometown, birthday) [35]. It is difficult for users to remember that this information is publicly available online, and so any communication that mentions the information seems personable and trustworthy. It is thus the responsibility of public access SIA systems to make users aware of the information that they make available to the world. National campaigns are already in place to create such awareness (cf. [93]).

3.3   Individual Differences Between Users

Section 2.4 discussed the technical solution of allowing users to remain anonymous, or to use a pseudonym in their interaction with the system. Interestingly, while pseudonyms and anonymity may reduce privacy concerns, social networks and public content systems increasingly require users to use their real name [306] (presumably to combat the increasing number of fake accounts), and even some governments require their citizens to verify their real name before signing up on certain popular websites (presumably to counter rumors and defamation of politicians during the election cycle) [54].

There is ample evidence that anonymous and pseudonymous users behave very differently from identifiable users [247], and these differences can have a profound effect on systems with a public audience. Specifically, the absence of a name allows users to produce content more freely, which increases creativity, but also induces a certain dissociation between the members of an online community [272]. On purely anonymous sites, this results in reduced political correctness and inhibition [55], while on pseudonymous sites such as Reddit, it increases the opportunity for intimacy and the sharing of secrets [294]. Indeed, Chen et al. [52] show that pseudonymity can increase users' ability to exercise their privacy rights, specifically in terms of "creativity" and "contemplation" (cf. Pedersen's [224] Privacy Function Rating Scale). In contrast, Cho et al. [54] show that real name requirements can reduce profanity and anti-normative expressions, especially among more-frequently participating users. Depending on the type of information to be collected, this difference is important for SIA system developers to understand: a SIA system based on users' creative expressions (e.g. social Q&A systems, social tagging systems) would benefit from a user base that interacts via pseudonyms, while a SIA system with materials for children may prefer to employ a real-name policy to keep profanity at bay.

3.4   Technical Solutions

A potential mitigation of the privacy problems that occur with public SIA systems is embedded in the European Court of Justice ruling on the "right to be forgotten" [79]. In this ruling, the court ruled that European citizens have the right to ask search engines to remove inaccurate, irrelevant, or excessive information about them from search results. The ruling has provisions to balance against the right to free speech and public interest, and does not require the information to be deleted altogether, just to be removed from search results. This makes the information harder—but not impossible—to find [200]. Critics were initially skeptical that the service would mainly be used by politicians and criminals to cover up serious crimes or scandals, but recent figures show that 95% of all requests come from private citizens out to protect their personal and private information [285].

While the Right to be Forgotten is more a legal solution than a technical solution, it relies on a technical limitation (i.e. removal from search results) to mitigate a privacy threat (i.e. the ease with which false or irrelevant private information can be found online, if available). Users can exercise their right to be forgotten by filing a request with a search engine. SIA system developers should be aware that if they operate a service that can be classified as a search engine, they are legally required to respond to such requests. Google responded to nearly 220,000 of such requests in the first year the ruling was in effect, and has granted 46% and rejected 38% of them [285]. Note that it is also not unthinkable that the European Union may expand the right to be forgotten to online aggregators and social networking systems, especially as these systems increasingly impact their users' social, professional, and financial lives.

3.5 Decision-Support Solutions

Another way to solve privacy problems that occur with public SIA systems is to try to prevent unwanted disclosures from happening at all. This involves a design intervention that "nudges" users away from disclosing too much information [2]. While such nudges have mainly been studied in the context of social networks (see Sect. 4.5), they are definitely applicable to public content systems such as social Q&A and tagging systems as well.

For example, default disclosure settings have been shown to have a very strong effect on users' disclosure decisions [120,158]. While many public content systems allow users to set their disclosures to private (e.g. Twitter, Instagram, Pinterest), most of them are public by default, and users have to actively change their settings if they prefer otherwise. This default setting is a typical "nudge": users are free to change the setting, but most tend to stay with the default setting [283] (see Sect. 4.5 for more details about defaults as nudges). However, while most users are unlikely to change the default, they may compensate for this in their subsequent disclosure behavior (i.e. by posting less personal information) [312]. SIA system developers thus have to make a tradeoff between having a higher *volume* of content available to the public (using a public-by-default setting) and having more *personal* content available to the user's private connections. One solution is to allow users to disclose some information in private while keeping their main posts public (e.g. Pinterests' "secret" pinboards).

Context-based hurdles are another nudge-based method that are particularly suitable for public content systems. Specifically, a system can create an additional barrier to disclosure when it detects a potentially regretful situation. GMail, for example provides a feature called "Mail Goggles" (see Fig. 4), which asks you to solve a few simple math problems before you can send an email late night on the weekend, when a user might be drunk [225]. Such barriers can give users just enough time to experience the disclosure regret *before* posting the information publicly.

4 Disclosure to Known Contacts: Social Network-Based Systems

"Social network-based systems" (SNSs) are SIA systems that depend on close interaction or collaboration between its members. In disclosing to these sys-

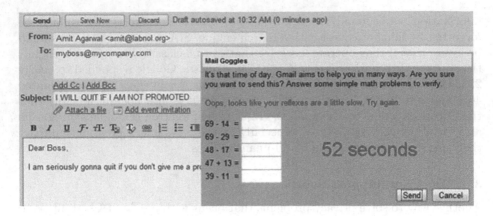

Fig. 4. GMail's "Mail Goggles" prevent drunk users from sending ill-advised emails.

tems, users have to consider three types of recipients: the system, their "friends" or "contacts", and (potentially) other users on the network. As a sizable percentage of Internet users use SNSs (in 2015, 72% of online adults actively used Facebook [70]), and as these systems make it increasingly easy to re-share content, the boundaries between SNSs and "public content systems" are steadily dissolving. As such, much of what was discussed in Sect. 3 applies to SNSs as well. Social network-based systems include collaborative systems (Chap. 4 of this book [323]) and location-sharing systems (Chap. 16 in this book [31]).

4.1   Motivations

Researchers have continuously been surprised by the fact that SNS users tend to share tremendous amounts of personal information, despite potential drawbacks [3]. Indeed, research shows that the social benefits of self-presentation far outweigh users' privacy concerns as an antecedent to the use of social networks [308]. Other benefits of using SNSs include increased social capital [39, 75, 76], social connectedness [153], self-esteem [76, 160], and personal well-being [75].

That said, users' concern with what they share online has grown over the past few years, as users increasingly worry about who can see what [28, 167, 219, 271]. Note, though, that SNS privacy concerns seem to be rather different from "traditional" privacy concerns: SNS users worry less about the confidentiality of their information, but more about information overload [74, 114, 260], social conventions (i.e., "netiquette" [213, 264, 289]), and leaving a wrong impression [150, 155, 276, 308, 320]. Strategies to deal with SNS privacy thus extend beyond limiting disclosure [77], and include things like selective sharing through customized friend lists, blocking people, blocking apps or event invitations, restricting chat availability, limiting access to or visibility of one's Timeline/Wall, untagging or asking a friend to take down an unwanted photo or post, and altering one's News Feed [310, 313].

It may seem that users' privacy concerns are at odds with the potential benefits of sharing on social networks: with each increase in privacy, users seem to give up benefits [164]. However, Wisniewski et al. [309] demonstrate that giving users the level of privacy they want increases their social connectedness, which in turn helps to build social capital. In other words, more openness does not lead to benefits if users' privacy desires are not met.

What influences users' SNS privacy preferences? Most social networking ties are with existing, offline relationships [32,159], and as a guiding principle for managing their SNS privacy, users usually attempt to mirror offline behavior. Indeed, Page et al. [213] found that many privacy concerns of users of location-sharing systems were rooted in the worry that the system will change their existing relationships with their friends on the network. Page et al. coined the term "boundary preservation" to describe this desire to preserve existing offline relationship boundaries, and argue that designers of SNSs should make sure that this desire is easy to accomplish. This objective could interfere with the goals of a SIA system; for example, a social visualization system could highlight certain aspects of social relationships (e.g. the amount or lack of interaction between users, an overlap or contrast in personal interests) that would not be apparent in the offline world, and therefore pertrude on an existing relationship boundary (e.g. by increasing the opportunity for "cyber-stalking"). An example from the area of location-sharing is the worry that the act of sharing a location update may be wrongly interpreted as an invitation to join the user at an inopportune time [213]. When designing SIA functionality that relates to or encourages social interaction, designers should make sure to respect existing social conventions, e.g. by explicitly clarifying the intent of an interaction [215]. This is especially true for collaborative systems, which rely on successful and meaningful interactions between users.

Special consideration should be given to *enterprise* social network-based systems. People are more likely to share their personal information on such systems, because colleagues are generally trusted [67]. However, corporations do monitor employees' social communications, and this may reduce sharing [72]. Investigating an enterprise social travel application, Aizenbud-Reshef et al. [9] found that people were less likely to share their travel information than they claimed they were comfortable with (a reversal of the privacy paradox), especially for upcoming travel. Smith et al. [258] note that privacy research at the organizational and supra-organizational (e.g. group) level is particularly sparse. They suggest that research on privacy within organizations and small groups could particularly focus on the development and negotiation of privacy-related norms.

4.2 Examples

4.2.1 Imagined Audiences and Context Collapse

As mentioned in Sect. 3.2, SIA system users may have trouble understanding the audience of their posts/disclosures [63,123,296]. The most straightforward examples of this problem are the numerous cases of an employee disparaging their job on an SNS, not realizing that their employer is part of the audience [19,

196,257]. Another example is the girl who unsuspectedly created a public event on Facebook for a small party, which somehow went viral, leading to a clash between police and 5,000 gatecrashers who decided to show up at the night of the party [243]. More subtle examples revolve around disagreements about what is deemed appropriate or relevant. In this sense, users often write their posts for an "imagined" audience [173], but researchers find that users' actual audience often reaches beyond this imagined audience [123]. Some users conceptualize their audience as an ideal person who shares the same interests and perspectives, and these users may be surprised to find that some of their audience members take offense with their posts. Marwick and Boyd [181] highlight an example of a person who got threatened with a lawsuit and loss of work because of one of their tweets. Some users try to prevent this problem by conceptualizing their audience as its most sensitive members, and adjust their postings accordingly [181,311]. In real life users can disclose selectively without heavy censorship, because audiences tend to differ depending on the social context. On social networks, this strategy is prevented by a phenomenon called "context collapse": on SNS users face a single audience that potentially receives *all* their disclosures. In effect, users end up disclosing to the lowest-common denominator.

The problem of imagined audiences is exacerbated by a SIA functionality that is found in many SNSs: the targeting of friends' content based on contextual cues. While this functionality is often seen as necessary to select relevant information from a seemingly endless stream of social content, Johnson et al. [123] argue that this functionality can inadvertently select an inappropriate audience based on the context of a post. For example, a cryptic post about a surprise birthday party may be featured in the news feed of the birthday person (even if they are never explicitly mentioned), due to the many mutual friends that are commenting on the post. Similarly, the news feed algorithm may decide to feature a personal post in which the user criticizes a certain politician in the news feed of family members who are supporters of that politician. Such unintended disclosures are impossible to prevent by restricting disclosures to one's social circle, because the social threat in this situation actually comes from a dynamically generated audience *inside* one's social circle. Content-sharing algorithms that avoid such potential conflict are an under-explored area of research.

4.2.2   Unwanted Friend Recommendations

Another SIA functionality that is often found in SNSs is "friend recommendation" (see Chap. 15 of this book [95]), in which an algorithm suggests whom to connect with on the social network. Unfortunately, these algorithms do not take into account extraneous social conditions that may make such a recommendation highly inappropriate. For example, when Facebook first implemented this feature in 2009, many users complained that Facebook recommended them to reconnect with ex-lovers and dead friends [45]. In one case Facebook has even recommended a rape victim to become friends with his rapist [130]. Facebook has introduced a feature that allows the accounts of the deceased to be memorialized, which prevents them from showing up in friend recommendations [48]. As for the other situations, more research into conflict-avoiding friend recommendation algorithms is needed.

4.2.3 Social Ads

A final SIA functionality that is causing privacy problems on SNSs is "social ads". In early 2011 both Facebook and LinkedIn rolled out a new feature: advertisements would show users which of their friends "liked" the advertised brand [104, 236, 292]. Importantly, both SNSs decided to automatically enable this feature for all users without much of an announcement, allowing them to opt out via a privacy setting buried deep inside the settings interface. LinkedIn has since removed this feature [236], but Facebook is to date still showing such ads to its users. In fact, it plans to even allow third party applications or ad networks to use your name and picture in their ads [189]. While social ads may be an interesting opportunity for SNSs to generate advertisement revenue, implementing this feature without obtaining full user consent can lead to very awkward social situations, especially given that Facebook "likes" have a habit of appearing on users' profiles without their own involvement [185].

4.3 Individual Differences Between Users

Whereas limiting one's disclosure is the primary method to regulate one's privacy in most "commercial" settings, SNS users have a plethora of strategies available [77, 311]. For example, Wisniewski et al. [310, 313] identify ten distinct privacy behaviors on Facebook: withholding basic or contact information, selective sharing through customized friend lists, blocking people, blocking apps or event invitations, restricting chat availability, limiting access to or visibility of one's Timeline/Wall, untagging or asking a friend to take down an unwanted photo or post, and altering one's News Feed. Moreover, they were able to classify participants into six categories (see Fig. 5) with distinct privacy management strategies:

- **Privacy Maximizers** use almost all of the available privacy features on the social network.
- **Self-Censors** use very few of the available privacy features, but primarily protect their privacy via the traditional method of withholding information.
- **Selective Sharers** share much more information, but they protect their privacy by sharing this content selectively, using custom friend lists.
- **Privacy Balancers** exhibit moderate levels of privacy management behaviors. Follow-up work shows that this class of SNS users contains both "informed balancers" (who carefully select the privacy mechanisms that suit their personal preferences) and "uninformed balancers" (who simply make do with the few mechanisms they are aware of) [313].
- **Time Savers/Consumers** use Facebook primarily for passively consuming other people's posts, and take precautions to limit or avoid direct interaction with other users (e.g. through chat).

– **Privacy Minimalist** use only a few common privacy features, but are generally very open in their disclosure.

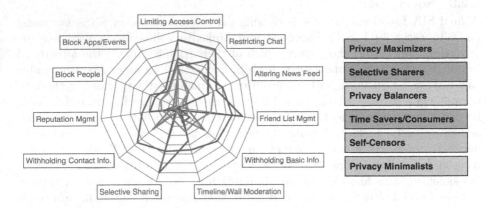

Fig. 5. The six privacy management strategies uncovered by Wisniewski et al. [310, 313]. See usabart.nl/chart for an interactive version.

Developers of social network-like SIA systems should realize that users expect such systems to have a wide variety of ways to manage one's privacy, and that different users will use these privacy mechanisms in different ways. This may also have consequences regarding the features offered to each user—while "Selective Sharers" may appreciate a SIA feature that allows them to better connect with subsets of their friends, such a feature is not useful for "Time Savers/Consumers". The latter group may be more appreciative of a time-saving feature that personalizes their News Feed to highlight the most relevant stories.

Beyond Facebook, Page et al. [211] suggests that the use of social networks— and indeed any social media—depends strongly on users' preferred communication style. Page et al. argue that users of services that broadcast implicit social signals (e.g. location-sharing SNS) are predominantly used by users who are predisposed to "FYI (For Your Information) communication". FYI communicators prefer to keep in touch with others through posting and reading status updates, i.e., without actually having to interact with them. They tend to benefit from the implicit social interaction mechanisms provided by location-sharing systems and social visualization systems. People who are not FYI communicators, on the other hand, would rather call others, or otherwise interact with them in a more direct manner, rather than passively reading about them on social media. They thus tend to benefit more from systems that promote more direct interaction, such as explicit collaboration features provided by some collaborative systems (see Chap. 4 of this book [323]). Page et al. find that older users are generally less likely to be FYI communicators, but that the tendency to FYI communication is somewhat higher for parents than for individuals without children.

4.4   Technical Solutions

With the increasing pervasiveness of SNSs, regulating one's privacy on these social networks is an inherently complex problem. For example, most Facebook users have over 200 contacts [256] and average seven new contacts a month [96]. To allow users to manage this vast number of connections, SNSs have to resort to "labyrinthian" privacy controls [58] that are incomprehensible [174, 270] and lead to disclosures that inconsistent with users' intentions [180].

To avoid the problems of "context collapse", SNS users seem to prefer to share their personal information with their social connections selectively [156, 167, 219]. Indeed, SNS users tend to restrict access to their profiles by sharing certain information with certain people only [129, 179, 322]. Facebook ("Friend Lists") and Google+ ("Circles") both allow users to share specific posts and profile items with specific categories of recipients [129, 303][4]. Note, though, that while many users seem to engage in categorizing their contacts (a privacy-related form of people tagging [233]), this feature is rarely used to make disclosure more selective [64, 270, 303, 310], because when people are prompted to categorize their friends into semantically meaningful categories, they often create categories that are inadequate for making privacy decisions [132]. While some researchers suggest that more granular categories are needed to foster information sharing in SNSs [22, 33, 241, 275], recent work has shown that five standard categories (Family members, Friends, Classmates, Colleagues, Acquaintances) are ultimately most convenient. Unexpectedly, more granular categorizations lead to *higher* rather than lower levels of over-disclosure threat [140].

Studies in computer-mediated interactions show that users sometimes *lie* as a privacy preservation tactic [98]. A user may for example tell a friend that she has fallen ill, rather than telling the friend that she does not want to go out with her that evening [212]. Researchers recommend that SNSs allow users to make white lies; a functionality that has been dubbed "plausible deniability" [14, 29, 166]. Page et al. [212] note, though, that users who lie may actually experience increased privacy concerns, for fear of being caught. Moreover, they acknowledge that the difficulty of maintaining a lie differs per medium (hence rates of lying vary significantly across media [99]). For example, in location-sharing systems, Page et al. [212] demonstrate that lying indeed *increases* users' boundary preservation concerns. The problem of lying on SNSs is thus a complex issue that involves balancing the opportunity for users to lie with the moral responsibility of creating honest social experiences. SIA system developers need to be acutely aware of this issue, since the SIA functionality may expose—or further exacerbate—users' lies.

Note also that there is a flipside to "plausible deniability", in that users dislike inaccuracies in their personal information, as they fear that such inaccuracies might be embarrassing. Take the following example: From an information-theoretic perspective, randomly perturbing a few of a user's data points gives

---

[4] Users also seem to prefer to share their personal information selectively with aggregators and public content systems [208], but mechanisms to automate selective sharing in these domains have not gained much traction [20].

users the ability to deny any data point, thereby effectively implementing plausible deniability. This method can also used to implement differential privacy in social networks (cf. [43,65,328] and Sect. 2.4). Users, however, may be uncomfortable with this approach, because it may be difficult for their audience to adjust their opinion after they have been told that a certain embarrassing data point is not real. Indeed, numerous studies in cognitive psychology have shown that people are unable to disregard false facts that they once thought were true [12,122,315]. Moreover, the occurrence of the embarrassing data point makes it *plausible* that this information is real, and the "accused" user's friend might simply not believe that it is not real. This type of plausible deniability thus thwarts what I call "deniable plausibility", which is arguably equally important to assure.

Finally, SIA system developers should acknowledge the variety of users' preferred communication styles, and adjust or adapt the functionality of their systems accordingly. For example, social visualization systems and location sharing systems establish implicit connections between users: this is great for FYI communicators, but the lack of explicit context and direct interaction may alienate users who are not apt to communicate this way. As a solution, Page [215] recommends for location-sharing SNSs to allow their users to communicate their location status in active, one-to-one communications, and/or to annotate their implicit location updates with explicit messages. Conversely, SIA systems that encourage direct interaction (as some collaborative systems may do) need to acknowledge that some readers prefer to communicate passively (e.g., through status updates) instead. These users should be given the option to limit direct interaction, lest they be overwhelmed by it. Most contemporary social media systems support both direct one-to-one and passively broadcasted interactions, but typically emphasize either the former (e.g. chat clients, video conferencing systems) or the latter (e.g. SNSs, blogs, location-sharing systems). More work could be done to investigate the opportunities for tailoring the interaction paradigm to the communication style of the user.

4.5 Decision-Support Solutions

As the privacy aspects on SNSs are often incredibly complex, traditional decision-support solutions—i.e., offering transparency and control—are deemed insufficient to help SNS users with their privacy decisions. As users often either misunderstand [174,180,270] or avoid the hassle [23,24,56,92,118,290] of exploiting privacy settings, extra care needs to be taken to use sensible default settings (see also Sect. 3.5). Even when users actively engage in setting their sharing preferences, this default setting may unconsciously influence their sharing decisions; a phenomenon that has ben dubbed the "default effect" [119,120,158].

There are several psychological explanations for this effect. First of all, people tend to stick with the default because it avoids the effort and stress of making an active decision ("status quo bias", cf. [18,121,127,242]. Moreover, people tend to cognitively regard the default option as a reference point in their decisions, and evaluate the alternative options in terms of losses and gains compared to this endowed option, which gives the default an advantage, because losses tend to loom larger than gains (endowment and loss aversion, cf. [127,128,218]). Finally, defaults act as an implied endorsement of the default value by the system, and users tend to comply with this endorsement [183,249].

How large is the effect of defaults? Both Johnson et al. [120] and Lai and Hui [158] find consistent differences in newsletter sign-up rates between opt-in and opt-out default settings of at least 25 percentage-points. Section 2.5 mentioned that the *order* in which information is requested can also influence users' disclosure decisions, with users usually disclosing more information when the most sensitive information is requested first. A similar effect has been demonstrated for the order in which contact categories are presented in SNSs [140]. Note that default effects are practically unavoidable; therefore, SIA system developers have a moral obligation to choose these defaults wisely [273].

Default privacy settings may "nudge" SNS users in the direction of more sharing or more privacy [283], and the general idea of "nudges" as a means to protect SNS users' privacy has recently gained a lot of attention in privacy research [2,17,112]. Nudges are subtle yet persuasive cues that makes people more likely to decide in one direction or the other [283]. Carefully designed nudges make it easier for people to make the right choice, without limiting their ability to choose freely. As such, nudges turn decision fallacies (like the default effect) into mechanisms that help them [2]: they exploit these fallacies to create a "choice architecture" that encourages wanted behavior and inhibits unwanted behavior [283].

Beyond default settings, "justifications" are another extensively implemented type of nudge (see also Sect. 2.5). A justification is a succinct reason to disclose or not disclose a certain piece of information that makes it easier to rationalize the decision and to minimize the regret associated with choosing the wrong option. Justifications include providing a *reason* for requesting the information [57], highlighting the *benefits* of disclosure [152,299], and appealing to the social norm [6,26,221]. The effect of justifications seems to vary, and an overview study shows that while users appreciate the support provided by justifications, they are arguably not subtle enough as a nudge, because they simply remind users of the privacy implications of using the system, which thwarts the positive effect of the nudge itself [139].

A related nudge is to give users feedback on the real or potential audience of a shared piece of information. For example, in location sharing services, researchers have experimented with giving users real-time feedback on who is requesting or viewing their location [117,288]. The results of these experiments are mixed: users appreciate the information, but it can easily become excessive and annoying. Similarly, Wang et al. [300,301] implemented and tested a tool that provides users with detailed feedback about the potential audience when posting a Facebook message. They find that at least some users consider this tool helpful, but they find no significant differences in posting behavior. Wang et al. [300,301] also consider "sentiment feedback" (a feature that tells users whether their SNS post is likely to be perceived as positive or negative) and a "post timer" (a feature that delays SNS posts by 10 s, allowing users to change their mind; see Fig. 6). While some of the participants in their study seemed to like these tools, others found them intrusive and annoying.

Fig. 6. Wang et al. [300,301]'s "sentiment feedback" and "post timer" features.

The use of nudges for privacy (and indeed, in general) is somewhat controversial. While nudges are supposed to help users navigate complex decisions, some researchers argue that they may threaten consumer autonomy, especially when they cause behavioral or cognitive biases [261,265]. These researchers argue for "smart default" settings that match the preferences of most users, or even "adaptive defaults" that take users' individual preferences into account [136]. The idea of adaptive defaults is covered in more detail in Sect. 5.

A final method to support users' privacy decisions in SNSs (one that can also be construed as a nudge) is to carefully plan the available options that users can use to set their privacy settings. As most users seem to have no fixed preference for their settings (i.e., their choice process is *constructive*, cf. [27,60]), they can the influenced by the available options. For example, if an "extreme" sharing option is introduced that is sufficiently distinct from the existing options, this not only causes some users to switch from the previously most extreme option to this new option, but it also causes some users to switch from a less extreme option to the previously most extreme option [143] due to the "compromise effect" [253]. In other words, such an extreme option may increase sharing across the board. Similarly, introducing or removing certain options can move users towards or away from the subjectively closest other options [143], in line with the "substitution effect" [109,291]. SIA system developers can thus selectively display privacy options in order to nudge users towards or away from certain behaviors. Again, this type of nudging can be personalized (cf. [137]), which will be discussed in more detail in Sect. 5.

5   SIA in Support of Privacy

While this chapter has primarily covered the privacy aspects of SIA systems, the current section will discuss the opposite angle: using SIA functionality to improve privacy decision-making. There are two ways in which social information can be used to support privacy decisions: The first way is to explicitly show users social information about privacy decisions that can help them navigate the privacy decision landscape. The second way is to use social information to provide "privacy recommendations". Both methods are discussed in more detail below.

5.1 Explicit Use of Social Information in Privacy Decision Support

Section 4.5 covered the idea of justifications as a method to support (or rather, nudge) users' privacy decisions. One of the justification methods, "appealing to the social norm", can be seen as a form of Social Navigation (see Chap. 5 of this book [81]), where users receive insight into how many other users have disclosed a certain piece of information, or made a certain privacy setting. Several researchers have used this method—in some occasions presenting the information as neutral as possible; at other times with an agenda of influencing users' decisions (i.e., as a nudge). Examples are Acumen (Fig. 7; a browser toolbar that tells users what percentage of other users had blocked certain tracking cookies) and Bonfire (Fig. 8; a desktop firewall that tells users what percentage of other users had blocked Internet access for certain applications) [91]. Note, though, that the effect of neither of these two tools on user behaviors have been formally evaluated. On the experimental side, Acquisti et al. [6] asked users to disclose embarrassing information about themselves, and found that users were about 27% more likely to disclose their personal information when they learned that many others decided to disclose the same information. In a study on Facebook privacy settings Besmer et al. [26] found that social cues had barely any effect on users' Facebook privacy settings: only the small subset of users who take the time to customize their settings may be influenced by strong negative social cues. Finally, Kobsa et al. [221] used social navigation cues to support users in setting their privacy settings in an IM client. They too rate social navigation as a secondary effect.

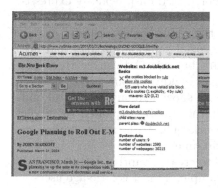

Fig. 7. The Acumen browser toolbar.

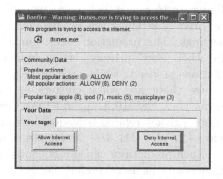

Fig. 8. The Bonfire desktop firewall.

An overview study of several justification methods showed that while none of the presented justifications really worked, users particularly dislike the "how many other users have made this decision" type information [139]. Specifically, this was the only information that was not even considered "helpful" by users. Interestingly enough, this was the only justification that actually had some effect on users (i.e., the displayed percentage actually influenced users' disclosure somewhat). Semi-structured interviews revealed that some users believe that social information, when presented as a justification, "feels like peer pressure", which makes them believe that this justification method is worse than having no justification at all. Future work should investigate the idea of using less conspicuous social navigation cues, such as the designs presented by DiGioia and Dourish [66].

5.2 Implicit Use of Social Information in Privacy Decision Support

Social information can also be used implicitly, by giving users "privacy recommendations". Privacy recommendations solve an important problem with privacy nudges (see Sect. 4.5): Nudges have a universal "direction" (i.e., they either nudge people to become more private, or to disclose more information), which means that they have to make implicit assumptions about the optimal balance between privacy and disclosure, and more importantly, that this optimal balance is the same for all users. But what if privacy preferences are *variable*? There is ample evidence that people vary in their disclosure behavior (see Sect. 2.3), and that this behavior depends on the context of the decision [22, 57, 167, 286]. Indeed, the variability and context-dependency of privacy preferences is at the core of many privacy theories such as Altman's privacy regulation theory [11], Nissenbaum's contextual integrity [201, 202], and Petronio's communication privacy management [226, 227].

Privacy recommendations solve this problem by *tailoring* nudges to the user and her context [87, 135, 148]. As with regular recommender systems (cf. Chaps. 10–15 of this book [30, 95, 116, 134, 168, 209]), privacy recommendations are predictions of liking and/or future behavior, based on previously expressed liking and/or past behaviors. In this case, users' future disclosure behaviors or privacy settings are predicted based on their past behaviors and privacy settings. The prediction algorithm leverages social information (i.e., patterns in other users' privacy settings and disclosure behaviors) to make these predictions. Indeed, while privacy behaviors differ wildly among users, research shows there are some clusters of users that show similar behavioral patterns [142, 208, 310, 313], and that these patterns can be exploited to predict users' privacy-related behaviors in social networks [69, 80], location-sharing services [137, 232, 241, 316], and smart home environments [216].

Such predictions can subsequently be used to create adaptive default settings that better fit users' preferences [136]. For example, if most college-age users do not disclose photos taken in the bar district on Friday evening to their parents, the privacy recommender may suggest the same to a new user who is about to start college. Similarly, predictions can be used to order information requests in a way that avoids asking for information that the user deems too private to disclose [136]. For example, a recommender system may avoid asking a user for their income or sexual preference (two decidedly sensitive pieces of information) if the user has already chosen not to disclose their race and education level (two medium-sensitive pieces of information), and ask the user for their age and gender (two decidedly non-sensitive pieces of information) instead. Another use of privacy recommendations would be to tailor justifications to the user, as some users are more susceptible of justifications that employ social information ("80% of other users decided to disclose their occupation to this Q&A system."), while others prefer justifications that provide more detailed explanations ("Disclosing your occupation will help others judge the quality of your answers based on your area of expertise."), and still others prefer to get no justification at all [138].

The bulk of the existing work on privacy recommendations is algorithmic, and focuses on how to predict users' privacy preferences and behaviors using existing machine learning techniques. Research on the effects of privacy recommendations on users' behavior and attitudes is still in its infancy, though. A first comprehensive evaluation of a system with built-in privacy recommendations showed promising results, but highlighted that more research is needed in this area [136]. The evaluation concerns a demographics-based health recommender system. The system asks demographics questions in a sequential order. When a user answer a demographics question, the recommendations are adapted to this answer on the fly. Alternatively, the user can decide to skip a question if they deem the question is too sensitive to answer. The evaluation tested several means of ordering the recommendations:

- **Most-sensitive first:** the demographic questions are ordered by decreasing sensitivity.
- **Least-sensitive-first:** the demographic questions are ordered by increasing sensitivity.
- **Most-useful-first:** the demographic questions are ordered by decreasing usefulness.
- **Static trade-off:** the demographic questions are ordered most-useful-first for items with a sensitivity below a certain threshold (manipulated to be either low or high), and least-sensitive-first for items above this threshold.
- **adaptive request order:** same as the static trade-off, but the threshold is adapted to users' previous disclosures.

Although the adaptive request order condition did not result in the hypothesized benefits, other versions that automatically traded off usefulness and sensitivity of the items to be disclosed did indeed improve users' experience. Particularly, the trade-off request order with a high sensitivity-to-usefulness threshold resulted in higher levels of trust and user satisfaction for people with some domain expertise or with low privacy concerns. Moreover, the trade-off and adaptive request orders resulted in better recommendations than "most-useful-first", arguably because users were more likely to answer more questions in the trade-off and adaptive versions since they avoided overly sensitive questions. This effect also crucially depended on users' privacy concerns: a higher trade-off threshold worked better for users with high concerns, while a lower threshold worked better for users with low concerns. As such, the study showed that automatic means to relieve some of the burden of controlling one's privacy settings is a promising endeavor. Future improvements to the adaptive request order may make this fully adaptive version more acceptable to users as well.

6   Recommendations for Future Work

Before concluding this chapter, I here reflect upon the information presented in this chapter, and make a number of recommendations for future research and practice regarding privacy aspects of Social Information Access systems.

Researchers and practitioners should **conduct careful privacy analyses** before embarking on projects with unintended data use, even if this type of secondary use is already permitted by the system's privacy policy. While the United States currently do not pose legal restrictions to unintended data use, but the European Union does [78], and the E.U. regulation applies to U.S. companies if they cater to E.U. users. Regardless of the legal ramifications, it is wise to address privacy problems *before* the widespread deployment of a system, because users who believe that their privacy is being violated may shy away from providing further content [16,147,248,274].

Academic researchers are not excused from conducting privacy analyses; researchers at U.S. academic institutions who perform research with human subjects are required to get permission for their study from their Institutional Review Board (IRB), and universities in many other countries have similar procedures. Even if the proposed study falls outside IRB restrictions, there is a moral obligation to minimize unintended harm to study participants. Failing to do so can damage the reputation of researchers, departments and institutions [86,331]. On a broader note, the ethics of using algorithms for information access are being discussed in the Human-Computer-Interaction field (cf. [177]) and I strongly encourage algorithms researchers to join this debate.

Researchers and practitioners should **conduct extensive online user experiments**. In developing privacy-enhancing features for their SIA systems, developers often assume that users will behave according to the various "attack models" defined by security researchers. User behavior may not follow such models, or even any model at all [4,5], and thus it is better not to make assumptions about users' decisions, but rather to test the privacy-enhancing features with real users. Conducting user experiments is a complicated endeavor, especially when privacy attitudes and behaviors are to be measured. It is important to strive for a realistic study setting (keeping users unaware of the purpose of the study, lest they behave differently), while keeping tight control over the experimental setup. I advise those who want to conduct a controlled experiment to consult Knijnenburg et al.'s user-centric evaluation framework [146] and methodology [145] for further guidance.

Researchers and practitioners should **integrate technical and decision-support related solutions** to privacy problems. Existing technical and decision-support related solutions come from different research fields, and integrated solutions are scarce because there is very little overlap between these fields, despite the existence of conferences like the Privacy Enhancing Technologies Symposium (PETS) and the Symposium On Usable Privacy and Security (SOUPS) trying to merge them. I encourage researchers and practitioners to form multidisciplinary teams that can attempt to tackle privacy issues in more integrative manners. This is absolutely necessary to advance the field of privacy research, because no privacy-enhancing technology is effective without user acceptance (cf. "client-side personalization" [149]), and conversely, decision-support related solutions seem to only be effective when tailored to the user (requiring privacy recommendations, see Sect. 5.2).

Researchers and practitioners should **make sure that that users *are* protected and *feel* protected** at the same time. This relates to the call to integrate technical and decision-support related solutions: many creators of technical solutions focus on protecting users against privacy threats, without addressing how users tend to *feel* about using the system. Unfortunately, privacy-enhancing features often remind users of the privacy implications of using the system, which ironically makes them more (rather than less) fearful [139]. A more integrative privacy-enhancing solution would not only make the system safer for the user, but also educate users about their privacy and/or convince them that their safety is improved.

The effect of decision-support related solutions, on the other hand, is often measured in terms of changed attitudes or behavioral intentions, without focusing on actual user behavior. Such attitudes and intentions are however not always directly related to actual behaviors [4,89,187,203,267], which means that seemingly successful decision-support interventions may not have a significant effect in real life.

It is also important to consider the *target behavior* that privacy-related interventions should aspire to attain. Convincing users to disclose *less* is not always a desirable goal, because that may simply preclude users from exploiting the benefits of SIA functionality [100,207,228]. A balance between privacy and disclosure is more desirable, so an alternative solution is to aspire to make users' disclosure more *consistent*. Users' privacy decision-making is often irrational though [4,5], which makes it difficult to decide which of any two inconsistent behaviors should be considered "correct". A third solution is therefore to take users' *personal preference* as the ground truth. But if the user is uninformed about privacy, this may be a suboptimal target as well. A final solution could be to offer users a "guided exploration", teaching them about their privacy and learning from their preferences in an iterative manner. This, however, brings us back to the same problem that plagues solutions involving transparency and control (see Sect. 1.4), in that users claim to want transparency and control, but often avoid the hassle of actually exploiting it. To resolve this conundrum, I suggest that researchers involved in discussing the ethics of using algorithms for information access should consider the target of privacy-related interventions as one of their main points of discussion.

Finally, I recommend that researchers and practitioners should **get involved in legislative initiatives** for privacy regulation. Widespread effective privacy protection in SIA systems depends on the forcing function provided by the regulatory environment in which these systems operate. Recent legal developments in Europe regarding the legal use of personal data have a profound impact on legal boundaries in which SIA systems are allowed to operate [78], and more extensive privacy regulations have been proposed in the United States as well [107]. The future of Social Information Access may very well depend on what legislators deem legitimate means of supporting users versus violating their right to privacy. I strongly encourage SIA system developers to get involved in the impending legal debate, lest they might see their field of work/research become illegal altogether!

7 Conclusion

User privacy is very relevant for SIA systems: on the one hand, these systems are dependent on high-quality user input; on the other hand, they have the potential of (mis)using user data in various privacy-sensitive ways. It is therefore important for SIA system developers to understand how users make disclosure decisions, and what factors influence these decisions. Privacy-preserving technical solutions can help assuage users' concerns, thereby increasing their disclosure. Technical solutions discussed in this chapter are:

- Pseudonyms (i.e., by forgoing a real name policy)
- Differential privacy (i.e., via random perturbation)
- Client-side computation (e.g., client-side recommendation algorithms)
- Giving users the right to be forgotten (as required by European law for search engines)
- Contact categorization (e.g., Google+ Circles and Facebook Friend Lists)
- Allowing for plausible deniability (e.g., by giving users the ability to lie about their data)
- Assuring deniable plausibility (e.g., by only displaying data about the user that is known to be correct)
- Tailoring the interaction method to various communication styles (i.e., by supporting both status updates and direct messages)

Note that these solutions have mainly been studied in isolation—very little work has been done on developing holistic and encompassing solutions that are likely to be required in actual SIA systems. Some of the discussed solutions are, in fact, inherently incompatible, and hence a tradeoff needs to be made between different privacy and security requirements. The challenge of trading off and integrating different privacy-enhancing solutions is still wide open for future research [87, 94].

Research shows that users often find it difficult to make sensible and consistent privacy decisions, it is therefore recommended that SIA system developers employ solutions to support users' privacy decision-making practices. Such decision support has traditionally come in the form of transparency and control, but recent work highlighted in this chapter shows that these methods are often insufficient in helping users navigate the complex privacy landscape. The privacy aspects of modern systems are simply too complex for most users to grasp, and users are typically not motivated to engage in difficult and time-consuming privacy setting behavior. Hence, this chapter has discussed a wide variety of decision support solutions:

- Explanations and feedback (i.e., telling users how their data will be used, or who will see their posts)
- Designs that promote more careful decision-making (e.g. add/remove buttons in Web form auto-completion tools)
- Carefully selected choice options that promote or discourage certain decisions (e.g., exploiting the compromise and substitution effect)
- Context-based hurdles (e.g. GMail's "Mail Goggles" to prevent drunk emailing)
- Smart or adaptive default disclosure settings (because most users never change their default settings)
- Smart or adaptive information request orders (i.e. prioritizing less sensitive and more useful personal information requests)

Above all, it is important for SIA system developers to realize that while the data that systems use to provide the SIA functionality is often deemed public, this does not mean that attention to privacy is irrelevant or unnecessary. In fact, in many cases users may view the SIA functionality as a secondary or unintended use of their data, which some may accept due to the added benefit the functionality provides, but others may perceive as an inexcusable privacy violation. This chapter has outlined several privacy "horror stories", and in most cases the secondary use of information for SIA purposes leads to the most severe privacy violations, especially when the SIA functionality is introduced after the fact as a new feature. It is thus advisable to conduct research to uncover users' privacy-related attitudes to proposed new functionalities, to communicate such new features extensively, and to give users ample opportunities to opt out of the functionality. And if possible, any new functionality that has privacy implications should probably be opt-in rather than opt-out!

References

1. Ackerman, M.S., Cranor, L.F., Reagle, J.: Privacy in e-commerce: examining user scenarios and privacy preferences. In: Proceedings of the 1st ACM Conference on Electronic Commerce, EC 1999, pp. 1–8. ACM, Denver (1999). https://doi.org/10.1145/336992.336995
2. Acquisti, A.: Nudging privacy: the behavioral economics of personal information. IEEE Secur. Priv. **7**, 82–85 (2009). https://doi.org/10.1109/MSP.2009.163
3. Acquisti, A., Gross, R.: Predicting social security numbers from public data. Proc. Natl. Acad. Sci. **106**(27), 10,975–10,980 (2009). https://doi.org/10.1073/pnas.0904891106

4. Acquisti, A., Grossklags, J.: Privacy and rationality in individual decision making. IEEE Secur. Priv. **3**(1), 26–33 (2005). https://doi.org/10.1109/MSP.2005.22
5. Acquisti, A., Grossklags, J.: What can behavioral economics teach us about privacy? In: Acquisti, A., De Capitani di Vimercati, S., Gritzalis, S., Lambrinoudakis, C. (eds.) Digital Privacy: Theory, Technologies, and Practices, pp. 363–377. Auerbach Publications, New York/London (2008). https://doi.org/10.1201/9781420052183.ch18
6. Acquisti, A., John, L.K., Loewenstein, G.: The impact of relative standards on the propensity to disclose. J. Mark. Res. **49**(2), 160–174 (2012). https://doi.org/10.1509/jmr.09.0215
7. Adjerid, I., Acquisti, A., Brandimarte, L., Loewenstein, G.: Sleights of privacy: framing, disclosures, and the limits of transparency. In: Proceedings of the Ninth Symposium on Usable Privacy and Security, SOUPS 2013, pp. 9:1–9:11. ACM, Newcastle (2013). https://doi.org/10.1145/2501604.2501613
8. Adomavicius, G., Tuzhilin, A.: Context-aware recommender systems. In: Ricci, F., Rokach, L., Shapira, B., Kantor, P.B. (eds.) Recommender Systems Handbook, pp. 217–253. Springer, Boston (2011). https://doi.org/10.1007/978-0-387-85820-3_7
9. Aizenbud-Reshef, N., Barger, A., Dubinsky, Y., Guy, I., Kremer-Davidson, S.: Privacy concerns in enterprise social travel: attitudes and actions. In: Campos, P., Graham, N., Jorge, J., Nunes, N., Palanque, P., Winckler, M. (eds.) INTERACT 2011. LNCS, vol. 6948, pp. 242–249. Springer, Heidelberg (2011). https://doi.org/10.1007/978-3-642-23765-2_17
10. Alluri, A.V.: Empirical study on key attributes of yelp dataset which account for susceptibility of a user to social influence. M.S. thesis, University of Cincinnati, Cincinnati, OH (2015)
11. Altman, I.: The Environment and Social Behavior: Privacy, Personal Space, Territory, and Crowding. Brooks/Cole Publishing Company, Monterey (1975)
12. Anderson, C.A., Lepper, M.R., Ross, L.: Perseverance of social theories: the role of explanation in the persistence of discredited information. J. Pers. Soc. Psychol. **39**(6), 1037–1049 (1980). https://doi.org/10.1037/h0077720
13. Angst, C.M., Agarwal, R.: Adoption of electronic health records in the presence of privacy concerns: the elaboration likelihood model and individual persuasion. MIS Q. **33**(2), 339–370 (2009)
14. Aoki, P., Woodruff, A.: Making space for stories. In: Proceedings of the SIGCHI Conference on Human Factors in Computing Systems, pp. 181–190 (2005). https://doi.org/10.1145/1054972.1054998
15. Arlein, R.M., Jai, B., Jakobsson, M., Monrose, F., Reiter, M.K.: Privacy-preserving global customization. In: 2nd ACM Conference on Electronic Commerce, EC 2000, pp. 176–184. ACM, Minneapolis (2000). https://doi.org/10.1145/352871.352891
16. Awad, N.F., Krishnan, M.S.: The personalization privacy paradox: an empirical evaluation of information transparency and the willingness to be profiled online for personalization. MIS Q. **30**(1), 13–28 (2006)
17. Balebako, R., Leon, P.G., Mugan, J., Acquisti, A., Cranor, L.F., Sadeh, N.: Nudging users towards privacy on mobile devices. In: CHI 2011 Workshop on Persuasion, Influence, Nudge and Coercion Through Mobile Devices, PINC 2011, Vancouver, Canada, pp. 23–26 (2011)
18. Baron, J., Ritov, I.: Reference points and omission bias. Organ. Behav. Hum. Decis. Process. **59**(3), 475–498 (1994). https://doi.org/10.1006/obhd.1994.1070

19. Bean, D.: 11 Brutal Reminders That You Can and Will Get Fired for What You Post on Facebook (2014). https://www.yahoo.com/tech/11-brutal-reminders-that-you-can-and-will-get-fired-for-84931050659.html

20. Beatty, P., Reay, I., Dick, S., Miller, J.: P3P adoption on e-commerce web sites: a survey and analysis. IEEE Internet Comput. **11**(2), 65–71 (2007). https://doi.org/10.1109/MIC.2007.45

21. Beckwith, R., Mainwaring, S.: Privacy: personal information, threats, and technologies. In: Proceedings of the Proceedings 2005 IEEE International Symposium on Technology and Society, ISTAS 2005, pp. 9–16. IEEE, Los Angeles (2005). https://doi.org/10.1109/ISTAS.2005.1452707

22. Benisch, M., Kelley, P.G., Sadeh, N., Cranor, L.F.: Capturing location-privacy preferences: quantifying accuracy and user-burden tradeoffs. Pers. Ubiquit. Comput. **15**(7), 679–694 (2011). https://doi.org/10.1007/s00779-010-0346-0

23. Berendt, B., Günther, O., Spiekermann, S.: Privacy in e-commerce: stated preferences vs. actual behavior. Commun. ACM **48**(4), 101–106 (2005). https://doi.org/10.1145/1053291.1053295

24. Bergmann, M.: Testing privacy awareness. In: Matyáš, V., Fischer-Hübner, S., Cvrček, D., Švenda, P. (eds.) Privacy and Identity 2008. IAICT, vol. 298, pp. 237–253. Springer, Heidelberg (2009). https://doi.org/10.1007/978-3-642-03315-5_18

25. Berry, M.J., Linoff, G.: Data Mining Techniques: For Marketing, Sales, and Customer Support. Wiley, New York (1997)

26. Besmer, A., Watson, J., Lipford, H.R.: The impact of social navigation on privacy policy configuration. In: Proceedings of the Sixth Symposium on Usable Privacy and Security, SOUPS 2010, pp. 7:1–7:10. ACM, Redmond (2010). https://doi.org/10.1145/1837110.1837120

27. Bettman, J.R., Luce, M.F., Payne, J.W.: Constructive consumer choice processes. J. Consum. Res. **25**(3), 187–217 (1998). https://doi.org/10.1086/209535

28. Binder, J., Howes, A., Sutcliffe, A.: The problem of conflicting social spheres: effects of network structure on experienced tension in social network sites. In: Proceedings of the 27th International Conference on Human Factors in Computing Systems, CHI 2009, pp. 965–974. ACM, Boston (2009). https://doi.org/10.1145/1518701.1518849

29. Birnholtz, J., Guillory, J., Hancock, J., Bazarova, N.: "On my way": deceptive texting and interpersonal awareness narratives. In: Proceedings of the 2010 ACM Conference on Computer Supported Cooperative Work, CSCW 2010, pp. 1–4. ACM, Savannah (2010). https://doi.org/10.1145/1718918.1718920

30. Bogers, T.: Tag-based recommendation. In: Brusilovsky, P., He, D. (eds.) Social Information Access. LNCS, vol. 10100, pp. 441–479. Springer, Cham (2018)

31. Bothorel, C., Lathia, N., Picot-Clemente, R., Noulas, A.: Location recommendation with social media data. In: Brusilovsky, P., He, D. (eds.) Social Information Access. LNCS, vol. 10100, pp. 624–653. Springer, Cham (2018)

32. Boyd, D.M., Ellison, N.B.: Social network sites: definition, history, and scholarship. J. Comput.-Mediat. Commun. **13**(1), 210–230 (2007). https://doi.org/10.1111/j.1083-6101.2007.00393.x

33. Brandimarte, L., Acquisti, A., Loewenstein, G.: Misplaced confidences: privacy and the control paradox. Soc. Psychol. Pers. Sci. **4**(3), 340–347 (2013). https://doi.org/10.1177/1948550612455931

34. Brody, R.G., Mulig, E., Kimball, V.: Phishing, pharming and identity theft. Acad. Account. Finan. Stud. J. **11**(3), 43–56 (2007)

35. Brown, G., Howe, T., Ihbe, M., Prakash, A., Borders, K.: Social networks and context-aware spam. In: Proceedings of the 2008 ACM Conference on Computer Supported Cooperative Work, CSCW 2008, pp. 403–412. ACM, San Diego (2008). https://doi.org/10.1145/1460563.1460628
36. Brusilovsky, P.: Social information access: the other side of the social web. In: Geffert, V., Karhumäki, J., Bertoni, A., Preneel, B., Návrat, P., Bieliková, M. (eds.) SOFSEM 2008. LNCS, vol. 4910, pp. 5–22. Springer, Heidelberg (2008). https://doi.org/10.1007/978-3-540-77566-9_2
37. Brusilovsky, P., He, D.: Social information access: definition and classification. In: Brusilovsky, P., He, D. (eds.) Social Information Access. LNCS, vol. 10100, pp. 1–18. Springer, Cham (2018)
38. Brusilovsky, P., Smyth, B., Shapira, B.: Social search. In: Brusilovsky, P., He, D. (eds.) Social Information Access. LNCS, vol. 10100, pp. 213–276. Springer, Cham (2018)
39. Burke, M., Marlow, C., Lento, T.: Social network activity and social well-being. In: Proceedings of the SIGCHI Conference on Human Factors in Computing Systems, CHI 2010, pp. 1909–1912. ACM, Atlanta (2010). https://doi.org/10.1145/1753326.1753613
40. Caine, K., Hanania, R.: Patients want granular privacy control over health information in electronic medical records. J. Am. Med. Inform. Assoc. **20**(1), 7–15 (2013). https://doi.org/10.1136/amiajnl-2012-001023
41. Caine, K., Kohn, S., Lawrence, C., Hanania, R., Meslin, E.M., Tierney, W.M.: Designing a patient-centered user interface for access decisions about EHR data: implications from patient interviews. J. Gen. Intern. Med. **30**(1), 7–16 (2015). https://doi.org/10.1007/s11606-014-3049-9
42. Calandrino, J.A., Kilzer, A., Narayanan, A., Felten, E.W., Shmatikov, V.: You might also like: privacy risks of collaborative filtering. In: Proceedings of the 2011 IEEE Symposium on Security and Privacy, pp. 231–246. IEEE, Oakland (2011). https://doi.org/10.1109/SP.2011.40
43. Campan, A., Truta, T.M.: Data and structural k-anonymity in social networks. In: Bonchi, F., Ferrari, E., Jiang, W., Malin, B. (eds.) PInKDD 2008. LNCS, vol. 5456, pp. 33–54. Springer, Heidelberg (2009). https://doi.org/10.1007/978-3-642-01718-6_4
44. Canny, J.F.: Collaborative filtering with privacy. In: Proceedings of the IEEE Symposium on Security and Privacy, pp. 45–57. IEEE Computer Society, Washington, DC (2002)
45. Cashmore, P.: Facebook Recommends Reconnecting with Ex-Lovers, Dead Friends (2009). http://mashable.com/2009/10/25/facebook-reconnect/
46. Cassel, L.N., Wolz, U.: Client side personalization. In: DELOS Workshop: Personalisation and Recommender Systems in Digital Libraries, Dublin, Ireland, pp. 8–12 (2001)
47. Cavusoglu, H., Phan, T., Cavusoglu, H.: Privacy controls and content sharing patterns of online social network users: a natural experiment. In: ICIS 2013 Proceedings, Milan, Italy (2013)
48. Chan, K.H.: Memories of Friends Departed Endure on Facebook (2009). https://www.facebook.com/notes/facebook/memories-of-friends-departed-endure-on-facebook/163091042130/
49. Chellappa, R.K., Sin, R.G.: Personalization versus privacy: an empirical examination of the online consumer's dilemma. Inf. Technol. Manag. **6**(2–3), 181–202 (2005). https://doi.org/10.1007/s10799-005-5879-y

50. Chen, D., Fraiberger, S.P., Moakler, R., Provost, F.: Enhancing transparency and control when drawing data-driven inferences about individuals. In: Working Paper (2015). http://archive.nyu.edu/handle/2451/33969

51. Chen, H., Chiang, R.H.L., Storey, V.C.: Business intelligence and analytics: from big data to big impact. MIS Q. **36**(4), 1165–1188 (2012)

52. Chen, H.G., Chen, C.C., Lo, L., Yang, S.C.: Online privacy control via anonymity and pseudonym: cross-cultural implications. Behav. Inf. Technol. **27**(3), 229–242 (2008). https://doi.org/10.1080/01449290601156817

53. Chen, Y.F.: Herd behavior in purchasing books online. Comput. Hum. Behav. **24**(5), 1977–1992 (2008). https://doi.org/10.1016/j.chb.2007.08.004

54. Cho, D., Kim, S., Acquisti, A.: Empirical analysis of online anonymity and user behaviors: the impact of real name policy. In: 2012 45th Hawaii International Conference on System Science, HICSS 2012, pp. 3041–3050. IEEE Computer Society, Maui (2012). https://doi.org/10.1109/HICSS.2012.241

55. Coleman, G.: Phreaks, hackers, and trolls: the politics of transgression and spectacle. In: Mandiberg, M. (ed.) The Social Media Reader, pp. 99–119. NYU Press, New York (2012)

56. Compañó, R., Lusoli, W.: The policy maker's anguish: regulating personal data behavior between paradoxes and dilemmas. In: Moore, T., Pym, D., Ioannidis, C. (eds.) Economics of Information Security and Privacy, pp. 169–185. Springer, New York (2010). https://doi.org/10.1007/978-1-4419-6967-5_9

57. Consolvo, S., Smith, I., Matthews, T., LaMarca, A., Tabert, J., Powledge, P.: Location disclosure to social relations: why, when, & what people want to share. In: Proceedings of the SIGCHI Conference on Human Factors in Computing Systems, CHI 2005, pp. 81–90. ACM, Portland (2005). https://doi.org/10.1145/1054972.1054985

58. Consumer Reports: Facebook & your privacy: who sees the data you share on the biggest social network? (2012). http://www.consumerreports.org/cro/magazine/2012/06/facebook-your-privacy

59. Cosley, D., Lam, S.K., Albert, I., Konstan, J.A., Riedl, J.: Is seeing believing?: how recommender system interfaces affect users' opinions. In: Proceedings of the SIGCHI Conference on Human Factors in Computing Systems, CHI 2003, pp. 585–592. ACM, Ft. Lauderdale (2003). https://doi.org/10.1145/642611.642713

60. Coupey, E., Irwin, J.R., Payne, J.W.: Product category familiarity and preference construction. J. Consum. Res. **24**(4), 459–468 (1998). https://doi.org/10.1086/209521

61. Culnan, M.J.: "How did they get my name?": an exploratory investigation of consumer attitudes toward secondary information use. MIS Q. **17**(3), 341–363 (1993). https://doi.org/10.2307/249775

62. Culnan, M.J.: Consumer awareness of name removal procedures: Implications for direct marketing. J. Dir. Mark. **9**(2), 10–19 (1995). https://doi.org/10.1002/dir.4000090204

63. De Wolf, R., Pierson, J.: Researching social privacy on SNS through developing and evaluating alternative privacy technologies. In: Proceedings of the CSCW 2013 Networked Social Privacy Workshop, San Antonio, TX (2013)

64. Deuker, A.: Friend-to-friend privacy protection on social networking sites: a grounded theory study. In: AMCIS 2012 Proceedings, Seattle, WA (2012)

65. Di Castro, D., Lewin-Eytan, L., Maarek, Y., Wolff, R., Zohar, E.: Enforcing K-anonymity in web mail auditing. In: Proceedings of the Ninth ACM International Conference on Web Search and Data Mining, WSDM 2016, pp. 327–336. ACM, New York (2016). https://doi.org/10.1145/2835776.2835803

66. DiGioia, P., Dourish, P.: Social navigation as a model for usable security. In: Proceedings of the 2005 Symposium on Usable Privacy and Security, SOUPS 2005, pp. 101–108. ACM, Pittsburgh (2005). https://doi.org/10.1145/1073001.1073011

67. DiMicco, J., Millen, D.R., Geyer, W., Dugan, C., Brownholtz, B., Muller, M.: Motivations for social networking at work. In: Proceedings of the 2008 ACM Conference on Computer Supported Cooperative Work, CSCW 2008, pp. 711–720. ACM, New York (2008). https://doi.org/10.1145/1460563.1460674

68. Dimitrov, D., Helic, D., Strohmaier, M.: Tag-based navigation and visualization. In: Brusilovsky, P., He, D. (eds.) Social Information Access. LNCS, vol. 10100, pp. 181–212. Springer, Cham (2018)

69. Dong, C., Jin, H., Knijnenburg, B.P.: Predicting privacy behavior on online social networks. In: Ninth International AAAI Conference on Web and Social Media, ICWSM 2015, pp. 91–100. AAAI Publications, Oxford (2015)

70. Duggan, M.: Mobile messaging and social media - 2015. Technical report, Pew Research Center (2015). http://www.pewinternet.org/2015/08/19/mobile-messaging-and-social-media-2015/

71. Duhigg, C.: How Companies Learn Your Secrets. The New York Times (2012). http://www.nytimes.com/2012/02/19/magazine/shopping-habits.html

72. D'Urso, S.C.: Who's watching us at work? Toward a structural-perceptual model of electronic monitoring and surveillance in organizations. Commun. Theory 16(3), 281–303 (2006). https://doi.org/10.1111/j.1468-2885.2006.00271.x

73. Egelman, S.: My profile is my password, verify me!: the privacy/convenience trade-off of Facebook connect. In: Proceedings of the SIGCHI Conference on Human Factors in Computing Systems, CHI 2013, pp. 2369–2378. ACM, Paris (2013). https://doi.org/10.1145/2470654.2481328

74. Ehrlich, K., Shami, N.S.: Microblogging inside and outside the workplace. In: Fourth International AAAI Conference on Weblogs and Social Media, ICWSM 2010, pp. 42–49. AAAI Publications, Washington, DC (2010)

75. Ellison, N.B., Steinfield, C., Lampe, C.: The benefits of Facebook "Friends:" social capital and college students' use of online social network sites. J. Comput.-Mediat. Commun. 12(4), 1143–1168 (2007). https://doi.org/10.1111/j.1083-6101.2007.00367.x

76. Ellison, N.B., Steinfield, C., Lampe, C.: Connection strategies: social capital implications of Facebook-enabled communication practices. New Media Soc. 13(6), 873–892 (2011). https://doi.org/10.1177/1461444810385389

77. Ellison, N.B., Vitak, J., Steinfield, C., Gray, R., Lampe, C.: Negotiating privacy concerns and social capital needs in a social media environment. In: Trepte, S., Reinecke, L. (eds.) Privacy Online: Perspectives on Privacy and Self-Disclosure in the Social Web, pp. 19–32. Springer, Heidelberg (2011). https://doi.org/10.1007/978-3-642-21521-6_3

78. EU: Regulation (EU) 2016/679 of the European Parliament and the Council of 27 April 2016 on the protection of individuals with regard to the processing of personal data and on the free movement of such data, and repealing Directive 95/46/EC (General Data Protection Regulation) (2016). http://eur-lex.europa.eu/legal-content/EN/TXT/PDF/?uri=CELEX:32016R0679

79. European Commission: Factsheet on the "Right to be Forgotten" ruling. Technical report C-131/12, European Commission, Brussels, Belgium (2014). http://ec.europa.eu/justice/data-protection/files/factsheets/factsheet_data_protection_en.pdf

80. Fang, L., LeFevre, K.: Privacy wizards for social networking sites. In: Proceedings of the 19th International Conference on World Wide Web, WWW 2010, pp. 351–360. ACM, Raleigh (2010). https://doi.org/10.1145/1772690.1772727

81. Farzan, R., Brusilovsky, P.: Social navigation. In: Brusilovsky, P., He, D. (eds.) Social Information Access. LNCS, vol. 10100, pp. 142–180. Springer, Cham (2018)

82. Featherman, M.S., Pavlou, P.A.: Predicting e-services adoption: a perceived risk facets perspective. Int. J. Hum.-Comput. Stud. **59**(4), 451–474 (2003). https://doi.org/10.1016/S1071-5819(03)00111-3

83. Federal Trade Committee: Privacy Online: Fair Information Practices in the Electronic Marketplace. A Report to Congress (2000). http://www.ftc.gov/reports/privacy2000/privacy2000.pdf

84. Federal Trade Committee: Protecting Consumer Privacy in an Era of Rapid Change: Recommendations for Businesses and Policymakers (2012). https://www.ftc.gov/sites/default/files/documents/reports/federal-trade-commission-report-protecting-consumer-privacy-era-rapid-change-recommendations/120326privacyreport.pdf

85. Fishbein, M., Ajzen, I.: Belief, Attitude, Intention, and Behavior: An Introduction to Theory and Research. Addison-Wesley Pub. Co., Reading (1975)

86. Fiske, S.T., Hauser, R.M.: Protecting human research participants in the age of big data. Proc. Natl. Acad. Sci. **111**(38), 13,675–13,676 (2014). https://doi.org/10.1073/pnas.1414626111

87. Friedman, A., Knijnenburg, B.P., Vanhecke, K., Martens, L., Berkovsky, S.: Privacy aspects of recommender systems. In: Ricci, F., Rokach, L., Shapira, B. (eds.) Recommender Systems Handbook, 2nd edn, pp. 649–688. Springer, Boston, MA (2015). https://doi.org/10.1007/978-1-4899-7637-6_19

88. Friedrich, G., Zanker, M.: A taxonomy for generating explanations in recommender systems. AI Mag. **32**(3), 90–98 (2011). https://doi.org/10.1609/aimag.v32i3.2365

89. van de Garde-Perik, E., Markopoulos, P., de Ruyter, B., Eggen, B., Ijsselsteijn, W.: Investigating privacy attitudes and behavior in relation to personalization. Soc. Sci. Comput. Rev. **26**(1), 20–43 (2008). https://doi.org/10.1177/0894439307307682

90. Gedikli, F., Jannach, D., Ge, M.: How should I explain? A comparison of different explanation types for recommender systems. Int. J. Hum.-Comput. Stud. **72**(4), 367–382 (2014). https://doi.org/10.1016/j.ijhcs.2013.12.007

91. Goecks, J., Edwards, W.K., Mynatt, E.D.: Challenges in supporting end-user privacy and security management with social navigation. In: Proceedings of the 5th Symposium on Usable Privacy and Security, SOUPS 2009, pp. 5:1–5:12. ACM, Mountain View (2009). https://doi.org/10.1145/1572532.1572539

92. Gross, R., Acquisti, A.: Information revelation and privacy in online social networks. In: Proceedings of the 2005 ACM Workshop on Privacy in the Electronic Society, WPES 2005, pp. 71–80. ACM, Alexandria (2005). https://doi.org/10.1145/1102199.1102214

93. Guillaume, D.: Amazing mind reader reveals his 'gift' (2012). https://www.youtube.com/watch?v=F7pYHN9iC9I

94. Gürses, S., Diaz, C.: Two tales of privacy in online social networks. IEEE Secur. Priv. **11**(3), 29–37 (2013). https://doi.org/10.1109/MSP.2013.47

95. Guy, I.: People recommendation on social media. In: Brusilovsky, P., He, D. (eds.) Social Information Access. LNCS, vol. 10100, pp. 570–623. Springer, Cham (2018)

96. Hampton, K., Goulet, L.S., Marlow, C., Rainie, L.: Why most Facebook users get more than they give. Technical report, Pew Internet & American Life Project (2012). http://www.pewinternet.org/2012/02/03/why-most-facebook-users-get-more-than-they-give/

97. Han, S., He, D.: Network-based social search. In: Brusilovsky, P., He, D. (eds.) Social Information Access. LNCS, vol. 10100, pp. 277–309. Springer, Cham (2018)

98. Hancock, J., Birnholtz, J., Bazarova, N., Guillory, J., Perlin, J., Amos, B.: Butler lies: awareness, deception and design. In: Proceedings of the 27th International Conference on Human Factors in Computing Systems, CHI 2009, pp. 517–526. ACM, Boston (2009). https://doi.org/10.1145/1518701.1518782

99. Hancock, J.T., Thom-Santelli, J., Ritchie, T.: Deception and design: the impact of communication technology on lying behavior. In: Proceedings of the SIGCHI Conference on Human Factors in Computing Systems, CHI 2004, pp. 129–134. ACM, Vienna (2004). https://doi.org/10.1145/985692.985709

100. Hann, I.H., Hui, K.L., Lee, S.Y., Png, I.: Overcoming online information privacy concerns: an information-processing theory approach. J. Manag. Inf. Syst. **24**(2), 13–42 (2007). https://doi.org/10.2753/MIS0742-1222240202

101. Harris Interactive Inc.: A Survey of Consumer Privacy Attitudes and Behaviors (2000). http://www.bbbonline.org/UnderstandingPrivacy/library/harrissummary.pdf

102. Hasler, L., Ruthven, I.: Escaping information poverty through internet newsgroups. In: Fifth International AAAI Conference on Weblogs and Social Media, ICWSM 2011, pp. 153–160. AAAI Publications, Barcelona (2011)

103. Hassing, L.: An exploratory study in the concerns for information privacy. Master thesis, University of Delft, Netherlands (2015)

104. Henry, A.: How Facebook Is Using You to Annoy Your Friends (and How to Stop It) (2013). http://lifehacker.com/5987248/how-facebook-is-using-you-to-annoy-your-friends-and-how-to-stop-it

105. Ho, S.Y., Tam, K.: Understanding the impact of web personalization on user information processing and decision outcomes. MIS Q. **30**(4), 865–890 (2006)

106. Hong, J.: The state of phishing attacks. Commun. ACM **55**(1), 74–81 (2012). https://doi.org/10.1145/2063176.2063197

107. House, W.: Consumer data privacy in a networked world: a framework for protecting privacy and promoting innovation in the global economy. Technical report, White House, Washington, D.C. (2012)

108. Huang, J.H., Chen, Y.F.: Herding in online product choices. Psychol. Market. **23**(5), 413–428 (2006). https://doi.org/10.1002/mar.20119

109. Huber, J., Puto, C.: Market boundaries and product choice: illustrating attraction and substitution effects. J. Consum. Res. **10**(1), 31–44 (1983)

110. Huberman, B.A., Adar, E., Fine, L.A.: Valuating privacy. IEEE Secur. Priv. **3**(5), 22–25 (2005). https://doi.org/10.1109/MSP.2005.137

111. Hui, K.L., Tan, B.C.Y., Goh, C.Y.: Online information disclosure: motivators and measurements. ACM Trans. Internet Technol. **6**(4), 415–441 (2006). https://doi.org/10.1145/1183463.1183467

112. Hull, G., Lipford, H.R., Latulipe, C.: Contextual gaps: privacy issues on Facebook. Ethics Inf. Technol. **13**(4), 289–302 (2011). https://doi.org/10.1007/s10676-010-9224-8

113. Iachello, G., Hong, J.: End-user privacy in human-computer interaction. Found. Trends® Hum.-Comput. Interact. **1**(1), 1–137 (2007). https://doi.org/10.1561/1100000004

114. Iachello, G., Smith, I., Consolvo, S., Chen, M., Abowd, G.D.: Developing privacy guidelines for social location disclosure applications and services. In: Proceedings of the 2005 Symposium on Usable Privacy and Security, SOUPS 2005, pp. 65–76. ACM, Philadelphia (2005). https://doi.org/10.1145/1073001.1073008

115. Iyilade, J.: Enforcing privacy in secondary user information sharing and usage. In: Carberry, S., Weibelzahl, S., Micarelli, A., Semeraro, G. (eds.) UMAP 2013. LNCS, vol. 7899, pp. 396–400. Springer, Heidelberg (2013). https://doi.org/10. 1007/978-3-642-38844-6_48

116. Jannach, D., Lerche, L., Zanker, M.: Recommending based on implicit feedback. In: Brusilovsky, P., He, D. (eds.) Social Information Access. LNCS, vol. 10100, pp. 510–569. Springer, Cham (2018)

117. Jedrzejczyk, L., Price, B.A., Bandara, A.K., Nuseibeh, B.: On the impact of real-time feedback on users' behaviour in mobile location-sharing applications. In: Proceedings of the Sixth Symposium on Usable Privacy and Security, SOUPS 2010, Redmond, Washington, pp. 14:1–14:12 (2010). https://doi.org/10.1145/1837110. 1837129

118. Jensen, C., Potts, C., Jensen, C.: Privacy practices of internet users: self-reports versus observed behavior. Int. J. Hum.-Comput. Stud. **63**(1–2), 203–227 (2005). https://doi.org/10.1016/j.ijhcs.2005.04.019

119. John, L.K., Acquisti, A., Loewenstein, G.: Strangers on a plane: context-dependent willingness to divulge sensitive information. J. Consum. Res. **37**(5), 858–873 (2011). https://doi.org/10.1086/656423

120. Johnson, E.J., Bellman, S., Lohse, G.L.: Defaults, framing and privacy: why opting in ≠ opting out. Mark. Lett. **13**(1), 5–15 (2002). https://doi.org/10.1023/A: 1015044207315

121. Johnson, E.J., Goldstein, D.: Do defaults save lives? Science **302**(5649), 1338–1339 (2003). https://doi.org/10.1126/science.1091721

122. Johnson, H.M., Seifert, C.M.: Sources of the continued influence effect: when misinformation in memory affects later inferences. J. Exp. Psychol.: Learn. Mem. Cogn. **20**(6), 1420–1436 (1994). https://doi.org/10.1037/0278-7393.20.6.1420

123. Johnson, M., Egelman, S., Bellovin, S.M.: Facebook and privacy: it's complicated. In: Proceedings of the 8th Symposium on Usable Privacy and Security, SOUPS 2012, pp. 9:1–9:15. ACM, Pittsburgh (2012). https://doi.org/10.1145/2335356. 2335369

124. Joinson, A.N., Paine, C., Buchanan, T., Reips, U.D.: Measuring self-disclosure online: blurring and non-response to sensitive items in web-based surveys. Comput. Hum. Behav. **24**(5), 2158–2171 (2008). https://doi.org/10.1016/j.chb.2007. 10.005

125. Joinson, A.N., Reips, U.D., Buchanan, T., Schofield, C.B.P.: Privacy, trust, and self-disclosure online. Hum.-Comput. Interact. **25**(1), 1–24 (2010). https://doi. org/10.1080/07370020903586662

126. Juels, A.: Targeted advertising ... and privacy too. In: Naccache, D. (ed.) CT-RSA 2001. LNCS, vol. 2020, pp. 408–424. Springer, Heidelberg (2001). https:// doi.org/10.1007/3-540-45353-9_30

127. Kahneman, D., Knetsch, J.L., Thaler, R.H.: Anomalies: the endowment effect, loss aversion, and status quo bias. J. Econ. Perspect. **5**(1), 193–206 (1991). https:// doi.org/10.1257/jep.5.1.193

128. Kahneman, D., Tversky, A.: Choices, Values, and Frames. Cambridge University Press, Cambridge (2000)

129. Kairam, S., Brzozowski, M., Huffaker, D., Chi, E.: Talking in circles: selective sharing in Google+. In: Proceedings of the SIGCHI Conference on Human Factors in Computing Systems, CHI 2012, pp. 1065–1074. ACM, Austin (2012). https://doi.org/10.1145/2207676.2208552

130. Kantor, K.: People You May Know (2015). https://www.youtube.com/watch?v=LoyfunmYIpU

131. Kay, J., Kummerfeld, B.: Creating personalized systems that people can scrutinize and control: drivers, principles and experience. ACM Trans. Interact. Intell. Syst. **2**(4), 24:1–24:42 (2013). https://doi.org/10.1145/2395123.2395129

132. Kelley, P.G., Brewer, R., Mayer, Y., Cranor, L.F., Sadeh, N.: An investigation into Facebook friend grouping. In: Campos, P., Graham, N., Jorge, J., Nunes, N., Palanque, P., Winckler, M. (eds.) INTERACT 2011. LNCS, vol. 6948, pp. 216–233. Springer, Heidelberg (2011). https://doi.org/10.1007/978-3-642-23765-2_15

133. Kim, D.J., Ferrin, D.L., Rao, H.R.: A trust-based consumer decision-making model in electronic commerce: the role of trust, perceived risk, and their antecedents. Decis. Support Syst. **44**(2), 544–564 (2008). https://doi.org/10.1016/j.dss.2007.07.001

134. Kluver, D., Ekstrand, M., Konstan, J.: Rating-based collaborative filtering: algorithms and evaluation. In: Brusilovsky, P., He, D. (eds.) Social Information Access. LNCS, vol. 10100, pp. 344–390. Springer, Cham (2018)

135. Knijnenburg, B.P.: Simplifying privacy decisions: towards interactive and adaptive solutions. In: Proceedings of the Recsys 2013 Workshop on Human Decision Making in Recommender Systems, Decisions@Recsys 2013, Hong Kong, China, pp. 40–41 (2013)

136. Knijnenburg, B.P.: A user-tailored approach to privacy decision support. Ph.D. thesis, University of California, Irvine, Irvine, CA (2015). http://search.proquest.com/docview/1725139739/abstract

137. Knijnenburg, B.P., Jin, H.: The persuasive effect of privacy recommendations. In: Twelfth Annual Workshop on HCI Research in MIS, Milan, Italy, pp. 16:1–16:5 (2013)

138. Knijnenburg, B.P., Kobsa, A.: Helping users with information disclosure decisions: potential for adaptation. In: Proceedings of the 2013 ACM International Conference on Intelligent User Interfaces, IUI 2013, pp. 407–416. ACM Press, Santa Monica (2013). https://doi.org/10.1145/2449396.2449448

139. Knijnenburg, B.P., Kobsa, A.: Making decisions about privacy: information disclosure in context-aware recommender systems. ACM Trans. Interact. Intell. Syst. **3**(3), 20:1–20:23 (2013). https://doi.org/10.1145/2499670

140. Knijnenburg, B.P., Kobsa, A.: Increasing sharing tendency without reducing satisfaction: finding the best privacy-settings user interface for social networks. In: ICIS 2014 Proceedings, Auckland, New Zealand (2014)

141. Knijnenburg, B.P., Kobsa, A., Jin, H.: Counteracting the negative effect of form auto-completion on the privacy calculus. In: ICIS 2013 Proceedings, Milan, Italy (2013)

142. Knijnenburg, B.P., Kobsa, A., Jin, H.: Dimensionality of information disclosure behavior. Int. J. Hum.-Comput. Stud. **71**(12), 1144–1162 (2013). https://doi.org/10.1016/j.ijhcs.2013.06.003

143. Knijnenburg, B.P., Kobsa, A., Jin, H.: Preference-based location sharing: are more privacy options really better? In: Proceedings of the SIGCHI Conference on Human Factors in Computing Systems, CHI 2013, pp. 2667–2676. ACM, Paris (2013). https://doi.org/10.1145/2470654.2481369

144. Knijnenburg, B.P., Sivakumar, S., Wilkinson, D.: Recommender systems for self-actualization. In: Proceedings of the 10th ACM Conference on Recommender Systems, RecSys 2016, pp. 11–14. ACM, Boston (2016). https://doi.org/10.1145/2959100.2959189

145. Knijnenburg, B.P., Willemsen, M.C.: Evaluating recommender systems with user experiments. In: Ricci, F., Rokach, L., Shapira, B. (eds.) Recommender Systems Handbook, 2nd edn, pp. 309–352. Springer, Boston (2015). https://doi.org/10.1007/978-1-4899-7637-6_9

146. Knijnenburg, B.P., Willemsen, M.C., Gantner, Z., Soncu, H., Newell, C.: Explaining the user experience of recommender systems. User Model. User-Adap. Interact. **22**(4–5), 441–504 (2012). https://doi.org/10.1007/s11257-011-9118-4

147. Knijnenburg, B.P., Willemsen, M.C., Hirtbach, S.: Receiving recommendations and providing feedback: the user-experience of a recommender system. In: Buccafurri, F., Semeraro, G. (eds.) EC-Web 2010. LNBIP, vol. 61, pp. 207–216. Springer, Heidelberg (2010). https://doi.org/10.1007/978-3-642-15208-5_19

148. Kobsa, A.: Tailoring privacy to users' needs (invited keynote). In: Bauer, M., Gmytrasiewicz, P.J., Vassileva, J. (eds.) UM 2001. LNCS (LNAI), vol. 2109, pp. 303–313. Springer, Heidelberg (2001). https://doi.org/10.1007/3-540-44566-8_52

149. Kobsa, A., Cho, H., Knijnenburg, B.P.: The effect of personalization provider characteristics on privacy attitudes and behaviors: an elaboration likelihood model approach. J. Assoc. Inf. Sci. Technol. **67**, 2587–2606 (2016). https://doi.org/10.1002/asi.23629

150. Kobsa, A., Patil, S., Meyer, B.: Privacy in instant messaging: an impression management model. Behav. Inf. Technol. **31**(4), 355–370 (2012). https://doi.org/10.1080/01449291003611326

151. Kobsa, A., Schreck, J.: Privacy through pseudonymity in user-adaptive systems. ACM Trans. Internet Technol. **3**(2), 149–183 (2003). https://doi.org/10.1145/767193.767196

152. Kobsa, A., Teltzrow, M.: Contextualized communication of privacy practices and personalization benefits: impacts on users' data sharing and purchase behavior. In: Martin, D., Serjantov, A. (eds.) PET 2004. LNCS, vol. 3424, pp. 329–343. Springer, Heidelberg (2005). https://doi.org/10.1007/11423409_21

153. Koroleva, K., Krasnova, H., Veltri, N., Günther, O.: It's all about networking! empirical investigation of social capital formation on social network sites. In: ICIS 2011 Proceedings, Shanghai, China (2011)

154. Kosinski, M., Stillwell, D., Graepel, T.: Private traits and attributes are predictable from digital records of human behavior. Proc. Natl. Acad. Sci. **110**(15), 5802–5805 (2013). https://doi.org/10.1073/pnas.1218772110

155. Krämer, N.C., Haferkamp, N.: Online self-presentation: balancing privacy concerns and impression construction on social networking sites. In: Trepte, S., Reinecke, L. (eds.) Privacy Online: Perspectives on Privacy and Self-Disclosure in the Social Web, pp. 127–141. Springer, Heidelberg (2011). https://doi.org/10.1007/978-3-642-21521-6_10

156. Krasnova, H., Hildebrand, T., Guenther, O.: Investigating the value of privacy in online social networks: conjoint analysis. In: ICIS 2009 Proceedings, Phoenix, AZ (2009)

157. Krishnamurthy, B., Wills, C.: Privacy diffusion on the web: a longitudinal perspective. In: Proceedings of the 18th International Conference on World Wide Web, WWW 2009, pp. 541–550. ACM, Madrid (2009). https://doi.org/10.1145/1526709.1526782

158. Lai, Y.L., Hui, K.L.: Internet opt-in and opt-out: investigating the roles of frames, defaults and privacy concerns. In: Proceedings of the 2006 ACM SIGMIS Conference on Computer Personnel Research, Claremont, CA, pp. 253–263 (2006). https://doi.org/10.1145/1125170.1125230

159. Lampe, C., Ellison, N., Steinfield, C.: A face(book) in the crowd: social searching vs. social browsing. In: Proceedings of the 2006 20th Anniversary Conference on Computer Supported Cooperative Work, CSCW 2006, pp. 167–170. ACM, Banff (2006). https://doi.org/10.1145/1180875.1180901

160. Lampinen, A., Lehtinen, V., Lehmuskallio, A., Tamminen, S.: We're in it together: interpersonal management of disclosure in social network services. In: Proceedings of the SIGCHI Conference on Human Factors in Computing Systems, CHI 2011, pp. 3217–3226. ACM, New York (2011). https://doi.org/10.1145/1978942.1979420

161. Lanier, J.: You Are Not a Gadget: A Manifesto, Large Print edn. Thorndike Press, Waterville (2010)

162. Lathia, N., Hailes, S., Capra, L.: Private distributed collaborative filtering using estimated concordance measures. In: Proceedings of the 2007 ACM Conference on Recommender Systems, RecSys 2007, pp. 1–8. ACM, New York (2007)

163. Laufer, R.S., Proshansky, H.M., Wolfe, M.: Some analytic dimensions of privacy. In: Küller, R. (ed.) Proceedings of the Lund Conference on Architectural Psychology. Dowden, Hutchinson & Ross, Lund (1973)

164. Laufer, R.S., Wolfe, M.: Privacy as a concept and a social issue: a multidimensional developmental theory. J. Soc. Issues **33**(3), 22–42 (1977). https://doi.org/10.1111/j.1540-4560.1977.tb01880.x

165. Leary, M.R.: Self-presentation: Impression Management and Interpersonal Behavior. Routledge, Boulder (1996). ISBN 978-0-8133-3004-4

166. Lederer, S., Hong, J.I., Dey, A.K., Landay, J.A.: Personal privacy through understanding and action: five pitfalls for designers. Pers. Ubiquit. Comput. **8**(6), 440–454 (2004). https://doi.org/10.1007/s00779-004-0304-9

167. Lederer, S., Mankoff, J., Dey, A.K.: Who wants to know what when? Privacy preference determinants in ubiquitous computing. In: Proceedings of the SIGCHI Conference on Human Factors in Computing Systems, CHI 2003, pp. 724–725. ACM, Ft. Lauderdale (2003). https://doi.org/10.1145/765891.765952

168. Lee, D., Brusilovsky, P.: Recommendations based on social links. In: Brusilovsky, P., He, D. (eds.) Social Information Access. LNCS, vol. 10100, pp. 391–440. Springer, Cham (2018)

169. Lee, H., Park, S.J.: MONERS: a news recommender for the mobile web. Expert Syst. Appl. **32**(1), 143–150 (2007). https://doi.org/10.1016/j.eswa.2005.11.010

170. Lee, J.S., Lee, J.C.: Context awareness by case-based reasoning in a music recommendation system. In: Ichikawa, H., Cho, W.-D., Satoh, I., Youn, H.Y. (eds.) UCS 2007. LNCS, vol. 4836, pp. 45–58. Springer, Heidelberg (2007). https://doi.org/10.1007/978-3-540-76772-5_4

171. Li, H., Sarathy, R., Xu, H.: The role of affect and cognition on online consumers' decision to disclose personal information to unfamiliar online vendors. Decis. Support Syst. **51**(3), 434–445 (2011). https://doi.org/10.1016/j.dss.2011.01.017

172. Li, Y.: Theories in online information privacy research: a critical review and an integrated framework. Decis. Support Syst. **54**(1), 471–481 (2012). https://doi.org/10.1016/j.dss.2012.06.010

173. Litt, E.: Knock, knock. who's there? The imagined audience. J. Broadcast. Electron. Media **56**(3), 330–345 (2012). https://doi.org/10.1080/08838151.2012.705195

174. Liu, Y., Gummadi, K.P., Krishnamurthy, B., Mislove, A.: Analyzing Facebook privacy settings: user expectations vs. reality. In: Proceedings of the 2011 ACM SIGCOMM Conference on Internet Measurement Conference, IMC 2011, pp. 61–70. ACM, Berlin (2011). https://doi.org/10.1145/2068816.2068823

175. Lou, J., Fang, Y., Lim, K.H., Peng, J.Z.: Contributing high quantity and quality knowledge to online Q&A communities. J. Am. Soc. Inf. Sci. Technol. **64**(2), 356–371 (2013). https://doi.org/10.1002/asi.22750

176. Lowry, P.B., Moody, G., Vance, A., Jensen, M., Jenkins, J., Wells, T.: Using an elaboration likelihood approach to better understand the persuasiveness of website privacy assurance cues for online consumers. J. Am. Soc. Inf. Sci. Technol. **63**(4), 755–776 (2012). https://doi.org/10.1002/asi.21705

177. Lustig, C., Pine, K., Nardi, B., Irani, L., Lee, M.K., Nafus, D., Sandvig, C.: Algorithmic authority: the ethics, politics, and economics of algorithms that interpret, decide, and manage. In: Proceedings of the 2016 CHI Conference Extended Abstracts on Human Factors in Computing Systems, CHI EA 2016, pp. 1057–1062. ACM, New York (2016). https://doi.org/10.1145/2851581.2886426

178. Machanavajjhala, A., Korolova, A., Sarma, A.D.: Personalized social recommendations: accurate or private. Proc. VLDB Endow. **4**(7), 440–450 (2011). https://doi.org/10.14778/1988776.1988780

179. Madden, M.: Privacy management on social media sites. Technical report, Pew Internet & American Life Project, Pew Research Center, Washington, DC (2012). http://www.pewinternet.org/2012/02/24/privacy-management-on-social-media-sites/

180. Madejski, M., Johnson, M., Bellovin, S.: A study of privacy settings errors in an online social network. In: 2012 IEEE International Conference on Pervasive Computing and Communications Workshops (PERCOM Workshops), SESOC 2012, Lugano, Switzerland, pp. 340–345 (2012). https://doi.org/10.1109/PerComW.2012.6197507

181. Marwick, A.E., Boyd, D.: I Tweet honestly, I Tweet passionately: Twitter users, context collapse, and the imagined audience. New Media Soc. **13**(1), 114–133 (2011). https://doi.org/10.1177/1461444810365313

182. McGinty, L., Smyth, B.: Adaptive selection: an analysis of critiquing and preference-based feedback in conversational recommender systems. Int. J. Electron. Commer. **11**(2), 35–57 (2006)

183. McKenzie, C.R.M., Liersch, M.J., Finkelstein, S.R.: Recommendations implicit in policy defaults. Psychol. Sci. **17**(5), 414–420 (2006). https://doi.org/10.1111/j.1467-9280.2006.01721.x

184. McSherry, F., Mironov, I.: Differentially private recommender systems: building privacy into the net. In: Proceedings of the 15th ACM SIGKDD International Conference on Knowledge Discovery and Data Mining, KDD 2009, pp. 627–636. ACM, Paris (2009). https://doi.org/10.1145/1557019.1557090

185. Meisler, B.: Why Are Dead People Liking Stuff on Facebook? (2012). http://readwrite.com/2012/12/11/why-are-dead-people-liking-stuff-on-facebook

186. Metzger, M.J.: Privacy, trust, and disclosure: exploring barriers to electronic commerce. J. Comput.-Mediat. Commun. **9**(4) (2004). https://doi.org/10.1111/j.1083-6101.2004.tb00292.x

187. Metzger, M.J.: Effects of site, vendor, and consumer characteristics on web site trust and disclosure. Commun. Res. **33**(3), 155–179 (2006). https://doi.org/10.1177/0093650206287076

188. Metzger, M.J.: Communication privacy management in electronic commerce. J. Comput.-Mediat. Commun. **12**(2), 335–361 (2007). https://doi.org/10.1111/j.1083-6101.2007.00328.x
189. Meyers, J.: How to Remove Your Name and Profile Picture from Facebook's Social Ads (2011). http://internet.wonderhowto.com/how-to/remove-your-name-and-profile-picture-from-facebooks-social-ads-0126187/
190. Mikians, J., Gyarmati, L., Erramilli, V., Laoutaris, N.: Detecting price and search discrimination on the internet. In: Proceedings of the 11th ACM Workshop on Hot Topics in Networks, HotNets-XI, pp. 79–84. ACM, Redmond (2012). https://doi.org/10.1145/2390231.2390245
191. Mikians, J., Gyarmati, L., Erramilli, V., Laoutaris, N.: Crowd-assisted search for price discrimination in e-commerce: first results. In: Proceedings of the Ninth ACM Conference on Emerging Networking Experiments and Technologies, CoNEXT 2013, pp. 1–6. ACM, Santa Barbara (2013). https://doi.org/10.1145/2535372.2535415
192. Milberg, S.J., Burke, S.J., Smith, H.J., Kallman, E.A.: Values, personal information, privacy and regulatory approaches. Commun. ACM **38**(12), 65–74 (1995). https://doi.org/10.1145/219663.219683
193. Milne, G.R.: Consumer participation in mailing lists: a field experiment. J. Public Policy Mark. **16**(2), 298–309 (1997)
194. Milne, G.R., Gordon, M.E.: Direct mail privacy-efficiency trade-offs within an implied social contract framework. J. Public Policy Mark. **12**(2), 206–215 (1993). https://doi.org/10.2307/30000091
195. Mirzadeh, N., Ricci, F., Bansal, M.: Feature selection methods for conversational recommender systems. In: Proceedings of the 2005 IEEE International Conference on e-Technology, e-Commerce and e-Service, EEE 2005, Hong Kong, China, pp. 772–777 (2005). https://doi.org/10.1109/EEE.2005.75
196. Moult, J.: Woman sacked on Facebook for complaining about her boss after forgetting she had added him as a friend (2009). http://www.dailymail.co.uk/news/article-1206491/Woman-sacked-Facebook-boss-insult-forgetting-added-friend.html
197. Mulligan, D., Schwartz, A.: Your place or mine?: privacy concerns and solutions for server and client-side storage of personal information. In: Tenth Conference on Computers, Freedom and Privacy, Toronto, Ontario, pp. 81–84 (2000). https://doi.org/10.1145/332186.332255
198. Narayanan, A., Shmatikov, V.: Robust de-anonymization of large sparse datasets. In: 2008 IEEE Symposium on Security and Privacy, pp. 111–125. IEEE (2008). https://doi.org/10.1109/SP.2008.33
199. Navarro Bullock, B., Hotho, A., Stumme, G.: Accessing information with tags: search and ranking. In: Brusilovsky, P., He, D. (eds.) Social Information Access. LNCS, vol. 10100, pp. 310–343. Springer, Cham (2018)
200. Newman, A.L.: What the "right to be forgotten" means for privacy in a digital age. Science **347**(6221), 507–508 (2015). https://doi.org/10.1126/science.aaa4603
201. Nissenbaum, H.: Privacy as contextual integrity. Wash. Law Rev. **79**, 119–157 (2004)
202. Nissenbaum, H.: Privacy in Context: Technology, Policy, and the Integrity of Social Life. Stanford University Press, Stanford (2009)
203. Norberg, P.A., Horne, D.R., Horne, D.A.: The privacy paradox: personal information disclosure intentions versus behaviors. J. Consum. Aff. **41**(1), 100–126 (2007). https://doi.org/10.1111/j.1745-6606.2006.00070.x

204. Norcie, G., Blythe, J., Caine, K., Camp, L.J.: Why Johnny can't blow the whistle: identifying and reducing usability issues in anonymity systems. In: NDSS 2014 Workshop on Usable Security, USEC 2014. Internet Society, San Diego (2014). https://doi.org/10.14722/usec.2014.23022
205. Oh, J.M., Moon, N.: User-selectable interactive recommendation system in mobile environment. Multimedia Tools Appl. **57**(2), 295–313 (2012). https://doi.org/10.1007/s11042-011-0737-x
206. Oh, S.: Social Q&A. In: Brusilovsky, P., He, D. (eds.) Social Information Access. LNCS, vol. 10100, pp. 75–107. Springer, Cham (2018)
207. Olivero, N., Lunt, P.: Privacy versus willingness to disclose in e-commerce exchanges: the effect of risk awareness on the relative role of trust and control. J. Econ. Psychol. **25**(2), 243–262 (2004). https://doi.org/10.1016/S0167-4870(02)00172-1
208. Olson, J.S., Grudin, J., Horvitz, E.: A study of preferences for sharing and privacy. In: CHI 2005 Extended Abstracts, pp. 1985–1988. ACM, Portland (2005). https://doi.org/10.1145/1056808.1057073
209. O'Mahoney, M., Smyth, B.: From opinions to recommendations. In: Brusilovsky, P., He, D. (eds.) Social Information Access. LNCS, vol. 10100, pp. 480–509. Springer, Cham (2018)
210. Orwell, G.: Nineteen Eighty-Four. A Novel. Secker & Warburg, London (1949)
211. Page, X., Knijnenburg, B.P., Kobsa, A.: FYI: communication style preferences underlie differences in location-sharing adoption and usage. In: Proceedings of the 2013 ACM International Joint Conference on Pervasive and Ubiquitous Computing, UbiComp 2013, pp. 153–162. ACM, Zurich (2013). https://doi.org/10.1145/2493432.2493487
212. Page, X., Knijnenburg, B.P., Kobsa, A.: What a tangled web we weave: lying backfires in location-sharing social media. In: Proceedings of the 2013 Conference on Computer Supported Cooperative Work, CSCW 2013, pp. 273–284. ACM, San Antonio (2013). https://doi.org/10.1145/2441776.2441808
213. Page, X., Kobsa, A., Knijnenburg, B.P.: Don't disturb my circles! boundary preservation is at the center of location-sharing concerns. In: Proceedings of the Sixth International AAAI Conference on Weblogs and Social Media, ICWSM 2012, pp. 266–273. AAAI Publications, Dublin (2012)
214. Page, X., Tang, K., Stutzman, F., Lampinen, A.: Measuring networked social privacy. In: Proceedings of the 2013 Conference on Computer Supported Cooperative Work Companion, CSCW 2013, pp. 315–320. ACM, San Antonio (2013). https://doi.org/10.1145/2441955.2442032
215. Page, X.W.: Factors that influence adoption and use of location-sharing social media. Ph.D. thesis, University of California, Irvine, Irvine, CA (2014). http://gradworks.umi.com/36/06/3606987.html
216. Pallapa, G., Das, S.K., Di Francesco, M., Aura, T.: Adaptive and context-aware privacy preservation exploiting user interactions in smart environments. Pervasive Mob. Comput. **12**, 232–243 (2014). https://doi.org/10.1016/j.pmcj.2013.12.004
217. Pariser, E.: The Filter Bubble: How the New Personalized Web is Changing What We Read and How We Think. Penguin Books, New York (2012)
218. Park, C.W., Jun, S.Y., MacInnis, D.J.: Choosing what I want versus rejecting what I do not want: an application of decision framing to product option choice decisions. J. Mark. Res. **37**(2), 187–202 (2000). https://doi.org/10.1509/jmkr.37.2.187.18731

219. Patil, S., Lai, J.: Who gets to know what when: configuring privacy permissions in an awareness application. In: Proceedings of the SIGCHI Conference on Human Factors in Computing Systems, CHI 2005, pp. 101–110. ACM, Portland (2005). https://doi.org/10.1145/1054972.1054987

220. Patil, S., Norcie, G., Kapadia, A., Lee, A.J.: Reasons, rewards, regrets: privacy considerations in location sharing as an interactive practice. In: Proceedings of the Eighth Symposium on Usable Privacy and Security, SOUPS 2012, pp. 5:1–5:15. ACM, Washington, DC (2012). https://doi.org/10.1145/2335356.2335363

221. Patil, S., Page, X., Kobsa, A.: With a little help from my friends: can social navigation inform interpersonal privacy preferences? In: Proceedings of the ACM 2011 Conference on Computer Supported Cooperative Work, CSCW 2011, pp. 391–394. ACM, Hangzhou (2011). https://doi.org/10.1145/1958824.1958885

222. Pavlou, P.A.: Consumer acceptance of electronic commerce: integrating trust and risk with the technology acceptance model. Int. J. Electron. Commer. 7(3), 101–134 (2003)

223. Pavlou, P.A.: State of the information privacy literature: where are we now and where should we go. MIS Q. 35(4), 977–988 (2011)

224. Pedersen, D.M.: Psychological functions of privacy. J. Environ. Psychol. 17(2), 147–156 (1997). https://doi.org/10.1006/jevp.1997.0049

225. Perlow, J.: New in labs: stop sending mail you later regret (2008). http://gmailblog.blogspot.com/2008/10/new-in-labs-stop-sending-mail-you-later.html

226. Petronio, S.: Communication boundary management: a theoretical model of managing disclosure of private information between marital couples. Commun. Theory 1(4), 311–335 (1991). https://doi.org/10.1111/j.1468-2885.1991.tb00023.x

227. Petronio, S.: Communication privacy management theory: what do we know about family privacy regulation? J. Fam. Theory Rev. 2(3), 175–196 (2010). https://doi.org/10.1111/j.1756-2589.2010.00052.x

228. Phelps, J., Nowak, G., Ferrell, E.: Privacy concerns and consumer willingness to provide personal information. J. Public Policy Mark. 19(1), 27–41 (2000). https://doi.org/10.1509/jppm.19.1.27.16941

229. Preibusch, S., Krol, K., Beresford, A.R.: The privacy economics of voluntary over-disclosure in web forms. In: Böhme, R. (ed.) WEIS 2012, pp. 183–203. Springer, Heidelberg (2013). https://doi.org/10.1007/978-3-642-39498-0_9

230. Raban, D.R., Danan, A., Ronen, I., Guy, I.: Impression management through people tagging in the enterprise: implications for social media sampling and design. J. Inf. Sci. 43, 295–315 (2016). https://doi.org/10.1177/0165551516636305

231. Rashid, A.M., Albert, I., Cosley, D., Lam, S.K., McNee, S.M., Konstan, J.A., Riedl, J.: Getting to know you: learning new user preferences in recommender systems. In: Proceedings of the 7th International Conference on Intelligent User Interfaces, IUI 2002, pp. 127–134. ACM, San Francisco (2002). https://doi.org/10.1145/502716.502737

232. Ravichandran, R., Benisch, M., Kelley, P.G., Sadeh, N.M.: Capturing social networking privacy preferences. In: Goldberg, I., Atallah, M.J. (eds.) PETS 2009. LNCS, vol. 5672, pp. 1–18. Springer, Heidelberg (2009). https://doi.org/10.1007/978-3-642-03168-7_1

233. Razavi, M.N., Iverson, L.: Improving personal privacy in social systems with people-tagging. In: Proceedings of the ACM 2009 International Conference on Supporting Group Work, GROUP 2009, pp. 11–20. ACM, New York (2009). https://doi.org/10.1145/1531674.1531677

234. Resnick, P., Kuwabara, K., Zeckhauser, R., Friedman, E.: Reputation systems. Commun. ACM 43(12), 45–48 (2000). https://doi.org/10.1145/355112.355122

235. Riboni, D., Bettini, C.: Private context-aware recommendation of points of interest: an initial investigation. In: Proceedings of the IEEE International Conference on Pervasive Computing and Communications Workshops, pp. 584–589. IEEE Computer Society, Los Alamitos (2012). https://doi.org/10.1109/PerComW.2012.6197582

236. Richmond, R.: LinkedIn's Social-Ad Misstep (2011). http://gadgetwise.blogs.nytimes.com/2011/08/17/linkedins-social-ad-misstep/

237. Riva, G.: The sociocognitive psychology of computer-mediated communication: the present and future of technology-based interactions. CyberPsychol. Behav. **5**(6), 581–598 (2002). https://doi.org/10.1089/109493102321018222

238. Ronson, J.: How One Stupid Tweet Blew Up Justine Sacco's Life. The New York Times (2015). http://www.nytimes.com/2015/02/15/magazine/how-one-stupid-tweet-ruined-justine-saccos-life.html

239. Rubinstein, I.S., Good, N.: Privacy by design: a counterfactual analysis of Google and Facebook privacy incidents. Berkeley Technol. Law J. **28**, 1333–1414 (2013). https://doi.org/10.15779/Z38G11N

240. Rust, R.T., Kannan, P.K., Peng, N.: The customer economics of internet privacy. J. Acad. Mark. Sci. **30**(4), 455–464 (2002). https://doi.org/10.1177/009207002236917

241. Sadeh, N., Hong, J., Cranor, L., Fette, I., Kelley, P., Prabaker, M., Rao, J.: Understanding and capturing people's privacy policies in a mobile social networking application. Pers. Ubiquit. Comput. **13**(6), 401–412 (2009). https://doi.org/10.1007/s00779-008-0214-3

242. Samuelson, W., Zeckhauser, R.: Status quo bias in decision making. J. Risk Uncertain. **1**(1), 7–59 (1988). https://doi.org/10.1007/BF00055564

243. Sawer, P.: Facebook party leads to riots in Dutch town (2012). http://www.telegraph.co.uk/news/worldnews/europe/netherlands/9559868/Facebook-party-leads-to-riots-in-Dutch-town.html

244. Schafer, J.B., Konstan, J.A., Riedl, J.: E-commerce recommendation applications. Data Min. Knowl. Disc. **5**, 115–153 (2001). https://doi.org/10.1023/A:1009804230409

245. Schau, H.J., Gilly, M.C.: We are what we post? Self-presentation in personal web space. J. Consum. Res. **30**(3), 385–404 (2003). https://doi.org/10.1086/378616

246. Sheehan, K.B.: Toward a typology of internet users and online privacy concerns. Inf. Soc. **18**(1), 21–32 (2002). https://doi.org/10.1080/01972240252818207

247. Shelton, M., Lo, K., Nardi, B.: Online media forums as separate social lives: a qualitative study of disclosure within and beyond Reddit. In: iConference 2015 Proceedings, Newport Beach, CA (2015)

248. Sheng, H., Nah, F.F.H., Siau, K.: An experimental study on ubiquitous commerce adoption: impact of personalization and privacy concerns. J. Assoc. Inf. Syst. **9**(6), 344–376 (2008)

249. Sher, S., McKenzie, C.R.M.: Information leakage from logically equivalent frames. Cognition **101**(3), 467–494 (2006). https://doi.org/10.1016/j.cognition.2005.11.001

250. Shi, P., Xu, H., Erickson, L., Zhang, C.: See friendship: interpersonal privacy management in a collective world. In: AMCIS 2012 Proceedings (2012)

251. Shklovski, I., Mainwaring, S.D., Skúladóttir, H.H., Borgthorsson, H.: Leakiness and creepiness in app space: perceptions of privacy and mobile app use. In: Proceedings of the 32nd Annual ACM Conference on Human Factors in Computing Systems, CHI 2014, pp. 2347–2356. ACM, Toronto (2014). https://doi.org/10.1145/2556288.2557421

252. Shokri, R., Pedarsani, P., Theodorakopoulos, G., Hubaux, J.P.: Preserving privacy in collaborative filtering through distributed aggregation of offline profiles. In: RecSys, pp. 157–164 (2009)

253. Simonson, I.: Choice based on reasons: the case of attraction and compromise effects. J. Consum. Res. **16**(2), 158–174 (1989)

254. Singel, R.: Netflix Spilled Your Brokeback Mountain Secret, Lawsuit Claims (2009). http://www.wired.com/2009/12/netflix-privacy-lawsuit

255. Sleeper, M., Balebako, R., Das, S., McConahy, A.L., Wiese, J., Cranor, L.F.: The post that wasn't: exploring self-censorship on Facebook. In: Proceedings of the 2013 Conference on Computer Supported Cooperative Work, CSCW 2013, pp. 793–802. ACM, San Antonio (2013). https://doi.org/10.1145/2441776.2441865

256. Smith, A.: 6 new facts about Facebook (2014)

257. Smith, C., Kanalley, C.: Fired Over Facebook: 13 Facebook Posts That Got People CANNED (2010). http://www.huffingtonpost.com/2010/07/26/fired-over-facebook-posts_n_659170.html

258. Smith, H.J., Dinev, T., Xu, H.: Information privacy research: an interdisciplinary review. MIS Q. **35**(4), 989–1016 (2011)

259. Smith, H.J., Milberg, S.J., Burke, S.J.: Information privacy: measuring individuals' concerns about organizational practices. MIS Q. **20**(2), 167–196 (1996). https://doi.org/10.2307/249477

260. Smith, H.J., Rogers, Y.: Managing one's social network: does age make a difference? In: Human Computer Interaction - INTERACT 2003, pp. 551–558. IOS Press (2003)

261. Smith, N.C., Goldstein, D.G., Johnson, E.J.: Choice without awareness: ethical and policy implications of defaults. J. Public Policy Mark. **32**(2), 159–172 (2013). https://doi.org/10.1509/jppm.10.114

262. Solove, D.J.: The Digital Person: Technology and Privacy in the Information Age. New York University Press, New York (2004)

263. Solove, D.J.: A taxonomy of privacy. Univ. Pa. Law Rev. **154**(3), 477–564 (2006). https://doi.org/10.2307/40041279

264. Solove, D.J.: Understanding Privacy. Harvard University Press, Cambridge (2008)

265. Solove, D.J.: Privacy self-management and the consent dilemma. Harv. Law Rev. **126**, 1880–1903 (2013)

266. Spiekermann, S.: User Control in Ubiquitous Computing: Design Alternatives and User Acceptance. Shaker Verlag, Aachen (2008)

267. Spiekermann, S., Grossklags, J., Berendt, B.: E-privacy in 2nd generation e-commerce: privacy preferences versus actual behavior. In: Proceedings of the 3rd ACM Conference on Electronic Commerce, Tampa, FL, pp. 38–47 (2001). https://doi.org/10.1145/501158.501163

268. Stilgherrian: Why big data evangelists need to be reprogrammed (2014). http://www.zdnet.com/article/why-big-data-evangelists-need-to-be-reprogrammed/

269. Stone, D.L.: The effects of the valence of outcomes for providing data and the perceived relevance of the data requested on privacy-related behaviors, beliefs, and attitudes. Ph.D. thesis, Purdue University (1981)

270. Strater, K., Lipford, H.R.: Strategies and struggles with privacy in an online social networking community. In: Proceedings of the 22nd British HCI Group Annual Conference on People and Computers, BCS-HCI 2008, pp. 111–119. British Computer Society, Swinton (2008)

271. Stutzman, F., Kramer-Duffield, J.: Friends only: examining a privacy-enhancing behavior in Facebook. In: Proceedings of the 28th International Conference on

Human Factors in Computing Systems, CHI 2010, pp. 1553–1562. ACM, Atlanta (2010). https://doi.org/10.1145/1753326.1753559

272. Suler, J.: The online disinhibition effect. CyberPsychol. Behav. **7**(3), 321–326 (2004). https://doi.org/10.1089/1094931041291295

273. Sunstein, C.R., Thaler, R.H.: Libertarian paternalism is not an oxymoron. Univ. Chicago Law Rev. **70**(4), 1159–1202 (2003). https://doi.org/10.2307/1600573

274. Sutanto, J., Palme, E., Tan, C.H., Phang, C.W.: Addressing the personalization-privacy paradox: an empirical assessment from a field experiment on smartphone users. MIS Q. **37**(4), 1141–1164 (2013)

275. Tang, K., Hong, J., Siewiorek, D.: The implications of offering more disclosure choices for social location sharing. In: Proceedings of the SIGCHI Conference on Human Factors in Computing Systems, CHI 2012, pp. 391–394. ACM, Austin (2012). https://doi.org/10.1145/2207676.2207730

276. Tang, K., Lin, J., Hong, J., Siewiorek, D., Sadeh, N.: Rethinking location sharing: exploring the implications of social-driven vs. purpose-driven location sharing. In: Proceedings of the 12th ACM International Conference Adjunct Papers on Ubiquitous Computing, UbiComp 2010, pp. 85–94. ACM, Copenhagen (2010). https://doi.org/10.1145/1864349.1864363

277. Taylor, C.: Twitter Turns Ugly Over PR Person's Idiotic Tweet (2013). http://mashable.com/2013/12/20/justine-sacco/

278. Taylor, D., Davis, D., Jillapalli, R.: Privacy concern and online personalization: the moderating effects of information control and compensation. Electron. Commer. Res. **9**(3), 203–223 (2009). https://doi.org/10.1007/s10660-009-9036-2

279. Taylor, H.: Most people are "Privacy Pragmatists" who, while concerned about privacy, will sometimes trade it off for other benefits. Technical report 17, Harris Interactive, Inc. (2003)

280. Teltzrow, M., Kobsa, A.: Impacts of user privacy preferences on personalized systems: a comparative study. In: Karat, C.M., Blom, J., Karat, J. (eds.) Designing Personalized User Experiences for eCommerce, pp. 315–332. Kluwer Academic Publishers, Dordrecht (2004). https://doi.org/10.1007/1-4020-2148-8_17

281. Tene, O., Polonetsky, J.: Big data for all: privacy and user control in the age of analytics. Northwest. J. Technol. Intellect. Prop. **11**(5), 239–272 (2012)

282. Tene, O., Polonetsky, J.: A theory of creepy: technology, privacy and shifting social norms. Yale J. Law Technol. **16**, 59–102 (2013)

283. Thaler, R.H., Sunstein, C.: Nudge: Improving Decisions About Health, Wealth, and Happiness. Yale University Press, New Haven, London (2008)

284. Tintarev, N., Masthoff, J.: Evaluating the effectiveness of explanations for recommender systems. User Model. User-Adap. Interact. **22**(4–5), 399–439 (2012). https://doi.org/10.1007/s11257-011-9117-5

285. Tippmann, S., Powles, J.: Google accidentally reveals data on 'right to be forgotten' requests. The Guardian (2015). http://www.theguardian.com/technology/2015/jul/14/google-accidentally-reveals-right-to-be-forgotten-requests

286. Toch, E., Cranshaw, J., Drielsma, P.H., Tsai, J.Y., Kelley, P.G., Springfield, J., Cranor, L., Hong, J., Sadeh, N.: Empirical models of privacy in location sharing. In: Proceedings of the 12th ACM International Conference on Ubiquitous Computing, UbiComp 2010, pp. 129–138. ACM, Copenhagen (2010). https://doi.org/10.1145/1864349.1864364

287. Treiblmaier, H., Pollach, I.: Users' perceptions of benefits and costs of personalization. In: ICIS 2007 Proceedings (2007)

288. Tsai, J.Y., Kelley, P., Drielsma, P., Cranor, L.F., Hong, J., Sadeh, N.: Who's viewed you?: the impact of feedback in a mobile location-sharing application. In: Proceedings of the 27th International Conference on Human Factors in Computing Systems, CHI 2009, pp. 2003–2012. ACM, Boston (2009). https://doi.org/10.1145/1518701.1519005

289. Tufekci, Z.: Can you see me now? Audience and disclosure regulation in online social network sites. Bull. Sci. Technol. Soc. **28**(1), 20–36 (2008). https://doi.org/10.1177/0270467607311484

290. Turner, M.A., Varghese, R.: Making sense of the privacy debate: a comparative analysis of leading consumer privacy surveys. Technical report, Privacy & American Business (2002)

291. Tversky, A.: Elimination by aspects: a theory of choice. Psychol. Rev. **79**(4), 281–299 (1972). https://doi.org/10.1037/h0032955

292. Tynan, D.: Facebook ads use your face for free (2011). http://www.itworld.com/article/2746556/networking-hardware/facebook-ads-use-your-face-for-free.html

293. Vallet, D., Friedman, A., Berkovsky, S.: Matrix factorization without user data retention. In: Tseng, V.S., Ho, T.B., Zhou, Z.-H., Chen, A.L.P., Kao, H.-Y. (eds.) PAKDD 2014. LNCS (LNAI), vol. 8443, pp. 569–580. Springer, Cham (2014). https://doi.org/10.1007/978-3-319-06608-0_47

294. Vickery, J.R.: The curious case of Confession Bear: the reappropriation of online macro-image memes. Inf. Commun. Soc. **17**(3), 301–325 (2014). https://doi.org/10.1080/1369118X.2013.871056

295. Vig, J., Sen, S., Riedl, J.: Tagsplanations: explaining recommendations using tags. In: Proceedings of the 14th International Conference on Intelligent User Interfaces, IUI 2009, pp. 47–56. ACM, Sanibel Island (2009). https://doi.org/10.1145/1502650.1502661

296. Vitak, J., Blasiola, S., Litt, E., Patil, S.: Balancing audience and privacy tensions on social network sites: strategies of highly engaged users. Int. J. Commun. **9**, 1485–1504 (2015)

297. Walther, J.B., Parks, M.R.: Cues filtered out, cues filtered in. In: Knapp, M.L., Daly, J.A. (eds.) Handbook of Interpersonal Communication, 4th edn, pp. 529–563. SAGE Publications Inc., Thousand Oaks (2002)

298. Wang, N., Grossklags, J., Xu, H.: An online experiment of privacy authorization dialogues for social applications. In: Proceedings of the 2013 Conference on Computer Supported Cooperative Work, CSCW 2013, pp. 261–272. ACM, San Antonio (2013). https://doi.org/10.1145/2441776.2441807

299. Wang, W., Benbasat, I.: Recommendation agents for electronic commerce: effects of explanation facilities on trusting beliefs. J. Manag. Inf. Syst. **23**(4), 217–246 (2007). https://doi.org/10.2753/MIS0742-1222230410

300. Wang, Y., Leon, P.G., Acquisti, A., Cranor, L.F., Forget, A., Sadeh, N.: A field trial of privacy nudges for Facebook. In: Proceedings of the 32nd Annual ACM Conference on Human Factors in Computing Systems, CHI 2014, pp. 2367–2376. ACM, Toronto (2014). https://doi.org/10.1145/2556288.2557413

301. Wang, Y., Leon, P.G., Scott, K., Chen, X., Acquisti, A., Cranor, L.F.: Privacy nudges for social media: an exploratory Facebook study. In: Second International Workshop on Privacy and Security in Online Social Media, PSOSM 2013, Rio De Janeiro, Brazil, pp. 763–770 (2013)

302. Wang, Y., Norcie, G., Komanduri, S., Acquisti, A., Leon, P.G., Cranor, L.F.: "I regretted the minute I pressed share": a qualitative study of regrets on Facebook. In: Proceedings of the Seventh Symposium on Usable Privacy and Security, pp. 10:1–10:16. ACM, Pittsburgh (2011). https://doi.org/10.1145/2078827.2078841

303. Watson, J., Besmer, A., Lipford, H.R.: +Your circles: sharing behavior on Google+. In: Proceedings of the 8th Symposium on Usable Privacy and Security, SOUPS 2012, pp. 12:1–12:10. ACM, Pittsburgh (2012). https://doi.org/10.1145/2335356.2335373

304. Westin, A.F., Harris (Louis) and Associates, Inc.: The Dimensions of Privacy: A National Opinion Research Survey of Attitudes Toward Privacy. Garland Publishing, New York (1981)

305. Westin, A.F., Maurici, D.: E-commerce & privacy: what the net users want. Technical report, Privacy & American Business, and PricewaterhouseCoopers LLP (1998). http://www.pwcglobal.com/gx/eng/svcs/privacy/images/E-Commerce.pdf

306. Wikipedia: Facebook real-name policy controversy. https://en.wikipedia.org/w/index.php?title=Facebook_real-name_policy_controversy

307. Wilkinson, D., Sivakumar, S., Cherry, D., Knijnenburg, B.P., Raybourn, E.M., Wisniewski, P., Sloan, H.: User-tailored privacy by design. In: Submitted to USEC 2017 (2017)

308. Wilson, D., Proudfoot, J., Valacich, J.: Saving face on Facebook: privacy concerns, social benefits, and impression management. In: ICIS 2014 Proceedings (2014)

309. Wisniewski, P., Islam, A.N., Knijnenburg, B.P., Patil, S.: Give social network users the privacy they want. In: Proceedings of the 18th ACM Conference on Computer Supported Cooperative Work & Social Computing, CSCW 2015, pp. 1427–1441. ACM, Vancouver (2015). https://doi.org/10.1145/2675133.2675256

310. Wisniewski, P., Knijnenburg, B.P., Richter Lipford, H.: Profiling Facebook users' privacy behaviors. In: SOUPS2014 Workshop on Privacy Personas and Segmentation, PPS 2014, Menlo Park, CA (2014)

311. Wisniewski, P., Lipford, H., Wilson, D.: Fighting for my space: coping mechanisms for SNS boundary regulation. In: Proceedings of the SIGCHI Conference on Human Factors in Computing Systems, CHI 2012, pp. 609–618. ACM, Austin (2012). https://doi.org/10.1145/2207676.2207761

312. Wisniewski, P., Xu, H., Chen, Y.: Understanding user adaptation strategies for the launching of Facebook timeline. In: Proceedings of the SIGCHI Conference on Human Factors in Computing Systems, CHI 2014, pp. 2421–2430. ACM, Toronto (2014). https://doi.org/10.1145/2556288.2557363

313. Wisniewski, P.J., Knijnenburg, B.P., Lipford, H.R.: Making privacy personal: profiling social network users to inform privacy education and nudging. Int. J. Hum.-Comput. Stud. 98, 95–108 (2017). https://doi.org/10.1016/j.ijhcs.2016.09.006

314. Woong Yun, G., Park, S.Y.: Selective posting: willingness to post a message online. J. Comput.-Mediat. Commun. 16(2), 201–227 (2011). https://doi.org/10.1111/j.1083-6101.2010.01533.x

315. Wyer, R.S., Budesheim, T.L.: Person memory and judgments: the impact of information that one is told to disregard. J. Pers. Soc. Psychol. 53(1), 14–29 (1987). https://doi.org/10.1037/0022-3514.53.1.14

316. Xie, J., Knijnenburg, B.P., Jin, H.: Location sharing privacy preference: analysis and personalized recommendation. In: Proceedings of the 19th International Conference on Intelligent User Interfaces, IUI 2014, pp. 189–198. ACM, Haifa (2014). https://doi.org/10.1145/2557500.2557504

317. Xu, H., Luo, X.R., Carroll, J.M., Rosson, M.B.: The personalization privacy paradox: an exploratory study of decision making process for location-aware marketing. Decis. Support Syst. 51(1), 42–52 (2011). https://doi.org/10.1016/j.dss.2010.11.017

318. Xu, H., Teo, H.H., Tan, B.C.Y., Agarwal, R.: The role of push-pull technology in privacy calculus: the case of location-based services. J. Manag. Inf. Syst. **26**(3), 135–174 (2009). https://doi.org/10.2753/MIS0742-1222260305

319. Xu, H., Wang, N., Grossklags, J.: Privacy-by-redesign: alleviating privacy concerns for third-party applications. In: ICIS 2012 Proceedings, Orlando, FL (2012)

320. Yang, J., Morris, M.R., Teevan, J., Adamic, L.A., Ackerman, M.S.: Culture matters: a survey study of social Q&A behavior. In: Fifth International AAAI Conference on Weblogs and Social Media, ICWSM 2011, pp. 409–416. AAAI Publications (2011)

321. Yang, S.C., Hung, W.C., Sung, K., Farn, C.K.: Investigating initial trust toward e-tailers from the elaboration likelihood model perspective. Psychol. Mark. **23**(5), 429–445 (2006). https://doi.org/10.1002/mar.20120

322. Young, A.L., Quan-Haase, A.: Information revelation and internet privacy concerns on social network sites: a case study of Facebook. In: Proceedings of the Fourth International Conference on Communities and Technologies, pp. 265–274. ACM, University Park (2009). https://doi.org/10.1145/1556460.1556499

323. Yue, Z., He, D.: Collaborative search. In: Brusilovsky, P., He, D. (eds.) Social Information Access. LNCS, vol. 10100, pp. 108–141. Springer, Cham (2018)

324. Zaslow, J.: If TiVo Thinks You Are Gay, Here's How to Set It Straight (2002). http://www.wsj.com/articles/SB1038261936872356908

325. Zhang, B., Wang, N., Jin, H.: Privacy concerns in online recommender systems: influences of control and user data input. In: Symposium on Usable Privacy and Security, SOUPS 2014, Menlo Park, CA, pp. 159–173 (2014)

326. Zheng, V.W., Zheng, Y., Xie, X., Yang, Q.: Towards mobile intelligence: learning from GPS history data for collaborative recommendation. Artif. Intell. **184–185**, 17–37 (2012). https://doi.org/10.1016/j.artint.2012.02.002

327. Zhong, L.: My pins are my dreams: pinterest, collective daydreams, and the aspirational gap. M.S. thesis, Massachusetts Institute of Technology (2014)

328. Zhou, B., Pei, J.: Preserving privacy in social networks against neighborhood attacks. In: 2008 IEEE 24th International Conference on Data Engineering, pp. 506–515 (2008). https://doi.org/10.1109/ICDE.2008.4497459

329. Zhou, T.: Understanding users' initial trust in mobile banking: an elaboration likelihood perspective. Comput. Hum. Behav. **28**(4), 1518–1525 (2012). https://doi.org/10.1016/j.chb.2012.03.021

330. Zhu, T., Ren, Y., Zhou, W., Rong, J., Xiong, P.: An effective privacy preserving algorithm for neighborhood-based collaborative filtering. Future Gener. Comput. Syst. **36**, 142–155 (2014). https://doi.org/10.1016/j.future.2013.07.019

331. Zimmer, M.: OkCupid Study Reveals the Perils of Big-Data Science (2016). https://www.wired.com/2016/05/okcupid-study-reveals-perils-big-data-science/

3
Social Q&A

Sanghee Oh$^{(\boxtimes)}$ ⓘ

Department of Library and Information Science,
Chungnam National University, Daejeon, South Korea
sanghee.oh@cnu.ac.kr

Abstract. Social questioning and answering (social Q&A or SQA) is
a community-based online service on which peer users ask and answer
questions to and for one another about various topics in everyday life.
Social Q&A has been labeled with several variations, such as community
Q&A, collaborative Q&A, and online Q&A, but it most often refers to a
free and open Q&A site with dedicated users who subscribe to the service
to ask and answer questions. This encourages people to bring up their
various issues, to actively seek solutions and suggestions, and to share
personal experiences as well as to give and receive social and emotional
support. This chapter provides a literature review of the recent social
Q&A research and explains the theories and methods that have been
applied to conducting social Q&A research with examples from previous
studies in order to show a range of diverse approaches to examining user
behaviors and interactions in social Q&A.

1 Introduction

Social Q&A is a venue for creating an extensive volume of user-generated contents on the web, empowered by social information access technologies. Like
other social media, such as Wikis, blogs, and resource sharing sites, all of the
questions, answers, and additional data generated from users' activities, such as
user comments and ratings/votes for questions and answers, are accumulated
and available freely and publicly for searching and browsing within social Q&A
sites. Social Q&A has merits in that users can describe and discuss their personal
inquiries and experiences in as much detail as they want with natural language
in their questions and obtain customized answers to their situations instantly
with information, suggestions, advice, and opinions from anonymous others who
share a topic of interests. Answers that have been generated by the user community in social Q&A becomes the "wisdom of crowds," guiding people to access
information that have been created for the purposes to satisfy their personal
needs presented in questions.

This chapter introduces the research trends on social Q&A briefly. Shah et
al. [95] and Gazan [33] provided literature reviews of social Q&A research two
years apart. This chapter presents the research development on social Q&A after

© Springer International Publishing AG, part of Springer Nature 2018
P. Brusilovsky and D. He (Eds.): Social Information Access, LNCS 10100, pp. 75–107, 2018.
https://doi.org/10.1007/978-3-319-90092-6_3

that. First, the examples of the social Q&A services were introduced with their common and unique features of allowing users to ask and answer questions in their interfaces. Then, the theories and methods that have been applied to conducting social Q&A research are explained with examples from previous studies in order to show a wide range of diverse approaches to examining user behaviors and interactions in social Q&A.

This book also covers the related topics to social Q&A. For example, social Q&A users collaborate one another to find solutions to a common problem by sharing their personal experiences and information. Chapter 5 "Collaborative Search" explains the factors affecting collaborative information search in the user communities and the technologies to support it [110]. Asking questions in social Q&A allows for users to perform "social searches" of seeking information with anonymous others collaboratively [48]. Chapter 8 "Social Search" describes the various strategies for improving the search process of accessing social information emerged from the communities of users [14]. Chapter 9 "Network-based Social Search" [38] and Chap. 10 "Tag-based Social Search" [70] discuss the specific approaches to enhance social searches in the community of users.

2   Common Features

The common features of interface design in social Q&A sites include questions and threads of corresponding answers to the questions. In most cases, users receive points or badges (e.g., top contributors or leaders) as social rewards for participating in asking and answering questions. Such user reputation building systems are perceived to encourage competition and increase productivity among users in the site [68]. This user information is viewed along with the question and corresponding answers. Other users can vote for questions and answers and these ratings are available to review.

Yahoo! Answers, one of the most popular social Q&A services with about five million monthly visitors [80], has been most frequently used as a test bed in social Q&A research due to its high volume of scales and uses [1,12,21,30,40, 49,56,62,73,106]. Figure 1 shows an example of a question and its corresponding answers listed along with user IDs. It is a question about politics. The asker can provide an additional explanation of his question to clarify what he wants in answers. A total of 97 answers are given. One of the answers is selected as the best answer by the asker. The asker's rating on the best answer is also available. Yahoo! Answers allow users to collect points when they ask or answer questions. Figure 1 also shows a Leaderboard with user IDs and their earned points. These leaders are highlighted because they earned many points from various activities in Yahoo! Answers. For example, Yahoo! Answers users earn 2 points when answering questions and 10 points when their answers are selected as the best answer among others.

WikiAnswers is a wiki-style Q&A service that enables multiple users to collaborate in creating and revising one answer over time. WikiAnswers encourage users submitting questions using words like "Who, What, Where, When, Why, How, etc." and creating one answer to the question. Figure 2 shows an example

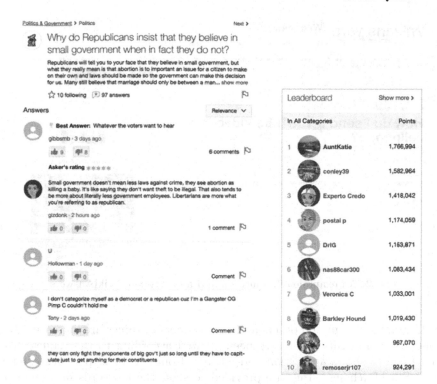

Fig. 1. An example of a question about politics and corresponding answers in Yahoo! Answers

of a question and an answer. There is an edit button so anyone can update information in the answer but it allows one answer only. The patterns of rephrasing or reformulating questions in WikiAnswers have been often studied [7,11].

Recently, several social-network-based Q&A services have emerged. Aardvark was launched in 2009, proposing a mechanism with which to create an expert community and enabling the experts to generate quick answers through instant messaging, emails, and Facebook [44]. Google purchased Aardvark in 2010 but discontinued it in 2011 due to low frequencies of use. Instead, Quora was launched in 2009 and has been popular with an average of 1.1 million monthly visits (Quantcast.com, 2015). Users ask and answer questions to their "friends" in social networks or acquaintances they get to know in Quora. User reputation or trust building has been important for users to gain popularity in Quora [78,103].

StackExchange, is another fast-growing social network-based Q&A. It was first launched as an online community for software developers and programmers to exchange questions and answers about technical programming languages or problems (named StackOverflow). There have been studies about computer programmers' or experts' question asking and answering behaviors and their contributions to StackOverflow or StackExchange [68,79,100,102]. The scope of topics has been expanded to such items as sports, languages, project management, pets,

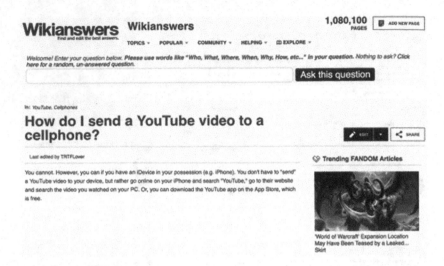

Fig. 2. An example of a question and an answer in WikiAnswers

health, graphic design and others and the numbers of topics are increasing [79]. About 160 community sites have been created according to the request from users. Figure 3 shows a question and answers from a User Experience community in StackExchange. Tags are provided to show the keywords in a question. The green check sign indicates that the answer is selected as a correct answer. There is about 8 comments on the question. Although it was not captured in the screen, there were about 2 or 3 comments attached to the 5 answers each. Users can vote on questions, answers, as well as comments.

There are non-English social Q&A services that are popular in use and research. *Naver Knowledge-iN* is a Korean social Q&A; it is the oldest one which has been successful in facilitating millions of visitors on a daily basis since its launch in 2002 [69]. Like Naver Knowledge-iN, *Baidu Knows* has been a huge success in China since its launch in 2005. It is known as the most popular and widely used Q&A site in China.

3   Research Trends

Social Q&A research has been carried out on various aspects of user interactions, information exchanges, and system designs to promote the activities within the sites. Shah et al. [95] described two streams of research related to users and contents in social Q&A. User-centered studies investigated various user roles of distributing knowledge and information [31,32] and locating authoritative users who provide good quality of answers in social Q&A [12,51,52]. Content-centered studies examined the quality of answers in social Q&A, proposing and testing a set of criteria with which to evaluate them [55,98] and worked on developing systematic algorithms with which to select high quality answers [63].

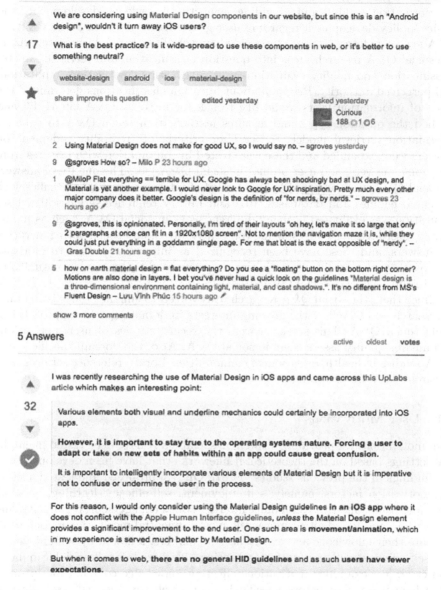

Fig. 3. An example of a question and answers from StackExchange (Color figure online)

Based on the review, Shah et al. [95] proposed a research agenda for future social Q&A, mainly illustrating three major areas, including users (investigating user needs, expectations, and motivations), information (evaluating cost and quality of questions and credibility of answers), and technology (developing better user interfaces, usability testing strategies, and business models of social Q&A), in addition to the potential areas of research regarding usage patterns and behaviors of information seeking and providing, collective knowledge development, and device/policy development regarding user-generated contents in social Q&A.

A few years later, Gazan [33] provided another literature review, categorizing the social Q&A research areas into question classification and retrieval, answer classification and quality evaluation, user satisfaction, motivation, reputation, and perceived authority. Research about question classifications determined the types of information users would like to ask for in social Q&A [39,41,47] and applied the question types and features learned from social Q&A to enhance information retrieval models [7,15]. User criteria of credibility judgment on answers were expanded and these were tested in combination with features from questions, answers, and user profiles in social Q&A [96]. The quality of answers in social Q&A was compared to the quality of information in other venues such as search engines and databases [101] and library reference services [91]. User reputation was evaluated in many different ways in social Q&A, such as points earned, user levels or badges, being top contributors, and social voting or rating on answers, and these have been recognized as important factors to motivate user participation in the site activities, to promote the sites, and, eventually, to lead to the success of social Q&A sites [94].

Since then, the social Q&A research has evolved to enhance the existing lines of research and to reflect the growing interests in domain-specific approaches of using social Q&A. This section covers the recent studies of user motivations and answer quality assessments in social Q&A. Also, the domain-specific social Q&A studies in health, open source communities, library reference services, and research Q&A are presented.

3.1   User Motivations

User motivations for providing answers in social Q&A have been tested in empirical settings based on the theoretical frameworks that have been developed from the findings in the previous studies [69,81]. Oh [73] proposed and tested a set of ten motivation factors, namely: self-enjoyment, self-efficacy, learning, personal gain, altruism, empathy, community interest, social engagement, reciprocity, and reputation, and found altruism was the most influential motivation among users to share their knowledge and experiences in answers, followed by self-enjoyment and self-efficacy, while reputation and reciprocity were less influential. Similarly, Lou et al. [66] examined motivations affecting knowledge contribution in social Q&A focusing on five factors, specifically: enjoy helping, knowledge self-efficacy, self-worth, learning, and rewards. Findings presented that intrinsic motivations such as learning, enjoying helping others, and self-efficacy have been significantly influential, while the effect of extrinsic motivations was relatively minimal. Jin et al. [50] proposed a research model of user motivations for continuing to answer questions in social Q&A, identifying the relationships among user motivations such as reputation enhancement, reciprocity, and enjoyment in helping others and user characteristics such as user satisfaction, user confirmation, and knowledge self-efficacy. The findings presented that user motivations are highly influenced by user satisfaction and knowledge self-efficacy.

Not only motivations for sharing, but also motivations for asking questions in social Q&A have been investigated. Zhang [111] reviewed the types of motivations that askers described in their questions on social Q&A sites and identified the following three factors: cognitive motivations for collecting information, social motivations for social supports, emotional motivations for sharing personal stories and feelings. Shah et al. [93] developed a framework of information seeking behaviors in online Q&A, consisting of modalities (sources and strategies), user motivations, and materials (contents) and emphasized the social aspects of collaborative information seeking in social Q&A. Choi et al. [18] defined five motivations pertaining to asking questions in social Q&A, including cognitive needs, affective needs, personal integrative needs, social integrative needs, and tension free needs, and tested the relationships between the motivations and user expectations from answers, including looking for quick responses, additional information, accurate or complete information, emotional support, verification of knowledge, and trustworthy sources.

The effect of a single motivation in social Q&A has been examined as well. Wu and Korfiatis [107] tested collective reciprocity as a motivation, which refers to behavioral patterns for allowing multilateral interactions among users by receiving benefits from exchanging questions and answers. They tested the association between users' efforts (i.e., answering questions) and benefits (i.e., posting questions) and the return from a social Q&A site (i.e., getting best answers ratings from Yahoo! Answers). Findings showed that the more users answer questions, the more likely the site would give back some favor. While the more users ask questions, however, the less likely the users would benefit from the site. Paul [78] investigated reputation building in a social network based Q&A, Quora. Users' reputation within a site is a critical cue for making judgments on their degrees of authority and, eventually, it influences evaluating the quality of answers given by the authoritative users. Findings showed that building reputations motivates users to participate since it could affect the development of their careers in real life due to the connections in social networks. Intrinsic motivations such as personal satisfaction and self-pleasure in researching new topics were influential for them, as well.

User motivations for asking and answering questions in social Q&A could vary across culture. Yang et al. [109] compared the motivations across four countries, the United States, the United Kingdom, China, and India. They found that Chinese are more likely ask questions not for fun but for social connectivity. It could be influenced by collectivity culture of people preferring to have rich and long-term social relationships with others and a prevention regulatory-focused culture of promoting safety and responsibility in a community. Both users in China and India are motivated to ask questions to those in their social networks. They also have commons in motivations for answering questions, affected by social reciprocity and social bonding.

The main reasons to not answer questions in social Q&A were also studied. Dearman and Truong [24] found that Yahoo! Answer users review the nature or content of a question and prefer not to answer if the question is not sincere, discriminates against a group or an individual, violates community rules, or involves illegal activities. Their time, effort and expertise required also matter. They do not answer a question with too many answers already given. Yang et al. [109] found other reasons such as not being interested, being too busy, or not knowing the answer. Privacy was an important reason for Chinese users of social Q&A. Users in the United States and the United Kingdom more likely hesitate to provide answers to those who do not know well.

Findings from the motivation studies shed lights on users' diverse intentions and perspectives on asking and answering questions in social Q&A and provide implications to enhance the interface design of social Q&A sites. Mamykina et al. [68] demonstrated the features that promote both intrinsic (i.e., altruism, learning) and extrinsic motivations (i.e., reputation building) could be integrated to the social Q&A interfaces using point systems, reward activities, or gamification mechanisms.

3.2 Answer Quality and Credibility Judgment

The studies about answer quality were expanded, covering various criteria used for the credibility assessment of answers. Social Q&A has been recognized as one of the frequently used social media platforms for seeking information from user generated contents [20, 53, 74] and users have applied various strategies to evaluate the content credibility. Kim et al. [53] noted that undergraduate students take actions to evaluate the quality of answers they find in social Q&A, checking out cited sources in answers, tone or style of answer writing, or reactions from other users (i.e., comments, votes, rates). Jeon and Rieh [49] found that users make credibility judgments on answers in social Q&A, considering answerers' involvement and efforts (attitude), answerers' intention or decency (trustworthiness), and answerers' expertise. There were also practical studies that assessed the quality of answers with a certain set of criteria. Fichman [30] evaluated the quality of answers across four social Q&A sites, i.e., Askville, WikiAnswers, Wikipedia Reference Desk and Yahoo! Answers, and found that the quality of answers across sites differs from one another significantly in terms of answer accuracy, completeness and verifiability and there was no correlation between the popularity and the answer quality in a social Q&A site. Oh and Worrall [75] identified ten user criteria for evaluating the quality of answers in social Q&A, in particular health answers, including accuracy, completeness, relevance, objectivity, readability, source credibility, empathy, politeness, confidence, and efforts, and then used them to evaluate the quality of health-related answers. There were significant differences in evaluating the quality of answers between the expert groups and the users in that the overall ratings of answer quality assessed by users were higher than those assessed by nurses and librarians across all of the criteria.

Answer quality in social Q&A is associated with external features as well as content credibility. Chua and Banerjee [21] examined the correlation between answer quality and the answer speed at which an answer is posted responding to a question. There was no significant relationships between answer quality and answer speed; the best answers are posted much later than the fastest answers, although the answer speed could differ across the types of questions asked. Li et al. [60] defined two types of features pertaining to answer quality assessment, web-captured and human-coded. Web-captured features include several auto-generated user point systems in a social Q&A site and the answer response time, and answer length. Human-coded features included social elements, consensus building, factual information, resource provision, references, opinions, and personal experiences. The relationships with these features and the answer quality across different degrees were examined.

3.3 Domain-Specific Studies

3.3.1 Health

Social Q&A has been an effective venue for those who have health concerns with a great potential for sharing information and experiences without exposing their personal identities. Social Q&A users can easily bring up their private and intimate health concerns and elaborate on their situations and experiences in questions and answers. Those who are suspicious about having a symptom can receive a quick answer if their condition is serious enough to go to see a physician. Or, those who are diagnosed and under treatment for a disease may want to seek additional information and social supports from others in social Q&A who have had experiences with similar diseases.

Health has been selected as one of the major topics about which to observe user interactions in social Q&A research [39]. The motivations for asking and answering health questions were investigated and the needs for affective and social supports have been observed from both askers and answerers [73,111]. The quality of health answers was evaluated by health experts (nurses), information experts (reference librarians), and Yahoo! Answers users. [75,105]. Health information needs presented in health questions of social Q&A were analyzed. Zhang [112] developed a layered model of contexts for consumer health information searching based on the findings from content analysis of health questions. Oh et al. [76] adapted the model to evaluate contexts associated with information seeking about cancer in social Q&A using the methods of text mining for analyzing cancer questions obtained from Yahoo! Answers. They verified and augmented the model to six layers, namely: demographic, cognitive, affective, social, situational, and technical layers.

3.3.2 Reference Services

Social Q&A was often compared to reference services in libraries. Both services allow users to obtain personalized answers to their information inquiries. In reference services, users can have one to one interactions with librarians who are trained professionals in searching information resources and in providing author-itative answers for their users. However, in social Q&A, users have one to many relationships with anonymous others who seek information collaboratively and may obtain multiple answers. Harper et al. [40] compared the quality and char-acteristics of answers across three different Q&A sites, including social Q&A, reference services, and expert Q&A (i.e., AllExperts) by deploying a set of ques-tions to the Q&A sites and evaluating and comparing the answers obtained from the sites. Shachaf [90] described a framework of "social references" which is extended from reference services but differs by involving social behaviors, that is, allowing a group of volunteers to produce answers collaboratively. The Wikipedia Reference Service is an example [91]. Shah and Kitzie [92] investigated the different points of views of reference librarians and end users toward virtual reference services and social Q&A and found that virtual reference services could be better than social Q&A in terms of answer customization and answer qual-ity, while social Q&A would be more cost-effective, speedy, voluminous, and socially-driven than reference services. There is a group of reference librarians who participate in the "Slam the Board" activities of posting answers in social Q&A to help users access authoritative sources of information and to promote reference services in the social contexts. Luo [67] interviewed these librarians to identify their strategies for providing answers in social Q&A.

3.3.3 Open Source Communities

Open source communities have used social Q&A as venues for collaboratively producing, sharing, and managing knowledge and support about software, pro-gramming, and other technology. Vasilescu et al. [102] found that the user rep-utation building mechanism in social Q&A, e.g., user point or badge earning, attracted people to social Q&A and encouraged them to answer quickly, com-paring it with sharing information in mailing lists. There were studies about the use of StackOverflow, a programing and technology Q&A site, exploring users' editing behaviors when formulating questions [108] and trends of question changes on a technical topic, for example, an API use in mobile programming [61]. Choi and Yi [19] investigated the publics' needs and motivations for sharing open source software knowledge and information in social Q&A.

3.3.4 Others

Additionally, there are other domain-specific studies of social Q&A. Savolainen [89] analyzed questions related to travel planning to examine the types of infor-mation discussed in social Q&A. Li et al. [60] examined the quality of answers given in an academic Q&A site hosted by ResearchGate, a social network site for academic researchers and scholars. Given the global popularity of social Q&A, Yang et al. [109] investigated the cultural differences associated with use of social Q&A across the United States, the United Kingdom, China, and India.

4 Theories

Many social Q&A studies are descriptive or exploratory, collecting and analyzing content and user data from a social Q&A site, in order to examine the new and popular phenomenon of social Q&A. These approaches are useful for building foundations for future development of research in social Q&A, but could be shallow in understanding the nature of human behaviors in social contexts. There are a few studies, however, which adapt social theories and theoretical models in information science to the emerging contexts of social Q&A. The definitions and applications of the theories in social Q&A are explained in this section.

4.1 Social Theories

Social theories have been used mainly to explain what motivates users to participate in the sites actively and continuously over a long term. The social exchange theory states that individuals intend to minimize costs and maximize benefits when they exchange goods with others [28,29]. Reciprocity is one of the main factors in this theory, encouraging social actions in "give-and-take" situations between individuals, but Blau [9] asserted that individuals do not expect to receive tangible rewards all the time when they interact with others. Instead, generalized reciprocity [27] may play an important role as it encourages individuals to exchange sources of information with the belief that someone else will offer similar help when they need it in the future. Wu and Korfiatis [107] redefined generalized reciprocity as collective reciprocity, indicating "a collective patterning in responding to kind or unkind intentions in multilateral interactions in a social network" (p. 2071) and applied it to explain askers' benefits and efforts (costs) related to activities of posting questions and answers to obtain best answers in Yahoo! Answers.

Similarly, social cognitive theory [5,6] was used to identify user characteristics of self-efficacy, which is referred to as "a form of self-evaluation regarding ones' capability of performing certain behaviors to attain certain goals." (p. 95). Jin et al. [50] tested the relationship between self-efficacy and user satisfaction in social Q&A with the assumption that users whose self-efficacy is high are satisfied with their question answering behaviors as sharing knowledge and it leads them to continuously participate in social Q&A. They also used the social exchange theory to test reciprocity as one of the factors for user satisfaction with Yahoo! Answers China. Oh [72] proposed both generalized reciprocity and self-efficacy as the main factors that motivate users to provide answers in Yahoo! Answers and tested it along with eight additional factors identified from a comprehensive review of motivation literature in online communities.

While social exchange and cognitive theories are useful explaining individuals' points of view when they join, seek, and share information in social Q&A, there are approaches used to view social Q&A as online communities and to examine users' social behaviors interacting with others and the technical supports within the structure of the communities. Rosenbaum and Shachaf [84] adapted the structuration theory [36] to explain both the social and the technical aspects of utilizing social Q&A for cultivating knowledge in communities of practice [59]. The structure of an online community affects the creation of social interactions within the community and these interactions reshape and develop the structure. The duality of structure indicates that users are highly involved in both enabling and constraining the evolution of structure by participating in various activities as social practices in using information and communication technology in social Q&A.

4.2 Relevance Judgment and Credibility

Although it is not a named theory, the concept of relevance judgment [85,87] has been widely used to develop the theoretical framework of evaluating answer quality in social Q&A. Relevance judgment stresses what users consider when evaluating the quality of information they obtain from various resources [86]. Kim and Oh [56] investigated askers' relevance criteria for selecting best answers by analyzing askers' comments in Yahoo! Answers and identified 23 criteria in six categories, namely: content, cognitive, utility, information source, extrinsic and socio-emotional criteria. Oh and Worrall [75] selected ten criteria for health answer evaluations, including accuracy, completeness, relevance, objectivity, readability, source credibility, politeness, confidence, empathy, and efforts and tested how users, nurses, and reference librarians perceive the quality of health answers in Yahoo! Answers differently. In a similar manner, credibility judgment was studied to investigate the criteria related to the credibility of information or sources of information. Kim [54] found that Yahoo! Answers users make credibility judgments of answers with criteria associated with the message features of the answers and the perceived expertise of the answerers. Recently, Jeon and Rieh [49] proposed a conceptual framework of credibility assessment in social Q&A, considering both askers' and answerers' perspectives on answer quality evaluation at the content, heuristics and interaction levels.

5 Research Methods and Techniques

The substantial amount of data generated from user activities in social Q&A has greatly attracted social Q&A researchers to investigate user behaviors and interactions in social contexts. Data used in social Q&A research include not only questions and answers but also user profiles, user comments and ratings on questions and answers, and other interactions that occur during exchanges of questions and answers. Researchers have adapted various approaches to collecting and analyzing data obtained from social Q&A depending on their theoretical and empirical interests of research.

A great deal of numeric data, such as question and answer lengths, number of questions or answers collected, time lapses between question and answer postings, user rating scores, page views, and transaction logs, have been automatically collected and used for statistical or cluster analyses [1,3,8,23,34,102].

The content of messages embedded in questions, answers or user comments are analyzed using the methods of content analysis and text mining [19,30,39,56, 88,112]. User attitudes and perceptions have also been investigated using the traditional methods of interviews and surveys [19,25,56,73,75,92,109].

In this section, the use of application program interfaces (APIs) was introduced as one of the effective methods for collecting massive amounts of contents and user-related data in social Q&A. The scales and types of social Q&A data collected from APIs could vary depending on research designs. Therefore, the research methods applied to analyze the data collected from APIs were explained, including social network analysis, content analysis, text mining, and surveys/interviews, with examples from the previous studies.

5.1   Data Collection Using Application Program Interfaces (APIs)

An application program interface (API) is a set of protocols, which allows computer programs to interact with one another. Social media companies often develop APIs of their services and make them available publicly, thus enabling third parties collect their data and use them for their purposes, for example, research. Extended Markup Languages (XMLs) and JavaScript Object Notations (JSON) are the common standards for the API scripts. Over 550 social APIs were made available for public use in 2011 [26], and the number has been increasing.

APIs enable researchers to obtain a large amount of data automatically and to carry out "big data" analysis regarding user behaviors and patterns of social media use. APIs could provide significant benefits to both qualitative and quantitative research of social media in that user interactions and patterns of Internet use can be revealed through "instantaneous" and "nonintrusive" methods of data collection and analysis [65]. There are a few limitations, however, when using data collected from APIs. The social media companies often set limits for collecting data through APIs and do not provide their strategies of random sampling for public data distribution [71]. Therefore, researchers must rely on analyzing a partial set of data collected using APIs without accessing the entire collection, and need to be careful when generalizing their study findings [42]. Also, the extensive amount of data collected using APIs could present a big picture of "what" users are doing in social contexts and this could be enhanced by understanding "why," implementing other qualitative methods, such as surveys and interviews [65].

About 60 Q&A APIs are currently available to researchers and the public according to ProgramableWeb (http://www.programmableweb.com/category/ qa/), including APIs for AnswerBag, Quora, and Stack Exchange (See Fig. 4). The social Q&A sites provide the API scripts and documentations for free, which is structured and often comprehensive. When questions arise regarding the scripts or application, program developers seek additional resources and participate in online communities in order to share their knowledge and experiences and develop crowd-sourced documents collaboratively [17].

Category: Q&A

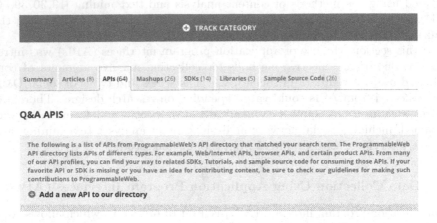

Fig. 4. ProgramableWeb Q&A API information page

Stack Exchange, for example, uses JSON as a standard of data structure for transmitting data through web programs and applications. The Stack Exchange API guideline provides information about authentication for data extraction and documents for script writing support. A number of data attributes available from Stack Exchange can be captured using APIs including data associated with answers, badges, comments, events, information, posts, privileges, questions, revisions, searches, suggested edits, tags, and users. More than 100 data types can be identified, collected, and used for analysis (A total set of data attributes is available from the Stack Exchange API site: https://api.stackexchange.com/docs/).

A number of textual and non-textual features can be identified from the extracted questions, answers, and user data using APIs and then utilized for analyzing contents and user interactions in social Q&A. For example, Bian et al. [8] conducted statistical analyses for better answer retrieval using a number of textual features such as question/answer lengths (i.e., number of words), question/answer lifetimes (i.e., how long a question or an answer has been posted), question/answer ranks, question popularity (i.e., number of corresponding answers), and question/answer votes, and a number of non-textual and user interaction features, such as users' total points, users' total number of answers provided, users' total number of best answers, users' total number of questions asked, and users' starts received in Yahoo! Answers.

Dalip et al. [23] developed systematical strategies ranking answers by quality, using measures obtained from Stack Overflow, such as users' features (i.e., number of questions, answers, or comments they posted and the associated ratings), users' graph features (i.e., users' expertise levels), review features (i.e., question and answer ages, number of edits in answers, number of users who suggest editing in answers or questions, number of suggestions approved or rejected), text structure features (i.e., image counts, section counts, paragraph counts, quoted text lengths and counts, number of links to external sources), length features (i.e., counts of characters, words, and sentences), style features (i.e., use of capitalization, punctuation, non-stop words in questions, answers, or other texts), and readability features using several readability indexes.

5.2 Social Network Analysis

Social network analysis (a.k.a., structure analysis) is known as an interdisciplinary research design comprised of strategies for identifying and explaining user behaviors in social contexts, named in social networks, in which a user develops relationships by interacting with their peers and other resources [77,104]. User behaviors and interactions in social networks are captured through a variety of data sources and are presented in graphs with a set of nodes (representing individual users, named actors) and links that connect nodes (representing relationships between actors). The volume and intensity of nodes and links are statistically measured and the density and centrality of the user relationships in social networks are visualized in order to interpret user dynamics and variations in the networks.

In social Q&A research, social network analyses were used to capture a variety of user roles and activities or to identify users with certain characteristics, e.g., user expertise or authority. For example, Adamic et al. [1] investigated knowledge sharing activities, such as the diversity of questions and answers, the breadth of answering, and the quality of answers, using the methods of social network analysis and non-network analysis, i.e., cluster analysis of topic categories. During a month long period, about one million questions, eight million answers, and associated user data were collected using Yahoo! Answers API. Through social network analysis, creating Q&A networks, the relationships between askers and answerers were examined based on the number of questions and answers users posted or received. Findings indicate that there is a separation of roles in asking and answering questions, and users provide answers in order to help others, or for fun, depending on the topics in social Q&A. The ego network analysis, which investigates the individual activities in the network, indicates that user connectivity differs according to topics; those who provide answers in the topic of "wrestling" are highly connected with others while those who provide answers in the topic of "programming" are not (See Fig. 5). Also, it was found that those who have high levels of expertise would like to offer answers to all levels of askers, but those who have lower levels of expertise would provide answers to those who have even less expertise.

Bouguessa et al. [12] assumed that askers in social Q&A would prefer to receive answers from their peer users who are authoritative. The more visibly present authoritative users are in social Q&A sites, the more high-quality and helpful answers could be distributed throughout the sites. They applied several different probability techniques driven from network analysis, such as PageRank, HITS, Z-score, and InDegree, to calculate the degree of user authority. In a social Q&A site setting, nodes are those who provide answers, and links represent

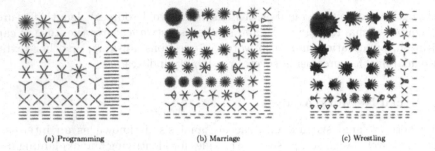

(a) Programming   (b) Marriage   (c) Wrestling

Fig. 5. A sample ego network generated from the social network analysis of the categories of programming, marriage, and wrestling; This figure is copied from the original article written by Adamic et al. [1]

answering behaviors. PageRank is used to decide the ranking of nodes (answerers) in a network, based on the calculation of the number of links attached to a node (the number of answers provided) and their associations with other linked nodes (who provides answers to whom). HITS algorithm is used to identify answerers who act as a hub. The degree of answerers' authorities can be identified through connections to many good nodes. Z-score calculates the probability of users having authority with the number of questions and answers. InDegree is measured by the number of nodes that are attached to a node. The answerers with higher InDegree are likely to be more authoritative. A number of questions and answers given by askers and best answerers were collected from the topic categories of engineering, biology, programing, mathematics, physics, and chemistry in Yahoo! Answers. About 65% of users ask questions only, about 30% of users answer questions only, and about 5% of them ask and answer across the topic categories. 70% of authoritative users across the categories show strong presence in the site, being very active in answering questions.

Harper et al. [39] developed systematic approaches to predict types of questions in social Q&A i.e., if they are "informational questions" (seeking for fact-based or advice-oriented answers) or "conversational questions" (stimulating discussion about a topic of interests). Social network analysis was used to specify user roles, if they are "answer people" who mainly provide information in answers, or "discussion people" who are encouraged to discuss about certain topics. A great deal of questions, answers, and user data have been collected from three social Q&A sites using their APIs; data include about 1 million users, 4 million questions, and 24 million answers from Yahoo! Answers, about 50,000 users, 140,000 questions and 800,000 answer data from AnswerBag, and about 10,000 users, 45, 000 questions and 650,000 answers Metafilter (the data of Metafilter had to be collected using customized programming scripts due to the lack of its API service.). The history of users asking and answering informational or conversational questions were collected and used to develop the ego network of users in social Q&A, defining the question asking and answering relationships among users. Findings revealed that those who ask conversational questions have greater numbers of relationships with high connectivity with others than those who ask informational questions.

5.3   Content Analysis

Content analysis is one of the widely used research methods used for analyzing the contents of questions and answers in social Q&A. Content analysis is "a research technique for making replicable and valid inferences from texts (or other meaningful matter) to the contexts of their use" (p. 18) [58]. Content analysis is known as a flexible method due to its open analysis of text data [16]. Hsieh and Shannon [45] defined qualitative content analysis as "a research method for the subjective interpretation of the content of text data through the systematic classification process of coding and identifying themes or patterns" (p. 1278). During content analysis, text data are coded into a set of categories. The contents of categories and the frequencies of the coded texts in the categories are used for interpreting the data. There are five practical steps of qualitative content analysis, as follows: (1) prepare text data by cleaning them up for review, (2) define the unit of analysis, which could be the basic unit of text to classify, (3) develop coding categories and apply coding categories that already exist, (3) test the coding schema on a sample of text, (4) code all of the text, (5) assess coding consistency, (6) draw conclusions from the coded data, (7) report methods and findings [113].

Hsieh and Shannon [45] clarified three approaches to qualitative content analysis and called them conventional, directed, and summative. Conventional content analysis is used to explore an emerging phenomenon, which is not yet fully explained with an existing theory. An open coding procedure is performed to identify themes within the text. Researchers thoroughly review the text, specify terms or sentences related to the themes, develop a coding schema, and apply it to code the text as a whole. This requires an iterative process of reviewing the text data and it develops a conceptual framework of the data analysis inductively. On the other hand, directed content analysis can be performed when there is an existing theory that researchers can refer to in order to develop their strategies of analyzing text data but would like to enhance the theory further. The coding can be initiated with a set of predetermined coding categories. The names and definitions of existing categories can be modified and new categories can emerge during the process. Researchers can expand or reshape the coding schema to comply with their purposes of research and contribute to further developing the existing theory. Summative content analysis is initiated by counting the number of specific words or explicit contents in the text to explore their usage. After that, latent content analysis, which is referred to as a process of interpreting the meaning of contents [43], is conducted to identify alternative terms of representation and to examine the contexts associated with the terms.

In social Q&A research, content analysis often involves a small set of data but the thorough and systematic examination of the data set yields meaningful findings for understanding user expressions during information seeking in questions and for understanding the contents of user-generated information presented in answers. For questions, content analysis was used to classify questions into certain types and to develop a framework of general discussions occurring in social Q&A. For example, Ignatova et al. [47] applied a traditional typology of questions developed by Graesser, McMahen, and Johnson [37] to develop a classification scheme of questions in social Q&A, including concept completion, definition, procedural, comparison, disjunctive, verification, quantification, casual, and general information needs (See Fig. 6).

Proposed Type	Examples from Yahoo! Answers
Concept Completion	*which r websites to learn web design?*
Definition	*what is KPO (knowledge processing and out sourcing)?*
Procedural	*How do I create a risk management database?*
Comparison	*what is the difference between retesting and regression testing?*
Disjunctive	*Which one is better to use for speech recognition and image processing..C,C++,VC++ or Matlab?*
Verification	*Does the linear regression of a data set pass through the centroid of the data set?*
Quantification	*how many bytes of storage are available just using the 6800's data registers?*
Causal	*why is 0.05 used as s significant value in data analysis?*
General Information Need	*i have a hard time dealing with database management can anyone please help me?*

Fig. 6. A scheme of questions types in social Q&A; This figure was copied from the original article written by Ignatova et al. [47]

A total of 805 questions about a topic of "data mining" were extracted using Yahoo! Answers API. Fifty of them were used as a training set for testing the classification schema and the rest of questions were used as a whole set to code. Three coders participated in classifying question types. The inter- and intra-annotator agreement was calculated using Kappa statistics [22]. Three types of agreement were measured, partial overlap (PO) and complete overlap (CO), among the three coders, and certainty attributes if the coders were sure of their coding results, labeling them as "sure." The kappa values across the three coders ranged from .800 to .947 in PO and POsure and .617 to 1.00 in Co and COsure, indicating that the annotation (the coding) is stable. Findings show that almost half of the questions (46.3%) in the data set were about concept completion, followed by definition (20.3%), procedural (17.1%), comparison (8.5%) casual, (3.2%), disjunctive (1.8%), verification (1.4%), quantification (0.9%), and general information needs (0.5%). This indicates that many questions in Yahoo! Answers are fact-finding questions. Among the questions, 3.8% were marked as opinion-based questions during coding, but the agreement value was pretty low, ranging from .267 to .493. The authors believe that the low agreement on opinion questions is due to the fact that coding requires domain knowledge in order to identify the nature of the opinion seeking or the opinion request.

Harper et al. [39] used mixed methods to identify types of questions and their usage in social Q&A and content analysis was used to identify the primary intent of questions, whether they are informational or conversational, and to evaluate the writing quality and to assign a value to archive. Thirty coders reviewed a total of 490 questions, randomly selected from Yahoo! Answers (163 questions), Answerbag (161 questions), and Metafilter (166 questions); each coder reviewed partial sets of questions. A question was examined by at least two coders. There is 87.1% agreement among the coders when they defined questions as being either informational or conversational, 74.4% agreement on the writing quality measure, and 70.4% agreement on archival value. Findings show that 63% of questions are informational while 32% are conversational. Yahoo! Answers contain greater numbers of informational questions (57%) than conversational questions (36%) while in Answerbag, there were greater numbers of conversational questions (57%) than informational questions (40%). The majority of questions in Metafilter were informational (91%), having the highest quality on the writing measure and on the archival values than the two other social Q&A sites.

Recently, questions were recognized as an explicit expression of information needs on certain topics of interest and content analysis was used to identify the kinds of information and the contexts askers presented in questions of social Q&A. For example, Bowler et al. [13] intended to ascertain teenagers' information needs on eating disorders as presented in Yahoo! Answers questions and develop a taxonomy of questions about the topic. They targeted question extraction, using Yahoo! Answers APIs, limiting their searches to specific keywords, such as teen+eating disorder, teen+anorexia, and teen+anorexic. After filtering the collected data with specific keywords such as "anorexia", "anorexic", "bulimia", "bulimic", and "eating disorder" in the question title, they used a set of 330 questions for content analysis and obtained a set of themes and subcategories of types of questions asked by teens in Yahoo! Answers. These include seeking information (factual, diagnosis, treatment or intervention), seeking emotional support (validation, and seeking comfort), seeking communication (conversation starters and deep talks), seeking self-expression (confession and reflection), and seeking help to complete tests (homework help and manuscript ideas). They further reviewed the expressions and presentations in questions and reported the specifics related to each type of question with narrative examples of questions.

Zhang [112] defined contexts as "a collection of factors that may influence consumer interactions with systems for health information" (p. 1159) and investigated the contextual factors that social Q&A users consider when asking questions in Yahoo! Answers. Instead of collecting data using API, Zhang obtained a pre-selected set of Yahoo! Answers data from Liu et al. [63], containing about 70,000 health-related questions across 23 subcategories. The coding unit for content analysis in this study is not a sentence or a question, but a questioning message with interrogative sentences that identify and interpret the factors directly related to askers' inquiries presented in their questions. A total of 600 messages were randomly selected from the data set; 540 messages were used for content analysis after discarding several messages due to the lack of authentic representations of consumer information needs. An open coding, which thoroughly reviewed the messages and identified themes and factors embedded in the messages, was carried out. The coding was mainly performed by the author but was also compared to the partial coding results (20% of all messages) from the second coder to test intercoder reliability. Findings from this study resulted in a layered model of context for consumer health information searching, composed of five layers, e.g., demographic, cognitive, affective, situational, and social & environmental layers.

Choi and Yi [19] investigated the general public's information needs utilizing open source software (OSS) by analyzing the content of questions posted in Yahoo! Answers. A total of 5,489 questions were collected from Yahoo! Answers using a customized Java web crawler that collected questions containing the keyword "open source." The authors developed an initial coding schema based on their expertise and experiences in OSS and tested the schema several times through an iterative process of reviewing sample questions and then modifying the schema; they ended up coding a total of 1,150 questions for content analysis. Multiple assignments were allowed to code one question; a total of 1,285 coding data were used for data interpretation. A total of nine categories of information needs on OSS were identified, specifically, software request (51.4%), general description (15.6%), technical issue (9.8%), advantage/disadvantage (7.0%), licensing (5.8%), future of OSS (3.6%), business model (2.4%), support availability (2.0%), and project management (1.7%).

Answers have been examined thoroughly using content analysis, mostly to evaluate the quality of the answers [2,10,30,40,60,75]. For example, Fichman [30] randomly collected 1,533 data (questions and corresponding answers) collected from four Q&A sites, Yahoo! Answers (N = 584), WikiAnswers (N = 605), Wikipedia Reference Desk (N = 77), and Askville (N = 256) by running several computer-programming scripts. Three codes were used to evaluate the quality of answers, accuracy (if the answer is correct), completeness (if an answer is thorough and includes enough information and provide responses to every inquiry in a question), and verifiability (if an answer provide sources of information). Three coders reviewed the answers and evaluated the presence of codes in the answers, noting 'yes.' A pair of coders reviewed answers. The intercoder reliability, the percentage of agreement among coders was .92 in general, indicating "acceptable" in the measure [64]. Findings show that there are variances among the Q&A sites; answers from Wikipedia Reference Desk are evaluated as the most accurate (56%) while answers from Yahoo! Answers were the least accurate (32%). There was no statistically significant difference in completeness across the four sites. In terms of verifiability, Wikipedia Reference Desk had the highest percentage (76%), while WikiAnswers had the lowest (6%), followed by Yahoo! Answers (25%). The authors infer that the quality differences across the four sites could have originated due to the differences of community sizes, user demographics, policies, motivators, and technical infrastructures and uses, and question types discussed in each community.

Li et al. [60] investigated answer quality and characteristics of a special topic-related Q&A site for academic communities, ResearchGate, a social network site for researchers and scholars to display their research profiles, share their publication records (and full-texts, if available), and make and develop social networks with their colleagues or researchers in their fields. The Q&A platform at ResearchGate is used as a venue for researchers and scholars to ask and answer questions regarding their career building in academia. A total of 1,128 posts (answers) responding to 107 question threads in the topics of library and information science, history of art, and astrophysics were used for content analysis. The contents of the answers and their posting time and date are collected using a programming-script. ResearchGate also allows users to vote for their favorable answers. Due to human-coding, seven types of evaluation criteria were identified, such as social elements, consensus building, factual information, providing resources, referring to other researchers, providing opinion, and providing personal experience. Three coders determined whether an element was presented in an answer and marked them as '1'. If not, it was marked as '0'. An initial set of 100 questions was coded by all three coders first to verify the coding schema, resulting in .83 coding agreement, and then three coders analyzed the rest of answers (1,021 answers) independently. The coding results indicate that high quality answers are more likely to have factual information, provide resources, refer to other researchers, provide opinions, and provide personal experiences than medium or low quality answers.

Both questions and answers are a set of data for describing the patterns of user communication and information exchange in social Q&A. Savolainen [88] proposed a framework of arguments, consisting of four patterns, i.e., failed opening (initiating a claim), nonoppositional (supporting the claim), oppositional (making a counterclaim and rebutting) and mixed (an argument combined with both oppositional and nonoppositional), based on the model by Toulmin [99]. He observed the argument patterns presented in between questions and answers on a debatable topic, global warming. A sample of 100 questions and associated answers were selected from the category of Environment/Global warming in Yahoo! Answers. An inductive approach to thoroughly review the contents and structures of questions and answers revealed that four types of discussions occurred in social contexts, facts, personal beliefs, opinions of other people, and emotional appeals. The findings revealed that the most frequently observed arguments in social Q&A was failed opening, indicating most discussions in social Q&A are broad, and not specific. Most users initiate their argument with their personal beliefs and provide evidence when they would like to counterclaim or support an argument.

Not only questions and answers but also user comments were analyzed. Kim and Oh [56] collected and analyzed about 7,366 askers' comments on answers that had been selected as the best among other answers (best answers) using Yahoo! Answers API. Both authors participated in content analysis during a two-step process. First, they reviewed about 10% of the comments (750 comments) using an initial coding schema of relevance criteria that they developed based on a comprehensive literature review of previous studies about relevance judgment in information seeking on the web. Their agreement with the coding was calculated using Cohen's kappa, yielding .836. Although this is an almost perfect level of agreement, they compared their coding results and refined the coding schema for better presentation of the criteria. The second step of coding includes 2,140 comments, resulting in a coding agreement level of .896. A question could include multiple statements assigned to different criteria. Thus, a total of 2,223 statements related to best answer selection criteria were identified. Findings show a comprehensive set of best answer selection criteria, including seven categories and their 23 associated criteria, namely: content category (accuracy, specificity, clarity, rationality, completeness, writing style, length), cognitive category (novelty, understandability), utility category (effectiveness, solution, feasibility), information source category (reference to external sources, answerers' expertise), extrinsic category (external verification, available alternatives, quickness), and socio-emotional category (emotional support, answerers' attitudes, answerers' efforts, answerers' experiences, agreement, taste, humor). In general, 79% of all of the selection criteria were fell into socio-emotional, content, and utility categories, although there were some variances in terms of distributions of criteria occurrence across topics.

5.4   Text Mining

Text mining is defined as "the discovery of previously unknown knowledge that can be found in text collections" [97]. Traditionally, there are three steps in text mining, (1) text preprocessing, (2) text representation, and (3) knowledge discovery. Text preprocessing means cleaning up data obtained from the text corpus by removing stop words or processing stemming, for example. Once the data are cleaned, major concepts and associated terms are extracted from the text utilizing several predicted models of text mining. In the end, the patterns and associations relevant to the extracted concepts and terms can be transformed into knowledge discovered from the text corpus [46].

A substantial amount of knowledge and information has been produced by the format of text in social media. Text mining can be an effective tool for extracting knowledge from massive unstructured text datasets drawn from social media environments [82], but it has been a challenge to analyze the data due to the distinct characteristics of the text in social media; it is time sensitive, short in length, unstructured, and contains abundant information with noise [46]. Therefore, a sophisticated approach to text mining should be designed and applied to analyze social media data, considering the nature and context of the data generated and exchanged in social media.

In social Q&A, users can retrieve questions or answers with a simple keyword search, but it is not easy to locate relevant information since information is scattered and buried in a large volume of questions and answers. Text mining of questions and answers can be a solution to systematically review the content, to identify the hidden value of the information, and to discover new patterns of knowledge among diverse questions and answers. For example, Kim et al. [57] collected 5,400 Influenza A Virus (H1N1)-related questions and answers from Yahoo! Answers. A keyword, "H1N1," was used to retrieve and identify questions related to the topic and the best answers associated with the selected questions were collected together. The major topics presented in H1N1 related questions and their corresponding best answers were analyzed using SPSS Clementine text mining software (SPSS was merged to IBM in 2009 and the software was renamed as IBM SPSS Modeler). The top 50 topics were extracted from the software and the authors grouped them into several categories such as general health, flu-specific topics, and nonmedical issues. The flu-specific topics were further analyzed by developing sub-categories. It was found that diseases/symptoms are the most frequently observed topics in questions, followed by special focus groups, influenza vaccines, influenza viruses, and more. SPSS Clementine software was also used to identify types of resources presented in answers by extracting URLs of the resources. An analysis of the types of organizations or locations revealed that many official resources from health organizations, governments, and universities were cited in answers.

Deng and Zhang [25] investigated social Q&A users' perceptions of library reference services by analyzing the contents of questions and answers related to the subject of library reference services, posted in Yahoo! Answers. Yahoo! Answers API was used to collect a total of 1,420 unique questions on the subjects of library reference, library reference services, reference services, virtual reference, and reference librarians and their responses (4,964 Q&A exchanges). IBM SPSS Modeler was used to identify words and phrases from questions and answers presenting key concepts related to reference services. Text mining enabled the creation of two category types and 12 associated categories, the service category (reference librarian, reference services, and Yahoo! Answers) and the attribute category (helpful, would recommend, accessible, like, satisfied, fast, reliable, meets needs, and intend to use). Reference librarian was the most frequently cited service category (68.9%) and "helpful" was the most frequently described expression of reference services in questions and answers (20.4%). IBM SPSS Modeler analyzed the associations among the categories and generated several concept maps, showing that reference librarians and their services are recognized as providing helpful services to Yahoo! Answers users and they would like to recommend reference services to others, although there are some users who are confused about the differences between reference services and other information services.

Oh et al. [76] conducted a large scale data analysis with Yahoo! Answers' cancer-related questions using IBM SPSS Modeler. A total of 81,434 cancer questions were randomly collected from Yahoo! Answers using its API. The software was used to pre-process the mined text, cleaning up term variations such as plural terms, synonyms, acronyms and typos, and to identify key concepts, which are both generic and medical terms cited in questions using the American English Dictionary and Medical Subject Headings (MeSH). A total of 534 terms were recognized as representations of cancer-related topics that users discuss in questions and were classified into six layers of contexts, namely, demographic, cognitive, affective, social, situational, and technical layers (See Fig. 7).

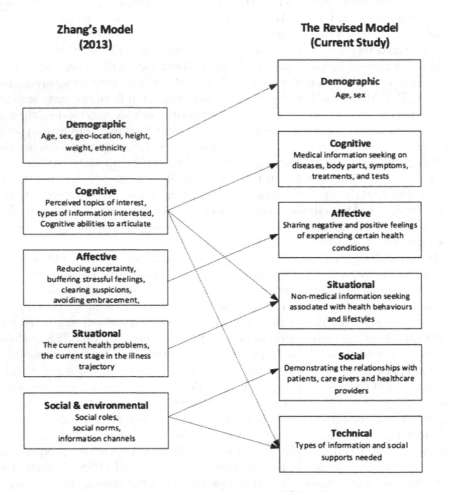

Fig. 7. A layered model of health questions in social Q&A; This figure is copied from the original article written by [76]

5.5   Interviews and Surveys

Social network analysis, content analysis, and text mining are good methods for observing what users have been doing during exchanges of questions and answers with their peers in social Q&A, but they may not be effective at fully discovering the reasons or the intentions of using social Q&A. To fill this gap, interviews and surveys were frequently used to obtain direct responses from users about their perceptions and activities in social Q&A. A semi-structured interview, which begins with a set of pre-determined questions with the flexibility of adding, deleting, and modifying the questions, or changing the question wording and the question order during the interviews [83], was often used to examine user perceptions on the use of social Q&A [49,54,69,78,92]. Online surveys were used to investigate user motivations [24,73,109]. One of the challenges to data sampling for interviews and surveys in social Q&A research is to reach users and recruit them to participate in the studies due to user anonymity in the sites. Thus, researchers often carry out purposive sampling by selecting a social Q&A site for a study and then developing strategies to recruit users online, considering the user contact options available from the sites, or convenient sampling by recruiting social Q&A users available in local areas, conducting user experiments, and then interviewing them face-to-face.

For example, Oh [73] investigated user motivations for sharing health information in Yahoo! Answers. She collected 124,926 user profiles of those who had provided answers in health categories of Yahoo! Answers, using Yahoo! Answers API. With the user data set, a stratified random sampling approach was used to select answerers according to their user levels of participation in Yahoo! Answers. A total of 1,800 answerers, 300 answerers at each level, from level two to level seven (The more users answer questions, the higher their levels. Level one was excluded because it is a starting level given to everyone without considering their contributions to the site), were initially included in the sample and invited to the survey by sending an invitation to those who enabled email contacts to be active in the Yahoo! Answers user profile page. The survey invitation was continued until collecting at least 30 responses at each level. The responses from users in lower levels were pretty low. Thus, an additional 150 answerers in level two and three were added to the sample and invited to the survey. Thus, a total of 2,139 invitations were sent out and 257 responded and participated in the online survey. In the survey, a total of ten motivation factors were provided to rate if they are important reasons for participants to provide answers in Yahoo! Answers. Each motivation factor was tested using five or six statements on a 5 point Likert scale. The internal consistency across the statement testing each motivation factor was tested using Cronbach's a, resulting in a range from .703 to .935, indicating acceptable to excellent [35]. It was found that altruism is the most influential factor, followed by enjoyment and efficacy.

Similarly, Paul [78] collected user profile data first and then invited users for interviews from the pre-defined set of user data in order to investigate users' perceptions on answer quality and reputation building in Quora. They first collected data related to 60 question topics and 3,917 associated information about

users using RSS feeds, selected users who have moderate to high activity levels in the site, and then randomly sent invitations to interview via a private
messaging system on Quora. A total of ten users responded and participated in
the 30 min interviews, including technologists, librarians, journalists, politicians,
and students. It was found that there are several unique characteristics of Quora
as compared to other social Q&A sites so that users prefer to search for authoritative answers and promote their reputations in that site. Users believe answers
with "primary sources of information" are authoritative, which is encouraged
by Quora across the site. Users need to use their real identities in Quora. This
helps them build their reputation within the site, which is a strong motivation
for them to participate in Quora. Social voting is recognized as a social signal
to build user reputation, but it may not be a good indicator of answer quality
since users may vote for best answers only from those in their social network.

Jeon and Rieh [49] conducted a semi-structured interview to ascertain people's credibility judgments on answers in social Q&A. Instead of inviting the
current users of social Q&A, they recruited undergraduate students in a hosting
university, trained them to post questions to Yahoo! Answers for a period of
one week and then interviewed them at the end of the experimental week. Their
questions and corresponding answers were collected for additional data analysis along with the interview transcriptions. Findings show that participants
make credibility judgments on answers considering answerers' involvement and
efforts (attitude), answerers' intention or decency (trustworthiness), and answerers' expertise.

Luo [67] interviewed a special group of social Q&A users, librarians who have
posted answers to Yahoo! Answers as members who participate in the "Slam the
Board" activity. This is an online activity of reference librarians to reach users
beyond their library services, to distribute good quality of information resources,
and to promote the presence of librarians in social Q&A sites. Any reference
librarian can participate in visits to social Q&A sites on the tenth of each month
and answer as many questions as they want. Due to their voluntary participation
and online anonymity, it was not possible to identify specific reference librarians
and reach individual members of populations through the social Q&A sites or the
Slam the Board website. Therefore, Luo [67] took a judgment sampling approach,
that is, selecting the study sample units based on researchers' judgments on usefulness and representativeness for the their research design [4], and recruited
participants from the Facebook group "Slam the Boards" and email listserves of
reference librarians (e.g., lib-ref, dig-ref, and ili-l). A total of 12 librarians participated in the interview over the phone, Skype, or email exchanges, responding to
questions about their motivations to participate in Slam the Boards, their time
commitment, their criteria for choosing a social Q&A site in which to participate
and for select questions to answer, the perceived differences between reference
services and social Q&A, and the benefits of their experiences with the Slam the
Board activities.

6   Conclusion

Social Q&A research has thrived over the past few years, developing vital streams of research in the field of user behaviors and data analytics in social contexts. Mixed methods have been used in combination with data/text mining, content analysis, user interviews, and surveys to examine the emerging and popular use of social Q&A from various angles. User-generated questions have been a good source for understanding user information needs in real lives. The anonymity of social Q&A encourages users to be honest and elaborative in describing their concerns and problems. User motivations for providing answers to the questions have been widely studied to promote knowledge and information sharing in social Q&A. Various strategies of assessing the quality of answers given by peer users were proposed and tested.

There is a great deal of potential to advance research of social Q&A in the future. In previous studies, contents and quality of questions and answers were thoroughly reviewed as separate data sets, but further research is needed to fully discover the relationships between questions and answers, regarding how effectively questions and answers are exchanged, how questions have been formulated to obtain good answers, and how effectively answers have provided responses to questions. User interactions associated with question asking and answering would need to be investigated further, identifying the roles of askers and answerers, as well as other users who read, make comments, vote, and rate questions and answers.

Data/text mining approaches could strengthen the ways to capture and analyze a variety of textual and non-textual features obtained from social Q&A. For example, text mining has been heavily used to analyze the contents of questions and answers, but it could be expanded to capture user comments on questions and answers. IBM SPSS Modeler is the software that has been mainly used for text mining. It is a popular and easy-to-use tool for analyzing texts in social Q&A, but could be limited when examining data using the pre-defined set of algorithms available from the software only. The use of software and associated algorithms could be diversified.

Social Q&A has been recognized as one of the important sources of information, but it is still unclear in what ways it has been useful for people to consult when making decisions in every day life. The use of social Q&A would need to be studied, and compared with other types of online, offline, and social media resources, in order to investigate how people consume a variety of resources in order to obtain information and the social supports they need as well as how important social Q&A are as sources of information.

References

1. Adamic, L., Zhang, J., Bakshy, E., Ackerman, M.: Knowledge sharing and Yahoo! Answers: everyone knows something. In: Proceedings of the 17th International Conference on World Wide Web, WWW 2008, pp. 665–674. ACM, New York (2008)

2. Agichtein, E., Castillo, C., Donato, D., Gionis, A., Mishne, G.: Finding high-quality content in social media. In: Proceedings of the 2008 International Conference on Web Search and Data Mining, WSDM 2008, pp. 183–194. ACM, New York (2008)

3. Ahn, J., Butler, B., Weng, C., Webster, S.: Learning to be a better q'er in social Q&A sites: social norms and information artifacts. Proc. Am. Soc. Inf. Sci. Technol. **50**(1), 1–10 (2013)

4. Babbie, E.R.: The Practice of Social Research. Wadsworth, Belmont (2012)

5. Bandura, A.: Social Foundations of Thought and Action: A Social Cognitive Theory. Prentice Hall, Englewood Cliffs (1986)

6. Bandura, A.: Self-efficacy: The Exercise of Control. Worth Publishers, New York (1997)

7. Bernhard, D., Gurevych, I.: Answering learners' questions by retrieving question paraphrases from social Q&A sites. In: Proceedings of the Third Workshop on Innovative Use of NLP for Building Educational Applications, EANL 2008, pp. 44–52. Association for Computational Linguistics, Stroudsburg (2008)

8. Bian, J., Liu, Y., Agichtein, E., Zha, H.: Finding the right facts in the crowd: factoid question answering over social media. In: Proceedings of the 17th International Conference on World Wide Web, WWW 2008, pp. 467–476. ACM, New York (2008)

9. Blau, P.: Exchange and Power in Social Life. Wiley, New York (1964)

10. Blooma, J., Chua, A.Y., Goh, D.: A predictive framework for retrieving the best answer. In: Proceedings of the 2008 ACM Symposium on Applied Computing, pp. 1107–1111. ACM, New York (2008)

11. Bordes, A., Chopra, S., Weston, J.: Question answering with subgraph embeddings. arXiv preprint arXiv:1406.3676 (2014)

12. Bouguessa, M., Doumoulin, B., Wang, S.: Identifying authoritative actors in question-answering forums: the case of Yahoo! Answers. In: Proceedings of the 14th ACM SIGKDD International Conference on Knowledge Discovery and Data Mining, pp. 866–874. ACM, New York (2008)

13. Bowler, L., Oh, J., He, D., Mattern, E., Jeng, W.: Eating disorder questions in Yahoo! Answers: information, conversation, or reflection? Proc. Am. Soc. Inf. Sci. Technol. **49**(1), 1–11 (2012)

14. Brusilovsky, P., Smyth, B., Shapira, B.: Social search. In: Brusilovsky, P., He, D. (eds.) Social Information Access. LNCS, vol. 10100, pp. 213–276. Springer, Cham (2018)

15. Cao, X., Cong, G., Cui, B., Jensen, C.: A generalized framework of exploring category information for question retrieval in community question answer archives. In: Proceedings of the 19th International Conference on World Wide Web, pp. 201–210. ACM, New York (2010)

16. Cavanagh, S.: Content analysis: concepts, methods and applications. Nurse Res. **4**(3), 5–13 (1997)

17. Chen, C., Zhang, K.: Who asked what: integrating crowdsourced FAQs into API documentation. In: Companion Proceedings of the 36th International Conference on Software Engineering, pp. 456–459. ACM, New York (2014)

18. Choi, E., Kitzie, V., Shah, C.: Investigating motivations and expectations of asking a question in social Q&A. First Monday **19**(3) (2014)

19. Choi, N., Yi, K.: Raising the general public's awareness and adoption of open source software through social Q&A interactions. Online Inf. Rev. **39**(1), 119–139 (2015)

20. Choi, W., Stvilia, B.: Web credibility assessment: conceptualization, operationalization, variability, and models. J. Assoc. Inf. Sci. Technol. **66**(12), 2399–2414 (2015)
21. Chua, A., Banerjee, S.: So fast so good: an analysis of answer quality and answer speed in community question-answering sites. J. Am. Soc. Inf. Sci. Technol. **64**(10), 2058–2068 (2013)
22. Cohen, J.: A coefficient of agreement for nominal scales. Educ. Psychol. Meas. **20**(1), 37–46 (1960)
23. Dalip, D., Gonçalves, M., Cristo, M., Calado, P.: Exploiting user feedback to learn to rank answers in Q&A forums: a case study with stack overflow. In: Proceedings of the 36th International ACM SIGIR Conference on Research and Development in Information Retrieval, pp. 543–552. ACM (2013)
24. Dearman, D., Truong, K.N.: Why users of Yahoo! Answers do not answer questions. In: Proceedings of the SIGCHI Conference on Human Factors in Computing Systems, pp. 329–332. ACM (2010)
25. Deng, S., Zhang, Y.: User perceptions of social questions and answer websites for library reference services: a content analysis. Electron. Libr. **33**(3), 386–399 (2015)
26. DuVander, A.: 550 Social APIs: over 35% added in 2011. Programable Web (2011)
27. Ekeh, P.P.: Social Exchange Theory: The Two Traditions. Heinemann, London (1974)
28. Emerson, R.M.: Power-dependence relations. Am. Sociol. Rev. 31–41 (1962)
29. Emerson, R.M.: Social exchange theory. Annu. Rev. Sociol. **2**(1), 335–362 (1976)
30. Fichman, P.: A comparative assessment of answer quality on four question answering sites. J. Inf. Sci. **37**(5), 476–486 (2011)
31. Gazan, R.: Specialists and synthesists in a question answering community. Proc. Am. Soc. Inf. Sci. Technol. **43**(1), 1–10 (2006)
32. Gazan, R.: Seekers sloths and social reference: homework questions submitted to a question answering community. N. Rev. Hypermed. Multimed. **13**(2), 239–248 (2007)
33. Gazan, R.: Social Q&A. J. Am. Soc. Inf. Sci. Technol. **62**(12), 2301–2312 (2011)
34. Gazan, R.: First-mover advantage in a social Q&A community. In: 2015 48th Hawaii International Conference on System Sciences (HICSS), pp. 1616–1623. IEEE (2015)
35. George, D., Mallery, P.: SPSS for Windows Step by Step: A Simple Guide and Reference, 4th edn. Allyn & Baconm, Boston (2003)
36. Giddens, A.: Central Problems in Social Theory: Action, Structure, and Contradiction in Social Analysis, vol. 241. University of California Press, Berkeley (1979)
37. Graesser, A., McMahen, C., Johnson, B.: Question asking and answering. In: Handbook of Psycholinguistics, pp. 517–538. Academic Press, San Diego (1994)
38. Han, S., He, D.: Network-based social search. In: Brusilovsky, P., He, D. (eds.) Social Information Access. LNCS, vol. 10100, pp. 277–309. Springer, Cham (2018)
39. Harper, F., Moy, D., Konstan, J.: Facts or friends? Distinguishing informational and conversational questions in social Q&A sites. In: Proceedings of the SIGCHI Conference on Human Factors in Computing Systems, pp. 759–768. ACM, New York (2009)
40. Harper, F., Raban, D., Rafaeli, S., Konstan, J.: Predictors of answer quality in online Q&A sites. In: Proceedings of the SIGCHI Conference on Human Factors in Computing Systems, pp. 865–874. ACM, New York (2008)

41. Harper, F., Weinberg, J., Logie, J., Konstan, J.: Question types in social Q&A sites. First Monday **15**(7) (2010)
42. Herring, S.C.: Web content analysis: expanding the paradigm. In: Hunsinger, J., Klastrup, L., Allen, M. (eds.) International Handbook of Internet Research. Springer, Dordrecht (2009). https://doi.org/10.1007/978-1-4020-9789-8_14
43. Holsti, O.R.: Content Analysis for the Social Sciences and Humanities. Addison-Wesley, Reading (1969)
44. Horowitz, D., Kamvar, S.D.: The anatomy of a large-scale social search engine. In: Proceedings of the 19th International Conference on World Wide Web, pp. 431–440. ACM, New York (2010)
45. Hsieh, H., Shannon, S.: Three approaches to qualitative content analysis. Qual. Health Res. **15**(9), 1277–1288 (2005)
46. Hu, X., Liu, H.: Text analytics in social media. In: Aggarwal, C., Zhai, C. (eds.) Mining Text Data. Springer, Boston (2012). https://doi.org/10.1007/978-1-4614-3223-4_12
47. Ignatova, K., Toprak, C., Bernhard, D., Gurevych, I.: Annotating question types in social Q&A sites. In: Hoeppner, W.(ed.) Tagungsband des GSCL Symposiums 'Sprachtechnologie und eHumanities', pp. 44–49 (2009)
48. Jeon, G., Rieh, S.: Do you trust answers?: credibility judgments in social search using social Q&A sites. Soc. Netw. **2**, 14 (2013)
49. Jeon, G., Rieh, S.: Answers from the crowd: how credible are strangers in social Q&A? In: iConference 2014 Proceedings (2014)
50. Jin, X., Zhou, Z., Lee, M., Cheung, C.: Why users keep answering questions in online question answering communities: a theoretical and empirical investigation. Int. J. Inf. Manag. **33**(1), 93–104 (2013)
51. Jurczyk, P., Agichtein, E.: Discovering authorities in question answer communities by using link analysis. In: Proceedings of the 16th ACM Conference on Conference on Information and Knowledge Management, pp. 919–922. ACM (2007)
52. Jurczyk, P., Agichtein, E.: Hits on question answer portals: exploration of link analysis for author ranking. In: Proceedings of the 30th Annual International ACM SIGIR Conference on Research and Development in Information Retrieval, pp. 845–846. ACM (2007)
53. Kim, K., Sin, S., Yoo-Lee, E.: Undergraduates' use of social media as information sources. College & Research Libraries, Chicago (2013). crl13-455
54. Kim, S.: Questioners' credibility judgments of answers in a social question and answer site. Inf. Res. **15**(2) (2010)
55. Kim, S., Oh, J.S., Oh, S.: Best-answer selection criteria in a social Q&A site from the user-oriented relevance perspective. Proc. Am. Soc. Inf. Sci. Technol. **44**(1), 1–15 (2007)
56. Kim, S., Oh, S.: Users' relevance criteria for evaluating answers in a social Q&A site. J. Am. Soc. Inf. Sci. Technol. **60**(4), 716–727 (2009)
57. Kim, S., Pinkerton, T., Ganesh, N.: Assessment of H1N1 questions and answers posted on the web. Am. J. Infect. Control **40**(3), 211–217 (2012)
58. Krippendorff, K.: Content Analysis: An Introduction to its Methodology. Sage, Thousand Oaks (2004)
59. Lave, J., Wenger, E.: Situated Learning: Legitimate Peripheral Participation. Cambridge University Press, Cambridge (1991)
60. Li, L., He, D., Jeng, W., Goodwin, S., Zhang, C.: Answer quality characteristics and prediction on an academic Q&A site: a case study on ResearchGate. In: Proceedings of the 24th International Conference on World Wide Web, WWW 2015 Companion, pp. 1453–1458. ACM, New York (2015)

61. Linares-Vásquez, M., Bavota, G., Di Penta, M., Oliveto, R., Poshyvanyk, D.: How do API changes trigger stack overflow discussions? A study on the Android SDK. In: Proceedings of the 22nd International Conference on Program Comprehension, ICPC 2014, pp. 83–94. ACM, New York (2014)

62. Liu, Y., Agichtein, E.: You've got answers: towards personalized models for predicting success in community question answering. In: Proceedings of the 46th Annual Meeting of the Association for Computational Linguistics on Human Language Technologies: Short Papers, pp. 97–100. Association for Computational Linguistics (2008)

63. Liu, Y., Bian, J., Agichtein, E.: Predicting information seeker satisfaction in community question answering. In: Proceedings of the 31st Annual International ACM SIGIR Conference on Research and Development in Information Retrieval, pp. 483–490. ACM (2008)

64. Lombard, M., Snyder-Duch, J., Bracken, C.: Content analysis in mass communication: assessment and reporting of intercoder reliability. Hum. Commun. Res. **28**(4), 587–604 (2002)

65. Lomborg, S., Bechmann, A.: Using APIs for data collection on social media. Inf. Soc. **30**(4), 256–265 (2014)

66. Lou, J., Fang, Y., Lim, K., Peng, J.: Contributing high quantity and quality knowledge to online Q&A communities. J. Am. Soc. Inf. Sci. Technol. **64**(2), 356–371 (2013)

67. Luo, L.: Slam the boards: librarians outreach into social Q&A sites. Internet Ref. Serv. Q. **19**(1), 33–47 (2014)

68. Mamykina, L., Manoim, B., Mittal, M., Hripcsak, G., Hartmann, B.: Design lessons from the fastest Q&A site in the west. In: Proceedings of the SIGCHI Conference on Human Factors in Computing Systems, pp. 2857–2866. ACM (2011)

69. Nam, K., Ackerman, M., Adamic, L.: Questions in, knowledge in? A study of Naver's question answering community. In: Proceedings of the SIGCHI Conference on Human Factors in Computing Systems, pp. 779–788. ACM (2009)

70. Navarro, B.B., Hotho, A., Stumme, G.: Accessing information with tags: search and ranking. In: Brusilovsky, P., He, D. (eds.) Social Information Access. LNCS, vol. 10100, pp. 310–343. Springer, Heidelberg (2018)

71. Neuhaus, F., Webmoor, T.: Agile ethics for massified research and visualization. Inf. Commun. Soc. **15**(1), 43–65 (2012)

72. Oh, S.: Answerers' motivations and strategies for providing information and social support in social Q&A: an investigation of health question answering. Ph.D. thesis, University of North Carolina, Chapel Hill (2010)

73. Oh, S.: The characteristics and motivations of health answerers for sharing information, knowledge, and experiences in online environments. J. Am. Soc. Inf. Sci. Technol. **63**(3), 543–557 (2012)

74. Oh, S., Kim, S.: College students' use of social media for health in the USA and Korea. Inf. Res. **19**(4), n4 (2014)

75. Oh, S., Worrall, A.: Health answer quality evaluation by librarians, nurses, and users in social Q&A. Libr. Inf. Sci. Res. **35**(4), 288–298 (2013)

76. Oh, S., Zhang, Y., Park, M.: Cancer information seeking in social question and answer services: identifying health-related topics in cancer questions on Yahoo! Answers. Inf. Res. **21**(3), 718 (2016)

77. Otte, E., Rousseau, R.: Social network analysis: a powerful strategy, also for the information sciences. J. Inf. Sci. **28**(6), 441–453 (2002)

78. Paul, S., Hong, L., Chi, E.: Who is authoritative? Understanding reputation mechanisms in quora. arXiv preprint arXiv:1204.3724 (2012)

79. Posnett, D., Warburg, E., Devanbu, P., Filkov, V.: Mining stack exchange: expertise is evident from initial contributions. In: 2012 International Conference on Social Informatics (SocialInformatics), pp. 199–204. IEEE (2012)
80. Quantcast: Top international websites & rankings. https://www.quantcast.com/top-sites/. Accessed 23 Sept 2017
81. Raban, D., Harper, F.: Motivations for answering questions online. N. Media Innovative Technol. **73** (2008)
82. Rajman, M., Besançon, R.: Text mining - knowledge extraction from unstructured textual data. In: Rizzi, A., Vichi, M., Bock, H.H. (eds.) Advances in Data Science and Classification. Studies in Classification, Data Analysis, and Knowledge Organization, pp. 473–480. Springer, Heidelberg (1998). https://doi.org/10.1007/978-3-642-72253-0_64
83. Robson, C.: Real World Research: A Resource for Social Scientists and Practitioner-Researchers, 2nd edn. Blackwell, Oxford (2002)
84. Rosenbaum, H., Shachaf, P.: A structuration approach to online communities of practice: the case of Q&A communities. J. Am. Soc. Inf. Sci. Technol. **61**(9), 1933–1944 (2010)
85. Saracevic, T.: Relevance: a review of an a framework for the thinking on the notion in information science. J. Am. Soc. Inf. Sci. **26**(6), 321–343 (1975)
86. Saracevic, T.: Relevance: a review of the literature and a framework for thinking on the notion in information science. Part iii: behavior and effects of relevance. J. Am. Soc. Inf. Sci. Technol. **58**(13), 2126–2144 (2007)
87. Saracevic, T.: Why is relevance still the basic notion in information science? (despite great advances in information technology). In: International Symposium on Information Science (ISI 2015), pp. 18–21 (2015)
88. Savolainen, R.: The structure of argument patterns on a social Q&A site. J. Am. Soc. Inf. Sci. Technol. **63**(12), 2536–2548 (2012)
89. Savolainen, R.: Providing informational support in an online discussion group and a Q&A site: the case of travel planning. J. Assoc. Inf. Sci. Technol. **66**(3), 450–461 (2015)
90. Shachaf, P.: The paradox of expertise: is the Wikipedia reference desk as good as your library? J. Doc. **65**(6), 977–996 (2009)
91. Shachaf, P., Rosenbaum, H.: Online social reference: a research agenda through a STIN framework. In: iConference (2009)
92. Shah, C., Kitzie, V.: Social Q&A and virtual reference: comparing apples and oranges with the help of experts and users. J. Am. Soc. Inf. Sci. Technol. **63**(10), 2020–2036 (2012)
93. Shah, C., Kitzie, V., Choi, E.: Modalities, motivations, and materials: investigating traditional and social online Q&A services. J. Inf. Sci. **40**(5), 669–687 (2014)
94. Shah, C., Oh, J.S., Oh, S.: Exploring characteristics and effects of user participation in online social Q&A sites. First Monday **13**(9) (2008)
95. Shah, C., Oh, S., Oh, J.S.: Research agenda for social Q&A. Libr. Inf. Sci. Res. **31**(4), 205–209 (2009)
96. Shah, C., Pomerantz, J.: Evaluating and predicting answer quality in community Q&A. In: Proceedings of the 33rd International ACM SIGIR Conference on Research and Development in Information Retrieval, pp. 411–418. ACM (2010)
97. Stavrianou, A., Andritsos, P., Nicoloyannis, N.: Overview and semantic issues of text mining. ACM Sigmod Rec. **36**(3), 23–34 (2007)
98. Su, Q., Pavlov, D., Chow, J., Baker, W.: Internet-scale collection of human-reviewed data. In: Proceedings of the 16th International Conference on World Wide Web, pp. 231–240. ACM (2007)

99. Toulmin, S.E.: The Uses of Argument. Cambridge University Press, Cambridge (2003)

100. Treude, C., Barzilay, O., Storey, M.: How do programmers ask and answer questions on the web? Nier track. In: 2011 33rd International Conference on Software Engineering (ICSE), pp. 804–807. IEEE (2011)

101. Tutos, A., Mollá, D.: A study on the use of search engines for answering clinical questions. In: Proceedings of the Fourth Australasian Workshop on Health Informatics and Knowledge Management, vol. 108, pp. 61–68. Australian Computer Society, Inc. (2010)

102. Vasilescu, B., Serebrenik, A., Devanbu, P., Filkov, V.: How social Q&A sites are changing knowledge sharing in open source software communities. In: Proceedings of the 17th ACM Conference on Computer Supported Cooperative Work & Social Computing, pp. 342–354. ACM (2014)

103. Wang, G., Gill, K., Mohanlal, M., Zheng, H., Zhao, B.: Wisdom in the social crowd: an analysis of quora. In: Proceedings of the 22nd International Conference on World Wide Web, pp. 1341–1352. ACM (2013)

104. Wellman, B., Berkowitz, S.: Social Structures: A Network Approach, vol. 2. CUP Archive, Cambridge (1988)

105. Worrall, A., Oh, S.: The place of health information and socio-emotional support in social questioning and answering. Inf. Res. **18**(3) (2013). n3

106. Wu, D., He, D.: Comparing IPL2 and Yahoo! Answers: a case study of digital reference and community based question answering. In: Proceedings of 2014 iConference, iSchools (2014)

107. Wu, P., Korfiatis, N.: You scratch someone's back and we'll scratch yours: collective reciprocity in social Q&A communities. J. Am. Soc. Inf. Sci. Technol. **64**(10), 2069–2077 (2013)

108. Yang, J., Hauff, C., Bozzon, A., Houben, G.: Asking the right question in collaborative Q&A systems. In: Proceedings of the 25th ACM Conference on Hypertext and Social Media, pp. 179–189. ACM (2014)

109. Yang, J., Morris, M., Teevan, J., Adamic, L., Ackerman, M.: Culture matters: a survey study of social Q&A behavior. In: International AAAI Conference on Web and Social Media Fifth International AAAI Conference on Weblogs and Social Media, vol. 11, pp. 409–416 (2011)

110. Yue, Z., He, D.: Collaborative information search. In: Brusilovsky, P., He, D. (eds.) Social Information Access. LNCS, vol. 10100, pp. 108–141. Springer, Heidelberg (2018)

111. Zhang, Y.: Contextualizing consumer health information searching: an analysis of questions in a social Q&A community. In: Proceedings of the 1st ACM International Health Informatics Symposium, pp. 210–219. ACM (2010)

112. Zhang, Y.: Toward a layered model of context for health information searching: an analysis of consumer-generated questions. J. Am. Soc. Inf. Sci. Technol. **64**(6), 1158–1172 (2013)

113. Zhang, Y., Wildemuth, B.: Qualitative analysis of content. Applications of Social Research Methods to Questions in Information and Library Science, pp. 308–319 (2016)

4
Collaborative Information Search

Zhen Yue[1] and Daqing He[2(✉)] (iD)

[1] Yahoo Research, Sunnyvale, CA, USA
zhenyue@yahoo-inc.com
[2] School of Computing and Information,
University of Pittsburgh, Pittsburgh, PA 15213, USA
dah44@pitt.edu

Abstract. In this chapter, we present one type of social information access called *Collaborative Information Search (CIS)*, where multiple people directly work as a team to collaborate explicitly to search relevant information for resolving a share information need. CIS integrates team collaboration with exploratory search, so that complex search tasks can be decomposed into simpler and smaller tasks for individual team members to resolve. In this chapter, we cover various factors that influence people's collaboration in search, and discuss the approaches that researchers have developed to support various forms of collaborative information search on the web, in academic setting, and in other environments. We will further talk about the evaluation of collaborative search systems, and then conclude with discussions on the remaining challenges and possible new directions on this topic.

1 Introduction

Social Information Access (SIA) can take various forms, one of which is called Collaborative Information Search (CIS). CIS concerns multiple people directly working as a team and collaborating explicitly to search for relevant information to satisfy a shared information need within the team. Therefore, CIS is a form of direct collaboration among members, and the members form a tight-knit team rather than a loose coordination as in many other forms of SIA. This makes CIS different from some other types of indirect collaborations in search discussed in this volume, such as social search in Chap. 7 [10], network-based social search in Chap. 8 [26], and tag-based social search in Chap. 9 [51].

The goal of this chapter is to introduce collaborative information search in the context of social information access. Particularly, as an introduction to this topic, this chapter fulfills two objectives. The first is to explain the concept of collaboration in the context of information search, which is related to human behaviors in the seeking process. The second objective is to introduce technologies in CIS, which includes the systems for conducting CIS, the techniques for supporting search in CIS, and the evaluation of CIS systems.

© Springer International Publishing AG, part of Springer Nature 2018
P. Brusilovsky and D. He (Eds.): Social Information Access, LNCS 10100, pp. 108–141, 2018.
https://doi.org/10.1007/978-3-319-90092-6_4

Consequently, the sections in this chapter are organized as follows. Section 2 examines collaboration as an independent concept both in a general environment and in the context of team work. Then the discussion focuses on the definition of collaborative information search and the studies of its appearance in academic and other settings in Sect. 3. Also in this section, we describe a model of the collaborative information search process so that major patterns and actions inside the search process can be quantitatively discussed. Through talking about factors affecting CIS in Sect. 4, we will introduce technologies that are designed to support users' search in CIS in Sect. 5. Finally, this chapter concludes with sections on the evaluation of CIS technologies and systems, and future directions in research.

Readers who are interested in exploring further about CIS can read survey papers about general collaborative information search [19, 20, 63] or collaborative web search [46, 47, 49].

2   Collaboration

2.1   Definition of Collaboration

In human society, collaboration is often natural and necessary to permit people to handle complex tasks or problems that cannot be handled by individuals. As Mattessich and Monsey [41] pointed out, *collaboration* is "a mutually beneficial and well-defined relationship entered into by two or more organizations to achieve common goals." Therefore, collaboration is a joint decision-making process, and the parties who see different aspects of a problem can constructively explore their differences and search for solutions that go beyond their own limited vision of what is possible [25].

Although people may use collaboration interchangeably with terms such as coordination and cooperation, Shah [61] stated that collaboration is a stronger form of people working together. For example, coordination is "a process of connecting different agents together for a harmonious action", and cooperation involves parties following some interaction rules in addition to coordination, which could include jointly planning actions, executing tasks and sharing resources. In contrast, collaboration emphasizes that the parties contribute their own individual expertise, take on different aspects of a problem, and complete a task or project together. Therefore, the final solution might "be more than the sum of each participant's contribution" [61].

To understand the concept of collaboration, we first need to understand the definition of a team because collaboration only exists in a team. When a number of people explicitly collaborate towards a common goal, they are often called a *work group* or *team* [3]. In order to uniquely define the team and set it apart from other types of groups, Paris et al. [54] summarizes the characteristics of a team as including multiple sources of information, task inter-dependencies, coordination among members, common and valued goals, specialized member roles and responsibilities, task-relevant knowledge, intensive communication, and adaptive strategies for responding to changes. Based on the analysis of characteristics that may be used to differentiate various type of groups, Andriessen [3] identified three types of groupings. The first one is called a *collection*, which refers to loosely coupled individuals that exchange information on an ad-hoc basis. The second is called a *community*, which is a group of people that have a common interest and therefore interact over a period of time. The last one is called a *team* because it is a group of people with a common goal, formality, and interdependence, which co-operates during a clearly delineated time period. Communities and teams are particularly interesting in relation to collaborative information seeking. This is because communities provide the context for the application of implicit collaborative technologies while teams are target users of collaborative exploratory search systems, which support explicit collaboration.

2.2 Collaboration and Teamwork Processes

Morgan et al. [45] discussed two categories of behaviors in a team, which are taskwork and teamwork. According to them, taskwork is performed by individual team members with the goal of executing the task while teamwork is related to team member interactions necessary to achieve team goals. There are many reasons to believe that teamwork is directly related to team performance [37]. Some research efforts have shown that teamwork is related to team functioning and task outcomes, and is fairly consistent across different task types [43]. Team research suggests that teamwork appears to be composed of a relatively stable set of behaviors and cognitive processes [59], and many researchers have devoted their efforts to identifying and organizing behaviors that define teamwork. Based on previous research efforts and review, Dickinson and McIntyre [15] proposed an influential teamwork model that highlights seven basic teamwork components: communication, team orientation, team leadership, monitoring, feed-back, back-up and coordination.

Another important and influential framework of teamwork was proposed by Marks et al. [40]. The framework is based on the argument that different teamwork processes are important at different phases of task execution. A phase is a distinguishable period of time that can be classified into an action phase and a transition phase. They claimed that a team focuses on a particular task in the action phase while teams in the transition phase review the previous efforts and prepare for future work. In addition, there is the third phase called interpersonal process, taking place in both the action and transition phases. They then provide further detail about three processes of teamwork. For example, the transition process involves mission analysis, goal specification and strategy formulation and planning. An action process reflects four types of activities including monitoring progress toward goals, system monitoring of team resources, team monitoring and back-up behavior, and coordination of activities of team members. Finally, the interpersonal process focuses on the management of interpersonal relationships, which includes conflict management, motivation and confidence building, and affection management. The benefit of having a teamwork process taxonomy is that it helps to understand how individuals collaborate in their interdependent efforts to achieve common goals.

2.3 Collaboration and Communication Media

In order to achieve collaboration, the members of the team have to communicate with each other. Among the stream of work investigating the media of communication, a well-known media richness theory [14] recognized four different types of communication medium according to the varying degree of richness. They are face-to-face, video, audio, and computer-mediated text transfers. Different tasks are best mediated by different media. For example, video is good for judgment tasks but too rich for generating ideas. Text messaging is good for generating ideas but not rich enough for negotiation [42]. Stone and Posey [73] examined the collaborative performance among team members through different communication media. They found that the perceived performance was lower in computer-mediated text groups than in face-to-face groups when the groups were not trained. But with training, there is no difference in perceived performance between the groups using the two different communication media.

Two types of communication styles were recognized in the literature: task-oriented versus socially-oriented [7]. Task-oriented communication focuses on fulfilling the responsibilities while socially-oriented communication focuses on satisfying the emotional needs of interpersonal relationships. In a study investigating communication in computer-supported collaborative learning environments [74], the researchers proposed a framework of coding communication messages, which can be used to distinguish social-oriented communication from task-oriented communication. There is also a line of work investigating the emotions involved in the text-based communication. For example, Brooks and colleagues [9] proposed a machine learning technique that can automatically detect and classify types of affections in chat logs.

The theories and methodologies from communication studies in general teamwork settings can be borrowed by researchers in collaborative search studies to investigate the role of communication in the collaborative search process.

2.4 Factors Affecting Collaboration

Collaboration is a complex process that requires multiple parties to contribute individually to resolving different aspects of a problem. Therefore, there are many factors that can affect collaboration, either by enabling people to communicate easier or completing the individual task more effectively.

Through reviewing 18 relevant studies in the literature, Mattessich and Monsey [41] identified 19 factors that influence the success of collaborations. These factors can be grouped into six categories: Environment, Membership, Process/Structure, Communications, Purpose, and Resources. Among the 19 factors, *Mutual respect, understanding and trust* and *Appropriate cross-section of members* are the two factors that were mentioned the most (11 out of 18 studies), and they both belong to the Membership category. This indicates that the characteristics of collaborators are probably the most important factors. The next commonly mentioned factor (9 out of 18 studies) is *Open and frequent communication*, which is in the Communication category. This shows that the information exchange plays a critical role as well. Collaboration is expensive, so *Sufficient funds* is the fourth commonly mentioned factor (8 out of 18 studies).

Collaboration involves multiple individuals or parties. It needs a person who can convene the collaborative group, who has organization and interpersonal skills, and who can carry out the role with fairness [41]. Therefore, *Skilled convener* is the fifth commonly mentioned factor (7 out of 18 studies).

Then there are several factors that share the same importance (6 out of 18 studies). *History of collaboration or cooperation in the community* is the factor in the Environment category, *Members view collaboration as in their self-interest* is the one in the Membership category, *Members share a stake in both process and outcome* and *Multiple layers of decision-making* are the two in the Process/Structure category.

Finally, the remaining factors include *Established informal and formal communication links* in the Communication category and *Concrete, attainable goals and objectives* in the Purpose category, which are mentioned in 5 out of 18 studies; *Shared vision* in the Purpose category, *Flexibility and Development of clear roles and policy guidelines* in the Process/Structure category, which are mentioned in 4 out of 18 studies; and *Unique purpose* in the Purpose category, *Collaborative group seen as a leader in community* and *Political/social climate favorable* in the Environment category, *Ability to compromise* in the Membership category, and *Adaptability* in the Process/Structure category, which are mentioned in three studies.

Within the context of online academic collaboration, Olson and colleagues [52] first proposed a set of factors in 2000, and then extended that work resulting in TORSC (Theory Of Remote Scientific Collaboration) in 2008 [53]. They listed five overarching factors that contribute the success of remote scientific collaboration: the nature of work, common ground, collaboration readiness, technology readiness, and management/planning/decision making [53].

All these research works demonstrate that collaboration can be affected by many factors from various categories. The most important factors are most likely related to the people involved in the collaboration as well as their organization and communication styles and capabilities. In addition, the work itself and the environment can play important roles as well.

3   Collaboration in Information Search

When people engage in search tasks, their needs can be as simple as fact-finding or a known-item search, but the needs can also be as complex as exploratory search. Given the complex nature of the information needs, people who conduct exploratory searches may decide to collaborate among themselves so that they form a search team and share the same search goals. For instance, students may work together to search for information for a collaborative course project; friends may search together while planning a vacation; healthcare providers might collaboratively search for information to diagnose a patients illness [58]; or family members might collaboratively search the web to buy a car [46]. In these cases, groups of people are engaging in *Collaborative Information Search (CIS)*, where they directly work as a team and collaborate explicitly to search for relevant information to satisfy a shared information need among them.

3.1 Collaborative Information Search in the Web Environment

It has been observed that CIS becomes increasingly common. Through conducting a survey among 204 information workers in 2006, Morris [46] found that many people conduct collaborative web search at least weekly, and the most common search tasks involving collaboration were travel planning, general shopping, and literature searches. A more recent survey by Morris [47] in 2013 reported that respondents engaged in collaborative web search on a daily basis has increased from 0.9% in 2006 to 11% in 2012. Morris suggested that the increased prevalence is a result of the significant change in the technology landscape particularly the increase usages of social networking sites and the growing usage of smart phones. Evans and Chi [16] also conducted a survey among 150 people using Mechanical Turk to investigate collaborative search strategies involved in the before-search, during-search and after-search stages. The surveys revealed that collaborative web search is a surprisingly common activity.

However, users who conduct CIS often encounter barriers that prevent them from achieving effective CIS using current search engines. For example, among the set of barriers identified by Morris [46], the two most common complaints are that the current search systems lack supports for parallelizing tasks without unnecessary duplication of effort and helping remote collaborators to navigate the shared context/focus. Her later study [47] showed again that users were still frustrated with the lack of awareness of collaborators' activities, which caused redundant work.

Consequently, systems that are specifically designed for collaborative information search have been constructed. For example, Han et al. [27] designed the CollabSearch system. As shown in Fig. 1, CollabSearch provides a common search interface that resembles generic web search engines such as Google (see the right part of the interface). In fact, its uses Google as the underlying search engine to retrieve web documents. After a query is entered in the query box at the top of this panel, Google is called to generate search results, and those results are displayed underneath the search box just like any generic search engine. To help the user remember his/her search process, a search history panel is located at the right side of the search results. This panel shows prior queries issued by the user as well as those issued by the other team members. At the left side of the interface, there is a chat box showing the chat history between the user and the other team members. This is the communication channel for the whole team to collaborate with each other while searching and sharing their search queries. Of course, this is just one implementation of a CIS system. There could be many other designs. We will talk more about various design considerations in Sects. 4 and 5.

In addition to the general Web environment, Collaborative Information Search can be found in many specific settings. In the next two sub-sections,

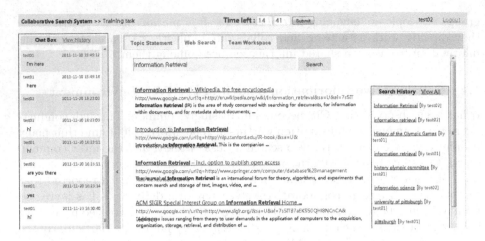

Fig. 1. CollabSearch Interface. (Adapted with permission from Yue et al. [87].)

we'll introduce CIS in the academic setting and other settings, which indicate the wide range of applications of collaborative information search across many domains.

3.2   Collaborative Information Search in Academic Settings

In the academic environment, studies of scholarly communication conducted in the 1960s and 1970s established that scholars social ties and networks profoundly affect their information gathering, reading, awareness and interpretation of documents and literature [76]. However, only recently did researchers start to focus on scholars' collaboration during the information search process.

In a combined ethnographic and experimental study of physicists, researchers discovered that successful scientific collaboration requires the collection and use of a range of awareness information to update team members on the current state of their team's activities [70]. The study investigated the types of information and knowledge that need to be shared to support situational awareness and the ways in which technology can be used to facilitate such information sharing.

Blake and Pratt [8] observed two groups of scientists in public health and bio-medicine conducting collaborative search for systematic literature reviews. They found that scientists actively collaborated as they refined the retrieval, extraction, and analysis phases of a process that the authors called information synthesis. Based on the characterizations of user behavior during information synthesis, they proposed the design and process to implement a tool METIS, which will support the collaborative, iterative, and interactive information synthesis processes of scientists.

Based on a comparative qualitative study of scholars across a range of humanistic, social-scientific, and scientific disciplines, Talja [75] found that existing search engines could not support collaboration among scholars in their searching and sharing information.

Students often collaborate in their academic tasks as well. Wu and Yu [84] explored the collaborative behavioral patterns of undergraduate students when they teamed together to write research competition proposals. They found that division of labor was the most popular collaboration strategy, and that the three key success factors are a strong team leader, clear division of labor, and active communication among team members. Leeder and Shah [36] also studied students' collaborative information seeking behaviors when they conducted authentic group work projects. Working with 41 participants in 10 groups, they found that students' performance during their search affected their CIS behaviors, and the students' pre-task attitudes and experiences toward group work can also strongly influence their CIS too. Finally, the students want collaborative search tools to be convenient, lightweight, and easy to use.

3.3   Collaborative Information Search in Other Settings

Researchers also studied collaborative information search behavior in several different settings in addition to academia. For example, in the patent domain, Hansen and Järvelin [28] studied collaborative information seeking activities in a real-life and information intensive setting. Their results showed that the patent search process involves highly collaborative aspects throughout the stages of the information seeking process. They categorized the activities into document-related collaborative activities and human-related activities. Finally, a refined IR framework involving collaborative aspects was proposed.

In the software design setting, Poltrock et al. [57] examined how members of two teams sought and shared with each other external information acquired within the team. They identified five collaborative information retrieval strategies: identifying needs collaboratively, formulating queries collaboratively, retrieving information collaboratively, communicating about information needs and sharing retrieved information, and coordinating information retrieval activities.

In the domain of healthcare, Reddy and Jansen [58] proposed a model for understanding collaborative search behavior in context based on their studies of two healthcare teams. They found that collaborative information behavior differs from individual information behavior with respect to how individuals interact with each other, the complexity of the information need, and the role of information technology. They also found triggers for collaboration, including a lack of domain expertise.

In the military command and control setting, Sonnenwald and Pierce [71] studied collaboration in dynamic situations with rapidly changing information and a need for continuous information exchange. They found that the commander played an important role in identifying critical information needs. Three types of collaborative information behavior were identified: information seeking by recommendation, direct questioning, and advertising information paths.

Within everyday life information seeking (ELIS) studies, McKenzie [44] found that people routinely assist each other in solving information problems. For example, in representing themselves as information seekers, participants gave accounts that showed them to be active and on guard, attentively receptive, and surrounded by a supportive network of others like them. The findings suggest that information seeking theories and models have limited insight into how information comes or goes due to the initiative or actions of another agent.

3.4   Collaborative Information Search Processes

Collaborative information search is complex and often involves multiple iterations of users communicating with each other and conducting searches. Therefore, it is important to consider the entire search process, and study the important stages, patterns, and activities inside the search process. This kind of study is referred to as information seeking behavior research [39].

The information seeking process is one of the major topics in information seeking behavior research. Researchers have employed two major approaches to investigate. One is modeling the macro-level information seeking process, which focuses on qualitative constructs such as stages and context in the information seeking process. A famous macro-level information seeking process model for individual search is Kuhlthau's ISP model [35]. The ISP model (see Fig. 2) presents a holistic view of information seeking from the users perspective consisting of six stages: task initiation, selection, exploration, focus formulation, collection and presentation. Based on empirical research, the model incorporates the physical, affective, and cognitive aspects of users' experience common to each stage. The other type of information seeking process model looks into the micro-level information seeking process by identifying descriptive categories such as user actions, search strategies or tactics and the transition relationships among them [33].

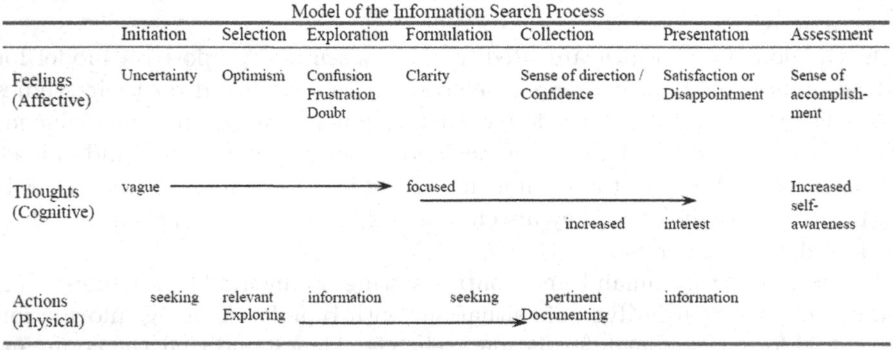

Model of the Information Search Process

	Initiation	Selection	Exploration	Formulation	Collection	Presentation	Assessment
Feelings (Affective)	Uncertainty	Optimism	Confusion Frustration Doubt	Clarity	Sense of direction / Confidence	Satisfaction or Disappointment	Sense of accomplishment
Thoughts (Cognitive)	vague ————————————→			focused —————————————→	increased	interest	Increased self-awareness
Actions (Physical)	seeking	relevant Exploring	information ———————————→	seeking	pertinent Documenting	information	

Fig. 2. Kuhlthau's information search process model. (Adapted from Kuhlthau's online information search process page [34])

In terms of macro-level collaborative search processes, there are several studies attempting to explore Kuhlthau's ISP model in a collaborative setting. For example, Hyldegård [31] utilized the ISP model in a group educational setting based on a qualitative preliminary case study. She found that a collaborative search process cannot be modeled the same way as an individual search process. Consequently, she suggested that the ISP model should be extended to incorporate the impact of social and contextual factors in relation to the collaborative information seeking process. Shah and Gonzalez-Ibanez [64] also attempted to map Kuhlthau's ISP model to collaborative information seeking. Through a laboratory study with 42 pairs of participants, they investigated the similarities and disparities between individual and collaborative information seeking processes. Similar to Hyldgard, they also declared that social elements should be added when applying the ISP model in a collaborative setting.

Yue et al. [87] conducted a thorough study of collaborative search processes at the micro-level. They created a novel approach using Hidden the Markov Model (HMM) to automatically model search process using hidden states. Different patterns of hidden states were identified and compared in both individual and collaborative search (see Table 1). In addition, the patterns of hidden states between two types of tasks were also compared, where T1 is an academic task that is a recall-oriented information-gathering task, and T2 is a leisure task that is a utility-based decision-making task.

Table 1. Hidden states in collaborative information seeking (extracted from Yue et al. [87])

Hidden state	Possible explanation
HQ	Formulate query, execute query
HV	Examine results
HS	Extract information
HD	Reflect/iterate/stop
HW	Check saved information in workspace
HC	Communicate with team members

Several important findings were drawn from Yue et al. [87]'s study (see both Table 1 and Fig. 3). First, two types of hidden states are identified in both individual and collaborative search processes: the search related hidden states and the sense-making related hidden states. Within the search related hidden states, users' interactions are focused directly on search activities, such as specifying a query (i.e., HQ), viewing a result (i.e., HV) or saving a result (i.e., HS), whereas the sense-making related hidden states tend to support search in terms of evaluating and defining search problems (i.e., HD), or making sense of the information through communications (i.e., HW and HC). Second, the search related hidden states are similar in both individual and collaborative search. However, the sense-making related hidden states are quite different. Individual searches only have one type of sense-making related hidden state (i.e., HD), but there are three different types in collaborative search (i.e., HD, HW, and HC). In addition, sense-making related hidden states have occurred significantly more in collaborative search than in individual search. Third, the percentage of sense-making is significantly higher in decision-making tasks than in information gathering tasks (see Fig. 3). These findings suggest that the demand for sense-making is higher in collaborative search and especially in decision-making tasks. Moreover, people are utilizing multiple approaches for sense-making in individual search. In particular, the cross-category transitions occurred more often in the decision-making task. These findings indicate that search and sense-making are more tightly connected with each other in collaborative search.

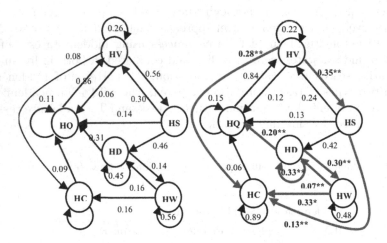

Fig. 3. Comparison of transition probabilities of hidden states in collaborative search for two tasks (left: T1, right: T2; red arrows indicate significant difference: *p < 0.05, **p < 0.01). (Adapted with permission from Yue et al. [87].) (Color figure online)

4   Factors Affecting Collaborative Information Search

In the field of individual information search, there have been several studies examining the factors that affect search. For example, Marchionini [39] pointed out that information search depends on six factors: information seeker herself, the task that requires search, the search system employed, the domain covering the task, the search setting, and search outcomes. It is through a careful manipulation of the interconnections of these factors that a search can be successfully completed. As an even more complex process, collaborative information search can be affected by many more factors [63]. In this section, we will mainly concentrate on those that are unique to collaborative information search, which can be roughly classified into people-side factors and search-side factors.

4.1   People-Side Factors

In individual search, people who conduct the search are the most important factor because their information needs trigger the search, and they have the final decision on the right search strategies and whether or not the search is successful [39]. In collaborative information search, the people who make up the team are still the most important factors, and they can be further classified into group size, age, collaboration style, affective signals, and communication style.

4.1.1   Group Size

There is a trade-off between the group size and the effectiveness of collaborative search. On the one hand, more people on the team means that each team member might work less and there is a higher likelihood of having the expertise to undertake the task. On the other hand, a bigger team means more collaboration interactions within the team, which could cause too much collaboration overhead. London [38] stated that collaboration works best in small groups, and it might break down in large groups. Morris [46] found that the groups involved in collaborative information search in her project are mostly small teams of two persons. Among the 109 self-identified users of collaborative information search, she found that 80.7% reported that they were involved in a two-person group including him/herself, while 19.3% participated in a three- or four-person group, with no respondents reporting a larger group size.

In a subsequent study on a similar research question regarding the group size for collaborative information search, Morris [47] found that small-group collaboration was still more common than large groups. For example, two-person groups were found to be around 31.2% of all such groups. However, between 2008 and 2013, the number of groups with three or four persons have increased greatly. Morris [47] found that the proportion of collaborative search with three and four person teams have increased to 22.9% and 23.9% respectively. Only groups with more than four members were still less common 9.2% reported working in groups of five; 4.6% in groups of six; and 8.3% in groups having seven or more members.

The possible reason for this change might be the adoption of remote collaboration technologies for simultaneous interaction among larger groups of users (e.g., video conference, group chat/message tools, and social networking sites), and mobile devices enabling co-located collaboration among each group member (e.g., smart phones) [47].

4.1.2   Age

Researchers also found that the age of the users who engage in collaborative information search is a significant factor, but it negatively correlates with the likelihood of participating in collaborative search [47]. This might be due to the fact that the new generation of technology-savvy users are more comfortable with new technologies (smart phones, social networking, etc.), and that they have more open attitudes toward online collaboration [47].

4.1.3   Collaborative Style and Roles of Users

When users engage in collaborative information search, they can form different styles of collaborative relationships. Some have tightly-coupled collaborations where the goal is focused and clearly defined. In contrast, some others are loosely-coupled collaborations and focused the means of taking different actions in a complex search process. The collaboration style thus can be classified into symmetric vs. asymmetric, based on the roles taken by the team members.

Team members who take the symmetric roles are called peers [20]. Peers have equal amounts of same control of the collaborative search process, and may work together or independently after a negotiation on the division of labors. The collaborators engaged in CollabSearch [87] and SearchTogether [48] are examples of peers involved in collaborative information seeking. A variation on the peer role mentioned by [20] is domain A expert working with domain B expert. Although they have different domain knowledge, which makes it natural for them to take leadership in their own domain, they still form an equal relationship in the overall collaborative search project.

The asymmetric relationship in collaborative search is often reflected by the fact that the team members can have different degrees of "expertise or familiarity with a domain and with search tools" [20]. Their roles can form around search expert vs. search novice, or domain expert vs. domain novice, or search expert vs. domain expert. No matter which type of relationship, the search expert could require more sophisticated functions in collaborative search interfaces, and the domain expert may form deeper and more complex information needs. Particularly in the pair of search expert and domain expert, the search expert could know "how to select collections and formulate queries against those collections, but can only make rudimentary relevance judgments based on the description of the information need provided by the domain expert." Therefore, the domain expert often is the main person to evaluate retrieved documents.

Another example of the asymmetric relationship is called prospector and miner, which was explored in [56]. Here, the prospector can search broadly, and generates many queries to explore the collection, but makes relatively few relevance judgments for each query. In contrast, the miner takes the inputs from the prospector, and makes detailed relevance judgments about the results returned to the queries issued by the prospector.

Although it seems that both symmetric and asymmetric relationships have their own suitable situations, Gray [25] warned that collaboration might not be the best approach when the power in a group is not evenly distributed. Shah [63] also stated that users in collaborative information search should be given freedom to choose their own way of collaboration, and the system should provide enough support for carrying out that collaboration.

4.1.4   Affective Signals

As the major driving force for the information search process, the cognitive and affective aspects of users are increasingly recognized as important factors influencing the interactions between users and search engines [4]. For example, affective processes may have direct impacts on users' strategies for processing information [18,32], and users' positive feelings could support their subsequent interactions with search engines, while negative feelings hindered that action [80].

In collaborative information search, researchers have started to examine affective signals' impacts on the search process. González-Ibáñez and Shah [24] proposed the notion of a Group's Affective Relevance (GAR), which refers to the overall emotional experience of each group member with regard to specific information objects that he/she shares with the group. Through analyzing more than 6000 chat messages produced by the CIR teams in their study, Shah and Marchioini [66] found an interesting correlation between members' expressed emotions and the performance of the group; that is, the closer the distance between the number of positive, negative, and neutral information judgments, the higher the performance of the teams as measured by precision.

In order to investigate a CIR process that has different affective states as initial conditions, and the implications of those different conditions on the CIR performance, González-Ibáñez and Shah [23] conducted a controlled experiment with affective induction to the team members. The pairs of team members were organized by positive-positive affective states, or positive-negative states, or negative-negative states. Then, the teams were asked to perform CIR tasks. Their results show that different combinations of initial affective states in the teams can affect the team's search and task performances. The negative-negative configuration produced more precision in solving fact-finding tasks than the other two, and the positive-negative configuration led to more efficient search processes than the other two. Their results are consistent with the general literature about affective states: people with negative affective states tend to employ more systematic and detailed actions, whereas people with positive affective states are less accurate but perform with higher efficiency [32,69]. Of course, CIR in reality would not have artificially infused affective states. According to the Affect Infusion Model (AIM), factors in the context of information processing (such as familiarity, complexity and typicality, personal relevance, specific motivation, cognitive capacity, and situational pragmatics) would produce differences in users' affective states.

4.1.5 Communication Style

Communication is the process of sending and receiving information. It is vital to the success of two or more individuals working as a team [15]. Although the study of communication in collaborative searches is a relatively new topic, researchers in other domains, such as Computer-Supported Cooperative Work (CSCW), Computer Mediated Communication (CMC) and Computer-Supported Collaborative Learning (CSCL), have studied communications for a long time.

The literature recognizes two types of communication styles: task oriented versus socially oriented [79]. Task oriented communication focuses on fulfilling the responsibilities while socially oriented communication focuses on satisfying the emotional needs of the collaborators. Both communication styles can be found in CIR. Through analyzing 1813 explicit communication records (chats) among team members in her studies of CIR, Yue et al. [87] found that between

64.46% to 74.46% of communication is task-oriented, 16.31% to 13.11% is task-related social communication, and 19.23% to 12.34% is purely social communication that is not task-related. This shows that people do not utilize one single communication style in their CIR.

Communication style is closely related to task types. For example, Yue et al. [87] found that information-gathering tasks trigger less task-oriented communication among team members than decision-making tasks. However users in information-gathering tasks engaged in much more coordination-related task communication whereas users in decision-making tasks performed more content-related task communication. The amount of their social-oriented communication is roughly the same.

Communication style is a factor affecting collaborative search strategies and tactics as well. Foley and Smeaton [17] proposed division of labor and sharing of knowledge as two important strategies of successful collaboration in search. Both strategies can be facilitated by the task-oriented communication between team members. In a study of library users, Twidale [78] identified a set of search tactics that may require task-oriented communication with others. For example, users may seek help from the reference librarian or brainstorm with others to generate new approaches to search. Reddy and colleagues [58] identified three reasons for task-oriented communication among team members while looking for information: consulting, brainstorming and team cognition.

Communication style greatly influences the selection of mechanisms for communication among team members. This is because each communication mechanism has its costs and benefits in the collaborative search process [21]. For example, greater demands for efficiency in task and social oriented communication styles caused most existing collaborative search systems to implement instant messaging as a function to support the communication among team members [48,60]. González-Ibáñez and colleagues [21] found that task-oriented communication and even task related social oriented communication favors face-to-face communications to allow team members to interact effortlessly, but there is also the risk of hurting search performance with more non-task related social communications. The users who engage in more task-related conversations would like to use texting, but they would also be disappointed with the limited capabilities of social communication in collaborative search. Therefore, González-Ibáñez and colleagues [21] thought that most users in collaborative search would probably like the audio plus text option because it provides the right level of social oriented communication and, at the same time, enables more focused task-oriented communication.

4.1.6 Knowledge Learning

Knowledge learning is recognized as an important component in the search process. Because prior search scenarios are relatively limited, existing studies often measure the knowledge growth before and after a search to study knowledge learning in search. However, this caused the lack of a fine-grained understanding of users' knowledge, changing patterns within a search process, and users' adoption of different sources for learning.

Knowledge learning is an even more important component in collaborative information search as the CIS tasks are usually exploratory, which triggers learning. A CIS task involves diverse learning resources such as self-explored search content, partners search content and explicit communication between them [13]. Through analyzing the data from a controlled laboratory user study with both collaborative and individual information seeking conditions, Chi et al. [13] demonstrated that users' knowledge keeps growing in both conditions, but significantly more diverse queries are issued in the collaborative condition. Their analysis of users queries also revealed that the rate of adoption of different learning resources varies at different information seeking stages, and the adoption is influenced by the nature of search tasks. These results motivated them to propose several insights for system design to enhance knowledge learning in CIS. For example, they discovered that the current knowledge sharing support through accessing teams workspace and explicit communication is probably enough to generate proper queries, but it still fails to facilitate a truly understanding of certain knowledge in a clicked document. Therefore, proper information visualization techniques might be useful here to summarize the knowledge states of the team members and/or the whole groups.

4.2   Search-Side Factors

Besides the people-side factors discussed above, various aspects of search and search environment can also affect collaborative information search.

4.2.1   Synchronized and Unsynchronized

Most collaborative information search in the traditional library setting was synchronous; only when searches are performed online can we see different models.

Collaboration in web search can also be synchronous or asynchronous in terms of concurrency. Both of these two types of collaboration are common in exploratory search tasks, yet they are very different from each other. In synchronous collaboration, team members can get instant feedback from each other while in asynchronous collaboration only those who search later can benefit from the work of earlier team members.

Synchronous collaboration was found to be more common than asynchronous, comprising roughly two-thirds (64.2%) of the incidents studied in [47]. Shah and Gonzlez-Ibez [64] also found that synchronous collaborative search requires the team to have an appropriate division of labor and mediated support for sharing knowledge.

Twidale and Nichols [77] designed the Ariadne system, which allows a user to collaborate with an information expert remotely and synchronously using a library catalog. It does not support asynchronous collaboration.

Yue et al. [86] presents a user study aiming to compare search processes in three different conditions: a pair of users working on the same Web search tasks

synchronously with explicit communication, pair of users working on the same
Web search tasks asynchronously without explicit communication, and individual users working separately. They stated that, in synchronous collaboration,
it is easier for collaborators to explicitly communicate with each other such as
verbal means or text chatting. However, in asynchronous collaboration, the formats of communication are more likely to be implicit as in sharing a document
or search history. In their study, chat represents explicit communication while
the Workspace activities represent implicit communication. Their results showed
that participants in synchronous search have fewer actions related to querying
and collecting relevant information than participants in asynchronous search.
However, pre-query analysis demonstrates the possible benefit of explicit communication in synchronous search condition to help users to generate queries.
They also found that both types of communication are common in synchronous
search. The fact that participants undertake a higher number of Workspace-
related actions than participants in synchronous search might indicate that the
explicit communication between participants promotes implicit communication.

Interestingly, social search (see Chap. 7 of this volume [10]) also has this
difference in concurrency. The information about other users' activities in the
community is often based on their past behaviors. Therefore, common social
search approaches can be viewed as asynchronous search. However, one type of
social search aims to connect experts in the community to the user so that there
could be direct communication between them. This type of social search would
be considered to be synchronous search.

4.2.2   Co-located and Remotely Located

In terms of location, collaborative web search can be co-located or remotely
located. Co-located means that team members are all at the same place, so they
may communicate directly without computer support. Collaboration remotely
implies the need for additional channels, such as chat, voice, or audio conferencing to coordinate searchers' activities.

In 2013, Morris [47] found that remote collaboration was more common than
co-located in 2013, characterizing 61.5% of the described searches as remote collaborations. This is a different result to the earlier survey [46], which found only
a slight prevalence of co-located search configurations. She also found that nearly
all of the smartphone owners (92.8%) reported using their phones to engage in
co-located collaborative searches in which several people simultaneously used
their smartphones to look up information. This suggests that co-located collaborative smartphone search may be a rich area for further investigation.

Shah and González-Ibáñez [65] compared team users and single user in five
different conditions (see Fig. 4). One of the conditions involves a single users
and the other condition artificially combines two single users. The remaining
three conditions are synchronous collaborative searches in which the location
types varies from each other, including co-located using same computer, co-
located using different computer and remotely located. They identified that two
collaborators working on remote collaborative searches could cover more unique

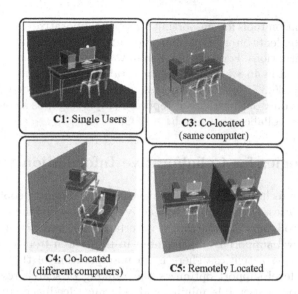

Fig. 4. Conditions for co-located and remotely collaborations. (Adapted with permission from Shah and González-Ibáñez [65].)

information, which shows that there is a value in remote collaboration when the task has clearly independent components. Interactions, although performed remotely, were able to avoid overlapping their explorations and take advantage of more sources of information. This is important for exploratory and recall-oriented search tasks.

Shah and Gonzlez-Ibez [65] also found benefits to co-located collaborating, which may influence the way in which users formulate their queries. Even though most teams split up the task, the physical closeness of users enabled them to hear what their peers thought aloud; or even have brief conversations (facilitated by face-to-face interaction). This may have influenced implicitly common queries between those participants.

4.2.3 Devices

With the rapid development of mobile technologies, we have seen more smartphones and tablets being used in collaborative search [47]. Morris found that about 30.3% of the collaborative searches she surveyed involved a smartphone and 11% involved a tablet. Technologies that might facilitate public sharing such as TVs and projectors were rarely employed, in only 1.8% and 0.9% of the searches, respectively. She also found that non-digital tools were an important part of collaborative searches; for instance, 11% of respondents reported using paper to support their collaborative search tasks.

It seems that collaborative search is increasingly occurring beyond the search engine, and people relied on multiple tools to successfully complete collaborative searches [47]. Because mainstream web browsers and search engines do not have communication tools for supporting remote collaborative search, Morris [47] found that respondents often employed communication tools that they have been using in other situations. For example, e-mail was the most common communication tool (involved in 46.8% of the searches), talking on the phone (27.5%), text messaging/SMS (30.3%), and instant messaging (12.8%). Videoconferencing was found to be rare; only one participant reported employing it as a communications channel during a collaborative search.

5   Technologies for Collaborative Information Search

Over time, many technologies have been developed to support people's collaborative search activities. Depending on how active the system is, there are two levels of support for collaborative search: (1) interface-level and (2) algorithm level. The interface-level support is implemented in the search front-end, facilitating users' collaboration activities. There is no manipulation of the search results. The algorithm level support optimizes the ranking of search results or query suggestions through user role mining and relevance feedback. In this section, we introduce different technologies and systems used in both levels to support collaborative search.

5.1   Interface-Level Support

Collaborative search should support both search activities and collaboration activities. Therefore, many collaborative search systems are designed based on traditional search systems with enhanced interface features for collaborative activities. These features are designed based on the understanding of various collaborative search strategies and tactics. Foley and Smeaton [17] proposed division of labor and sharing of knowledge as two important strategies of successful collaboration in search. Through a study of library users, Twidale [78] identified process-related and product-related collaborative search tactics, such as asking someone else for help, coordinated searching, or sharing the search products. Facilitating explicit communication, maintaining awareness and supporting information sharing among team members can enhance most of these collaborative search strategies and tactics.

5.1.1   Communication Support

Based on the identified importance of communication in collaborative search, most existing collaborative search systems have implemented features to support the explicit communication among team members [48, 60, 87].

When users are co-located, it is very convenient for them to communicate face-to-face with each other. CoSearch [2] is a tool that provides explicit support for groups of co-located people to search the web when gathered around a single computer. The primary design goal for CoSearch was to enhance the experience of co-located collaborative web search in settings where computing resources are limited, by enabling distributed control and division of labor while maintaining group communication and awareness levels. Instant message is the simplest way of supporting communication for remote users and it offers a great deal of user freedom. Instant message can take various forms, including text, audio and video. Text messages can apply to both synchronous and asynchronous collaboration while audio and video are more applicable to synchronous collaboration.

The design of advanced support for communication should be based upon the understanding of costs and benefits of communications in the collaborative search process [21]. Hertzum [29] found that communications could be effective in establishing common ground between team members. However, other researchers also reported that communication could introduce extra workload or distract users from their search tasks [12]. González-Ibáñez and colleagues [21] investigated the costs and benefits of three different communication mediums: face-to-face, computer-mediated text, and audio plus text. They found that the face-to-face medium allows users to interact effortlessly, but it also generated more non-task related communications, which may hurt the search performance. The communication through text medium was more focused on task-related conversations but also limited the social aspects of communication in collaborative search. The audio plus text medium was able to provide the right level of social presence and at the same time it did not distract team members from the task.

Yue [85] further studied the content of communication and the timing of communication in different task types. She found that the before search stage communication is more focused on task coordination. In the during search stage, team members are more involved in task content communication. Task social communication is more common in the before search and after search stages than in the during search stage. She also analyzed the correlation between communication patterns and search outcomes, which reveals the costs and benefits of communications. The results suggested that communication could encourage participants to explore a wider range of vocabularies for the queries. However, communication also takes time and additional effort on the part of the participants, thereby decreasing the recall and increasing the cognitive load. An interesting finding is that task social communication actually has a positive correlation with the recall and satisfaction, suggesting that the social interaction may engage participants to the search task. However, there is indeed a cost to other types of communications. Task coordination, task content and non-task related communications have negative relationships with recall, and task content also has a positive correlation with the cognitive load. Therefore, there are both benefits and costs to communications in the collaborative search processes. This study suggests that the key for success in collaborative search might be the interpersonal social interactions among the team members, which provide social support and increase team members' engagement with the search. Therefore, the collaborative search system should not take over all the collaboration mediation, which results in removing the personal interactions among team members. Instead, the collaborative search system should be designed to support team members in providing social support for each other.

5.1.2 Awareness Support

Researchers (e.g., Yue [85]) found that users in collaborative search often need to make a division of labor in terms of search topics. The system should be able to provide support for them to divide the topics in the search task and take ownership of sub-topics. Also, the team members need to be aware of everyone's progress on the sub-topics so that they can make adjustments to the coordination as the search is going on. The system should provide support for such awareness and the mechanism for adjustment. Yue [85] pointed out that it's important to design interface-mediated support for coordination among team members because coordination through communication is costly. Another study [62] presented the effects of three different awareness conditions on coordination through chat messages. The findings showed that a lower level of awareness support increases the cost of coordination in the collaborative search process. Morris [47] also suggests that lack of awareness of collaborators' activities result in redundant work and frustration on the part of users.

Fig. 5. The SearchTogether system: (a) integrating messaging, (b) query awareness, (c) current results, (d) recommendation queue, (e)–(g) search buttons, (h) page-specific metadata, (i) toolbar, (j) browser. (Adapted with permission from Morris and Horvitz [48].)

Existing collaborative search systems have implemented various functions to support awareness among users. As shown in Fig. 5, SearchTogether [48] is a prototype that enables remote users to synchronously or asynchronously collaborate when searching the web. The system aims to support collaboration with several mechanisms for awareness, including shared search histories, split searching, peek-and-follow browsing, and integrated chat. An updated version of SearchTogether, called CoSense [55] added several new features for collaborative information sense-making, including search strategies view, timeline view, workspace view and chat-centric view. All the features are designed to enhance awareness among users.

It may be helpful for the system to visualize the information space that has been explored by the team members so that they can evaluate the status of the search task.

Coagmento [60], which is now a free open-source product[1], is a system supporting multiple people working together to conduct online information seeking tasks. As shown in Fig. 6, it provides a set of action histories, including queries submitted and page/snippets saved to support the users' peripheral awareness. These awareness supports are designed for users to keep track of the search status. González-Ibáñez and Shah [22] studied the effect of different peripheral awareness levels (none, personal, group) on collaborative search. They found that support for group awareness is significant for effective collaboration.

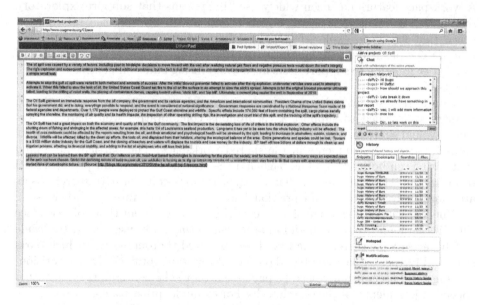

Fig. 6. The awareness design in the Coagmento system [22]. (Adapted with permission from an image supplied by Prof. Shirag Shah.)

Shared query history is a commonly used feature to support group awareness. Capra et al. [11] studied the impact of shared query history on query reformulation in asynchronous collaborative search. They conducted an asynchronous collaborative search user study using a system called Results Space. In the study, subjects who did the search later were provided with query histories

---

[1] http://coagmento.org/.

of the participants who did the search earlier. They found that although only four participants actually clicked on the provided queries from previous participants, 10 out of 11 participants reported that they indeed looked at the query history and made use of it. Four motivations for using the query history were summarized from the interviews. The first motivation is to write different queries from what the previous participants had already done. The second one is to get an overall sense of topics on which the previous participants had searched. The third motivation is trying to figure out where to start their search by examining the train of thought of previous participants through query history. The last motivation is that the query history can inspire participants with new ideas for their queries.

5.1.3 Information Sharing Support

Sharing search products is an important collaborative search tactic. The collaborative search system should have a good mechanism for sharing and organizing search results, which can help the users to assess the information obtained [85]. A workspace feature has been widely used in systems that support exploratory search. It allows users to save, organize and make sense of the information space that has been explored [81]. In collaborative search, the sense-making is more complex. Many collaborative search systems implemented the shared workspace function to support users in sharing, rating and commenting on search results [55, 66, 87].

Capra [11] studied the document view and rating behaviors in shared workspace designed to support asynchronous collaborative web search. Users can view and share ratings on search results. They found that all 11 participants viewed and rated documents in the shared workspace. The participants found that ratings on both the landing page and the search engine result page (SERP) to be useful. They also found that in asynchronous collaborative search, users are more likely to rate documents that had been previous rated by other users.

Yue [85] studied the usage of shared workspace in different types of tasks (see Fig. 7). She found that in the information-gathering task, users need to decide which parts of the search topic have been explored thoroughly and which parts still need further exploration based on the information obtained. In the decision-making task, users usually need to pay attention to the constraints on, or other special requirements for the search tasks. For example, price is a constraint in the travel-planning task. The system should allow users to specify their price criteria, organizing their saved items using the price, and making the price for the each saved item easily seen to all the team members to facilitate the decision-making.

5.2 Algorithm-Level Support

Beyond the interface-level support, researchers also developed various algorithm-level support for collaborative search, such as balancing user roles in collaborative search, query recommendation based on collaborative activity and re-ranking search results based on the community's preferences.

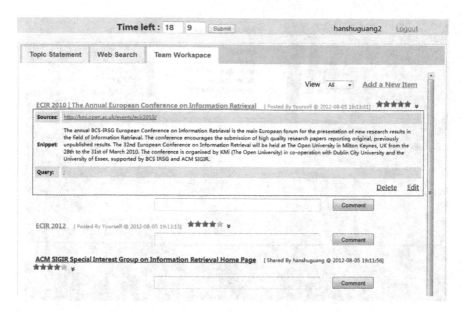

Fig. 7. Shared workspace in CollabSearch interface. (Adapted with permission from Yue et al. [87].)

5.2.1   Role-Based Mediation

In collaborative search, the users can take symmetric roles or different roles [56]. For symmetric roles, the users are peers and they usually split the search task into sub-tasks, with each user in charge of a set of sub-tasks [85]. In this case, there is no further difference in supporting users in their search tasks.

However, users in collaborative search can also take different roles. For example, in a collaborative search system called Cerchiamo [56], one member of the two-person team acts as the Prospector, whose job is to discover potentially relevant documents, whereas the other member acts as the Miner who selects relevant documents based on the documents collected by the prospector. In order to support the team's collaboration, the system provides three different user interfaces. The one for the prospector is a rich query interface developed based on MediaMagic [1] so that powerful online search can be performed and returned documents can be quickly examined. The second interface is for the miner, which is just a browsing interface called RSVP with the function of rapid serial visualization of results. To provide shared context for the team, the third interface is a shared display showing the progress of the search session (see Fig. 8).

In addition to the interface support, Cerchiamo also configures algorithms to support the prospector and the miner, respectively. For example, on the miner's side, the sequence ordering of returned documents for the miner to judge is based on a score assigned to each document. The score is based on the rank of the document in a returned ranked list collected by the prospector, the relevance weight the system estimates for the ranked list, and the freshness weight the

Fig. 8. Prospector and Minor role-based medication in Cerchiamo interface. Each user's UI is suited to their role: prospector (left) and Miner (right). Center screen shows the shared query state, and the two large screens on the side show sample relevant shots for the current topic. (Adapted with permission from an image shared by Dr. Pickens.)

system estimates for the ranked list. Each time, the miner judges a document to be relevant, the relevance weight for the ranked lists containing the document increases, but the freshness weight of the ranked lists decreases. Thus, the system constantly adjusts the sequence ranking of the documents that the miner is judging.

Similarly, Shah et al. [67] proposed an algorithm to redistribute documents between two team members who act as a Gatherer and a Surveyor. The Gatherer focuses on quickly finding as much relevant information as possible, while the Surveyor works on exploring more diverse information. Both members would issue queries to look for relevant documents, and the system generates initial results based on the queries, then fuses the two results together, and splits the fused results based on the role of the two members. That is, the Gatherer receives one list that is optimized for effectiveness (e.g., high precision), and the Surveyor receives the other list that is optimized for exploration (e.g., high diversity). These experiment results demonstrated that the technique could provide support to satisfy the needs of both members.

Soulier et al. [72] proposed a framework of techniques to monitor users activities in collaborative search and predict the roles that team members can perform using a supervised learning method. Through simulations on two different user study datasets, the authors demonstrated that users could achieve better performance if the suggested roles were followed.

5.2.2   Collaborative Querying

Collaborative querying refers to how users of an information retrieval system, during their query formulation and reformulation stages, can draw help from previous query preferences of other users [19]. This means that previously-learned queries and relevant documents are reused in new and similar search sessions to improve the overall quality of the queries [30].

For example, Yue et al. [88] studied how to utilize the user activities in explicit collaborative search for query reformulation. In particular, they studied the influences of different collaborative activities on how users generate new terms for query reformulation. Through log analysis of data collected from a user study, they compared possible sources for query terms. The results show that both search and collaborative actions are possible sources of new query terms. Traditional resources for query expansion, such as previous search histories and relevant documents, are still important sources of new query terms. The content in chat and workspace generated by participants themselves seems more likely to be the source for new query terms than that of their partners. Task types also affect the influences on query reformulations. For the academic task, previously saved relevance documents are the most important resources for new query terms while chat histories are the most important resources for the leisure task.

5.2.3   Context-Sensitive Reranking

Contextual-sensitive reranking aims to use users' recent search history (past queries and clicked documents) to generate a better ranking of the returned documents [68]. However, the context in collaborative web search is more complicated, and thus provides some interesting opportunities to draw contextual supports from multiple sources [27]. As shown in Fig. 9, the context information available in collaborative search would include the search history of the user himself/herself, the search history of the partner, and the chat messages between them. Compared to the context in individual search, the last two are unique in collaborative search.

Based on a user study data collection with 54 participants, Han and his colleagues [27] compared the effectiveness of contextual support using the user's own search history, the partners' history and the chat messages. Within each search history, they also explored the performance differences between using the past queries and the clicked documents underneath each query. Their results confirmed that contextual supports can significantly improve retrieval effectiveness, either using the user's own search history, which essentially becomes a contextual-sensitive individual search, or combining the user's and the partner's search history. In particular, adding the partner's search history on top of the user's own search history could further improve the retrieval effectiveness, which demonstrates the usefulness of the partner's search history. The team's explicit collaboration behaviors (i.e., chat messages) can significantly improve the retrieval performance as well, and the improvement is even more significant than that of using the user's own search history. More interestingly, although the chat messages contain a great deal of noisy and irrelevant information (such

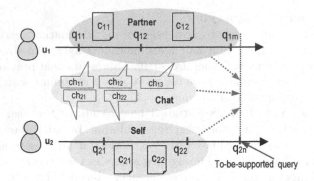

Fig. 9. Context information available in collaborative search, in which self-history, partner-history and chat messages are all part of the context. (Adapted with permission from Han et al. [27].)

as social greetings and off-topic chats), those noises are so off the topic that they do not affect the ranking of relevant documents at all. Therefore, there is no need to clean the chat messages for the purpose of improving retrieval.

6   Evaluating Collaborative Search Systems

6.1   System-Oriented Evaluation

Evaluation in a collaborative information seeking environment can be a huge challenge due to the variety of interactions between systems and users. A few efforts had been made to evaluate various parameters in a collaborative information seeking environment by using traditional information retrieval (IR) or human-computer interaction (HCI) measures [61]. Baeza-Yates and Pino [5] presented an initial attempt to evaluate performance measures in collaborative IR. They tried to extend the performance measure in a single-user IR system and treat the performance of a group as the summation of performance of individuals. In a later work [6], they evaluated the relationships among quality of the outcomes, number of people involved, time spent on the overall task, and total work done. As both efforts only used measures for evaluating performance, how well the system can support users in their collaboration process was not evaluated.

Capra et al. [11] used the TREC Robust corpus for a collaborative search user study so that standard recall and precision measures could be computed. However, when the collection is the open web, there is no ground truth to be used to calculate recall and precision. Shah and Gonzlez-Ibez [65] proposed precision and recall measures that can be used in an open-web collection context. Recall is defined as the ratio of relevant web pages collected by a single team to the relevant web pages collected by all of the teams. Precision is defined as the ratio of relevant web pages collected by a single team to all the web pages viewed by that team. In addition, the authors also proposed other measurements such as query diversity, useful webpages, and likelihood discovery. Lavenshtein distance is used to compute the distance between pairs of queries for each team to measure the query diversity. Useful webpages are defined as webpages on which a user spends at least 30 s. Likelihood of discovery is used to measure hard-to-find information, which is measured by the inverted frequency that each webpage is visited by all the teams.

6.2   User-Oriented Evaluation

There were several studies focused on the usability of the collaborative interface. Wilson and Schraefel [82] proposed an analytical inspection evaluation for information seeking interfaces which incorporated information seeking models in HCI usability evaluation method. And later, Wilson and Schreafel [83] extended the framework to evaluate collaborative search interface. This method was designed for HCI experts to evaluate the usability of the interface; no real users were involved in the evaluation. Morris and Horvitz [48] evaluated their SearchTogether system with a user study of 14 subjects in 7 pairs. They collected log, observation and questionnaire data from the study. The evaluation revealed the effectiveness of their interface by analyzing the usage of certain features and asking users how they assessed the effectiveness of the features in helping them accomplish the task. In the evaluation of the CoSearch system, Amershi and Morris [2] recruited 36 subjects in 12 groups to use the system. Subjects were asked to comment on the usability of Co-Search by answering 5-point Likert scale questions. Shah [60] evaluated the Coagmento system using a set of objective and subjective measures in a user study involving 42 pairs of subjects. Objective measures included effectiveness and efficiency which are based on analyzing search outcome of individual and group. Subjective measures such as awareness, effort, ease of use, satisfaction, and engagement were evaluated through questionnaire.

7   Discussion and Conclusion

7.1   Discussion

Collaborative information search is an active research topic, and many interesting achievements have been made over the past decade. However, as seen in this chapter, there are still many significant challenges to be addressed.

7.1.1   User Groups

Evidence shows that populations such as the elderly, recent immigrants, and people in developing countries may engage in collaborative search activities, but very little detailed data is currently available on the search needs and practices of these groups.

Smart Splitting [50] considers the expertise areas of each group member, users' roles and their impacts on collaborative search, which demonstrates the potential for role-tailored group search systems, but this is still a rich and valuable area for further investigation.

Existing studies mostly concentrate on two-person teams. But collaboration can happen in groups of different sizes. It is therefore important to explore the optimal size of groups for collaborative search in various tasks, and to examine the influence and limits of current technologies on those groups. It is also an open question to understand how the performance of proposed collaborative search systems can scale up with larger group sizes.

7.1.2   New Devices

We have discussed the impacts of devices on collaborative search. But it is still unknown how the high precision touch input, high resolution output of those modern mobile devices can affect people's collaborative search.

Further more, it is a challenge to design and develop collaborative search systems that enable collaboration among different group members using different devices with different capabilities.

7.1.3   Collaborative Search in Specific Contexts

We know that information needs are generated from more and broader contexts that merely search. This is also true in collaborative information search. Although the literature has studied collaborative search behaviors in various settings, the research has yet to identify the specific influence of context factors, as well as the detailed support required during collaborative search in various contexts. For example, collaborative search can happen in an educational setting, where students might work together or they collaborate with their tutors. It can also happen in health care, where patients work with caregivers or with doctors, and caregivers work with doctors. Collaborative search can happen in military agencies, where commanders need collaboration with teams of intelligence analysts; or collaboration can happen in e-discovery, where plaintiff lawyers work with defendant lawyers, or lead lawyers work with junior lawyers. All these contexts impose their unique contributions and constraints to collaborative search in terms of group size, team member relationships, information needs, and system support.

7.1.4   Integrated Collaborative Search Systems

Researchers in the field of collaborative search have been working on designing and implementing integrated collaborative search systems so that users can conduct collaborative search within one interface. However, as Morris [47] showed many users do not use integrated collaborative search system for their collaborative search at all.

The main question, therefore, is whether or not there is a need for integrated collaborative search systems. Should we instead try to couple the communication and collaboration tools that are part of the users' everyday routines (e.g., e-mail, texting, instant messaging, phone calls, and social networking) with search technologies to build a collaborative search system on-the-go? If this should be the route, what does this mean for the existing web browser and search engine technologies, as well as current communication and collaboration technologies?

7.2 Conclusion

In this chapter, we discussed collaborative information search, where a team of people work together to achieve a search goal that is shared by all team members. Collaborative search consists of search activities and collaboration activities, but it is not just simple search part plus collaboration part, because studies show that collaborative search can achieve better users' satisfaction and lower cognitive loads than individual search [85]. This demonstrates the usefulness of collaborative search, and also indicates that much more powerful support is needed to further improve the design of collaborative search systems.

References

1. Adcock, J., Cooper, M., Girgensohn, A., Wilcox, L.: Interactive video search using multilevel indexing. In: Leow, W.-K., Lew, M.S., Chua, T.-S., Ma, W.-Y., Chaisorn, L., Bakker, E.M. (eds.) CIVR 2005. LNCS, vol. 3568, pp. 205–214. Springer, Heidelberg (2005). https://doi.org/10.1007/11526346_24
2. Amershi, S., Morris, M.R.: CoSearch: a system for co-located collaborative web search. In: Proceedings of the SIGCHI Conference on Human Factors in Computing Systems, pp. 1647–1656. ACM (2008)
3. Andriessen, J.H.E.: Group processes. In: Andriessen, J.H.E. (ed.) Working with Groupware, pp. 89–124. Springer, London (2003). https://doi.org/10.1007/978-1-4471-0067-6_6
4. Arapakis, I., Jose, J.M., Gray, P.D.: Affective feedback: an investigation into the role of emotions in the information seeking process. In: Proceedings of the 31st Annual International ACM SIGIR Conference on Research and Development in Information Retrieval, pp. 395–402. ACM (2008)
5. Baeza-Yates, R., Pino, J.A.: A first step to formally evaluate collaborative work. In: Proceedings of the International ACM SIGGROUP Conference on Supporting Group Work: The Integration Challenge, pp. 56–60. ACM (1997)
6. Baeza-Yates, R., Pino, J.A.: Towards formal evaluation of collaborative work. Inf. Res. **11**(4), 11–14 (2006)
7. Bass, B.M., Stogdill, R.M.: Handbook of Leadership, vol. 11. Free Press, New York (1990)
8. Blake, C., Pratt, W.: Collaborative information synthesis. Proc. Am. Soc. Inf. Sci. Technol. **39**(1), 44–56 (2002)
9. Brooks, M., Kuksenok, K., Torkildson, M.K., Perry, D., Robinson, J.J., Scott, T.J., Anicello, O., Zukowski, A., Harris, P., Aragon, C.R.: Statistical affect detection in collaborative chat. In: Proceedings of the 2013 Conference on Computer Supported Cooperative Work, pp. 317–328. ACM (2013)
10. Brusilovsky, P., Smyth, B., Shapira, B.: Social search. In: Brusilovsky, P., He, D. (eds.) Social Information Access. LNCS, vol. 10100, pp. 213–276. Springer, Heidelberg (2017)
11. Capra, R., Chen, A.T., Hawthorne, K., Arguello, J., Shaw, L., Marchionini, G.: Design and evaluation of a system to support collaborative search. Proc. Am. Soc. Inf. Sci. Technol. **49**(1), 1–10 (2012)

12. Carroll, J.M., Rosson, M.B., Convertino, G., Ganoe, C.H.: Awareness and team-work in computer-supported collaborations. Interact. Comput. **18**(1), 21–46 (2006)
13. Chi, Y., Han, S., He, D., Meng, R.: Exploring knowledge learning in collaborative information seeking process. In: Proceedings of the 2016 ACM SIGIR 2016 Workshop: Searching as Learning (SAL), pp. 1–5. ACM (2016)
14. Daft, R.L., Weick, K.E.: Toward a model of organizations as interpretation systems. Acad. Manag. Rev. **9**(2), 284–295 (1984)
15. Dickinson, T.L., McIntyre, R.M.: A conceptual framework for teamwork measurement. In: Team Performance Assessment and Measurement, pp. 19–43 (1997)
16. Evans, B.M., Chi, E.H.: Towards a model of understanding social search. In: Proceedings of the 2008 ACM Conference on Computer Supported Cooperative Work, pp. 485–494. ACM (2008)
17. Foley, C., Smeaton, A.F.: Division of labour and sharing of knowledge for synchronous collaborative information retrieval. Inf. Process. Manag. **46**(6), 762–772 (2010)
18. Forgas, J.P.: The affect infusion model (AIM): an integrative theory of mood effects on cognition and judgments (2001)
19. Foster, J.: Collaborative information seeking and retrieval. Annu. Rev. Inf. Sci. Technol. **40**(1), 329–356 (2006)
20. Golovchinsky, G., Qvarfordt, P., Pickens, J.: Collaborative information seeking. Computer **3**, 47–51 (2009)
21. González-Ibáñez, R., Haseki, M., Shah, C.: Lets search together, but not too close! An analysis of communication and performance in collaborative information seeking. Inf. Process. Manag. **49**(5), 1165–1179 (2013)
22. González-Ibáñez, R., Shah, C.: Coagmento: a system for supporting collaborative information seeking. Proc. Am. Soc. Inf. Sci. Technol. **48**(1), 1–4 (2011)
23. González-Ibáñez, R., Shah, C.: Performance effects of positive and negative affective states in a collaborative information seeking task. In: Baloian, N., Burstein, F., Ogata, H., Santoro, F., Zurita, G. (eds.) CRIWG 2014. LNCS, vol. 8658, pp. 153–168. Springer, Cham (2014). https://doi.org/10.1007/978-3-319-10166-8_14
24. González-Ibáñez, R.I., Shah, C.: Group's affective relevance: a proposal for studying affective relevance in collaborative information seeking. In: Proceedings of the 16th ACM International Conference on Supporting Group Work, pp. 317–318. ACM (2010)
25. Gray, B.: Collaborating: Finding Common Ground for Multiparty Problems. Jossey-Bass, San Francisco (1989)
26. Han, S., He, D.: Network-based social search. In: Brusilovsky, P., He, D. (eds.) Social Information Access. LNCS, vol. 10100, pp. 277–309. Springer, Heidelberg (2017)
27. Han, S., He, D., Yue, Z., Jiang, J.: Contextual support for collaborative information retrieval. In: Proceedings of the 2016 ACM on Conference on Human Information Interaction and Retrieval, pp. 33–42. ACM (2016)
28. Hansen, P., Järvelin, K.: Collaborative information retrieval in an information-intensive domain. Inf. Process. Manag. **41**(5), 1101–1119 (2005)
29. Hertzum, M.: Collaborative information seeking: the combined activity of information seeking and collaborative grounding. Inf. Process. Manag. **44**(2), 957–962 (2008)
30. Hust, A.: Introducing query expansion methods for collaborative information retrieval. In: Dengel, A., Junker, M., Weisbecker, A. (eds.) Reading and Learning. LNCS, vol. 2956, pp. 252–280. Springer, Heidelberg (2004). https://doi.org/10.1007/978-3-540-24642-8_15

31. Hyldegård, J.: Collaborative information behaviour–exploring Kuhlthaus information search process model in a group-based educational setting. Inf. Process. Manag. **42**(1), 276–298 (2006)
32. Isen, A.M., Means, B.: The influence of positive affect on decision-making strategy. Soc. Cogn. **2**(1), 18–31 (1983)
33. Kim, J.: Describing and predicting information-seeking behavior on the web. J. Am. Soc. Inform. Sci. Technol. **60**(4), 679–693 (2009)
34. Kuhlthau, C.: Information search process. http://wp.comminfo.rutgers.edu/ckuhlthau/information-search-process/. Accessed 23 May 2017
35. Kuhlthau, C.C.: Inside the search process: information seeking from the user's perspective. J. Am. Soc. Inf. Sci. **42**(5), 361 (1991)
36. Leeder, C., Shah, C.: Collaborative information seeking in student group projects. Aslib J. Inf. Manag. **68**(5), 526–544 (2016)
37. LePine, J.A., Piccolo, R.F., Jackson, C.L., Mathieu, J.E., Saul, J.R.: A meta-analysis of teamwork processes: tests of a multidimensional model and relationships with team effectiveness criteria. Pers. Psychol. **61**(2), 273–307 (2008)
38. London, S.: Collaboration and community. Pew partnership for civic change (1995). http://www.scottlondon.com/articles/oncollaboration.html. Accessed 30 Nov 2015
39. Marchionini, G.: Information Seeking in Electronic Environments, vol. 9. Cambridge University Press, Cambridge (1997)
40. Marks, M.A., Mathieu, J.E., Zaccaro, S.J.: A temporally based framework and taxonomy of team processes. Acad. Manag. Rev. **26**(3), 356–376 (2001)
41. Mattessich, P.W., Monsey, B.R.: Collaboration: What Makes it Work. A Review of Research Literature on Factors Influencing Successful Collaboration. ERIC, Amherst H. Wilder Foundation, Saint Paul, MN (1992)
42. McGrath, J.E., Hollingshead, A.B.: Putting the group back in group support systems: some theoretical issues about dynamic processes in groups with technological enhancements. In: Group Support Systems: New Perspectives, pp. 78–96 (1993)
43. McIntyre, R.M., Salas, E.: Measuring and managing for team performance: emerging principles from complex environments. In: Team Effectiveness and Decision Making in Organizations, pp. 9–45 (1995)
44. McKenzie, P.J.: A model of information practices in accounts of everyday-life information seeking. J. Doc. **59**(1), 19–40 (2003)
45. Morgan Jr., B.B., Glickman, A.S., Woodard, E.A., Blaiwes, A.S., Salas, E.: Measurement of team behaviors in a Navy environment. Technical report, DTIC Document (1986)
46. Morris, M.R.: A survey of collaborative web search practices. In: Proceedings of the SIGCHI Conference on Human Factors in Computing Systems, pp. 1657–1660. ACM (2008)
47. Morris, M.R.: Collaborative search revisited. In: Proceedings of the 2013 Conference on Computer Supported Cooperative Work, pp. 1181–1192. ACM (2013)
48. Morris, M.R., Horvitz, E.: SearchTogether: an interface for collaborative web search. In: Proceedings of the 20th Annual ACM Symposium on User Interface Software and Technology, pp. 3–12. ACM (2007)
49. Morris, M.R., Teevan, J.: Collaborative web search: who, what, where, when, and why. Synth. Lect. Inf. Concepts Retr. Serv. **1**(1), 1–99 (2009)
50. Morris, M.R., Teevan, J., Bush, S.: Enhancing collaborative web search with personalization: groupization, smart splitting, and group hit-highlighting. In: Proceedings of the 2008 ACM Conference on Computer Supported Cooperative Work, pp. 481–484. ACM (2008)

51. Navarro Bullock, B., Hotho, A., Stumme, G.: Tag-based social search. In: Brusilovsky, P., He, D. (eds.) Social Information Access. LNCS, vol. 10100, pp. 310–343. Springer, Heidelberg (2017)
52. Olson, G.M., Olson, J.S.: Distance matters. Hum.-Comput. Interact. **15**(2), 139–178 (2000)
53. Olson, J.S., Hofer, E.C., Bos, N., Zimmerman, A., Olson, G.M., Cooney, D., Faniel, I.: A theory of remote scientific collaboration. Sci. Collab. Internet **1**, 73 (2008)
54. Paris, C.R., Salas, E., Cannon-Bowers, J.A.: Teamwork in multi-person systems: a review and analysis. Ergonomics **43**(8), 1052–1075 (2000)
55. Paul, S.A., Morris, M.R.: Sensemaking in collaborative web search. Hum.-Comput. Interact. **26**(1–2), 72–122 (2011)
56. Pickens, J., Golovchinsky, G., Shah, C., Qvarfordt, P., Back, M.: Algorithmic mediation for collaborative exploratory search. In: Proceedings of the 31st Annual International ACM SIGIR Conference on Research and Development in Information Retrieval, pp. 315–322. ACM (2008)
57. Poltrock, S., Grudin, J., Dumais, S., Fidel, R., Bruce, H., Pejtersen, A.M.: Information seeking and sharing in design teams. In: Proceedings of the 2003 International ACM SIGGROUP Conference on Supporting Group Work, pp. 239–247. ACM (2003)
58. Reddy, M.C., Jansen, B.J., Krishnappa, R.: The role of communication in collaborative information searching. Proc. Am. Soc. Inf. Sci. Technol. **45**(1), 1–10 (2008)
59. Salas, E., Prince, C., Baker, D.P., Shrestha, L.: Situation awareness in team performance: implications for measurement and training. Hum. Factors: J. Hum. Factors Ergon. Soc. **37**(1), 123–136 (1995)
60. Shah, C.: Coagmento-a collaborative information seeking, synthesis and sensemaking framework. In: Integrated Demo at CSCW, pp. 6–11 (2010)
61. Shah, C.: Collaborative information seeking: a literature review. Adv. Librariansh. **32**(2010), 3–33 (2010)
62. Shah, C.: Effects of awareness on coordination in collaborative information seeking. J. Am. Soc. Inform. Sci. Technol. **64**(6), 1122–1143 (2013)
63. Shah, C.: Collaborative information seeking. J. Assoc. Inf. Sci. Technol. **65**(2), 215–236 (2014)
64. Shah, C., González-Ibáñez, R.: Exploring information seeking processes in collaborative search tasks. Proc. Am. Soc. Inf. Sci. Technol. **47**(1), 1–7 (2010)
65. Shah, C., González-Ibáñez, R.: Evaluating the synergic effect of collaboration in information seeking. In: Proceedings of the 34th International ACM SIGIR Conference on Research and Development in Information Retrieval, pp. 913–922. ACM (2011)
66. Shah, C., Marchionini, G.: Awareness in collaborative information seeking. J. Am. Soc. Inform. Sci. Technol. **61**(10), 1970–1986 (2010)
67. Shah, C., Pickens, J., Golovchinsky, G.: Role-based results redistribution for collaborative information retrieval. Inf. Process. Manag. **46**(6), 773–781 (2010)
68. Shen, X., Tan, B., Zhai, C.: Context-sensitive information retrieval using implicit feedback. In: Proceedings of the 28th Annual International ACM SIGIR Conference on Research and Development in Information Retrieval, pp. 43–50. ACM (2005)
69. Sinclair, R.C., Mark, M.M.: The effects of mood state on judgemental accuracy: processing strategy as a mechanism. Cogn. Emot. **9**(5), 417–438 (1995)
70. Sonnenwald, D.H., Maglaughlin, K.L., Whitton, M.C.: Designing to support situation awareness across distances: an example from a scientific collaboratory. Inf. Process. Manag. **40**(6), 989–1011 (2004)

71. Sonnenwald, D.H., Pierce, L.G.: Information behavior in dynamic group work contexts: interwoven situational awareness, dense social networks and contested collaboration in command and control. Inf. Process. Manag. **36**(3), 461–479 (2000)

72. Soulier, L., Shah, C., Tamine, L.: User-driven system-mediated collaborative information retrieval. In: Proceedings of the 37th International ACM SIGIR Conference on Research and Development in Information Retrieval, pp. 485–494. ACM (2014)

73. Stone, N.J., Posey, M.: Understanding coordination in computer-mediated versus face-to-face groups. Comput. Hum. Behav. **24**(3), 827–851 (2008)

74. Strijbos, J.W., Martens, R.L., Jochems, W.M.: Designing for interaction: six steps to designing computer-supported group-based learning. Comput. Educ. **42**(4), 403–424 (2004)

75. Talja, S.: Supporting scholars' collaboration in document seeking, retrieval, and filtering. In: Proceedings of the Workshop on Computer Supported Scientific Collaboration, pp. 49–55 (2003)

76. Talja, S., Hansen, P.: Information sharing. In: Spink, A., Cole, C. (eds.) New Directions in Human Information Behavior, pp. 113–134. Springer, Dordrecht (2006). https://doi.org/10.1007/1-4020-3670-1_7

77. Twidale, M., Nichols, D.: Collaborative browsing and visualisation of the search process. Aslib Proc. **48**, 177–182 (1996)

78. Twidale, M.B., Nichols, D.M., Paice, C.D.: Browsing is a collaborative process. Inf. Process. Manag. **33**(6), 761–783 (1997)

79. Van Dolen, W.M., Dabholkar, P.A., De Ruyter, K.: Satisfaction with online commercial group chat: the influence of perceived technology attributes, chat group characteristics, and advisor communication style. J. Retail. **83**(3), 339–358 (2007)

80. Wang, P., Hawk, W.B., Tenopir, C.: Users interaction with world wide web resources: an exploratory study using a holistic approach. Inf. Process. Manag. **36**(2), 229–251 (2000)

81. White, R.W., Roth, R.A.: Exploratory search: beyond the query-response paradigm. Synth. Lect. Inf. Concepts Retr. Serv. **1**(1), 1–98 (2009)

82. Wilson, M.L., Schraefel, M.C.: Evaluating collaborative search interfaces with information seeking theory. arXiv preprint arXiv:0908.0703 (2009)

83. Wilson, M.L., et al.: Evaluating collaborative information seeking interfaces with a search-oriented inspection method and re-framed information seeking theory. Inf. Process. Manag. **46**(6), 718–732 (2010)

84. Wu, D., Yu, W.: Undergraduates' team work strategies in writing research proposals. In: Proceedings of the 18th ACM Conference Companion on Computer Supported Cooperative Work and Social Computing, pp. 199–202. ACM (2015)

85. Yue, Z.: Investigating search processes in collaborative exploratory web search. Ph.D. thesis, School of Information Sciences, University of Pittsburgh, June 2014. http://d-scholarship.pitt.edu/21711/

86. Yue, Z., Han, S., He, D.: A comparison of action transitions in individual and collaborative exploratory web search. In: Hou, Y., Nie, J.-Y., Sun, L., Wang, B., Zhang, P. (eds.) AIRS 2012. LNCS, vol. 7675, pp. 52–63. Springer, Heidelberg (2012). https://doi.org/10.1007/978-3-642-35341-3_5

87. Yue, Z., Han, S., He, D.: Modeling search processes using hidden states in collaborative exploratory web search. In: Proceedings of the 17th ACM Conference on Computer Supported Cooperative Work and Social Computing, pp. 820–830. ACM (2014)

88. Yue, Z., Han, S., He, D., Jiang, J.: Influences on query reformulation in collaborative web search. Computer **3**, 46–53 (2014)

5

Social Navigation

Rosta Farzan(✉) and Peter Brusilovsky

School of Computing and Information, University of Pittsburgh,
Pittsburgh, PA 15260, USA
{rfarzan,peterb}@pitt.edu

Abstract. In this chapter we present one of the pioneer approaches in supporting users in navigating the complex information spaces, *social navigation support*. Social navigation support is inspired by natural tendencies of individuals to follow traces of each other in exploring the world, especially when dealing with uncertainties. In this chapter, we cover details on various approaches in implementing social navigation support in the information space as we also connect the concept to supporting theories. The first part of this chapter reviews related theories and introduces the design space of social navigation support through a series of example applications. The second part of the chapter discusses the common challenges in design and implementation of social navigation support, demonstrates how these challenges have been addressed, and reviews more recent direction of social navigation support. Furthermore, as social navigation support has been an inspirational approach to various other social information access approaches we discuss how social navigation support can be integrated with those approaches. We conclude with a review of evaluation methods for social navigation support and remarks about its current state.

1 Introduction

Navigation through the ever-changing information space is becoming increasingly difficult. Recent research efforts have highlighted the interactive nature of information access behavior and promoted the potential value of harnessing user activity patterns to drive navigation in information space. "Social Navigation", defined as "moving towards cluster of people" has been introduced for Web as a response to the problem of disorientation in information space [31]. The idea of social navigation in information space stems from the natural tendency of humans to follow direct and indirect cues of each other when feeling lost [5]. Social navigation in information space as well as the term social navigation was introduced by Dourish and Chalmers [31]; however, the idea of social navigation is frequently traced back to the pioneer Edit Wear and Read Wear systems [55,56]. In this system, Hill and Hollan introduced the idea of physical wear in the domain of document processing as "computational wear." Computational wear

© Springer International Publishing AG, part of Springer Nature 2018
P. Brusilovsky and D. He (Eds.): Social Information Access, LNCS 10100, pp. 142–180, 2018.
https://doi.org/10.1007/978-3-319-90092-6_5

is the visualization of the history of authors' and readers' interactions with a document. The visualization of the history enables the new users to quickly locate most viewed or edited parts of the document. As suggested by Dieberger [26], social navigation support does not necessarily change users' navigation behaviors but it increases their awareness inside the information space. Social navigation support is offered by utilizing traces of activities of latent users to guide newer users; for example, which links have been traversed by majority of users [26,115] or which pages are being explored by other users at the moment [72,108].

Introduced in few pioneer projects in 1990es in the context of navigation in information space, the ideas of social navigation attracted a lot of followers from other areas of information access. In a number of follow-up papers and books [27] the term "social navigation" was used to refer to other kinds of social information access, such as collaborative filtering. For example, Wong et al., defined social navigation is a mechanism to "enable actions not based on spatial or semantic information, but on social information" [117]. This chapter, however, focuses on social navigation in its original meaning, as an approach to help users navigating in information space by utilizing traces of behavior left by previous users. We attempt to provide a comprehensive view of social navigation by discussing how it supports users' navigation in the information space, theoretical support, original approaches in implementing it in the information space, and evaluation methods of the existing implementations. Furthermore, we have tried to discuss how advancement of social computing fields has advanced the implementation approaches in social navigation. We end the chapter with highlighting challenges for researchers and practitioners interested in social navigation in information space.

2   Supporting Theories

Social navigation is inspired from principles that have been discovered in nature. People have observed a variety of interesting behaviors among insects or animals in nature. Animals and insects such as birds, fish, ants, or termites engage in collective or swarm behavior [78]. A swarm is a collection of unsophisticated agents that are cooperating to achieve a goal. Each agent follows simple local rules from their environment in a relatively independent manner but collectively they achieve the swarm's objectives. This emergent collective intelligence is known as "Swarm Intelligence (SI) [7]." "SI is the property of a system whereby the collective behaviors of (unsophisticated) agents interacting locally with their environment cause coherent functional global patterns to emerge" [7]. An example of SI in nature is the food foraging behavior of ants. Ants use their pheromone to mark trails connecting the nest to food sources. The pheromone gets richer and richer as more ants follow the trail to carry food to the nest. At each point the trail with the highest density of the pheromone has the highest chance of being chosen by the ants.

While interacting with complex information spaces, humans behave similar to animals in trying to achieve collective intelligence. Information seeking tasks on the Web can be mapped to a biological society. The Web represent the society, and the surfer represent the animal which is an autonomous agent with limited knowledge given the available information abundance. Desired information is food for which the surfer is browsing. Click-stream and other browsing behavior is the Web pheromone and the popularity of the Web page represents the density of the pheromone. Wu and Aberer [118] conducted a "Quest for Treasure" experiment to evaluate the collective intelligence behavior of humans in information space. The experiment involved 12 rooms that visitors could navigate to. Two of the rooms had a chest treasure in them. For each link they presented the raw visit click and pheromone density. Pheromone density was calculated taking into account positive and negative feedback. Positive feedback includes accumulation of visits and spreading of pheromone from other links. Negative feedback includes diffusion of the popularity of a link and was modeled by a half-life time function. Following the link pheromone, one could quickly find the treasure chests. The result of their experiment showed a simple form of self-organization and demonstrated the value of "swarm of Internet surfers."

Effect of social navigation in information space can be explained by the *information foraging theory*. Related to SI, the information foraging theory [95] is an analogy to food foraging strategies among animals which states that "when feasible, natural information systems evolve toward stable states that maximizes gains of valuable information per unit cost." Information foraging is the result of human adaptation to the explosive information growth. The central problem the theory tries to address is allocation of attention to the most useful information. The goal is to maximize profitability of information resources by increasing information gained per unit cost. Information scent is used to assess the profitability of information resources. Information scent is the "perception of the value, cost, or access path of information sources obtained from proximal cues, such as bibliographic citations, WWW links, or icons representing the sources."

Information foraging has mainly focused on explaining information seeking behavior of individual users. Pirolli introduced the idea of "Social Information Foraging" (SIF) [94]. SIF is based on the idea that information foragers engage in social exchange of information. Connected to the idea of swarm intelligence, information foragers cooperate to increase the likelihood of high-value information discoveries. The basic SIF model assumes existence of hints from the group of information forager about the likely location of useful information patches. It attempts to model the benefit of cooperation and social capital in information seeking tasks. Recent social Web technologies such as Weblogs, collaborative tagging, and recommender systems have emerged to exploit or enhance SIF. The success of those technologies implies the effectiveness of social information foraging.

SIF connects social navigation and information foraging. Social navigation support (SNS) can enrich information scent and assist in scent detection to judge the potential relevance of information resources. Information foragers have to navigate through information patches to find what they need. SNS can decrease the cost of information gain by both enriching between-patch and within patch foraging gains. Figure 1 depicts the possible effect of SNS on information gain. To satisfy information needs, first, information foragers should find the relevant patches. As they go through the information patches they gain information as represented by the information gain function up to the point that they reach the information gain threshold. Social navigation cues can enrich between-patch information gain by highlighting the patches with useful information and decreasing the time needed to assess different patches. While navigating inside a patch, social navigation support can improve the return from a patch by highlighting the useful resources inside the patch; e.g. highlighting the part of the document that received the most attention by previous users.

Lunich et al. have proposed a theoretical framework to explain social navigation process in information space from communication perspective [79]. They explain social navigation in terms of users' decision to generate traces and to follow traces as well as the attributes of the content. The model proposes that users' decisions can be influenced by personal traits, interpersonal relationships, contextual factors, and content.

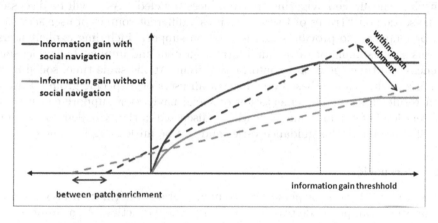

Fig. 1. Information foraging model with social navigation support

3   Influencing Users' Experiences

Supporting social navigation in information space has the potential to improve user experiences through four main mechanisms: guidance, persuasion, engagement, and social presentation. Below, we describe each mechanism along with supporting theories.

3.1   Guidance

Social navigation support has been initially motivated by the challenge of information overload in information space. While navigating the Web, users often are faced with large amount of information and overwhelming set of options to follow in search of their desired information. It is commonly documented that users on the Web often experience information overload and anxiety dealing with too much information and too many choices [8,53] To address the challenge of information overload, researchers have been studying ways to provide guidance to users in information space. Motivated by the natural tendency of human to follow traces of each other, especially when feel lost, researchers have studied social navigation support in information space as an approach to visualize and highlight information traces of users in the space. The ability to relatively easily track user activities and traces in information space provides the opportunity to make use of these traces to guide individuals about what information others have been accessing, or seeking. As argued by Wong et al. [117], social navigation support can be employed to support information discovery and guidance in information space in three major ways: (1) aid navigation to most popular content by highlighting what resources everyone else is accessing; (2) support serendipitous discovery by highlighting important resources that have not drawn attention of a large group of people; (3) diverge attention from resources with highest popularity and encourage navigating the "road less traveled". As it will be discussed in the section on "Traces of Users Activities", different sources of user activities can be employed to provide social navigation support, including explicit users' actions such as liking or rating an information item and implicit behaviors such as clicks and time spent on an information item. At the same time, social navigation guidance can be based on traces of all users or a specific group of users. As a result, different implementation of social navigation support can provide different level of guidance and as it will be discussed in the "Challenges" section, the effectiveness of this guidance varies and can be misleading in some cases.

3.2 Persuasion

Supported by theories of persuasive communication, information about activities of others can persuade people to take a particular action. As a result, social navigation support has the power of persuasion by relying on and presenting information about actions of others. The strength of persuasion interacts with the source of information [21]. For example, people are more likely to follow figures of authority, others similar to them, or those whom they have a strong relationship with. Therefore, depending on the source of social navigation support, its power of persuasion can vary. Kulakarni and Chi [71] in two experimental studies of social navigation in the form of augmented annotation in the context of news articles, showed that users are likely to follow recommendation of others as long as they are not total strangers for whom they have no basis for assessing the reliability of their actions.

3.3 Engagement

In addition to power of social navigation support in providing guidance and pursuing users to follow a particular path and access specific information, social navigation support can increase users' engagement within the information space by adding social affordances to the space. It has been shown that activities even the ones not intrinsically very engaging can become more engaging through integration with social interactions [70]. For example, Farzan et al., showed that individuals are more likely to be engaged with even a solitary game if the game is integrated with a social context and in association with teams [40]. Social navigation support can turn information seeking that has been traditionally thought of as solitary action into social interactions through direct and indirect communications with other users. Observing footprints of others or an ability to directly communicate with others can serve as a social mechanism encouraging further engagement within the information space.

3.4 Social Presentation

As discussed earlier, social navigation support adds social dimension into information space and information seeking tasks. At the same time, information about activities of others is an indicator that their actions have been recorded by the system and will be presented to others. As a result, users can perceive that any action they take in the information space contributes to the way they have been presented to others. As suggested by Goffman's [51], individuals alter their behavior and performance based on their audience to mange their self-presentation. The presentation of self in age of social media and online sites has been the focus of many studies [57,83,91]. It has been shown that users of social networking site employ various strategies to manage their self presentation and the presentation of the identity through the nature and amount of information they share with others [113,114]. In turn, this perception of social navigation as a way of social presentation and self-presentation can influence their information seeking behavior [71].

4   Pioneer Examples of Social Navigation

Following the ideas introduced in the seminal Edit Wear and Read Wear system [55] and an early attempt to conceptualize social navigation in [31], two pioneer systems played an important role in the development of social navigation research stream. These systems, Juggler [26] and Footprints [115] implemented ideas of social navigation in two meaningful contexts and demonstrated how it could help users navigating in two kinds of informations spaces, a Web site and a text-based virtual environment.

Footprints [115,116] introduces the idea of interaction history for digital information which is taken from extensive human use of history traces in the physical world. Footprints provides contextualized navigation through the use of several interface features such as maps, path views, annotations, and sign posts. The system tracks all transitions from different sources such as selecting a link, typing a URL, or selecting a bookmark. It visualizes the interaction history by presenting the traffic through a Web site, percentage of users following each link, and popular paths to the Web sites. Additionally it allows the users to provide direct guidance by adding signposts expressing their opinions about different resources and the path to reach the resource. Figure 2 shows different views of the documents and navigating through the documents in Footprints and how they are augmented with social navigation support such as coloring the nodes representing the popular documents in the site map interface or showing what percentage of users have followed each link on the page by annotating the links with the percentage as shown in the right bottom side of the figure. Footprints does not present any identifiable information and social navigation support is offered based on aggregated and anonymous users' activities.

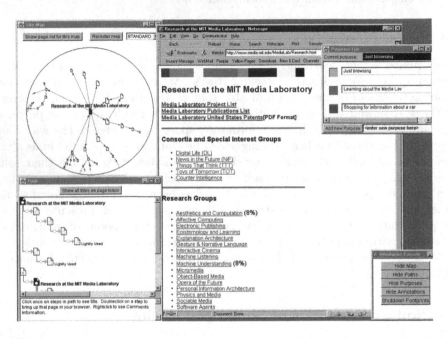

Fig. 2. Social navigation support on different views of footprints systems. Image retrieved from http://alumni.media.mit.edu/~wex/Screenshots/final-fullscreen.gif

Juggler [26] was designed to support interaction between a teacher and students in a remote teaching support system. By its nature, it is a text-based virtual environment (known as MOO) enhanced with a Web browser for displaying Web pages. Juggler provides an example of implementing a history enriched environment in a MOO context. It highlights major navigation paths through different textual bulletin boards (rooms) and adds the computational wear to each bulletin boards by showing the number of times it was accessed. Juggler also supports an intentional form of social navigation by encouraging users to directly recommend useful resources (such as URLs) to each other.

Another pioneer system to acknowledge is EFOL, an online food store developed by Kristina Höök and her colleagues in the PERSONA project [27,108,109]. Unlike Juggler and Footprints that were inspired by the ideas of history enriched space of Edit Wear and Read Wear, The PERSONA team was motivated by the recognized need to support users navigating in information spaces [4] and the idea of adaptive navigation support introduced by adaptive hypermedia [10]. However, in contrast to traditional adaptive hypermedia where navigation supported was based on knowledge engineering provided by system creators, the PERSONA team called for "Edited Adaptive Hypermedia" [58] where navigation support could be offered on the basis of explicit and implicit activities of earlier system users. EFOL also implemented the idea of a populated information space where synchronous presence of other users in different parts of information space (recipe clubs) was indicated by their avatars encouraging other users to navigate to a populated place. Once in the same "club" users were able to chat just like in a real information space.

While these pioneer systems were more proof of concept than practical systems highly utilized by regular users, they played an important role in defining the design space for social navigation. Using these systems as motivating examples, their authors promoted social navigation in a series of workshops and edited books [27,61,87]. Altogether, this work and established social navigation as research direction and defined its research agenda.

5   Exploring the Design Space of Social Navigation

Social navigation augments the information space with traces of activities of others. In design of such augmentation, one can observe three main foci: (1) history-enriched environments that attempt to enrich users' experiences by visualizing history of users' interactions; (2) co-presence enriched environments that aim to enrich users' experiences by visualizing the presence of others and to increase users' awareness of others in the information space; (3) organized guided information seeking that aim to guide users' navigation through the information space through explicit cues provided by other users.

5.1   Users' Activities and Their Traces

Independent of design focus, various tracers of users' activities can be leveraged to offer social navigation support. We classify these traces along two dimensions – intention and synchrony. *Intention* dimension represent whether the users are leaving traces with the explicit intention of providing feedback to the system and to others or whether they are just performing their activities on the system and those can be used as implicit indicator of feedback to the system [109]. *Synchrony* indicates whether users are communicating the feedback to each other synchronously and directly or the feedback is communicated to others asynchronously [27]. This is an extension to the original classification suggested by Dieberget et al. to distinguish social navigation based on the communication mode between the actors, into "direct social navigation" when the actors are in direct communication with each other and "indirect social navigation" when contacts between the actors are anonymous and indirect [29]. Examples of each kinds of traces have been presented in Table 1.

Table 1. Classification of users' activities and traces of the activities

		Synchrony	
		Asynchronous	Synchronous
Intention	Implicit	Clicks, time spent downloads, highlighting text, scrolling, bookmarking, mouse movements	Editing a shared document such as Google documents, browsing a Web page
	Explicit	Likes, Ratings, Recommendations, Comments, actions	Web page recommendation in a chat message

Independent of the source of user traces, in implementation of social naviga-tion support, one can employ traces of all users of the system or a specific group of users. At the same time, the anonymity of social navigation support traces can range from aggregated and anonymous to individual anonymous or individual and non-anonymous. Each of this decisions influence how social navigation sup-port affect the users' decisions in the information space and they have been topics of interest in various research studies as we discuss in the section "Evaluation Methods". While protecting user privacy is important and necessary, visibility and translucency can be beneficial to increase trust and awareness [33].

Mapping the design space to the classification of users' activities described in Table 1, synchronous explicit approaches are more in the form of recommen-dation that are less strongly considered as social navigation and other chapters of this on recommender system provide more details on that. Below, we discuss each approach in details and provide examples for each approach.

5.2 History Enriched Environments via Implicit Asynchronous Traces

In search of solutions for the challenge of information overload and difficulty in finding the most desirable information, researchers explored the idea of enrich-ing information space with the navigation history of the latent users. These approaches often rely on *asynchronous and implicit* traces of those who have already navigated the information space. These traces, such as click-through or download history, can be employed to provide social navigation support.

The Jugler and Footprints systems reviewed above provide two early exam-ples of history enriched environments that leverage implicit asynchronous traces of user navigation. This work motivated a number of follow-up projects that attempted to expand this approach in several direction. Social Navigation swiki or CoWeb [28] provide an interesting example of implementing the ideas of social navigation in the context of a Wiki system, i.e., a user-expandable hyperspace [74]. Unlike a regular Web site where that the end users can only browse leaving their clickstreams, Wiki allows all users to update existing pages and create new ones. In this context, page creation and update activities form another stream of implicit traces. Social navigation Swiki provides a history enriched page view that shows recency of user page updates and browsing by attaching two kinds of visual cues to Swiki page links (Fig. 3): one to show the recency of page update ("new" sign) and another to show the recency of page use (pair of footprints). The color of each visual cue (red-hot, yellow, gray) reflects three levels of recency.

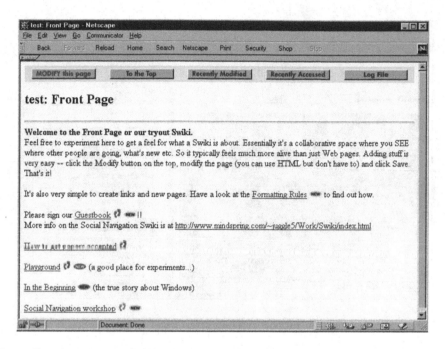

Fig. 3. Social navigation Swiki page showing recency footprints for browsing and page updating activities. Used with permision from [28] (Color figure online)

KnowledgeSea II [13], an educational information system, was designed to help students find relevant information among hundreds of online tutorial pages distributed over the Web by augmenting the interface for accessing educational resources with information about the collective behavior of students in a class. It provided social navigation support based on prior students' interactions with the online resources and pages they have been visiting every week as the course had been progressing. More specifically, it used the number of clicks made by all students in the class on a specific page or topics as a sign of its importance in the context of the class and used blue color of different intensity to visualize this social importance to the users. Figure 4 shows the main interface of the system which includes a grid of course topics annotated with background color and other social cues based on students' activities in a particular class and the content of each topic cell as the list of resources in that cell. KnowledgeSea II introduced two extensions of the original idea of history enriched hyperspace introduced by Footprints [115]. First, it offered two-level social navigation that starts by leading readers to valuable topic cells by visualizing a cumulative importance of its resources and then allows to select valuable resources within the topic cell.

Second, it offered social comparison by contrasting user own navigation (shown as the intensity of human figure color) with the navigation of the whole class (shown as the intensity of background color).

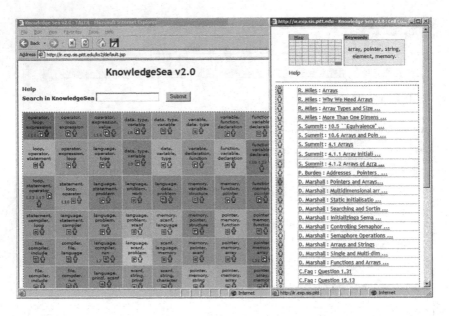

Fig. 4. Social navigation support in main interface of KnowledgeSea II system

5.3   Co-presence Enriched Environments via Implicit Synchronous Traces

While approaches in implementing history-enriched environments rely on asynchronous and implicit traces of users' activities, another set of approaches aim at enriching users' information navigation experiences by presenting a live and social image of the information space and where other users are at the moment. These approaches still rely on implicit traces of users; i.e. users do not explicitly communicate with each other but they are aware of presence of each other *synchronously*. The value of *awareness* of presence of others has been traditionally studied and highlighted within the computer-mediated communication field [17,92]. Research in the field of social navigation followed these ideas to extend the values of co-presence to *information navigation support*.

A classic example in this context is EDUCO [72]. EDUCO is a collaborative learning environment, which implemented social navigation support to enrich learners' experiences in Web-based learning. EDUCO supports synchronous social navigation by visualizing the presence of others in the learning environment. As users of the system are accessing the educational Web documents, others can view their presence as dots next to the documents, as shown in Fig. 5. The color of the documents represent the popularity of the document among the users based on how many times they have been clicked. Furthermore, users can leave comments associated with documents that are visible to others navigating to the document.

Fig. 5. Representation of documents and users within EDUCO learning environment

5.4 Sharing Destinations and Paths via Explicit Asynchronous Traces

While Dourish and Chalmers [31] originally defined social navigation as navigation towards a cluster of people or navigation because other people have looked at something, Dieberger [26] argued that various kinds of direct information sharing (i.e., sharing a web page in a bulletin board post, sharing it on a "pointer" page such as a list of bookmarks, or a list of favorite links on one's home page) should be considered as examples of social navigation. In a classification of social navigation approaches introduced in [27], this kind of direct information sharing is considered as *direct asynchronous* social navigation.

By its complexity, direct information sharing could be classified into sharing individual destinations and sharing sequential paths. Sharing destinations (i.e., Web URLs) is a simpler kind of explicit information sharing. In early days of the Web, when search engines have not yet reached their current power, research teams explored a range of ideas for explicit sharing of URLs in- and out-of-context of a specific page. At that time, various kinds of bulletin boards such as USENET newsgroups provided an easy mechanism for explicit sharing of "out-of-context", i.e., generally useful links. The original bulletin board format, however, offered no useful interface for funding and re-using this information.

The need to improve mechanism for direct sharing of USENET information motivated several interesting projects [52,81,111] A classic example of leveraging USENET information to support more convenient direct social navigation interface is offered by PHOAKS system [111]. PHOAKS used a set of rules to extract useful links shared by the users in their posts to USENET newsgroups and listed extracted links for each group as recommendation to its users. Links were ranked by its social support, i.e., number of users recommending the link. At the same time, a few educational hypermedia systems offered their users the ability to share useful links "in-context", i.e., adding a new useful link on a specific hypertext page [44,86]. The ability to add a new link to the existing page, also became a part of the core functionality of Wiki systems.

In the second part of 1990s, collaborative bookmarking systems gradually emerged as a more efficient platform for explicit sharing of Web links. The idea of collaboratively sharing and using bookmarks that were originally meant to be personal collections of valuable Web links appeared to be very productive. Between 1997 and 2005, researchers and practitioners explored multiple approaches for organizing shared bookmarks [32,65,75]. Gradually, an approach to characterise each link with multiple tags originally introduced by WebTagger [65] become dominant. With the introduction of collaborative tagging, social bookmarking systems, which started as a specific kind of social navigation, emerged into a new kind of social information access that can support both search and navigation. Since collaborative tagging and bookmarking are analyzed in details in other chapters [30,89] we will not discuss it further in this chapter.

Systems for sharing paths and trails could be considered as a more advanced case of explicit social navigation. In this case users share not just a single resource or destination, but a whole sequential navigation path. In some sense, this kind of social navigation could be also considered as the oldest since the idea or sharing paths was introduced by Vannevar Bush as a key component of his visionary system Memex [18]. The inspiration provided by Memex ideas certainly contributed to the development of several practical "guided path" (or guided tours) systems at the end of 1980s in the context of Hypertext research [82,112,119]. The original guided tours have not fully implemented Memex vision of sharing paths between users, serving rather as another tool in the hands of the original hypertext authors to enhance the usability of hypertext systems [82,112]. However, just 10 years after the debut of guided paths in classic hypertext, the fast growth of the Web and the increasing engagement of end users as contributors lead to re-emergence of guided paths as true social navigation tools. The systems for sharing Web paths (or trails) appeared at the second half of 1990s in parallel with many other kinds of social information access systems [46,54,90]. These systems were directly influenced by Memex and earlier work on guided tours rather than by the early work on social navigation [31]. A classic example of a system for sharing Web navigation paths is Walden's Paths [46]. As a number of other early social navigation systems [26,32,72], Walden's Paths system was developed for educational context. The key idea of the system was to separate path authoring from content authoring. In contrast to the common approach,

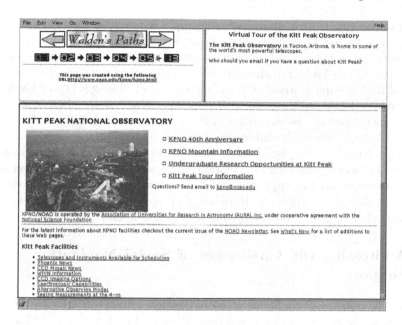

Fig. 6. Following a shared trail in Walden's Paths system

the original paper declared that "in general the author of the path is not the author of the supporting documents" [46]. A "path" in the system was defined simply as a sequence of Web pages (URLs) where each page cane be extended with annotation commenting on the page and its role in the path. The Walden's Paths provided a powerful interface for any interested Web users to define and share "paths" and an interface for navigating shared paths (Fig. 6). The navigation interface included current page in the path along with authored comments and overview of the whole path showing the position of the current page. The users were encouraged to explore pages around the path by following links from the current page. However, the position in the path was preserved even when the user wandered off-path and a "lost" user could return back to the path with a click of "return back to the path" button. Walden's Paths has been evaluated in several contexts, some lessons learned were summarized in [103].

The success of Walden Paths and early trail-sharing system encouraged a range of similar projects that explored tools and infrastructures for authoring and sharing guided paths for the Web such as Ariadne [63], Ethemeral Paths [45], TRAILGUIDE [97], TrailTRECer [47], or HATS [69]. It is important to note that in contrast to early work on shared guided paths that was inspired by Memex and was not positioned in the context of research on social navigation, more recent works in the stream [47,49,97] clearly articulated the place of shared paths in the context of social navigation and other kinds of social information access. In turn, it helped to generalize the idea of shared trails as navigation support tools moving it from its Web origin to other kinds of electronic environments. An early example of this generalization is trail-based navigation in shared directories [49]. A more recent example is provided by systems for collecting and sharing physical trails such as pedestrian walks, cycling paths, or travel itineraries. While modern online physical trail sharing systems look quite different from the Web trail-sharing systems, the early motivating examples of physical trail sharing systems such as Salzburg Trail Manager [48] or Cyclopath [96] were developed within team with solid experience in social navigation and directly motivated by the earlier research on social navigation in digital world. This example is especially interesting because it demonstrates how the ideas of social navigation completed a full circle between physical and digital word. Motivated originally by social navigation in physical world, the work on social navigation explored the application of these ideas to help users in navigating in various digital environmets. Enriched and elaborated, these ideas are now coming back to improve our navigation in physical world.

6   Addressing the Challenges of Social Navigation Support

Despite the potential benefits of social navigation support in information space highlighted above, researchers and practitioners faced various challenges in design and implementation of those ideas and in enriching users' information navigation experiences. These challenges were gradually identified and extensively discussed. The need to address these challenges encouraged a number of projects that could be classified as the third generation of research on social navigation. The majority of these systems were developed between 2005 and 2010 and represent a considerably more mature endeavors. Many of these systems have been used in real life context with hundreds and thousands of users. This section attempts to provide a representative review of this work. We start with discussing major challenges of social navigation support. Following that, we review some most representative systems of the third generation stressing specific approaches that these system used to address some of these identified challenges. Not all challenges have been addressed in these systems and some stay as open challenges.

The major challenges in implementation of the social navigation support can be categorized as below:

Tracking Users' Traces: Privacy, Information Efficiency, and Effectiveness:
Implementation and evaluation of social navigation mechanism have included various sources of user traces as a basis for social navigation support, including anonymous individual traces [107], traces of identified individuals, or aggregated traces; however, it stays as open research question which navigation trails should be logged and visualized to support an effective social navigation. Each approach include advantages and disadvantages. On one hand, more information can be beneficial in providing richer and more accurate social navigation support; however, there are privacy and social representation issues associated with collecting detailed and identifiable information. Being aware that each action is being recorded by the system and is going to be presented can lead into users'

change of behavior to present their navigation behavior in more desirable way. At the same time, such behavior of the system can raise users' concern about their privacy that not only their navigation in the system is logged by the system, it can be visible to other users' of the system. Moreover, more information is not always more beneficial. At times, abundance of information can cause information overload for users, especially if it is difficult for users to assess the relevance of information. At the same time, visualizing large amount information can introduce technical challenges [117]. Similarly, in terms of information efficiency and effectiveness, trace aggregation faces challenges in terms of the level of aggregation. Aggregation can be done at the group level by defining groups of similar users, collaborating users, or competing users [35,59]. However, the current research lacks conclusive results on effectiveness of different approaches.

Reliability of User Traces: Snowball Effect and Cognitive Biases

Social navigation relies on recommending the path traveled by others; however, users' reaction to social navigation support can be influenced by different cognitive biases. Several researchers have attempted to study the significance and degree of such biases experimentally. Salganik et al. [100] studied the impact of social influence on users' decision in an artificially created online music market. They showed that social influence, presented and prior number of downloads, can persuade individuals independent of the actual quality of the songs. Following on these experiment, in a series of experiment Lerman distinguished the position versus social influence cognitive bias in individuals' information access behavior [73]. She presented that independent of the quality of information and in addition to social influence, the position of information on the screen can significantly influence users' decision to access it.

As a result of such cognitive biases, social navigation systems often are challenged by snowball effect: if the first user heads in the wrong direction, all other users of the system enhanced with social navigation can be attracted to the same wrong path. This "snowball effect" is a special concern for systems that rely mostly on implicit feedback that could be frequently unreliable, especially considered in isolation. For example, a click on a page link might indicate a true interest in a page content or a mistake caused by an unclear link anchor. Therefore, it is important to be able to detect these paths and to prevent the system from directing users on to them.

Combining several types of implicit feedback can partially address this problem; for example, combining time spent reading with clickstream data [22]. If a user has gone through a page by mistake, the chance that they spend only a very short amount of time on the page is high. As a result, considering the time can help to eliminate some of the misleading pitfalls. In addition, different kinds of user traces carry different reliability in registering user true interests. While low-commitment actions such as clicking on a link are inherently unreliable, such actions as leaving a comment, downloading, or purchasing indicate a higher commitment and could be used for providing more reliable navigation support and minimizing the snowball effect.

Drift of Interest

A known challenge in implementation of social navigation is the concept of drift of interest [107]. Over time, the interest of people and the importance of information are changing. What is very important to a community of users today might not have much value in several months. This is especially important for highly dynamic context such as educational context in which the interest of students is dependent on the specific topic they are studying at the moment.

This problem can be addressed by weighting more recent visits, providing social navigation support based on the data from a specific period of time, or showing the recency of social guidance [28,105]. Often it is important to preserve old data in addition to recent ones. For example, in educational contexts, students might be interested in the currently discussed information to work on the latest assignment, and, at the same time, they might be interested in previously discussed materials to prepare for the midterm exam.

Bootstrapping and Engaging Users

A very important and well identified challenge in developing social navigation systems is how to get the system started. This is known as the "cold start" problem in collaborative filtering based recommender systems. Social navigation heavily relies on feedback provided by users - implicitly or explicitly. Early users will not have many navigational aids which might get disappointed by the system. On the other hand, as a result of not having navigational aid, they might head in the wrong direction which will affect the whole functionality of the system by accumulating a trail on the wrong path. Therefore, guiding and motivating early on users is a key challenge in effectiveness of social navigation systems.

A study of social navigation in educational context demonstrated that students with better knowledge of the subject are usually the first to explore "uncharted" territory where social navigation support is not yet available [59]. These students have the highest chance to locate most appropriate resources thus "blazing trails" for less knowledgeable students to follow. This results suggest that this group of users can be specifically encouraged to bootstrap a new system. It is not evident, however, that the situation with most prepared users blazing trails for the rest of the community will assure the proper bootstrapping in other contexts. Combining content based navigation support approaches with social navigation [101] could be recommended as a more general way of addressing the cold start problem.

At the same time, extrinsic reward can be introduced to encourage participation of early-on users, such as gamification approaches in providing points and badges for encouraging contribution that has been shown to be very effective [42]. However, such extrinsic approaches can also face challenges, especially with regards to undermining the quality of contribution and intrinsic motivation in those who have already been motivated to participate [20,41]. Other works have investigated approaches in introducing alternative mechanism on the system to allow the users to benefits from their contribution early-on when the user cannot yet benefit from the social aspects of the system [38].

6.1   AnnotatEd and KALAS: Exploring More Reliable Traces

The reliability of social traces was among the first challenges addressed by the third-generation social navigation systems. Many of these systems tried to avoid snowball effect by providing more reliable sources of user traces through implicit and explicit actions. Two good examples of transitioning to more reliable traces can be provided by AnnotatED and KALAS systems developed as extensions of earlier social navigation projects.

AnnotatEd [35], an educational hypertext reading support system, was designed as an extension of the KnowledgeSea II system [13] mentioned above to address several challenges faced by the classic implementation of social navigation support in KnowledgeSea II. Main focus of this extension was improving the quality of social navigation support by using more reliable evidence of user interest in a page (such as leaving an annotation rather than just clicking on a page) or a smarter processing of unreliable click traces. As presented in Fig. 7, AnnotatEd allowed users to add public or private comments to the section of online tutorials and textbooks they visited and classify their comments as a praise, a problem, or a general note. This information then was used to augment links to reading resources with with social and personal visual cues to represent presence, type, and density of students' annotations associated. Annotations are known as very reliable signs of user interes and page relevance [9] and a study of AnnotatEd [35] confirmed the ability of annotation-based navigation support to direct users to important relevant pages. Furthermore, AnnotatEd tracked the time each user spent on each page and determine a "depth" or each "footprint" taking into account time spent and the length of the text in each page [34]. As a result, a click could be considered as leaving only a half-deep "footprint" or no footprint at all depending on the time spend reading the page. AnnotatEd also extended the visibility of social navigation support. While KnowledgeSea II focused on social augmentation of Web links on specially created navigation maps, AnnotatEd added social visual cues to all regular within-page links. Note also that both AnnotatEd [35] and KnowledgeSea II system [13] addressed the global-level aggregation problem since it used traces of student behavior from the same class to provide social navigation. This filtered out the behavior of users who might used the same information with a different need or from a different prospect.

KALAS [107], an extension of the pioneer EFOL food recipe system [108], attempted to address some of the above-mentioned challenges by synthesizing a group of social navigation support features. It provides social navigation support by visualizing the aggregated trail of users through the environment. The trail includes the comments left by the users as well as information about the number of users who have downloaded a recipe. To provide social navigation support, KALAS collected users' feedback in an implicit and explicit format. For implicit feedback, KALAS focuses on reliable evidence of interests such as downloading, printing, or saving a recipe. Any of these actions leaves a positive vote for that recipe. Explicit feedback is collected by allowing users to click on a "good recipe" button or to check the thumbs-up/thumbs-down option in the recipe list. This provides an explicit positive or negative vote for the recipe. KALAS also supports synchronous social navigation by displaying currently logged on

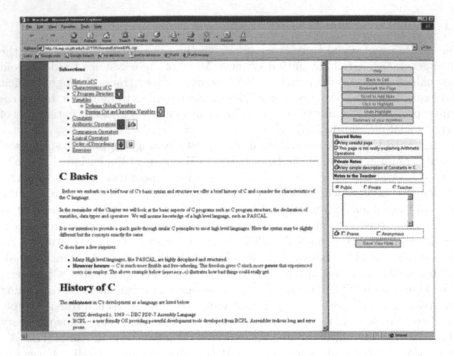

Fig. 7. AnnotatEd: annotation based social navigation support with Web page resources

users in each section of the system and allowing real-time chat among the users. Such implementation of social navigation support can often be observed in large scale commercial systems, such as in Amazon.com that the aggregate purchasing and browsing information of all customers or specific group of customers are presented to individuals to assist their shopping decisions.

6.2 Conference Navigator: Reliable Privacy-Protected Traces

Conference Navigator (CN) [36], a community-based conference support system, was designed to explore the value of social navigation in the context of planning a conference attendance. Conference attendees in multiple parallel-session conferences often have a difficult time deciding which talk to attend. The CN system explored the value of Social Navigation support to assist the conference attendees with finding the most relevant talk in each session of the conference to their research interests. CN system addressed two critical issues in implementation of social navigation support: reliability of traces and users' privacy. To address users' privacy concerns and their concerns with social presentation, the CN system allows users to join sub-communities defined in the system. Each sub-community represents a specific research interest. As shown in Fig. 8, while the users browse the schedule of the conference, they can look at it from a prospect of their a sub-community (e.g. "Social Learning" community in Fig. 8). Each user can belong to as many sub-communities as they desire but only one

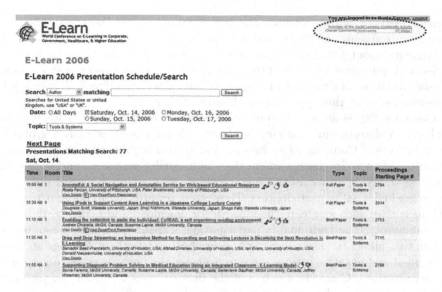

Fig. 8. Conference schedule browser with social visual cues

sub-community is selected as active at each time. As they browse the conference schedule, they can indicate interest in a specific talk by "scheduling to attend the talk" or by explicitly up-voting or down-voting the talk as it relates to the interests of their active community. This information is then used to provide social navigation support for the sub-community by guiding users to the talks that are most relevant to the interests of the community. As in many other social navigation systems, the navigation support was implemented by augmenting links to relevant talks with social visual cues.

6.3   Comtella and CourseAgent: Engaging Users

While reliability of users' traces is essential to provide meaningful social navigation support, *encouraging* users to leave traces, especially traces based on explicit actions is even more essential to systems relying on social navigation functionality. Various systems have tried different approaches to increase users' engagement with the system. In this section, we review two examples that show how user engagement can be increased by using two alternative approaches - "intrinsic" and "extrinsic" motivation to participate.

CourseAgent [38] is a course recommendation systems that is based on students' explicit feedback about the difficulty level of courses as well as course relevance to specific career goals. The systems uses this feedback to provide social navigation support to future students in making decisions about what courses to take. Encouraging students to provide feedback about courses they have taken is a key challenge for such systems, especially when students who have already taken a number of courses might not directly benefit from the navigation support. To do so, the CourseAgent system transforms the action to provide feedback into an *intrinsically* beneficial action for the users. This was done by introducing the study progress dashboard where the feedback provided by students about taken courses is used to calculate how far along they are in terms of progress towards each of their career goals. This approach is an example of using intrinsic motivation to increase student explicit feedback. A user study demonstrated that this approach was highly efficient [38].

Comtella [20] is an social information system designed for researchers and students to share useful academic and educational resources with a group of users. Success of Comtella as an information system highly relies on active participation of users in sharing interesting high quality resources and voting on resources shared by other users. Comtella employs an adaptive reward system to encourage high quality participation. The system rewards more cooperative users with incentives such as greater bandwidth for download and higher visibility in the community. The high quality participation is ensured through a reputation system that allows the users to rate the contributions of others. Ratings are then aggregated and negative ratings serve to decrease the rewards given to low quality contributions. Comtella was one of the first systems to explore engagement based on rewards and reputation that form the foundation of an increasingly more popular extrinsic motivation approach to increase participation. Morover, as shown in Fig. 9, users are visualized as stars in the system with different size and level of brightness based on their participation in the system. Visualization

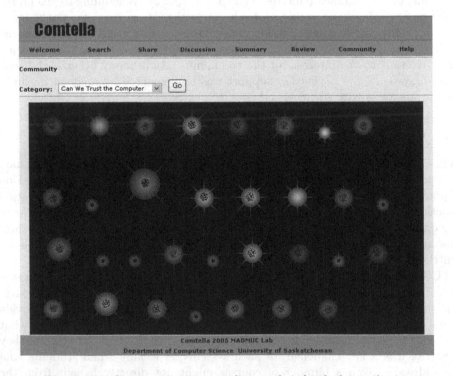

Fig. 9. Visualization of users according to their level of contribution

is also designed as an approach in encouraging participation by increasing users' awareness about their participation as compared to others and by enforcing a sense of social responsibility and social comparison.

6.4   Progressor: Social Navigation and Engagement with Social Comparison

An interesting approach that combines the benefits of social navigation support and user engagement is social comparison. Social comparison is known as strong factor encouraging user participation [19]. KnowledgeSea II [13] mentioned above was the first system to introduce social comparison in the context of social navigation, however, in this system it was based on less reliable navigation footprints and its effect was relatively small. A more elaborated example of extending traditional social navigation with social comparison using more reliable traces of user behavior is provided by Progressor [59], an educational practice system in the domain of computer programming. By its nature, a practice system provides access to various kinds of educational *practice content*. The work with this content is not mandatory and it doesn't carry credit points, it is, however, an opportunity to practice knowledge gained in a regular class and improve target skills. The use of practice content has two known problems. First, good practice systems offer an abundance of practice content of different difficulty levels to address the needs of students with different level of knowledge. In turn, it makes it hard for students to select the most appropriate content to practice. Second, despite their educational effectiveness, practice systems are usually under-used by students who prefer to focus on credit-bearing activities.

Progressor attempts to address both problems using a combination of social navigation support and social comparison. The system arranges practice problems into topics that are visualized as segments of a circle as shown on Fig. 10. The color of each segment represents the amount of knowledge gained by a student by working on practice problems for this topic, from red (no knowledge) to green (mastery). This kind of knowledge representation is known as an *open learner model*. The student could view in parallel his own model (left) or either a model of class peer or a group knowledge model of the whole class (right). The models shown on the right, especially the cumulative class model, offer social navigation support. Here students can see which topics have been already successfully mastered by the whole class, which topics were only attempted by a few advanced peers, and which are not yet practiced by anyone. Comparing his or her current knowledge level with the knowledge of the class or specific peers, the student can easily select most appropriate topics to practice while also getting a strong motivation to work on bridging the gap between her knowledge and class knowledge. Clicking on a topic brings a list of practice problems for this topic that uses the same color-coding knowledge representation to help choosing most appropriate problems to practice. As a study of Progressor [59] shows, both social navigation and social comparison were highly effective significantly improving student success rate with practice problems and increasing the amount

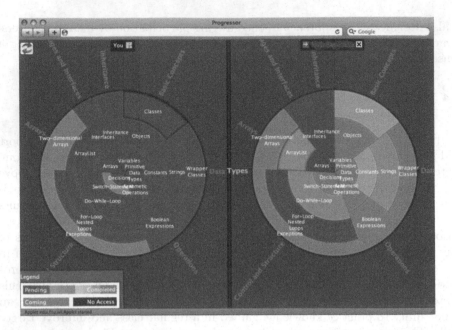

Fig. 10. Social navigation support and social comparison in Progressor (Color figure online)

of student work with non-mandatory content by more than 150%. Studies performed with similar systems Progressor+ [60] and Mastery Grids [16] confirmed this remarkable double effect of social navigation and social comparison.

The systems presented have been successful to address some of the identified challenges at various level; however, researchers and practitioners are still inspired to find ways in improving social navigation support by tackling these challenges and some of these challenges such as "drift of interest" or concerns with "social presentation of users' activities" are less frequently addressed within the existing implementations.

7   Social Navigation Beyond Hypertext and Hyperlinks

The early research on social navigation focused on assisting users in hypertext-style browsing, i.e., traversing the hyperlink space and identifying links to desirable resources. However, challenges of information access does not stop at the link level and vast amount of information as well challenge users once they arrive at a specific resource. As a result, social navigation support needs to also consider within resource support; i.e. tracking users' traces as they go through a particular page. For example, allowing the users to highlight specific parts of text within a page or associate comments with specific section of the page. Within-resource social navigation support becomes more challenging when considering large number of resources on the Web that are in multimedia and other continuous media formats with temporal dimensions. However, a range of recent projects demonstrated that the ideas of social navigation could be creativly apply to help the users finding the right place within a page or in continuous media. Moreover, several pioneed projects demonstrated that social navigation could be used to enhance other kinds of information access, beyond its original focus on "browsing". In this section, we review a sample of projects that explored social navigation ideas beyond hypertext and hyperlinks.

7.1 Spatial Social Navigation

While most implementation approaches of Web-based social navigation support has focused on facilitating navigation between Web pages, the original idea of social navigation support as it was imagined by the Edit Wear and Read Wear focused on helping a user to navigate within a single document space. Unfortunately, the idea of fine-grained tracing of user behavior that Edit Wear and Read Wear implemented in the context of a text editor was not easy to replicate in a hypertext and Web context. In a regular hypertext or Web system users leave nothing but page-level clicks. However, a Web system enhanced with annotation functionality opens opportunities for within-page social navigation based on user annotation behavior.

Web annotation technology became quite popular with various Web annotation systems created in the peak of its work between 1995 and 2005 [25,64,99,106]. While many of these systems supported only page-level annotations (just like AnnotatEd system reviewed above), several systems including popular Annotea project from WWW Consortium [64] allowed adding comments for any HTML fragment or simply mark-up most valuable fragments. Some of these systems limited access to this information to the original users, others allowed sharing annotations (this stream of work contributed to modern social tagging systems). The majority of these annotation systems also allowed users to share their annotations with all users of the system offering some kind of within-page social navigation.

In parallel to the research on Web-based annotation systems, Schilit et al. [102] explored the use of annotations in the context of a pioneer tablet-based reading tool XLibris. Unlike the Web annotation tools, which focused on page-level and "linear" within-page text annotation, XLibris pioneered spatial annotation, that enables XLibris users to manipulate the position of the annotation in addition to the text of the annotation. XLibris offered a pen-based, free form annotation tool that supports highlighting, underlining, and commenting. XLibris also pioneered some form of annotation-based social navigation such as a skimming mode, which highlights only the most important parts of a document, based on other users' annotations.

The ideas of Web page annotation and spatial document annotations were integrated in a Spatial Annotation system developed by Kim et al. [67]. The system was designed as an extension of AnnotatEd [35] to support Web-based access to digitized scanned books produced by large-scale book digitization projects, such as the Carnegie Mellon Million Book project [23]. Unlike the original AnnotatEd that supported only page-level annotations, the Spatial Annotation system allowed users to mark any rectangular page fragment (that might include a figure, a paragraph, or just a few words) and add any kind of comments. Spacial marks and comments might be visible to other users of the system who might add their own comments to any annotation creating a localized discussion. Further, to guide the readers to the most commented and appreciated fragments, the Spatial Annotation system provided within-page social navigation support based on prior users' annotations through visualizing traces of users' activities related to page fragments. To represent prior users' activities, the system extensively used visualize cues. As shown in Fig. 11 the thickness of the border of an annotated fragment indicates the volume of associated annotations while and the color of the border and and the background color indicates whether an annotations was created by the target users or someone else, is it public or private, positive or neutral.

A more recent example of spatial social navigation support within a Web-based document space was provided by Wong et al. [117] who focused on supporting sense-making and exploration of visual information. They implemented social navigation support as annotations to online maps such as Bing Maps by adding information about which parts of the map users had explored in response to a particular geo-location search task.

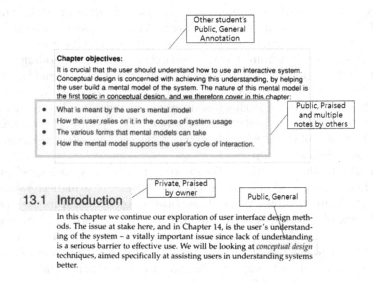

Fig. 11. Visual cues based on spatial annotations provide withing-page social navigation support

7.2   Social Navigation in Continuous Media

Social navigation in continuous media such as video is similar in several aspects to within-page social navigation reviewed above. While the visionary Edit Wear and Read Wear interface offered some ideas of continuous social navigation, this topic has not been addressed in early social navigation research. However, with the increasing popularity of online video, especially video-based Web lectures [11] encouraged the application of social navigation ideas in this context. A traditional (1–3 h) Web lecture contains many mundane parts such as course logistics, but also many important fragments explaining core domain concepts. However, a regular Web lecture interface, even extended with special video navigation tools such as sliders and scrolling, provides no hint on the importance of various fragments. Mertens et al. [84, 85] described VirtPresenter system that attempted to address this problem using an extension of a classic footprint-based approach to continuous media. VirtPresenter considered each viewing one video frame by a user as a social footprint indicating possible importance of this frame and displayed a cumulative history of frame-level lecture viewing in a graphic form next to the video scrolling bar (see Fig. 12). This approach made it easy to identify (and not to miss) most watched parts of the lecture. To address students drift of interest that is natural in a semester-long course, VirtPresenter introduced week-based filtering: the students were able to choose which social data are used to construct the social viewing graph, the amount of data gathered during the whole term or just the interaction recorded during specific weeks. VirtPresenter also enabled explicit social navigation allowing students to bookmark specific parts of the video and send these bookmarks as Web links by e-mail to their friends and peers.

It is important to observe that by its use of less reliable implicit "footprint" data, VirtPresenter was similar to the first generation "click-based" social navigation for the Web. While the simple approach pioneered by VirtPresenter has been later used with variations in other systems [66, 68], several follow-up projects focused on improving the reliability of social navigation for Web lectures. The set of explored ideas was mostly similar to those explored in research on Web-based social navigation reviewed above. For example, the CLAS system [98] attempted to use explicit vs. implicit footprints to identify most important lecture fragments. The idea of the CLAS approach is really simple – it

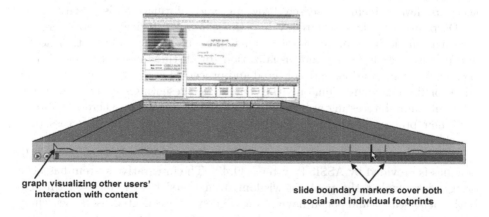

graph visualizing other users'
interaction with content

slide boundary markers cover both
social and individual footprints

Fig. 12. Social navigation interface for a video lecture in VirtPresenter

encourages students to mark important parts of the lecture while watching, by simply pressing the spacebar. In return, all watched lectures are enhanced with the visually annotated timeline showing important spots. Another project [50] explored a smarter use of several kinds of of implicit social feedback (such as the use of pause, play, skip, and rewind) to identify most important fragments. The DIVER platform [93] offered students the ability to create "dives" by marking and commenting video fragments and share these dives with other students. This approach enabled annotation-based social navigation in video context. The Video Colaboratory [104] made annotation-based social navigation more transparent by visualizing comments and marks of participating students as *signposts* attached to the video navigation bar.

7.3 Integrating Social Navigation with Other Social Information Access Approaches

Social navigation could be naturally combined with other information access approaches. Wherever the link to an information object is displayed, be it among other links on a Web page, in the list of search results, or in the information visualization space, it could be augmented with visual cues expressing various kinds of socially-produced information associated with an object. In fact, Knowledge-Sea (Fig. 4), Educo (Fig. 5), Comtella (Fig. 9), and Progressor (Fig. 10) reviewed in this chapter present social navigation in the context of different information visualizations displaying correspondingly the volume of traffic and annotations, co-presence, activity, and performance associated with elements of visualization. Two other examples of more advanced "social visualization" displaying traffic and annotations associated with information items can be found in [3,80]. Similarly, a typical example of using social navigation in search context is social annotation of search results in the ranked list with associated traffic [2] and social link [88] information. These examples are reviewed in more details in the Social Search chapter of this book [15]. A study presented in [14] has shown that it is more influential to provide social navigation support across multiple information access pathways, including search, browsing, and information visualization.

Despite of its demonstrated value, the examples reviewed above present a rather simple integration of social navigation into other information access approaches such as search and visualization. In all these cases, the social data (i.e., clicks or annotations) are collected and processed in the same way as for the traditional social navigation, only the context for presenting social visual cues is different. More interesting are cases of more tight integration where social data and their processing approach traditionally used for one type of access (i.e., search) are used for social navigation.

An example of a tight integration of *social search and social navigation* technologies is provided by ASSIST system [39,43]. The integrative system has been designed to exploit the pools of wisdom from users' traces collected through both social search and social navigation. The system collected users' searching traces such as the search queries and clicks on search results as well users' browsing traces such as time spent on each page, page annotations, and navigation

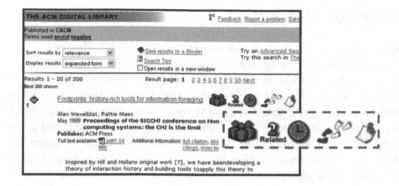

Fig. 13. An integration of social navigation and social search in ASSIST

from search results to other Web pages. Both kinds of traces were used then to augment user search and browsing interfaces with social visual cues (Fig. 13). An evaluation of the integrative system in the context of research paper access in ACM digital library suggested the potential for integration to provide information access support beyond just the sum of two approaches [43]. A similar attempt to use traces of both search and navigation behavior in a context of supporting user access to YouTube videos is presented in [24].

As an example of integration of *social navigation and collaborative recommendation* approaches, we can consider social link generation based on a broader picture of navigation behavior. Link generation is considered to be one of the major types of adaptive navigation support [12], yet almost all social navigation approaches focus on social augmentation of links that are already present on a page rather than generate additional links that would benefit users browsing this page. This helps the users to select possibly best navigation step, but doesn't bring them sufficiently close to their possible navigation destination. By taking into account user navigation behavior beyond this single page, it might be possible to deduce more distant or event ultimate destinations of user navigation and generate links to these destination. This idea has been first implemented by Bollen and Heylighen [6] who demonstrated how multi-step social navigation links could be generated by a transitive closure approach (i.e., $A{\rightarrow}B$ & $B{\rightarrow}C \implies A{\rightarrow}C$). The result of this "distant links" generation – a list of recommended link added to the page – combines the features of social navigation and collaborative filtering and can be generated using data collection and processing technologies from either area. Indeed, one stream of work on "distant link" generation including Bollen and Heylighen's was motivated by swarm intelligence ideas and used social navigation approaches [110,118] while another stream was associated with the field of recommender systems and used item-to-item [76,77], graph-based [62], and contextual recommendation approaches [1]. Probably the best known example of generated social navigation link is provided by Amazon.com recommendations *"Customers Who Bought This Item Also Bought..."* or *"What Other Items Do Customers Buy After Viewing This Item?"* on a specific product page.

8   Evaluation Methods

Evaluation of social navigation technology is particularly challenging. On one hand, to accurately evaluate the impact of social navigation support, it is required to study a natural system with a large number of users who can generate data as sources of social navigation support and to allow users to perform information seeking tasks and navigate through the information space as naturally as possible. There is very little, however, that can be controlled in the field studies with natural settings and as a result only the overall impact of social navigation can be observed in this kind of studies. Details about how various aspects influence the impact of social navigation support cannot be studied in such setting. On the other hand, the manipulated nature of controlled lab studies can be very obvious to study participants and as result their behavior can be significantly altered compared to the natural conditions. Therefore, researchers in this area have been employing mixed methodologies and pseudo-experiments in an attempt to evaluate different aspects of social navigation support. Evaluation of social navigation technology has been focused on the following aspects: overall impact of social navigation support, presentation of social navigation, and circumstances under which social navigation support is positively effective.

8.1   Overall Impact of Social Navigation on Users' Behavior

Studies that examine the overall impact of social navigation use both natural settings and experimental conditions to understand how social navigation support changes users' behavior and what kind of "additional value" it can bring by affecting this behavior. The studies focused on behavior change compare user behavior with social navigation enabled or disabled as well as access to information items enhanced or not enhanced with different social visual cues. In particular, studies evaluating aforementioned systems such as KnowledgeSea II, CourseAgent, Progressor, and Educo show that users' behavior are significantly influenced by social navigation cues. Users' frequently notice the navigation cues and make use of the cues to access information they seek more effectively. Results of such evaluations showed that resources with navigation cues were accessed at significantly higher rates and users of the systems followed footprints of each other creating a clear path across resources. The studies focused on "additional value" attempt to register various kind of benefits that the presence of social interaction could deliver. For example, the study of Progressor [59] demonstrated that social navigation significantly increases user motivation to work with practice problem while also improving user success rate. KALAS [107] has been evaluated by 302 users. The result of the evaluation shows that users make use of the recommendation feature very often and are very likely to be attracted to the most populated sections of the system; however, they were less influenced by the implicit trail left by other users and made little use of leaving comments.

Other studies have documented mixed results on the impact of social navigation support on users' performance. While a group of users have exhibited benefiting from social navigation cues to access relevant information more effectively, others, especially those with high level of interpersonal trust were likely to be lead to less relevant resources as a result of being highly influenced by social navigation cues [37]. In another work, an evaluation of social navigation cues in geographical maps [117], confirms similar results that users' performance in finding geographical spaces are improved with social navigation cues only if the cues come from users who have been guided too and are reliable sources of cues; otherwise, presence of social navigation cues does not affect users' performance. Connected to these results, a study in the context of news search has shown that users are highly persuaded by navigation cues on which news article to read as well as more satisfied with their choice as long as such cues are generated by others they know and they are not persuaded by navigation cues produced by strangers [71].

8.2   Presentation of Social Navigation

Evaluation of presentation of social navigation has focused on studying ways to visualize and highlight social navigation support. It is an important research question to understand how different presentation approaches of social navigation support affect users' decision in adherence to the cues. Similarly, it is important to understand how different presentation approaches vary in terms of attracting users' attention to social navigation cues. In evaluation of social navigation presentation, researchers most often employed log analysis that has been complemented by eye-tracking and qualitative evaluations as well as conducting controlled lab experiments [39]. Their results show that the location of social navigation cues influences how much users notice those cues. The results suggest that the visibility of social cues highly interacts with users' visual parsing behavior. Social annotations draw more attention when places on top of search result snippets, especially when the snippet is shorter.

8.3   Circumstances Under Which Social Navigation Support is Effective

Majority of studies of social navigation have focused on field studies; however, there has been a few studies that attempted at assessing the impact of social navigation on information seeking behavior in lab experiments under controlled settings. One such study was done in the context of fact finding and generating informational reports [37]. The participants in that study were required to find factual information in response to a set of questions from a very large corpus of relevant and irrelevant news articles. The experiment was conducted as within-subject experiment, manipulating task difficulty and time available to complete the task along with availability of social navigation support. The experiment interface followed typical search engine look and feel. However, as shown in Fig. 14, in the conditions with social navigation support, the search results were augmented by two kinds of social navigation support that were presented to participants as other participants' footprints but in reality were pre-planned by

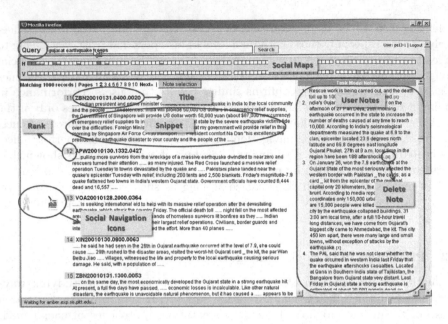

Fig. 14. Social navigation support for fact finding in a large corpus of news articles

the study and were the same for all the participants. The results of the study indicated that participants are more likely to make use of social navigation cues under time pressure.

9   Concluding Remarks

As a field of research, social navigation is now 20 years old. Over these 20 years, the field make a large transition from a narrow topic investigated by a few like-minded researchers to a relatively large direction of work that influenced many kinds if interactive systems and affected all kinds of information access. Most importantly, with the growing popularity of social Web applications, there has been a large adoption of the ideas of social navigation support by real-world systems. Many Web applications such as News websites integrate social information about how many other people have read a news article, or have liked it, and even extract information from users' social networks about the articles. These information often appear on the sites as "most read", "most forwarded", or "most downloaded" items. Many Web-based information-oriented systems have been transformed into "populated places" imagined by the early research on KALAT. In these systems users become first-class citizens that can leave feedback, reviews, and communicate with each other. It is now also a standard practice to engage users in rating of products and information items and display the overall rating alongside the product in every context it is being displayed. Moreover, social navigation, originally motivated by real world navigation and later enriched by the experience of information navigation, were brought back to help us navigate the real word through location-based systems (such as Yelp.com or Foursquare.com) and trail-sharing systems (such as Cyclopath.org or trailrunproject.com).

With all that real world success, it is important to mention that the majority of practical application of social navigation use it in its simplest form, most often asynchronous and indirect navigation cues that can be implemented as an overlay of social information on the existing interface. In some sense, we can say that the majority of practical application of social navigation use techniques that are about 10 years old. While most of these applications are affected by social navigation problems reviewed in this chapter, very few apply more recent and more advanced techniques that allow to deal with these problems. We think that more recearch on advanced social navigation is required as well as more work on integrating the results of new research into practical systems. We hope that this chapter will help both researchers and practitioners in their work on social navigation.

References

1. Adomavicius, G., Mobasher, B., Ricci, F., Tuzhilin, A.: Context-aware recommender systems. AI Mag. **32**(3), 67–80 (2011)
2. Ahn, J.W., Farzan, R., Brusilovsky, P.: Social search in the context of social navigation. J. Korean Soc. Inf. Manage. **23**(2), 147–165 (2006)
3. Ahn, J.W., Farzan, R., Brusilovsky, P.: A two-level adaptive visualization for information access to open-corpus educational resources. In: Brusilovsky, P., Dron, J., Kurhila, J. (eds.) Workshop on the Social Navigation and Community-Based Adaptation Technologies at the 4th International Conference on Adaptive Hypermedia and Adaptive Web-Based Systems, pp. 497–505 (2006). http://www.sis.pitt.edu/%7epaws/SNC_BAT06/crc/ahn.pdf
4. Benyon, D., Höök, K.: Navigation in information spaces: supporting the individual. In: Howard, S., Hammond, J., Lindgaard, G. (eds.) Human-Computer Interaction INTERACT '97. ITIFIP, pp. 39–46. Springer, Boston, MA (1997). https://doi.org/10.1007/978-0-387-35175-9_7
5. Bilandzic, M., Foth, M., De Luca, A.: CityFlocks: designing social navigation for urban mobile information systems. In: Proceedings of the 7th ACM Conference on Designing Interactive Systems, pp. 174–183. ACM (2008)
6. Bollen, J., Heylighen, F.: A system to restructure hypertext networks into valid user models. New Rev. Multimed. Hypermed. **4**, 189–213 (1998)
7. Bonabeau, E., Dorigo, M., Theraulaz, G.: Swarm Intelligence: From Natural to Artificial Systems. Oxford University Press, Oxford (1999)
8. Borchers, A., Herlocker, J., Konstan, J., Reidl, J.: Ganging up on information overload. Computer **31**(4), 106–108 (1998)
9. Bradshaw, S., Light, M.: Annotation consensus: implications for passage recommendation in scientific literature. In: Proceedings of Eighteenth ACM Conference on Hypertext and Hypermedia, Hypertext 2007, pp. 209–216. ACM Press (2007)
10. Brusilovsky, P.: Methods and techniques of adaptive hypermedia. User Model. User-Adap. Interact. **6**(2–3), 87–129 (1996)
11. Brusilovsky, P.: Web lectures: electronic presentations in web-based instruction. Syllabus **13**(5), 18–23 (2000)

12. Brusilovsky, P.: Adaptive navigation support. In: Brusilovsky, P., Kobsa, A., Nejdl, W. (eds.) The Adaptive Web. LNCS, vol. 4321, pp. 263–290. Springer, Heidelberg (2007). https://doi.org/10.1007/978-3-540-72079-9_8

13. Brusilovsky, P., Chavan, G., Farzan, R.: Social adaptive navigation support for open corpus electronic textbooks. In: De Bra, P.M.E., Nejdl, W. (eds.) AH 2004. LNCS, vol. 3137, pp. 24–33. Springer, Heidelberg (2004). https://doi.org/10.1007/978-3-540-27780-4_6

14. Brusilovsky, P., Farzan, R., Ahn, J.W.: Comprehensive personalized information access in an educational digital library. In: Proceedings of the 5th ACM/IEEE-CS Joint Conference on Digital Libraries, JCDL 2005, pp. 9–18. IEEE (2005)

15. Brusilovsky, P., Smyth, B., Shapira, B.: Social search. In: Brusilovsky, P., He, D. (eds.) Social Information Access. LNCS, vol. 10100, pp. 213–276. Springer, Heidelberg (2018)

16. Brusilovsky, P., Somyürek, S., Guerra, J., Hosseini, R., Zadorozhny, V.: The value of social: comparing open student modeling and open social student modeling. In: Ricci, F., Bontcheva, K., Conlan, O., Lawless, S. (eds.) UMAP 2015. LNCS, vol. 9146, pp. 44–55. Springer, Cham (2015). https://doi.org/10.1007/978-3-319-20267-9_4

17. Budzik, J., Fu, X., Hammond, K.J.: Facilitating opportunistic communication by tracking the documents people use. In: Workshop on Awareness and the WWW, held at the ACM Conference on Computer Supported Cooperative Work (CSCW 2000) (2000). https://pdfs.semanticscholar.org/4a8b/b1ff02aef04d40f951db1e52424af0822ecd.pdf

18. Bush, V.: As we may think. Atlantic 176, 101–108 (1945)

19. Buunk, A.P., Gibbons, F.X.: Social comparison: the end of a theory and the emergence of a field. Organ. Behav. Hum. Decis. Process. 102(1), 3–21 (2007)

20. Cheng, R., Vassileva, J.: Adaptive reward mechanism for sustainable online learning community. In: AIED, pp. 152–159 (2005)

21. Cialdini, R.B.: Harnessing the science of persuasion. Harvard Bus. Rev. 79(9), 72–81 (2001)

22. Claypool, M., Le, P., Wased, M., Brown, D.: Implicit interest indicators. In: 6th International Conference on Intelligent User Interfaces, pp. 33–40. ACM Press (2001)

23. Coyle, K.: Mass digitization of books. J. Acad. Librariansh. 32(6), 641–645 (2006)

24. Coyle, M., Freyne, J., Brusilovsky, P., Smyth, B.: Social information access for the rest of us: an exploration of social YouTube. In: Nejdl, W., Kay, J., Pu, P., Herder, E. (eds.) AH 2008. LNCS, vol. 5149, pp. 93–102. Springer, Heidelberg (2008). https://doi.org/10.1007/978-3-540-70987-9_12

25. Davis, R.C.: Annotate the web: four ways to mark up web content. Behav. Soc. Sci. Librar. 35(1), 46–49 (2016)

26. Dieberger, A.: Supporting social navigation on the world wide web. Int. J. Hum. Comput. Stud. 46(6), 805–825 (1997)

27. Dieberger, A., Dourish, P., Höök, K., Resnick, P., Wexelblat, A.: Social navigation: techniques for building more usable systems. Interactions 7(6), 36–45 (2000)

28. Dieberger, A., Guzdial, M.: CoWeb - experiences with collaborative web spaces. In: Lueg, C., Fisher, D. (eds.) From Usenet to CoWebs: Interacting with Social Information Spaces, pp. 155–166. Springer, New York (2003). https://doi.org/10.1007/978-1-4471-0057-7_8

29. Dieberger, A., Höök, K., Svensson, M., Lönnqvist, P.: Social navigation research agenda. In: Extended Abstracts of the SIGCHI Conference on Human Factors in Computing Systems, CHI 2001, pp. 107–108. ACM (2001)

30. Dimitrov, D., Helic, D., Strohmaier, M.: Tag-based navigation and visualization. In: Brusilovsky, P., He, D. (eds.) Social Information Access. LNCS, vol. 10100, pp. 181–212. Springer, Heidelberg (2018)

31. Dourish, P., Chalmers, M.: Running out of space: models of information navigation. Short paper present. HCI **94**, 23–26 (1994)

32. Dron, J., Boyne, C., Mitchell, R., Siviter, P.: CoFIND: steps towards a self-organising learning environment. In: Davies, G., Owen, C. (eds.) World Conference of the WWW and Internet, WebNet 2000, pp. 75–80. AACE (2000)

33. Erickson, T., Kellogg, W.A.: Social translucence: using minimalist visualisations of social activity to support collective interaction. In: Höök, K., Benyon, D., Munro, A.J. (eds.) Designing Information Spaces: The Social Navigation Approach. Computer Supported Cooperative Work, pp. 17–41. Springer, London (2003). https://doi.org/10.1007/978-1-4471-0035-5_2

34. Farzan, R., Brusilovsky, P.: Social navigation support in E-learning: what are the real footprints. Intell. Tech. Web Personal. **8**(1), 49–80 (2005)

35. Farzan, R., Brusilovsky, P.: Annotated: a social navigation and annotation service for web-based educational resources. New Rev. Hypermed. Multimed. **14**(1), 3–32 (2008)

36. Farzan, R., Brusilovsky, P.: Where did the researchers go?: supporting social navigation at a large academic. In: Proceedings of the Nineteenth ACM conference on Hypertext and Hypermedia, Hypertext 2008, pp. 203–212. ACM (2008)

37. Farzan, R., Brusilovsky, P.: Social navigation support for information seeking: if you build it, will they come? In: Houben, G.-J., McCalla, G., Pianesi, F., Zancanaro, M. (eds.) UMAP 2009. LNCS, vol. 5535, pp. 66–77. Springer, Heidelberg (2009). https://doi.org/10.1007/978-3-642-02247-0_9

38. Farzan, R., Brusilovsky, P.: Encouraging user participation in a course recommender system: an impact on user behavior. Comput. Hum. Behav. **27**(1), 276–284 (2011)

39. Farzan, R., Coyle, M., Freyne, J., Brusilovsky, P., Smyth, B.: ASSIST: adaptive social support for information space traversal. In: Proceedings of Eighteenth ACM Conference on Hypertext and Hypermedia, Hypertext 2007, pp. 199–208. ACM (2007)

40. Farzan, R., Dabbish, L.A., Kraut, R.E., Postmes, T.: Increasing commitment to online communities by designing for social presence. In: Proceedings of the 2011 ACM Conference on Computer Supported Cooperative work, pp. 321–330. ACM (2011)

41. Farzan, R., DiMicco, J.M., Millen, D.R., Brownholtz, B., Geyer, W., Dugan, C.: When the experiment is over: deploying an incentive system to all the users. In: Symposium on Persuasive Technology, in Conjunction with the AISB 2008 Convention (2008)

42. Farzan, R., DiMicco, J.M., Millen, D.R., Dugan, C., Geyer, W., Brownholtz, E.A.: Results from deploying a participation incentive mechanism within the enterprise. In: Proceedings of the SIGCHI Conference on Human Factors in Computing Systems, CHI 2008, pp. 563–572. ACM (2008)

43. Freyne, J., Farzan, R., Brusilovsky, P., Smyth, B., Coyle, M.: Collecting community wisdom: integrating social search and social navigation. In: Proceedings of the 12th International Conference on Intelligent user Interfaces, pp. 52–61. ACM (2007)

44. Fulantelli, G., Corrao, R., Munna, G.: Enhancing user interaction on the web. In: Bra, P.D., Leggett, J. (eds.) WebNet 1999, World Conference of the WWW and Internet, pp. 403–408. AACE (1999)

45. Furuta, R., III, F.M.S., Francisco-Revilla, L., Hsieh, H., Karadkar, U., Hu, S.C.: Ephemeral paths on the WWW: the Walden's paths lightweight path mechanism. In: Bra, P.D., Leggett, J. (eds.) WebNet 1999, World Conference of the WWW and Internet, pp. 409–414. AACE (1999)
46. Furuta, R., Shipman III, F.M., Marshall, C.C., Brenner, D., Hsieh, H.w.: Hypertext paths and the world-wide web: experience with Walden's paths. In: Bernstein, M., Carr, L., Østerbye, K. (eds.) Proceedings of Eight ACM International Hypertext Conference, Hypertext 1997, pp. 167–176. ACM (1997)
47. Gams, E., Berka, T., Reich, S.: The TrailTRECer framework - a platform for trail-enabled recommender applications. In: Hameurlain, A., Cicchetti, R., Traunmüller, R. (eds.) DEXA 2002. LNCS, vol. 2453, pp. 638–647. Springer, Heidelberg (2002). https://doi.org/10.1007/3-540-46146-9_63
48. Gams, E., Rehrl, K., Kaschl, D.: Providing other people's trails for navigation assistance in physical environments. Int. J. Spat. Data Infrastruct. Res. 3(1), 3–19 (2008)
49. Gams, E., Reich, S.: Following your colleagues' footprints: navigation support with trails in shared directories. In: Proceedings of Fifteenth ACM Conference on Hypertext and Hypermedia, Hypertext 2004, pp. 89–90. ACM Press (2004)
50. Gkonela, C., Chorianopoulos, K.: VideoSkip: event detection in social web videos with an implicit user heuristic. Multimed. Tools Appl. 69(2), 383–396 (2014)
51. Goffman, E., et al.: The Presentation of Self in Everyday Life. Anchor Books, Harmondsworth (1978). https://en.wikipedia.org/wiki/The_Presentation_of_Self_in_Everyday_Life
52. Goldberg, D., Nichols, D., Oki, B.M., Terry, D.: Using collaborative filtering to weave an information tapestry. Commun. ACM 35(2), 61–70 (1992)
53. Hanani, U., Shapira, B., Shoval, P.: Information filtering: overview of issues, research and systems. User Model. User-Adap. Interact. 11(3), 203–259 (2001)
54. Hauck, F.J.: Supporting hierarchical guided tours in the world wide web. Comput. Netw. ISDN Syst. 28(7), 1233–1242 (1996)
55. Hill, W.C., Hollan, J.D.: History-enriched digital objects: prototypes and policy issues. Inf. Soc. 10(2), 139–145 (1994)
56. Hill, W.C., Hollan, J.D., Wroblewski, D., McCandless, T.: Edit wear and read wear. In: Proceedings of the SIGCHI Conference on Human Factors in Computing Systems, CHI 1992, pp. 3–9. ACM Press (1992)
57. Hogan, B.: The presentation of self in the age of social media: distinguishing performances and exhibitions online. Bull. Sci. Technol. Soc. 30(6), 377–386 (2010)
58. Höök, K., Rudstrom, A., Waern, A.: Edited adaptive hypermedia: combining human and machine intelligence to achieve filtered information. In: Flexible Hypertext Workshop at the Eight ACM International Hypertext Conference, Hypertext 1997, pp. 54–58 (1997)
59. Hsiao, I.H., Bakalov, F., Brusilovsky, P., König-Ries, B.: Progressor: social navigation support through open social student modeling. New Rev. Hypermed. Multimed. 19(2), 112–131 (2013)
60. Hsiao, I.-H., Brusilovsky, P.: Motivational social visualizations for personalized E-learning. In: Ravenscroft, A., Lindstaedt, S., Kloos, C.D., Hernández-Leo, D. (eds.) EC-TEL 2012. LNCS, vol. 7563, pp. 153–165. Springer, Heidelberg (2012). https://doi.org/10.1007/978-3-642-33263-0_13
61. Höök, K., Benyon, D., Munro, A.J.: Designing Information Spaces: The Social Navigation Approach. Springer, Berlin (2003). https://doi.org/10.1007/978-1-4471-0035-5

62. Jäschke, R., Marinho, L., Hotho, A., Schmidt-Thieme, L., Stumme, G.: Tag recommendations in folksonomies. In: Kok, J.N., Koronacki, J., Lopez de Mantaras, R., Matwin, S., Mladenič, D., Skowron, A. (eds.) PKDD 2007. LNCS (LNAI), vol. 4702, pp. 506–514. Springer, Heidelberg (2007). https://doi.org/10.1007/978-3-540-74976-9_52

63. Jühne, J., Jensen, A.T., Grønbæk, K.: Ariadne: a Java-based guided tour system for the world wide web. In: Ashman, H., Thistewaite, P. (eds.) Proceedings of Seventh International World Wide Web Conference, vol. 30, pp. 131–139. Elsevier (1998)

64. Kahan, J., Koivunen, M.R., Prud'Hommeaux, E., Swick, R.R.: Annotea: an open RDF infrastructure for shared web annotations. Comput. Netw. **39**(5), 589–608 (2002)

65. Keller, R.M., Wolfe, S.R., Chen, J.R., Rabinowitz, J.L., Mathe, N.: A bookmarking service for organizing and sharing URLs. In: Proceedings of Sixth International World Wide Web Conference, pp. 1103–1114 (1997)

66. Ketterl, M., Mertens, R., Vornberger, O.: Bringing web 2.0 to web lectures. Interact. Technol. Smart Educ. **6**(2), 82–96 (2009)

67. Kim, J.K., Farzan, R., Brusilovsky, P.: Social navigation and annotation for electronic books. In: Proceedings of BooksOnline 2008 Workshop at the 17th ACM Conference on Information and Knowledge Management, CIKM 2008, pp. 25–28. ACM (2008). https://doi.org/10.1145/1458412.1458421

68. Kim, J., Guo, P., Cai, C., Li, S.W.D., Gajos, K., Miller, R.: Data-driven interaction techniques for improving navigation of educational videos. In: Proceedings of the 27th Annual ACM Symposium on User Interface Software and Technology, pp. 563–572. ACM (2014)

69. Kim, S., Slater, M., Whitehead Jr., E.J.: WebDAV-based hypertext annotation and trail system. In: Proceedings of Fifteenth ACM Conference on Hypertext and Hypermedia, Hypertext 2004, pp. 87–88. ACM Press (2004)

70. Kraut, R.E., Resnick, P.: Encouraging contribution to online communities. In: Building Successful Online Communities: Evidence-Based Social Design, pp. 21–76. MIT Press, Cambridge (2011)

71. Kulkarni, C., Chi, E.: All the news that's fit to read: a study of social annotations for news reading. In: Proceedings of the SIGCHI Conference on Human Factors in Computing Systems, CHI 2013, pp. 2407–2416. ACM (2013)

72. Kurhila, J., Miettinen, M., Nokelainen, P., Tirri, H.: EDUCO - a collaborative learning environment based on social navigation. In: De Bra, P., Brusilovsky, P., Conejo, R. (eds.) AH 2002. LNCS, vol. 2347, pp. 242–252. Springer, Heidelberg (2002). https://doi.org/10.1007/3-540-47952-X_26

73. Lerman, K.: Information is not a virus, and other consequences of human cognitive limits. Future Internet **8**(2), 21 (2016)

74. Leuf, B., Cunningham, W.: The Wiki Way: Quick Collaboration on the Web. Addison-Wesley Longman Publishing Co., Inc., Boston (2001)

75. Li, W.S., Vu, Q., Agrawal, D., Hara, Y., Takano, H.: PowerBookmarks: a system for perzonalizable web information organization, sharing, and management. In: Proceedings of Eight International World Wide Web Conference, pp. 297–311 (1999)

76. Linden, G.: People who read this article also read... IEEE Spectrum, March 2008. About news personalization including Findory

77. Linden, G., Smith, B., York, J.: Amazon.com recommendations: item-to-item collaborative filtering. IEEE Internet Comput. **7**(1), 76–80 (2003)

78. Liu, Y., Passino, K.M.: Swarm intelligence: literature overview. Department of Electrical Engineering, the Ohio State University (2000)

79. Lünich, M., Rössler, P., Hautzer, L.: Social navigation on the internet: a framework for the analysis of communication processes. J. Technol. Hum. Serv. **30**(3–4), 232–249 (2012)

80. Ma, K.L., Wang, C.: Social-aware collaborative visualization for large scientific projects. In: International Symposium on Collaborative Technologies and Systems, pp. 190–195. IEEE (2008)

81. Maltz, D., Ehrlich, K.: Pointing the way: active collaborative filtering. In: Proceedings of the SIGCHI Conference on Human Factors in Computing Systems, CHI 1995, pp. 202–209. ACM (1995)

82. Marshall, C.C., Irish, P.M.: Guided tours and on-line presentations: how authors make existing hypertext intelligible for readers. In: Proceedings of the Second ACM Conference on Hypertext, Hypertext 1989, pp. 15–26. ACM (1989)

83. Marshall, P.D.: The promotion and presentation of the self: celebrity as marker of presentational media. Celebr. Stud. **1**(1), 35–48 (2010)

84. Mertens, R., Farzan, R., Brusilovsky, P.: Social navigation in web lectures. In: Wiil, U.K., Nürnberg, P.J., Rubart, J. (eds.) Proceedings of Seventeenth ACM Conference on Hypertext and Hypermedia, Hypertext 2006, pp. 41–44. ACM Pres (2006)

85. Mertens, R., Ketterl, M., Brusilovsky, P.: Social navigation in web lectures: a study of virtPresenter. Interact. Technol. Smart Educ. **7**(3), 181–196 (2010)

86. Mitsuhara, H., Kanenishi, K., Yano, Y.: Learning process sharing for educational modification of the web. In: Cantoni, L., McLoughlin, C. (eds.) ED-MEDIA 2004 - World Conference on Educational Multimedia, Hypermedia and Telecommunications, pp. 1187–1192. AACE (2004)

87. Munro, A.J., Höök, K., Benyon, D.: Social Navigation of Information Space. Springer, Berlin (1999). https://doi.org/10.1007/978-1-4471-0837-5

88. Muralidharan, A., Gyongyi, Z., Chi, E.: Social annotations in web search. In: Proceedings of the SIGCHI Conference on Human Factors in Computing Systems, CHI 2012, pp. 1085–1094. ACM (2012)

89. Navarro Bullock, B., Hotho, A., Stumme, G.: Accessing information with tags: search and ranking. In: Brusilovsky, P., He, D. (eds.) Social Information Access. LNCS, vol. 10100, pp. 310–343. Springer, Cham (2018)

90. Nicol, D., Smeaton, C., Slater, A.F.: Footsteps: trail-blaizing the web. In: Kroemker, D. (ed.) Proceedings of Third International World-Wide Web Conference, pp. 879–885 (2005)

91. Ong, E.Y., Ang, R.P., Ho, J.C., Lim, J.C., Goh, D.H., Lee, C.S., Chua, A.Y.: Narcissism, extraversion and adolescents' self-presentation on facebook. Personality Individ. Differ. **50**(2), 180–185 (2011)

92. Palfreyman, K., Rodden, T.: A protocol for user awareness on the world wide web. In: Proceedings of the 1996 ACM Conference on Computer Supported Cooperative Work, pp. 130–139. ACM, New York (1996)

93. Pea, R., Lindgren, R.: Video collaboratories for research and education: an analysis of collaboration design patterns. IEEE Trans. Learn. Technol. **1**(4), 235–247 (2008)

94. Pirolli, P.: An elementary social information foraging model. In: Proceedings of the SIGCHI Conference on Human Factors in Computing Systems, CHI 2009, pp. 605–614. ACM (2009)

95. Pirolli, P., Card, S.: Information foraging. Psychol. Rev. **106**(4), 643 (1999)

96. Priedhorsky, R.: The value of geographic Wikis. Ph.D. thesis (2010). http://conservancy.umn.edu/handle/11299/97898

97. Riedl, M.: A computational model and classification framework for social navigation. In: 6th International Conference on Intelligent User Interfaces, IUI 2001, pp. 137–144. ACM (2001)

98. Risko, E., Foulsham, T., Dawson, S., Kingstone, A.: The collaborative lecture annotation system (CLAS): a new tool for distributed learning. IEEE Trans. Learn. Technol. **6**(1), 4–13 (2013)

99. Röscheisen, M., Mogensen, C., Winograd, T.: Shared web annotations as a platform for third-party value-added information providers: architecture, protocols, and usage examples. Technical report, Computer Science Department, Stanford University (1996). http://www-diglib.stanford.edu/rmr/TR/TR.html

100. Salganik, M.J., Dodds, P.S., Watts, D.J.: Experimental study of inequality and unpredictability in an artificial cultural market. Science **311**(5762), 854–856 (2006)

101. Schein, A.I., Popescul, A., Ungar, L.H., Pennock, D.M.: Methods and metrics for cold-start recommendations. In: Proceedings of the 25th Annual International ACM SIGIR Conference on Research and Development in Information Retrieval, pp. 253–260. ACM (2002)

102. Schilit, B.N., Golovchinsky, G., Price, M.N.: Beyong paper: supporting active reading with free form digital ink annotations. In: Karat, C.M., Lund, A., Coutaz, J., Karat, J. (eds.) Proceedings of the SIGCHI Conference on Human Factors in Computing Systems, CHI 1998. ACM Press (1998)

103. Shipman, F., Furuta, R., Brenner, D., Chung, C., Hsieh, H.: Guided paths through web-based collections: design, experiences, and adaptations. J. Am. Soc. Inf. Sci. **51**(3), 260–272 (2000)

104. Singh, V., Abdellahi, S., Maher, M.L., Latulipe, C.: The video collaboratory as a learning environment. In: The 47th ACM Technical Symposium on Computer Science Education, SIGCSE 2016, pp. 352–357. ACM Press (2016)

105. Smyth, B., Balfe, E., Freyne, J., Briggs, P., Coyle, M., Boydell, O.: Exploiting query repetition and regularity in an adaptive community-based web search engine. User Model. User-Adap. Interact. **14**(5), 383–423 (2004)

106. Su, A.Y., Yang, S.J., Hwang, W.Y., Zhang, J.: A web 2.0-based collaborative annotation system for enhancing knowledge sharing in collaborative learning environments. Comput. Educ. **55**(2), 752–766 (2010)

107. Svensson, M., Höök, K., Cöster, R.: Designing and evaluating KALAS: a social navigation system for food recipes. ACM Trans. Comput. Hum. Interact. **12**(3), 374–400 (2005)

108. Svensson, M., Höök, K., Laaksolahti, J., Waern, A.: Social navigation of food recipes. In: Proceedings of the SIGCHI Conference on Human Factors in Computing Systems, CHI 2001, pp. 341–348. ACM (2001)

109. Svensson, M., Laaksolahti, J., Höök, K., Waern, A.: A recipe based on-line food store. In: Proceedings of the 5th International Conference on Intelligent User Interfaces, pp. 260–263. ACM (2000)

110. Tattersall, C., van den Berg, B., van Es, R., Janssen, J., Manderveld, J., Koper, R.: Swarm-based adaptation: wayfinding support for lifelong learners. In: De Bra, P.M.E., Nejdl, W. (eds.) AH 2004. LNCS, vol. 3137, pp. 336–339. Springer, Heidelberg (2004). https://doi.org/10.1007/978-3-540-27780-4_46

111. Terveen, L., Hill, W., Amento, B., McDonald, D., Creter, J.: Phoaks: a system for sharing recommendations. Commun. ACM **40**(3), 59–62 (1997)

112. Trigg, R.H.: Guided tours and tabletops: tools for communicating in a hypertext environment. ACM Trans. Inf. Syst. **6**(4), 398–414 (1988)
113. Tufekci, Z.: Can you see me now? audience and disclosure regulation in online social network sites. Bull. Sci. Technol. Soc. **28**(1), 20–36 (2008)
114. Vitak, J.: Balancing privacy concerns and impression management strategies on Facebook. In: Symposium on Usable Privacy and Security (SOUPS) (2015). https://cups.cs.cmu.edu/soups/2015/papers/ppsVitak.pdf
115. Wexelblat, A., Maes, P.: Footprints: history-rich tools for information foraging. In: Proceedings of the SIGCHI Conference on Human Factors in Computing Systems, CHI 1999, pp. 270–277. ACM (1999)
116. Wexelblat, A., Mayes, P.: Footprints: History rich web browsing. In: Conference on Computer-Assisted Information Retrieval, RIAO 1997, pp. 75–84 (1997)
117. Wong, Y.L., Zhao, J., Elmqvist, N.: Evaluating social navigation visualization in online geographic maps. Int. J. Hum. Comput. Interact. **31**(2), 118–127 (2015)
118. Wu, J., Aberer, K.: Swarm intelligent surfing in the web. In: Lovelle, J.M.C., Rodríguez, B.M.G., Gayo, J.E.L., del Puerto Paule Ruiz, M., Aguilar, L.J. (eds.) ICWE 2003. LNCS, vol. 2722, pp. 431–440. Springer, Heidelberg (2003). https://doi.org/10.1007/3-540-45068-8_80
119. Zellweger, P.T.: Active paths through multimedia documents. In: Proceedings of the International Conference on Electronic Publishing on Document Manipulation and Typography, pp. 19–34. Cambridge University Press, New York (1988)

6
Tag-Based Navigation and Visualization

Dimitar Dimitrov[1]([✉]) [iD], Denis Helic[2] [iD], and Markus Strohmaier[3] [iD]

[1] GESIS – Leibniz Institute for the Social Sciences,
University of Koblenz-Landau, Cologne, Germany
`dimitar.dimitrov@gesis.org`
[2] Graz University of Technology, Graz, Austria
`dhelic@tugraz.at`
[3] GESIS – Leibniz Institute for the Social Sciences,
RWTH Aachen University, Cologne, Germany
`markus.strohmaier@gesis.org`

1 Introduction

Online information seeking has become an everyday task in lives of modern people. In principle, we distinguish between two strategies to explore and discover information spaces: search and navigation. Search implies a query formulation, whereas navigation is the process of finding a way to a given target by following hyperlinks. Navigation without a specific target is also referred to as browsing.

One of the main advantages of navigation as compared to search, which is tightly related to our cognitive abilities as humans—is that recognizing what we are looking for is much easier than formulating and describing our information need in a couple of keywords [34]. In literature, the formulation of an information need is also referred to as the *vocabulary problem* [29]. To overcome this problem in the early days of the Web, the information space has been structured by hand using predefined *controlled vocabulary* terms. As the Web continued to rapidly grow, the biggest disadvantage of this approach—the static structure—became more and more visible. Together with the rise of search engines, this led to the vanishing of even famous websites using controlled vocabularies such as for example DMOZ[1].

A new way of organizing a set of resources emerged with the introduction of social tagging systems. Prominent instances of social tagging systems on the Web are, e.g., BibSonomy[2], CiteULike[3], Delicious[4], and Flickr[5], where BibSonomy offers sharing of literature and bookmarks, CiteULike the sharing of citations, Delicious the sharing of bookmarks, and Flickr the sharing of photos. These

---

[1] DMOZ has been closed as of Mar 17, 2017, and it is no longer available under https://www.dmoz.org. The editors have set up a static mirror under http://dmoztools.net/.
[2] http://www.bibsonomy.org.
[3] http://www.citeulike.org.
[4] http://del.icio.us.
[5] https://www.flickr.com.

© Springer International Publishing AG, part of Springer Nature 2018
P. Brusilovsky and D. He (Eds.): Social Information Access, LNCS 10100, pp. 181–212, 2018.
https://doi.org/10.1007/978-3-319-90092-6_6

systems allow users to annotate a set of resources according to their needs with freely chosen words also called *tags*. This free-form annotation approached the vocabulary problem from a *social* angle and introduced new research directions, i.e., for structuring and visualizing the information space. Moreover, new models and theories for tag-based navigation have been developed and helped to establish it as a novel way of information access. *Tag-based navigation* is defined as the process of finding a way between two resources of a social tagging system following user assigned tags [38]. This way of exploring the information space is usually supported by a *tag cloud*. The tag cloud is a user interface that visualizes the tags describing a given set of resources. In that sense, a tag cloud is a textual representation of the topic or subject of the resource set and it captures its *aboutness* [24]. Navigation and browsing in a social tagging system are commonly initiated, e.g., by a system-wide tag cloud, by traversing tag hierarchy or by executing a search query typed into a traditional search box (see Sect. 3.1). Using tags as search terms (see "Tag-based social search" in this book [67]) or following recommended tags (see "Tag-based Recommendation" in this book [8]) are user activities very similar to tag-based navigation.

We organize our chapter in the following way. In Sect. 2, we describe the fundamental social tagging process for shaping the information space of a social tagging system. In Sect. 3, we discuss the tag cloud-based user interaction schema in a social tagging system, layouts, usefulness and evaluation of tag clouds, and visualization of trends in tagging data. We also give an overview of more complex interfaces that integrate tag clouds or expose tag hierarchies to users. We discuss clustering of tagging data in Sect. 4, as it is a way of dealing with one of the main problems with tagging data—the lack of structure. In this section, we give an overview over flat and hierarchical tag clustering. The flat tag clustering produces groups of similar tags, whereas the hierarchical tag clustering produces a tag hierarchy in which tags occupy a given hierarchy level based on, e.g., their generality. We show how tag-based navigation is modeled in Sect. 5. In general, models of user navigational behavior are used for providing navigational support such as recommending or highlighting links, adapting the navigational hierarchy, or even removing particular navigational links. In a particular case of tag-based navigation user models can be used to, for instance, adapt a given tag-cloud, include additional tags into the tag-cloud, or for ranking of resources whenever a given tag from a tag-cloud is selected. Finally, we discuss the navigability of social tagging systems from different theoretic perspectives in Sect. 6. Analyzing tag-based navigation with a plethora of network-theoretic tools allows us to evaluate and assess the quality, efficiency, and usefulness of the navigational structures imposed by various social tagging systems. We can use this information to further adapt and improve tag-based navigational constructs. For example, by measuring the average distance between resources in a social-tagging system we obtain a lower bound on the average number of clicks that a user needs to make to traverse between any two given resources. In the cases where the average distance exceeds a typical number of clicks that users make on the Web we have a strong indication for a poorly designed navigational interface that, consequently, we need to improve.

2   Social Tagging

In a social tagging system, users assign tags to resources. This process shapes the structure of the social tagging system and is called *social tagging*. The result of such a human-based annotation of resources is referred to as a *folksonomy*—a folk-generated taxonomy. A folksonomy is defined as a tuple $F := (U, T, R, Y)$ where U, T and R are finite sets, whose elements are called users, tags and resources, respectively, and Y is a ternary relation between them, i.e., $Y \subseteq U \times T \times R$, called tag assignments. A folksonomy can also be seen as a tri-partite hypergraph where the node set is divided into three disjoint sets - $V = T \cup U \cup R$ with *hyperedges* expressed by one tag, one user and one resource - t, u, r. The presented definition follows the notion of Hotho et al. [44]. For further formal definitions of folksonomies the interested reader may consult "Tag-based social search" in this book [67].

As pointed out by Furnas et al., the social tagging process is the collective effort of solving the vocabulary problem [28]. In this sense, tags are beneficial for navigating the information space since they provide useful hints about a resource collection. Understanding how users create tags is important for: (i) designing user interfaces (see Sect. 3), (ii) designing clustering algorithms (see Sect. 4), (iii) modeling tag-based navigation (see Sect. 5), (iv) studying the theoretic navigability of social tagging systems (see Sect. 6). Steps towards gaining such understanding have been made by Golder and Huberman who studied the regularities in the users' activities and the tag frequencies in social tagging systems [31]. They also identified some of the problems that arise when users create tags such as synonymy (multiple tags that share the same meaning, e.g., little/small), polysemy (a tag that has many related meanings, e.g., wood (a piece of a tree)/wood (an area with many trees)), or homonymy (a tag that has different not related meanings, e.g., band (a musical group)/band (a ring)). Körner et al. presented a different view on the social tagging process by characterizing the users and their tagging motivations (see Table 1) [52]. The authors split the users into at least two main groups depending on their motivation:

1. Categorizers—users who try to divide the resources into categories by assigning tags sound with some personal or shared conceptualization.
2. Describers—users who try to assign tags that describe the resource best.

The categorizers assign tags to use them as a navigational aid and try to develop a consistent taxonomy. Resources are tagged according to a common characteristic important to the mental model of the user (e.g., "pictures", "projects", "drafts", or "archive"). The describers typically assign tags to support indexing of resources, and thus support search and retrieval tasks. The assigned tags are mainly descriptive and stemming from a dynamically changing open set of tags. Although, this user separation is very nice from a theoretic point of view, in the real world, social tagging system users probably belong to these two groups simultaneously. For example, a user can use a very small categorization schema while she is assigning a lot of descriptive tags. At the same time, a tag can be part

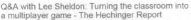

Tags

(a) Resource tag list in BibSonomy

(b) System-wide tag cloud in BibSonomy

(c) Tag hierarchy in tagFlake

(d) Tag hierarchy in ELSABer

Fig. 1. Tag-based user interfaces. BibSonomy (see footnote 2) provides a resource-specific tag list (a) to navigate between resources and a system-wide tag cloud (b) to initiate browsing. Browsing can also be initiated by a tag hierarchy as implemented in tagFlake (c) [21] and ELSABer (d) [56].

of a categorization schema and still be used as a descriptive tag. A more detailed description of the different tag types, their intended usage and classification is provided in "Tag-based social search" in this book [67].

3   User Interfaces and Visualization

Using tags to organize content introduced new research problems, i.e., exposing the content through user interfaces that leverage the advantages of the free-form annotation. In this section, we present interfaces developed to navigate the content of a social tagging system, i.e., the tag cloud and other interfaces that integrate tag clouds or facilitate tag-based browsing, e.g, through tag hierarchies.

Table 1. Tagging motivations as identified by Körner et al. in [52].

	Categorizers	Describers
Goal of tagging	Later browsing	Later retrieval
Change of vocabulary	Costly	Cheap
Size of vocabulary	Limited	Open
Tags	Subjective	Objective

We also review research literature dealing with problems naturally arising with the introduction of these interfaces, e.g., tag selection, tag cloud layouts and usefulness, tag cloud evaluation, and trend visualization using tag clouds. Table 2 shows a brief summary of the contributions along the research lines discussed in this section.

3.1 Tag Clouds

Tag clouds are widely adopted across many social tagging systems because they visualize the information space in an intuitive way. A tag cloud is a textual representation of the topic or subject of a resource as collectively seen by the users and it captures the *aboutness* of the resource [24]. There are three possible visualizations of the relationship between users, tags and resources in a social tagging system. The first presents users and their connection to tags, the second shows users connections to resources and the last presents tags and how they connect to resources. From network theoretic perspective, these are all possible combinations related to the bipartite projections of the tripartite tagging hypergraph. When using tag clouds, however, it is up to the operators of a given social tagging system to decide which one of these combinations to offer.

Let us exemplify a tag cloud with the interaction schema of a user navigating a tag-resource bipartite network [39, 40]:

1. The system presents a tag cloud to the user for a given resource.
2. The user chooses a tag from the tag cloud.
3. The system delivers a list of resources tagged with the selected tag.
4. The user selects a resource from the list.
5. The resource is displayed and the process starts anew.

Table 2. User interfaces literature

Research line	Research work
Tag selection	Venetis et al. in [84], Skoutas and Alrifai in [80], Helic et al. in [39, 40]
Tag cloud layouts	Gambette and Véronis in [30], Jafee et al. in [46], Bielenberg and Zacher in [7], Kaser and Lemire in [48], Seifert et al. in [77], Viegas et al. in [85], Eda et al. in [23]
Tag cloud usefulness	Rivadeneira et al. in [73], Sinclair and Cardew-Hall in [78], Lohmann et al. in [60], Bateman et al. in [4], Zubiaga in [94], Kuo et al. in [53], Halvey and Keane in [32], Millen and Feinberg in [65]
Tag clouds over time	Lee et al. in [55], Collins et al. in [17], Dubinko et al. [22], Russell in [74], Wagner et al. in [89]
Tag cloud evaluation	Skoutas and Alrifai in [80], Venetis et al. in [84], Trattner et al. in [82], Helic et al. in [39], Aouiche et al. in [2]
Integrated interfaces	Kammerer et al. in [47], Lin et al. in [58], Helic and Strohmaier in [36], Li et al. in [56], Di Caro et al. in [21] Vig et al. in [86–88]

With this interaction schema, a user navigates on a tag-resource bipartite network using resource-specific tag clouds (see Fig. 1(a)). In step three of the interaction schema, the system may also provide a tag cloud that captures the aboutness of the currently presented list of resources. To initiate tag-based navigation, a social tagging system may present a system-wide tag cloud capturing the aboutness of the whole social tagging system (see Fig. 1(b)) or a user-wide tag cloud covering the tags assigned by the currently logged-in user (see Fig. 2(c)). Tag-based browsing can be also initiated by a tag hierarchy. For example, tagFlake by Di Caro et al. [21] and ELSABer by Li et al. [56] are user interfaces that offer top down tag hierarchy browsing. Both interfaces work in a similar manner. First, a tag hierarchy is presented on the left side of the screen. After a tag is selected, the associated resources are displayed on the right side of the screen (see Figure 1(c) and (d)).

Tag Selection. One of the first questions arising when designing a tag cloud refers to the tag cloud size, i.e., which and how many tags should be displayed. For example, there are systems presenting only twenty tags, whereas others offer a much bigger number (see Fig. 2). Proposed by Venetis et al., the simple TopN tag selection algorithm is very widely adopted [84]. To create a tag cloud, this algorithm considers only tags assigned to a specific resource. The algorithm selects the top n tags with the highest resource-specific frequency to present in the tag cloud. If there are less than n tags available for a resource, the remaining tag positions in the cloud are left empty. In the same work, Venetis et al. proposed also algorithms for selecting top tags based on standard text features and maximum resource coverage [84]. The tag selection problem has also been studied by Skoutas and Alrifai who introduced a tag selection framework based on frequency, diversity and rank aggregation [80].

Tag Could Layouts and Functionality. Figure 2(a), (b) and (c) shows the basic tag cloud layout. However, more sophisticated approaches exist. For example, Gambette and Veronis arranged tags in a tree structure (TreeCloud) so that their semantic proximity is reflected (see Fig. 2(d)) [30]. Proposed by Jafee et al., TagMaps is a unique layout using real geographical space to create a tag cloud for large collections of geo-referenced photographs [46]. Bielenberg and Zacher introduced a circular tag cloud layout and compared it to the typical rectangular layout [7]. In the proposed circular layout, the distance to the center and the font size of the tag represent its importance. In this layout, the distance between tags in the cloud does not reflect their similarity. Different researchers concentrated on the aesthetic issues regarding the tag cloud layouts. Kaser and Lemire arranged tags in nested HTML tables in which tag relationships are considered. To tackle white spaces in tag clouds, emerging due to different font size usage, they proposed to use the *min-cut placement* Electronic Design Automation algorithm [48]. In another work, Seifert et al. concentrated on the visual issues in layouts and proposed a new algorithm utilizing arbitrary convex polygons to bound tags and reduce white spaces [77]. Viegas et al. introduced Wordle,

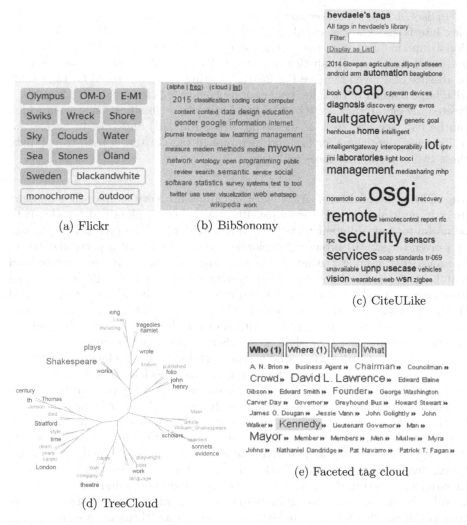

(a) Flickr

(b) BibSonomy

(c) CiteULike

(d) TreeCloud

(e) Faceted tag cloud

Fig. 2. Tag cloud layouts and functionality. Flickr (see footnote 5) (a) uses different colors in the tag cloud to distinguish between automatically and user assigned tags. BibSonomy (see footnote 2) (b) and CiteULike (see footnote 3) (c) use different font sizes to indicate tag importance and popularity. Changing the layout from a cloud to an alphabetically or frequency sorted list is offered by BibSonomy (b), whereas CiteULike (c) allows tag filtering. TreeCloud (d) uses a tree structure to reflect semantic proximity between tags [30]. In a faceted tag cloud (e), tags are classified into "Who", "Where", "When" and "What" facets [82].

a distinctive layout that concentrates on the balance of colors, typography and other visual features [85]. Eda et al. concentrated on experienced emotions when using tag clouds and proposed a layout in which the font size is determined by the tag's entropy and not by its content popularity [23].

In Sect. 3.2, we will show how tag clouds have been integrated into more complex interfaces. The tag cloud interface itself, however, can also provide additional functionality, e.g., tag sorting, filtering or faceting. Rearranging tags to present them as an alphabetically sorted list is helpful for finding the presence or absence of a given tag (see Fig. 2(b)). Filtering of tags is useful for tag clouds with large number of tags, i.e. system or user-wide tag clouds (see Fig. 2(c)). In a faceted tag cloud, tags are structured according to a classification schema which can be flat (see Fig. 2(e)) or hierarchical (see Fig. 1(d) and (c)). In Sect. 4, we will discuss algorithms for flat and hierarchical tag clustering used to populate these two types of interfaces with tags.

Tag Cloud Usefulness. Although tag clouds are very simple, they support users in multiple ways. For example, Rivadeneira et al. identified tag clouds to be useful for four different tasks [73]: (i) search: finding the presence or absence of a given target, (ii) browsing: exploring the cloud without a particular target in mind, (iii) gaining (visual) impression about a topic, (iv) recognition and matching: recognizing the tag cloud as data describing a specific topic. In an experiment on gaining impressions and recognition, Rivadeneira et al. studied the tag font size and the cloud layout. The results suggested that the font size has a strong effect on recognition. Although the different layouts changed the accuracy of impression, they had no significant effect on recognition.

Sinclair and Cardew-Hall studied the usefulness of tag clouds for different information retrieval tasks and found tag clouds especially useful for browsing scenarios [78]. In such scenarios, tag clouds support discovery of items (resources, users, or other tags) that a user might not have thought of or known about.

Halvey and Keane examined the usefulness of tag clouds for finding a specific target by comparing them to horizontal and vertical alphabetically sorted lists [32]. Their results showed that tag clouds are outperformed by both list types, suggesting that alphabetization aids users for orientation. They also experimented with the tag cloud typography, i.e., font sizes and found out that targets with larger font sizes are found more quickly. Regarding tag font size and position in the cloud, they came to similar conclusion as Rivadeneira et al. in [73], namely, that the font size strongly contributes to recall, whereas proximity to the largest tag has no effect.

Bateman et al. concentrated on the visual features of tag clouds and how they affect the visual search of a tag in the cloud [4]. They concluded that font size has a more significant impact on finding a tag than other visual features such as, e.g., color, tag string length and tag location.

Kuo et al. compared the usefulness of tag clouds and lists for summarizing search results from the biomedical domain [53]. They considered tag clouds superior to search result lists with respect to the presentation of descriptive information. However, tag clouds performed significantly worse when presenting relationships between concepts.

Lohmann et al. studied different tag cloud layouts, i.e., sequential layout (alphabetical sorting), clustered layout (thematic clusters), circular layout

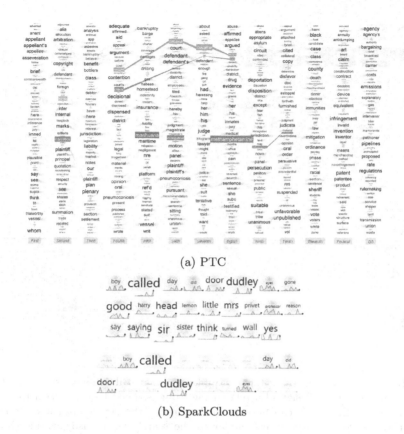

(a) PTC

(b) SparkClouds

Fig. 3. Tag clouds over time. PTC (a) compares the popularity of a given tag between multiple time periods [17]. In SparkClouds (b) a sparkline shows, e.g., if a tag is new or if it has experienced high popularity in the system over a time period [55].

(decreasing popularity), and their ability to support typical information seeking tasks [60]. For finding a specific tag, they suggest the sequential layout with alphabetical sorting. The thematically clustered layout performed best for finding tags that belong to a certain topic, whereas the circular layout with decreasing popularity is more appropriate for finding the most popular tags.

Overall, literature suggests that visualizing tags, i.e., their font size and layout of the cloud have significant effect on the tag cloud usefulness for tag-based navigation. Furthermore, the presented findings highlight the intrinsic connection of tag-based navigation and the way tagging data is visualized.

Tag Clouds Over Time. Tag clouds are also useful for visualizing trends and comparing resource collections over time. Research in this direction has been conducted by Lee et al. who introduced SparkClouds [55] and by Collins et al. who presented Parallel Tag Clouds (PTC) [17]. PTC is designed to con-

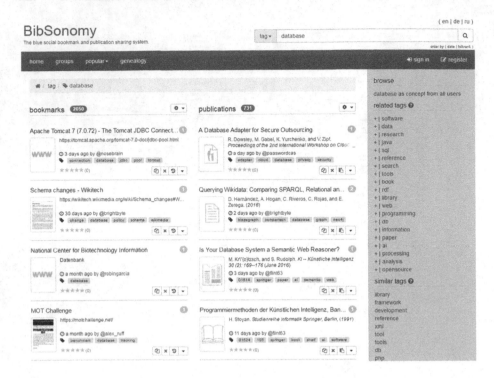

Fig. 4. Query expansion using tags as implemented in BibSonomy (see footnote 2). By clicking on the "+" sign of a related tag—tag assigned together with the search tag (top left)—the query can be expanded to refine the set of presented resources. Exploring the content using tags similar to the search tag is also possible (bottom left).

sider and understand changes across multiple resource collections by presenting their tag clouds simultaneously. It combines parallel coordinate plots visualization [45] with tag clouds and can highlight the underuse and overuse of a tag (see Fig. 3(a)). SparkClouds unifies sparklines [83] with typical tag cloud features to visualize evidence of change across multiple tag clouds (see Fig. 3(b)).

As discussed later on, the maturity of a social tagging system influences its navigability. Wagner et al. compared different methods for estimating the system's maturity, i.e., with respect to its semantic stability [89]. Taglines and Cloudalicious are two prominent tools for visualizing the usage and semantic stability of tags. Introduced by Dubinko et al., Taglines is a visualization working with the *river metaphor*—tags flow from left to right—and the *waterfall metaphor*—tags are presented in fixed slots through which they can "travel" over time [22]. With these metaphors, Taglines presents tags that possess a significantly high occurrence frequency inside a given time period, compared to outside this period. Proposed by Russel, Cloudalicious visualizes the evolution of tags over time [74]. The tool works with Delicious tagging data and produces

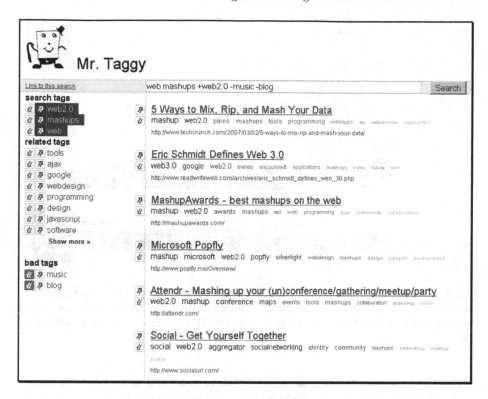

Fig. 5. MrTaggy as presented by Kammerer et al. in [47]. The MrTaggy browser allows a user to specify a query and then rate both presented resources (right) and related tags (left) by clicking on the arrows on the left. Clicking a tag arrow up or down refines the query by adding or excluding the tag from the query. Clicking a resource arrow up or down highlights similar results or excludes the resource from the result list.

a graph of the collective tagging activity for a given URL. It shows the relative weights of the most popular tags for the URL. Indications of stabilization are observed as the lines of the graph move from left to right. This pattern expresses the collective opinion of the users with respect to the URL. An even more interesting pattern in the graph are diagonal lines, as such lines suggest that users changed the URL describing tags.

Tag Cloud Evaluation. Skoutas and Alrifai, and Venetis et al. conducted research on tag cloud evaluation [80, 84]. Both author groups propose very similar evaluation metrics for tag clouds with respect to coverage, overlap and selectivity. Apart from the evaluation metrics, Skoutas and Alrifai introduced a user navigation model that combined with the evaluation metrics allows tag cloud evaluation with respect to navigation. Aouiche et al. proposed an entropy-based metric for evaluating the informativeness of a tag cloud [2].

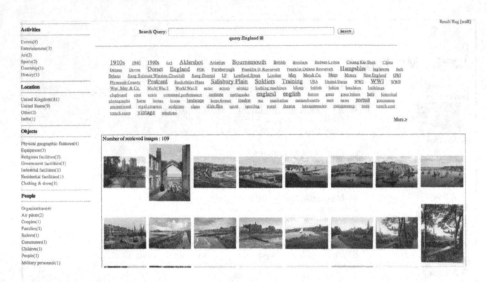

Fig. 6. The DPNF interface as proposed by Lin et al. in [58]. The DPNF interface integrates a search box to query the system and start navigation (top), controlled vocabularies as facets (left) and a tag cloud (middle) to explore an image collection (bottom).

Another method for evaluating tag clouds has been followed by Trattner et al. who performed a user study to evaluate tag-based information access in image collections [82]. In the study, they compared traditional and faceted tag clouds with a baseline (search-only) interface. Both tag cloud types performed better than the search-only interface with respect to a predefined search task. Additionally, the authors observed that the faceted tag cloud is more difficult to use initially, but it is considered as more powerful by the study participants in the long run.

Helic et al. showed that the navigability assumption—the widely adopted belief that tag clouds are useful for navigation—does not hold for every social tagging system [39]. Furthermore, they showed that the usefulness of tag clouds is sensitive to the adoption phase of the system, i.e., its maturity and that the navigability assumption may only hold for more mature systems. One very useful finding by Helic et al. is that the limitation of the tag cloud size to a practically more feasible size, e.g., five, ten or more tags does not influence the navigability. Depending on the maturity of the social tagging system and the tag cloud type (e.g., system-wide or resource-specific), however, a tag could covers hundreds or even thousands of resources. In such cases, the resources displayed after a tag selection are often sorted by their reverse chronological order and paginated which reduces the navigability. To tackle this problem, Helic et al. introduced a generalized pagination algorithm and experimented with different context preservation functions.

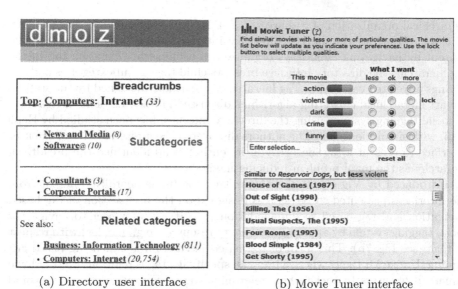

<table>
</table>

(a) Directory user interface   (b) Movie Tuner interface

Fig. 7. The directory user interface (a) as implemented in DMOZ (see footnote 1). The typical interface elements include breadcrumbs, subcategories and related categories. The Movie Tuner interface (b) as proposed by Vig et al. in [87]. The Movie Tuner interface offers users to critique a resource with respect to tags, e.g., "less violent" critique is applied to the movie *Reservoir Dogs*.

3.2 Integrated Interfaces

Very often tag clouds are integrated into more complex user interfaces that allow a user to start, e.g., with a query formulation and then narrow down the presented results using the tag cloud. Some social tagging systems, e.g., BibSonomy also allow users to expand or narrow down the query using related or similar tags (see Fig. 4). Presented by Kammerer et al., MrTaggy is a very similar interface that allows searching and browsing of resources and tags by exploiting the relationships between them and the collected relevance feedback (see Fig. 5).

The Dual-Perspective Navigation Framework (DPNF) by Lin et al. introduced an interface seamlessly combining controlled vocabularies (metadata) and free vocabularies (social tags) [58]. By combining both vocabulary types, DPNF aims to provide better resource findability at each navigation step. In a user study, the authors compared the DPNF interface with a tag-only and metadata-only interface and concluded that the DPNF interface preforms best with respect to lookup and exploratory search tasks.

Another more complex interface than the tag cloud is the directory interface. It offers the following user interface elements: (i) breadcrumbs—provide a complete path to the root category, (ii) subcategories—deliver a list of links to more fine-grained categories, (iii) related categories—provide links to related categories. The directory interface is usually adopted in information systems with hierarchical organization of resources, e.g, DMOZ (see footnote 1) (see Fig. 7a). In a social tagging system, the directory interface allows users to explore the content by navigating along tags organized in a hierarchy. For each tag in the tag hierarchy, the directory interface presents child tags as subcategories and tag siblings as related categories. Tag hierarchies are usually obtained by the state of the art hierarchical clustering algorithms discussed in Sect. 4.2. The navigability of social tagging systems using the directory interface has been studied by Helic et al. with simulations [36]. The authors discovered the limited ability of the user interface to present a tag hierarchy in its entirety and identified the breadth of hierarchy as the main problem reducing navigability.

Introduced by Vig et al., the Movie Tuner is fundamentally different from the interfaces presented so far as it is designed on the intersection of tag-based navigation and tag-based recommendation [87]. It allows a user to navigate a social tagging system by applying *critique* to resources (e.g., movies) with respect to tags (see Fig. 7b). This form of tag-based navigation is called tag-based critiquing. Unlike MrTaggy where a user can specify if a tag is relevant or not, Movie Tuner allows users to specify how relevant a tag is. For example, a user could explore the system for movies that are "less violent" than the movie *Reservoir Dogs*. Similar to the directory interface, Movie Tuner also operates on a special data structure called *tag genome* [86,88]. The tag genome captures the relevance of tags to resources and addresses three limitations of the social tagging process: (i) binary tag-resource relationships (the strength of the relationship is not reflected), (ii) tag sparsity (not all relevant tags may be assigned) and (iii) only positive tag-resource relationships (irrelevance of tags cannot be indicated as they are not assigned). The Movie tuner interface uses a multi-objective tag selection algorithm to choose the tags displayed for a given resource. The optimized tag selection objectives are: critique value, popularity and diversity. After applying a critique across one or multiple tags, the systems recommends resources that satisfy the user's critique. To respond to a user critique, the Movie Tuner interface resorts to an algorithm that selects resources based on the difference to the critiqued tags and similarity to the original resource.

The presented ways of integrating tag clouds into more complex interfaces, aim to achieve better overall user experience and to provide even better support for tag-based navigation.

4   Tag Clustering

As pointed out in Sect. 2, synonymy, polysemy and homonymy have been identified as problematic regrading the semantic of tags. Another crucial issue with social tagging data is the lack of structure. The efficiency of tag-based navigation and browsing, however, depends on the structure of the information space. To this end, creating groups of resources meaningful to users by exploiting, i.e., semantic relationships between tags is of special importance for tag-based navigation. Moreover, grouping tags that are semantically related to each other with additional taxonomy relations between tag groups allows us to come up with hierarchical tag-based interfaces. As various previous studies have shown users are able to efficiently navigate hierarchical interfaces—hence providing such interfaces in a social tagging system is particularly important for supporting users in their explorations of the information space.

In this section, we present an overview of the state of the art algorithms for clustering tagging data. They tackle the above problems by organizing tags according to a classification schema. Depending on the classification schema, there are flat (see Table 3) and hierarchical (see Table 4) clustering algorithms. In general, the discussed tag clustering algorithms represent adaptations of existing state of the art clustering algorithms, e.g., K-Means, or Affinity Propagation. Unlike the algorithms for tag selection focusing on resource-specific tag clouds, the algorithms presented here create tag clouds for resource collections. For example, the flat clustering algorithms organize tags for the presentation through a faceted tag cloud or a system-wide tag cloud. The hierarchical clustering algorithms produce tag hierarchies suitable, e.g., for the directory interface or for interfaces such as ELSABer and tagFlake. We pay special attention to hierarchical clustering due to its importance for modeling tag-based navigation (see Sect. 5). The algorithms reviewed in this section are divided into three different classes: content-based, graph-based and machine learning.

4.1 Flat Tag Clustering

The content-based approach has been followed by Specia and Motta who proposed an algorithm for creating semantically related clusters of tags based on their co-occurrence [81]. The algorithm performs statistical analysis of the tag space and constructs a co-occurrence vector for each tag. Clusters are then created using cosine similarity between tags given their co-occurrence vectors. Zubiaga et al. introduced a content-based algorithm using unsupervised neural networks to obtain flat tag clusters [95]. Using language modeling techniques, the clusters are then labeled with the most discriminative tag in a cluster.

The graph-based approach has been adopted by Begelman et al. who proposed a recursive algorithm that uses spectral bisection to split a graph of connected tags into two clusters [5]. Similar to Begelman et al., Au Yeung et al. also introduced a graph-based clustering algorithm using a modularity function to evaluate the quality of division [3]. The authors evaluated their algorithm on three different networks based on users, co-occurrence of tags and context of tags. Hereby, the best results have been achieved using context tags networks.

The machine learning approach has been followed by Remage et al. and by Hassan-Montero and Herrero-Solana. Hassan-Montero and Herrero-Solana proposed an algorithm for tag clustering that considers the semantic relationships between tags [33]. Under a predefined number of clusters and a number of selected relevant tags, the proposed algorithm resorts to K-Means clustering on a tag similarity matrix estimated by means of the Jaccard coefficient. Ramage et al. studied the usage of K-Means clustering in an extended vector space model that contains not only tags but also texts from web pages [72]. They also proposed a novel generative algorithm based on latent Dirichlet Allocation (LDA) that uses tags and web pages texts. They found that the usage of tags in combination with web pages texts improves the cluster quality.

4.2   Hierarchical Tag Clustering

In this section, we cover the following three hierarchical clustering algorithms in a more detailed fashion: Hierarchical K-Means [20], Affinity Propagation [71] and Generality in Tag Similarity Graph [41]. Among many other algorithms presented here, these three algorithms are commonly used, simple to implement and exist in different variations.

Hierarchical K-Means. K-Means is probably the most prominent clustering algorithm [25,59]. The K-Means versions that we present here complement the flat clustering version from the previous section. For example, Zhong introduced a spherical online version of K-Means [93]. Dhillon et al. adapted the algorithm to work with textual data by replacing Euclidian distance with cosine similarity [20]. A combination of these two K-Means version creates a tag hierarchy in a top-down manner. The algorithm starts by splitting the whole input data into ten clusters. Clusters with more than ten samples are processed iteratively in the same manner, whereas clusters with less than ten samples are considered as leaf clusters. A special case is introduced to handle clusters with eleven samples which initially would have been also split into ten clusters. This special case gives freedom to the partitioning as it allows the division of clusters with eleven samples not into ten but into three clusters. Each node in the hierarchy is represented by the nearest tag to the centroid. This tag is removed from the actual tags contained in a cluster if the cluster is further partitioned.

Affinity Propagation. Affinity propagation has been originally proposed by Frey and Dueck [26]. The input of the algorithm is a set of similarities between data samples provided in a matrix. The diagonal of the matrix contains the self-similarity values representing the suitability of the data sample to serve as a cluster center. They are also called preferences. Specifying a number of desired clusters is not needed, however, there is a correlation between the preference values and the number of clusters (lower preference values imply a low number of clusters and vice versa). Affinity propagation characterizes each data sample according to its "responsibility" and its "availability" values. The responsibility expresses the ability of the sample to serve as an exemplar for other samples, whereas the availability shows the suitability of other data samples to be the exemplars for a specific data sample. Affinity propagation exchanges messages

Table 3. Flat clustering

Algorithm type	Example approaches
Machine learning	Ramage et al. in [72], Hassan-Montero and Herrero-Solana in [33]
Graph-based	Begelman et al. in [5], Au Yeung et al. in [3]
Content-based	Specia and Motta in [81], Zubiaga et al. in [95]

between data samples and iteratively updates the responsibility and availability values of each sample with a parameter λ as an update factor. By adding structural constraints into the global objective function of affinity propagation, Plangprasopchok et al. adapted the algorithm to create a taxonomy [71]. Another approach to induce a hierarchy is to use the original version of affinity propagation recursively in a bottom-up manner. The algorithm starts with a matrix containing the top ten cosine similarities between the tags in a given dataset. The minimum of those similarities acts as preference for all data samples. The clusters are produced by selecting examples with associated data samples. Depending on an adjustable parameter specifying the ratio between the desired number of clusters and the data samples, the results are returned or another iteration starts. If the number of selected clusters in the previous run was too high, the preference values are lowered. Otherwise, they are increased. The sum of the connected data samples normalized to unit length represents the centroid of the cluster, while cosine similarity between the centroids of the clusters serve as input matrix for the next iteration. This process is repeated until the top-level is reached. As the output of the algorithm should be a hierarchy, each node in the hierarchy needs to represent a unique tag. To this end, the nearest tag to the centroid is selected as the tag representing the node. Additionally, the selected tag is removed from the actual tags contained in the leaf cluster and it cannot be used in lower hierarchy levels. The update factor λ can be dynamically adjusted in each iteration. The algorithm terminates when a given number of iterations is reached or if the clusters are stable for at least ten iterations.

Generality in Tag Similarity Graph. Introduced by Heymann and Garcia-Molina, this algorithm receives a *tag similarity graph* as input [41]. The tag similarity graph is an unweighted graph in which each tag is represented by a node and two nodes have an edge between them if the similarity between

Table 4. Hierarchical clustering

Algorithm type	Example approaches
Machine learning	Dhillon et al. in [20], Zhong in [93] (K-Means), Schmitz et al. in [75] (Association rules), Di Caro et al. in [21] (LSA), Candan et al. in [12] (LSA), Li et al. in [56] (Decision trees)
Graph-based	
- Graph clustering	Muchnik et al. in [66], Lancichinetti et al. in [54]
- Affinity propagation	Frey and Dueck in [26], Plangprasopchok et al. in [71]
Content-based	
- Generality in tag similarity	Heymann and Garcia-Molina in [41], Benz et al. in [6], Helic and Strohmaier in [36]
- Other	Schmitz in [76], Brooks and Montanez in [11]

their respective tags is above some threshold. The algorithm starts by setting the most general node (central node in the similarity graph) as root of the hierarchy. All other nodes are added to the hierarchy in descending order of their centrality in the similarity graph. For each candidate node, the similarity between all currently present nodes in the hierarchy and the candidate node is calculated. The candidate node is added as a child of the most similar node in the hierarchy if their similarity is above a given threshold. Otherwise, the candidate node is added as a child of the root. The algorithm makes three main assumptions:

1. Hierarchy representation assumption—the edges representing a given hierarchy are also present in the similarity graph.
2. Noise assumption—there are noisy connections between unrelated tags (mainly due to spamming activities).
3. General-general assumption—the noisy connections between tags occur more often in the higher levels of a given hierarchy.

According to Heymann and Garcia-Molina, the hierarchy representation assumption is essential for detecting hierarchies based on similarity measures. Since tagging data exhibits a lot of noise [13], noisy tags would be of high degree in the tag similarity graph. Thus, they would occupy high hierarchy levels which would eventually reduce the ability of the produced hierarchy to guide navigation. This makes the second assumption also fundamental as it accounts for noisy tag connections. The general - general assumption is based on the intuition that higher level (more general) tags are likely to co-occur by chance. Inserting the more general (central) nodes in the similarity graph in the top of a hierarchy assures short hierarchy distances between the most general tags.

The authors mention the possibility to use different similarity measures as well as different centrality measures. Typical versions of the algorithm are degree centrality as centrality measure and co-occurrence as similarity measure (Deg-Cen/Cooc) and closeness centrality and Cosine similarity (CloCen/Cos). As pointed out by Heymann and Garcia-Molina, more control over the properties of the hierarchy is possible by dynamically adjusting the similarity threshold.

Other Algorithms. Muchnic et al. discussed an algorithm for condensing a hierarchy based on metrics for estimating the hierarchy level of single nodes in a network [66]. Clauset et al. presented a general approach for extracting hierarchies from network data demonstrating that the existence of a hierarchy can simultaneously explain and quantitatively reproduce several commonly observed topological properties of networks, e.g., right-skewed degree distributions, high clustering coefficients and short path lengths [16]. Lancichinetti et al. proposed an approach for discovering hierarchies based on overlapping network community structures [54]. They introduced a fitness function for estimating the quality of cover and used it to find the most appropriate community for each network node.

Benz et al. computed the generality in tag similarity graph algorithm by Heymann and Garcia-Molina by using co-occurrence as similarity measure and degree centrality as centrality measure [6]. Additionally, they introduced an extensive preprocessing of the data to remove synonyms and resolve ambiguous tags. Helic and Strohmaier also adapted the generality in tag similarity graph algorithm to control for the breadth in the top levels of the created hierarchy as they identified it as a navigability reducing factor [36].

Li et al. presented the Effective Large Scale Annotation Browser (ELSABer) to browse social annotation data [56]. The algorithm creates a hierarchy using a decision tree and tag features containing, e.g., tag coverage, inverse coverage rate and intersection rate. Schmitz et al. applied association rule mining to extract hierarchies from tagging data concentrating on ontology learning and emergent semantics [75]. Another algorithm for creating tag hierarchies has been presented by Candan et al. who constructed a hierarchy by transforming the tag space into a tag graph and then minimizing its spanning tree [12]. The algorithm uses a similarity lower bound to prevent a context drifting of the tags in the hierarchy. Di Caro et al. described an algorithm that extracts the most significant tags from text documents (not from tagging data) and maps them to a hierarchy so that descendant tags are contextually dependant on their ancestors within a given document corpus [21]. Both algorithms by Di Caro et al. and by Candan et al. applied latent Semantic Analysis (LSA).

Brooks and Montanez presented an agglomerative clustering algorithm to induce a tag hierarchy using abstract tags and abstract tag clusters [11]. Schmitz introduced a subsumption-based algorithm for inducing tag hierarchies [76]. The algorithm uses co-occurrence statistics and builds a graph of possible parent-child relationships. For each node, the best path to a root is calculated under the consideration of reinforced possible parents. The paths are then composed into a tree.

5 Modeling Navigation in Social Tagging Systems

As shortly mentioned in Introduction models of user navigational behavior have been extensively used to improve information retrieval capabilities of the Web-based information systems—the most famous example being the Google's random surfer model used to improve rankings of search results. Similarly to the random surfer model various other navigational models have been applied for providing further navigational improvements such as adaptations of links and navigational interfaces by, for example, inserting, highlighting or removing links. For exactly these reasons navigational models have been also applied and adapted for social tagging systems.

In this section, we present two main frameworks for modeling tag-based navigation: Markov chains and decentralized search. Modeling navigation in tagging systems has been recognized as an important step towards better understanding of user navigation behavior [35], which in turn has major practical implications such as implementing more efficient user interfaces [36].

Markov chains have been regularly applied for modeling navigation on the Web, i.e., on information networks [92]. On the other hand, decentralized search approaches has been applied to study the navigational efficiency of broad and narrow folksonomies [35], to evaluate a folksonomy from a pragmatic point of view with respect to tag-based navigation [38] and to build directories for social tagging systems [36].

As shown in Sect. 3.1, tag-based navigation is facilitated either by traversing a tag hierarchy or by navigating between tags connecting resources on a tag-resource network. For pragmatic reasons, however, tag-based navigation is often modeled on a tag-tag network projected over a tag-resource network. Such network mappings reduce complexity and are shown to be effective, e.g., in the field of ontology learning [64].

5.1 Markov Chain Models

Navigation on the Web is the process of following links between web pages. Markov chains model navigation on the Web by assigning transition probabilities between web pages also called *states* [9,19,57,79]. Although Markov chain models can also be of higher order (the transition probability between two states depends on several previous states), first order Markov chains (the transition probability depends only on the current state) are more commonly used due to their simplicity. Navigation on the tag-tag network of a social tagging system is modeled with Markov chains by representing each tag as a state. Transition probabilities between states are then assigned according to the distance between the tags in a tag hierarchy induced, e.g., by the algorithms presented in the previous section.

5.2 Decentralized Search

Decentralized search is an algorithm designed by Kleinberg to explain the ability of humans to efficiently search other people in huge social networks [49,50]. The algorithm has been since its invention also used to model navigation in information networks. To model navigation, the *decentralized search* algorithm passes messages between network nodes. In decentralized search, the message holder forwards a message to one of its immediate neighbor nodes until the intended message recipient (the target node) is found. For selecting the next step in the navigation process, decentralized search resorts to a *background knowledge* to rank the neighbors of the current node (also called candidate nodes) and forward the message to one of them.

When modeling tag-based navigation, the background knowledge is a tag hierarchy and the message is passed to the neighbor j with the shortest hierarchy distance $d(j,t)$ to the target node t.

In Fig. 8, we see an example of decentralized search in a tag-tag network using a tag hierarchy as hierarchical background knowledge. The goal in this example is to find a path between the start node 13 (marked yellow) and the target node 33 (marked red). To select the next node, the algorithm looks up

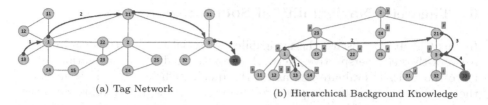

(a) Tag Network   (b) Hierarchical Background Knowledge

Fig. 8. Decentralized search using hierarchy as background knowledge as shown by Helic et al. in [38]. (Color figure online)

the distance of the neighbors of the current node to the target node in the hierarchical background knowledge. The neighbor with the shortest hierarchy distance is then selected. For the first step, node 1 is selected since it is the only adjacent node to 13. At step two, the set of neighbors of node 1 contains 11, 12, 13, 14, 21, 22 and 23 and the node with the shortest distance to the target node is node 21 (number in boxes in (b) provides the distance of each node to the target node). The procedure is repeated until the target node is reached. The red arrows show the resulting path.

In the given example, the message is always passed to the node with the shortest hierarchy distance. Thus, a distance greedy action selection is used to model a confidently navigating user which is a plausible scenario in a social tagging system. The action selections presented next are also applicable although originally developed for modeling navigation on information networks [37]:

e-**greedy:** The e-greedy action selection chooses the candidate node j with the shortest distance to the target node t with a probability $1-e$. With a probability e, another candidate node is chosen uniformly at random.

Softmax Rule: The softmax rule [10,18] chooses a candidate node with shortest distance to the target node with a probability $p(j) \propto e^{cf(j)}$. Hereby, $f(j)$ represents the fitness function calculated from the distances $d(j,t)$, and c is the user's confidence in her intuition. For high values of c, the softmax rule selects the candidate node with the shortest distance to the target nodes, thus, reduces to greedy selection. For small values of c, the softmax rule is tuned to select other candidate nodes based on $f(j)$, thus, it models a user with low confidence.

Inverse Distance Rule: The inverse distance rule [63] is very similar to the softmax rule as it selects the candidate node with a probability $p(j) \propto f(j)^{-c}$. The parameter c expresses again the confidence. The main difference to the softmax rule is the different probability distribution.

Dacaying e-greedy: The decaying e-greedy rule [37] is based on the idea that humans do not possess sufficient intuition in the beginning of the navigation process, but their intuition becomes better and better during the process. The rule is based on a decay function that adapts e at every step of the navigation. Different decay functions are possible, but normally $e(t) = e_0 \lambda^{-t}$ is used. Hereby, e_0 is the initial value of e, and λ is a decaying factor at step t.

6   Theoretic Navigability of Social Tagging Systems

In this Chapter we discuss the navigability of social tagging systems from the network-theoretic perspective. Network-theoretic analysis of navigation allows us to theoretically evaluate and assess the quality, efficiency, and usefulness of the navigational structures imposed by various social tagging systems. Such theoretical analysis provides us with theoretical bounds on various aspects of social tagging systems and provides the first evaluation results and first indications for potential navigational bottlenecks and problems. Subsequently, we may remedy the problems even before performing expensive usability tests with real users. For example, network-theoretic tools allows us to study connectivity of various parts of our information systems—in this way we can identify completely disconnected or poorly connected groups of resources in our system.

So far, the theoretic navigability of social tagging systems has been studied from four different perspectives on which we want to shed light in this section: network theoretic, information theoretic, information foraging and tagging vs. library approach. Each perspective emphasizes that the navigability of a social tagging systems depends on the ability of the users to assign tags to resources, i.e., to solve the vocabulary problem.

6.1   Network Theoretic Perspective

Adamic et al. studied navigation in power-law degree distributed networks and showed that random walks naturally tend to select nodes with a high degree [1]. Based on this observation, they proposed a version of the decentralized search algorithm that exploits the degree distribution for finding the target node by passing the message to the candidate node with the highest degree. Such an algorithm makes each power-law degree distributed network theoretically searchable. Folksonomies—the data structures of social tagging systems—possess power-law degree distributions (see "Tag-based social search" in this book [67]), thus, they are easily navigable with Adamic's algorithm. In the case of tag-based navigation, however, the navigability of a folksonomy cannot be measured in this way as the algorithm does not exploit the semantic relationships between tags but only the network topology. Furthermore, tag hierarchies created by the algorithms presented in Sect. 4 are structures capturing not only semantic relationships between tags but also other useful properties of the social tagging process which cannot be neglected when looking at the network theoretic perspective of tag-based navigation. To this end, in this section we concentrate on two aspects: (i) the general navigability of a folksonomy as a graph and (ii) the ability of tag hierarchies to guide navigation in such a graph.

Navigability of a Folksonomy as a Graph. Cattuto et al. described the navigability of a folksonomy in terms of its "small world" network properties [13]. Small world networks are easy to navigate as all network nodes are reachable within few steps. In a pioneer work, Watts and Strogatz defined the class of "small world" networks based on the characteristic path length and the clustering coefficient [91]. The characteristic path length is a global network topology measure specifying the average shortest path distance for all possible node pairs. The clustering coefficient is a local measure and specifies the extent to which the neighbors of a given node form a clique. Since folksonomies are tri-partite graphs, the above measures cannot be directly applied to study their the network properties. To this end, Cattuto et al. redefined the characteristic path length and the clustering coefficient for three mode data. After comparing observed folksonomies with two randomly generated folksonomies of equal size with respect to both measures, they found that the observed folksonomies have extremely high clustering coefficients and comparable to lower characteristic path lengths. Cattuto et al. also noticed the small characteristic path length values (about 3.5) which did not change significantly as observed folksonomies grew. This is a very important observation since it implies that on average every resource, user and tag is reachable from any other resource, user or tag within a couple of clicks. This high reachability explains also why folksonomies support serendipitous discovery [62].

Navigation Supported by Tag Hierarchies. The ability of tag hierarchies to guide tag-based navigation has been studied by Helic et al. using the small world network models [38]. In general, there are two types of small world network models—lattice-based (ring lattice model by Watts and Strogatz [91] and 2D-lattice model by Kleinberg [50]) and hierarchy-based (single hierarchy model by Kleinberg [51] and multiple hierarchies model by Watts et al. [90]). Essentially, those small world models generate networks in which the balance between the local network structure (short range links) and the global network structure (long range links) is used to guide navigation modeled, e.g., as decentralized search. In the hierarchy-based models, this balance is regulated through the distance distribution between the nodes of the hierarchies generating the network. Inspired by the above observations and models, Helic et al. proposed a theoretic evaluation of the suitability of tag hierarchies to support tag-based navigation in a tag-tag network. For all connected node pairs in the network, they suggested to measure the distance between the pair nodes in a given tag hierarchy and to create a distance distribution. The theoretic suitability of the tag hierarchy to support navigation is then estimated using this distance distribution. More precisely, the authors introduced an indirect comparison of the distance distribution of a tag hierarchy and the class of theoretically searchable networks according to Watts' model. A direct comparison is not possible due to the following differences: (i) in Watts' model, the degree distribution is uniform, whereas the tag degree distribution has been shown to follow a power-law; (ii) in a tag hierarchy, tags could be potentially attached everywhere in a hierarchy, which is not the case in the model of Watts where they would be attached only to leaves. To tackle these differences, Helic et al. adapted Watts' model to tagging networks and discussed the distance distributions of two synthetic tag hierarchies—the random and the homophily-based hierarchy. The homophily distance distribution

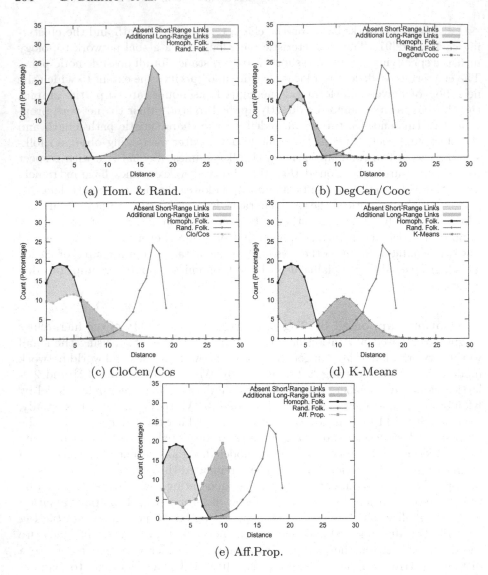

(a) Hom. & Rand.

(b) DegCen/Cooc

(c) CloCen/Cos

(d) K-Means

(e) Aff.Prop.

Fig. 9. Comparison of distance distributions for four hierarchical tag clustering algorithms as shown by Helic et al. in [38]. (Color figure online)

describes a hierarchy that only supports short range connections in the tag-tag network, whereas the random distance distribution mimics a tag hierarchy with both short and long range connections. None of the distributions is optimal (see Fig. 9(a)). The homophily distance distribution is dominated by short range links, whereas the random distance distribution is dominated by long range links. However, the homophilous tag hierarchy, which is lacking some long range links, is theoretically more suitable to guide tag-based navigation.

In Fig. 9, we also see a comparison between the distance distributions of tag hierarchies created by the algorithms in Sect. 4.2 on a BibSonomy dataset[6] and the two synthetic distance distributions—random and homophily-based distributions. The gray and the yellow areas represent the differences in the number of short range links and long range links, respectively. The gray area is called the Absent Short-Range Links area, whereas the yellow area is called the Additional Long-Range Links area. Theoretically, both areas need to be greater than zero but still rather small. Otherwise, the distance distribution will incorporate too many long range links and will become similar to the random distance distribution. The tag hierarchies created by DegCen/Cooc and CloCen/Cos versions of the generality in tag similarity graph algorithm should perform best from a theoretic point of view (see Fig. 9(b) and (c)) as they exhibit distance distributions with many short range links mixed with a few long range links. The distance distributions of the tag hierarchies induced by K-Means and Affinity Propagation seem theoretically less suitable since they possess too many long range links and too few short range links (see Fig. 9(d) and (e)).

6.2   Information Theoretic Perspective

The navigability of social tagging systems has also been studied by Chi and Mytkowicz from an information theoretic perspective [14,15]. In their work, Chi and Mytkowicz see social tagging as the collective effort of creating a mental map summarizing an information space. They suggest that users can benefit from this map as a navigational aid for efficiently exploring the information space. This idea is very similar to the idea of using hierarchies induced from folksonomies as background knowledge for modeling tag-based navigation with decentralized search or Markov models. With their work, the authors address the vocabulary problem— the ability of users to efficiently assign tags to resources, thus, to create mental maps of the information space. In their analysis, Chi and Mytkowicz calculated three information-theoretic measures: (i) entropy—a measure of uncertainty in a random variable, (ii) conditional entropy—a measure of the remaining entropy of a random variable given the that the value of the second random variable is known and (iii) mutual information—a symmetric measure of the independence of two random variables. Chi and Mytkowicz applied these measures to a Delicious folksonomy and found out that over time as the social tagging systems mature (i) the tagging efficiency is decreasing, thus, tags lose their descriptiveness, (ii) tags lose their ability to deliver conspicuous navigability and (iii) there is a decaying ability of the users to navigate between tags and resources.

6.3   Information Foraging Perspective

Pirolli and Card proposed the information foraging theory to describe the human information seeking in a digital environment [69]. In subsequent work, the original theory has been adapted to model user navigation on the Web [27,70] and to an elementary social information foraging model (SIF) [68].

---

[6] http://www.kde.cs.uni-kassel.de/ws/dc09/.

In their work, the authors establish the SIF model as a useful mathematical tool for studying how social collaboration influences information foraging. SIF assumes that in the process of information foraging, *hints* (tags) are created and shared between individuals which makes navigation easier and improves the organization of the information space. The model presents a perspective that goes hand in hand and complements the perspectives presented earlier in this section. It also captures different aspects of the collective information foraging, i.e., time, diversity of the hints, cost of cooperation and social capital of the collaborators. SIF considers the effectiveness of the hints—tightly related to the entropy and mutual information analysis from the information theoretic perspective—in terms of their amount, validity, and interpretability by the forager in a certain step during navigation. According to the model, lowering the cost of effort associated with creating and sharing tags leads to higher productivity of the users of a social tagging systems. Click2Tag [43], a tagging technique that follows this idea, is shown to have lower tagging costs compared to the widely adopted "type-to-tag" approach.

6.4   Tagging vs. Library Approach

Macgregor and McCulloch discussed the advantages and disadvantages of the "tagging approach" (resources are annotated by users with freely chosen tags) and the "library approach" (resources are annotated by users with predefined controlled vocabulary) [61]. They proposed a definition of a controlled vocabulary and compared unrestricted free-form vocabularies emerged in social tagging systems to controlled vocabularies. Macgregor and McCulloch pointed out that controlled vocabularies have advantages in dealing with synonyms and homonyms, thus, provide good semantic clues. Compared to free-form vocabularies that exhibit a lot of noise introduced by the users, controlled vocabularies can handle lexical anomalies. Macgregor and McCulloch concluded that the precision and the recall of free-form vocabularies depend on the distribution of the tags. Consequently, general tags exhibit high recall and suffer precision, whereas specific tags suffer recall and enjoy precision. Heymann et al. made also a comparison of the navigational characteristics of the "tagging approach" and the "library approach" represented by tagging distributions and library terms distributions, respectively [42]. In their comparison, the authors focused on three major large scale organizational features of the tagging and library approaches: consistency, i.e., ability to deal with synonyms, quality, i.e., with respect to tag distributions and completeness, i.e., correspondence between tag and library terms. These organizational features give a different perspective to the "vocabulary problem" addressed from the information theoretic perspective by Chi and Mytkowicz. Their results suggest that tagging systems tend to be at least to some extent consistent, of high quality and complete. They found that: (i) synonyms are not problematic, (ii) moderately common user tags are perceived as even more helpful than library annotations assigned by an expert and (iii) top tags correspond to library terms.

7   Conclusion

In this chapter, we discussed the challenges researchers faced with the emergence of social tagging systems offering a free-form annotation of resources using tags. First, we presented the user interfaces that allow tag-based navigation, i.e., tag clouds and tag hierarchies. We paid special attention to the tag clouds as the most common interface that accounts for the unstructured nature of tagging data. We reviewed literature focusing on tag cloud usefulness, layouts and evaluation. Furthermore, we discussed trend visualizations with tag clouds. We also presented how tag clouds have been integrated into more complex user interfaces. We summarized the most popular state of the art algorithms for tag clustering used to populate, e.g., a tag cloud or to create a tag hierarchy. Lastly, we showed how tag-based navigation have been modeled and provided an overview of the different theoretic perspectives regarding the ability of folksonomies to support tag-based navigation, i.e., the network theoretic, the information foraging and entropy of information perspectives, and the "tagging approach" vs. "library approach".

References

1. Adamic, L.A., Lukose, R.M., Puniyani, A.R., Huberman, B.A.: Search in power-law networks. Phys. Rev. E **64**(4), 046135 (2001)
2. Aouiche, K., Lemire, D., Godin, R.: Web 2.0 OLAP: from data cubes to tag clouds. CoRR abs/0905.2657 (2009). http://arxiv.org/abs/0905.2657
3. Au Yeung, C., Gibbins, N., Shadbolt, N.: Contextualising tags in collaborative tagging systems. In: Proceedings of the 20th ACM Conference on Hypertext and Hypermedia, HT 2009, pp. 251–260. ACM, New York (2009)
4. Bateman, S., Gutwin, C., Nacenta, M.: Seeing things in the clouds: the effect of visual features on tag cloud selections. In: Proceedings of the Nineteenth ACM Conference on Hypertext and Hypermedia, HT 2008, pp. 193–202. ACM, New York (2008)
5. Begelman, G., Keller, P., Smadja, F.: Automated tag clustering: improving search and exploration in the tag space. In: Collaborative web tagging workshop at WWW 2006, vol. 50, Edinburgh, Scotland (2006)
6. Benz, D., Hotho, A., Stumme, G., Sttzcr, S.: Scmantics made by you and me: self-emerging ontologies can capture the diversity of shared knowledge. In: Proceedings of the 2nd Web Science Conference, WebSci 2010 (2010)
7. Bielenberg, K., Zacher, M.: Groups in social software: utilizing tagging to integrate individual contexts for social navigation (2006)
8. Bogers, T.: Tag-based recommendation. In: Brusilovsky, P., He, D. (eds.) Social Information Access. LNCS, vol. 10100, pp. 441–479. Springer, Cham (2018)
9. Borges, J., Levene, M.: Evaluating variable-length markov chain models for analysis of user web navigation sessions. IEEE Trans. Knowl. Data Eng. **19**(4), 441–452 (2007)
10. Bridle, J.S.: Probabilistic interpretation of feedforward classification network outputs, with relationships to statistical pattern recognition. In: Soulié, F.F., Hérault, J. (eds.) Neurocomputing, vol. 68, pp. 227–236. Springer, Heidelberg (1990). https://doi.org/10.1007/978-3-642-76153-9_28

11. Brooks, C.H., Montanez, N.: Improved annotation of the blogosphere via autotagging and hierarchical clustering. In: Proceedings of the 15th International Conference on World Wide Web, WWW 2006, pp. 625–632. ACM, New York (2006)
12. Candan, K.S., Di Caro, L., Sapino, M.L.: Creating tag hierarchies for effective navigation in social media. In: Proceedings of the 2008 ACM Workshop on Search in Social Media, SSM 2008, pp. 75–82. ACM, New York (2008)
13. Cattuto, C., Schmitz, C., Baldassarri, A., Servedio, V.D., Loreto, V., Hotho, A., Grahl, M., Stumme, G.: Network properties of folksonomies. AI Commun. **20**(4), 245–262 (2007)
14. Chi, E.H., Mytkowicz, T.: Understanding navigability of social tagging systems. In: Proceedings of CHI, vol. 7 (2007)
15. Chi, E.H., Mytkowicz, T.: Understanding the efficiency of social tagging systems using information theory. In: Proceedings of the Nineteenth ACM Conference on Hypertext and Hypermedia, HT 2008, pp. 81–88. ACM, New York (2008)
16. Clauset, A., Moore, C., Newman, M.E.: Hierarchical structure and the prediction of missing links in networks. Nature **453**(7191), 98–101 (2008)
17. Collins, C., Viegas, F.B., Wattenberg, M.: Parallel tag clouds to explore and analyze faceted text corpora. In: IEEE Symposium on Visual Analytics Science and Technology, VAST 2009, pp. 91–98. IEEE (2009)
18. Daw, N.D., O'Doherty, J.P., Dayan, P., Seymour, B., Dolan, R.J.: Cortical substrates for exploratory decisions in humans. Nature **441**(7095), 876–879 (2006)
19. Deshpande, M., Karypis, G.: Selective markov models for predicting web page accesses. ACM Trans. Internet Technol. (TOIT) **4**(2), 163–184 (2004)
20. Dhillon, I.S., Fan, J., Guan, Y.: Efficient clustering of very large document collections. In: Grossman, R.L., Kamath, C., Kegelmeyer, P., Kumar, V., Namburu, R.R. (eds.) Data Mining for Scientific and Engineering Applications. MC, vol. 2, pp. 357–381. Springer, Boston, MA (2001). https://doi.org/10.1007/978-1-4615-1733-7_20
21. Di Caro, L., Candan, K.S., Sapino, M.L.: Using tagflake for condensing navigable tag hierarchies from tag clouds. In: Proceedings of the 14th ACM SIGKDD International Conference on Knowledge Discovery and Data Mining, KDD 2008, pp. 1069–1072. ACM, New York (2008)
22. Dubinko, M., Kumar, R., Magnani, J., Novak, J., Raghavan, P., Tomkins, A.: Visualizing tags over time. ACM Trans. Web **1**(2), 7 (2007)
23. Eda, T., Uchiyama, T., Uchiyama, T., Yoshikawa, M.: Signaling emotion in tagclouds. In: Proceedings of the 18th International Conference on World Wide Web, WWW 2009, pp. 1199–1200. ACM, New York (2009)
24. Fairthorne, R.A.: Content analysis, specification and control. Annu. Rev. Inf. Sci. Technol. **4**, 73–109 (1969)
25. Forgy, E.W.: Cluster analysis of multivariate data: efficiency versus interpretability of classifications. Biometrics **21**, 768–769 (1965)
26. Frey, B.J., Dueck, D.: Clustering by passing messages between data points. Science **315**(5814), 972–976 (2007)
27. Fu, W.T., Pirolli, P.: SNIF-ACT: a cognitive model of user navigation on the world wide web. Hum.-Comput. Interact. **22**(4), 355–412 (2007)
28. Furnas, G.W., Fake, C., von Ahn, L., Schachter, J., Golder, S., Fox, K., Davis, M., Marlow, C., Naaman, M.: Why do tagging systems work? In: CHI 2006 Extended Abstracts on Human Factors in Computing Systems, CHI EA 2006, pp. 36–39. ACM, New York (2006)
29. Furnas, G.W., Landauer, T.K., Gomez, L.M., Dumais, S.T.: The vocabulary problem in human-system communication. Commun. ACM **30**(11), 964–971 (1987)

30. Gambette, P., Véronis, J.: Visualising a text with a tree cloud. In: Locarek-Junge, H., Weihs, C. (eds.) International Federation of Classification Societies Conference, IFCS 2009, pp. 561–569. Springer, Heidelberg (2009). https://doi.org/10.1007/978-3-642-10745-0_61

31. Golder, S.A., Huberman, B.A.: Usage patterns of collaborative tagging systems. J. Inf. Sci. **32**(2), 198–208 (2006)

32. Halvey, M.J., Keane, M.T.: An assessment of tag presentation techniques. In: Proceedings of the 16th International Conference on World Wide Web, WWW 2007, pp. 1313–1314. ACM, New York (2007)

33. Hassan-Montero, Y., Herrero-Solana, V.: Improving tag-clouds as visual information retrieval interfaces. In: International Conference on Multidisciplinary Information Sciences and Technologies, InSciT 2006 (2006)

34. Hearst, M.: Search User Interfaces. Cambridge University Press, Cambridge (2009)

35. Helic, D., Körner, C., Granitzer, M., Strohmaier, M., Trattner, C.: Navigational efficiency of broad vs. narrow folksonomies. In: Proceedings of the 23rd ACM Conference on Hypertext and Social Media, HT 2012, pp. 63–72. ACM, New York (2012)

36. Helic, D., Strohmaier, M.: Building directories for social tagging systems. In: Proceedings of the 20th ACM International Conference on Information and Knowledge Management, CIKM 2011, pp. 525–534. ACM, New York (2011)

37. Helic, D., Strohmaier, M., Granitzer, M., Scherer, R.: Models of human navigation in information networks based on decentralized search. In: Proceedings of the 24th ACM Conference on Hypertext and Social Media, HT 2013, pp. 89–98. ACM, New York (2013). https://doi.org/10.1145/2481492.2481502

38. Helic, D., Strohmaier, M., Trattner, C., Muhr, M., Lerman, K.: Pragmatic evaluation of folksonomies. In: Proceedings of the 20th International Conference on World Wide Web, WWW 2011, pp. 417–426. ACM, New York (2011)

39. Helic, D., Trattner, C., Strohmaier, M., Andrews, K.: On the navigability of social tagging systems. In: 2010 IEEE Second International Conference on Social Computing (SocialCom), pp. 161–168. IEEE (2010)

40. Helic, D., Trattner, C., Strohmaier, M., Andrews, K.: Are tag clouds useful for navigation? A network-theoretic analysis. Int. J. Soc. Comput. Cyber-Phys. Syst. **1**(1), 33–55 (2011)

41. Heymann, P., Garcia-Molina, H.: Collaborative creation of communal hierarchical taxonomies in social tagging systems. Technical report 2006–2010, Stanford University, April 2006

42. Heymann, P., Paepcke, A., Garcia-Molina, H.: Tagging human knowledge. In: Proceedings of the Third ACM International Conference on Web Search and Data Mining, WSDM 2010, pp. 51–60. ACM, New York (2010)

43. Hong, L., Chi, E.H., Budiu, R., Pirolli, P., Nelson, L.: SparTag.us: a low cost tagging system for foraging of web content. In: Proceedings of the Working Conference on Advanced Visual Interfaces, AVI 2008, pp. 65–72. ACM, New York (2008)

44. Hotho, A., Jäschke, R., Schmitz, C., Stumme, G.: Information retrieval in folksonomies: search and ranking. In: Sure, Y., Domingue, J. (eds.) ESWC 2006. LNCS, vol. 4011, pp. 411–426. Springer, Heidelberg (2006). https://doi.org/10.1007/11762256_31

45. Inselberg, A.: The plane with parallel coordinates. Vis. Comput. **1**(2), 69–91 (1985)

46. Jaffe, A., Naaman, M., Tassa, T., Davis, M.: Generating summaries and visualization for large collections of geo-referenced photographs. In: Proceedings of the 8th ACM International Workshop on Multimedia Information Retrieval, MIR 2006, pp. 89–98. ACM, New York (2006)

47. Kammerer, Y., Nairn, R., Pirolli, P., Chi, E.H.: Signpost from the masses: learning effects in an exploratory social tag search browser. In: Proceedings of the SIGCHI Conference on Human Factors in Computing Systems, CHI 2009, pp. 625–634. ACM, New York (2009)

48. Kaser, O., Lemire, D.: Tag-cloud drawing: algorithms for cloud visualization. arXiv preprint arXiv:0703109 (2007)

49. Kleinberg, J.: Navigation in a small world. Nature 406(6798), 845 (2000)

50. Kleinberg, J.: The small-world phenomenon: an algorithmic perspective. In: Proceedings of the Thirty-second Annual ACM Symposium on Theory of Computing, pp. 163–170. ACM (2000)

51. Kleinberg, J.: Small-world phenomena and the dynamics of information. Adv. Neural Inf. Process. syst. 1, 431–438 (2002)

52. Körner, C., Benz, D., Hotho, A., Strohmaier, M., Stumme, G.: Stop thinking, start tagging: tag semantics emerge from collaborative verbosity. In: Proceedings of the 19th International Conference on World Wide Web, WWW 2010, pp. 521–530. ACM, New York (2010)

53. Kuo, B.Y.L., Hentrich, T., Good, B.M., Wilkinson, M.D.: Tag clouds for summarizing web search results. In: Proceedings of the 16th International Conference on World Wide Web, WWW 2007, pp. 1203–1204. ACM, New York (2007)

54. Lancichinetti, A., Fortunato, S., Kertész, J.: Detecting the overlapping and hierarchical community structure in complex networks. New J. Phys. 11(3), 033015 (2009)

55. Lee, B., Riche, N.H., Karlson, A.K., Carpendale, S.: Sparkclouds: visualizing trends in tag clouds. IEEE Trans. Vis. Comput. Graph. 16(6), 1182–1189 (2010)

56. Li, R., Bao, S., Yu, Y., Fei, B., Su, Z.: Towards effective browsing of large scale social annotations. In: Proceedings of the 16th International Conference on World Wide Web, WWW 2007, pp. 943–952. ACM, New York (2007)

57. Li, Z., Tian, J.: Testing the suitability of Markov chains as web usage models. In: Proceedings 27th Annual International Computer Software and Applications Conference, COMPSAC 2003, pp. 356–361. IEEE (2003)

58. Lin, Y.L., Brusilovsky, P., He, D.: Finding cultural heritage images through a dual-perspective navigation framework. Inf. Process. Manag. 52(5), 820–839 (2016)

59. Lloyd, S.: Least squares quantization in PCM. IEEE Trans. Inf. Theory 28(2), 129–137 (1982)

60. Lohmann, S., Ziegler, J., Tetzlaff, L.: Comparison of tag cloud layouts: task-related performance and visual exploration. In: Gross, T., Gulliksen, J., Kotzé, P., Oestreicher, L., Palanque, P., Prates, R.O., Winckler, M. (eds.) INTERACT 2009. LNCS, vol. 5726, pp. 392–404. Springer, Heidelberg (2009). https://doi.org/10.1007/978-3-642-03655-2_43

61. Macgregor, G., McCulloch, E.: Collaborative tagging as a knowledge organisation and resource discovery tool. Libr. Rev. 55(5), 291–300 (2006)

62. Mathes, A.: Folksonomies-cooperative classification and communication through shared metadata (2004). http://www.adammathes.com/academic/computer-mediated-communication/folksonomies.html

63. Miao, G., Tao, S., Cheng, W., Moulic, R., Moser, L.E., Lo, D., Yan, X.: Understanding task-driven information flow in collaborative networks. In: Proceedings of the 21st International Conference on World Wide Web, WWW 2012, pp. 849–858. ACM, New York (2012)

64. Mika, P.: Ontologies are us: a unified model of social networks and semantics. Web Semant. Sci. Serv. Agents World Wide Web 5(1), 5–15 (2007)

65. Millen, D.R., Feinberg, J.: Using social tagging to improve social navigation. In: Workshop on the Social Navigation and Community based Adaptation Technologies (2006)
66. Muchnik, L., Itzhack, R., Solomon, S., Louzoun, Y.: Self-emergence of knowledge trees: extraction of the Wikipedia hierarchies. Phys. Rev. E **76**(1), 016106 (2007)
67. Navarro Bullock, B., Hotho, A., Stumme, G.: Accessing information with tags: search and ranking. In: Brusilovsky, P., He, D. (eds.) Social Information Access. LNCS, vol. 10100, pp. 310–343. Springer, Cham (2018)
68. Pirolli, P.: An elementary social information foraging model. In: Proceedings of the SIGCHI Conference on Human Factors in Computing Systems, CHI 2009, pp. 605–614. ACM, New York (2009)
69. Pirolli, P., Card, S.: Information foraging. Psychol. Rev. **106**(4), 643 (1999)
70. Pirolli, P., Fu, W.-T.: SNIF-ACT: a model of information foraging on the world wide web. In: Brusilovsky, P., Corbett, A., de Rosis, F. (eds.) UM 2003. LNCS (LNAI), vol. 2702, pp. 45–54. Springer, Heidelberg (2003). https://doi.org/10.1007/3-540-44963-9_8
71. Plangprasopchok, A., Lerman, K., Getoor, L.: From saplings to a tree: integrating structured metadata via relational affinity propagation. In: Proceedings of the AAAI Workshop on Statistical Relational AI, July 2010
72. Ramage, D., Heymann, P., Manning, C.D., Garcia-Molina, H.: Clustering the tagged web. In: Proceedings of the Second ACM International Conference on Web Search and Data Mining, WSDM 2009, pp. 54–63. ACM, New York (2009)
73. Rivadeneira, A.W., Gruen, D.M., Muller, M.J., Millen, D.R.: Getting our head in the clouds: toward evaluation studies of tagclouds. In: Proceedings of the SIGCHI Conference on Human Factors in Computing Systems, CHI 2007, pp. 995–998. ACM, New York (2007)
74. Russell, T.: Cloudalicious: Folksonomy over time. In: Proceedings of the 6th ACM/IEEE-CS Joint Conference on Digital Libraries, JCDL 2006, p. 364. ACM, New York (2006)
75. Schmitz, C., Hotho, A., Jäschke, R., Stumme, G.: Mining association rules in folksonomies. In: Batagelj, V., Bock, H.H., Ferligoj, A., Žiberna, A. (eds.) Data Science and Classification, pp. 261–270. Springer, Heidelberg (2006). https://doi.org/10.1007/3-540-34416-0_28
76. Schmitz, P.: Inducing ontology from flickr tags. In: Collaborative Web Tagging Workshop at WWW 2006, vol. 50, Edinburgh, Scotland (2006)
77. Seifert, C., Kump, B., Kienreich, W., Granitzer, G., Granitzer, M.: On the beauty and usability of tag clouds. In: 12th International Conference Information Visualisation, 2008, IV 2008, pp. 17–25. IEEE (2008)
78. Sinclair, J., Cardew-Hall, M.: The folksonomy tag cloud: when is it useful? J. Inf. Sci. **34**(1), 15–29 (2008)
79. Singer, P., Helic, D., Hotho, A., Strohmaier, M.: Hyptrails: a bayesian approach for comparing hypotheses about human trails on the web. In: Proceedings of the 24th International Conference on World Wide Web, International World Wide Web Conferences Steering Committee, Republic and Canton of Geneva, Switzerland WWW 2015, pp. 1003–1013 (2015)
80. Skoutas, D., Alrifai, M.: Tag clouds revisited. In: Proceedings of the 20th ACM International Conference on Information and Knowledge Management, CIKM 2011, pp. 221–230. ACM, New York (2011)
81. Specia, L., Motta, E.: Integrating folksonomies with the semantic web. In: Franconi, E., Kifer, M., May, W. (eds.) ESWC 2007. LNCS, vol. 4519, pp. 624–639. Springer, Heidelberg (2007). https://doi.org/10.1007/978-3-540-72667-8_44

82. Trattner, C., Lin, Y., Parra, D., Yue, Z., Real, W., Brusilovsky, P.: Evaluating tag-based information access in image collections. In: Proceedings of the 23rd ACM Conference on Hypertext and Social Media, HT 2012, pp. 113–122. ACM, New York (2012)
83. Tufte, E.R.: Beautiful Evidence. Graphis Press, New York City (2006)
84. Venetis, P., Koutrika, G., Garcia-Molina, H.: On the selection of tags for tag clouds. In: Proceedings of the Fourth ACM International Conference on Web Search and Data Mining, WSDM 2011, pp. 835–844. ACM, New York (2011)
85. Viegas, F.B., Wattenberg, M., Feinberg, J.: Participatory visualization with wordle. IEEE Trans. Vis. Comput. Graph. **15**(6), 1137–1144 (2009)
86. Vig, J., Sen, S.: Computing the tag genome. Technical report, 10 September 2010
87. Vig, J., Sen, S., Riedl, J.: Navigating the tag genome. In: Proceedings of the 16th International Conference on Intelligent User Interfaces, IUI 2011, pp. 93–102. ACM, New York (2011)
88. Vig, J., Sen, S., Riedl, J.: The tag genome: encoding community knowledge to support novel interaction. ACM Trans. Interact. Intell. Syst. (TiiS) **2**(3), 13 (2012)
89. Wagner, C., Singer, P., Strohmaier, M., Huberman, B.A.: Semantic stability in social tagging streams. In: Proceedings of the 23rd International Conference on World Wide Web, WWW 2014, pp. 735–746. ACM, New York (2014). https://doi.org/10.1145/2566486.2567979
90. Watts, D.J., Dodds, P.S., Newman, M.E.: Identity and search in social networks. Science **296**(5571), 1302–1305 (2002)
91. Watts, D.J., Strogatz, S.H.: Collective dynamics of small-worldnetworks. Nature **393**(6684), 440–442 (1998)
92. West, R., Leskovec, J.: Human wayfinding in information networks. In: Proceedings of the 21st International Conference on World Wide Web, WWW 2012, pp. 619–628. ACM, New York, (2012)
93. Zhong, S.: Efficient online spherical k-means clustering. In: Proceedings, 2005 IEEE International Joint Conference on Neural Networks, IJCNN 2005, vol. 5, pp. 3180–3185. IEEE (2005)
94. Zubiaga, A.: Enhancing navigation on wikipedia with social tags. CoRR abs/1202.5469 (2012). http://arxiv.org/abs/1202.5469
95. Zubiaga, A., García-Plaza, A.P., Fresno, V., Martínez, R.: Content-based clustering for tag cloud visualization. In: Proceedings of the 2009 International Conference on Advances in Social Network Analysis and Mining, ASONAM 2009, pp. 316–319. IEEE Computer Society, Washington (2009)

7
Social Search

Peter Brusilovsky[1]([⊠])(iD), Barry Smyth[2], and Bracha Shapira[3]

[1] School of Computing and Information, University of Pittsburgh,
Pittsburgh, PA 15260, USA
peterb@pitt.edu
[2] Insight Centre for Data Analytics School of Computer Science,
University College Dublin, Dublin, Ireland
barry.smyth@ucd.ie
[3] Department of software and Information Systems Engineering,
Ben-Gurion University of the Negev, 84105 Beer-Sheva, Israel
bshapira@bgu.ac.il

Abstract. Today, most people find what they are looking for online by using search engines such as Google, Bing, or Baidu. Modern web search engines have evolved from their roots in information retrieval to developing new ways to cope with the unique nature of web search. In this chapter, we review recent research that aims to make search a more social activity by combining readily available social signals with various strategies for using these signals to influence or adapt more conventional search results. The chapter begins by framing the social search landscape in terms of the sources of data available and the ways in which this can be leveraged before, during, and after search. This includes a number of detailed case studies that serve to mark important milestones in the evolution of social search research and practice.

1 Introduction

Search-based information access is the most popular way for people to locate information online, and the text-based query-box and search button have become a ubiquitous user interface component across the gamut of operating systems and apps. Usually when we have a general information need, we start with a query and a search engine, such as Google or Bing, and simple browsing or recommendation engines play a secondary role. Search-based information access has a long research heritage that began with early work on information retrieval during the early 1970s [169]. Somewhat surprisingly, despite at least 40 years of active research and practice, the world of information retrieval and web search has largely focused on a single-user embodiment that still informs modern search engines. By and large, modern search is framed as a single-user activity in which a lone searcher conducts their search in isolation from others, despite the fact that other users may experience similar information needs, or may have satisfied the same or similar needs in the past. Given this, why shouldn't search be viewed as

© Springer International Publishing AG, part of Springer Nature 2018
P. Brusilovsky and D. He (Eds.): Social Information Access, LNCS 10100, pp. 213–276, 2018.
https://doi.org/10.1007/978-3-319-90092-6_7

a more social or collaborative enterprise by harnessing the past work or present needs of others? While a few visionary projects from Memex [39] to Super-book [165] have suggested some ideas about how to leverage the power of community for information finding, it was not until the second part of 1990 that this idea of a more social approach to search—what we will refer to as "social search"—enjoyed a sustained and systematic interest from the research community.

In this chapter, we adopt the major theme of this book by defining social search as any approach to information search that harnesses the information access patterns of other users, whether past or present. One way to achieve this is to leverage the query and selection traces of past searchers to help future searchers to find what they are looking for, either by helping them to formulate better queries or to identify more promising leads by promoting more relevant content. Before the age of the web, the opportunities for collecting sufficient volume of user "search traces" were limited. This changed as the world of the web became the world of search. Today, the volume of queries and selections that feed modern search engines provides a unique resource for social search endeavors. In fact, as search engines continuously look for new ways to improve their performance, such data doesn't just enable social search, but provides an appealing opportunity to radically transform how we search. By 1997, the search engines of the day were already collecting increasing volumes of queries and clicks from their users. Even at this early stage, the Web had outgrown traditional query-based, term-matching, information-retrieval techniques as a viable solution to the challenges of general web search. This led to a explosion of creative ideas, by exploring several kinds of novel search data and user traces to address pressing problems such as query formulation [58,74,85,163] and document ranking [36,53,57,217].

Indeed, the "big idea" for modern search emerged from this early work, as several researchers independently highlighted the power of various forms of search traces – what we might now refer to as social, community, or crowd data – to fundamentally change web search from its information-retrieval origins. Most influential were two novel search engines: Google [36], which leveraged community page linking for better ranking; and DirectHit [202], which leveraged user queries and page exploration behavior for query term suggestion [58] and ranking [57]. Over the next 15 years, this stream of work gradually became an important and active research direction at the crossroads of information retrieval, social computing, and intelligent interfaces. The work produced over these years explored a much larger variety of social traces and suggested many novel ideas for using social information to enhance search.

This chapter attempts to provide a systematic overview of social search as a research stream, with a comprehensive account of the core ideas that underpin modern social search approaches and by highlighting social search in practice through a variety of case studies. It complements two other chapters in this book that offer a deeper analysis of two specific areas of social search: Chap. 8 on network-based social search [92] and Chap. 9 on using tags for social search [149]. The chapter is structured as follows. The next section attempts to frame social search to define key dimensions for its classification. We position social search in a broader context, compare and contrast various definitions of social search, identify key sources of information for social search, and elucidate the main stages of the information retrieval process where social information can be leveraged. The subsequent sections go on to review the main approaches for using social data in each of the identified stages. Following that, we examine the problem of making social search more personalized with a review of a number of cases studies to demonstrate different ways of addressing the personalization challenge. We conclude with a final review of two case studies that attempt to go beyond the usual borders of social search by connecting social search with other forms of social information access approaches reviewed in this book.

2   Framing Social Search

In this chapter, we attempt to frame social search as a unique approach to web search and information discovery, which sits at the intersection between information retrieval, social computing, and intelligent interfaces. We pay particular attention to the sources of information that drive social search and the social search process. But first, it is useful to review the evolution of search and, in particular, to distinguish between two common viewpoints: using sources of social data to support search versus searching social data sources.

2.1   Defining Social Search

A useful working definition for *social search*, albeit a very broad one, is that it refers to *a group of approaches that use past user behavior to assist current users in finding information that satisfies a specific information need.* Such an information need is typically represented by a query or question and is satisfied by the retrieval of some unit of information, which is typically (but not limited to) a web page. This definition follows the original understanding of the term social search, which emerged from several early papers between 1997 and 2007 during the first decade of research on this topic [8,37,74,76,77,102,140]. This perspective can be also considered as a useful match with terminology used in other well-defined types of social information access, such as social navigation and social bookmarking, although with a different emphasis on how information needs are expressed or satisfied.

It is important to note, however, how the increased popularity of social media and social networking in the decade that followed has encouraged a range of other definitions of social search, many of which differ from one another as well as from its original meaning. Most of these recent definitions framed social search as search within social media content or data. For example, Carmel et al. [41] use the term social search "to describe the search process over 'social' data gathered from Web 2.0 applications, such as social bookmarking systems, wikis, blogs, forums, social network sites (SNSs), and many others". Similarly, Wikipedia defines social search as "a behavior of retrieving and searching on a social searching engine that mainly searches user-generated content such as news, videos, and images related search queries on social media like Facebook, Twitter, Instagram, and Flickr" [203].

On the other hand, Bao et al. [21], who were among the first to use the term "social search" in the context of social media, defined it as "utilizing social annotations for better web search". Amitay et al. [13] also understand social search as "leveraging social information to enhance search results" where by social information authors mean information collected by social systems such social bookmarking and networking services. McDonnell and Shiri [137] define social search as "the use of social media to assist in finding information on the internet". Evans and Chi [67,68] expand this understanding of social search further: "Social search is an umbrella term used to describe search acts that make use of social interactions with others. These interactions may be explicit or implicit, co-located or remote, synchronous or asynchronous."

The first group of these definitions focus on *searching social data sources* and, as such, are somewhat orthogonal to our understanding of social search, since they distinguish social search by the *type of content* to be searched (i.e., social media contents). The second group emphasizes *the use of social data during search*, regardless of what is being searched, and is fully consistent with the meaning of social search used in this paper. At the same time, this definition narrows the scope of the "social data" that can be used to improve search to those data collected by social systems. In contrast, for the purpose of this work, our definition of social search refers to approaches that can improve search using *all* kinds of information traces left by past users/searchers, both inside and outside social media systems. As will be explained below, this chapter considers a variety of past information traces that range from queries and clicks, to votes and tags, to comments and social links, and everything in between. The common denominator is the use of any and all information left, or contributed by, users as a side effect of their natural information-seeking behaviors.

2.2 Sources of Social Information

Following the goal of the book, this chapter attempts to show how search processes can be improved using various kinds of explicit and implicit social information, and specifically traces left by previous users. The introductory chapter of this book [38] offered a classification of such traces and where they might be sourced. Here we focus on a subset of these traces: those that have been found to be useful for users and that have been used to explore search systems. Below, we review the main sources of social information for social search, along with the types of social information that is usually available within these sources. These sources provide one of the two primary dimensions in our classification of social search techniques.

2.2.1 Search Engine Logs

Search engines routinely collect query session logs. A typical query session log includes queries issued by the users and the result pages selected, or "clicked", by the user on the search engine result page (SERP). More sophisticated logging may include additional data, such as dwell time, within-page behavior, and even post-query browsing. Either way, queries and clicks are perhaps the most important and popular kinds of social information, which can be used alone or in concert to guide social search.

2.2.2 Information Links

User-generated links, created by content authors between information items, such as Web pages, are another important source of community wisdom. Web page authors make extensive use of hyperlinks to create the webs of content that underpin the world wide web. And these links have served as one of the earliest sources of social information used to improve the search process [36,84]. Moreover, beyond traditional "open Web" and its hyperlinks, social search systems now use links created by users of the many and varied Web-based systems the exist today, such as wikis, blogs, and micro-blogs, for example.

2.2.3 Browsing Trails

Surfers of the "open Web" and other Web-based systems leave extensive browsing trails as they travel from content to content. Within a single Web site, clicked links are also routinely collected in web logs and provide yet another source of information-seeking behavior. Beyond these routine browsing traces, traces of browsing behavior can also be accumulated by using various browsing agents [129], proxy-based intermediaries [23] or browser plugins, such as Alexa or HeyStaks [181]. For the interested reader, a comprehensive review of user tracking approaches beyond single-site borders can be found in [83].

2.2.4 Annotation and Comments

User annotations and comments are examples of the "secondary" information content that can accumulate to augment primary content such as Web pages or product descriptions. In a Web context, annotation functionality is provided by various Web annotation systems including the original Annotea project from the WWW Consortium [108]. Indeed, today user-generated comments and reviews are commonplace across most e-commerce sites and have proven to be a powerful source of collective user opinion, helping to guide many users as they consider their choices among an ocean of products and services, from Amazon[1] to Yelp[2].

2.2.5 Curation Systems

We use the term *curation system* to distinguish several types of systems where users collectively collaborate to collect and organize content for future consumption. This includes hierarchically organized directories, such as the Open Directory Project (dmoz.org), guided path systems [81], and page grouping systems [2,177]. Most important among the curation systems, from the perspective of Web search, are social bookmarking systems, which allow their users to openly

---

[1] http://amazon.com.
[2] http://yelp.com.

share various information items (Web pages, photographs, research papers) while providing textual *comments* and *tag-based* annotations.

2.2.6   Blogs and Microblogs

Conventional blogs, as well as micro-blog systems such as Twitter, have emerged as important contemporary sources of primary user-generated content, and are now increasingly used by search systems. Just like in other contexts, an explicit reference to a Web page (i.e, a Web link) or another item (i.e., a movie) in a blog post or a tweet is a signal of its social value, with surrounding text providing a context for mentioning the link.

2.2.7   Social Links

Social links—such as the friendship connections between social network users—carry important information about relationships between users. There are several examples of social search systems that use direct links of various kinds (trust, friendship, etc.), as well as indirect links, such as those formed by the users who have joined the same social group. Social links originate in the social linking systems—originally systems like Friendster and MySpace, and today, platforms like Facebook and LinkedIn—but today the idea of a social link has evolved beyond a simple friendship connection. For example, Twitter allows user to follow and mention other users, while social bookmarking systems usually support "watch" links that help to keep interested users informed about changes and updates.

2.3   The Social Search Process

While conventional search might appear to be an atomic, mostly single-shot process—"submit a query, get results"—it is naturally comprised of several distinct stages. These various all stand to benefit in different ways from different kinds of social information. To better organize our review of social search, this section presents an "anatomy" of the social search process as a sequence of actions to be performed by both the user and the search system.

A sensible and straightforward decomposition of search identifies three obvious stages: "before search", "during search", and "after search". This decomposition was used in the past for the analysis of personalized [139] and social search [68]. It remains useful here because it serves to separate social search enhancements into those that might be provided as *external* supports (before or after search) and therefore potentially provided as third-party enhancements, versus those than are contrived as *internal* supports (during search), and therefore more naturally provided by search engines themselves.

2.3.1   Before Search

This stage includes all actions that must be performed before a query meets the search engine. On the search engine side, it includes resource discovery (known as *crawling* in Web search context) and indexing. On the user side, it includes query formulation and query expansion. The reason to separate query formulation from query expansion is to emphasize how these processes differ from a user engagement perspective. Query formulation is a user-centered stage during which a search support system can offer valuable assistance. Query expansion is conceived of as a fully automatic process that attempts to improve the query after it is formulated by the user, but before it is received by the search engine.

2.3.2 During Search

This stage includes two related processes performed by the search engine: matching items to the query (and retrieving relevant items) and ranking these items. To distinguish the ranking produced by the search engine from various external or secondary re-ranking approaches, we refer to the former as *primary item ranking*; the latter will usually be referred to as *re-ranking*.

2.3.3 After Search

This stage includes several types of search enhancements that can be implemented after obtaining an original ranked list of search results in order to produce a better search engine results page. Historically, these enhancements were produced by "third party" systems, which took the list of results produced by a search engine and returned it in an enhanced form. However, nowadays, some of these enhancements are done by the search engines themselves, as shown in Table 1. The most popular of these enhancements are various kinds of item *re-ranking and recommendation* that use social or personal information to change or augment the primary ranking; e.g. [50,77,180,183]. Once the order of presentation is determined, a final *item presentation* needs to be generated, including an item summary or *snippet* to provide the user with a synopsis of the item.

Table 1. Stages of (social) search process

	Add-ons	Engines
I Before search		
1. Resource discovery and indexing		✓
2. Query formulation	✓	✓
3. Query expansion	✓	
II During search		✓
4. Matching items to query		✓
5. Primary item ranking		✓
III After search (SERP)		
6. Item re-ranking and recommendation	✓	
7. Item presentation (snippet building)	✓	✓
8. Item augmentation	✓	✓
9. After SERP support	✓	

The presentation of individual items may be also *augmented* to emphasize their social or personal relevance. While this concludes the SERP presentation, some projects have explored opportunities to use social information *after SERP*, i.e., to support the user after a specific SERP item has been accessed, as we discuss later in this chapter.

2.4   The Big Picture View

To summarize, these core stages of the search process are a useful way to organize our review of social search technologies. The following seven sections (3 to 9) review the use of social information to support each step of the search process listed in Table 1. The unique needs of each step help to frame the approaches taken and the types of social information used. Moreover, by grouping social search approaches by stages, we can clarify their commonalities and connections with their underlying social data sources. To better understand this connection, we also offer Table 2 that groups together approaches, which support the same step of the search process while using the same social information source.

Table 2. Stages of social search and their support using various sources of social information

Source of social information	Query formulation	Query expansion	Indexing	Primary ranking	Re-ranking and recommendation	Presentation	Augmentation	After SERP
Search engine logs	[19, 22, 58, 60, 75, 85, 101, 107, 121, 127, 132, 184, 192]	[25, 56, 74, 101, 120, 196]	[14, 61, 80, 82, 87, 114, 151, 159, 170, 171, 209]	[5, 28, 36, 46, 54, 55, 57, 63, 64, 82, 89, 104, 106, 112, 158, 160, 162, 176, 193, 209]	[32–35, 138, 176, 178–180, 183, 188, 200, 201]	[9, 30, 31, 170]	[50, 51, 124, 170, 195]	[9, 48, 50, 51, 72, 76]
Browsing trails		[120]	[215]	[16, 24, 65, 90, 130, 131, 190, 210, 216]			[8, 37, 48, 70, 72, 76]	[48, 72, 76]
Information links		[15, 52, 59, 119]	[44, 53, 78, 87, 117, 151, 153, 198, 215]	[36, 43, 44, 53, 78, 84, 115, 117, 153, 160, 198, 205, 215]				
Annotations and comments			[61, 165]		[145, 146]		[8, 37, 70, 72, 76]	
Curation systems		[4, 128, 207, 213]	[21, 26, 40, 47, 87, 97, 150, 151, 164, 208, 214]	[1, 21, 40, 100, 126, 206, 212, 214]	[103, 150, 208, 211]		[13, 109, 142, 211]	
Blogs and microblogs			[11, 45, 62, 66, 87, 123, 144, 157, 168, 191]	[7, 87, 136]	[7, 157, 168]		[13, 157]	
Social links	[20]			[189]	[77, 95, 176, 178–180]	[9, 30, 31]	[72, 73, 76, 122, 148, 155, 173]	

In the context of this chapter, Table 2 serves as a "big picture" that provides an overview of the whole social search area and highlighting both well-explored areas and "white spots" on the map of social search. In our review and the organization of Table 2 we mostly follow the order of the search process listed Table 1. One exception is the Sect. 5, where we review together tightly interrelated social indexing and matching, which belong to different stages in our classification. This section is placed between query expansion and ranking, where its matching aspect belongs.

3 Query Formulation, Elaboration, and Recommendation

Web search is challenging, not only because of the sheer scale of information that exists online, but also because of the simple lack of information retrieval expertise that most searchers exhibit. For example, it is well documented that most searchers begin their searches with vague or under-specified queries. Early work on web search established that most queries contain only 2–3 terms [186, 187, 204] and rarely offer a clear account of the searcher's real information needs. At the same time, the advanced search features offered by modern search engines (e.g. boolean operators, positional switches, filters etc.) are seldom if ever used; see, for example, [185, 187]. In short, a typical search query is usually woefully incomplete when it comes to the specification of a searcher's real information needs.

In response to this, there has been a significant amount of work undertaken on how to better support searchers by, for example, elaborating vague queries or recommending better ones. As a source of social information for query elaboration and recommendation, most systems use search engine logs. The idea of using such logs for accumulating successful queries was originally suggested by [163], and the first simple approach for using search logs to suggest new terms to narrow a search was patented by DirectHit [58]. The first example of a full-scale query recommendation service based on accumulated queries was described in [85] and was followed by a large body of work on query recommendation [19, 22, 60, 75, 101, 121, 132, 184, 192].

One of the early approaches to query recommendation was the mining of query logs to identify clusters of related queries [17, 19, 156, 197]. For example, in [19] the clustering process was based on the queries, and the terms contained in selected URLs and term-based similarity can be used to rank and recommend a set of queries that are similar to a particular target query. Another popular approach was based on building query networks using various techniques for query association [18, 85, 127, 132, 156, 184]. Yet another approach [22] borrows ideas from recommender systems research to apply collaborative filtering techniques for query recommendations, based on search sessions that contain successful queries.

The approaches reviewed above, and many further variations that exist, rely solely on submitted queries and result selections, and as a result, are less satisfactory as examples of social search, because they lack any real social context other than the set of users who happened to use a particular search engine. The work of Balfe and Smyth [20] serves as a more tangible social counterpoint, in this regard, by applying query-log analysis and recommendation techniques in an early collaborative search setting. They describe a novel query recommendation technique that suggest a queries that have proven to be successful during past search sessions *within a community of related searchers.* In particular, they propose a novel recommendation ranking metric that prioritizes queries for recommendation that are based on their relevance and coverage characteristics, preferring queries that are not only relevant to the user's likely information need—but that also have the potential to cover a broad set of variations of these needs.

4   Query Expansion

One of the most natural and successful techniques to resolve the query-mismatch problem is to expand the original query with other query terms to better capture the actual user intent or to produce a query that is more likely to retrieve relevant documents. Automatic query expansion (AQE) has been studied since 1960, when it was suggested by Maron and Kuhns [133]. AQE is applied to the original query, as submitted by the user. This query is expanded to include additional terms and is then submitted to the search engine to drive search, and hopefully to retrieve better results. Thus, from the perspective of the user, AQE is a type of query formulation technique that is transparent, in that it is performed in the background in a manner that is typically unseen by the user.

Many techniques have been examined for AQE, such as relevance feedback [167] and term distribution analysis [161]. For many years, until relatively recently, the practical benefits of AQE were questioned in web search, as results mainly benefited from improved recall but with a cost to precision [94]; not an ideal combination for most web search tasks.

Recently, and largely due to the huge amount of available data and the low quality of user queries, AQE has been revisited. New AQE techniques have been presented that use new data sources and employ more sophisticated methods for finding new features that can correlate with the query terms. AQE is now regaining popularity as a promising method and has been adopted in various commercial applications, such as intranet and desktop search or domain-specific search engines [42]. However, it is still not commonly employed in major Web search engines, mainly due to their emphasis on precision, but also due to the need for fast response times, which preclude the costly computations that most AQE methods require.

A recent survey [42] classifies AQE techniques into five main groups according to the conceptual paradigm used for finding the expansion features: linguistic methods, corpus-specific statistical approaches, query-specific statistical approaches, search log analysis, and Web data. The linguistic approaches examine global language properties of the query, while the corpus-specific and query-specific techniques analyze the entire database or the query context to identify correlated word features to use for expansion. The search log and web data analysis approaches are more social in nature and are deserving of further discussion.

The use of search logs for AQE was originally suggested in [74]. This paper was one of the first to recognize search improvement based on social data as a type of "social searching". Search logs typically contain the queries submitted by users, alongwith result URLs that were selected by the searchers. By mining these query-result associations, AQE techniques can, for example, predict the relevance of selected URLs to a given query, as in [175]. Expansion terms can be extracted from the top results, or the clicked results, or from documents that are frequently returned for similar queries [56,74]. The underlying assumption is that users who click on results usually do not make their choice randomly, but rather select results rationally, based on their perceived relevance to their information need. Even if some of the clicks are erroneous or mistaken, query logs remain a valuable source for automatic query expansion, due to the sheer volume of query and click data that mainstream search engines can provide.

Some systems, such as [56], investigate the association between query terms and the documents that were retrieved for the query during a specific search interaction. Their method is based on probabilistic term correlations that are computed by linking query terms to terms in documents that were retrieved in the same session. When a new query is submitted, for every query term, all correlated document terms are selected based on the probability that was pre-computed between them. For every query, the top-ranked terms, by their probability, can be selected as expansion terms. Another example is described in [25]. The authors suggest and compare some variations of associating queries extracted from a freely available large query log with documents of the corpus that is being searched. The association between pairs of documents from the corpus, and a query from the log, is based on close statistical matches between them. For each document in the corpus, query associations are maintained as document surrogates. Then, the terms of the associated queries are used in the search process, either as expansion terms, or as surrogates for document ranking. The authors describe experiments with almost one million prior query associations, which improved the effectiveness of queries by 26%29%, and showed that their method outperformed a conventional query expansion approach using the corpus of past queries.

Other query log analysis methods for AQE look at the association within query terms in the log. For example [196] analyzes term co-occurrence in queries, to discover association patterns for terms; more specifically, term substitution and addition within multi-word queries. The authors look for terms that can syntactically substitute for other terms (e.g. "auto" - "car"; or "yahoo" - "google") and terms that often occur together (e.g. "car" - "insurance"). They frame the term association pattern mining as a probability estimation problem, so that high-probability patterns are used for query expansion and reformulation. Similarly, [101] also look at the association between terms that appear in queries, but rather than analyzing their co-occurrence in a single query, they consider the entire user's query session to infer term associations across sequences of related queries. Query logs can be also combined with other social data for the purpose of AQE. For example, Kramár et al. [120] built communities of like-minded users by mining user browsing behavior and then used query analysis within a user community to produce more user-sensitive query expansion.

Another AQE technique, based on social data, harnesses *Web data sources* – such as web pages, wikis, blogs, and others – for term expansion, especially since query logs are not always available for the developers of AQE approaches, who may be external to the operators of a given search engine. A large variety of external linguistic sources can be used for AQE. One type (not related to social search) is expert-generated data, such as WordNet [143,213]. However, various kinds of user-generated content, from regular Web pages to Wikipedia, is also a popular choice for AQE. Such data presumably reflects the perception of users about relevant terms and the semantic relationships between their terms.

The anchor text inside Web pages have also been used for AQE and specifically for query refinement (via expansion or substitution of terms) as suggested by Kraft and Zien [119], and as an alternative to query log analysis. Kraft and Zien [119] developed a simple and elegant approach to automatic query refinement using anchor text and median rank aggregation. Briefly, the query is matched with similar anchor texts and additional terms from the matching anchor texts are then used for query expansion. This is based on the notion that anchor text is an strong surrogate for the document to which a corresponding link connects. While this is a fairly sound assumption to make, a number of challenges exist when it comes to using this approach in practice. For example, popular queries may match too many anchor texts, and some anchor texts might be useless if they are automatically generate – rather than hand-coded by the Web page creator – a ranking algorithm is required to assess the utility of any matching anchor texts, to focus on those that are most likely to act as a source of high-quality additional terms. Kraft and Zien [119] used factors such as the weighted number of occurrences of an anchor text, the number of terms in the anchor text, and the number of characters in the anchor text, to rank the anchor texts. Their experiments successfully demonstrated how mining anchor text for query refinement in this way was capable of outperforming similar methods using document collections.

Following on from this work, Dang and Croft [59] constructed an anchor log from the anchor texts in a web test collection. They compared their log-based query reformulation techniques, using both the anchor log and a real query log, and showed that the former produces results that are at least as effective as the latter. Their anchor log was built from the TREC Gov-2 web collection, which contains 25 million web pages crawled from the .gov domain during early 2004. The integration of this algorithm into a search process was based on providing the user with a list of ranked query refinements, from which the user selected one that best suited their needs.

A more recent study by Craswell et al. [52] extended the work of Dang and Croft [59] by fully automating the formulation process. The authors again used a large anchor graph as a linguistic resource for the query rewriting, but introduced new algorithms based on a random-walk approach, which proved to be capable of offering statistically significant improvements in precision.

Wikipedia, DBpedia, FreeBase, and the like, are also considered as effective sources for AQE [4]. Wikipedia is the largest and most popular collaborative effort of volunteer contributors, and is routinely ranked among the ten most popular websites. It is the leading, and most updated open encyclopedia, and offers wide coverage of a diverse range of topics, events, entities, and more. It is considered to be a reliable data source and has found application in many AQEs. DBpedia [125] is a large-scale multilingual, semantic, knowledge base extracted from Wikipedia info-boxes, and is connected to several other linked data repositories on the Semantic Web. FreeBase was a large collaborative knowledge base consisting of data composed mainly by its community members. The data was collected from various sources, including user-submitted wiki contributions. FreeBase data is now being moved to WikiData. All of these sources have been used to enrich and expand queries with new terms that are inferred as relevant based on the content and/or structure of these other resources. Given a query string, the search engine may identify relevant pages or entities within the source that it uses, such as relevant Wikipedia concepts, and use their titles, categories, or other components to infer relatedness between terms to expand the query. Different approaches rely on different sources of query terms, use different methods to identify relevant information within a source, and rely on different approaches when it comes to selecting and ranking terms for use in AQE.

For example, Arguello et al. [15] used the links that appear in Wikipedia pages, which are relevant to the query string, and the anchor text associated with these links to expand queries. Li et al. [128] used the categories from Wikipedia pages for AQE. Briefly, the starting query is run against a Wikipedia collection; each category is assigned a weight that is proportional to the number of top-ranked articles assigned to it and articles are re-ranked based on the sum of the weights of their corresponding categories.

Elsewhere, Xiong and Callan [207] used FreeBase as their source. They investigated two methods for identifying FreeBase entities that are relevant to a query, and two methods of using these entities for query expansion, by utilizing information derived from Freebase linked entity's description and categories, to select the terms for expansion. Like most of the systems that utilized user generated content for expansion, results are superior to more conventional state of the art AQE methods. According to Xiong and Callan [207], using experiments conducted on the ClueWeb09 dataset with TREC Web Track queries, these methods are almost 30% more effective than state-of-the-art query expansion algorithms. This makes sense: query expansion aims at expanding queries with terms that users would consider relevant in order to enhance the queries; therefore AQE should benefit from harnessing user-generated content for expansion, since it reflects how people view the relationships between terms. A user who assigns a page to a category is signaling that they believe that the topic of the page is related to that category.

5 Social Indexing and Matching

An inherent challenge for all search systems is the surprising degree of variety with which people refer to the same thing. Frequently referred as *the vocabulary problem* [79], it means that people with the same information need are liable to describe it using very different query terms. Sometimes people with different information needs will formulate similar queries. This makes it exceedingly difficult to index content for future retrieval. This problem has been recognized in the early days of automated search based on manual document indexing. Back then, professional indexers were often blamed for conceptualizing a document incorrectly, or at least in a way that was at odds with searchers of the document. The switch to automatic content-based document indexing, which was expected to resolve this problem, served only to highlight the potential for mismatch between the vocabulary used by content producers (i.e, the authors of the item text) and content consumers (i.e., the search system users who are looking for this item).

As early as 1993, Bollmann-Sdorra and Raghavan [27] argued that the structure of the query space is very different from the structure of the document space, and that it makes sense to consider documents and queries to be elements from different term spaces. Cui et al. citecui2002probabilistic attempted to quantify the differences between the document-space and the query-space by measuring the typical similarities between document vectors and query vectors by using two months of query logs from the Microsoft Encarta search engine. They found very low average similarities between queries and document vectors, with the vast majority of similarities at less than 0.4 (using a standard a cosine similarity metric) and a mean similarity value of only 0.28 across the collection as a whole.

The lack of correspondence between the query space and the document space in traditional information retrieval settings is only amplified in Web search, for two basic reasons. First, as mentioned above, Web search queries tend to be short and vague, often including no more than 2 or 3 query terms [204]. This limits the opportunity for overlap between queries and documents. Second, Web content is unusually sensitive to changing trends and fashions, such that documents that were originally produced for one purpose (and indexed accordingly) might later become more relevant in an entirely different context, which may not be captured by their existing index terms.

Fortunately, while exacerbating the potential for vocabulary mismatches, the Web also offers a potential cure: the social information that started to accumulate rapidly on the Web has emerged as a viable bridge capable of connecting content (producer) and query (consumer) spaces. Indeed, certain unique types of social information are contributed mostly by document consumers who typically refer to a document from the consumer perspective, which may change over time. For example, the query terms used by users seeking a particular item may be useful as a means to augment automatically index terms or extracted item content during query-item matching; other forms of social information in addition to query terms may also be used. These approaches are often referred to as social *document expansion* or *social indexing*. A more flexible and, arguably, safer way to achieve this is *social matching*, by separating content indexing from social information related to a document while taking both kinds of information into account at the query/matching time. Over the last 15 years, an increased variety of social information has been used for social indexing and matching. The remaining part of this section briefly reviews the most popular approaches, organizing them by the type of social information they endeavor to leverage.

5.1   Link Anchors

The first kind of social information explored as a source for enhanced document representation was the anchor text associated with document links, which we have discussed in the previous section on query formulation, elaboration, and recommendation. When a link is created from one document to some target document, the link creator will typically use anchor text that references the target document in a particular way, revealing their own perspectives and biases. The intuition is that this anchor text will likely contain terms that will appear in queries for the target document that are submitted by future searchers who share a similar perspective. These terms may be absent from the original target document and therefore absent from its index; for example the linking document may be created at a much later date and serves to place the target document in an entirely new context that was unknown when it was originally created. These new anchor text terms may provide useful and novel indexing terms for the target document.

The use of anchor text in this way was pioneered by McBryan [135] in the World Wide Web Worm. A few years later, Brin and Page [36] used it as one of the foundations of their Google search engine, along with Web link structure, and Chakrabarti et al. [44] used it for compiling a list of authoritative web resources on a given topic. Following that, several research teams demonstrated the value of anchor text as a source of information in the context of site search and known-item Web search [53, 153, 198]. This work attracted broad attention to the use of anchor text and helped the approach to emerge as a mainstream information source for document expansion, which is today used by many (if not all) of the major Web search engines.

The study of anchor text also continued as a research stream. For example, Zhou et al. [215] compared two ways to combine anchor text with Web page content for site-level search: a combined representation with a single ranking and a separate representation with a combined ranking. Fujii [78] separately assessed the value of anchor text for answering informational and navigational queries and showed that anchor-based retrieval is especially effective for navigational queries. Koolen and Kamps [117] demonstrated the value of anchor text during ad-hoc search in sufficiently large Web collections. Ritchie et al. [166] expanded this line of work for research literature search by using the text associated with citations as a form of anchor text to expand document representations. Currently, anchor text is considered to be a gold standard for the social expansion of document representations, and it is typically used as a baseline against which to evaluate alternative social expansion approaches [14, 87, 123, 151].

5.2   Query Logs

The second primary source for social document expansion, indexing, and matching is search engine logs. The idea to use queries for social indexing is a natural one, since queries are, by definition, an expression of a user's information needs, albeit a partial one that uses a particular vocabulary. It is therefore an obvious step to enhance a document representation using query terms [14,114,170,171,209]; or the terms could be used to create an alternative query-based representation [159].

The main problem with this idea is that it is not always straightforward to create a reliable association between documents and queries by using query logs alone. First-generation research on query-document association focused on two natural ideas: associate a query with each of the top-N retrieved documents [163, 170], or with each of the selected documents on a SERP [82,114,151,159,197, 209]. Both approaches were found to be satisfactory for such tasks as document labeling or clustering, but were not able to match the quality of anchor-based document expansion.

A breakthrough in the use of query-based document expansion came once researchers recognized that query logs served to accumulate social feedback across multiple queries and even across multiple search sessions. For instance, users frequently have to go through a sequence of query reformulation, starting with one or more failed queries that provided no satisfactory results, and ending with successful queries that return some satisfactory results. Somewhat surprisingly, it is the failed queries that turned out to provide some of the best terms for document expansion, since these terms indicate the elements of the user's conceptualization of the document that are missing from the document's default representation. Moreover, another insight was that not every selected document truly matches user information needs, since clicks are made on the basis of SERP snippets that rarely give a complete picture of the document in question. Documents accessed at the end of the query session or extensively analyzed documents are more likely to represent the information that the searcher was looking for. The idea of using failed queries for document expansion was originally suggested by Furnas [80] and was based on explicit user feedback. The use of automatic query-based document expansion based on "smart" session mining was suggested by Amitay et al. [14] and independently explored in a few other projects [61,87]. Currently, "smart" mining of query sessions serves as one of the main sources for both social indexing and document ranking. It will be discussed in more detail below.

5.3   Annotations and Tags

While link anchors and queries can be considered as two indirect ways for information consumers to describe their conceptualization of documents, various page annotations and comments offer users a more direct way. The idea to use page annotations for better document indexing was pioneered by the Superbook system [165]. Dmitriev et al. [61] explored the use of annotation for document indexing in a much larger context of intranet search, where document annotations were collected using a special corporate browser, Trevi. This project is notable because it also attempted to incorporate all kinds of available social sources for document indexing, including anchors, query chains, and annotations.

The new stage of work on the use of social annotations for document indexing and search was associated with the rapid rise of social bookmarking systems

[86, 91] in the middle of 2000. Social bookmarking systems, such as Del.icio.us[3] and Flickr[4], allowed users to collect and organize various kinds of resources, such as Web pages and images. Most of the systems allowed their users to add free-text annotations and more structured social tags to the bookmarked resources. The use of both kinds of annotations for document indexing and search within these systems was then quite natural. Indeed, in at least some social systems, such as Flickr, it was the only indexable information.

It was also shown how social annotations could be useful in the broader context of Web search. As early as in 2007, two seminal papers [21, 150] proposed social search engines that used social annotations, accumulated by Del.icio.us, to build enhanced document representations and offer better retrieval and ranking. A direct comparison of queries, anchors, and annotations as a source of index data demonstrated that tags outperform anchors in several retrieval tasks [151]. This work was expanded in a sequence of follow-up studies that examined the idea using larger datasets [47, 97] and experimented with building better tag-based document representations using more sophisticated tag analysis techniques [26, 40, 87, 164, 208, 214]. Since this book considers tag-based social search as a separate social information access approach, we refer the reader to Chap. 9 for a more extensive discussion of this topic [149].

5.4   Blogs and Microblogs

Among other sources of social information that could be used for social indexing, it is also worth mentioning various kinds of blogs, and specifically microblogs, such as Twitter. Many blog and microblog posts include links to Web pages or mentions of other kinds of indexable objects (such as people, events, or products). Naturally, the content of these posts could be used as a source of information for enriching representations of these objects, just as link anchors can be used for regular Web pages. Lee and Croft [123] explored several kinds of social media in the context of Web search and referred to this kind of data as *social anchors* (similarly, Mishne and Lin [144] referred to Twitter text as *twanchor*).

Over the last five years, this source of information has been extensively explored in several contexts and using several kinds of blogs and microblogs. For example, short posts in Blippr [66] were used to build representations of users and movies, while tweets and Facebook posts were used as a source of terms for Web document representation [11, 87, 144, 157, 168, 191] and user profiling [3]. In addition to their text content, which is similar in nature to anchor text and annotations, blog and microblog posts offer an important temporal dimension. The natural tendency of blog posts to refer to recent events and actual documents could be used for just-in-time crawling and document expansion with reference to time [45, 62], which in turn, is valuable for recency-based ranking.

---

[3] https://del.icio.us.
[4] https://flickr.com.

6 Ranking

When information retrieval systems – from classic search systems to modern web search engines – receive a user query, they have to perform two important tasks: (1) retrieve documents that are relevant to a query (this is also known as query-document matching); and (2) rank these documents based on their relevance to the query. While different models of information retrieval offer a variety of approaches to query-document matching, all these approaches are based on some form of document indexing. As a result, from the perspective of using traces of past users, social matching is equivalent to social indexing, which was reviewed above. Ranking, however, and the subject of this section, uses social traces in significantly different ways.

Ranking search results has been always an important problem in information retrieval. The ability to correctly rank search results has been long advocated as a benefit or modern vector and probabilistic models of information retrieval, in comparison with alternative set-based Boolean models [118]. The goal of ranking is to sort results in descending order of their relevance to the user's true information need. Getting this right is critically important, given that users have a limited practical ability, or willingness, to examine and access many results; if the right result isn't near the top of the ranking, then a user is unlikely to be satisfied by the results of their search, and may even abandon their search session altogether. Eye-tracking studies [88,105] have shown that results located near the top of the ranked list have a much higher probability of being noticed and examined by the user. This probability rapidly decreases as we move down the ranking. Indeed, [113] demonstrated how the judgments of searchers are significantly influenced by the rank position of a result, as users tend to inherently trust in the ability of a search engine to correctly rank results by their relevance; so much so that manipulating lists of results to position relevant results at the end of the ranking failed to attract user attention. In other words, it is not sufficient for a search engine just to have good results somewhere in the returned list (as ensured by matching). It is equally important to position the best results at the top of the ranking if the user is to notice and attend to them.

Traditional ranking approaches in both vector and probabilistic models of information retrieval are based solely on the content of documents (e.g. web pages). Similarly, primary ranking in early Web search engines and search tools was fully based on document content. While this approach works relatively well for smaller-scale search systems, the world of web search quickly outgrew this approach. With the rapid growth of the Web, almost every query produced an increasingly large number of matched Web pages. In this situation, content-based ranking routinely failed to identify the most relevant documents. It was a perfect context to start harnessing other sources of information for ranking, including the social information that naturally accumulates from traces left by past Web searchers. As mentioned in the introduction, the idea of using social traces for ranking was pioneered in parallel by two novel search engines of the day: Google [36], which leveraged page links, and DirectHit [202], which leveraged user page exploration behavior [57]. This seminal work encouraged a large volume of follow-up research on "social ranking".

The first two subsections of this section review the use of social traces for the "primary rankings" produced by retrieval systems and engines, as well as search components of various Web systems with the use of data available to these systems. In this respect, primary ranking approaches differ from various re-ranking and other result-promotion approaches produced by third-party systems, which have no access to search engine data, but which leverage other valuable sources of ranking data. These re-ranking approaches are reviewed separately in the following sections.

6.1   Web Links for Primary Ranking

Web links were the first kind of social information leveraged for ranking purposes. The authors of link-based ranking approaches argue that the availability of links is a feature that distinguishes valuable pages and suggest different ways to promote pages with more links in the ranked list. The first approach, suggested by Carriere and Kazman [43], didn't differentiate incoming and outgoing links and used only local link analysis to re-rank results based on the total number of links (both incoming and outgoing). However, it was the other two link-based ranking approaches—PageRank [36] and HITS [84,115]—that demonstrated the potential power of link-based ranking. PageRank and HITS are frequently referred to together because they were introduced almost at the same time and inspired by similar ideas. Both approaches focused on incoming links and argued that incoming links created by multiple authors signify page value or "authority" [115].

The key difference between PageRank and HITS was the kind of link analysis that was performed. PageRank adopted ideas from citation analysis: papers with a larger number of citations (i.e., incoming links) are more important, but citations by more important papers weight more than citations from less important papers. To combine these two basic ideas, PageRank introduced a recursive calculation of page importance (called PageRank) based on a random walk in the whole Web linking graph [36].

HITS was also based on the results of an analysis of the Web hyperlink structure [44] to identify so-called hubs (sources of outgoing links) and authorities (sources of incoming links) in the web graph. However, the network analysis of HITS is query oriented. It is focused on the top ranked search results and their parent and children pages. Both approaches are explained in more detail in Chap. 8 of this book [92], but suffice it to say that PageRank and HITS inspired a generation of network-based ranking approaches whose reach often extended well beyond web search. Some of these approaches, as they relate to social search, are reviewed below with additional coverage in Chaps. 8 [92] and 9 [149] of this book.

Link anchors, the "other side" of Web links, emerged as a popular source of information for ranking at the same time as link topology [36,44]. For example, the Clever system [44] enhanced the original HITS approach by using an analysis of anchor text to differentially weight the contribution of individual links.

The simplest ranking, using anchor text, can be implemented by re-using the regular ranking approach, but using the anchor text instead of the original content [53,117] or the anchor text merged with the original content [215]. Such approaches can be effective in certain settings. For example, Craswell et al. [53] demonstrated how re-ranking based on link anchor text is twice as effective as ranking based on document content for site finding.

More advanced ranking approaches use separate content-based and anchor-based document representations. The main challenge that remains is to find an appropriate way to combine ranking information produced by these two independent sources. One approach is to blend together the independent rankings; see [198,205,215]. Another approach is to incorporate both sources of information into a single probabilistic ranking model; see [78,153,198].

6.2   Query Logs for Primary Ranking

The idea of using query logs for ranking is based on an assumption that a click on a specific link in a ranked list of results, generated by a specific query, is an indication of the *general importance* of a page behind the link as well as its *relevance* to the query. With this assumption in mind, every click can be considered as a vote in favor of the clicked page, and a query log can be treated as a collection of votes that can be used for ranking. The problem is that this voting data is very noisy, for several different reasons. First, an isolated click on a link is not a reliable vote. As mentioned earlier, link selection is usually based on a small result snippet that might be misleading. Second, as shown by the studies mentioned above, a chance to click on a link depends not only of the true relevance of the document, but also to a large extent on its position of the link in the ranked list; as shown in [113], even highly relevant links tend to be ignored if they are ranked at the bottom of a result list. Third, different users might need different information in response to the same query. Thus, using "click votes" from one group of users could harm the ranking for another group.

Due to the noise in the data, the use of query logs for ranking present a considerable practical challenge, as demonstrated by the rather limited success of Direct Hit [202], the first major search engine to use query session clicks as relevance votes for ranking. Direct Hit did attempt to handle the noise in click-through data by accounting for time spent by a user on the actual page, as well as the position of the clicked link in the ranked list [57]. However, its use of query log data did not prevail and could not compete with the link-based ranking system popularized by Google. More recent research using query log-based ranking focused on addressing the problem of noisy click-through data. For example, clustering can be used (query-centered, document-centered, and user-centered clustering) [17,63,82,176,209] to reduce the noise from isolated "votes". A switch from absolute to the relative treatment of click-through votes [104,106,162], and later, more advanced click models [46,89,158] consider the noise associated with different positions of clicked documents in the ranked lists displayed to users. Using a broader set of features in combination with more advanced machine-learning approaches also led to considerable quality improvements in log-based ranking approaches [5].

The post-Direct Hit research on log-based ranking can be divided into several strands of research. One popular approach explored various graph-based methods to better leverage connections between the queries and pages accumulated in the log. Another popular strand of research harnessed machine-learning techniques by treating search ranking as a type of *learning-to-rank* task. The benefits of this approach include its ability to consider richer feature sets extracted from query logs. Yet another strand of research looked at grouping users into more coherent clusters to generate social ranking within a cluster; namely, using only the click data of like-minded users. This led to good results without the need to use more advanced document representations or ranking approaches.

Graph-based approaches to query log analysis condense the social wisdom represented in the click log in a concentrated graph form that can be explored in several ways. For example, [82, 209] focused on using social data as an additional source of *features* (known as click-though features) to improve document representation for improving matching and ranking. Both groups of authors attempted to address two known problems with click-through indexing: (1) the overall *reliability* of click-based evidence mentioned above; and (2) the *sparsity* of click-through data. While popular queries and pages appear in many query-click pairs, many queries and pages receive too few associations in the logs to be used reliably. To solve both problems, Xue et al. [209] suggested moving from the level of isolated click "votes" to clusters of similar queries and clicked pages. By representing query log data as a bipartite graph, they applied an iterative reinforcement algorithm to compute the similarity between web pages and queries, which fully explore the relationships between web pages and queries. After the similarity between web pages are computed, two similar web pages can "share" queries; for example, queries associated with one page can be assigned to similar pages as extra metadata.

Gao et al. [82] followed the same idea for more reliable query-based indexing based on the expansion of sparse click-through data. Defining this approach as "smoothing", this paper explored two such smoothing techniques: (1) query clustering via Random Walk on click graphs; and (2) a discounting method inspired by the Good-Turing estimator.

Elsewhere, Craswell and Szummer [54] used a random walk approach on a graph-based representation of a click log to produce a probabilistic ranking of documents for a given query. They explored several types of Markov random walk models to select the most effective combination of parameters. Poblete et al. [160] attempted to combine the benefits of link-based and log-based ranking by also applying a random walk model to the integrated graph, which included both static hypertext links between pages and query-page links reconstructed from the log.

The idea of considering log-based ranking as a machine-learning problem was introduced by Joachims [104] as a learning-to-rank problem. While learning-to-rank has been used in the past with explicit feedback data, the intrinsic noise of log data, based on implicit feedback, made it a challenge to use conventional learning-to-rank approaches in this context. To overcome this challenge, Joachims [104] considered more reliable *relative* implicit feedback that could be obtained by analyzing user clickthrough patterns in the log; for example, a click on the third link in SERP while ignoring the second link indicates that the third document is more relevant to the query than the second, from the user's point of view. In follow-up work, Radlinski and Joachims [162] expanded the original approach by considering multi-query search sessions and using more complex log patterns that allowed for engaging more broadly across sessions with more reliable social wisdom not available within a single query.

Following this pioneering work, learning-to-rank gradually emerged as one of the key approaches to incorporate implicit feedback accumulated in query logs into the ranking of search results. More recent work in this area advanced earlier research by exploring different learning-to-rank approaches and alternative ideas to reduce the impact of log noise. For example, Agichtein et al. [5] re-examined the use of absolute implicit feedback, originally explored in Direct Hit [57]. To deal with the noise, the authors extracted a much larger set of feedback-related features from the log and fused them with a RankNet supervised algorithm to learn a ranking function that best predicts relevance judgments. In particular, they aggregated click frequencies to filter out noisy clicks and used a deviation from expected click probability to more accurately consider the value of a click at different positions in the SERP. Dou et al. [63] followed suit [104] by using the relative feedback approach, but adopted the idea of feedback aggregation and RankNet fusion suggested in [5]. Instead of relying on single feedback cases, the authors aggregated large numbers of user clicks for each query-document pair and extracted pairwise preferences, based on the aggregated click frequencies of documents.

Another important advance in the use of machine-learning techniques for ranking was the introduction of formal *click models*. Click models attempt to model user search behaviors to address rank position bias in a more holistic way than pairwise preferences. Once learned from data, click models can produce a data-informed ranking of web pages for a given query. The idea of learning a model of user search behavior was introduced by Agichtein et al. [6]. Early exploration of several click models was performed in 2008 by Craswell et al. [55] as well as Dupret and Piwowarski [64]. In the following year, several papers [46,89,158] pioneered the use of probabilistic graphical models (i.e., Bayesian networks) to represent and learn click models. In a broad stream of follow-up work, the use of click models and learning-to-rank approaches for log-based ranking have been expanded and refined in many different ways. For example, recently, Katarya et al. [112] extended the approach of [89] using interactive learning. Wang et al. [193] captured non-sequential SERP examination behavior in a click model. Borisov et al. [28] experimented with neural networks as an alternative basis for click models, and Wang et al. [194] explored learning-to-rank approaches in the sparse context of personalized search.

The use of user groups to cluster social wisdom accumulated in query logs into smaller and more coherent communities was pioneered in several projects [10,77,111,179,180]. Some of these projects suggested using existing social groups, such as the implicit communities harnessed by [179,180] based on the origin of search queries. Others attempted to match users and form groups dynamically [10,111]. For example, Almeida and Almeida [10] suggested a "community-aware" search engine, which used a graph-based representation of a query log to identify multiple communities of interest and associate these communities with documents. The engine then helped users in the search process by matching them to existing communities and providing community-biased ranking by fusing community-based relevance and content relevance. The use of static groups based on demography and other characteristics has also been extensively explored. For example, Teevan et al. [189] demonstrated how the value of "groupization" depends on the type of the group and query category. Work-related groups were found to have no cohesion on social queries, while demographic groups have no cohesion on work-related queries. The author also suggested an alternative approach to group-based ranking, by applying ideas from personalized ranking at the group level. More details on several group-based social search approaches can be found in the Sect. 10.

6.3   Using Browsing Trails and Page Behavior for Primary Ranking

Browsing trails can offer several types of social wisdom that can be used to improve rankings. First, browsing trails provide some evidence of page value and importance through various implicit indicators, such as reading-time, or, with appropriate instrumentation, within-page scrolling, and mousing (for desktops) or "fingering" for mobile browsing [93].

Second, assuming that users have coherent goals when browsing, the sequence of pages followed can help to create associations between pages. This information can be used to enhance network-based algorithms for estimating page importance, such as PageRank. For example, while the fixed links between pages, used by PageRank, already influence navigation pathways, without real user information, PageRank has to assume an equal probability of navigation from one page to all outgoing links. In reality, some pages may be much more important than others, and browsing data could help to capture this information.

The easiest way to collect browsing information is through browsing logs, which are maintained by most Web sites and information systems. However, due to its localized nature, this information can be used only for improving ranking at the site level or for system-level searching. To be useful for Web search, user browsing needs to be captured beyond a single site, which requires some client-side or server-side instrumentation. Client-site instrumentation, such as user agents or browser plug-ins, originally developed for personalized search [83] can be used to collect user browsing data across multiple sites, and usually on a deeper level than site logs. On the other hand, server-side instrumentation, such as the link-following technology used by most modern search engines, can register a user's continuous browsing session starting from an instrumented page or SERP.

Each of these approaches has its own pros and cons. Browser-level data facilitates navigation capture in a much broader context, but because browser-based instrumentation must be installed, tits adoption has been limited. In contrast, link following is now commonplace, but it typically includes only short, truncated browsing segments.

Site-level browsing data, which is the easiest data source to collect, was also the first to be explored for ranking. Xue et al. [210] attempted to improve the PageRank approach by calculating page importance using "implicit links" instead of explicit hypertext links between site pages. The authors argued that implicit links exist between pairs of pages if there are frequent transitions between them (as mined from browsing logs), and that such associations are more valuable in a "small web" context, where explicit links might not reliably reflect main navigation pathways. Their study demonstrated that a PageRank-style algorithm, based on implicit links, provided the best ranking for site-level search, as it outperformed regular PageRank, HITS, and other approaches. This work was followed by a number of like-minded attempts to improve PageRank by using site-level browsing data. For example, Eirinaki and Vazirgiannis [65] suggested an approach to bias the PageRank calculation to rank pages more highly that previous users visited more often. Guo et al. [90] took time spent on a page into consideration and suggested an approach to bias the PageRank calculation so as to give a higher ranking to pages that previous users visited more frequently and for a longer period of time.

The use of post-search browsing trails, collected by the search engines and through the link-following approach, has been investigated by Bilenko and White [24] in the context of their research on recommending search destinations [200]. The nature of post-search trails is different from site-level browsing trails, since each trail originates from a specific query. As a result, the use of this data is more similar to the use of query sessions, rather than the use of general browsing trails. Each page selection and its dwell time indicates a page value for the original query, rather than its general importance. In their work, Bilenko and White [24] explored several approaches to process and use trail data and confirmed that post-search browsing behavior logs provided a strong signal for inferring document relevance for future queries. In particular, their work indicated that using full trails can lead to better results and that using the logarithm of dwell time is the best approach when using this source of information.

More recent work has attempted to go beyond site-level and post-search trails by using server-side and client side instrumentation to collect a broader set of trails that could be used to improve regular Web searching. The most well-known of the these approaches is BrowseRank, which was introduced in [130]. The authors suggest using a navigation graph, augmented with time data in place of a "timeless" link graph, used by PageRank. Instead of the usual approach, which applies PageRank on the top of browsing data, they used a continuous-time Markov process on the user browsing graph as a model for computing the stationary probability distribution of the process as page importance. The authors argued that this approach could leverage this new kind of data better than the original PageRank, a discrete-time Markov process on the web link graph as a model. In a follow-up paper, [131] suggested and evaluated several other approaches for browsing-based estimation of page importance. Using a large "link following" dataset collected by a major search engine, the authors demonstrated that BrowseRank significantly outperforms PageRank and other simple algorithms. While this type of data, as discussed above, might not be the best match to evaluate BrowseRank, a follow-up work [190] demonstrated that BrowseRank also outperforms PageRank in the context of site-level search. Zhu and Mishne [216] suggested an alternative approach to calculate browsing-based importance using both click order and page dwell time. The authors evaluated ClickRank using a large volume of user browsing logs collected from the Yahoo! toolbar and demonstrated that this method outperforms both PageRank and BrowseRank; see also [16].

6.4   Reranking and Recommending Web Search Results

In this section, we consider how social signals (which might not be available to the search engines and thus cannot be used for primary ranking) might be used to better bring search results to the attention of the end users; for example, understanding the searcher's community may help to influence search results. We will look at how such information can be used to re-rank an original set of results or used to insert new results into an original result list.

To begin with, the I-SPY [178–180] system provides an early example of a form of social ranking or, more correctly, social *re-ranking*, since I-SPY re-ranks a set of results provided by an underlying search engine. We discuss I-SPY in further detail as a later case study, but for now, it is sufficient to say that I-SPY uses click-through data re-ranking based on community interests. Briefly, result pages that have been frequently selected in the past for a query that is the same or similar to the target query are considered to be more relevant than result pages that have been selected less frequently. I-SPY calculates a relevance score based on click-through rates and weighted by query similarity, and uses this score to promote and re-rank results. Moreover, I-SPY leverages click-through data that originates from a community of like-minded searchers by sourcing its search queries from search boxes placed on topical websites. For instance, an I-SPY search box on a motoring web site is likely to attract queries from motoring enthusiasts, and their click patterns will help to differentiate and promote pictures of cars for a query like "jaguar photo", instead of pictures of large cats.

This combination of click-through data and community focus is closely linked to the idea of trust, expertise, and reputation as a ranking signal, which has become increasingly important in recommender systems. Intuitively, not all searchers/users are created equally. Some will have more or less expertise on certain topics and may make better recommendation sources as a result. This idea was first explored in the work on *trust-based recommender-systems* [134,152], in which they explicitly modeled the trustworthiness of users based on how often their past predictions were deemed to be correct. By prioritizing users who are more trustworthy in general or in respect to a particular topic, it was possible to significantly improve recommendation quality. Similar ideas were explored by expertise-based recommender systems that attempted to model user expertise [12]. In parallel, these ideas were adapted for community-based web search in the work of [32, 35] with search results ranked based on the *reputation* of the users who previously selected them for similar queries.

In this sense, we can view reputation as yet another type of social signal that arises from the result selection behavior of searchers, and how this selection behavior helps others in the future. Users who search frequently on a given topic, and whose selections are recommended to, and re-selected by, other searchers can be usefully considered to have a higher reputation on such topics, in comparison to other users who rarely search on these topics, or whose selections are rarely re-selected when recommended. This idea informed the work of [138, 183], which proposed an explicit model of search collaboration and considered a variety of graph-based reputation models for use in social search. Briefly, different ways to distribute and aggregate reputation were evaluated as ranking signals, with significant precision benefits accruing to reputation-based ranking compared to alternative ranking approaches. Similar ideas are explored by the ExpertRec system [188], which modeled user expertise by observing search sessions and subsequently promoted search results preferred by experts.

The previous examples are all examples of ranking/re-ranking an original result list using social signals, but we are not limited to the original list of results. Many researchers have looked at ways to make new recommendations by adding novel results into an original result list. For example, White et al. [200, 201] describes a technique to exploit the searching and browsing behavior of many users to augment regular search results with popular *destination pages*, which may be located many clicks beyond the original search results. This offers the promise of a significant time-saver for searchers by short-circuiting long navigation trails from SERP results to a final destination. They focused on navigation trails that began with a search query submitted to a popular search engine. Each trail represented a single navigation path (that is, a sequence of clicked hyperlinks) from a search results page to a terminating page, at which point the user went no further. These terminating or *destination* pages are then indexed using their original query terms so that they can be recommended and added to result lists for similar queries in the future. In this way, this approach is related to some ideas presented earlier on the topic of social indexing, in the sense that these destination pages are effectively being socially indexed based on the queries that have led to them.

In fact, authors describe a number of different ways to harness these destination data. For example, in one interface, for a given search query, a set of destinations are presented that have been frequently navigated to after similar queries. In another interface, searchers have recommended destinations that have been navigated to, not for the current query, but for typical follow-on queries (from the same search session). The study found that systems that offer popular destinations lead to more successful and efficient searching compared to query suggestion and unaided Web search but that destinations recommended based on follow-on queries had the potential to be less relevant to searcher needs.

Likewise, the HeyStaks collaborative search system [181, 182] also attempts to add new pages to an existing SERP; it too is described in more detail as a later case study. Briefly, like I-SPY, it organizes searchers and searches into communities, but this time, they are based on user-defined search topics and interests. These interests are captured in so-called *search staks*, which act as repositories of past search histories that encompass a variety of search-related data, including queries, result selections, result ratings, and tags, among others. Searchers continue to search using their favorite search engine, but by using the HeyStaks browser app, their searches are recorded and aggregated in staks that they have created or joined—and the search results that they receive from an underlying search engine are enriched with additional recommendations from relevant staks. For instance, if a user is searching for "ski chalets" on Google, as part of a "ski vacation" stak, then, in addition to Google's results, they may see other results that have been found by other members of the stak, either for similar queries or from different search engines. These recommendations are selected and ranked based on a weighted scoring metric that combines query similarity, selection popularity, and other social signals (rating, tags, and so on).

6.5   Ranking and Re-ranking with Social Media Data

Social media data plays a dual role in social search. In its *local* role, socially posted data are used to perform search within the very social system where these data were posted; for example, searching for images or bookmarks in a social sharing system or searching for people in a social network, where comments, tags, and other user-contributed information are vital for finding relevant information. However, more and more frequently, social data are used more *globally* – to improve search ranking beyond the host system. For example, posts in a collaborative Web bookmarking system like Del.icio.us or tweets of Web pages could improve Web search, while posts in a research paper bookmarking systems such as CiteULike[5] or blog posts mentioning research papers could improve searching in academic search systems, such as Semantic Scholar[6]. The use of social data for searching within a social system is naturally a component of primary ranking, since these data are directly available to the social system search component. In contrast, the majority of work for using social data in a more global search context is to perform a social re-ranking of results that are returned by a general search engine. The use of social data could considerably improve both local and global rankings in both primary ranking and re-ranking settings. Moreover, the approaches to improve ranking using social data are quite similar in all these contexts. Here, we review the use of social data for both primary ranking and re-ranking with a focus on the approaches taken, rather than their application contexts.

Using social data could improve ranking in internal and external search systems in three ways. First, as reviewed in Sects. 5.3 and 5.4, a resource (i.e., a paper, a Web page) shared in a social system is usually augmented with comments, tag, or at least some surrounding content (like in tweets and blogs).

---

[5] http://www.citeulike.org.
[6] http://www.semanticscholar.org/.

This content complements the content of the original resource and can be used to improve both matching and ranking in the same way as it can be done by using query texts or anchors.

Second, the very fact that a resource was shared on social media and the scale of this sharing (i.e., the number of tweeting or bookmarking users) could be interpreted as the sign of value, and used for promoting shared content for both internal and external search. Simple approaches based on this idea follow the DirectHit path that treats every sharing event as a vote. For example, in an internal Twitter search, the number of re-tweets is treated as a sign of a tweet's importance, which can be used for tweet ranking [7]. For external ranking, blog posts could be treated as votes for various news articles and leveraged to produce better news article rankings [136]. Similarly, sharing a research paper in social bookmarking system could be treated as a vote that the paper is worth reading [103]. More sophisticated approaches adopt ideas from PageRank to extract importance data from the complex network of connections between users, shared items, and tags. For example, a network of tweets and tweeters could be used to estimate the *authority* of each tweeter and give a higher weight to more authoritative tweets or re-tweets [157, 168].

Finally, resources and posts explicitly shared by individual users provide a much more reliable indication of user interests, as compared with traces left while searching or browsing. A model of user interests extracted from shared social data could be used to further improve ranking by making it more personalized; a considerable share of work on tag-based search was at least partially focused on personalization [1, 40, 126, 150, 208].

Among all kinds of social media data, tags were both the first and the most popular source of information for improving search ranking. The work on tag-based ranking technologies has started with research on improving ranking within social tagging systems. Several advanced approaches for tag-based ranking were reported as early as 2006 [100, 206]. Subsequently, several papers adopted different approaches for using social tagging data to improve the ranking of Web search results [21, 150, 211]. While most early papers explored straightforward approaches for integrating tags into rankings, such as simple "vote counting" to estimate page importance [211] or traditional vector model to measure query-document similarity [150], a increasing share of work focused on more sophisticated uses of tags for ranking.

Network-based ranking approaches could be considered to be the most popular group of such advanced approaches. Most are motivated by Google's PageRank and try to adapt it for social systems. While PageRank and HITS tried to leverage the information encapsulated in a complex network of interconnecting Web pages, the ranking approaches in social tagging systems attempted to leverage information hidden in the even more complex network that is formed by users, tags, and resources. While each Web link establishes a new edge in a network of pages, each tagging event (a user U tags item I with a tah T) creates a new *hyperedge* that connects U, I and T, which could be alternately represented with three edges $U - I$, $U - T$, and $I - T$.

Hotho et al. [100] were the first to suggest an advanced network-based ranking approach, *FolkRank*, to extract information from a tagging network for the internal ranking of social bookmarks in BibSonomy; see Chap. 9 of this book [149]. Bao et al. [21] used similar ideas to leverage the information encapsulated in a social bookmarking system to improve external result ranking. The authors suggested two network-based ranking approaches to optimize Web search. First, they proposed a novel similarity-based ranking algorithm *SocialSimRank* based on the idea that social tags can act as a type of metadata for the similarity calculation between a query and a web page. Second, they described an *importance ranking* approach, *SocialPageRank*, inspired by PageRank, to estimate the popularity of a page based on its tag history. In a similar attempt, Abel et al. [1] suggested a graph-based ranking approach that was inspired by HITS, rather than PageRank.

Other early approaches focused on exploiting tags to build a better semantic representation of the document space (and the user space), which could be leveraged for better ranking. A range of formalisms were used for the included Semantic Web [206], such as topic models [126,208] and language models [215]. Some radically different approaches were also introduced by researchers from other communities. For example, Zanardi and Capra [212] used ideas from collaborative filtering to combine both social and personal data for tag-based search ranking, di Sciascio et al. suggested a user-driven approach to fuse traditional query-based relevance with tag-based relevance [172]. Over the last 10 years, the research on tag-based ranking expanded into a distinct research direction with papers exploring different ways to leverage graph information, extract semantics, or combine these with other approaches. An extensive review of tag-based ranking for both search in social systems and Web search is provided in Chap. 9 of this book [149].

7   Resource Presentation – Social Summarization

Almost all web search engines follow a similar pattern of result presentation, the so-called *10-blue links* approach. That is to say, results are presented as a simple list of URL links, with each result made up of a title, a URL, and some suitable *snippet* to summarize the result. These snippets play a vital *sense-making* role, as previously discussed. They help searchers to efficiently make sense of a collection of search results, as well as to determine the likely relevance of individual results. Snippets are typically extracts of content from the corresponding web page. They are also typically query-sensitive, in the sense that the selected text is chosen because it is relevant to the current query. For example, the most popular approach is to select text in the web page that contains some or all of the query terms; however, it is not easy to decide which parts of the page and which keywords will be most helpful for the users to recognize a page as a true match to the query. This is one resource presentation task where social data can help.

The idea of using social data to generate page snippets was first suggested in [170]. The authors proposed to use past queries as a component of page snippets by arguing that they offered the best characterization of a document from an end-user perspective. To generate these kinds of snippets, they used query logs to incrementally build associations between queries and their selected documents. However, these past queries were not truly integrated into a presentation snippet, but instead were added after the snippet summary, and served more as an augmentation rather than a summary. It also provided a one-size-fits-all approach to snippet generation that was not adapted to a specific user or context.

The work of [30,31] resolved both of the shortcomings mentioned above and demonstrated an example of using social signals to guide the generation of "smooth" snippets in a *community-based search context*. They describe an approach to personalizing snippets for the needs of a community of like-minded searchers. Their approach, called *community-focused social summarization*, uses community search behavior – query repetition and result selection regularity – as the basis for generating community-focused snippets. It uses the standard, query-focused snippet generation technique of the underlying search engine, but each time a result is selected, for some query, the corresponding snippet is recorded so that, over time, popular results within a community of searchers come to be associated with a growing set of (possibly overlapping) snippets, based on the queries for which they were selected. Sentence fragments can be scored based on how frequently they recur in snippets, and a social summary can then be generated from the most popular fragments and weighted according to how similar the current query is to the query that resulted in a particular snippet fragment. In this way, highly personalized, query-focused snippets can be generated at a given length specification. In tests, these snippets prove to be superior to those produced by alternative summarization techniques, including those that involve sophisticated natural language processing techniques; see also Alhindi et al. [9] for related ideas on generating group-adapted page snippets.

8   Augmenting Search Results: Annotations and Explanations

In the previous section, we focused on the presentation of search results – generating snippets that present each search result to the user – but that is not the only way to improve the presentation of results for the benefit of searchers. In this section, we consider SERP *augmentation*: the different ways that the SERP can be *decorated* with additional information to assist the searcher. SERP augmentation is, in some aspects, similar to the link augmentation in social navigation reviewed in Chap. 5 of this book [71].

There has been a long history of research into how search results might be presented to users so that they can better understand their relevance. For instance, TileBars [96] introduced a visualization paradigm that offers an explanation of not just the strength of the match between the users query and a given result page, but also the frequency of each term, how each term is distributed in the text, and where the terms overlap within the document. This approach provides the user with additional explanatory information that can help them to come to a decision as to whether or not each page is relevant to their information needs. In a social search context, explanations can be derived from social data, such as the interactions of other users as they search or explicit social links.

8.1 Query Logs

Earlier, in Sect. 6.2, we discussed the role of query logs for primary ranking, but sequences of queries can also be used for the purposes of result annotation and explanation. As mentioned in Sect. 7, in 2002, Scholer and Williams [170] suggested augmenting document presentation with past queries to result in this document selection. The I-SPY collaborative search engine [178–180], mentioned in Sect. 6.2, and discussed later in Sect. 10.2, is another early example of the use of social annotations, and query annotations in particular. For example, Fig. 1 shows a sample SERP annotated with additional information that is based on how relevant a result is to the searcher's community, the recency of result selections, and the availability of related queries that have caused a result to be selected.

This work was subsequently extended by the *SearchGuide* project [50,51], which built on-top of I-SPY to provide a even richer augmentation interface [50,51], and which will be discussed in further detail in Sect. 9. Briefly, for now, SearchGuide (see Fig. 3) provides an enhanced interface to brings similar types of social annotations to bear on regular content pages, in addition to SERPs.

Using query logs to augment search results has also been explored in several other projects. For example, Lehikoinen et al. [124] demonstrated how past users interacted with search results in the context of meta-search in P2P networks. Elsewhere, Wang et al. [195] used query log data to build a topic map of a search space and used this map to augment search results with a set of related topics as the basis for further exploration.

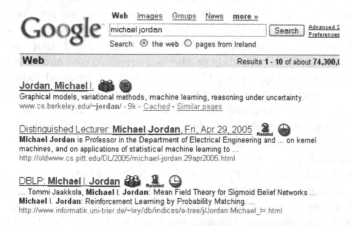

Fig. 1. An example of the augmented SERP used by a version of the I-SPY system that shows a set of results that are augmented with additional icons to reflect how relevant a result is to the searcher's community, the recency of result selections, and the availability of one or more related queries for a result.

8.2   Browsing and Annotation

Research on using navigation and annotation activity of past users to augment search results was performed mostly in parallel with the work on using query sessions, though it was originally motivated by the ideas of social navigation rather than social search. The social navigation research stream in early 2000, as reviewed in Chap. 5 of this book [71], aimed to guide users through an information space by augmenting navigation links with "social wisdom" extracted from the history of past navigation. The main impediment to using this approach in a search context is selecting a meaningful subset of users, such that their browsing data will be helpful in selecting query results. While the use of query sessions for augmentation facilitates focusing on users who issued the same (or similar) queries, and thus are likely to have a high probability of similar information needs, browsing traces can come from all kinds of users. The majority of browsers are likely to have highly different needs from those who issued a specific query. As a result "everyones" browsing behavior will hardly help in a specific search context. However, if a community of like-minded users who share similar goals can be identified, then their browsing behavior may help other community users with their searches.

This basic idea was originally developed in the Knowledge Sea project [37], which leveraged the browsing behavior of users taking the same course in an e-learning context. Knowledge Sea attempted to use social navigation to support several kinds of information access to educational content, in the form of a collection of online textbooks, including browsing, search, and information visualization. While the collection of textbooks can be accessed by students who take different college classes (and thus have different information needs and priorities), Knowledge Sea considered students within the same class as sufficiently like-minded to apply social navigation support. In the context of search, social navigation support was provided by augmenting SERP links with visual cues, which reflected how much each search result had been read and annotated by students *of the same class* (Fig. 2). The browsing-based visual cue was shown as a blue human icon on a blue background. The density of color indicated the cumulative amount of page reading by the user (figure color) and the class (background color). For example, a light icon on a dark background indicated pages frequently explored by other students in class, but so far ignored by the current user. The annotation-based cue used a yellow background color to indicate how many annotations made by students in class each SERP page has, and also indicated how positive were these annotations.

This approach made it easy to recognize pages that members of the class found to be useful, especially if the user had so far paid little attention to them. Several rounds of studies with Knowledge Sea demonstrated that pages with high levels of class browsing and annotation behavior were especially appealing to the students: the presence of annotation considerably affected their navigation and reading choices [37]. Moreover, in the context of search, user result-selection and reading-time data demonstrated how higher levels of browsing behavior, indicated by visual cues, offered a stronger signal of search result relevance, compared to being among the top three results in the ranked list [8].

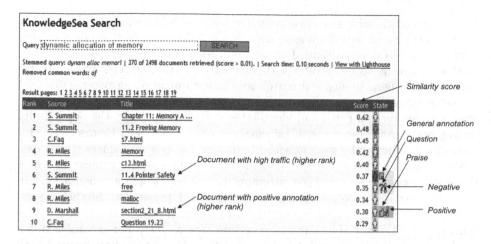

Fig. 2. Augmentation of search results with visual cues to indicate browsing and annotation behavior of students in the same class. Darker blue color indicated pages frequently selected and read by the class, while yellow-orange color indicates pages annotated by users in class (Color figure online)

The idea of group-level augmentation of search results with browsing data was further explored in the ASSIST project [72,76]. Using group-level data collection, ASSIST integrated the ideas of SearchGuide and Knowledge Sea and used both search and browsing data to augment the presentation of search results. This project is reviewed in more details in Sect. 11.1. The use of annotation data to augment search results has been further explored in an exploratory search context, in [70]. In this work, browsing and annotation behavior produced by a group of people working on a set of similar information analysis tasks was used to augment each search result to stress the extent to which the retrieved document has been read and annotated by other users in a group. In addition, the SERP was also augmented with a social "map" icon to show the location of documents with the group footprint among the top 100 retrieved documents.

8.3   Social Media and Social Links

Augmenting search results with data from various social media systems (i.e., bookmarking, tagging, blogging, and microblogging data) may help with SERP augmentation in several different ways. First, since document sharing in social media is a sign of that document's importance, augmenting a SERP document with information about the number of shares (i.e., posted to a bookmarking system, shared in a microblog) could help to guide users to prominent documents. Of course, users who bookmarked or otherwise shared documents may have done so for a variety of reasons, which may have little connection to the goals of their search or the goals of the new searcher. Consequently, to make this idea practical, social data should be carefully collected from a subcommunity of like-minded users; for example, users within the same enterprise, as suggested in [13].

Moreover, additional information provided by users while sharing could also be used to augment shared documents with "social" descriptions that could further help during selection decisions. For example, social bookmarking data could be used to augment search results with related tags [13,109,211], while Twitter data could be used to augment search results with mentioning tweets [157].

Since almost every instance of information sharing in social media is associated with an authenticated and identifiable user, this opens up the opportunity to augment search results with information about the users themselves. This approach was originally used in social bookmarking systems, where it was natural to show who bookmarked a document, regardless of the context in which the document was shown. First attempts to apply "people augmentation" in a broader setting were performed by several IBM researchers in the context of enterprise search, which leveraged information from enterprise social systems such as IBM social bookmarking system Dogear [141]. For example, Millen et al. [142] explicitly injected data retrieved from Dogear (and augmented with some people information) into all search results. Amitay et al. [13] used Dogear and the IBM blog system BlogCentral for a more elaborated augmentation of search results, using a list of people who shared the retrieved documents.

As with other types of social media augmentation, the main challenge with "people" augmentation is ensuring that there is a match to the current searcher's needs: the reasons for one person to share a result might be completely irrelevant to a future search context, and while simple people augmentation can work well in a narrow enterprise search context [13,142,147] its value quickly decreases in broader settings. To address these issue, IBM researchers explored the use of social networks by focusing on the social connections of the searcher [173]. In an enterprise context, social connections are usually professionally oriented and documents shared by connections have a much higher chance to be relevant to future searchers. In addition, the availability of social connections can act as important signals of authenticity and credibility of the shared content. A study of a file sharing system at IBM demonstrated that users are more likely to download a file when the file author is in their social network [173].

By 2011, "people" augmentation for search results had become widespread, reaching major search engines like Google and Bing [148]. It was natural to expect this approach to be beneficial for searchers. However, a sequence of studies of social augmentation of Web search, from 2012–2013, demonstrated the situation to be more complex. An eyetracking study by Muralidharan et al. [148] demonstrated how the social augmentation of search results remained unnoticed in the vast majority of cases. In post-study interviews subjects indicated this kind of annotation to be useful for only a subset of search topics, which they classified as "social" and "subjective". A follow-up study by Fernquist and Chi [73] confirmed that searchers often simply did not notice result annotations, mostly because we have evolved fairly rigid attention patterns when it comes to parsing search result-lists and these patterns tend to focus exclusively on titles and URLs, a form of *intentional blindness*. Moreover, when searchers did notice annotations, they tended to disregard those from strangers or unfamiliar people with uncertain expertise.

In a related study, Pantel et al. [155] performed a utility analysis of social annotations. They produced a taxonomy social relevance to capture and model the different types of features (query features, content features, social connection features) that can influence social relevance in search. Their findings corroborate some of those above. They also established that close social connections and experts on a given search topic provided the most utility, as compared to more distant contacts or those with uncertain expertise. This study also demonstrated how the value of different types of connections (i.e., a work colleague, a personal friend, an expert) is not universal, but depends on the topic of the query. For example, the presence of a friend in augmentation of movie search results increases the value of the result, while the presence of work colleague reduces it. Moreover, [155] described how their approach can be used to predict whether a given social annotation is likely to be relevant to a given query-page pair, which may have applications when it comes to a more selective, and possibly personalized, approach to automatically annotate individual search results.

The idea of selective annotation was further explored by [122], in the context of news reading. The authors consider how different types of annotations affect peoples' news selection behavior, and report on results from experiments looking at social annotations in two different news reading contexts. Although not strictly search-focused the results are relevant because they confirm, unsurprisingly, that the annotations of strangers have no persuasive effects, while the annotation of personal friends do have a positive impact, on article selection and reader engagement. Intriguingly, the results also suggest that annotations do more than simply influence selection: they can make (social) content more interesting by their presence, at least in part, by providing additional context to the annotated content.

9 Beyond the SERP

Finally, in this section, we consider the opportunity to support searchers beyond the SERP, which we have only touched on briefly in what has come before. In a conventional web search setting, once a searcher selects a result, they are redirected to the appropriate URL, where they are effectively left to their own devices. In other words, once they select a result, they leave behind the search engine and any ability for it to further support their search needs, which may or may not be satisfied by the selected page. At the very least, this is a missed opportunity when it comes to helping the searcher to find what they are looking for.

For instance, many search results, depending on the query, will be for high-level landing pages. They may bring the search close to the information they are looking for, but the searcher may have to engage in additional browsing to locate the specific page they need. When planning a vacation, and looking for a hotel, a search engine like Google might bring us to a travel site, or a city-level page, but often not to a specific hotel – even if it does bring the searcher to a hotel page. It is likely that users will want to search further using the hotel site's own search interface. This begs the question as to whether the primary search engine might be able to further support searchers as they continue to search and browse, with the added benefit that the primary search engine can then learn from these off-SERP interactions. These ideas are related in spirit to some discussed earlier on the topic of augmenting search results, but instead of adding new information to the SERP, it is all about augmenting non-SERP pages, but with search-related information.

Earlier, we referred to the SearchGuide system [50,51], and how it went one step further than SERP augmentation by also supporting searchers as they navigated beyond the SERP. SearchGuide uses a browser plug-in to augment regular web pages with search information that is relevant to the current session (see Fig. 3). For example, SearchGuides navigation bar provides a visualization of a pages "computational wear"; see also [98]. This navigation bar is calibrated to the length of the page and visualizes the distribution of query terms within the page. Each icon acts as a hyperlink to a query term occurrence within the page,

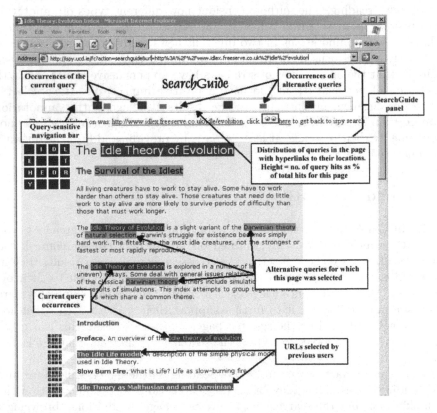

Fig. 3. An example of SearchGuide in operation, showing: the query-sensitive navigation bar; annotated page content, to emphasize occurrences of the current query; related queries within the main page content; and highlighted links that were selected by previous users. (Color figure online)

so the user can jump rapidly to regions of the page which appear to interesting, but without having to read or scan through the other (possibly irrelevant) content.

Page content is also augmented: *virtual signposts* (see also [199]) bring the user's attention to key areas of the page that relate to their query or to related queries. Finally, SearchGuide provides additional navigation support by highlighting hyperlinks within the page that have been frequently selected by users for *similar queries*, thereby serving to identify common navigation trails that past searchers have followed. The ideas in SearchGuide were further extended as part of the ASSIST project, which considered an even broader set of social feedback. This project is reviewed in more detail in Sect. 11.1.

10   Personalizing Social Search

One of the challenges faced by social search is where to draw the line between social data integration and the resulting adaptation of the search experience. Social search traces can come from a wide variety of people with many different types of information needs, but blindly using the search traces from all users may not help a particular user who has specialized needs.

This differentiates social search from other social recommendation techniques, where users are matched by a profile of their interests (see, for example, Chap. 10 of this book [116]), rather than a time-sensitive slice of current needs. In this chapter, we offer a deeper review of several case studies of practical systems, which have been developed to support more personalized social search experiences, chosen to convey the evolution of personalized search. The interested reader is encouraged to follow the citations provided for more detailed information on each case study.

10.1   Antworld

The AntWorld system [29, 110, 111, 174] was one of the first ad-hoc search engines to implement the sharing of community knowledge in order to improve the accuracy and speed of finding information on the Web. AntWorld supports users in resolving information "quests" rather than simple queries, as it attempts to understand the context of the user's information need. Following the world of ants as a computational metaphor, the system implements an asynchronous collaboration mode, where information trails from user quests are "deposited" for other community members to follow, just as ants leave pheromone trails to food sources. The AntWorld system accommodates the posting and sharing of communal knowledge as community members share their gained knowledge with the communal repository by providing feedback on how well specific search results answer their particular information needs.

For each user "quest" (Fig. 4), formulated as keywords (short description), and a longer natural language text (long description), the system computes and stores a summarized quest profile. In addition to the text of the quest, the profile contains the pages that the user browsed after receiving the system's response to the quest, as well as their judgment about the relevance of each of these pages. This additional information reflects, to some degree, the contextual information about the user's need and their relevant level of knowledge about the domain of the quest. The quest profile is analyzed and stored, and is used to guide users who search for similar quests in the future.

During a user's interaction with the system, their quest profile is dynamically built and matched against stored quests. The system presents the user with a list of pages that other users judged to be useful for similar quests. The system also puts an ant icon next to pages that were found to be useful for similar quests. As the user provides more feedback, the system's confidence in the quest profile increases, and it is able to identify similar quests with higher accuracy. Quest profiles are represented as vectors, using the vector space model with several variations of the *TF-IDF* scheme. The terms included in the short and long description of the quest are assigned higher weights than the text of the documents that the users judged as relevant. The similarity between the user quest and the quests stored in the system is computed using cosine similarity. For further technical details, see [29].

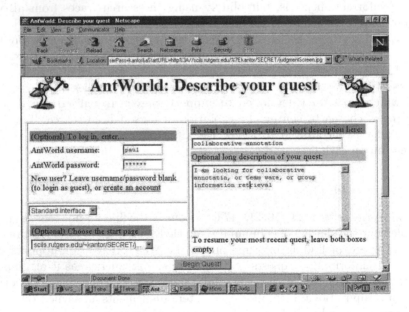

Fig. 4. AntWorld: quest description interface

A user study performed to evaluate AntWorld's potential effectiveness, and especially the extent to which users would make the effort and contribute knowledge to the community was presented in [174]. The experiment was conducted with students who used the AntWorld system to find specific information for their course assignments. In one session, users were not extrinsically motivated, while another session included an extrinsic incentive for providing evaluations (pizza coupons for the most contributing user). The results show that the extrinsically motivated group exhibited a more significant contributing behavior than the less active group without the extrinsic motivations. A clear conclusion was drawn about the need for some type of extrinsic motivation, in order to encourage users to provide feedback, since the productivity benefits of the system did not prove motivation enough. It might well be that if a similar study were conducted today, when social networks are popular and sharing knowledge and feedback has become habitual to users, results would be different.

Following AntWorld, the SERF system implemented a similar idea in a library search context [107]. In a manner similar to AntWorld's "quests", SERF encouraged users to submit extended and informative queries and collected feedback from users as to whether search results met their information needs. The system used the feedback to provide recommendations to later users with similar needs. Over time, the SERF system learned from the users about which documents were valuable for which information needs. One difference between AntWorld and SERF is that AntWorld builds a dynamic quest profile that is adapted during the user's interaction in relation to an information need, and updates the list of similar quests and related documents as the search goes on. SERF provides only one list of similar queries and their relevant documents after the user submits their query. Initial user studies to evaluate SERF concluded that recommendations based on prior users with similar queries could increase the efficiency, and potentially the effectiveness, of library website search at Oregon State University, where an experiment was conducted. In respect to user's willingness to provide feedback, the results followed the findings of AntWorld, with relatively low participation observed.

10.2 I-SPY

In this section and the next, we summarize a pair of related collaborative/social search approaches – I-Spy and HeyStaks – which have both been mentioned earlier in this chapter. They have been chosen as early and influential examples of collaborative search and social search with the aim of making traditional web search more personalised with respects to the needs and interests of groups, or communities, of like-minded searchers.

I-Spy is an early example of a *collaborative* web search engine [178–180]. It was developed as a meta-search engine which drew its results from an number of underlying search engines: queries to I-SPY were dispatched to a variety of underlying (third-party) search engines, such as Google and Bing, and their result lists were normalised and aggregated to provide I-SPY with an initial set of results. These combined results were then ranked and returned to the search using a variety of social signals; see Fig. 5.

I-SPY used an implicit model of a search community, by using the source of search queries as a proxy for topically related searches and searchers. Thus, for example, by hosting an I-SPY searchbox on a wildlife site one would expect queries for "Jaguar photos" to result in clicks for pages with photographs of the wildcat, rather than the motor car, or Apple operating system. In contrast,

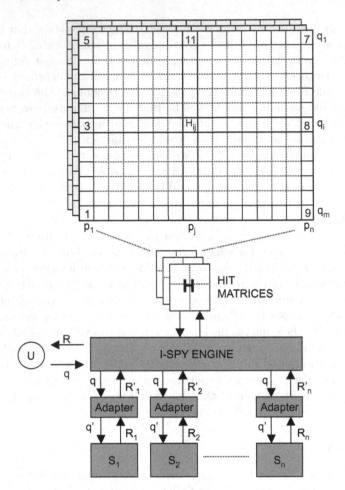

Fig. 5. The I-Spy system architecture showing the I-SPY collaborative search engine, search engine adaptors, and core hit-matrix as the basis for result relevance.

an I-SPY search box hosted on an antique automobile site might also attract "jaguar photos" queries but their selection histories would, presumably, link to car related pictures.

I-SPY records the past queries of users from a given community and their corresponding result selections. These data are stored in a data structure called a *hit matrix* (see Fig. 5). Then, for a new target query, q_t, the relevance of a page p_j is calculated as the proportion of selections for p_j given q_t. This simple relevance metric was extended in [178–180] to accommodate page selections for queries that were *similar* (based on term-overlap) to q_t. Thus, the relevance of a page p_j depends on a weighted-sum of its selections for similar queries. If the page was selected for many similar queries, then it received a higher relevance score than if it was selected for fewer, less similar queries. In this way I-SPY leveraged

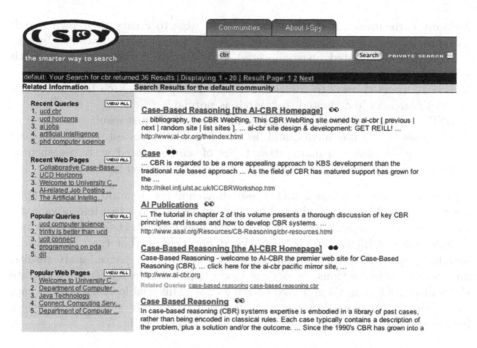

Fig. 6. An example I-Spy result-list showing a set of results retrieved for a given query. These results originate from underling search engines such as Google and Bing but are re-ranked based on social search data, such as past selections for similar queries.

a combination of conventional web search approaches (used of its underlying search engines) to identify a set of candidate results while harnessing community selection behavior in order to rank results.

An example of I-SPY in operation is presented in Fig. 6, which shows the result-list returned for the rather ambiguous query "cbr". In this example, the query originated from an I-SPY search box hosted on a AI research site, and therefore "cbr" referred to *case-based reasoning*, a form of machine learning. As a result, the results returned are all examples of case-based reasoning or AI related results, as reflected by the past selections of other searchers in this community.

10.3   HeyStaks

More recently, HeyStaks [181, 182] built on many of the ideas developed as part of I-SPY, but provided a more flexible social search experience and was motivated by a number of problems with the original I-SPY approach:

1. I-SPY used an *implicit* form of community, based on the origin of the search queries, as discussed above. At the time, it was common to host mainstream search-engine query-fields on third-party sites, but in the end this did not gain traction—especially when browsers implemented more dynamic "navigation fields" which allowed users to freely enter queries or web addresses – and it became obvious that most people interacted with search engines via their favorite search engine interface.

2. Many early users expressed a desire to be able to create their own search communities, based on different interests that they might have (e.g. work-related, travel-related, personal, etc.), but without the need to host search boxes on sites that were beyond their control.
3. I-SPY expected users to transition to an entirely new search interface when most searchers just wanted to "search as normal."
4. I-SPY used a limited set of social signals (essentially just result selections) as the basis for its judgments, and, as the social web evolved, it became clear that users engaged in many other types of search-related activities such as the tagging or sharing of results, as previously discussed in this chapter.

HeyStaks was developed with these shortcomings in mind. It was implemented using a browser plugin and toolbar, which carried a two-fold advantage. First, the toolbar was always available to the user through their browser, which allowed users to interact with HeyStaks at any time, rather than only during search. They could create or join staks (see below), tag pages, share results, and perform other actions. Second, it made it possible to seamlessly integrate HeyStaks with the user's preferred search engine. This enabled HeyStaks to capture queries and page selections and allowed it to directly augment the search interface of a search engine, such as Google or Bing; see Fig. 7.

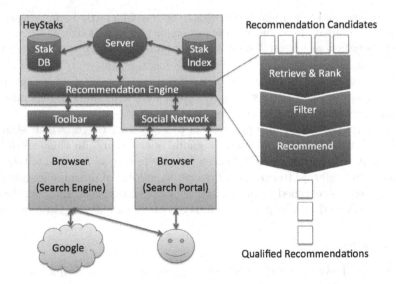

Fig. 7. HeyStaks is implemented as a back-end social search and recommendation system with a user-facing browser toolbar, thereby providing a seamless integration between HeyStaks and an existing search engine, such as Google.

The social and collaborative focus of HeyStaks was based on the ability of users to create *search staks* as types of folders for their search experiences. For example, a searcher might create a stak called "Canada Trip" as a repository for search information generated as they researched an upcoming trip to Canada;

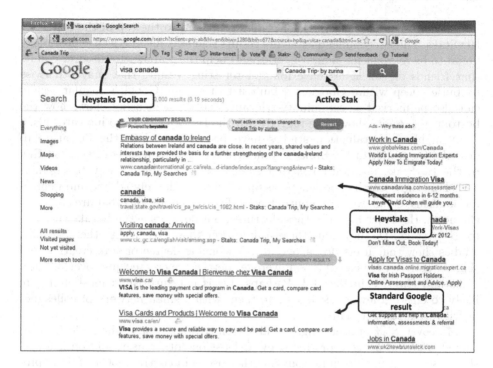

Fig. 8. An example of HeyStaks in action with Google. The HeyStaks toolbar is shown as a browser toolbar and provides the user with access to various features such as stak creation and sharing, and feedback options, such as voting and tagging. The screenshot shows a Google SERP that has been augmented with HeyStaks various augmentations, including a set of top-3 community promotions at the head of the Google result-list and annotations on regular Google results to indicate other community-relevant results.

creating a stak is a simple matter of completing a short pop-up form by using an option on the HeyStaks toolbar.

Next, as stak members search, their queries and selections are associated with a particular stak. Staks can be "shared" with others so that their searches will also be added to the staks they choose to join. Like I-Spy, HeyStaks tracks queries and result selections, but it also records other forms of search-related actions. For example, users can explicitly tag or share pages and can provide explicit relevance feedback in the form of positive and negative votes. These signals are integrated with queries and result selections in order to determine the relevance of a page for a new query, with a greater weighting given to explicit indicators of relevance, such as tagging and sharing results, while less weighting is given to implicit signals, such as a result selection. In this way, HeyStaks implements a version of a hit-matrix with each stak and uses this at search time to generate and rank recommendations.

An example, is shown in Fig. 8 which shows the results returned for the query "canada visa" based on our searcher's "Canada Trip" stak. The screenshot shows the regular results returned by Google as part of the normal Google SERP, but in addition, there are a number of results promoted to the top of the SERP by HeyStaks. These recommendations are results that stak members have previously found to be relevant for similar queries and help the searcher to discover results that friends or colleagues have found interesting, results that may otherwise be buried deep within Google's default list of results. Google's regular results can also be marked as community-relevant, and the screenshot shows how the bottom two results shown are tagged with the HeyStaks icon to indicate this.

A number of evaluation studies have been reported in the literature to describe the utility of HeyStaks in practical search settings; see [181–183]. Key to the HeyStaks proposition is that searchers need a better way to organize and share their search experiences, as opposed to the largely ad-hoc and manual mechanisms (email, word of mouth, face-to-face collaboration) that are currently the norm. HeyStaks provides these features, but do users actually take the time to create staks? Do they share these staks or join those created by others? Briefly, studies show that users do engage in a reasonable degree of stak creation and sharing activity; for example, on average, beta users created just over 3.2 new staks and joined a further 1.4. Perhaps this is not surprising: most users are likely to create a few staks and share them with a small network of colleagues or friends, at least initially.

Moreover, 85% of users engaged in search collaborations. The majority consumed results that were produced by at least one other user, and on average, these users consumed results from 7.5 other users. In contrast, 50% of users produced knowledge that was consumed by at least one other user, and each of these producers created search knowledge that was consumed by more than 12 other users on average. While users often re-selected promotions that stemmed from their own past search histories in a stak, 33% of the time, they selected results that had been contributed by other stak members. Thus, there is evidence that many users were helping other users and many users were helped by other users.

10.4   Social Search Engine – Search with Social Links

SSE [176] is a social search engine that uses both the collaborative analysis of search logs (similar to the AntWorld collaborative Quest idea) and the data obtained from the user's social network to personalize search results. Unlike other collaborative/social engines (e.g., I-SPY and HeyStaks), the user is not required to explicitly form search communities for various search topics for which collaboration is desired, but rather the system searches for relevant social ties. SSE looks for queries that were submitted by the user's friends, based on their social network, that are relevant to the user's current need. SSE merges results obtained from the collaborative analysis and the social network analysis, with results obtained by implementing standard search engines, to produce personalized and more accurate results for users. SSE integrates existing social network data (users friendships) and network metrics to rank documents. It considers the opinions of close friends about similar topics to the user's query topics as the more important metric for estimating document relevance. In addition, SSE integrates the socially sourced results with standard search results to better balance precision and recall. This balance is achieved by including results based on the opinions of friends and, at the same time, including results that were not previously identified by the these friends.

SSE consists of a standard, underlying search engine (based on Lucene.Net 2.4.0) that realizes standard keyword-based retrieval and which is expanded with modules to implement the two algorithms, based on query logs and social links. SSE also includes a merging algorithm that integrates the search results from all algorithms into a unified ranked list. The first personalization algorithm is a collaborative algorithm that looks for documents that received positive feedback from users for similar queries. The similarity between queries is based on the query terms, as well as on the documents that were returned for the query in a similar process to the AntWorld algorithm. The second personalization algorithm uses users' social links and follows the intuition that a document that was considered relevant by a close friend (from the user's social network) is more relevant than a document suggested by a more distant friend. The system, therefore, maintains a Friendship Value $FV(U_i, F_j)$ for each member of the social network and their friends in the network. The $FV(U_i, F_j)$ - friend value between user (U_i and another member on the network, F_j, is the centrality of the friend in the network normalized by the geodesic distance between them (i.e., the shortest path between (U_i and F_j). SSE builds on the idea of sharing knowledge between users of a community in order to enhance a user's search results. However, SSE is unique in integrating knowledge from two sources, both from the set of users' friends and from any other users that had the same need in the past and rated relevant documents for that need. SSE uses these sources, along with the user's personal profile, to personalize the search results; i.e., the system re-ranks results obtained from a standard search engine according to both of the above-mentioned sources.

Figure 9 presents the SSE's main processes. As soon as the user submits a query through the GUI, it is sent to a traditional SE whose results are returned to the ranker module. The query is also submitted to the Social Filter, which consults both the Collaborative knowledge base (KB) and the Social KB. The Social Filter returns a set of ranked documents from the set of documents that were previously seen by other users. As a last step, the ranker merges the two lists of documents (from the traditional SE and the Social Filter) and returns the merged list to the SSE GUI that displays the final set of ranked documents to the user. The user may evaluate any of the retrieved documents via the SSE GUI to enrich the system with additional feedback. For further results, the reader is referred to [176].

SSE demonstrated the benefit of personalizing search results by making use of collaborative knowledge and data from social networks. While sharing knowledge between users with similar needs improves search results, the integration of social information contributes further improvements. The SSE algorithm used the social network metric to indicate the strength of friendship between a current user and the user who is the source of the document to be returned as an indicator of the document's relevance for the current user. Specifically, a combination

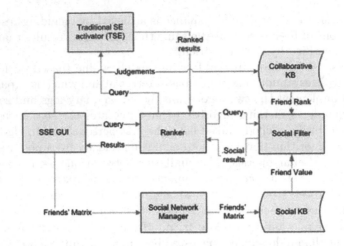

Fig. 9. The SSE architecture and outline recommendation model.

of the centrality of members in the network and the geodesic distance between the two users was used. As opposed to other community-based engines, SSE does not require users to join communities or to look for relevant searches that were performed by other users, but rather computes and identifies user communities, on the fly, that are based on network metrics. Hence, SSE could be used for short-term information needs; that is, ad hoc queries which constitute the most common usage of SEs. Although somewhat limited in scale, user studies were conducted to evaluate SSE (60 students in a lab and 7 search topics) that did not allow for a broad generalization of results, the trend was clearly towards better result accuracy when social information was used. Thus, given the popularity of social networks today, the integration of friendship data from several social networks has become very important, and recent publications [7] follow a similar approach, including a recent patent by Google [95].

11 Expanding the Borders of Social Search

As the chapter shows, social search technologies have demonstrated their ability to support various steps of the search process and leverage a range of social traces for this purpose. Yet, the majority of existing social search projects are very narrow in their coverage. Quite typically, a single social search system supports just one aspect of the search process and uses one kind of social data. We believe that overcoming this limitation is another important challenge of social search as a research field. It means developing approaches and systems that can use multiple kinds of data, support a wider set of search steps, and even support other kinds of information access. In the final section of this chapter, we showcase two further projects that go beyond the usual borders of social search systems to connect social search technologies with other types of social information access. The ASSIST platform, which was discussed earlier, demonstrates how social search can be integrated with social navigation, another social information access technology reviewed in Chap. 5 of this book [71]. ASSIST shows how these technologies can collaborate by using community browsing data to improve search, but also by using social data collected during search to improve navigation. We also consider the Aardvark [99] system, which crosses the boundary between traditional social searching and social Q&A technology, reviewed elsewhere in Chap. 3 of this book [154].

11.1 ASSIST – From Social Search to Social Navigation

The ASSIST platform [48, 72, 76] is a general-purpose approach to incorporating social visual cues into existing information access systems. ASSIST was designed to integrate elements of social search and social navigation into a single platform to assist users in both searching and browsing by using both active and passive social guidance. The motivation for ASSIST included I-SPY [179, 180] and Knowledge Sea [69]. While both systems used link annotation with social visual cues, I-SPY used search log data exclusively and focused on supporting user search, while Knowledge Sea used browsing data exclusively and focused on supporting navigation. The first attempts to use I-SPY to support browsing [30], and Knowledge Sea to support search [8] demonstrated how these approaches could be integrated, and led to the joint work on ASSIST.

ASSIST collects search and browsing data on a group level, and uses it in an information-exploration context (during search and browsing). The first implementation of this idea, called ASSIST-ACM, was developed for exploring research papers in the ACM Digital Library [72, 76]. It combined the hit matrix of I-SPY (Fig. 5) with a similar browsing-based hit matrix to offer I-SPY visual cues (Fig. 10) to the user as they used the ACM Digital Library.

To explain the work of the ASSIST platform in more detail, we will refer to a more advanced version of the system, ASSIST-YouTube, which was designed

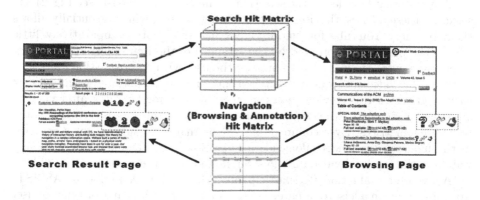

Fig. 10. ASSIST-ACM systems collects past user search and browsing traces and used them to help future users in both search and browsing.

to provide enhanced social supports, but for YouTube users [48,49]. ASSIST-YouTube offers a number of enhancements to the standard YouTube interface to improve search and browsing capabilities. In terms of active recommendations, ASSIST re-ranks lists of videos, offered by YouTube, to reflect accumulated community preferences.

In a search context, ASSIST re-ranks YouTube's search results in response to a user query Q and according to their relevance to Q. ASSIST leverages similar search history data as that used by I-SPY to assign relevance scores to videos, based on past search interactions. The relevance of video item I to query Q is estimated by the number of times I has been selected in response to Q as a fraction of the total number of selections across all items for Q. As in I-SPY, ASSIST also identifies videos that have been selected for similar queries (using a simple term-overlap similarity metric) and weights their relevance to their associated query Q_i by the similarity of Q_i to Q. These promotion candidates are ranked according to their weighted relevance score and are placed at the top of the result list for query Q.

In a browsing context, ASSIST re-ranks YouTube's *related videos*, which are displayed alongside the video that is currently being watched. This list is a valuable source of complementary content for engaging in browsing activities, and thus the position of videos within this list is important. ASSIST re-ranks the related video list, according to the items' contextual browsing popularity.

To provide passive recommendations, ASSIST augments content hyperlinks with visual social cues throughout the interface, highlighting areas of interest and suggesting paths through the wider information space. The presence of these cues signals previous encounters by community members with the content of the link. For example, when a user mouses over a cue icon, they are presented with the items search and browsing history by community members (Fig. 2). The search history presented conveys to users how the associated content has been chosen by a community member in relation to a query, as well as the strength of the item-query relationship (i.e. the relevance score). The mouse-over also includes a list of all queries that have led to the selection of the video in the past (see Fig. 2). By selecting these queries, the user can begin a new search, which essentially allows them to query YouTube for "more videos like this" with comparatively little effort. The query list is ordered by the strength of the item-query relationship. The mouse-overs are also used to provide the user with Amazon-style users who watched this video subsequently watched these recommendations. As mentioned earlier, in Sect. 2.1, if previous users engaged in browsing behavior after viewing a particular video (i.e., they selected a related video), this fact is recorded in the browse-hit matrices. By recommending videos that were subsequently watched in the mouseover provided alongside a hyperlink, the user may choose to skip watching the top-level video and go straight to one of the recommended videos.

A live-user evaluation (21 participants over a 14-week period of ASSIST-YouTube usage in a leisure-oriented context) uncovered three major types of user behavior in YouTube: traditional goal-directed search; direct browsing, following an externally recommended link; and casual browsing by watching interesting, but not search-specified videos. While the ASSIST-YouTube social recommendation engine was designed to assist only the first type of activity, the nature of its browsing support component also makes it useful for the social support of casual browsing. However, the social support of casual browsing turned out to be more challenging than the social support of goal-directed browsing. While the search goals of the users of a specific community have some reasonable overlap, their casual browsing is driven by their general interests, rather than focused goals. In particular, the evaluation highlighted how the effects and types of social enhancements should be engineered to match the user task with the target content repository. The original social support in ASSIST was engineered for a goal-driven search task, which needs to be done with minimal effort, versus an entertaining exploratory task with fewer time constraints.

11.2   Aardvark – From Social Search to Social Q&A

Aardvark [99] was yet another take on social search. Although it was subsequently acquired by Google (and subsequently shut down), it remains as a useful case-study in an alternative vision of social search. Aardvark is not a conventional web search engine, like Google or Bing: it is not designed to find web pages in response to user queries. Instead, it is closer to a Q&A service where users provide 'queries' in the form of questions, and the 'results' are answers returned by other users. We view this as a form of search, and include Aardvark here because of the central role that social information plays in the sourcing and ranking of its results (answers).

Briefly, the central idea behind Aardvark was to harness the knowledge of individuals to directly answer questions and queries posed by users. Its key contribution was its role as a social search platform, which attempted to capture and index the expertise and interests and social networks of users in order to match these users with incoming queries. Users ask a question using a variety of channels; for example, by instant message, email, web input, text message, or even voice. Aardvark routes the question to people in the user's extended social network who are likely to be able, and available, to answer it. Accordingly, there were four key features that distinguished Aardvark from more conventional search engines:

1. *Social Indexing.* Conventional search engines focus on indexing documents, but Aardvark focuses on indexing people and, in particular, on modeling user-topic and user-user relationships, rather than query-document relationships.
2. *Question Classification.* On receiving a new question (query) Aardvark must classify the topic of the question so as to relate it to users with the right type of expertise and interests.
3. *Question Routing.* Aardvark must route a new question to the right subset of users. It does this by using a variety of information, including an aspect model that captures the topical content of the question (and candidate answerers), social network information in order to connect the questioner with people they may know, and availability information based on historical responsiveness/availability data.

4. *Answer Ranking.* Once answers come to be generated, they must be ranked so that, ideally, the most helpful answer appears at the top of the ranking. Aardvark performed this ranking based on a number of factors, including topic expertise and availability, for example.

Compared to a traditional web search engine, where the challenge lies in finding the right document to satisfy a user's information need, the challenge in a social search engine like Aardvark lies in finding the right person to satisfy a user's information need. Moreover, while traditional search engines emphasize the authority of pages when it comes to ranking, with Aardvark, trust, which is based on intimacy and reputation, is critical. Moreover, we can expect that the type of questions that a user might submit to a search engine like Aardvark to be different from those entered into a search engine like Google. The latter are optimized for information retrieval and information discovery, whereas we might expect the former to be better suited for uncovering insightful user opinions on topics that are more challenging to convey as a simple search query.

The early evidence from Aardvark was promising. For example, [99] described the results of a preliminary trial to evaluate Aardvark against Google. This trial found, for example, that Aardvark was capable of successfully answering 71.5% of questions submitted with a mean answer rating of 3.93, as compared to 70.5% of questions answered by Google with a mean rating of 3.07. Admittedly, this trial was somewhat biased towards Aardvark in the sense that the 200 or so test questions originated with Aardvark and thus were not necessarily indicative of a typical Google search query. Nevertheless, it shows that for at least these type of questions Aardvark's human-powered social search has the potential to reliably deliver high-quality answers.

As a final, case study in this chapter, Aardvark serves as a useful indication that many opportunities remain for delivering more social search experiences beyond traditional search engines. This is likely to be especially important as the world of the web continues to move towards a mobile-first future.

12   Conclusions

Today, search is among the most popular ways that people access information and the search box has become an ever-present user interface component across all operating systems and most applications where information access and discovery is relevant. This is no more obvious than in the world of Web search, where search engines like Google, Bing, and Baidu continue to serve the information needs of millions of searchers, every hour of every day.

In the quest to consistently provide the best search experience to users, the world of Web search, in particular, has been evolving, and in this chapter we consider how the increasingly social world of the web has come to bear on web search. Specifically, this chapter attempts to review how web search has become (and will continue to become) more and more social, as new social signals come to be leveraged to make web search more relevant and personal for end users and communities of like-minded users.

To do this, we framed our treatment of social search along two key dimensions. First, we considered the sources and types of social information available, from links and logs, to tags and trails, annotations, and comments. Second, we considered three key stages of the search process and the opportunities to influence searching *before*, *during*, and *after* search. Accordingly, the main contribution of this chapter included a comprehensive review of how these sources of social information can be used throughout the traditional steps of the search process, including enhancing query formation, content indexing, result ranking, and final result presentation. Throughout this chapter, we provided concrete examples from the literature of the many and varied systems that have implemented different approaches to social search, and have concluded with a number of detailed case studies to highlight a number of seminal systems that served as important milestones in the evolution of social search to address the important challenges of this field.

References

1. Abel, F., Baldoni, M., Baroglio, C., Henze, N., Krause, D., Patti, V.: Context-based ranking in folksonomies. In: Proceedings of the 20th ACM Conference on Hypertext and Hypermedia, pp. 209–218. ACM (2009)
2. Abel, F., Frank, M., Henze, N., Krause, D., Plappert, D., Siehndel, P.: GroupMe! - where semantic web meets web 2.0. In: Aberer, K., et al. (eds.) ASWC/ISWC -2007. LNCS, vol. 4825, pp. 871–878. Springer, Heidelberg (2007). https://doi.org/10.1007/978-3-540-76298-0_63
3. Abel, F., Gao, Q., Houben, G.-J., Tao, K.: Analyzing user modeling on Twitter for personalized news recommendations. In: Konstan, J.A., Conejo, R., Marzo, J.L., Oliver, N. (eds.) UMAP 2011. LNCS, vol. 6787, pp. 1–12. Springer, Heidelberg (2011). https://doi.org/10.1007/978-3-642-22362-4_1
4. Aggarwal, N., Buitelaar, P.: Query expansion using Wikipedia and DBpedia. In: Forner, P., Karlgren, J., Womser-Hacker, C. (eds.) CLEF (Online Working Notes/Labs/Workshop) (2012)
5. Agichtein, E., Brill, E., Dumais, S.: Improving web search ranking by incorporating user behavior information. In: The 29th Annual International ACM SIGIR Conference on Research and Development in Information Retrieval, pp. 19–26. ACM Press (2006)
6. Agichtein, E., Brill, E., Dumais, S., Ragno, R.: Learning user interaction models for predicting web search result preferences. In: Proceedings of the 29th Annual International ACM SIGIR Conference on Research and Development in Information Retrieval, pp. 3–10. ACM (2006)
7. Agrawal, R., Golshan, B., Papalexakis, E.: Whither social networks for web search? In: Proceedings of the 21th ACM SIGKDD International Conference on Knowledge Discovery and Data Mining, pp. 1661–1670. ACM (2015)
8. Ahn, J.W., Farzan, R., Brusilovsky, P.: Social search in the context of social navigation. J. Korean Soc. Inf. Manag. **23**(2), 147–165 (2006)
9. Alhindi, A., Kruschwitz, U., Fox, C., Albakour, D.: Profile-based summarisation for web site navigation. ACM Trans. Inf. Syst. **33**(1), 4 (2015)
10. Almeida, R.B., Almeida, V.A.F.: A community-aware search engine. In: The Thirteenth International World Wide Web Conference, WWW 2004, pp. 413–421. ACM Press (2004)

11. Alonso, O., Bannur, S., Khandelwal, K., Kalyanaraman, S.: The world conversation: web page metadata generation from social sources. In: Proceedings of the 24th International Conference on World Wide Web, pp. 385–395. ACM (2015)
12. Amatriain, X., Lathia, N., Pujol, J.M., Kwak, H., Oliver, N.: The wisdom of the few: a collaborative filtering approach based on expert opinions from the web. In: Proceedings of the 32nd International ACM SIGIR Conference on Research and Development in Information Retrieval, pp. 532–539. ACM (2009)
13. Amitay, E., Carmel, D., Har'El, N., Koifman, S.O., Soffer, A., Yogev, S., Golbandi, N.: Social search and discovery using a unified approach. In: Proceedings of the 20th ACM Conference on Hypertext and Hypermedia, pp. 199–208. ACM (2009)
14. Amitay, E., Darlow, A., Konopnicki, D., Weiss, U.: Queries as anchors: selection by association. In: Proceedings of the 16th ACM Conference on Hypertext and Hypermedia, pp. 193–201 (2005)
15. Arguello, J., Elsas, J.L., Callan, J., Carbonell, J.: Document representation and query expansion models for blog recommendation. In: 2nd International Conference on Weblogs and Social Media, ICWSM 2008, pp. 10–18 (2008)
16. Ashyralyyev, S., Cambazoglu, B.B., Aykanat, C.: Incorporating the surfing behavior of web users into PageRank. In: 22nd ACM International Conference on Information and Knowledge Management, pp. 2351–2356. ACM (2013)
17. Baeza-Yates, R., Hurtado, C., Mendoza, M.: Improving search engines by query clustering. J. Am. Soc. Inf. Sci. Technol. 58(12), 1793–1804 (2007)
18. Baeza-Yates, R., Tiberi, A.: Extracting semantic relations from query logs. In: KDD 2007: Proceedings of the 13th ACM SIGKDD International Conference on Knowledge Discovery and Data Mining, pp. 76–85. ACM (2007)
19. Baeza-Yates, R.A., Hurtado, C.A., Mendoza, M.: Query recommendation using query logs in search engines. In: Current Trends in Database Technology - EDBT 2004 Workshops, Heraklion, Crete, Greece, 14–18 March 2004, Revised Selected Papers, pp. 588–596 (2004)
20. Balfe, E., Smyth, B.: Improving web search through collaborative query recommendation. In: Proceedings of the 16th European Conference on Artificial Intelligence, pp. 268–272 (2004)
21. Bao, S., Xue, G., Wu, X., Yu, Y., Fei, B., Su, Z.: Optimizing web search using social annotations. In: Proceedings of the 16th International Conference on World Wide Web, Banff, Alberta, Canada, 8–12 May 2007, pp. 501–510 (2007)
22. Baraglia, R., Cacheda, F., Carneiro, V., Fernández, D., Formoso, V., Perego, R., Silvestri, F.: Search shortcuts: a new approach to the recommendation of queries. In: Proceedings of the 2009 ACM Conference on Recommender Systems, pp. 77–84 (2009)
23. Barrett, R., Maglio, P.P.: Intermediaries: an approach to manipulating information streams. IBM Syst. J. 38(4), 629–641 (1999)
24. Bilenko, M., White, R.W.: Mining the search trails of surfing crowds: identifying relevant websites from user activity. In: The 17th International Conference on World Wide Web, pp. 51–60. ACM (2008)
25. Billerbeck, B., Scholer, F., Williams, H.E., Zobel, J.: Query expansion using associated queries. In: Proceedings of the Twelfth International Conference on Information and Knowledge Management, pp. 2–9. ACM, New York (2003)
26. Bischoff, K., Firan, C.S., Nejdl, W., Paiu, R.: Can all tags be used for search? In: Shanahan, J.G., Amer-Yahia, S., Zhang, Y., Kolcz, A., Chowdhury, A., Kelly, D. (eds.) The 17th ACM Conference on Conference on Information and Knowledge Management, pp. 203–212. ACM Press (2008)

27. Bollmann-Sdorra, P., Raghavan, V.V.: On the delusiveness of adopting a common space for modeling IR objects: are queries documents? J. Am. Soc. Inf. Sci. **44**(10), 579–587 (1993)

28. Borisov, A., Markov, I., de Rijke, M., Serdyukov, P.: A neural click model for web search. In: Proceedings of the 25th International Conference on World Wide Web, pp. 531–541. International World Wide Web Conferences Steering Committee (2016)

29. Boros, E., Kantor, P.B., Neu, D.J.: Pheromonic representation of user quests by digital structures. In: Proceedings of the 62nd Annual Meeting of the American Society for Information Science, pp. 633–642 (1999)

30. Boydell, O., Smyth, B.: From social bookmarking to social summarization: an experiment in community-based summary generation. In: Proceedings of the 2007 International Conference on Intelligent User Interfaces, 28–31 January 2007, Honolulu, Hawaii, USA, pp. 42–51 (2007)

31. Boydell, O., Smyth, B.: Social summarization in collaborative web search. Inf. Process. Manag. **46**(6), 782–798 (2010)

32. Briggs, P., Smyth, B.: On the role of trust in collaborative web search. Artif. Intell. Rev. **25**(1–2), 97–117 (2006)

33. Briggs, P., Smyth, B.: Harnessing trust in social search. In: Advances in Information Retrieval, 29th European Conference on IR Research, Rome, Italy, 2–5 April 2007, Proceedings, pp. 525–532 (2007)

34. Briggs, P., Smyth, B.: Trusted search communities. In: Proceedings of the 2007 International Conference on Intelligent User Interfaces, 28–31 January 2007, Honolulu, Hawaii, USA, pp. 337–340 (2007)

35. Briggs, P., Smyth, B.: Provenance, trust, and sharing in peer-to-peer case-based web search. In: Althoff, K.-D., Bergmann, R., Minor, M., Hanft, A. (eds.) ECCBR 2008. LNCS (LNAI), vol. 5239, pp. 89–103. Springer, Heidelberg (2008). https://doi.org/10.1007/978-3-540-85502-6_6

36. Brin, S., Page, L.: The anatomy of a large-scale hypertextual (web) search engine. In: Ashman, H., Thistewaite, P. (eds.) Seventh International World Wide Web Conference, vol. 30, pp. 107–117. Elsevier Science B.V, Amsterdam (1998)

37. Brusilovsky, P., Farzan, R., Ahn, J.W.: Comprehensive personalized information access in an educational digital library. In: The 5th ACM/IEEE-CS Joint Conference on Digital Libraries, pp. 9–18. ACM Press (2005)

38. Brusilovsky, P., He, D.: Introduction to social information access. In: Brusilovsky, P., He, D. (eds.) Social Information Access. LNCS, vol. 10100, pp. 1–18. Springer, Cham (2018)

39. Bush, V.: As we may think. The Atlantic, July 1945

40. Cai, Y., Li, Q.: Personalized search by tag-based user profile and resource profile in collaborative tagging systems. In: The 19th ACM Conference on Information and Knowledge Management (CIKM 2010), pp. 969–978. ACM (2010)

41. Carmel, D., Zwerdling, N., Guy, I., Koifman, S.O., Har'el, N., Ronen, I., Uziel, E., Yogev, S., Chernov, S.: Personalized social search based on the user's social network. In: Proceedings of the 18th ACM Conference on Information and Knowledge Management, pp. 1227–1236. ACM (2009)

42. Carpineto, C., Romano, G.: A survey of automatic query expansion in information retrieval. ACM Comput. Surv. **44**(1), 1:1–1:50 (2012)

43. Carriere, S.J., Kazman, R.: WebQuery: searching and visualizing the web through connectivity. In: Sixth International World Wide Web Conference, pp. 1257–1267. Elsevier (1997)

44. Chakrabarti, S., Dom, B., Raghavan, P., Rajagopalan, S., Gibson, D., Kleinberg, J.: Automatic resource compilation by analyzing hyperlink structure and associated text. In: Proceedings of the Seventh International Conference on World Wide Web 7, vol. 30, pp. 65–74. Elsevier Science Publishers B.V. (1998)

45. Chang, Y., Dong, A., Kolari, P., Zhang, R., Inagaki, Y., Diaz, F., Zha, H., Liu, Y.: Improving recency ranking using Twitter data. ACM Trans. Intell. Syst. Technol. 4(1), Article No. 4 (2013)

46. Chapelle, O., Zhang, Y.: A dynamic Bayesian network click model for web search ranking. In: Proceedings of the 18th International Conference on World Wide Web, pp. 1–10. ACM (2009)

47. Choochaiwattana, W.: Using social annotation to improve web search. Ph.D. thesis (2008). http://d-scholarship.pitt.edu/7832/

48. Coyle, M., Freyne, J., Brusilovsky, P., Smyth, B.: Social information access for the rest of us: an exploration of social YouTube. In: Nejdl, W., Kay, J., Pu, P., Herder, E. (eds.) AH 2008. LNCS, vol. 5149, pp. 93–102. Springer, Heidelberg (2008). https://doi.org/10.1007/978-3-540-70987-9_12

49. Coyle, M., Freyne, J., Farzan, R., Smyth, B., Brusilovsky, P.: Reducing click distance through social adaptive interfacing. In: ReColl 2008, International Workshop on Recommendation and Collaboration at 2008 International Conference on Intelligent User Interfaces (2008)

50. Coyle, M., Smyth, B.: SearchGuide: beyond the results page. In: De Bra, P.M.E., Nejdl, W. (eds.) AH 2004. LNCS, vol. 3137, pp. 296–299. Springer, Heidelberg (2004). https://doi.org/10.1007/978-3-540-27780-4_36

51. Coyle, M., Smyth, B.: Supporting intelligent web search. ACM Trans. Internet Techn. 7(4), 20 (2007)

52. Craswell, N., Billerbeck, B., Fetterly, D., Najork, M.: Robust query rewriting using anchor data. In: Proceedings of the Sixth ACM International Conference on Web Search and Data Mining, pp. 335–344. ACM, New York (2013)

53. Craswell, N., Hawking, D., Robertson, S.: Effective site finding using link anchor information. In: 24th Annual International ACM SIGIR Conference on Research and Development in Information Retrieval, pp. 250–257. ACM Press (2001)

54. Craswell, N., Szummer, M.: Random walks on the click graph. In: 30th Annual International ACM SIGIR Conference on Research and Development in Information Retrieval, pp. 239–246 (2007)

55. Craswell, N., Zoeter, O., Taylor, M., Ramsey, B.: An experimental comparison of click position-bias models. In: Proceedings of the 2008 International Conference on Web Search and Data Mining, pp. 87–94. ACM (2008)

56. Cui, H., Wen, J.R., Nie, J.Y., Ma, W.Y.: Probabilistic query expansion using query logs. In: Proceedings of the 11th International Conference on World Wide Web, pp. 325–332. ACM (2002)

57. Culliss, G.: Method for organizing information, 20 June 2000. http://www.google.com/patents/US6078916, US Patent 6,078,916

58. Culliss, G.: Method for organizing information, 11 January 11 2000. http://www.google.com/patents/US6014665

59. Dang, V., Croft, B.W.: Query reformulation using anchor text. In: Proceedings of the Third ACM International Conference on Web Search and Data Mining, pp. 41–50. ACM, New York (2010)

60. Dennis, S., Bruza, P., McArthur, R.: Web searching: a process-oriented experimental study of three interactive search paradigms. J. Am. Soc. Inf. Sci. Technol. 52(2), 120–133 (2002)

61. Dmitriev, P., Eiron, N., Fontoura, M., Shekita, E.: Using annotations in enterprise search. In: 15th International Conference on World Wide Web, pp. 811–817. ACM (2006)

62. Dong, A., Zhang, R., Kolari, P., Bai, J., Diaz, F., Chang, Y., Zheng, Z., Zha, H.: Time is of the essence: improving recency ranking using Twitter data. In: Proceedings of the 19th International Conference on World Wide Web, pp. 331–340. ACM (2010)

63. Dou, Z., Song, R., Yuan, X., Wen, J.: Are click-through data adequate for learning web search rankings? In: Proceedings of the 17th ACM Conference on Information and Knowledge Management, pp. 73–82. ACM (2008)

64. Dupret, G., Piwowarski, B.: A user browsing model to predict search engine click data from past observations. In: Proceedings of the 31st Annual International ACM SIGIR Conference on Research and Development in Information Retrieval, pp. 331–338. ACM (2008)

65. Eirinaki, M., Vazirgiannis, M.: Usage-based PageRank for web personalization. In: The Fifth IEEE International Conference on Data Mining, pp. 130–137 (2005)

66. Esparza, S.G., O'Mahony, M.P., Smyth, B.: On the real-time web as a source of recommendation knowledge. In: Proceedings of the Fourth ACM Conference on Recommender Systems, pp. 305–308. ACM (2010)

67. Evans, B.M., Chi, E.H.: Towards a model of understanding social search. In: 2008 ACM Conference on Computer Supported Cooperative Work, pp. 485–494. ACM (2008)

68. Evans, B.M., Chi, E.H.: An elaborated model of social search. Inf. Process. Manag. **46**(6), 656–678 (2010)

69. Farzan, R., Brusilovsky, P.: Social navigation support through annotation-based group modeling. In: Ardissono, L., Brna, P., Mitrovic, A. (eds.) UM 2005. LNCS (LNAI), vol. 3538, pp. 463–472. Springer, Heidelberg (2005). https://doi.org/10.1007/11527886_64

70. Farzan, R., Brusilovsky, P.: Social navigation support for information seeking: if you build it, will they come? In: Houben, G.-J., McCalla, G., Pianesi, F., Zancanaro, M. (eds.) UMAP 2009. LNCS, vol. 5535, pp. 66–77. Springer, Heidelberg (2009). https://doi.org/10.1007/978-3-642-02247-0_9

71. Farzan, R., Brusilovsky, P.: Social navigation. In: Brusilovsky, P., He, D. (eds.) Social Information Access. LNCS, vol. 10100, pp. 142–180. Springer, Cham (2018)

72. Farzan, R., Coyle, M., Freyne, J., Brusilovsky, P., Smyth, B.: ASSIST: adaptive social support for information space traversal. In: 18th Conference on Hypertext and Hypermedia, pp. 199–208. ACM Press (2007)

73. Fernquist, J., Chi, E.H.: Perception and understanding of social annotations in web search. In: 22nd International World Wide Web Conference, WWW 2013, Rio de Janeiro, Brazil, pp. 403–412, 13–17 May 2013

74. Fitzpatrick, L., Dent, M.: Automatic feedback using past queries: social searching? In: Proceedings of the 20th Annual International ACM SIGIR Conference on Research and Development in Information Retrieval, pp. 306–313 (1997)

75. Fonseca, B., Golgher, P., De Moura, E., Pôssas, B., Ziviani, N.: Discovering search engine related queries using association rules. J. Web Eng. **2**(4), 215–227 (2003)

76. Freyne, J., Farzan, R., Brusilovsky, P., Smyth, B., Coyle, M.: Collecting community wisdom: integrating social search and social navigation. In: International Conference on Intelligent User Interfaces, pp. 52–61. ACM Press (2007)

77. Freyne, J., Smyth, B.: An experiment in social search. In: De Bra, P.M.E., Nejdl, W. (eds.) AH 2004. LNCS, vol. 3137, pp. 95–103. Springer, Heidelberg (2004). https://doi.org/10.1007/978-3-540-27780-4_13

78. Fujii, A.: Modeling anchor text and classifying queries to enhance web document retrieval. In: The 17th International Conference on World Wide Web, pp. 337–346. ACM (2008)

79. Furnas, G.W., Landauer, T.K., Gomez, L.M., Dumais, S.T.: The vocabulary problem in human-system communication. Commun. ACM **30**(11), 964–971 (1987)

80. Furnas, G.W.: Experience with an adaptive indexing scheme. SIGCHI Bull. **16**(4), 131–135 (1985)

81. Furuta, R., Shipman III, F.M., Marshall, C.C., Brenner, D., Hsieh, H.W.: Hypertext paths and the world-wide web: experience with Walden's paths. In: Bernstein, M., Carr, L., Østerbye, K. (eds.) Eight ACM International Hypertext Conference (Hypertext 1997), pp. 167–176. ACM (1997)

82. Gao, J., Yuan, W., Li, X., Deng, K., Nie, J.: Smoothing clickthrough data for web search ranking. In: Proceedings of the 32nd International ACM SIGIR Conference on Research and Development in Information Retrieval, pp. 355–362. ACM (2009)

83. Gauch, S., Speretta, M., Chandramouli, A., Micarelli, A.: User profiles for personalized information access. In: Brusilovsky, P., Kobsa, A., Nejdl, W. (eds.) The Adaptive Web. LNCS, vol. 4321, pp. 54–89. Springer, Heidelberg (2007). https://doi.org/10.1007/978-3-540-72079-9_2

84. Gibson, D., Kleinberg, J., Raghavan, P.: Inferring web communities from link topology. In: Ninth ACM International Hypertext Conference, pp. 50–57. ACM Press (1998)

85. Glance, N.: Community search assistant. In: Proceedings of the 6th International Conference on Intelligent User Interfaces, pp. 91–96. ACM (2001)

86. Golder, S.A., Huberman, B.A.: Usage patterns of collaborative tagging systems. J. Inf. Sci. **32**(2), 198–208 (2006)

87. Graus, D., Tsagkias, M., Weerkamp, W., Meij, E., de Rijke, M.: Dynamic collective entity representations for entity ranking. In: Proceedings of the Ninth ACM International Conference on Web Search and Data Mining, pp. 595–604. ACM (2016)

88. Guan, Z., Cutrell, E.: What are you looking for?: an eye-tracking study of information usage in web search. In: ACM SIGCHI Conference on Human Factors in Computing Systems, pp. 407–416. ACM Press (2007)

89. Guo, F., Liu, C., Kannan, A., Minka, T., Taylor, M., Wang, Y., Faloutsos, C.: Click chain model in web search. In: Proceedings of the 18th International Conference on World Wide Web, pp. 11–20. ACM (2009)

90. Guo, Y.Z., Ramamohanarao, K., Park, L.A.F.: Personalized PageRank for web page prediction based on access time-length and frequency. In: Lin, T.Y., Haas, L., Kacprzyk, J., Motwani, R., Broder, A., Ho, H. (eds.) The 2007 International Conference on Web Intelligence, pp. 687–690. IEEE (2007)

91. Hammond, T., Hannay, T., Lund, B., Scott, J.: Social bookmarking tools (i): a general review. D-Lib Mag. **11**(4) (2005)

92. Han, S., He, D.: Network-based social search. In: Brusilovsky, P., He, D. (eds.) Social Information Access. LNCS, vol. 10100, pp. 277–309. Springer, Cham (2018)

93. Han, S., He, D., Yue, Z., Brusilovsky, P.: Supporting cross-device web search with social navigation-based mobile touch interactions. In: Ricci, F., Bontcheva, K., Conlan, O., Lawless, S. (eds.) UMAP 2015. LNCS, vol. 9146, pp. 143–155. Springer, Cham (2015). https://doi.org/10.1007/978-3-319-20267-9_12

94. Harman, D.: Information retrieval. In: Relevance Feedback and Other Query Modification Techniques, pp. 241–263. Prentice-Hall, Inc. (1992)

95. Harrington, T., Shenoy, R., Najork, M., Panigrahy, R.: Social network recommended content and recommending members for personalized search results, 3 February 2015. https://www.google.com/patents/US8949232, US Patent 8,949,232

96. Hearst, M.A.: TileBars: visualization of term distribution information in full text information access. In: Human Factors in Computing Systems, CHI 1995 Conference Proceedings, Denver, Colorado, USA, 7–11 May 1995, pp. 59–66 (1995)

97. Heymann, P., Koutrika, G., Garcia-Molina, H.: Can social bookmarking improve web search? In: The International Conference on Web Search and Web Data Mining (WSDM 2008), pp. 195–206 (2008)

98. Hill, W.C., Hollan, J.D., Wroblewski, D., McCandless, T.: Edit wear and read wear. In: Proceedings of the SIGCHI Conference on Human Factors in Computing Systems, pp. 3–9. ACM, New York (1992)

99. Horowitz, D., Kamvar, S.: The anatomy of a large-scale social search engine. In: The 19th International Conference on World Wide Web, pp. 431–440. ACM (2010)

100. Hotho, A., Jäschke, R., Schmitz, C., Stumme, G.: Information retrieval in folksonomies: search and ranking. In: Sure, Y., Domingue, J. (eds.) ESWC 2006. LNCS, vol. 4011, pp. 411–426. Springer, Heidelberg (2006). https://doi.org/10.1007/11762256_31

101. Huang, C.K., Chien, L.F., Oyang, Y.J.: Relevant term suggestion in interactive web search based on contextual information in query session logs. J. Am. Soc. Inf. Sci. Technol. 54(7), 638–649 (2003)

102. Huberman, B.A., Adamic, L.A.: Novelty and social search in the world wide web. CoRR cs.MA/9809025 (1998). http://arxiv.org/abs/cs.MA/9809025

103. Jiang, J., Yue, Z., Han, S., He, D.: Finding readings for scientists from social websites. In: 35th International ACM SIGIR Conference on Research and Development in Information Retrieval, pp. 1075–1076 (2012)

104. Joachims, T.: Optimizing search engines using clickthrough data. In: Proceedings of the Eighth ACM SIGKDD International Conference on Knowledge Discovery and Data Mining (KDD 2002), pp. 133–142. ACM (2002)

105. Joachims, T., Granka, L., Pan, B., Hembrooke, H., Gay, G.: Accurately interpreting clickthrough data as implicit feedback. In: 28th Annual International ACM SIGIR Conference, pp. 154–161. ACM Press (2005)

106. Joachims, T., Radlinski, F.: Search engines that learn from implicit feedback. Computer 40(8), 34–40 (2007)

107. Jung, S., Harris, K., Webster, J., Herlocker, J.L.: Serf: integrating human recommendations with search. In: Proceedings of the Thirteenth ACM International Conference on Information and Knowledge Management, pp. 571–580. ACM (2004)

108. Kahan, J., Koivunen, M.R., Prud'Hommeaux, E., Swick, R.R.: Annotea: an open RDF infrastructure for shared web annotations. Comput. Netw. 39(5), 589–608 (2002)

109. Kammerer, Y., Nairn, R., Pirolli, P., Chi, E.H.: Signpost from the masses: learning effects in an exploratory social tag search browser. In: 27th International Conference on Human Factors in Computing Systems, pp. 625–634 (2009)

110. Kantor, P., Boros, E., Melamed, B., Neu, D., Menkov, V., Shi, Q., Kim, M.H.: Ant world (demonstration abstract). In: Proceedings of the 22nd Annual International ACM SIGIR Conference on Research and Development in Information Retrieval, p. 323. ACM (1999)

111. Kantor, P.B., Boros, E., Melamed, B., Meñkov, V., Shapira, B., Neu, D.J.: Capturing human intelligence in the net. Commun. ACM **43**(8), 112–115 (2000)
112. Katariya, S., Kveton, B., Szepesvári, C., Wen, Z.: DCM bandits: learning to rank with multiple clicks. In: Proceedings of the 33rd International Conference on International Conference on Machine Learning, vol. 48, pp. 1215–1224. JMLR.org (2016)
113. Keane, M., O'Brien, M., Smyth, B.: Are people biased in their use of search engines? Commun. ACM **51**(2), 49–52 (2008)
114. Kemp, C., Ramamohanarao, K.: Long-term learning for web search engines. In: Elomaa, T., Mannila, H., Toivonen, H. (eds.) PKDD 2002. LNCS, vol. 2431, pp. 263–274. Springer, Heidelberg (2002). https://doi.org/10.1007/3-540-45681-3_22
115. Kleinberg, J.: Authoritative sources in a hyperlinked environment. J. ACM **46**(5), 604–632 (1999)
116. Kluver, D., Ekstrand, M., Konstan, J.: Rating-based collaborative filtering: algorithms and evaluation. In: Brusilovsky, P., He, D. (eds.) Social Information Access. LNCS, vol. 10100, pp. 344–390. Springer, Cham (2018)
117. Koolen, M., Kamps, J.: The importance of anchor text for ad hoc search revisited. In: Proceedings of the 33rd International ACM SIGIR Conference on Research and Development in Information Retrieval, pp. 122–129. ACM (2010)
118. Korfhage, R.R.: Information Storage and Retrieval. Wiley Computer Publishing, Hoboken (1997)
119. Kraft, R., Zien, J.: Mining anchor text for query refinement. In: Proceedings of the 13th International Conference on World Wide Web, pp. 666–674. ACM, New York (2004)
120. Kramár, T., Barla, M., Bieliková, M.: Disambiguating search by leveraging a social context based on the stream of user's activity. In: De Bra, P., Kobsa, A., Chin, D. (eds.) UMAP 2010. LNCS, vol. 6075, pp. 387–392. Springer, Heidelberg (2010). https://doi.org/10.1007/978-3-642-13470-8_37
121. Kruschwitz, U., Lungley, D., Albakour, M.D., Song, D.: Deriving query suggestions for site search. J. Am. Soc. Inf. Sci. Technol. **64**(10), 1975–1994 (2013)
122. Kulkarni, C., Chi, E.: All the news that's fit to read: a study of social annotations for news reading. In: 2013 ACM SIGCHI Conference on Human Factors in Computing Systems, CHI 2013, Paris, France, 27 April–2 May 2013, pp. 2407–2416 (2013)
123. Lee, C., Croft, B.: Incorporating social anchors for ad hoc retrieval. In: Proceedings of the 10th Conference on Open Research Areas in Information Retrieval, pp. 181–188 (2013)
124. Lehikoinen, J., Salminen, I., Aaltonen, A., Huuskonen, P., Kaario, J.: Metasearches in peer-to-peer networks. Pers. Ubiquitous Comput. **10**(6), 357–367 (2006)
125. Lehmann, J., Isele, R., Jakob, M., Jentzsch, A., Kontokostas, D., Mendes, P.N., Hellmann, S., Morsey, M., van Kleef, P., Auer, S., et al.: DBpedia-a large-scale, multilingual knowledge base extracted from Wikipedia. Seman. Web **6**(2), 167–195 (2015)
126. Lerman, K., Plangprasopchok, A., Wong, C.: Personalizing image search results on flickr. In: AAAI07 workshop on Intelligent Information Personalization (2007). http://arxiv.org/abs/0704.1676
127. Li, L., Otsuka, S., Kitsuregawa, M.: Finding related search engine queries by web community based query enrichment. World Wide Web **13**(1), 121–142 (2010)

128. Li, Y., Luk, W.P.R., Ho, K.S.E., Chung, F.L.K.: Improving weak ad-hoc queries using Wikipedia as external corpus. In: Proceedings of the 30th Annual International ACM SIGIR Conference on Research and Development in Information Retrieval, pp. 797–798. ACM (2007)

129. Lieberman, H.: Letizia: an agent that assists web browsing. In: The Fourteenth International Joint Conference on Artificial Intelligence, pp. 924–929 (1995)

130. Liu, Y., Gao, B., Liu, T.Y., Zhang, Y., Ma, Z., He, S., Li, H.: BrowseRank: letting web users vote for page importance. In: 31st Annual International ACM SIGIR Conference on Research and Development in Information Retrieval, pp. 451–458. ACM (2008)

131. Liu, Y., Liu, T.Y., Gao, B., Ma, Z., Li, H.: A framework to compute page importance based on user behaviors. Inf. Retr. 13(1), 22–45 (2010)

132. Ma, H., Yang, H., King, I., Lyu, M.R.: Learning latent semantic relations from clickthrough data for query suggestion. In: Shanahan, J.G., Amer-Yahia, S., Zhang, Y., Kolcz, A., Chowdhury, A., Kelly, D. (eds.) The 17th ACM Conference on Conference on Information and Knowledge Management: CIKM 2008, pp. 709–718. ACM Press (2008)

133. Maron, M.E., Kuhns, J.L.: On relevance, probabilistic indexing and information retrieval. J. ACM 7(3), 216–244 (1960)

134. Massa, P., Avesani, P.: Trust-aware recommender systems. In: 2007 ACM Conference on Recommender Systems, pp. 17–24. ACM (2007)

135. McBryan, O.A.: GENVL and WWWW: tools for taming the web. In: The 1st International World Wide Web Conference, pp. 79–90 (1994)

136. McCreadie, R., Macdonald, C., Ounis, I.: News article ranking: leveraging the wisdom of bloggers. In: RIAO 2010 - 9th RIAO Conference (2010)

137. McDonnell, M., Shiri, A.: Social search: a taxonomy of, and a user-centred approach to, social web search. Program 45(1), 6–28 (2011)

138. McNally, K., O'Mahony, M.P., Smyth, B.: A comparative study of collaboration-based reputation models for social recommender systems. User Model. User-Adapt. Interact. 24(3), 219–260 (2014)

139. Micarelli, A., Gasparetti, F., Sciarrone, F., Gauch, S.: Personalized search on the world wide web. In: Brusilovsky, P., Kobsa, A., Nejdl, W. (eds.) The Adaptive Web. LNCS, vol. 4321, pp. 195–230. Springer, Heidelberg (2007). https://doi.org/10.1007/978-3-540-72079-9_6

140. Millen, D., Whittaker, S., Yang, M., Feinberg, J.: Supporting social search with social bookmarking. In: The HCIC 2007 Winter Workshop (2007). http://www.hcic.org/hcic2007/papers.phtml

141. Millen, D.R., Feinberg, J., Kerr, B.: Dogear: social bookmarking in the enterprise. In: SIGCHI Conference on Human Factors in Computing Systems, CHI 2006, pp. 111–120 (2006)

142. Millen, D.R., Yang, M., Whittaker, S., Feinberg, J.: Social bookmarking and exploratory search. In: Bannon, L.J., Wagner, I., Gutwin, C., Harper, R.H.R., Schmidt, K. (eds.) ECSCW 2007. Springer, London (2007). https://doi.org/10.1007/978-1-84800-031-5_2

143. Miller, G.A.: WordNet: a lexical database for English. Commun. ACM 38(11), 39–41 (1995)

144. Mishne, G., Lin, J.: Twanchor text: a preliminary study of the value of tweets as anchor text. In: Proceedings of the 35th International ACM SIGIR Conference on Research and Development in Information Retrieval, pp. 1159–1160. ACM (2012)

145. Muhammad, K., Lawlor, A., Rafter, R., Smyth, B.: Great explanations: opinionated explanations for recommendations. In: Hüllermeier, E., Minor, M. (eds.) ICCBR 2015. LNCS (LNAI), vol. 9343, pp. 244–258. Springer, Cham (2015). https://doi.org/10.1007/978-3-319-24586-7_17

146. Muhammad, K.I., Lawlor, A., Smyth, B.: A live-user study of opinionated explanations for recommender systems. In: Proceedings of the 21st International Conference on Intelligent User Interfaces, IUI 2016, Sonoma, CA, USA, 07–10 March 2016, pp. 256–260 (2016)

147. Muller, M., Millen, D., Feinberg, J.: Information curators in an enterprise file sharing system. In: 11th European Conference on Computer Supported Cooperative Work, ECSCW 2009 (2009)

148. Muralidharan, A., Gyongyi, Z., Chi, E.: Social annotations in web search. In: Proceedings of the SIGCHI Conference on Human Factors in Computing Systems, pp. 1085–1094. ACM (2012)

149. Navarro Bullock, B., Hotho, A., Stumme, G.: Accessing information with tags: search and ranking. In: Brusilovsky, P., He, D. (eds.) Social Information Access. LNCS, vol. 10100, pp. 310–343. Springer, Cham (2018)

150. Noll, M.G., Meinel, C.: Web search personalization via social bookmarking and tagging. In: Aberer, K., et al. (eds.) ASWC/ISWC -2007. LNCS, vol. 4825, pp. 367–380. Springer, Heidelberg (2007). https://doi.org/10.1007/978-3-540-76298-0_27

151. Noll, M., Meinel, C.: The metadata triumvirate: social annotations, anchor texts and search queries. In: IEEE/WIC/ACM International Conference on Web Intelligence and Intelligent Agent Technology, WI-IAT 2008, vol. 1, pp. 640–647. IEEE (2008)

152. O'Donovan, J., Smyth, B.: Trust in recommender systems. In: IUI 2005: Proceedings of the 10th International Conference on Intelligent User Interfaces, pp. 167–174. ACM Press (2005)

153. Ogilvie, P., Callan, J.: Combining document representations for known-item search. In: Proceedings of the 26th Annual International ACM SIGIR Conference on Research and Development in Information Retrieval, pp. 143–150. ACM (2003)

154. Oh, S.: Social Q&A. In: Brusilovsky, P., He, D. (eds.) Social Information Access. LNCS, vol. 10100, pp. 75–107. Springer, Cham (2018)

155. Pantel, P., Gamon, M., Alonso, O., Haas, K.: Social annotations: utility and prediction modeling. In: The 35th International ACM SIGIR Conference on Research and Development in Information Retrieval, Portland, OR, USA, 12–16 August 2012, pp. 285–294 (2012)

156. Parikh, N., Sundaresan, N.: Inferring semantic query relations from collective user behavior. In: Shanahan, J.G., Amer-Yahia, S., Zhang, Y., Kolcz, A., Chowdhury, A., Kelly, D. (eds.) The 17th ACM Conference on Conference on Information and Knowledge Management: CIKM 2008, pp. 349–358. ACM Press (2008)

157. Phelan, O., Mccarthy, K., Smyth, B.: Yokie - a curated, real-time search and discovery system using Twitter. In: RSWEB 2011: 3rd Workshop on Recommender Systems and the Social Web at RecSys 2011 (2011)

158. Piwowarski, B., Dupret, G., Jones, R.: Mining user web search activity with layered Bayesian networks or how to capture a click in its context. In: Proceedings of the Second ACM International Conference on Web Search and Data Mining, pp. 162–171. ACM (2009)

159. Poblete, B., Baeza-Yates, R.: Query-sets: using implicit feedback and query patterns to organize web documents. In: The 17th International Conference on World Wide Web, pp. 41–50. ACM (2008)

160. Poblete, B., Castillo, C., Gionis, A.: Dr. Searcher and Mr. Browser: a unified hyperlink-click graph. In: Shanahan, J.G., Amer-Yahia, S., Zhang, Y., Kolcz, A., Chowdhury, A., Kelly, D. (eds.) The 17th ACM Conference on Conference on Information and Knowledge Management: CIKM 2008, pp. 1123–1132. ACM Press (2008)

161. Porter, M.F.: Implementing a probabilistic information retrieval system. Inf. Technol.: Res. Dev. 1(2), 131–156 (1982)

162. Radlinski, F., Joachims, T.: Query chains: learning to rank from implicit feedback. In: Proceedings of the 11th ACM SIGKDD International Conference on Knowledge Discovery in Data Mining, pp. 239–248. ACM (2005)

163. Raghavan, V.V., Sever, H.: On the reuse of past optimal queries. In: 18th Annual International ACM SIGIR Conference on Research and Development in Information Retrieval, pp. 344–350. ACM (1995)

164. Ramage, D., Heymann, P., Manning, C., Garcia-Molina, H.: Clustering the tagged web. In: Proceedings of the Second ACM International Conference on Web Search and Data Mining, pp. 54–63 (2009)

165. Remde, J.R., Gomez, L.M., Landauer, T.K.: SuperBook: an automatic tool for information exploration—hypertext? In: The ACM Conference on Hypertext, Hypertext 1987, pp. 175–188 (1987)

166. Ritchie, A., Robertson, S., Teufel, S.: Comparing citation contexts for information retrieval. In: Shanahan, J.G., Amer-Yahia, S., Zhang, Y., Kolcz, A., Chowdhury, A., Kelly, D. (eds.) The 17th ACM Conference on Information and Knowledge Management: CIKM 2008, pp. 213–222. ACM Press (2008)

167. Rocchio, J.J.: Relevance feedback in information retrieval (Chap. 14). In: Salton, G. (ed.) The SMART Retrieval System: Experiments in Automatic Document Processing. Prentice-Hall Series in Automatic Computation, pp. 313–323. Prentice-Hall, Englewood Cliffs (1971)

168. Rowlands, T., Hawking, D., Sankaranarayana, R.: New-web search with microblog annotations. In: Proceedings of the 19th International Conference on World Wide Web, pp. 1293–1296. ACM (2010)

169. Salton, G.: The SMART Retrieval System: Experiments in Automatic Document Processing. Prentice-Hall, Englewood Cliffs (1971)

170. Scholer, F., Williams, H.E.: Query association for effective retrieval. In: ACM 11th Conference on Information and Knowledge Management, CIKM 2002, pp. 324–331 (2002)

171. Scholer, F., Williams, H., Turpin, A.: Query association surrogates for web search. J. Am. Soc. Inf. Sci. 55(7), 637–650 (2004)

172. di Sciascio, C., Brusilovsky, P., Veas, E.: A study on user-controllable social exploratory search. In: 23rd International Conference on Intelligent User Interfaces. ACM (2018)

173. Shami, N.S., Muller, M., Millen, D.: Social search and metadata in predicting file discovery. In: Fifth International AAAI Conference on Weblogs and Social Media, pp. 337–344. AAAI Publications (2011)

174. Shapira, B., Kantor, P.B., Melamed, B.: The effect of extrinsic motivation on user behavior in a collaborative information finding system. J. Am. Soc. Inf. Sci. Technol. 52(11), 879–887 (2001)

175. Shapira, B., Taieb-Maimon, M., Moskowitz, A.: Study of the usefulness of known and new implicit indicators and their optimal combination for accurate inference of users interests. In: Proceedings of the 2006 ACM Symposium on Applied Computing, pp. 1118–1119. ACM (2006)

176. Shapira, B., Zabar, B.: Personalized search: Integrating collaboration and social networks. J. Am. Soc. Inf. Sci. Technol. 62(1), 146–160 (2011)

177. Shipman, F.M., Marshall, C.C.: Spatial hypertext: an alternative to navigational and semantic links. ACM Comput. Surv. 31(4es), Article No. 14 (1999)

178. Smyth, B., Balfe, E., Boydell, O., Bradley, K., Briggs, P., Coyle, M., Freyne, J.: A live-user evaluation of collaborative web search. In: IJCAI-2005, Proceedings of the Nineteenth International Joint Conference on Artificial Intelligence, Edinburgh, Scotland, UK, 30 July–5 August 2005, pp. 1419–1424 (2005)

179. Smyth, B., Balfe, E., Briggs, P., Coyle, M., Freyne, J.: Collaborative web search. In: IJCAI-2003, Proceedings of the Eighteenth International Joint Conference on Artificial Intelligence, Acapulco, Mexico, 9–15 August 2003, pp. 1417–1419 (2003)

180. Smyth, B., Balfe, E., Freyne, J., Briggs, P., Coyle, M., Boydell, O.: Exploiting query repetition and regularity in an adaptive community-based web search engine. User Model. User-Adapt. Interact. 14(5), 383–423 (2004)

181. Smyth, B., Briggs, P., Coyle, M., O'Mahony, M.: Google shared. A case-study in social search. In: Houben, G.-J., McCalla, G., Pianesi, F., Zancanaro, M. (eds.) UMAP 2009. LNCS, vol. 5535, pp. 283–294. Springer, Heidelberg (2009). https://doi.org/10.1007/978-3-642-02247-0_27

182. Smyth, B., Coyle, M., Briggs, P.: Heystaks: a real-world deployment of social search. In: Sixth ACM Conference on Recommender Systems, Dublin, Ireland, 9–13 September 2012, pp. 289–292 (2012)

183. Smyth, B., Coyle, M., Briggs, P., McNally, K., O'Mahony, M.P.: Collaboration, reputation and recommender systems in social web search. In: Ricci, F., Rokach, L., Shapira, B. (eds.) Recommender Systems Handbook, pp. 569–608. Springer, Boston (2015). https://doi.org/10.1007/978-1-4899-7637-6_17

184. Song, Y., Zhou, D., He, L.: Query suggestion by constructing term-transition graphs. In: Proceedings of the Fifth ACM International Conference on Web Search and Data Mining, pp. 353–362. ACM (2012)

185. Spink, A.: Web search: emerging patterns. Libr. Trends 52(2), 299–306 (2003)

186. Spink, A., Bateman, J., Jansen, B.J.: Users' searching behavior on the excite web search engine. In: Proceedings of WebNet 1998 - World Conference on the WWW and Internet, Orlando, Florida, USA, 7–12 November 1998

187. Spink, A., Jansen, B.J., Wolfram, D., Saracevic, T.: From e-sex to e-commerce: web search changes. IEEE Comput. 35(3), 107–109 (2002)

188. Sun, J., Zhong, N., Yu, X.: Collaborative web search utilizing experts' experiences. In: 2010 IEEE/WIC/ACM International Conference on Web Intelligence and Intelligent Agent Technology, pp. 120–127 (2010)

189. Teevan, J., Morris, M., Bush, S.: Discovering and using groups to improve personalized search. In: Proceedings of the Second ACM International Conference on Web Search and Data Mining, pp. 15–24. ACM (2009)

190. Trevisiol, M., Chiarandini, L., Aiello, L.M., Jaimes, A.: Image ranking based on user browsing behavior. In: 35th International ACM SIGIR Conference on Research and Development in Information Retrieval, pp. 445–454. ACM (2012)

191. Uherčík, T., Šimko, M., Bieliková, M.: Utilizing microblogs for web page relevant term acquisition. In: van Emde Boas, P., Groen, F., Italiano, G., Nawrocki, J., Sack, H. (eds.) SOFSEM 2013: Theory and Practice of Computer Science. LNCS, vol. 7741, pp. 457–468. Springer, Heidelberg (2013)

192. Vahabi, H., Ackerman, M., Loker, D., Baeza-Yates, R., Ortiz, A.: Orthogonal query recommendation. In: Proceedings of the 7th ACM Conference on Recommender Systems, pp. 33–40. ACM (2013)

193. Wang, C., Liu, Y., Wang, M., Zhou, K., Nie, J.Y., Ma, S.: Incorporating nonsequential behavior into click models. In: Proceedings of the 38th International ACM SIGIR Conference on Research and Development in Information Retrieval, pp. 283–292. ACM (2015)

194. Wang, X., Bendersky, M., Metzler, D., Najork, M.: Learning to rank with selection bias in personal search. In: Proceedings of the 39th International ACM SIGIR Conference on Research and Development in Information Retrieval, pp. 115–124. ACM (2016)

195. Wang, X., Tan, B., Shakery, A., Zhai, C.: Beyond hyperlinks: organizing information footprints in search logs to support effective browsing. In: ACM International Conference on Information and Knowledge Management 2009 (CIKM 2009), pp. 1237–1246 (2009)

196. Wang, X., Zhai, C.: Mining term association patterns from search logs for effective query reformulation. In: Proceedings of the 17th ACM Conference on Information and Knowledge Management, pp. 479–488. ACM (2008)

197. Wen, J.R., Nie, J.Y., Zhang, H.J.: Query clustering using user logs. ACM Trans. Inf. Syst. **20**(1), 59–81 (2002)

198. Westerveld, T., Kraaij, W., Hiemstra, D.: Retrieving web pages using content, links, URLs and anchors. In: Tenth Text REtrieval Conference, TREC 2001, pp. 663–672 (2002)

199. Wexelblat, A., Maes, P.: Footprints: history-rich tools for information foraging. In: Proceeding of the CHI 1999 Conference on Human Factors in Computing Systems: The CHI is the Limit, Pittsburgh, PA, USA, 15–20 May 1999, pp. 270–277 (1999)

200. White, R., Bilenko, M., Cucerzan, S.: Leveraging popular destinations to enhance web search interaction. ACM Trans. Web **2**(3), 1–30 (2008)

201. White, R.W., Bilenko, M., Cucerzan, S.: Studying the use of popular destinations to enhance web search interaction. In: Proceedings of the 30th Annual International ACM SIGIR Conference on Research and Development in Information Retrieval, Amsterdam, The Netherlands, 23–27 July 2007, pp. 159–166 (2007)

202. Wikipedia: Direct Hit Technologies, September 2015. https://en.wikipedia.org/wiki/Direct_Hit_Technologies

203. Wikipedia: Social search, November 2016. https://en.wikipedia.org/wiki/Social_search

204. Wolfram, D., Spink, A., Jansen, B.J., Saracevic, T.: Vox populi: the public searching of the web. JASIST **52**(12), 1073–1074 (2001)

205. Wu, M., Hawking, D., Turpin, A., Scholer, F.: Using anchor text for homepage and topic distillation search tasks. J. Assoc. Inf. Sci. Technol. **63**(6), 1235–1255 (2012)

206. Wu, X., Zhang, L., Yu, Y.: Exploring social annotations for the semantic web. In: 15th International Conference on World Wide Web, pp. 417–426. ACM Press (2006)

207. Xiong, C., Callan, J.: Query expansion with freebase. In: Proceedings of the 2015 International Conference on the Theory of Information Retrieval, pp. 111–120. ACM (2015)

208. Xu, S., Bao, S., Fei, B., Su, Z., Yu, Y.: Exploring folksonomy for personalized search. In: Proceedings of the 31st Annual International ACM SIGIR Conference on Research and Development in Information Retrieval, pp. 155–162. ACM (2008)

209. Xue, G., Zeng, H., Chen, Z., Yu, Y., Ma, W., Xi, W., Fan, W.: Optimizing web search using web click-through data. In: Proceedings of the Thirteenth ACM International Conference on Information and Knowledge Management, pp. 118–126. ACM (2004)

210. Xue, G.R., Zeng, H.J., Chen, Z., Ma, W.Y., Zhang, H.J., Lu, C.J.: Implicit link analysis for small web search. In: Proceedings of the 26th Annual International ACM SIGIR Conference on Research and Development in Informtaion Retrieval, pp. 56–63. ACM (2003)

211. Yanbe, Y., Jatowt, A., Nakamura, S., Tanaka, K.: Can social bookmarking enhance search in the web? In: JCDL 2007: 2007 Conference on Digital Libraries, pp. 107–116. ACM Press (2007)

212. Zanardi, V., Capra, L.: Social ranking: uncovering relevant content using tag-based recommender systems. In: Proceedings of the 2008 ACM Conference on Recommender Systems, pp. 51–58. ACM (2008)

213. Zhang, J., Deng, B., Li, X.: Concept based query expansion using WordNet. In: Proceedings of the 2009 International e-Conference on Advanced Science and Technology, pp. 52–55. IEEE (2009)

214. Zhou, D., Bian, J., Zheng, S., Zha, H., Giles, C.L.: Exploring social annotations for information retrieval. In: The 17th International Conference on World Wide Web, pp. 715–724. ACM (2008)

215. Zhou, J., Ding, C., Androutsos, D.: Improving web site search using web server logs. In: CASCON 2006: Proceedings of the 2006 Conference of the Center for Advanced Studies on Collaborative Research. ACM (2006)

216. Zhu, G., Mishne, G.: ClickRank: learning session-context models to enrich web search ranking. ACM Trans. Web $6(1)$, 1:1–1:22 (2012)

217. Zhu, J., Hong, J., Hughes, J.: PageRate: counting web users' votes. In: Proceedings of the Twelfth ACM Conference on Hypertext and Hypermedia, pp. 131–132. ACM (2001)

8
Network-Based Social Search

Shuguang Han and Daqing He[⊠] [iD]

School of Computing and Information, University of Pittsburgh,
Pittsburgh, PA 15213, USA
{shh69,dah44}@pitt.edu

Abstract. With the wide adoption of social media in recent years, researchers on social information access are gaining more interests on applying various of social interactions (e.g., friendship, bookmarking, tagging) for satisfying people's information needs. In this chapter, we focus on methods and technologies to boost information retrieval performance based on the idea of representing social information as networks. We study three different types of networks: people-centric networks, document-centric networks and heterogeneous networks combining both. Information from these networks has been utilized to compute vertex similarity (at the individual level), identify network clusters (at the community level) and calculate entire network measurements (at the network level), which are further applied to help search problems not only for seeking documents but also when searching for people. This chapter provides an extensive reviews of existing methods and technologies for performing such two search topics using networks. Through this chapter, our goal is to provide readers with introductory review of the existing work, and provide concrete presentations of relevant technologies for designing and developing network-based social search systems. Finally, we also point out potential remaining challenges on this topic.

1 Introduction

Search is a major part of people's daily lives. With the wide adoption of social media, people's information needs can be satisfied through adopting the information generated from social media. This so called social search can take researchers to focus on different types of social information. For example, Chap. 7 in this book discusses the general idea of social search [12], Chap. 9 reviews issues related social search using social tag information [94]. Different from these two chapters that only target on the understanding of one type of social media content, this chapter aims to provide an extensive review of the methods that utilized social information for search. Particularly, we focus on one specific school of approaches for utilizing social information, where various types of networks are created to represent the social information so that social network analysis technologies [135] can be applied to improve retrieval performance. This is what we called *Network-based Social Search*.

© Springer International Publishing AG, part of Springer Nature 2018
P. Brusilovsky and D. He (Eds.): Social Information Access, LNCS 10100, pp. 277–309, 2018.
https://doi.org/10.1007/978-3-319-90092-6_8

Network-based social search integrates information retrieval, social media and social network analysis into one topic. Therefore, to present this topic well to the readers, we develop an outline that decomposes this topic into several small components, and through presenting each component and then the integration of them, we hope to establish foundations for readers to master the main design and technologies related to network-based social search.

Firstly in Sect. 2, we briefly introduce the concepts of networks and social search to establish the idea that social media information can be represented as networks for achieving social search. Then we propose a general scheme for social search, in which three types of networks, including people-centric networks, document-centric networks and heterogeneous networks are considered. Furthermore, because both documents and people are involved in social search, it is natural to think about the support of both document search and people search needs [40,55]. In addition, we also define three different levels (individual level, community level and entire network level) to indicate the type of network information we used for search. A more detailed discussion of such division can be found in Sect. 2.3 and Fig. 4.

Secondly, Sect. 3 describes methods for performing various network computation tasks, which are all important tasks and useful when applying network information at three different levels. Example computation tasks include computing vertex similarity (at the individual level), identifying network clusters (at the community level) and calculating entire network measures (at the network level). Computation methods of these tasks are easily generalizable to different types of networks so that they are applicable to any type of networks (people-centric, document-centric and heterogeneous networks).

Thirdly in Sect. 4, we provide a comprehensive review for methods used for network-based document search. This section contains three subsections, each of which examines one type of networks (Sect. 4.1 for people-centric network, Sect. 4.2 for document-centric network, and Sect. 4.3 for heterogeneous network). Through the review of existing studies, we want the readers to know that most of the current studies only either focused on document-centric network or targeted people-centric network, we expect more and more future studies on applying heterogeneous networks for document search.

Fourthly in Sect. 5, we provide a detailed analysis for people search, where we follow the same content organization scheme as document search. Our objective is to review the current studies regarding to network-based people search, and establish the understanding that the major paradigm for people search is that *people search for people*, in which documents are only intermediates. This may account for a relatively few number of people search studies that are performed on top of the document-centric networks. As for the utilization of heterogeneous networks, there are indeed a few number of existing studies working on this topic; however, there is still much room to improve in the future.

2   Network-Based Social Search

2.1   Networks

People live in a connected world. They maintain relationships with family members, engage in activities with friends, and collaborate on professional work with colleagues. A **network** (also referred to as a graph) is often adopted to model these relations among different entities [25]. Formally, a network G represents real-world relations as two major components: nodes V and edges E. The nodes denote entities and the edges indicate entity-entity relations. The edge can be either defined as any real-world relation such as co-authors [97], colleagues [49], neighbors [138], or an indirect relation such as people who shared the same interests, bought the same product, or checked-in the same location. To further differentiate the strength of different types of relations, a weight W is introduced to measure the connectedness of two nodes [135]. Therefore, a network can be represented as G = (V, E, W) with W denoting the strength of different edges.

Many related studies have been devoted to understanding the underlying network patterns and their applications to other domains. For example, there are studies on investigating network evolution [73,74], predicting future connections [75] and discovering latent modular network structures [98]. Methods developed in these studies can be further applied in real-world applications such as recommending the best people to follow in LinkedIn and Twitter [38,130], and supporting people's daily web search activities with enriched network-based social contexts [15,113,139,146]. In this chapter, we are interested in applying network information in search-related applications, which are often referred as social search.

2.2   Social Search

Despite being an active research topic, social search does not own a clear, unified definition [17]. According to the Social Search Chapter in this volume [12], most of the existing studies view it as a type of web search that integrates the searcher's social networks [13,52]. Under this definition, search results returned by a social search engine are not only affected by the search query but also by the behavioral information of other users in the searcher's social graph. A social graph can be defined as the searcher's explicit connections such as friendship, coauthorship or any other relationship [19]. It can also be constructed using weaker connections such as people undertaken the same search task or shared the same social community [30,40,113,139].

The above-mentioned studies of social search only affect the information seeking process when a query is issued, while Chi [15] argued that social interactions can play important roles not only when people are examining search results but also in forming information needs and reflecting on the obtained results. For example, an information need may come from discussion with friends (in this case, the social interactions happen before the search), or people sometimes distribute/share their search results with colleagues (in this case, the social interactions happen after the search) [91,150]. A holistic view of the social search should therefore consider the social interactions in the whole search process [15]. In addition, social search can also go beyond the traditional query-result paradigm. For instance, recent studies [48,128] discovered that people sometimes ask (i.e., ask questions through social media posts) for information in social media websites such as Facebook and Twitter. However, due to the difficulty of obtaining such resources, existing studies preferred using information without privacy concerns.

Social search studies mostly focused on applying social interactions from the searcher, whereas the social information at the document-side[1] is often missing. As an important component in a search system, document acts as an equally important role as the searcher. Social information from documents usually looks more implicitly. For example, people's voting on document importance is one such type and has been widely adopted for *Link Analysis* in modern search engines. Examples include two widely-cite algorithms - PageRank [100] and HITS [68], for which the importance of a document is measured by the quantity and quality of hyperlinks that point to it. Since hyperlinks in a webpage are created by the owner of the page, the document importance essentially reflects people's collective wisdom.

To summarize, social search in this chapter focuses on applying social interaction information for supporting a search query. The applied social information not only involves the social graphs related to searchers but also includes the social information regarding to documents. However, we do not consider supporting the process of forming information needs before issuing a search query and the reflection on information obtained after the search query. Readers who want to know more about social search should read the Social Search chapter in this volume [12].

2.3   Applying Networks for Social Search

To apply social interactions in a social search system, we firstly represent them as networks [25]. In this section, we will discuss different types of networks and information needs for social search, then a simple outline summarizing our plan to discuss network-based social search.

2.3.1   Networks for Social Search

A typical search system has two major components: a data collection (e.g., a large amount of web pages) and a set of information seekers. As stated above, social interactions can be represented explicitly based on the communication between information seekers as well as their social connections, or be measured implicitly by people's co-voting on documents. These two types of social interactions were referred to as **people-centric networks** and **document-centric networks** in this Chapter. In people-centric networks, nodes are people and edges are their social connections, whereas nodes in document-centric networks are documents and edges represent people's co-voting information on document importance.

---

[1] Depending on the type of information needs, a document may refer to a web page in a web search system, a person in a people search system, or some other forms.

Here co-voting is in a general sense, and can indicate voting actions in social media, citing articles, constructing hyperlinks, and etc.

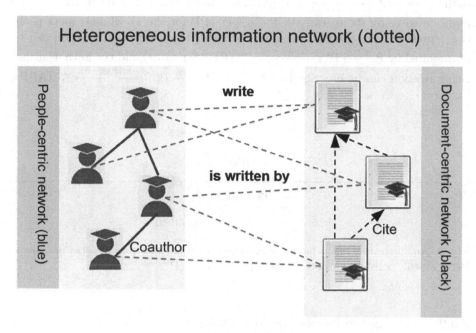

Fig. 1. An illustration of heterogeneous bibliographic information network. The blue lines denote the people-centric network, the black dotted lines denote the document-centric networks and the dotted lines (both in red and black) denote the heterogeneous information network. (Color figure online)

Either document-centric networks or people-centric networks only take into account one type of nodes and one type of relations, whereas many real-world networks are usually *heterogeneous* containing nodes and relations of different types [117]. As illustrated in Fig. 1, a bibliographic information network includes two types of entities: authors and papers. Two or more authors may co-write a paper (co-authorship relation) and one paper may cite many other papers (citation relation). Document-centric networks only consider the citation relation while people-centric networks only focus on the co-authorship. A heterogeneous information network bridges both networks through a new type of relation (i.e., authorship) indicating whether an author writes a paper. We refer this network as the **heterogeneous information network**. Note that it is different from a *k-partite network* [84] since edges in a k-partite network only connect disjoint sets of nodes (nodes of different types) while the example in Fig. 1 also connects nodes with the same type (e.g., document-document connection and author-author connection). In our example, network schema of the heterogeneous network can be illustrated with Fig. 2, where the coauthor relation and citation relation can be described using two **meta paths** [118]. Here, a meta path refers to a network path that connects different entities (with the same or different types). For example, a coauthor relation can be represented through a meta path $A \xrightarrow{\text{writes}} P \xrightarrow{\text{is-written-by}} A'$ (APA for short) and a citation relation can be represented using the meta path $P \xrightarrow{\text{cites}} P'$ (PP for short). Similar meta paths can represent more complex network relations. The co-citation relation can be represented by $P \xrightarrow{\text{cites}} P' \xrightarrow{\text{is-cited-by}} P''$ (PPP for short) and the author-author citing relation can be described as $A \xrightarrow{\text{writes}} P \xrightarrow{\text{cites}} P' \xrightarrow{\text{is-written-by}} A'$ (APPA for short).

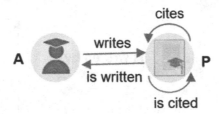

Fig. 2. An illustration of the heterogeneous bibliographic information network schema. Here, we only consider authors (**A**) and publications (**P**).

2.3.2 Information Needs for Social Search

Web search has always been the dominating search format even dated back to decades ago. Therefore, social search studies have mainly focused on extracting and applying social information for supporting web search performance [4,31,115]. The recent quick expansion of Social Network Services (SNS) such as Facebook and LinkedIn drives the formation of alternative information needs to search for people [55,116]. Although people search has been studied in academia [42,136], existing research mostly focused on searching for experts and using purely textual information [5]. We move forward to the topic of applying social information for people search.

To summarize, we will take into account two types of information needs for social search: social-based **document search** and social-based **people search**. It is worth noting that users might have many other vertical information needs such as searching for products, images, locations and news articles, which will not be considered in this chapter.

2.3.3 Outline for Network-Based Social Search

Given the above-mentioned three types of networks and two types of social search needs, this chapter will adopt the following outline (as shown in Fig. 3) with six different combinations of network type and information need.

Depending on the amount and the type of network information used in a social search system, we divide the use of social information into three different levels—the individual social connection level [13,52], the community level

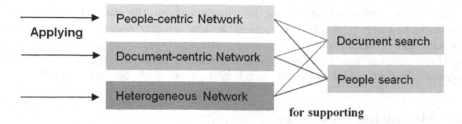

Fig. 3. Outline of this chapter for network-based social search

[113, 139] and the entire network level [42]. Figure 4 provides a simple illustration of employing people-centric networks for document search at such three levels. Similar framework can also be applied for document-centric networks and heterogeneous information networks, except that connections in document-centric networks are hyperlinks among documents and heterogeneous networks contain both documents and people. Besides document search, similar approaches can also be applied for people search. More details are explained in the following paragraph.

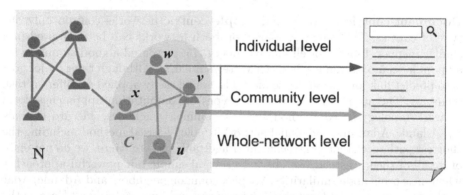

Fig. 4. An illustration of employing network information for social search. Here, we only simulate ways of employing people-centric networks for document search.

As shown in Fig. 4, suppose that we have a searcher u connecting with both v and w, and all of them belong to the same social community C. Here, a network community refers to a group of nodes who are densely connected with each other and sparsely interacted with external nodes [28]. All of these nodes and communities come from the same network $\mathcal{N}$. At the individual level, a social search engine ranks search results not only based on u's search query but also utilized the search information (e.g., issued queries and clicked documents) from her connections v and w. The community-level use of social information refers to leveraging the latent knowledge created in the social community C (e.g., the common search queries, search patterns), which involves u, v, w and

x. At the network level, statistical information about the entire people-centric networks can also be applied for understanding popular nodes and dominating information needs. Similarly, document-centric networks can also be exploited at three different levels for improving document search performance, and both networks will be useful in people search systems [6, 39, 40, 68, 100, 144, 153].

3 Measuring Networks

To effectively apply the social network information at the individual, community and network levels as shown in Fig. 4, we need to understand ways of computing node-node similarity (i.e., vertex similarity [14]), extracting network communities and modeling node authority at the global network level. This section provides more details on these topics for all three types of networks mentioned above including document-centric networks, people-centric networks and heterogeneous networks. The techniques mentioned in this section will be eventually applied in people search and document search algorithms for improving model performances. More details will be presented in the next two Chapters.

3.1 Measuring Node Similarity in Networks

Document-centric networks and **people-centric** networks contain only one type of entities. The vertex similarity on both networks can be computed in a similar manner. Liben-Nowell and Kleinberg [75] presented a good summary for all popular computation methods for vertex similarity. Although their major goal is to predict link formation, these methods can be easily applied at different task scenarios. The authors introduced two types of computation approaches: local network topology-based method such as common neighbors, Jaccard similarity, Adamic/Adar, and the global network topology-based method including the shortest path, preferential attachment, PageRank and etc. Here, we do not elaborate all of them. Instead, we illustrate several simple but powerful approaches when measuring node similarities. We pick common neighbors and Adamic/Adar as the representative of local network feature-based approaches, and Katz as the representative of global network feature based approach.

Formally, suppose that we want to compute vertex similarity between node x and node y. Let $\Gamma(x)$ and $\Gamma(y)$ represent a set of direct network neighbors for x and y, respectively. The common neighbor approach computes vertex similarity based on the number of common connections between x and y. This can be illustrated as Eq. 1. The common neighbor approach has been widely adopted in existing studies due to its good performance and simple implementation [14, 39]. The Adamic/Adar approach further takes into account the *degree* (degree refers to the number of nodes that are directly connected to a given node) of each common neighbor z. Let $\Gamma(z)$ denotes a node's direct connections, Adamic/Adar metric can then be represented using Eq. 2. Different from the above two methods, Katz index [64] takes into account the global network structure. It is defined as the ensemble of all paths with different length l (which lies between $[1, \infty])$

between x and y. It can be further computed using Eq. 3, where l indicates the path length, $\text{Path}_{x,y}^{(l)}$ denotes the set of all length-l paths between x and y and β is the damping factor that controls the importance of length-l paths. In terms of the setting of damping factor, one can refers to Bonchi et al. [9].

$$|\Gamma(x) \cap \Gamma(y)| \tag{1}$$

$$\sum_{z \in \Gamma(x) \cap \Gamma(y)} \frac{1}{\log |\Gamma(z)|} \tag{2}$$

$$\sum_{l=1}^{\infty} \beta^l \cdot |\text{Path}_{x,y}^{(l)}| \tag{3}$$

The above approaches only work on homogeneous networks, which cannot be applied for similarity computation in **heterogeneous networks** since connections in a heterogeneous network might have different semantic implications (e.g., connections between two persons vs. connections between person and document). To deal with this problem, meta path based vertex similarity measure named *PathSim* was proposed to retain the semantic meaning of each meta path [117,118]. Given a meta path $\mathcal{P}$, *PathSim* computes the vertex similarity between x and y based on Eq. 4, where s(x, y) denotes the path-based vertex similarity and $|\text{Path}_{x \to y}|$ indicates the number of path instances from x to y for the given meta path (similar for computing $|\text{Path}_{x \to x}|$ and $|\text{Path}_{y \to y}|$).

$$s(x, y) = \frac{2 \times |\text{Path}_{x \to y} : \text{Path}_{x \to y} \in \mathcal{P}|}{|\text{Path}_{x \to x} : \text{Path}_{x \to x} \in \mathcal{P}| + |\text{Path}_{y \to y} : \text{Path}_{y \to y} \in \mathcal{P}|} \tag{4}$$

Let's illustrate this idea with a simple example (adapted from Sun et al. [118]) on bibliographic networks. Suppose that we have authors x and y, x published 5 papers on conference venue SIGIR and 3 on KDD (we use V to denote conference venues) and y has 1 paper on SIGIR but 10 on KDD. Vertex similarity between x and y for meta path AVA (author - venue - author, i.e., the co-conference relations) can be computed via Eq. 5, where we can obtain 5 × 1 meta path instances for A-SIGIR-A and 3 × 10 meta path instances for A-KDD-A. The two parts in denominators can be calculated in the same way. As a result, the final node similarity is 0.56 for meta path AVA.

$$s(x, y) = \frac{2 \times (5 \times 1 + 3 \times 10)}{(5 \times 5 + 3 \times 3) + (1 \times 1 + 10 \times 10)} = 0.56 \tag{5}$$

3.2 Modular Structure in Networks

A common approach of utilizing social information in search is to treat searchers' social connections as search context, and then apply it for boosting the search result ranking performance [15]. However, this approach might encounter the data sparseness problem, particularly when the searcher is isolated in a network. Recent studies attempted to resolve this problem through estimating user context from other users of similar interests (also referred as social community)

[113,139,146]. Among the existing studies, social communities are either predefined or detected with heuristic rules. We believe that recent developments of the group discovery research from network science domain can help automatically discover latent network communities, and we will provide more details in the following paragraphs. Here, we firstly discuss community detection in homogeneous document-centric and/or people-centric networks, and then move to heterogeneous networks.

Many network analysis studies focus on understanding statistical characteristics of networks [7,96], where they usually assume a uniform network generation mechanism while more and more studies find that network structure is not even and it has obvious modular structure [37,98]. Some nodes are well-connected with each other whereas there are fewer connections with another group of nodes. This refers to *community* in a network. Community detection algorithms, extensively surveyed in Fortunato [28], allow us to uncover community structures (or clusters) in networks. Example methods include modularity maximization [98], information-theoretic based algorithm [108] and generative model based approach [104].

Modularity maximization is a widely-adopted community detection algorithm, whose basic idea is to traverse all possible community structures and choose the one that maximizes network modularity (referred to as Q). Q is computed in the following way: let e_{ij} be the fraction of edges that connect community i and community j, and $a_i = \sum_j e_{ij}$. Then, Q is defined as the fraction of edges that fall within communities, minus the expected value of the same quantity if edges are connected at random, as shown in Eq. 6. A larger value of Q means that nodes are more densely connected within communities but more sparsely connected outside of the same communities. The whole procedure of modularity maximization algorithm can be illustrated using Fig. 5. In the beginning, each node is initialized as a community. Then, we merge two communities at each step based on the criteria whether such merging can produce a better modularity score. This process is then repeated multiple iterations until all nodes are merged into one community. Finally, we choose a cutoff step with the maximized modularity value. Note that this algorithm requires a hard assignment for each node to a community while not supporting overlapping community detection. Later extensions of this algorithm resolve this problem by representing the node-community assignment as a probabilistic distribution so that one node does not have a hard assignment for only one community [101,104,145]. In addition, modularity maximization algorithm was also found to encounter the resolution limit problem which cannot effectively handle small communities in large networks [29,71]. Recent studies are still working for better solutions for this problem.

$$Q = \sum_i (e_{ii} - a_i^2) \tag{6}$$

A large number of community discovery algorithms only worked with **homogeneous networks**, which cannot effectively handle **heterogeneous networks**. This drives recent studies to develop new algorithms for community

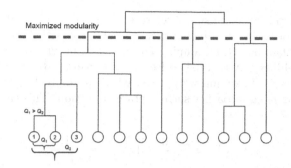

Fig. 5. An illustration of modularity maximization method for community detection. This figure is adapted from Newman and Girvan [99]. In this example, merging community 1 and 2 can produce a greater modularity than merging community 1 and 3, so we prefer to combine 1 and 2 in the second step.

detection on multi-typed nodes with multi-relations. For instance, several recent studies have tried to extend modularity maximization algorithm to heterogeneous networks, and proposed bipartite modularity [36, 92], tripartite modularity [93] and k-partite modularity [95]. Besides optimizing network modularity, other researchers have also examined the probabilistic generative models. Lin et al. [77] developed Metafac, a matrix factorization-based approach for automatic community discovery from multi-dimensional social contexts and interactions. Sun et al. proposed RankClus [119] and NetClus [120] for community detection from bi-typed and multi-typed heterogeneous information networks, respectively.

3.3 Measuring Node Importance in Networks

In social search systems, network information is often used to identify highly important nodes (people or documents) so that the low-quality nodes can be filtered in search results [6, 40, 100, 153]. Although existing studies have developed various algorithms to compute node importance [46, 57, 59, 61, 120, 153], most are developed on top of PageRank [100] and HITS [68], two widely-cited link analysis algorithms. In the following section, we illustrate the basic idea of PageRank. Readers can refer to [68] if want to know more details about HITS. We will firstly describe PageRank on homogeneous networks and then discuss its extensions to heterogeneous networks.

Proposed by Google co-founders Larry Page and Sergey Brin in late 90s, PageRank is a well-known algorithm for measuring web page importance through the analysis of hyperlinks among web pages [100]. Its basic assumption is that a web page is more important if it has more votes (hyperlinks) from other important web pages. Formally, let's assume that we want to compute PageRank score for page A, and this page has m web pages $(T_1, T_2, \ldots, T_m)$ point to it. The PageRank computation steps can be formalized using Eq. 7, where $PR(X)$ denotes the PageRank for page X, N represents the total number of web pages in collection, $C(X)$ stands for the total number of out links for page X and d is the damping factor that is usually set to 0.85. Page et al. [100] explained the computation process as an *imaginary surfer* who keeps clicking next web page at a certain probability (the damping factor d can be viewed as the continuation click probability). According to Eq. 7, computing PR(A) requires knowing PageRank values for incoming web pages, which needs to be obtained through iterative or recursive approaches such as power iteration. This can be illustrated with Algorithm 1.

$$PR(A) = \frac{1-d}{N} + d \times \sum_{x=1}^{m} \frac{PR(T_x)}{C(T_x)} \tag{7}$$

Algorithm 1. Power iteration method for PageRank computation

Input: A web graph with nodes **V** indicating web pages and edges **E** revealing hyperlinks; the maximized number of iterations T.

Output: PageRank values for all web pages **PR(V)**

1: Initializes PageRank value for each web page uniformly as $PR_{x \in V}(X; 0) = 1.0/|V|$

2: **for** t in [1, T]:

3:    **if** $\sum_{x \in V} |PR(x; t) - PR(x; t - 1)| < \varepsilon$

4:      break

5:    **for** v in **V**:

6:        compute PageRank value for each v based on Eq. 7

7:    **end for**

8: **end for**

To better understand the computation process, we provide a simple example. Suppose that we only have four web pages with hyperlinks as shown in Fig. 6. In the initialization phase, we assign PageRank values uniformly for four web pages, each has a value of 0.25. Then, based on Algorithm 1, we iterate our computation multiple times till it converges. Empirical studies usually find that PageRank values are quite stable after a few number of iterations (e.g., 5 times), the values usually do not change dramatically even if we iterate a lot more times.

So far, we mainly discuss approaches of identifying important nodes from **document-centric networks**. Later studies find that the same methods can be applied at **people-centric networks** to identify important people [39,40,79,144]. Furthermore, recent studies have also explored ways of identifying important documents and people simultaneously [6,23,153]. Sun et al. [119,120] provided a generalized framework that can propagate node importance mutually across networks with multiple nodes and relations. Experimental results show that document-centric and people-centric networks can mutually reinforce each other and generate much better results. In this case, node authority is essentially computed based on **heterogeneous information networks**.

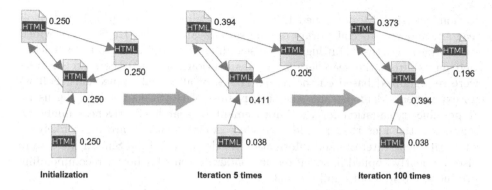

Fig. 6. An illustration of PageRank score change with different number of iterations.

4   Network-Based Document Search

Compared to people search, searching for documents remains to be the dominating information need in modern search engines. Therefore, this section starts with investigating ways of applying different levels of network information (including individual, community and entire network levels, as shown in Fig. 4) for document search engines. Here, three different types of network information, including people-centric, document-centric and heterogeneous networks as shown in Fig. 3, is taken into account. The following subsections elaborate existing studies on each of them.

4.1   Applying People-Centric Network for Document Search

Various of prior studies have been devoted to applying people-centric networks for improving document search [4,13,30,52,113,115,149], among which the dominating approach is to convert such network information into additional search result ranking signals. The document relevance, therefore, not only depends on the match between the document content and user queries, but also relies on the relation between the document and the searcher's social networks. This section describes how the people-centric network information can be applied for document ranking at three different levels.

The *individual level* use of people-centric networks refers to applying an individual's social connection information for better ranking of relevant documents. Recent researchers have examined quite a few approaches regarding to this topic [13,115,149]. Carmel et al. [13] compared the effectiveness of employing three different types of social networks for search personalization: (1) familiarity-based network that measured people's social relations based on familiarity (e.g., friendship); (2) similarity-based network that computed people's social relations based on their behavior similarity; and (3) the combined networks that included both types. Their experiments discovered that these three network-based personalization systems all significantly outperformed the one without search personalization. Smyth [113] developed HeyStaks that allowed a group of friends, colleagues, and etc. to collaborate on the same search tasks. HeyStaks provided two major functions: (1) it highlighted search traces (e.g., issued queries, clicked web pages) from the searcher's social connections; and (2) the search results were re-organized based on the search traces of all social connections. Followup experiments demonstrated the effectiveness of HeyStaks for engaging users. A possible explanation for the improvement of search effectiveness might be related to the fact that people with close social relations are more likely to share the same interests and information needs [13,115,149]. Similar ideas were also frequently applied in social recommender systems for better recommending products, music, books and etc. [53,85,146].

Applying the individual level network information is likely to encounter data sparsity when the individual has few social connections. This motivates researchers to explore alternative resources for expanding the network context. Observing that users from the same community (e.g., members of the same forum topic, users who search similar queries) usually read similar search content, developed similar interests and formed similar information needs [114], recent studies have attempted to employ users' behavioral information from a community of like-minded users (i.e., *community-level* network information) to improve search performance [3,63,113,114,127,141]. It is worth noting that the communities, among the above-mentioned studies, were either predefined [113–115], or determined based on simple heuristic rules such as whether people shared the same user traits (e.g., gender, location) [127] or performed the same search tasks [127,139], whereas automatic community detection algorithms (see Sect. 3.2) were seldom considered. We do think that these automatic algorithms can be beneficial and should be properly considered in the future. Empirical experiments and analysis from the above-mentioned studies all demonstrated that the community-level network information can effectively solve the cold start problem and augment users' search performances.

Despite the entire *network level* information was frequently exploited for inferring document importance in document-centric networks (see Sect. 4.2), there are rare studies on the direct usage of people-centric networks for document search. However, several recent attempts on employing the entire people-centric network information could potentially be useful for enhancing document search. For example, Zhou et al. [153] tried to identify important authors and important documents simultaneously based on the mutual reinforcement of coauthor networks and paper citation networks. Though they did not perform follow-up experiments on examining document search performance, their results can help filter low-quality documents in returned search results. Bao et al. [6] and Xu et al. [142] examined the utilities of applying networks of social annotations for improving web search performance. Here, the people-centric network about annotators (researchers sometimes define a social relation of co-tagging [132] among annotators) acted as an important component for identifying important annotations and their corresponding web documents, which were further demonstrated to be effective in improving document search performance. It is worth noting that the entire people-centric network was usually not employed as a stand-alone module, which was often tightly connected with document-centric networks. This is referred as the heterogeneous network in existing studies and we will discuss it in Sect. 4.3.

4.2 Applying Document-Centric Networks for Document Search

A document-centric network connects documents through their relations. Existing literature often adopted two types of document relations [68,90,100]. The first type defined a connection between two documents if one had a hyperlink to the other. This definition has been widely adopted in the research of *link analysis* [68,100]. The second type measured document connectedness through the content similarity between two documents [90]. Related text mining techniques such as vector space model [20,109] or latent topic modeling [8] were commonly adopted to compute document similarity based on the corresponding textual content. Other definitions of document relations, such as determining a connection if two documents were co-clicked under the same query [16] or if two documents were co-visited by the same users [83], were also employed and were particularly useful depending on the corresponding task contexts. However, instead of studying the differences among multiple relation measures, this section focuses on investigating ways of applying different document-centric networks for improving retrieval techniques.

The *individual level* use of document-centric networks refers to favoring documents that are better connected with individuals' previously-interacted documents when responding users' information requests. This topic has been extensively studied as relevance feedback in the information retrieval community [88], whose underlying idea is to rank documents that are not only matched with user queries but also associated with their feedbacks of relevant documents. Depending on the ways how users provided such feedback information, two types of feedbacks are commonly adopted in existing literature: explicit feedback and implicit feedback [66]. Explicit feedback is obtained through users' explicit judgment of document relevance/usefulness. However, due to the high cost of acquiring such information in live search systems, researchers have explored ways of utilizing *pseudo* relevance feedback, where the top-k retrieved documents of a given query were treated as relevant documents [88] for search results re-ranking. The performance of pseudo relevance feedback highly relies on the quality of the initial round of ranked results (i.e., before the re-ranking), and usually lead to the homogeneity of search results. Other researchers then experimented alternative approaches to infer document relevance through implicit user behaviors such as dwell time, clicking, saving and bookmarking [60,66,111,140]. For example, clicking a document with a long-dwell time usually indicated that the corresponding document is relevant, whereas no click or document click with short dwell time might reflect non-relevance. These ideas have been extensively explored in many related studies [44,67,72,111,121,126,134].

Existing literatures applied document-centric networks at the *community level* for enhancing document search performance. Here, a document community refers to a set of documents that are similar enough to each other while they are dissimilar to the documents outside of the community, which shares similar definition as cluster(s) in document clustering algorithms [143]. Prior studies mainly developed two types of approaches. In the first approach, the document community information was exploited to locate relevant documents based on the assumption that similar documents are more likely to satisfy the same/similar information needs [56,106]. Document communities were automatically discovered through standard clustering algorithms such as hierarchical clustering [131] and recently Latent Dirichilet Allocation (i.e., LDA) [137]. At the retrieval stage, a document is returned not only based on the document-query relevance but also depending cluster-query relevance. A detailed explanation of the whole procedure can be found in Liu and Croft [80]. The second approach emphasized on providing a better presentation of search engine result pages (SERP) for facilitating people's search experience. One typical example is the Scatter/Gather representation of SERPs [20,47]. In addition to providing a list of ranked search results, the Scatter/Gather mechanism further provided a set of clusters and summaries regarding to the result space for end users [35]. This can be illustrated as Fig. 7 – a Scatter/Gather example in a live search engine[2]. As illustrated, after typing a query, Vivisimo displays its search results on the main panel, and the corresponding topics (i.e., clusters) on the left panel. Clicking a cluster can help narrow down the searcher's information need. It is worth noting that although document clusters can be automatically detected either through the document content similarity [131,137] or hyperlinks [27,34], most approaches utilized the content information. This is probably due to that the information retrieval mainly focuses on searching for content whereas a hyperlink usually overlooks the content information.

The entire *network level* information of document-centric networks was frequently applied to infer important documents through *link analysis* algorithms [32]. World Wide Web environment is full of freedom. Different from books, journals and news articles, webpages can be created or written by anyone with any purpose. Although many of them are of high quality, some are poorly written or even deliberately misleading [148]. To develop a better quality control procedure with the capability of identifying high-quality web pages, past studies developed a set of link analysis algorithms such as PageRank [100] and HITs [68]. The high-quality documents are then preferred over low-quality ones once their document relevance scores are at the same level. Although the webpage quality can be applied for information retrieval applications in many different ways, the most common approach is to use it as an additional feature for search result ranking [78]. Particularly, when adopting statistical language model for information retrieval [18], such Webpage quality information can be exploited as the document prior in a more principled way. Specifically, the statistical language model [18,151] assumes that each document d is generated by a document language model (noted as ϑ_d, which is often estimated through word distribution probability), and the retrieval of relevant documents is to estimate the probability

---

[2] It was acquire by IBM in 2012, see https://en.wikipedia.org/wiki/Vivisimo.

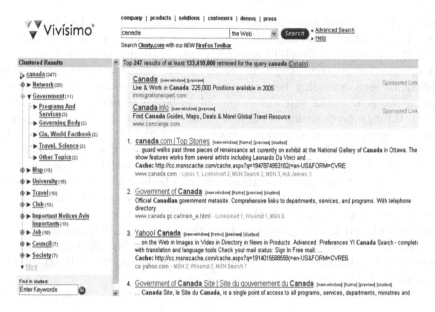

Fig. 7. An illustration of the Scatter/Gather-based search engine result representation on Vivisimo. The figure is adopted from Koshman et al. [69].

of d is relevant for a given query q, i.e., $P(\vartheta_q|d)$, which can be computed with Eq. 8. Under the Bayesian rule and the assumption of uniform query prior $p(q)$, $P(\vartheta_q|d)$ can be computed based on Eq. 9. Here, $P(q|\vartheta_d)$ measures the relevance between q and d, and $P(d)$ denotes the document prior. The document quality values such as PageRank value can be utilized to represent $P(d)$ [46,151]. More information about the statistical language model can be found in Zhai [151].

$$P(\vartheta_d|q) = \frac{P(d)P(q|\vartheta_d)}{P(q)} \tag{8}$$

$$\propto P(d)P(q|\vartheta_d) \tag{9}$$

4.3 Applying Heterogeneous Networks for Document Search

Heterogeneous networks take into account people-centric networks and document-centric networks at the same time. Applying such information for document search is a new topic since many recent studies still worked on understanding and extending traditional single-relation network mining algorithms onto this new type of multi-relation networks [117,118], in which the document content was usually ignored. For example, Huang et al. [54] proposed a Tri-HITs algorithm that automatically propagated document quality scores in heterogeneous networks among all three types of entities including users, tweets and web documents mentioned in a tweet. Sun et al. [118] discussed the similarity computation using meta-path in the context of heterogeneous information networks.

Besides, recent studies have also started to exploring the ways of modeling document content information and further improving web search performance. Deng et al. [23] developed a Co-HITs algorithm that combines both the content and link information (at the entire *network level*). The algorithm outputs were then applied for recommending better search queries. The social annotation-based heterogeneous networks were also employed for better estimating web document quality through propagating document importance among annotators, annotations and webpages (at the entire *network level*) [6, 142]. Also, Liu et al. [82] attempted to rank relevant publications for a given search query based on pseudo relevance feedback (PRF). Different from the traditional PRF that only included document content similarity, the authors further included the obtained feedback information from sixteen different meta-paths on the heterogeneous bibliographic networks (see Sect. 1.3 in Liu et al. [82], such network information was exploited at the **individual level**). Later on, the authors took a similar recommendation approach for open education resource recommendation [81]. Although with rare research and publications on this topic at this moment, we still expect more publications in the future.

5   Network-Based People Search

5.1   People Search

Modern search engines such as Google, Bing and Yahoo! provide a convenient way for people to access billions of web documents. However, there are many occasions where people would like to find the right persons rather than accessing the right documents [1]. For example, a company may need to find the right people to hire [55], researchers may need to find the right collaborators [39] and Ph.D. students may want to find the right committee members for their doctoral thesis [40]. This drives lots of recent efforts on studying the people search problem [40]. People search covers several research topics such as finding experts [5], discovering potential collaborators [14] and assigning reviewers for paper manuscript [24]. Other topics such as finding keynote speakers and locating conference program committee members were also mentioned in prior studies, but they are not well-investigated because of the difficulty of evaluation. The above-mentioned ideas were also implemented in several working systems. For example, AMiner[3] developed by Tang et al. [124] can support multiple people search tasks such as finding experts and recommending reviewers. PeopleExplorer[4] designed by Han et al. [40] can support complex people search tasks with users' interactive adjustments on facet importance.

More recently, the expert-finding problem was studied in the context of social medias such as Facebook and Twitter. For example, the SearchBuddies system automatically provided responders to questions posted on Facebook through

---

[3] https://aminer.org/.
[4] http://crystal.exp.sis.pitt.edu:8080/PeopleExplorer/.

matching the most relevant social contacts [48]. The Cognos system [33] recommends experts on Twitter by considering multiple types of metadata information. User-generated metadata was discovered as an important type of resource when modeling people's expertise. Similar conclusions were also found in Jiang et al. [58], in which Weibo (the Chinese version of Twitter) was used. The recent wide expansion of social media services such as Facebook and LinkedIn significantly increases the needs of searching for people, and they drive many further studies regarding to this topic [55, 107, 116, 133].

Previous studies [14, 41, 76, 102, 122, 125] found that combining multiple factors such as content and social relations can help locate appropriate candidates. The algorithms for combining different factors usually assume uniform user preference on different factors, and try to learn globally optimized parameters for all users. However, users may have significantly different preferences about the importance of each factor under different scenarios [147]. To resolve this problem, related work [10, 11] has experimented the ways of involving users in the algorithm process, and allowing them to explicitly express their preferences over different factors. The transparency of algorithm process were identified to be helpful in providing recommendations for items (such as books, music). The problem gets more and more complicated when it comes to recommend or search for people [40] since the access for people is way more complex. It is unclear whether or not users can really optimize the usage of the system [105].

A recent study from Han et al. [40] built PeopleExplorer, an interactive people search system with configurable user preferences (see Fig. 8), and examined system usability with lab controlled user study. Three different types of factors were taken into account for locating relevant candidates: the social similarity between the searcher and each to-be-searched candidate, the candidate authority and the content relevance between query and candidate. With 24 participants with four different people search tasks, the study showed that participants indeed prefer the systems with configurable search interface over the baseline system without the configuration feature and they behave more efficient and effective in this interface. A follow-up study [45] shows that users indeed reach reasonable configurations but there are still rooms to reach the best configurations.

In terms of the people search task, although most of the current research on people search focused on the expert-finding problem in academia and in enterprise setting [5], we do not limit our study only on these two topics and target our study to a more general people search context. People search differs from expert-finding problem in the target entity selection criteria: the goal for expert finding is to locate the people with the right expertise while people search emphasizes on satisfying users' multiple search objectives. In addition, our interests in people search are different from finding similar entities only based on network topology analysis [2, 118]. We are particularly interested in satisfying users' textual people search queries.

5.2   Applying People-Centric Networks for People Search

A major difference between people search and document search is the accessibility of the target entity (i.e., a document or a person) [40]. As Terveen and McDonald [129] pointed out, people search fundamentally changes the retrieval problem because the returned results are people instead of documents. People are social creatures; "assessing" people is more complex than assessing web documents. They proposed the concept of "social matching" to emphasize the "social" dimension of people search. Systems such as Referral Web [65] and Expertise Recommender [89] are able to return highly relevant candidates who are also socially related to the information seekers. As a result, people-centric networks play important roles in developing people search systems.

When incorporating the **individual level** people-centric network for people search, researchers have utilized network node similarity (i.e., social similarity between the searcher and a candidate) [55,76] or combine it with content relevance [14,40,52] when developing people search algorithms. The social similarity is computed using the methods we introduced in Sect. 3.1. The approaches of integrating social similarity with content relevance can either be a simple linear combination [14,41], a regularization based combination [125], or a random walk based method [122]. These algorithms assume uniform preference for every user; however, they are hard to be personalized for diverse user preferences. To accommodate the high variability of user preference, an interactive display allowing users to adjust preferences over multiple factors was examined [40,41,52,76,103,124]. The people search algorithm can pick up user preference and retrieve different candidates according to the configurations. For example, PeopleExplorer developed by Han et al. [40] (Fig. 8) designed sliders for users to customize the importance of social similarity, content relevance and candidate authority, and also provided the shortest path to reach each candidate. By doing this, the authors discovered that users can quickly locate the right candidates. For the same purpose, SaNDVIS and SmallBlue, developed by IBM researchers [76,102], provide visualizations of people's social networks, and allow users to customize different types of relationships of their interests (such as friendship, co-authorship, co-organization and etc.).

When providing visualization assistance for people search, recent studies also tried the ways of displaying people who might come from the same social community as the searcher (e.g., the same affiliation, the same country). The **community level** network provides extra information to model user contexts from the perspective of social connectedness for both searcher and candidates [22,62]. When measuring candidate relevance for a given people search query, Karimzadehgan et al. [62] considered both the content match and the relevance of all colleagues within the same organization. Deng et al. [22] measured the community-aware authority for each candidate on top of the AuthorRank, a node authority computation approach proposed by Liu et al. [79]. Experimental results from both studies have demonstrated the usefulness of such information for enhancing people search performance.

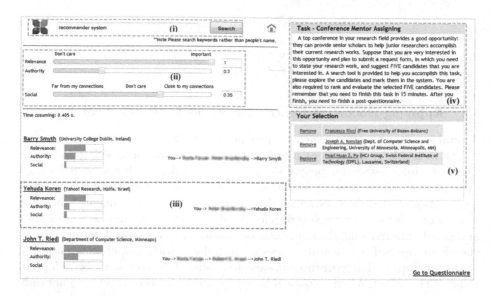

Fig. 8. A screen shot of the PeopleExplorer system (adopted from Han et al. [40]): (i) a search box; (ii) three factors: *relevance, authoritativeness* and *social*; the sliders are used to tune the weights of each factor. (iii) Candidate representation, including a visualized score for each factor and the connection(s). Eight candidates are displayed in each page. The author name hyperlinks to the ACM author profile page. (iv) the task description and (v) the selected candidates.

Information needs for different people search tasks vary extensively. In many real-world tasks, candidate authority, credibility and availability are important factors [51,147]. Particularly, the candidate authority, measuring the importance of a candidate based on the entire **network level** information, has been widely applied in existing studies [22,26,40,41,50,51,110]. Candidate authority can be employed in different ways. For example, IRIS IPS [41] and PeopleExplorer [40] treated candidate authority as an important user-controllable facet in people search interfaces so that users can manually configure their preferences (see Fig. 8). In addition to the interface-level usage, candidate authority can also be integrated into candidate ranking algorithms. To be specific, similar to the statistical language modeling approach for document retrieval (see Eq. 8), the expert finding model proposed by Balog et al. [5] modeled the candidate-query relevance P(ca|q) as Eq. 10, where P(q) and P(ca) denote the query and candidate prior probabilities, respectively. Balog et al. [5] further developed two approaches to compute P(q|ca): the document-centric expert finding model in 11 and the people-centric expert finding model in 12. The **network level** candidate authority can be used as P(ca) in both Eqs. 10 and 11.

$$P(ca|q) = \frac{P(ca)P(q|ca)}{P(q)} \qquad (10)$$

$$P(q|ca) \propto \sum_d P(q|d)P(d|ca)P(ca) \qquad (11)$$

$$P(q|ca) \propto \sum_d P(q|d)P(ca|d)P(d) \qquad (12)$$

5.3   Applying Document-Centric Networks for People Search

Several studies utilized document-centric networks for people search in literature. At the **individual document level**, researchers focused on applying document similarity information for people search. Macdonald and Ounis [86,87] tested the effectiveness of relevance feedback for people search. For each people search query, they treated the top-ranked documents as relevant documents, and such documents were further employed to locate similar and relevant documents. The authors discovered that the document-centric relevance feedback model can be effectively applied for locating relevant candidates with proper differentiation on search topics. The **community-level** document cluster was also adopted. Deng et al. [22] estimated statistical language model for each candidate not only based on the directly-related documents (for each candidate) but also based on the communities that those documents belong to. At the **network level**, document-centric networks are used for computing document importance, which was further employed in people-centric expert finding models as shown in Eq. 12. Document importance can be computed either through a modified PageRank algorithm [42,79,144]. So far, we haven't found any related studies on this topic, which we think is mainly due to the nature of people search. People search systems allow people to search for people, so the roles of document is more implicit.

5.4   Applying Heterogeneous Networks for People Search

Besides employing either document-centric networks or people-centric networks, there are also a large number of studies working on combining both for people search. As mentioned above (see Sect. 2.3.1), networks with multi-typed entities are referred as heterogeneous information networks. At the initial phase of this research topic, most of them focused on studying similarity search in such networks [112,117–120,152]. For example, Sun et al. [118] proposed a novel similarity metric named PathSim (as shown in Eq. 4) to measure entity-entity relations in heterogeneous networks based on the ensemble of meta paths. However, existing studies are usually query-independent, in which the similarity is computed without user-issued queries. However, this paper is more interested in the query-dependent people search problem.

We do not observe studies that explicitly mention using **individual-level** heterogeneous networks for people search. The closest work comes from Liu et al. [82], who experimented a set of 18 heterogeneous network-based features for document ranking. Although their major focus is on document search, we expect that the proposed approach can be easily generalized to people search. As for community-level usage, Deng et al. [22] developed two community-aware strategies for enhancing the people search performance - one is to enrich document representation with all documents within the same community, and the other is to model candidates' authorities based on their social communities instead of the whole networks. The authors discovered that both strategies can consistently improve people search performance. The entire **network level** information were adopted in a unified topic model by Tang et al. [123] when analyzing topical aspects of multiple academic entities such as publications, authors and conferences, simultaneously. Then, the model outputs about topical information was applied together with traditional approach for better ranking of relevant people. Similarly, Deng et al. [21] applied the entire network information for expertise retrieval based on a regularization based framework, in which the authors considered three constraints for regularization: the *document consistency* constraint implying that similar documents should have similar relevance to a given query, the *co-author consistency* constraint meaning that coauthors should have similar relevance to a given query, and the *document-author consistency* constraint denoting that authors' expertise should be consistent with the paper topics. Experimental results demonstrate the effectiveness of the regularization-based approach.

6   Conclusions and Challenges

6.1   Conclusions

Nowadays, people are exposed to a wide range of social media websites in their daily lives. Social information from different websites are universally recognized as an important resource for understanding, modeling and supporting different levels of information needs. This chapter provides an extensive review of the techniques about utilizing social information for supporting two types of information needs—searching for documents (document search) and searching for people (people search).

6.2   Challenges

Although there has been a great number of studies for network-based social search, there are still several challenges that need to be properly addressed in the future. We believe that the following three topics might attract more attentions.

The first challenge is to tackle data heterogeneity [22]. In both Sects. 4 and 5, we have observed a large number of studies on applying homogeneous networks for both document search and people search. However, the research of utilizing heterogeneous networks remains relatively few, and most of them are query-independent which cannot be customized for different user queries and thus may not be able to provide personalized search results for different users. Social search itself implies personalized information needs under certain social context, and we believe that there will have increasing research efforts on this topic in the future.

The second challenge is to deal with data incompleteness [70]. Many network-based approaches, particularly the individual-level network information, heavily rely on the completeness of network information. For example, Han et al. [43] tested ways of hiding certain network information for people search. They found that local network features (i.e., the individual level network information) of the information seeker can heavily affect the performance of people search, while the global network features (i.e., entire network level information) do not impact too much. Data incompleteness might be due to multiple reasons. First, social network services often allow users to configure privacy settings, which can significantly change the information visibility for different groups of users. Second, social information for a user might also scatter around multiple social networking platforms and the user may have different levels of activities on each platform. This further hinders us to collect a complete set of information.

The third challenge is to involve users in the people search process (human-in-the-loop). Most of existing studies on people search are barely off-line, not requiring/allowing user interactions. Few of them such as SanDVIS [102], Small-Blue [76] and PeopleExplorer [40] have taken the interactive approach, either through lab-controlled user studies or via field studies. They all find the effectiveness of interactive people search interfaces. However, these studies were all performed at a small scale with limited number of users and participants, and the included controllable factors are usually preset and task-specific. We expect more comprehensive studies on this topic in the future.

References

1. Ackerman, M.S., Pipek, V., Wulf, V.: Sharing Expertise: Beyond Knowledge Management. MIT Press, Cambridge (2003)
2. Adamic, L., Adar, E.: How to search a social network. Soc. Netw. **27**(3), 187–203 (2005)
3. Almeida, R.B., Almeida, V.A.: A community-aware search engine. In: Proceedings of the 13th International Conference on World Wide Web, pp. 413–421. ACM (2004)
4. Amitay, E., Carmel, D., Har'El, N., Ofek-Koifman, S., Soffer, A., Yogev, S., Golbandi, N.: Social search and discovery using a unified approach. In: Proceedings of the 20th ACM Conference on Hypertext and Hypermedia, pp. 199–208. ACM (2009)
5. Balog, K., Azzopardi, L., De Rijke, M.: Formal models for expert finding in enterprise corpora. In: Proceedings of the 29th Annual International ACM SIGIR Conference on Research and Development in Information Retrieval, pp. 43–50. ACM (2006)
6. Bao, S., Xue, G., Wu, X., Yu, Y., Fei, B., Su, Z.: Optimizing web search using social annotations. In: Proceedings of the 16th International Conference on World Wide Web, pp. 501–510. ACM (2007)
7. Barabási, A.L., Albert, R.: Emergence of scaling in random networks. Science **286**(5439), 509–512 (1999)
8. Blei, D.M., Ng, A.Y., Jordan, M.I.: Latent Dirichlet allocation. J. Mach. Learn. Res. **3**(Jan), 993–1022 (2003)

9. Bonchi, F., Esfandiar, P., Gleich, D.F., Greif, C., Lakshmanan, L.V.: Fast matrix computations for pairwise and columnwise commute times and Katz scores. Internet Math. **8**(1–2), 73–112 (2012)

10. Bostandjiev, S., O'Donovan, J., Höllerer, T.: TasteWeights: a visual interactive hybrid recommender system. In: Proceedings of the Sixth ACM Conference on Recommender Systems, pp. 35–42. ACM (2012)

11. Bostandjiev, S., O'Donovan, J., Höllerer, T.: LinkedVis: exploring social and semantic career recommendations. In: Proceedings of the 2013 International Conference on Intelligent User Interfaces, pp. 107–116. ACM (2013)

12. Brusilovsky, P., Smyth, B., Shapira, B.: Social search. In: Brusilovsky, P., He, D. (eds.) Social Information Access. LNCS, vol. 10100, pp. 213–276. Springer, Heidelberg (2018)

13. Carmel, D., Zwerdling, N., Guy, I., Ofek-Koifman, S., Har'El, N., Ronen, I., Uziel, E., Yogev, S., Chernov, S.: Personalized social search based on the user's social network. In: Proceedings of the 18th ACM Conference on Information and Knowledge Management, pp. 1227–1236. ACM (2009)

14. Chen, H.H., Gou, L., Zhang, X., Giles, C.L.: CollabSeer: a search engine for collaboration discovery. In: Proceedings of the 11th Annual International ACM/IEEE Joint Conference on Digital Libraries, pp. 231–240. ACM (2011)

15. Chi, E.H.: Information seeking can be social. Computer **42**(3), 42–46 (2009)

16. Craswell, N., Szummer, M.: Random walks on the click graph. In: Proceedings of the 30th Annual International ACM SIGIR Conference on Research and Development in Information Retrieval, pp. 239–246. ACM (2007)

17. Croft, W.B., Metzler, D., Strohman, T.: Search Engines: Information Retrieval in Practice. Addison-Wesley Reading, Menlo Park (2010)

18. Croft, W.B., Turtle, H.: A retrieval model incorporating hypertext links. In: Proceedings of the Second Annual ACM Conference on Hypertext, pp. 213–224. ACM (1989)

19. Curtiss, M., Becker, I., Bosman, T., Doroshenko, S., Grijincu, L., Jackson, T., Kunnatur, S., Lassen, S., Pronin, P., Sankar, S., et al.: Unicorn: a system for searching the social graph. Proc. VLDB Endow. **6**(11), 1150–1161 (2013)

20. Cutting, D.R., Karger, D.R., Pedersen, J.O., Tukey, J.W.: Scatter/Gather: a cluster-based approach to browsing large document collections. In: Proceedings of the 15th Annual International ACM SIGIR Conference on Research and Development in Information Retrieval, pp. 318–329. ACM (1992)

21. Deng, H., Han, J., Lyu, M.R., King, I.: Modeling and exploiting heterogeneous bibliographic networks for expertise ranking. In: Proceedings of the 12th ACM/IEEE-CS Joint Conference on Digital Libraries, pp. 71–80. ACM (2012)

22. Deng, H., King, I., Lyu, M.R.: Enhanced models for expertise retrieval using community-aware strategies. IEEE Trans. Syst. Man Cybern. Part B Cybern. **42**(1), 93–106 (2012)

23. Deng, H., Lyu, M.R., King, I.: A generalized Co-HITS algorithm and its application to bipartite graphs. In: Proceedings of the 15th ACM SIGKDD International Conference on Knowledge Discovery and Data Mining, pp. 239–248. ACM (2009)

24. Dumais, S.T., Nielsen, J.: Automating the assignment of submitted manuscripts to reviewers. In: Proceedings of the 15th Annual International ACM SIGIR Conference on Research and Development in Information Retrieval, pp. 233–244. ACM (1992)

25. Easley, D., Kleinberg, J.: Networks, Crowds, and Markets: Reasoning About a Highly Connected World. Cambridge University Press, Cambridge (2010)

26. Fang, H., Zhai, C.X.: Probabilistic models for expert finding. In: Amati, G., Carpineto, C., Romano, G. (eds.) ECIR 2007. LNCS, vol. 4425, pp. 418–430. Springer, Heidelberg (2007). https://doi.org/10.1007/978-3-540-71496-5_38

27. Flake, G.W., Lawrence, S., Giles, C.L., Coetzee, F.M.: Self-organization and identification of web communities. Computer **35**(3), 66–70 (2002)

28. Fortunato, S.: Community detection in graphs. Phys. Rep. **486**(3), 75–174 (2010)

29. Fortunato, S., Barthélemy, M.: Resolution limit in community detection. Proc. Natl. Acad. Sci. **104**(1), 36–41 (2007)

30. Freyne, J., Farzan, R., Brusilovsky, P., Smyth, B., Coyle, M.: Collecting community wisdom: integrating social search & social navigation. In: Proceedings of the 12th International Conference on Intelligent User Interfaces, pp. 52–61. ACM (2007)

31. Freyne, J., Smyth, B.: An experiment in social search. In: De Bra, P.M.E., Nejdl, W. (eds.) AH 2004. LNCS, vol. 3137, pp. 95–103. Springer, Heidelberg (2004). https://doi.org/10.1007/978-3-540-27780-4_13

32. Getoor, L., Diehl, C.P.: Link mining: a survey. ACM SIGKDD Explor. Newsl. **7**(2), 3–12 (2005)

33. Ghosh, S., Sharma, N., Benevenuto, F., Ganguly, N., Gummadi, K.: Cognos: crowdsourcing search for topic experts in microblogs. In: Proceedings of the 35th International ACM SIGIR Conference on Research and Development in Information Retrieval, pp. 575–590. ACM (2012)

34. Gibson, D., Kleinberg, J., Raghavan, P.: Inferring web communities from link topology. In: Proceedings of the Ninth ACM Conference on Hypertext and Hypermedia: Links, Objects, Time and Space–Structure in Hypermedia Systems: Links, Objects, Time and Space–Structure in Hypermedia Systems, pp. 225–234. ACM (1998)

35. Gong, X., Ke, W., Zhang, Y., Broussard, R.: Interactive search result clustering: a study of user behavior and retrieval effectiveness. In: Proceedings of the 13th ACM/IEEE-CS Joint Conference on Digital Libraries, pp. 167–170. ACM (2013)

36. Guimerà, R., Sales-Pardo, M., Amaral, L.A.N.: Module identification in bipartite and directed networks. Phys. Rev. E **76**(3), 036102 (2007)

37. Guimera, R., Sales-Pardo, M., Amaral, L.A.: Classes of complex networks defined by role-to-role connectivity profiles. Nat. Phys. **3**(1), 63–69 (2007)

38. Gupta, P., Goel, A., Lin, J., Sharma, A., Wang, D., Zadeh, R.: WTF: the who to follow service at Twitter. In: Proceedings of the 22nd International Conference on World Wide Web, pp. 505–514. International World Wide Web Conferences Steering Committee (2013)

39. Han, S., He, D., Brusilovsky, P., Yue, Z.: Coauthor prediction for junior researchers. In: Greenberg, A.M., Kennedy, W.G., Bos, N.D. (eds.) SBP 2013. LNCS, vol. 7812, pp. 274–283. Springer, Heidelberg (2013). https://doi.org/10.1007/978-3-642-37210-0_30

40. Han, S., He, D., Jiang, J., Yue, Z.: Supporting exploratory people search: a study of factor transparency and user control. In: Proceedings of the 22nd ACM International Conference on Information & Knowledge Management, pp. 449–458. ACM (2013)

41. Han, S., He, D., Yue, Z., Jiang, J., Jeng, W.: IRIS-IPS: an interactive people search system for HCIR challenge. In: The Proceedings of HCIR (2012)

42. Han, S., He, D., Yue, Z., Brusilovsky, P.: Supporting cross-device web search with social navigation-based mobile touch interactions. In: Ricci, F., Bontcheva, K., Conlan, O., Lawless, S. (eds.) UMAP 2015. LNCS, vol. 9146, pp. 143–155. Springer, Cham (2015). https://doi.org/10.1007/978-3-319-20267-9_12

43. Han, S., He, D., Yue, Z.: Benchmarking the privacy-preserving people search. arXiv preprint arXiv:1409.5524 (2014)

44. Han, S., Yi, X., Yue, Z., Geng, Z., Glass, A.: Framing mobile information needs: an investigation of hierarchical query sequence structure. In: Proceedings of the 25th ACM International on Conference on Information and Knowledge Management, pp. 2131–2136. ACM (2016)

45. Han, S., Zhang, D., He, D., Cheng, Q.: User exploration of slider facets in interactive people search system. In: IConference 2016 Proceedings (2016)

46. Haveliwala, T.H.: Topic-sensitive PageRank. In: Proceedings of the 11th International Conference on World Wide Web, pp. 517–526. ACM (2002)

47. Hearst, M.A., Pedersen, J.O.: Reexamining the cluster hypothesis: Scatter/Gather on retrieval results. In: Proceedings of the 19th Annual International ACM SIGIR Conference on Research and Development in Information Retrieval, pp. 76–84. ACM (1996)

48. Hecht, B., Teevan, J., Morris, M.R., Liebling, D.J.: SearchBuddies: bringing search engines into the conversation. ICWSM **12**, 138–145 (2012)

49. Hitchcock, M.A., Bland, C.J., Hekelman, F.P., Blumenthal, M.G.: Professional networks: the influence of colleagues on the academic success of faculty. Acad. Med. **70**(12), 1108–1116 (1995)

50. Hofmann, K., Balog, K., Bogers, T., De Rijke, M.: Integrating contextual factors into topic-centric retrieval models for finding similar experts. In: Proceedings of ACM SIGIR 2008 Workshop on Future Challenges in Expert Retrieval, pp. 29–36 (2008)

51. Hofmann, K., Balog, K., Bogers, T., De Rijke, M.: Contextual factors for finding similar experts. J. Am. Soc. Inform. Sci. Technol. **61**(5), 994–1014 (2010)

52. Horowitz, D., Kamvar, S.D.: The anatomy of a large-scale social search engine. In: Proceedings of the 19th International Conference on World Wide Web, pp. 431–440. ACM (2010)

53. Hsiao, K.J., Kulesza, A., Hero, A.O.: Social collaborative retrieval. IEEE J. Sel. Top. Signal Process. **8**(4), 680–689 (2014)

54. Huang, H., Zubiaga, A., Ji, H., Deng, H., Wang, D., Le, H.K., Abdelzaher, T.F., Han, J., Leung, A., Hancock, J.P., et al.: Tweet ranking based on heterogeneous networks. In: COLING, pp. 1239–1256 (2012)

55. Huang, S.W., Tunkelang, D., Karahalios, K.: The role of network distance in LinkedIn people search. In: Proceedings of the 37th International ACM SIGIR Conference on Research & Development in Information Retrieval, pp. 867–870. ACM (2014)

56. Jardine, N., van Rijsbergen, C.J.: The use of hierarchic clustering in information retrieval. Inf. Storage Retr. **7**(5), 217–240 (1971)

57. Jeh, G., Widom, J.: SimRank: a measure of structural-context similarity. In: Proceedings of the Eighth ACM SIGKDD International Conference on Knowledge Discovery and Data Mining, pp. 538–543. ACM (2002)

58. Jiang, M., Cui, P., Wang, F., Yang, Q., Zhu, W., Yang, S.: Social recommendation across multiple relational domains. In: Proceedings of the 21st ACM International Conference on Information and Knowledge Management, pp. 1422–1431. ACM (2012)

59. Jing, Y., Baluja, S.: VisualRank: applying PageRank to large-scale image search. IEEE Trans. Pattern Anal. Mach. Intell. **30**(11), 1877–1890 (2008)

60. Joachims, T., Granka, L., Pan, B., Hembrooke, H., Gay, G.: Accurately interpreting clickthrough data as implicit feedback. In: Proceedings of the 28th Annual International ACM SIGIR Conference on Research and Development in Information Retrieval, pp. 154–161. ACM (2005)

61. Kamvar, S.D., Schlosser, M.T., Garcia-Molina, H.: The eigentrust algorithm for reputation management in P2P networks. In: Proceedings of the 12th International Conference on World Wide Web, pp. 640–651. ACM (2003)

62. Karimzadehgan, M., White, R.W., Richardson, M.: Enhancing expert finding using organizational hierarchies. In: Boughanem, M., Berrut, C., Mothe, J., Soule-Dupuy, C. (eds.) ECIR 2009. LNCS, vol. 5478, pp. 177–188. Springer, Heidelberg (2009). https://doi.org/10.1007/978-3-642-00958-7_18

63. Kashyap, A., Amini, R., Hristidis, V.: SonetRank: leveraging social networks to personalize search. In: Proceedings of the 21st ACM International Conference on Information and Knowledge Management, pp. 2045–2049. ACM (2012)

64. Katz, L.: A new status index derived from sociometric analysis. Psychometrika **18**(1), 39–43 (1953)

65. Kautz, H., Selman, B., Shah, M.: Referral web: combining social networks and collaborative filtering. Commun. ACM **40**(3), 63–65 (1997)

66. Kelly, D., Teevan, J.: Implicit feedback for inferring user preference: a bibliography. In: ACM SIGIR Forum, vol. 37, pp. 18–28. ACM (2003)

67. Kim, Y., Hassan, A., White, R.W., Zitouni, I.: Modeling dwell time to predict click-level satisfaction. In: Proceedings of the 7th ACM International Conference on Web Search and Data Mining, pp. 193–202. ACM (2014)

68. Kleinberg, J.M.: Authoritative sources in a hyperlinked environment. J. ACM (JACM) **46**(5), 604–632 (1999)

69. Koshman, S., Spink, A., Jansen, B.J.: Web searching on the Vivisimo search engine. J. Am. Soc. Inform. Sci. Technol. **57**(14), 1875–1887 (2006)

70. Kossinets, G.: Effects of missing data in social networks. Soc. Netw. **28**(3), 247–268 (2006)

71. Lancichinetti, A., Fortunato, S.: Limits of modularity maximization in community detection. Phys. Rev. E **84**(6), 066122 (2011)

72. Leskovec, J., Dumais, S., Horvitz, E.: Web projections: learning from contextual subgraphs of the web. In: Proceedings of the 16th International Conference on World Wide Web, pp. 471–480. ACM (2007)

73. Leskovec, J., Kleinberg, J., Faloutsos, C.: Graphs over time: densification laws, shrinking diameters and possible explanations. In: Proceedings of the Eleventh ACM SIGKDD International Conference on Knowledge Discovery in Data Mining, pp. 177–187. ACM (2005)

74. Leskovec, J., Kleinberg, J., Faloutsos, C.: Graph evolution: densification and shrinking diameters. ACM Trans. Knowl. Discov. Data (TKDD) **1**(1), 2 (2007)

75. Liben-Nowell, D., Kleinberg, J.: The link-prediction problem for social networks. J. Am. Soc. Inform. Sci. Technol. **58**(7), 1019–1031 (2007)

76. Lin, C.Y., Cao, N., Liu, S.X., Papadimitriou, S., Sun, J., Yan, X.: SmallBlue: social network analysis for expertise search and collective intelligence. In: IEEE 25th International Conference on Data Engineering, ICDE 2009, pp. 1483–1486. IEEE (2009)

77. Lin, Y.R., Sun, J., Castro, P., Konuru, R., Sundaram, H., Kelliher, A.: MetaFac: community discovery via relational hypergraph factorization. In: Proceedings of the 15th ACM SIGKDD International Conference on Knowledge Discovery and Data Mining, pp. 527–536. ACM (2009)

78. Liu, T.Y., et al.: Learning to rank for information retrieval. Found. Trends® Inf. Retr. **3**(3), 225–331 (2009)
79. Liu, X., Bollen, J., Nelson, M.L., Van de Sompel, H.: Co-authorship networks in the digital library research community. Inf. Process. Manag. **41**(6), 1462–1480 (2005)
80. Liu, X., Croft, W.B.: Cluster-based retrieval using language models. In: Proceedings of the 27th Annual International ACM SIGIR Conference on Research and Development in Information Retrieval, pp. 186–193. ACM (2004)
81. Liu, X., Jiang, Z., Gao, L.: Scientific information understanding via open educational resources (OER). In: Proceedings of the 38th International ACM SIGIR Conference on Research and Development in Information Retrieval, pp. 645–654. ACM (2015)
82. Liu, X., Yu, Y., Guo, C., Sun, Y.: Meta-path-based ranking with pseudo relevance feedback on heterogeneous graph for citation recommendation. In: Proceedings of the 23rd ACM International Conference on Information and Knowledge Management, pp. 121–130. ACM (2014)
83. Liu, Y., Gao, B., Liu, T.Y., Zhang, Y., Ma, Z., He, S., Li, H.: BrowseRank: letting web users vote for page importance. In: Proceedings of the 31st Annual International ACM SIGIR Conference on Research and Development in Information Retrieval, pp. 451–458. ACM (2008)
84. Long, B., Wu, X., Zhang, Z.M., Yu, P.S.: Unsupervised learning on k-partite graphs. In: Proceedings of the 12th ACM SIGKDD International Conference on Knowledge Discovery and Data Mining, pp. 317–326. ACM (2006)
85. Ma, H., Zhou, D., Liu, C., Lyu, M.R., King, I.: Recommender systems with social regularization. In: Proceedings of the Fourth ACM International Conference on Web Search and Data Mining, pp. 287–296. ACM (2011)
86. Macdonald, C., Ounis, I.: Expertise drift and query expansion in expert search. In: Proceedings of the Sixteenth ACM Conference on Information and Knowledge Management, pp. 341–350. ACM (2007)
87. Macdonald, C., Ounis, I.: Using relevance feedback in expert search. In: Amati, G., Carpineto, C., Romano, G. (eds.) ECIR 2007. LNCS, vol. 4425, pp. 431–443. Springer, Heidelberg (2007). https://doi.org/10.1007/978-3-540-71496-5_39
88. Manning, C.D., Raghavan, P., Schütze, H., et al.: Introduction to Information Retrieval, vol. 1. Cambridge University Press, Cambridge (2008)
89. McDonald, D.W., Ackerman, M.S.: Just talk to me: a field study of expertise location. In: Proceedings of the 1998 ACM Conference on Computer Supported Cooperative Work, pp. 315–324. ACM (1998)
90. Menczer, F.: Evolution of document networks. Proc. Natl. Acad. Sci. **101**(suppl 1), 5261–5265 (2004)
91. Morris, M.R.: A survey of collaborative web search practices. In: Proceedings of the SIGCHI Conference on Human Factors in Computing Systems, pp. 1657–1660. ACM (2008)
92. Murata, T.: Detecting communities from tripartite networks. In: Proceedings of the 19th International Conference on World Wide Web, pp. 1159–1160. ACM (2010)
93. Murata, T., Ikeya, T.: A new modularity for detecting one-to-many correspondence of communities in bipartite networks. Adv. Complex Syst. **13**(01), 19–31 (2010)
94. Navarro Bullock, B., Hotho, A., Stumme, G.: Accessing information with tags: search and ranking. In: Brusilovsky, P., He, D. (eds.) Social Information Access. LNCS, vol. 10100, pp. 310–343. Springer, Heidelberg (2018)

95. Neubauer, N., Obermayer, K.: Towards community detection in k-partite k-uniform hypergraphs. In: Proceedings of the NIPS 2009 Workshop on Analyzing Networks and Learning with Graphs, pp. 1–9 (2009)

96. Newman, M.E.: The structure and function of complex networks. SIAM Rev. **45**(2), 167–256 (2003)

97. Newman, M.E.: Coauthorship networks and patterns of scientific collaboration. Proc. Natl. Acad. Sci. **101**(suppl 1), 5200–5205 (2004)

98. Newman, M.E.: Modularity and community structure in networks. Proc. Natl. Acad. Sci. **103**(23), 8577–8582 (2006)

99. Newman, M.E., Girvan, M.: Finding and evaluating community structure in networks. Phys. Rev. E **69**(2), 026113 (2004)

100. Page, L., Brin, S., Motwani, R., Winograd, T.: The PageRank citation ranking: bringing order to the web (1999)

101. Palla, G., Derényi, I., Farkas, I., Vicsek, T.: Uncovering the overlapping community structure of complex networks in nature and society. Nature **435**(7043), 814–818 (2005)

102. Perer, A., Guy, I.: SaNDVis: visual social network analytics for the enterprise. In: Proceedings of the ACM 2012 Conference on Computer Supported Cooperative Work Companion, pp. 275–276. ACM (2012)

103. Perer, A., Guy, I., Uziel, E., Ronen, I., Jacovi, M.: Visual social network analytics for relationship discovery in the enterprise. In: 2011 IEEE Conference on Visual Analytics Science and Technology, VAST, pp. 71–79. IEEE (2011)

104. Psorakis, I., Roberts, S., Ebden, M., Sheldon, B.: Overlapping community detection using Bayesian non-negative matrix factorization. Phys. Rev. E **83**(6), 066114 (2011)

105. Reichling, T., Wulf, V.: Expert recommender systems in practice: evaluating semi-automatic profile generation. In: Proceedings of the SIGCHI Conference on Human Factors in Computing Systems, pp. 59–68. ACM (2009)

106. Rijsbergen, C.J.V.: Information Retrieval, 2nd edn. Butterworth-Heinemann, Newton (1979)

107. Rodriguez, M., Posse, C., Zhang, E.: Multiple objective optimization in recommender systems. In: Proceedings of the Sixth ACM Conference on Recommender Systems, pp. 11–18. ACM (2012)

108. Rosvall, M., Bergstrom, C.T.: An information-theoretic framework for resolving community structure in complex networks. Proc. Natl. Acad. Sci. **104**(18), 7327–7331 (2007)

109. Salton, G., Wong, A., Yang, C.S.: A vector space model for automatic indexing. Commun. ACM **18**(11), 613–620 (1975)

110. Serdyukov, P., Hiemstra, D.: Modeling documents as mixtures of persons for expert finding. In: Macdonald, C., Ounis, I., Plachouras, V., Ruthven, I., White, R.W. (eds.) ECIR 2008. LNCS, vol. 4956, pp. 309–320. Springer, Heidelberg (2008). https://doi.org/10.1007/978-3-540-78646-7_29

111. Shen, X., Tan, B., Zhai, C.: Context-sensitive information retrieval using implicit feedback. In: Proceedings of the 28th Annual International ACM SIGIR Conference on Research and Development in Information Retrieval, pp. 43–50. ACM (2005)

112. Shi, C., Kong, X., Yu, P.S., Xie, S., Wu, B.: Relevance search in heterogeneous networks. In: Proceedings of the 15th International Conference on Extending Database Technology, pp. 180–191. ACM (2012)

113. Smyth, B.: A community-based approach to personalizing web search. Computer **40**(8), 42–50 (2007)

114. Smyth, B., Balfe, E., Freyne, J., Briggs, P., Coyle, M., Boydell, O.: Exploiting query repetition and regularity in an adaptive community-based web search engine. User Model. User-Adapt. Interact. **14**(5), 383–423 (2004)

115. Smyth, B., Briggs, P., Coyle, M., O'Mahony, M.: Google shared. A case-study in social search. In: Houben, G.-J., McCalla, G., Pianesi, F., Zancanaro, M. (eds.) UMAP 2009. LNCS, vol. 5535, pp. 283–294. Springer, Heidelberg (2009). https://doi.org/10.1007/978-3-642-02247-0_27

116. Spirin, N.V., He, J., Develin, M., Karahalios, K.G., Boucher, M.: People search within an online social network: large scale analysis of Facebook graph search query logs. In: Proceedings of the 23rd ACM International Conference on Information and Knowledge Management, pp. 1009–1018. ACM (2014)

117. Sun, Y., Han, J.: Mining heterogeneous information networks: principles and methodologies. Synth. Lect. Data Min. Knowl. Discov. **3**(2), 1–159 (2012)

118. Sun, Y., Han, J., Yan, X., Yu, P.S., Wu, T.: PathSim: meta path-based top-k similarity search in heterogeneous information networks. In: VLDB 2011 (2011)

119. Sun, Y., Han, J., Zhao, P., Yin, Z., Cheng, H., Wu, T.: RankClus: integrating clustering with ranking for heterogeneous information network analysis. In: Proceedings of the 12th International Conference on Extending Database Technology: Advances in Database Technology, pp. 565–576. ACM (2009)

120. Sun, Y., Yu, Y., Han, J.: Ranking-based clustering of heterogeneous information networks with star network schema. In: Proceedings of the 15th ACM SIGKDD International Conference on Knowledge Discovery and Data Mining, pp. 797–806. ACM (2009)

121. Tan, B., Shen, X., Zhai, C.: Mining long-term search history to improve search accuracy. In: Proceedings of the 12th ACM SIGKDD International Conference on Knowledge Discovery and Data Mining, pp. 718–723. ACM (2006)

122. Tang, J., Wu, S., Sun, J., Su, H.: Cross-domain collaboration recommendation. In: Proceedings of the 18th ACM SIGKDD International Conference on Knowledge Discovery and Data Mining, pp. 1285–1293. ACM (2012)

123. Tang, J., Zhang, J., Jin, R., Yang, Z., Cai, K., Zhang, L., Su, Z.: Topic level expertise search over heterogeneous networks. Mach. Learn. **82**(2), 211–237 (2011)

124. Tang, J., Zhang, J., Yao, L., Li, J., Zhang, L., Su, Z.: ArnetMiner: extraction and mining of academic social networks. In: Proceedings of the 14th ACM SIGKDD International Conference on Knowledge Discovery and Data Mining, pp. 990–998. ACM (2008)

125. Tang, W., Tang, J., Lei, T., Tan, C., Gao, B., Li, T.: On optimization of expertise matching with various constraints. Neurocomputing **76**(1), 71–83 (2012)

126. Teevan, J., Dumais, S.T., Horvitz, E.: Personalizing search via automated analysis of interests and activities. In: Proceedings of the 28th Annual International ACM SIGIR Conference on Research and Development in Information Retrieval, pp. 449–456. ACM (2005)

127. Teevan, J., Morris, M.R., Bush, S.: Discovering and using groups to improve personalized search. In: Proceedings of the Second ACM International Conference on Web Search and Data Mining, pp. 15–24. ACM (2009)

128. Teevan, J., Ramage, D., Morris, M.R.: #TwitterSearch: a comparison of microblog search and web search. In: Proceedings of the Fourth ACM International Conference on Web Search and Data Mining, pp. 35–44. ACM (2011)

129. Terveen, L., McDonald, D.W.: Social matching: a framework and research agenda. ACM Trans. Comput.-Hum. Interact. (TOCHI) **12**(3), 401–434 (2005)

130. Tiwari, M.: Large-scale social recommender systems: challenges and opportuni-
     ties. In: Proceedings of the 22nd International Conference on World Wide Web
     Companion, pp. 939–940. International World Wide Web Conferences Steering
     Committee (2013)
131. Tombros, A., Villa, R., Van Rijsbergen, C.J.: The effectiveness of query-specific
     hierarchic clustering in information retrieval. Inf. Process. Manag. **38**(4), 559–582
     (2002)
132. Ugander, J., Backstrom, L., Marlow, C., Kleinberg, J.: Structural diversity in
     social contagion. Proc. Natl. Acad. Sci. **109**(16), 5962–5966 (2012)
133. Ugander, J., Karrer, B., Backstrom, L., Marlow, C.: The anatomy of the Facebook
     social graph. arXiv preprint arXiv:1111.4503 (2011)
134. Vassilvitskii, S., Brill, E.: Using web-graph distance for relevance feedback in web
     search. In: Proceedings of the 29th Annual International ACM SIGIR Conference
     on Research and Development in Information Retrieval, pp. 147–153. ACM (2006)
135. Wasserman, S., Faust, K.: Social Network Analysis: Methods and Applications,
     vol. 8. Cambridge University Press, Cambridge (1994)
136. Weerkamp, W., Berendsen, R., Kovachev, B., Meij, E., Balog, K., De Rijke, M.:
     People searching for people: analysis of a people search engine log. In: Proceedings
     of the 34th International ACM SIGIR Conference on Research and Development
     in Information Retrieval, pp. 45–54. ACM (2011)
137. Wei, X., Croft, W.B.: LDA-based document models for ad-hoc retrieval. In: Pro-
     ceedings of the 29th Annual International ACM SIGIR Conference on Research
     and Development in Information Retrieval, pp. 178–185. ACM (2006)
138. Wellman, B.: Community: from neighborhood to network. Commun. ACM
     48(10), 53–55 (2005)
139. White, R.W., Chu, W., Hassan, A., He, X., Song, Y., Wang, H.: Enhancing per-
     sonalized search by mining and modeling task behavior. In: Proceedings of the
     22nd International Conference on World Wide Web, pp. 1411–1420. International
     World Wide Web Conferences Steering Committee (2013)
140. White, R.W., Jose, J.M., Ruthven, I.: An implicit feedback approach for interac-
     tive information retrieval. Inf. Process. Manag. **42**(1), 166–190 (2006)
141. Xie, H.R., Li, Q., Cai, Y.: Community-aware resource profiling for personalized
     search in folksonomy. J. Comput. Sci. Technol. **27**(3), 599–610 (2012)
142. Xu, S., Bao, S., Fei, B., Su, Z., Yu, Y.: Exploring folksonomy for personalized
     search. In: Proceedings of the 31st Annual International ACM SIGIR Conference
     on Research and Development in Information Retrieval, pp. 155–162. ACM (2008)
143. Xu, W., Liu, X., Gong, Y.: Document clustering based on non-negative matrix fac-
     torization. In: Proceedings of the 26th Annual International ACM SIGIR Confer-
     ence on Research and Development in Information Retrieval, pp. 267–273. ACM
     (2003)
144. Yan, E., Ding, Y.: Discovering author impact: a PageRank perspective. Inf. Pro-
     cess. Manag. **47**(1), 125–134 (2011)
145. Yang, J., Leskovec, J.: Overlapping community detection at scale: a nonnegative
     matrix factorization approach. In: Proceedings of the Sixth ACM International
     Conference on Web Search and Data Mining, pp. 587–596. ACM (2013)
146. Yang, X., Steck, H., Liu, Y.: Circle-based recommendation in online social net-
     works. In: Proceedings of the 18th ACM SIGKDD International Conference on
     Knowledge Discovery and Data Mining, pp. 1267–1275. ACM (2012)
147. Yarosh, S., Matthews, T., Zhou, M.: Asking the right person: supporting expertise
     selection in the enterprise. In: Proceedings of the SIGCHI Conference on Human
     Factors in Computing Systems, pp. 2247–2256. ACM (2012)

148. Yi, L., Liu, B., Li, X.: Eliminating noisy information in web pages for data mining. In: Proceedings of the Ninth ACM SIGKDD International Conference on Knowledge Discovery and Data Mining, pp. 296–305. ACM (2003)
149. Yin, P., Lee, W.C., Lee, K.C.: On top-k social web search. In: Proceedings of the 19th ACM International Conference on Information and Knowledge Management, pp. 1313–1316. ACM (2010)
150. Yue, Z., Han, S., He, D.: Modeling search processes using hidden states in collaborative exploratory web search. In: Proceedings of the 17th ACM Conference on Computer Supported Cooperative Work & Social Computing, pp. 820–830. ACM (2014)
151. Zhai, C.: Statistical language models for information retrieval. Synth. Lect. Hum. Lang. Technol. 1(1), 1–141 (2008)
152. Zhao, P., Han, J., Sun, Y.: P-Rank: a comprehensive structural similarity measure over information networks. In: Proceedings of the 18th ACM Conference on Information and Knowledge Management, pp. 553–562. ACM (2009)
153. Zhou, D., Orshanskiy, S., Zha, H., Giles, C.L., et al.: Co-ranking authors and documents in a heterogeneous network. In: Seventh IEEE International Conference on Data Mining, ICDM 2007, pp. 739–744. IEEE (2007)

9
Accessing Information with Tags: Search and Ranking

Beate Navarro Bullock[1], Andreas Hotho[1]◉, and Gerd Stumme[2(✉)]◉

[1] DMIR Research Group at the Computer Science Institute,
University of Würzburg, Am Hubland, 97074 Würzburg, Germany
[2] Knowledge and Data Engineering Group, Research Center for Information System
Design, University of Kassel, Wilhelmshöher Allee 73, 34121 Kassel, Germany
stumme@cs.uni-kassel.de
http://www.dmir.uni-wuerzburg.de/, http://www.kde.cs.uni-kassel.de/

Abstract. With the growth of the Social Web, a variety of new web-based services arose and changed the way users interact with the internet and consume information. One central phenomenon was and is *tagging* which allows to manage, organize and access information in social systems. Tagging helps to manage all kinds of resources, making their access much easier. The first type of social tagging systems were *social bookmarking systems*, i.e., platforms for storing and sharing bookmarks on the web rather than just in the browser. Meanwhile, (hash-)tagging is central in many other Social Media systems such as social networking sites and micro-blogging platforms. To allow for efficient information access, special algorithms have been developed to guide the user, to search for information and to rank the content based on tagging information contributed by the users.

In this article we review several aspects of the tagging process and its role for accessing information using search and ranking in tagging systems. A literature review of existing work in this area will be complemented by case studies which showcase findings of our own research. We will start with discussing typical properties of tagging systems, present example systems and their typical functionality, their strengths and weaknesses, the users' motivations, and different types of tags and annotators. To get an understanding of search and ranking methods, we use the formalization of tagging systems as a tripartite graph of users, tags, and resources – known as *folksonomy* – and discuss its network properties.

Ranking in folksonomies is a core component of information access in such systems. We review two central algorithms, FolkRank and Adjusted Hits, before focussing on a tighter integration of Web search and folksonomies. For this, we compare search in standard search engines with tag-based search, review Social PageRank, a method for ranking web pages that is using the information of tagging systems, and discuss learning-to-rank methods which also utilize tags to improve the ranking of web pages. Finally, we present the concept of *logsonomies* which provide a unified view on search and tagging by considering clicks on

© Springer International Publishing AG, part of Springer Nature 2018
P. Brusilovsky and D. He (Eds.): Social Information Access, LNCS 10100, pp. 310–343, 2018.
https://doi.org/10.1007/978-3-319-90092-6_9

search results as an implicit tagging process. Concluding, we discuss future options for a tighter integration of tagging and search with the goal of improving information access based on user provided content.

1   Introduction

At the start of the Internet era, people predominantly used the Web to consume mostly static digital content. A paradigm shift occurred with the advent of the *Social Web* where user-generated content, user participation and the creation of virtual social relationships became integral parts of Internet usage. Typical applications comprise online marketplaces such as eBay or Amazon, content sharing sites such as Flickr or Delicious, online social networks such as Facebook or Twitter, blogs and the online encyclopedia Wikipedia. Nowadays, social services are part of the day-to-day digital experience of most users'. Many activities of everyday life have moved partially or fully into the virtual world. Friendships are maintained via social networks, online product review and online shopping saves time and energy, and news or opinions are communicated via text messaging services. All these developments are summarized under the term *Social Media* which better reflects the broad range of different applications and services available in the Web today.

The growing popularity of the Social Web has led to a wealth of digital user data which can be collected and analysed. The exploration of such data helps to gain knowledge about user interests and preferences as well as the connection between them. This information can be harnessed by applications, for example by providing personalized recommendations and personal search algorithms or by tailoring commercial advertisements for individual users (see also the introduction of this book [21]).

One of the first types of Social Media applications were social bookmarking systems, where users share online resources in form of bookmarks. To manage bookmarks and make the bookmarks retrievable for themselves and others, they add tags – i.e., descriptive keywords – to their resources. This kind of crowd-sourced classification of (web) documents by tags soon became very popular and is now commonly used in many other systems, e.g.,in video and photo sharing platforms, social networking sites, and micro-blogging platforms – in the latter mostly as so-called *hashtags*.

With many different users sharing objects and tagging them, a common information structure is created which is called *folksonomy*. The structure emerges over time, influenced by the numerous users and the possibility to interact with each other. In general, it reflects a common knowledge, a form of *collective intelligence*, among the system's users. Social bookmarking systems provide large document collections covering the interests and topics of their active users. People can therefore use social bookmarking systems to find information – in a similar way as they use other existing online information retrieval systems such as search engines.

One major difference between information retrieval in search engines and social bookmarking systems is the way in which the document collection is created [59]. Search engine providers automatically crawl the Web. New sources are retrieved by following the hyperlinks of an already processed website. In social bookmarking systems, the document collection is created manually by the system's users who post and annotate bookmarks they find interesting.

While the popularity of dedicated social bookmarking systems is decreasing, tagging is still very popular and is an essential feature in many other social media systems, like Twitter or Youtube. The analysis of Twitter's hashtags shows comparable emerging structures as in dedicated tagging systems (cf. [101]). Additionally to annotating content in these systems, tags are also used to represents its own content and used as conversation streams [65]. A comparison of web search and search in Twitter showed that behaviour and search goals are quite different within both environments (cf. [98]). (Hash)tags play a central role to drive the search in Twitter. But unlike a search engine, Twitter does not need to crawl the content and is thus able to provide more information more timely than search engines.

The key algorithmic aspect of any (social) search application—including systems relying on tags—is its ranking functionality, as it ensures that important content is included in the top results. Other facets of search are discussed in the chapters on Collaborative Search [114] and Social Search [22] of this book. Beyond search, the integration of user created content and the collection of user generated data like tags opens a wide field for exploration: From gaining an understanding about the dynamic structures evolving in tagging systems, over building user-tailored functionalities based on their personal data, to solving specific system problems such as spam.

This chapter is structured as follows: The first part introduces the phenomena of tagging, example systems, typical functionality and discusses aspects like users' motivations, types of tags and types of annotators in Sect. 2. Then, we introduce a formal data model – folksonomies – and present typical systems, analyze the properties of the underlying network and investigate the emergence of semantics within such a system in Sect. 3. In Sect. 4, we will focus on ranking methods in tagging systems and discuss standard approaches like FolkRank or adjusted HITS and its application. We continue in Sect. 5 by comparing search in social bookmarking systems with existing web search for understanding commonalities and differences in both systems and discuss simple and more advanced ways of integrating tag in search web search. Finally, a more unified view on search and tagging is introduced in Sect. 6 before the future of search and ranking in tagging systems is discussed in Sect. 7. Additionally to all this general discussion of search and ranking with tags, we provide, in a case study, in more depth some specific results of our own previous work on data of the bookmarking system Delicious and search engine logs of the Microsoft Search Engine MSN.

2   Tagging Systems

Tagging systems enable users to store, organize and share resources such as bookmarks, publication metadata or photos over the Web. The resources can be labeled with any keyword (tag) a user can think of [86]. The tags serve as a representation of the document, summarizing its content, describing the relation to the resource's owner or expressing some kind of emotion the resource owner feels. The process itself is coined *tagging*. It is a kind of manual annotation process of tagged resources. Tags can be used for "future navigation, filtering or search" [28,39]. In particular, tags can play the role of index terms in search engines and can – taking the role of meta data/content description – improve the ranking of tagged resources.

The notion of adding keywords or annotations to digital resources (mostly to documents) existed before the arrival of tagging systems either in form of manual indexing or by automatically extracting keywords from text. In the scientific community the approach has been discussed under the name of *document indexing*, also called subject indexing [112]. The process of (manual) document indexing is twofold: A *conceptual analysis* of the item helps to understand the meaning and importance of it and then a index term has to be chosen. The result can be different for different people, depending on their background, interests or intentions [64]. In most classical document categorization and indexing schemes resources have to be classified into predefined categories which does not hold for tagging as keywords (tags) are personal and can be freely chosen (details on tags later in this section). A key difference between indexing and tagging is that the former typically result in one – collaboratively agreed – category per object whereas a social tagging structure (cf. folksonomies in 3) preserves the individual – potentially conflicting – multiple classifications of the users. Nevertheless, tagging can be seen as a kind of indexing as the process is mostly characterizing the tagged resource.

On the Web, the standard way of organizing documents is by automatic indexing. Manual approaches such as Yahoo!'s directory became more and more difficult to maintain due to the Web's rapid growth and changing nature. In contrast, search engines such as Google[1] or Bing[2], automatically index documents. They create an inverted index where documents are ordered according to specific automatically extracted keywords. Such keywords can then be used in the retrieval process. Since its introduction, automatic keyword extraction has long been the predominant way of preparing data for retrieval.

With the advent of tagging systems, manual indexing became popular again. In contrast to pre-defined indexing schemes, tagging allows spontaneous annotation. Keywords can be selected as they come up in one's mind without having to conform to predefined rules or standards. Hence, not only field experts are able to annotate resources, but web participants themselves freely categorize content. The properties and architecture of tagging systems which enable the simple annotation of resources will be discussed in the next section. Once the resources are annotated, they can be exploited for search and ranking, both within the tagging systems and in addition to standard web search. We discuss different approaches for both in the subsequent sections.

---

[1] http://www.google.com/.
[2] http://www.bing.com/.

2.1   Example Systems

The first tagging systems emerged in 2003 and became popular in the next years. Prominent examples are Delicious[3] for bookmarks, Connotea[4] and CiteULike[5] for publication metadata, BibSonomy[6] for bookmarks and publication metadata, Flickr[7] for photos or YouTube[8] for videos. Meanwhile, the process of tagging is an integral part of many websites – for instance in Technorati[9] (weblog posts) or Twitter[10] (micromessaging posts). Here, we present two example systems in more detail, which cover all core features of tagging systems, and whose data have been used frequently in research on different types of algorithms, including ranking, recommendations and spam detection: Delicious and BibSonomy. These systems are also used in the case studies of this chapter. Most of the research findings will hold for other systems as well even if they have slightly different properties or additional influencing factors (in Twitter, for instance, the inclusion of (hash)tags in the content or the existence of followers). Therefore, we will not focus on other popular systems using tags like Flickr, Twitter or YouTube and refer the reader to, e.g., [26, 36, 76].

2.1.1   BibSonomy

BibSonomy was introduced in 2006 [49]. The social bookmarking system is hosted by the Data Mining and Information Retrieval Group at the University of Würzburg and the Knowledge Engineering Group at the University of Kassel. The target user group are university users including students, teachers and scientists. As their work requires both the collection of web links and publications, BibSonomy combines the management of both types of resources (see Fig. 1).

As of September 2017 the system has about 10,000 active users, who share half a million bookmarks and one million publication metadata entries. Additionally, the system contains—mirroring the computer science library DBLP[11]— about four million publications and 51,000 homepages of research conferences, workshops, and persons, as well as the metadata of one and a half million dissertations[12], which have been imported from the dissertation catalogue of the German National Library (Deutsche Nationalbibliothek).

Further system features were developed to support researchers in their daily work, e.g.,finding relevant information, storing and structuring information, managing references and creating publication lists be it for a diploma thesis,

---

[3] http://del.icio.us (as of June 2017, the service stopped).
[4] http://www.connotea.org/ (as of March 2013, the service stopped).
[5] http://www.citeulike.org.
[6] http://www.bibsonomy.org.
[7] http://www.flickr.com.
[8] http://www.youtube.com.
[9] http://technorati.com/.
[10] https://twitter.com/.
[11] http://www.dblp.uni-trier.de.
[12] https://www.bibsonomy.org/persons.

Fig. 1. Search result for the tags 'ranking' and 'folksonomy' within the collection of user 'hotho' in BibSonomy

a research paper or the website of the research group. BibSonomy also promotes social interactions between users by offering friend connections and the possibility to follow the posts of other users. A more complete description of the system's features can be found in [13,49].

In [14], BibSonomy was used as a research platform for the research group of the Knowledge and Data Engineering team of Kassel. The team conducted experiments concerning different aspects of data mining and analysis including network properties, semantic characteristics, recommender systems, search and spam detection.

Finally, BibSonomy offers system snapshots to other researchers in order to support investigations about tagging data. In two challenges (ECML/PKDD discovery challenge 2008 and 2009) BibSonomy data was used. Several papers were published using those datasets, among them [89] exploring the semantics of tagging systems, [90] analysing communities, [12,29,52,91,94,113] building recommender services, [51,75,81,82,111] creating features and algorithms for spam detection. [30] used BibSonomy data for analysing how the evaluation of recommender algorithms is systematically biased by reducing the data to so-called cores – a procedure frequently performed in scientific publications dealing with long tailed data. Recently, a series of studies began to exploit the access logs of BibSonomy, giving new insights into the user behavior and usage of such tagging systems [32,84,116]. One major finding is the observation that the system is used above all for organizing one's own content, which contradicts somehow the naming of a "social" system. Nevertheless, social interaction can be observed, although not very prominently. Overall, BibSonomy provides an ecosystem for researchers which allows them to collect and manage their own publication data but also to share them with others and to benefit from the collection of the whole community.

2.1.2 Delicious

One of the first social bookmarking systems to become popular was *Delicious*. The system, founded by Joshua Schachter, went online in September 2003[13]. It arouse from a system called Memepool in which Schachter simply collected interesting bookmarks. Over time, users sent him more and more interesting links so that he wrote an application (Muxway) which allowed him to organise his links with short labels—tags. He then realised that not only him, but other internet users might be interested in organizing and sharing their internet links with the help of tags—and rewrote Muxway so that it became the website Delicious. Soon, Delicious became very popular. From December 2005 it was operated by Yahoo! Inc. As the system did not provide financial benefits for the company, it was sold to the company AVOS which was started by the founders of YouTube, Chad Hurley and Steve Chen[14] in 2011. Since then, several aspects of the system have been redesigned in order to introduce more social features into the system[15]. Afterwards, the ownership of the platform changed several times[16], before it was finally turned into a read only modus in June 2017 and its users were encouraged to move to Pinboard[17]. Since Delicious was one of the first highly popular bookmarking systems, its data have been heavily used by researchers. Data are still available, e.g., at http://konect.uni-koblenz.de/ and http://socialcomputing.asu.edu/datasets/Delicious and can be further used for research.

2.2 Design and Functionality of Tagging Systems

Tagging systems differ in their design and the functionalities they offer. [76] proposed several dimensions which allow the classification of tagging systems. According to the authors, the applications vary according to the kinds of objects (web bookmarks, photos or videos) they provide storage space for. The source of such objects also differs: In user-centric systems such as BibSonomy or Flickr, the users are collecting the material. In other systems, the providers themselves present the material, which can then be annotated by their users (for example Last.fm, which provides music). Concerning the process of tagging, systems support and restrict their users in different ways. Many systems, for example,

---

[13] http://www.technologyreview.com/tr35/profile.aspx?trid=432.

[14] http://techcrunch.com/2011/04/27/yahoo-sells-delicious-to-youtube-founders/.

[15] http://mashable.com/2012/10/04/delicious-redesign/ and https://blog.pinboard. in/2017/06/pinboard_acquires_delicious/.

[16] see e.g.,http://mashable.com/2014/05/08/delicious-acquired-science-inc/.

[17] http://pinboard.in/.

have implemented recommender systems to suggest tags and help users finding appropriate vocabulary (cf. chapter on tag-based recommendations in this book (cf. [17]). In some systems users can annotate all resources (Delicious), while other systems allow their users to decide explicitly whether other users are allowed to tag their resources or not. Also, the aggregation model of tag-resource assignments is different. While different users in BibSonomy are allowed to assign the same tag to a resource, Flickr prohibits the same tag-resource assignments among different users. [102] is referring to these two cases as *broad* and *narrow folksonomies*. Finally, many systems provide additional functionalities concerning social connections among users (for example: joining groups) and their resources (for example: organizing photos in an album).

2.3 Strengths and Weaknesses of Tagging Systems

To understand the success of tagging systems, it is beneficial to consider the strengths and weaknesses of social tagging. Several aspects were discussed in [41, 76,93,110] and also in the dissertations of [85,112]. Such properties have a strong influence on the use of the data for ranking and any other algorithmic tasks. We will briefly summarize the key points in the following.

2.3.1 Strengths

Low cognitive cost and entry barriers. While more formal classification systems such as catalogues or ontologies require the consideration of the domain and specific vocabulary or rules, no prior knowledge or training is required when starting social tagging [110].

Serendipity. One of the fascinating features of tagging systems is their ability to guide users to unknown, unexpected, but nonetheless useful resources. This ability is triggered by the system's small world property: with only a few clicks one can reach totally different resources, tags and users in the system.

Adaptability. In contrast to top-down approaches where a pre-defined classification system is given and experts or at least people knowing the system are required to classify resources according to the classification scheme, social tagging systems allow a bottom-up approach where users can add keywords without having to adhere to a pre-defined vocabulary, authority or fixed classification. The liberty of using arbitrary annotations allows a flexible adaptation to a changing environment where new terms and concepts are introduced [96,109]. However, as the majority of users annotates their resources with similar or the same tags, a classification system can still evolve.

Long tail. As everyone can participate and no pre-requisites have to be met, "every voice gains its space" [93]. Consequently, the systems do not only contain mainstream contents, but also original and individual items which might turn into popular ideas.

2.3.2 Weaknesses

Ambiguity of tags. The missing control of what kind of vocabulary is used in tagging systems entails the typical challenges which a natural language environment provides: ambiguity, polysemy, synonyms, acronyms or basic level variation [40, 96].

Multiple languages. Since people with different cultural backgrounds use tagging systems, multiple terms from different languages with the same meaning can be encountered.

Syntax issues. In most of the social tagging systems users can enter different tags by using the space as delimiter. Problems arise when users want to add tags consisting of more than one term. Often, the underscore character or hyphens are used to combine such terms.

No formal semantics. Tags as they are entered into the system have no relation among each other. One needs to apply further techniques to discover inherent patterns.

2.4   Users' Motivations for Tagging

Why do people manually label items? [43] distinguished eight motives in their survey of tagging techniques: Most users annotate resources in order to facilitate their *future retrieval*, which is confirmed to some extent by findings in [31, 32, 84]. By making a resource public, categorizing it and sometimes even adding it to a specific interest community, the resource becomes available for a system's audience (*contribution and sharing*). Often, annotators use popular tags to make people aware of their resources (*attracting attention*) or they use tags to express some part of their identity (*self presentation*). The tag *myown* in the social tagging system BibSonomy, for example, states that the annotating user is an author of the referenced paper. Furthermore, with the help of tagging, users can demonstrate their opinion about certain resources (*opinion expression*). Tags such as *funny, helpful* or *elicit* are examples of such value judgements. Some tags reflect organisational purposes of the annotator. Often used examples are *toread* or *jobsearch* (*task categorization*). For some users, tagging is a *game or competition*, triggered by some pre-defined rules: Playing the ESP game [100], one user has to guess labels another user would also choose to describe images displayed to both users. Other users earn money with tagging (*financial benefits*): There are websites such as Squidoo[18] and its successor Hubpages[19] which pay users a small amount of money for annotating resources.

2.5   Classification of Tags

In order to get a better understanding of the nature of tags, various authors [6, 16, 40, 87, 95, 103] identified different types of tags.

Most useful for search, ranking or more general analysis tasks are *factual tags* indicating facts about the resource. Three kinds of factual tags (cf. [6, 95]) can be listed:

---

[18] www.squidoo.com.
[19] www.hubpages.com.

- Content-based tags describe a resource's actual content such as *ranking-algorithm, java, subaru* or *parental-guide.*
- Context-based tags can be used to identify an items context in which it was created or can be located. Examples are *San Francisco, christmas* or *www-conf.*
- (Objective) attribute tags describe an object's specific characteristics explicitly For example, the blog of the hypertext conference can be tagged with *hypertext-blog.*

Attribute tags can also be part of the category of *personal tags*, when they serve to express an opinion or a specific feeling such as *funny* or *interesting.* Personal tags are often more difficult to use for inferring general, descriptive knowledge. They can be used, however, for specific tasks such as sentiment analysis. Personal tags include:

- Subjective tags state an annotator's opinion.
- Ownership tags express who owns the object, e.g., *mypaper, myblog* or *mywork.*
- Organisational tags denote what to do with the specific resource. They are often time-sensitive. For example, the *to-do,read,listen* tags loose significance if the task has been carried out.
- Purpose tags describe a certain objective the user has in mind considering the specific resource. Often, this relates to information seeking tasks such as learning about java (*learning_java*) or collecting resources for a chapter of the dissertation (*review_spam*).

Authors (such as [16,87,103]) intend to automatically identify tag types in order to better explore the semantic structure of a tagging system. Categories can then be used for tag recommendation, categorization, faceted browsing or information retrieval.

2.6 Types of Annotators

Another way to look at the annotation process is to describe the nature of taggers. [62,63] identified two types of taggers: *categorizers* and *describers.*

- A categorizer annotates a resource using terms of a systematic shared or personal vocabulary, often some kind of taxonomy. Their size of vocabulary is limited and terms are often reused. A categorizer aims at using tags for his or her personal retrieval [62].
- Describers annotate resources having their later retrieval in mind. They consider tags as descriptive labels which characterize the resource and can be searched for. The size of a describer's vocabulary can be large. Often, tags are not reused [62].

In [62] it could be shown, that the collaborative verbosity of describers is more useful to extract the semantic structure from tags. Most users show a mixed behaviour but the main type can be identified by the users' behavior. If users own many tags only applied once, they tend to be describers. Additionally, a vocabulary growing quickly hints towards a describer. Categorizers can be identified by their low tag entropy as they apply their tags in a balanced way. [62] restrict their findings to moderate verbose taggers excluding spammers, which negatively influence the semantic accuracy.[20]

Like for all other properties, the type of annotator has impact on the tagged information and the emerging social tagging structure, and should thus taken into account when designing appropriate search and ranking algorithms.

3   Folksonomies

In order to understand how search and ranking in tagging systems work, we need to understand their structure. The resulting structure of tagging systems has been termed *folksonomy*, being composed of *folk* and *taxonomy*. *Folk* refers to people, i.e., the users of the tagging system. A *taxonomy* can be considered as a hierarchical structure used to classify different concepts [57]. Such a hierarchy is built from "is-a" relationships, i.e., subsumptions of more specific concepts under more specific ones. Taxonomies are normally designed by an expert of the domain. An example of a taxonomy is the Dewey Decimal Classification system [OCL], introduced by librarians to organize their collections [19]. Web documents have also been categorized with the help of taxonomies (examples here are the Yahoo! Directory[21] and the Open Directory Project (ODP)[22]).

The composition of *folk* and *taxonomy* describes the creation of a lightweight taxonomy which emerges from the fact that many people ("folk") with a similar conceptual and linguistic background as well as common interests annotate and organize resources [76].

In this section, we present the formal definition of folksonomies and discuss some of their structural properties before diving into the subject of ranking algorithms for folksonomies. The understanding of the underlying structural properties and processes in folksonomies forms the basis to formulate effective ranking and search approaches as discussed in Sect. 4. This information can also be exploited in standard web search as discussed in Sect. 5.

3.1   Formal Model for Tagging

Formal definitions of a folksonomy have been presented, among others, by [45, 50,78]. They all have in common that they describe the connections between users, tags and resources. We follow the notion of [50], which is formalized in Definition 1. The definitions further down (Definitions 2 and 3) are also based on [50].

---

[20] See also Sect. 1 of Chap. 6 of this book [28].
[21] http://dir.yahoo.com/.
[22] http://dmoz.org/.

Definition 1. A *folksonomy* is a tuple $\mathbb{F} := (U, T, R, Y, \prec)$ where

- U, T, and R are finite sets, whose elements are called *users, tags* and *resources*, respectively, and
- Y is a ternary relation between them, i.e., $Y \subseteq U \times T \times R$, whose elements are called tag assignments (TAS for short).
- $\prec$ is a user-specific subtag/supertag-relation, i.e., $\prec \subseteq U \times T \times T$, called *is-a relation*.

Figure 2 illustrates Definition 1 except for subtag/supertag-relation. Elements of one of the three sets are connected to elements of the remaining sets through the ternary relation Y. For example, (u_1, t_1, r_1) is a TAS of the depicted folksonomy. In terms of graph theory, a folksonomy can be considered to be a *hypergraph*, as each edge in Y connects three (and thus more than two) vertices. It can be called a *tripartite* hypergraph, as each edge ends at exactly one vertice in each of the three disjoint subsets U, T, and R of the vertex set.

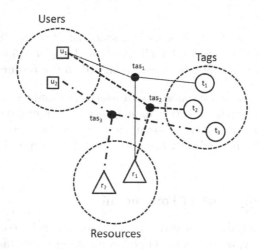

Fig. 2. Illustration of the folksonomy $\mathbb{F} := (U, T, R, Y)$ incl. the three finite sets users U, tags T and resources R connected through the ternary relation Y. For example, user u_1, tag t_1 and resource r_1 is the tag assignment tas_1 represented by a hyperedge.

Users of a bookmarking system are normally identified by a unique name they selected when registering. Tags are arbitrary strings. In most of the systems, they are divided by empty spaces. A folksonomy's resource can vary from URLs (for example Delicious, BibSonomy) to photos (Flickr) or videos (YouTube).

The *is-a relation* described in the definition classifies tags in form of super-/sub-concept relationships [50]. Not all tagging systems realize this functionality. In BibSonomy, such relations can be defined by the system's users. Delicious allows the creation of so-called tag *bundles*: users can define a set of tags and assign a group name to them. One can ignore the is-a relation, and simply define a folksonomy as a quadruple $\mathbb{F} := (U, T, R, Y)$. This structure is known in Formal Concept Analysis [37,107] as a *triadic context* [67,97].

Definition 2 describes a folksonomy for one user – a personomy. It basically considers only the tags and resources which the user in question submitted to the system. Figure 2 depicts two personomies. Both users have tagged one resource. However, the personomy of user u_1 contains one TAS more since the user assigned two tags to the resource.

Definition 2. The *personomy* $\mathbb{P}_u$ of a given user $u \in U$ is the restriction of $\mathbb{F}$ to u, i.e., $\mathbb{P}_u := (T_u, R_u, I_u, \prec_u)$ with $I_u := \{(t, r) \in T \times R \mid (u, t, r) \in Y\}$, $T_u := \pi_1(I_u)$, $R_u := \pi_2(I_u)$, and $\prec_u := \{(t_1, t_2) \in T \times T \mid (u, t_1, t_2) \in \prec\}$, where π_i denotes the projection on the ith dimension.

An important concept in the world of folksonomies is a *post*, which is presented in Definition 3. A post basically represents the set of tag assignments of one user for one resource. Figure 2 depicts two posts. The post of user u_1 is composed of two tag assignments, while the post of user u_2 contains one tag assignment.

Definition 3. The set P of all *posts* is defined as $P := \{(u, S, r) \mid u \in U, r \in R, S = \text{tags}(u, r), S \neq \emptyset\}$ where, for all $u \in U$ and $r \in R$, $\text{tags}(u, r) := \{t \in T \mid (u, t, r) \in Y\}$ denotes all tags the user u assigned to the resource r.

Though we focus on a folksonomy with users, tags and resources as elements, it is possible to enhance the structure to include more dimensions [1]. The authors of [110], for instance, consider a fourth set of elements: timestamps which are assigned to tag-resource pairs in order to consider temporal aspects in their analysis.

3.2 Network Properties of Folksonomies

From a network analysis perspective, a folksonomy can be considered as a tri-partite undirected hypergraph $G = (V, E)$ connecting users, tags and resources. $V = U \dot\cup T \dot\cup R$ is the set of nodes, and $E = \{\{u, t, r\} \mid (u, t, r) \in Y\}$ is the set of hyperedges. Those hypergraphs show interesting properties which help to understand a folksonomy's structure and the underlying dynamics.

In order to gain a better understanding of the basic properties of complex networks, especially scale-free and small-world networks, this section will present the main concepts and characteristics. A discussion of folksonomy properties with respect to navigation can be found the in the Chapter "Tag-based navigation and visualization" in this book (cf. [28]).

3.2.1 Power Law Distributions and Scale-Free Networks

A *power law* describes the phenomena where highly connected nodes in a network are rare while less connected nodes are common [4]. It indicates that the probability $P(k)$ that a vertex in the network is connected to k other vertices decays according to $P(k) \sim k^{-\gamma}$ [9] where $\gamma > 0$ is a constant and $\sim$ means asymptotically proportional to as $k \to \infty$. The distribution of a power-law is highly skewed having a *long tail*, which means that the probability of selecting a node which has less connections than the average is high. According to [108] "most nodes have small degree and few nodes have very high degree, with the result that the average node degree is essentially non-informative". Plotted on a log-log scale, power-law distributions will appear as a straight line with the gradient $-\gamma$.

Scale-free networks refer to networks which have a power-law degree distribution. No matter how many nodes the network consists of, the characteristic constant (γ) does not change, which makes such networks "scale-free" [83].

Various network structures induced from the structure of folksonomies exhibit power-law distributions. Especially the distribution of tag usage and tag co-occurrence have been carefully examined in respect to power laws. Based on such findings, different statements about user behaviour and the overall network dynamics have been made.

- [93] mentions the power law distribution of tag usage in broad folksonomies. He states that the "power law reveals that many people agree on using a few popular tags but also that smaller groups often prefer less known terms to describe their items of interest."
- [44] show, that for a small dataset of 100 URLs (which were tagged at least 1000 times) and their 25 most popular tags, the tag usage frequency follows a power law. The authors conclude that "the distribution of tag frequencies stabilizes [into power laws]", which indicates that users tend to agree about which tags optimally describe a particular resource.
- [24] examine tag co-occurrence using a set of preselected tags. They found a power law decay of the frequency-rank curves for high ranks while the curve for lower-ranked parts was flatter.
- [69, 105] analyse the distributions of tags per post, users per post and bookmarks per post in the social bookmarking system Delicious. Their results are similar to our results in Sect. 6 where we compare folksonomies to the tripartite structure of clickdata. While the occurrence of tags and URLs follow a power law distribution, the distribution of users is less straight (see Fig. 3). Still, one can observe that many users have only few posts, while a few users hold many posts.

As in many other real-word networks, e.g. for the topology of the World Wide Web [10, 35], the genetic network of proteins, or industrial networks and their trade relationships [9], a power law can be observed in the degree distributions of various network structures of folksonomies as well. The power law property of both the frequency and the degree distributions is something one would expect for a socially emerging structure, and has to be taken into account when designing search and ranking algorithms [92].

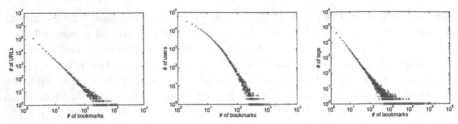

(a) Frequency distributions of [71]. The figures show the frequency distributions of URLs, users and tags.

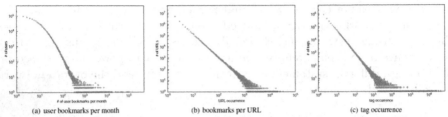

(a) user bookmarks per month   (b) bookmarks per URL   (c) tag occurrence

(b) Frequency distributions of [107] in a slightly different order. The figures show the frequency distributions of users, URLs and tags.

Fig. 3. Frequency distributions taken from [69, 105]. Please note that, while the order of the figures and the scale of the datasets used for the experiments is slightly different, the distributions themselves are similar for different datasets.

3.2.2 Small World Networks

Human activities in online media frequently results in so-called small world networks. It is therefore of interest to analyze if folksonomies are also of this type. Small world networks are networks with low average shortest path lengths that exhibit at the same time significantly higher clustering coefficients than random networks where nodes are connected randomly [104]. Here, the average shortest path length is the average of the shortest distances between any two nodes in the graph. The clustering coefficient quantifies the extent to which a node's immediate neighborhood is completely connected. [25] analyse the network properties of folksonomies. They adjust various measures (i.e., the average shortest path length and the clustering coefficient defined above) to the tripartite structure of folksonomies and demonstrate small world characteristics on the Delicious graph. The small world properties explains to some degree the advantages of folksonomies, like their support of serendipitous browsing as reported by users (cf. [77]) and therefore are another important property of folksonomies.

3.2.3 Preferential Attachment

Some models exist that can explain the emergence of such small world networks [7] and the generation of tags in folksonomies. A key question is which are the underlying processes in the model explaining the stabilization of tag distributions (into a power law)? Different phenomena have been discussed, the most dominant one being the *preferential attachment*, where users imitate each other either directly by looking at each other's resources or indirectly by accepting tags suggested by a tag recommender. In this "rich get richer" effect [11], a newly added node preferentially connects to a node which already has a high network degree. In a tagging system this implies that a tag that has already been used to describe a resource is more likely to be added again [45].

Various tag generation models have been introduced since the first model of Halpin [45] to describe the tagging process (for a survey see [43]). For example, users do not only imitate each other. They seem to have a similar understanding of a resource due to sharing a similar background. Based on that, [27] present a generative tagging model integrating the two perspectives: a new tag is assigned to a resource either by imitating a previous tag assignment or by selecting a tag from the user's active vocabulary. When choosing tags, users can also be influenced by external factors. [71], for example, analysed the influence of a resource's title on tagging behaviour and found that tags which appear in a resource's title are more often used than other tags with the same meaning.

Research about network structure and tagging dynamics has helped to understand the creation of folksonomies. It can be used to improve mining algorithms such as tag recommender systems (see Sect. 4.3) and to develop methods to make the implicit knowlegde and structure of a folksonomy explicit.

4   Ranking in Folksonomies

In the past, tagging systems were able to attract a large number of users who created huge amounts of information. But with the growing number of resources stored within each user's personomy and thus in the folksonomy, it became more and more difficult for the user to find and retrieve the saved resources. A first step to support search in folksonomy-based systems – complementing the browsing interface of the system – is to employ standard information retrieval techniques which are also used in web search engines. Since users are accustomed to web search engines, they likely will accept a similar interface for search in folksonomy-based systems.

The open challenge is how to provide suitable ranking mechanisms, similar to those based on the web graph structure, but exploiting the properties and structure of folksonomies. Traditional IR-techniques like the simple vector space model [73] cannot handle those challenges adequately. Due to their reliance on the occurrence of terms in the documents, they tend to "fall for" spam pages artificially augmented by large amounts of keywords.

As a start, web search engines provide a variety of possible technical solutions. Two algorithms (and many variations developed afterwards) proposed in the 1990s deal with the challenge of Web page ranking by considering the hyperlink structure of the Web: PageRank and the Hyperlink-Induced Topic Search (HITS) algorithm. Both algorithms model the Web as a network where nodes correspond to web pages and a directed edge between two nodes exists if one page has a hyperlink to the second one.

The PageRank algorithm [20,88] is the foundation of the popular search engine Google[23]. The algorithm models the behaviour of a random Web surfer who randomly follows a link without showing any preferences for specific pages. Consequently, all links on a page have equal probability of being followed. Periodically, the random surfer does not follow the offered links but jumps to a randomly selected page.

The other suggestion how to make use of the link structure of the Web in the calculation of rankings is Kleinberg's Hyperlink-Induced Topic Search (HITS) algorithm [56]. Kleinberg identified two different entities which can be found in the Web: hubs and authorities. According to the author, the two items "exhibit what could be called a mutually reinforcing relationship: a good hub is a page that points to many good authorities; a good authority is a page that is pointed to by many good hubs." More details about HITS and PageRank can be found in the chapter on collaborative search [114] in this volume.

Based on PageRank and HITS, several ranking schemes for folksonomies have been developed in the last years which allow to rank any of the three folksonomy dimensions: users, tags and resources. We will recap the most influential ones, namely FolkRank and an adjusted versions of HITS. These ranking mechanisms are also useful in other settings than search; e.g. in recommender systems, as described in the chapter on tag-based recommendations in this book (cf. [17]).

4.1   The FolkRank Algorithm

Several adaptations of the PageRank algorithm to the folksonomy structure exist. The FolkRank algorithm was first presented in [14,50].

The FolkRank algorithm operates on an undirected, tripartite graph $G_{\mathbb{F}} = (V, E)$, where $V = U \dot\cup T \dot\cup R$ (elements) and the set of edges E results from splitting tag assignments into three undirected edges each, i.e., $E = \{\{u,t\}, \{t,r\}, \{u,r\} \mid (u,t,r) \in Y\}$. Then, the PageRank formula as introduced in the chapter on collaborative search [114] of this book can then be analogously applied to the folksonomy graph:

$$\mathbf{w_{t+1}} = dA^T\mathbf{w_t} + (1-d)\mathbf{p},$$

where w_t is weight vector at iteration t, $\mathbf{p}$ is used as preference vector and $d \in [0,1]$ is a constant which controls the influence of the preference vector. A is the row-stochastic version of the adjacency matrix of $G_{\mathbb{F}}$.

As for the original PageRank, one can specify preference weights which are set in the preference vector $\mathbf{p}$ in order to compute a ranking of tags, resources and/or users tailored to the preferred element from V. For web-search-like ranking in folksonomies, the tags representing search terms receive a higher weight in the preference vector compared to the remaining elements (i.e., remaining tags, all users and all resources) whose weight scores follow an equal distribution. The FolkRank algorithm is outlined in Algorithm 1 and needs to be applied for each search term separately.

---

[23] http://www.google.de.

Algorithm 1. FolkRank

Input: Undirected, tripartite graph G_F, the set of preferred elements as vector $\mathbf{p}$, a
randomly chosen baseline vector $\mathbf{w_0}$ and a randomly chosen FolkRank vector $\mathbf{w_1}$.
1. Compute baseline vector $\mathbf{w_0}$ with $\hat{\mathbf{p}} = \mathbf{1}$ and $\mathbf{1} = [1, \ldots, 1]^T$.
2. Compute topic specific vector $\mathbf{w_1}$ with specific preference vector $\mathbf{p}$.
3. $\mathbf{w} := \mathbf{w_1} - \mathbf{w_0}$ is the final weight vector.
Output: FolkRank vector $\mathbf{w}$.

As can be seen in Algorithm 1, the computation of the ranking consists of
two main steps: First, a baseline with a uniform preference vector needs to be
computed (step 1 in Algorithm 1). The result of this computation is the fixed
point $\mathbf{w_0}$. Second, the fixed point $\mathbf{w_1}$ (step 2) is computed by using the preference
vector $\mathbf{p}$. The final weight vector for a specific search term is then $\mathbf{w} := \mathbf{w_1} - \mathbf{w_0}$.
The subtraction of the baseline reinforces the elements which are close to the
preferred elements, while it degrades elements which are popular in general.

FolkRank has a number of limitations. The most prominent one is its runtime.
For each query, one needs to run the full computation with the query term set as
preference, which cannot be done online. One workaround is the pre-computation
of rankings for all single tags. The result list for queries with more than one
tag can then be calculated by adding their ranking vectors. Other follow-up
works addressing the speed-up too, like [61], show comparable results but are
considerably faster. The FolkRank algorithm was not only successful in search
but also in other applications like recommendation, semantic similarity, trend
detection (see Subsect. 4.3 for details).

4.2 Adjusted Versions of the HITS Algorithm to a Folksonomy

The HITS algorithm [56] has also been adjusted to rank items in a folksonomy.
Two versions exist which have been called *Naive* HITS and *Social* HITS by [2].
The challenge of using HITS to rank resources in a folksonomy is the transformation of the undirected tripartite graph of an Folksonomy into a directed
graph needed by HITS. The two algorithms above are based on the transformation proposed by [110]: A tag assignment $(u, t, r) \in Y$ is split into two edges
$u \rightarrow t$ and $t \rightarrow r$. The resulting structure is a directed graph where hubs are
users and authorities are resources (as resources have no outgoing links their
hub weights become 0). While the *naive* HITS implementation uses this structure, *Social* HITS extends the graph by allowing for authority users and hub
resources. Given a tag assignment $(u, t, r) \in Y$ they derive two directed edges
from user actions: $u \rightarrow t$ and $u \rightarrow r$. Additionally, they create an edge $u_h \rightarrow u_a$,
whenever an arbitrary user u_h annotated a resource after it had already been
tagged by user u_a.

In experiments, Social HITS was tested, although not on a large scale. It could
be shown that the Social HITS algorithm is better than the simple Hits [110],
but the context of a user group (introduced in [3]) leads to even better results.

4.3   Beyond Search: Other Applications of FolkRank

All these graph-based ranking approaches which allow to rank not only resources but user and tags as well, can be used in other tasks like trend detection and tag or item recommendation. Here, we review these two application areas to shortly illustrate the applicability of the ranking approaches.

Trend detection. FolkRank can be applied for analyzing the emergence of common semantics by exploring trends in the folksonomy [48]. Since the structure of a folksonomy is symmetric with respect to the dimensions "user', 'tag', and 'resource' (all of them can be used as preference for ranking), one can apply the same approach to study upcoming users, upcoming tags, and upcoming resources over time. In [48], FolkRank is used to compute topic-specific rankings on users, tags, and resources. In a second step, these rankings for snapshots of the system are calculated at different points in time. One can discover both the absolute rankings (who is in the Top Ten?) and winners and losers (who rose/fell most?). A technique for analyzing the evolution of topic-specific trends is presented as well that could be used to automatically detect interesting developments in a tagging system. Another approach for detecting trends in folksonomies has been presented in [106]. It is a probabilistic generative model combined with smoothing used applied on Delicious.

Recommendation in Tagging Systems. Recommender systems aim at identifying items which match the interests of a specific user. To find those items, a variety of information sources related to both the user and the content items are considered, for example history of purchases, ratings, clickdata from logfiles or demographic information [54,74]. Ranking methods developed to improve search results like FolkRank can also be used to compute recommendations. As FolkRank and related approaches allow for tag and resource (or items) ranking even under the condition of a special user, one of the first works focused on the adoption of the method for this task [74]. A detailed discussion of recommender approaches around tagging is given in Chap. 12 in this book (cf. [17]).

5   Web Search with Folksonomies

Both folksonomy systems and search engines support users in retrieving resources from the web. The major differences between them are related to interface and content-creation aspects. Folksonomies allow users to organize, index and share web content ranging from simple web pages over images and videos to more complex meta data like publication records. By contrast, classical search engines automatically index the Web and offer a simple user interface to search in this index, relying on an advanced ranking scheme. The index itself is created by constantly crawling the Web, while the content of a folksonomy emerges from explicit tagging, a manual annotation and indexing process, by its users. As a consequence, users, not a (search) algorithm, decide about relevance in a folksonomy and by its contribution. This kind of user information contains the perception of users about the importance of the collected resources.

Section 4 was focusing on the adoption of web ranking and search mechanisms to folksonomies. Now, wego the other way: As folksonomies contain valuable human judgments about the importance of resources (including web pages), we will review work making use of tagging information for improving web search. Before discussing different ways for folksonomy integration into web search, we aim at a better understanding of traditional search approaches and their differences to tag-based search. Then we will discuss the SocialPageRank approach, which computes a global ranking on web resources; and a more advanced integration of folksonomies by adopting learning-to-rank approaches (cf. [53]) such that a user's personomy is interpreted as his set of preferences.

Social tagging is not the only aspect of the Social Web which has stimulated the development of classical search engines in the past. The new generation of search engines is heavily influenced by other social activities of users, as discussed in more detail in Chap. 7 of this book [22].

5.1   Comparing Traditional and Tag-Based Search

First experiments investigating bookmarks for information retrieval were conducted by [46]. The authors created a dataset of the social bookmarking system Delicious to run different analyses considering the system's tags and bookmarks. The authors found that the set of social bookmarks contains URLs which are often updated and also appear to be prominent in the result lists of search engines. A weak point is the fact that the URL collection of a social bookmarking system is likely to be too small to significantly impact the crawl ordering and the ranking of a major search engine.

[58] analyse the suitability of bookmarks for web search. They used Delicious as well as randomly selected URLs to feed a crawler. The authors found that the average external outdegree of Delicious URLs close to the seed was more than three times larger that for the neighbors of random URL seeds. Based on this finding, they conclude that Delicious URLs are a good source for discovering new content. Furthermore, the click rate of Delicious URLs is higher compared to a random selection of examples meaning that users tend to click on search results which also have been tagged in Delicious. This finding could be used for influencing the rank score of a page.

[79] performed a user study to compare rankings from social bookmarking sites against rankings of search engines and subject directories. Participants had to rate results from both systems after having submitted a query. The authors found that search results of both systems are overlapping. Furthermore, hits appearing in both search lists have a higher probability of being relevant than those returned by only one of the two systems.

In [59,60,80], we compared search and tagging behavior in bookmarking systems with ranking structures of web search systems. We describe our findings in the following case study, which is composed of three parts. (Parts 2 and 3 will follow in Sects. 5.3 and 6.)

Case Study (Part 1). We analyzed data from Delicious and MSN search logs provided by Microsoft. We discovered both similar and diverging behaviour in both kinds of systems [59]. The collection of Delicious tags is only about a quarter of the size of the MSN queries; and the overlap is rather small, due to a long tail of infrequent items in both systems. Once the sets are reduced to the frequent items, the relative overlap is higher. The remaining differences are due to different usage, e.g.,to the composition of multi-word lexems to single terms in Delicious, and the use of (parts of) URLs as query terms in MSN.

We have seen that, for a relatively high number of items, the search and tagging time series were significantly correlated. We have also observed that important events trigger both search and tagging without significant time delay, and that this behaviour is correlated over time.

Considering the fact that both the available search engine data and the folksonomy data cover only a minor part of the WWW, the overlaps of the sets of URLs of the different systems are rather high. This indicates that users of social bookmarking systems are likely to tag web pages that are also ranked highly by traditional search engines. The URLs of the social bookmarking system over-proportionally match the top results of the search engine rankings. A likely explanation is that taggers use search engines to find interesting bookmarks.

We observe that a comparison of rankings is difficult due to sparse overlaps of the data sets. It turns out that the top hits of the rankings produced by FolkRank are closer to the top hits of the search engines than the top hits of a simple vector based method. Furthermore, we could observe that the overlap between Delicious and the search engine results is larger in the top parts of the search engine rankings.

We also observe that the folksonomy rankings are more strongly correlated to the Google rankings than to those of MSN and AOL, whereby the graph-based FolkRank is closer to the Google rankings than TF and TF-IDF. Again, we assume that taggers preferably use search engines (and most of all Google) to find information they then proceed to tag. A qualitative analysis showed that the correlations were higher for specific IT topics, where Delicious has a particularly good coverage.

5.2   Integration of Tags in Search

After comparing the properties of classic search and ranking approaches with folksonomy-based methods, we will start to discuss the use of folksonomy information as a kind of social annotation in classical search methods. The goal is to improve the quality of the web search ranking. One of the early methods is the SocialPageRank (SPR) algorithm introduced in [8] which relies on the hypergraph of the folksonomy as the FolkRank algorithm. Both algorithms (SPR and FolkRank) are based on spreading weights along the link structure of the folksonomy graph by simulating a random surfer, but differ in the paths a random surfer can follow. While FolkRank allows all sorts of paths through the tripartite network, SocialPageRank restricts possible paths to a number of bi-partite subgraphs (described by so called association matrices A) and restricts the computation to a single type of paths, i.e., resource $\rightarrow$ user $\rightarrow$ tag $\rightarrow$ resource $\rightarrow$ tag $\rightarrow$ user [3]. We use the notation of [3] to present the spreading scheme which results in the global ranking $\mathbf{w_r}$ of the pages as presented in Algorithm 2.

Algorithm 2. Social PageRank

---

input : Association matrices A_{TR}, A_{RU}, A_{UT}, and a randomly chosen
            SocialPageRank vector $\mathbf{w_{r_0}}$

until $\mathbf{w_{r_0}}$ converges do:

$\mathbf{w_{u_i}} = A_{RU}^T * \mathbf{w_{r_i}}$

$\mathbf{w_{t_i}} = A_{UT}^T * \mathbf{w_{u_i}}$

$\mathbf{w'_{r_i}} = A_{TR}^T * \mathbf{w_{t_i}}$

$\mathbf{w'_{t_i}} = A_{TR} * \mathbf{w'_{t_i}}$

$\mathbf{w'_{u_i}} = A_{UT} * \mathbf{w'_{u_i}}$

$\mathbf{w_{r_{i+1}}} = A_{RU} * \mathbf{w'_{r_i}}$

output: SocialPageRank vector $\mathbf{w_r}$.

In the same paper, the SocialSimRank algorithm [8] is proposed, which is used to calculate similarity between a search query and web pages, again using the hypergraph. The main idea is to compute a query expansion by extending the query terms with tags which are used to annotate the same resource by the users of the folksonomy. In this way, the similarity takes the graph structure into account. Bao et al. [8] showed an improvement over standard BM25 ranking for both their SocialPageRank and for the combination of SocialSimRank and PageRank.

This first works on SPR and SSP showed that social annotations are helpful for improving web search. The intention of the proposed SocialPageRank is to compute a ranking for all web pages. As for FolkRank, the method is limited to the pages annotated by the user community as the SPR score can only be computed for these pages. Both approaches (SPR and FolkRank) differ mainly in the way to compute the ranking. While FolkRank uses the damping factor of the standard PageRank to compute a ranking for a given query (or for a given user) and needs to be run for each ranking calculation separately, SocialPageRank just computes one global ranking and uses this information in the same way as PageRank to come up with rankings for a given query, i.e., by restricting the global ranking to the result set of the query. When the SocialSimRank component is used for ranking, one needs to run the calculation for each term again, just as for FolkRank. In any case, FolkRank can also be used to calculate a ranking for web pages in a similar way as SocialPageRank and has the same limitations.

Due to these runtime and coverage limitations, most of the direct follow-up works use graph-based rankings only for personalization of web search, e.g., [99], or for query expansion in web search, e.g., [15], but not directly to rank Web search results. To make use of the tag resources, a more exploratory way to access the content of tagging systems is proposed in [55]. After crawling several tagging systems the collected content is indexed by the standard search system Lucene, and a simple user interface for searching for tags (including related and "bad" tags) is built. The easy access and suitability for explorative search is shown in several user studies, but the content is still limited to the manually annotated and collected resources. Another potential of tagging for web search is its exploitability whenever the content of the resources cannot be easily accessed or represented directly as a bag of words (as in standard text retrieval). In particular, this is the case for multi-media resources. For instance, [38] show how tagging can be used for relevance learning for image search.

In the next section, we will discuss how to overcome these limitations and integrate social annotations directly into search engines ranking, regardless whether the content of page of interest is accessible or annotated.

5.3   Advanced Integration of Tags in Search

Until now, all discussed approaches directly use the graph structure of the folksonomy to compute a ranking. Another way to exploit user feedback in terms of tags for search is to learn a ranking function in order to (re-)sort a given web search result. The main idea of such a learning-to-rank approach is to utilize the user feedback in form of clicked search results. Such information is easy to get, as search engines usually track their users and utilize this implicit information for a long while now. Unfortunately, the feedback is noisy, as a click does not always indicate relevance. On the other hand, the process of storing and annotating web links in a social bookmarking system can be seen as a – more explicit – expression of relevance. The process of annotating a resource (a web page) with a specific word or tag attaches the word to it and expresses a preference of the user to relate this resource to the tag compared to other resources. One could use this tagging information in a similar way as click log information for learning a new ranking function.

Mostly, tags describe a topic, the resource's context or the user's reason for tagging the resource, as discussed in Sect. 2.5 above which makes them useful as user feedback. For instance, the tag *learning-to-rank* describes the web link http://en.wikipedia.org/wiki/Learning_to_rank, while the tag *lecture* provides the context and the tag *toread* the intention what to do with the specific resource. It looks like, tagging data seems to be more explicit and thus better suitable for learning–to-rank, as the process of tagging involves finding a relevant resource and then storing and categorizing it in the bookmarking system. Nevertheless, they are still noisy with respect to content description because of the existence of tags with other purposes.

In general, the learning-to-rank method works as follows: Given the preferences of users for queries, the training data for the learning approach consists of queries and documents matching the query ordered according to their relevance score to the query. A query-document pair is usually represented by feature vectors with features such as the frequency of the query term in the document's summary or the length of the document. Then, different approaches exist for solving the sketched learning-to-rank task: pointwise, pairwise and list-wise approaches [33]. The relevance scores for the training data are derived by exploiting the user search behavior such as click data as described in [34,53,72].

In our case study, we have analyzed if folksonomy data can be used an additional source of knowledge about user behavior with more explicit feedback and could be helpful for learning-to-rank approaches:

Case Study (Part 2). In order to explore the usefulness of tags for search, we compared feedback generated from the Delicious tagging data to implicit feedback generated from the MSN clickdata [80]. To this end, we investigated different ways of interpreting tagging data as click data: Given a search query and the ranking of a search engine, we match the query and the URLs of the ranking with tags and resources of a social bookmarking system. We thereby assume that a URL in a ranking list is important if it has been tagged in the folksonomy with the query terms (or similar tags). At the same time, we assume that the URL is relevant for a search query if it has been clicked on after the submission of that query. Both types of information have been shown to be comparable, and can be fed into a learning-to-rank algorithm. A comparison of different strategies to infer implicit feedback for a learning-to-rank scenario has shown that strategies tend to perform better when the same data (either tagging or click data) is used to generate feedback and to predict feedback. The best results when mixing tagging and click data for learning and evaluation are obtained from a strategy based on the *FolkRank* algorithm (cf. [80]).

In the case study, it was shown that tagging data can be used for improving search results. Ranking functions can be further enhanced by social information either by re-ranking the documents of a result list or by personalizing a result list. [68] use tagging information to re-rank documents. They assume that documents with high similarity score between document terms and tags should retrieve a similar retrieval score. After a preliminary ranking, they compute similarities between documents in the ranking list using matrix factorization methods and utilize the similarity degree to re-rank documents. The authors of [66] propose the construction of a social inverted index taking not only the document and its terms but also the user tagging the document and its tags into account. [18] propose a linear weighting function which integrates a vector representing the social representation (i.e., tags) of a document into the Vector Space Model. Additionally, they account for a user's personal interests by computing the similarity between a user profile and the social document representation.

Several authors explore the use of social annotations for query expansion [15,42,70,115]. They enhance existing expansion techniques with tagging information. For example, the authors of [42] extend the co-occurrence matrix to measure how often tags and query terms appear together. Overall, the different studies show that tags can serve as a knowledge base for information retrieval tasks.

To summarize, while search queries express a specific information need and there is no evidence as to whether a clicked URL fulfills this need or not, tags serve as a description or categorization for the specific resource and can be useful. Search engine companies have started to follow the trend of integrating users information into the search process. The search engine Google, for example, released its own social platform where users can register and connect to friends and other associates. The ranking results of a specific search in this system also include content published or liked by a user's friends [47] (see also Chap. 7 of this book a more detailed discussion, [22]). Social bookmarking systems and tagging information can be a valuable resource for search. On the other hand, tagging systems profit from the technologies and methods of search algorithms.

6   Logsonomies: A Unified View on Search and Tagging

As discussed in Sect. 5.3, the click behavior of users in search engines contains information about the importance of web pages which can be used to improve ranking. A detailed analysis of click log data can provide a better understanding of the users, its behavior and the hidden information of such a log. Each click of a user on a result of a search query reflects his mental model about the relationship of the query terms to the search result. Hence when user u clicks on entry r of the resulting list of a search query which contains a search term t, this can be understood as an implicit expression of u's understanding of term t being relevant to resource r. Click-log data have thus the same tri-partite structure as folksonomies.

We call a *logsonomy* the representation of click-log data in form of a folksonomy $\mathbb{F}$ [60]. To map the click-log data to the three dimensions of the folksonomy (cf. Definition 1 of a folksonomy), we set

- U to be the set of users of the search engine. Depending on how users in logs are tracked, a user is represented either by an anonymized user ID, or by a session ID.
- T to be the set of queries the users gave to the search engine (where one query either results in one tag, or will be split at whitespaces into several tags).
- R to be the set of URLs which have been clicked by the search engine users.

In a logsonomy, we assume an association between t, u and r when user u clicked on resource r of a result set after having submitted a query term t (eventually with other query terms). The resulting relation $Y \times T \times R$ corresponds to the tag assignments in a folksonomy.

Case Study (Part 3). In order to compare the structure and semantics of folksonomies and search engines, we transformed the click data file of a search engine into a logsonomy. As in Sect. 3.2.2, using shortest path lengths and clustering coefficients in order to compare small world properties, we demonstrated that logsonomies do present a folksonomy like structure. Differences consist in the notion of users: while folksonomies store bookmarks from registered users, logsonomies track the interests in form of SessionIDs.

We analyzed the graph structure and semantic aspects of the logsonomies MSN and AOL (cf. [60]). We observed similar user, resource and tag distributions, whereby the split by query word datasets are closer to the original folksonomy than the complete query datasets, considering the complete query as a single tag. We could show that both graph structures have small world properties in that they exhibit relatively short shortest path length and high clustering coefficients. In general, the differences between the folksonomy and logsonomy model did not affect the graph structure of the logsonomies. Minor differences are triggered by the session IDs, which do not have the same thematic overlap as user IDs. Also, full queries show less inherent semantics in the graph structure than the split datasets do. To analyse semantic aspects, we used different relatedness measures and WordNet. Due to the fact that queries were split up into single terms, we observed that most co-occurrence related measures restored compound expressions. Interestingly, applying the resource-context-relatedness to logsonomies is much less precise for discovering semantically-close terms when compared to a folksonomy. We attribute this mainly to the incomplete user knowledge about the content of a page link they click on, leading e. g., to "erroneous" clicks. The behaviour of the tag context measure is more similar to the folksonomy case, which recommends it as a candidate for synonym and "sibling" term identification.

To summarize, in terms of emergent semantics as found in folksonomy systems [23], logsonomies show slightly different characteristics. However, the structure of the underlying graph of a logsonomy shows only minor differences compared to folksonomies. This indicates that it is worth investigating if findings made for folksonomies could the transfered to logsonomies. It further explains the success of the learning-to-rank method featuring the click log information as feedback for web search, since both click-logs and logsonomies are rich sources of information.

7 Future of Search and Ranking in Tagging

Over the last years, the use of tagging functionality has been shifted from being the core feature of stand-alone bookmarking systems to an additional component of all kinds of social media. Meanwhile it is implemented in almost all platforms, frequently in the form of hash-tags. This wide-spread usage of tagging bears a large potential for improving web search in the near future. In particular, we see the following upcoming trends.

- Tagging can help spot trends in society. People who are tagging can often be viewed as trend setters or early adopters of innovative ideas—their data is valuable for improving a search engine's diversity and novelty.
- Users could enrich visited URLs with their own tags (besides the automatically added words from the query) and the search engine could use these tags to consider such URLs for later queries—also from other users. Thus, those tags could improve the general quality of the search engine's results.

- Search engines typically have the problem of finding new, unlinked web pages. Assuming, users store new pages in the folksonomy, the search engine could better direct its crawlers to new pages [46]. Additionally, those URLs would have been already annotated by the user's tags. Therefore, even without crawling the pages it would be possible to present them in result sets.
- Bookmarked URLs of the user may include pages the search engine can not reach (intranet, password-protected pages, etc.). These pages can then be integrated into personalized search results. Giving search engines access to more private information can result in similar privacy issues as for search engines click logs which often allow the identification of the users themselves [5]. Certainly, this issue requires attention when integrating web search and tagging more tightly.

References

1. Abel, F.: Contextualization, user modeling and personalization in the social web: from social tagging via context to cross-system user modeling and personalization. Ph.D. thesis, University of Hanover (2011). http://d-nb.info/1014252423
2. Abel, F., Baldoni, M., Baroglio, C., Henze, N., Kawase, R., Krause, D., Patti, V.: Leveraging search and content exploration by exploiting context in folksonomy systems. New Rev. Hypermedia Multimed. 16(1–2), 33–70 (2010)
3. Abel, F., Henze, N., Krause, D.: Ranking in folksonomy systems: can context help? In: Shanahan, J.G., Amer-Yahia, S., Manolescu, I., Zhang, Y., Evans, D.A., Kolcz, A., Choi, K.-S., Chowdhury, A. (eds.) Proceedings of the 17th ACM Conference on Information and Knowledge Management, CIKM 2008, Napa Valley, California, USA, 26–30 October 2008, pp. 1429–1430. ACM (2008)
4. Adamic, L.: Zipf, power-laws, and pareto - a ranking tutorial (2002). http://www.hpl.hp.com/research/idl/papers/ranking/ranking.html
5. Adar, E.: User 4xxxxx9: anonymizing query logs. In: Query Logs Workshop at WWW 2006 (2007)
6. Al-Khalifa, H.S., Davis, H.C.: Towards better understanding of folksonomic patterns. In: Proceedings of the Eighteenth Conference on Hypertext and Hypermedia, HT 2007, pp. 163–166. ACM, New York (2007)
7. Amaral, L.A.N., Scala, A., Barthelemy, M., Stanley, H.: Classes of small-world networks. Proc. Nat. Acad. Sci. USA 97, 11149–11152 (2000)
8. Bao, S., Xue, G., Wu, X., Yu, Y., Fei, B., Su, Z.: Optimizing web search using social annotations. In: Proceedings of the WWW 2007, Banff, Canada, pp. 501–510 (2007)
9. Barabasi, A.-L., Albert, R.: Emergence of scaling in random networks. Science 286, 509–512 (1999)
10. Barabási, A.-L., Albert, R., Jeong, H.: Scale-free characteristics of random networks: the topology of the world-wide web. Phys. A Stat. Mech. Appl. 281(1–4), 69–77 (2000)
11. Barabasi, A.-L., Bonabeau, E.: Scale-free networks. Sci. Am. 288, 60–69 (2003)
12. Belém, F.M., Martins, E.F., Almeida, J.M., Gonçalves, M.A.: Personalized and object-centered tag recommendation methods for web 2.0 applications. Inf. Proc. Manag. 50(4), 524–553 (2014)

13. Benz, D., Eisterlehner, F., Hotho, A., Jäschke, R., Krause, B., Stumme, G.: Managing publications and bookmarks with bibsonomy. In: Cattuto, C., Ruffo, G., Menczer, F. (eds.) Proceedings of the 20th ACM Conference on Hypertext and Hypermedia, HT 2009, pp. 323–324. ACM, New York , June 2009

14. Benz, D., Hotho, A., Jäschke, R., Krause, B., Mitzlaff, F., Schmitz, C., Stumme, G.: The social bookmark and publication management system bibsonomy. VLDB J. **19**(6), 849–875 (2010)

15. Biancalana, C., Gasparetti, F., Micarelli, A., Sansonetti, G.: Social semantic query expansion. ACM Trans. Intell. Syst. Technol. (TIST) **4**(4), 60 (2013)

16. Bischoff, K., Firan, C.S., Kadar, C., Nejdl, W., Paiu, R.: Automatically identifying tag types. In: Huang, R., Yang, Q., Pei, J., Gama, J., Meng, X., Li, X. (eds.) ADMA 2009. LNCS (LNAI), vol. 5678, pp. 31–42. Springer, Heidelberg (2009). https://doi.org/10.1007/978-3-642-03348-3_7

17. Bogers, T.: Tag-based recommendation. In: Brusilovsky, P., He, D. (eds.) Social Information Access, LNCS, 10100, pp. 441–479. Springer, Heidelberg (2018)

18. Bouadjenck, M.R., Hacid, H., Bouzeghoub, M.: SoPRa: a new social personalized ranking function for improving web search. In: Proceedings of the 36th International ACM SIGIR Conference on Research and Development in Information Retrieval, SIGIR 2013, pp. 861–864. ACM, New York (2013)

19. Breitman, K., Casanova, M.A.: Semantic Web: Concepts, Technologies and Applications. Springer-Verlag London Limited, New York (2007). https://doi.org/10.1007/978-1-84628-710-7

20. Brin, S., Page, L.: The anatomy of a large-scale hypertextual web search engine. Comput. Netw. ISDN Syst. **30**(1–7), 107–117 (1998)

21. Brusilovsky, P., He, D.: Introduction to social information access. In: Brusilovsky, P., He, D. (eds.) Social Information Access. LNCS, vol. 10100, pp. 1–18. Springer, Cham (2018)

22. Brusilovsky, P., Smyth, B., Shapira, B.: Social search. In: Brusilovsky, P., He, D. (eds.) Social Information Access. LNCS, vol. 10100, pp. 213–276. Springer, Cham (2018)

23. Cattuto, C., Benz, D., Hotho, A., Stumme, G.: Semantic grounding of tag relatedness in social bookmarking systems. In: Sheth, A., Staab, S., Dean, M., Paolucci, M., Maynard, D., Finin, T., Thirunarayan, K. (eds.) ISWC 2008. LNCS, vol. 5318, pp. 615–631. Springer, Heidelberg (2008). https://doi.org/10.1007/978-3-540-88564-1_39

24. Cattuto, C., Loreto, V., Pietronero, L.: Collaborative tagging and semiotic dynamics. CoRR, abs/cs/0605015 (2006)

25. Cattuto, C., Schmitz, C., Baldassarri, A., Servedio, V.D.P., Loreto, V., Hotho, A., Grahl, M., Stumme, G.: Network properties of folksonomies. AI Commun. J. **20**(4), 245–262 (2007). Special Issue on "Network Analysis in Natural Sciences and Engineering"

26. Cunha, E., Magno, G., Comarela, G., Almeida, V., Gonçalves, M.A., Benevenuto, F.: Analyzing the dynamic evolution of hashtags on twitter: a language-based approach. In: Proceedings of the Workshop on Languages in Social Media, pp. 58–65. Association for Computational Linguistics (2011)

27. Dellschaft, K., Staab, S.: An epistemic dynamic model for tagging systems. In: Proceedings of the Nineteenth ACM Conference on Hypertext and hypermedia, HT 2008, pp. 71–80. ACM, New York (2008)

28. Dimitrov, D., Helic, D., Strohmaier, M.: Tag-based navigation and visualization. In: Brusilovsky, P., He, D. (eds.) Social Information Access. LNCS, vol. 10100, pp. 181–212. Springer, Cham (2018)

29. Djuana, E., Xu, Y., Li, Y., Jøsang, A.: A combined method for mitigating sparsity problem in tag recommendation. In: 47th Hawaii International Conference on System Sciences, HICSS 2014, Waikoloa, HI, USA, 6–9 January 2014, pp. 906–915 (2014)

30. Doerfel, S., Jäschke, R., Stumme, G.: The role of cores in recommender benchmarking for social bookmarking systems. ACM Trans. Intell. Syst. Technol. **7**(3), 40:1–40:33 (2016)

31. Doerfel, S., Zöller, D., Singer, P., Niebler, T., Hotho, A., Strohmaier, M.: Of course we share! testing assumptions about social tagging systems. CoRR, abs/1401.0629 (2014)

32. Doerfel, S., Zoller, D., Singer, P., Niebler, T., Hotho, A., Strohmaier, M.: What users actually do in a social tagging system: a study of user behavior in bibsonomy. ACM Trans. Web **10**(2), 14:1–14:32 (2016)

33. Dong, X., Chen, X., Guan, Y., Yu, Z., Li, S.: An overview of learning to rank for information retrieval. In: Burgin, M., Chowdhury, M.H., Ham, C.H., Ludwig, S.A., Su, W., Yenduri, S. (eds.) CSIE (3), pp. 600 606. IEEE Computer Society (2009)

34. Dou, Z., Song, R., Yuan, X., Wen, J.-R.: Are click-through data adequate for learning web search rankings? In: Proceeding of the 17th ACM Conference on Information and Knowledge Management, CIKM 2008, pp. 73–82. ACM, New York (2008)

35. Faloutsos, M., Faloutsos, P., Faloutsos, C.: On power-law relationships of the internet topology. In: Proceedings of the Conference on Applications, Technologies, Architectures, and Protocols for Computer Communication, SIGCOMM 1999, pp. 251–262. ACM, New York (1999)

36. Ferragina, P., Piccinno, F., Santoro, R.: On analyzing hashtags in Twitter. In: Ninth International AAAI Conference on Web and Social Media (2015)

37. Ganter, B., Wille, R.: Formal Concept Analysis: Mathematical Foundations. Springer, Heidelberg (1999). https://doi.org/10.1007/978-3-642-59830-2

38. Gao, Y., Wang, M., Zha, Z.J., Shen, J., Li, X., Wu, X.: Visual-textual joint relevance learning for tag-based social image search. IEEE Trans. Image Process. **22**(1), 363–376 (2013)

39. Golder, S., Huberman, B.A.: The structure of collaborative tagging systems, August 2005

40. Golder, S., Huberman, B.A.: The structure of collaborative tagging systems. J. Inf. Sci. **32**(2), 198–208 (2006)

41. Golder, S.A., Huberman, B.A.: Usage patterns of collaborative tagging systems. J. Inf. Sci. **32**, 198–208 (2006)

42. Guo, Q., Liu, W., Lin, Y., Lin, H.: Query expansion based on user quality in folksonomy. In: Hou, Y., Nie, J.-Y., Sun, L., Wang, B., Zhang, P. (eds.) AIRS 2012. LNCS, vol. 7675, pp. 396–405. Springer, Heidelberg (2012). https://doi.org/10.1007/978-3-642-35341-3_35

43. Gupta, M., Li, R., Yin, Z., Han, J.: Survey on social tagging techniques. SIGKDD Explor. Newsl. **12**, 58–72 (2010)

44. Halpin, H., Robu, V., Shepard, H.: The dynamics and semantics of collaborative tagging. In: Proceedings of the 1st Semantic Authoring and Annotation Workshop (SAAW 2006), pp. 211–220 (2006)

45. Halpin, H., Robu, V., Shepherd, H.: The complex dynamics of collaborative tagging. In: Proceedings of the 16th International Conference on World Wide Web, WWW 2007, pp. 211–220. ACM, New York (2007)

46. Heymann, P., Koutrika, G., Molina, H.: Can social bookmarking improve web search? In: Proceedings of the International Conference on Web Search and Web Data Mining, WSDM 2008, pp. 195–206. ACM, Palo Alto (2008)

47. Heymans, M.: Introducing Google social search: i finally found my friend's New York blog! (2009)

48. Hotho, A., Jäschke, R., Schmitz, C., Stumme, G.: Trend detection in folksonomies. In: Avrithis, Y., Kompatsiaris, Y., Staab, S., O'Connor, N.E. (eds.) SAMT 2006. LNCS, vol. 4306, pp. 56–70. Springer, Heidelberg (2006). https://doi.org/10.1007/11930334_5

49. Hotho, A., Jäschke, R., Schmitz, C., Stumme, G.: BibSonomy: a social bookmark and publication sharing system. In: de Moor, A., Polovina, S., Delugach, H. (eds.) Proceedings of the Conceptual Structures Tool Interoperability Workshop at the 14th International Conference on Conceptual Structures, Aalborg, Denmark. Aalborg University Press, July 2006

50. Hotho, A., Jäschke, R., Schmitz, C., Stumme, G.: Information retrieval in folksonomies: search and ranking. In: Sure, Y., Domingue, J. (eds.) ESWC 2006. LNCS, vol. 4011, pp. 411–426. Springer, Heidelberg (2006). https://doi.org/10.1007/11762256_31

51. Ignatov, D., Zhuk, R., Konstantinova, N.: Learning hypotheses from triadic labeled data. In: 2014 IEEE/WIC/ACM International Joint Conferences on Web Intelligence (WI) and Intelligent Agent Technologies (IAT), vol. 2, pp. 474–480, August 2014

52. Jin, Y., Li, R., Cai, Y., Li, Q., Daud, A., Li, Y.: Semantic grounding of hybridization for tag recommendation. In: Chen, L., Tang, C., Yang, J., Gao, Y. (eds.) semantic grounding of hybridization for tag recommendation. LNCS, vol. 6184, pp. 139–150. Springer, Heidelberg (2010). https://doi.org/10.1007/978-3-642-14246-8_16

53. Joachims, T.: Optimizing search engines using clickthrough data. In: ACM SIGKDD Conference on Knowledge Discovery and Data Mining (KDD), pp. 133–142 (2002)

54. Jäschke, R., Eisterlehner, F., Hotho, A., Stumme, G.: Testing and evaluating tag recommenders in a live system. In: Proceedings of the Third ACM Conference on Recommender Systems, RecSys 2009, pp. 369–372. ACM, New York (2009)

55. Kammerer, Y., Nairn, R., Pirolli, P., Chi, E.H.: Signpost from the masses: learning effects in an exploratory social tag search browser. In: Proceedings of the SIGCHI Conference on Human Factors in Computing Systems, CHI 2009, pp. 625–634. ACM, New York (2009)

56. Kleinberg, J.M.: Authoritative sources in a hyperlinked environment. J. ACM 46(5), 604–632 (1999)

57. Knerr, T.: Tagging ontology - towards a common ontology for folksonomies (2006). http://tagont.googlecode.com/files/TagOntPaper.pdf

58. Kolay, S., Dasdan, A.: The value of socially tagged URLs for a search engine. In: Quemada, J., Leon, G., Maarek, Y.S., Nejdl, W. (eds.) WWW, pp. 1203–1204. ACM (2009)

59. Krause, B., Hotho, A., Stumme, G.: A comparison of social bookmarking with traditional search. In: Macdonald, C., Ounis, I., Plachouras, V., Ruthven, I., White, R.W. (eds.) ECIR 2008. LNCS, vol. 4956, pp. 101–113. Springer, Heidelberg (2008). https://doi.org/10.1007/978-3-540-78646-7_12

60. Krause, B., Jäschke, R., Hotho, A., Stumme, G.: Logsonomy - social information retrieval with logdata. In: Proceedings of the Nineteenth ACM Conference on Hypertext and Hypermedia, HT 2008, pp. 157–166. ACM, New York (2008)

61. Kubatz, M., Gedikli, F., Jannach, D.: Localrank - neighborhood-based, fast computation of tag recommendations. In: Huemer, C., Setzer, T. (eds.) EC-Web 2011. LNBIP, vol. 85, pp. 258–269. Springer, Heidelberg (2011). https://doi.org/10.1007/978-3-642-23014-1_22

62. Körner, C., Benz, D., Strohmaier, M., Hotho, A., Stumme, G.: Stop thinking, start tagging - tag semantics emerge from collaborative verbosity. In: Proceedings of the 19th International World Wide Web Conference (WWW 2010). ACM, Raleigh, April 2010

63. Körner, C., Kern, R., Grahsl, H.-P., Strohmaier, M.: Of categorizers and describers: an evaluation of quantitative measures for tagging motivation. In: Proceedings of the 21st ACM Conference on Hypertext and Hypermedia, HT 2010, pp. 157–166. ACM, New York (2010)

64. Lancaster, F.W.: Indexing and Abstracting in Theory and Practice. University of Illinois, Chicago, Graduate School of Library and Information Science (2003)

65. Laniado, D., Mika, P.: Making sense of Twitter. In: Patel-Schneider, P.F., Pan, Y., Hitzler, P., Mika, P., Zhang, L., Pan, J.Z., Horrocks, I., Glimm, B. (eds.) ISWC 2010. LNCS, vol. 6496, pp. 470–485. Springer, Heidelberg (2010). https://doi.org/10.1007/978-3-642-17746-0_30

66. Lee, K.-P., Kim, H.-G., Kim, H.-J.: A social inverted index for social-tagging-based information retrieval. J. Inf. Sci. **38**(4), 313–332 (2012)

67. Lehmann, F., Wille, R.: A triadic approach to formal concept analysis. In: Ellis, G., Levinson, R., Rich, W., Sowa, J.F. (eds.) ICCS-ConceptStruct 1995. LNCS, vol. 954, pp. 32–43. Springer, Heidelberg (1995). https://doi.org/10.1007/3-540-60161-9_27

68. Li, P., Nie, J.-Y., Wang, B., He, J.: Document re-ranking using partial social tagging. In: 2012 IEEE/WIC/ACM International Conferences on Web Intelligence and Intelligent Agent Technology (WI-IAT), vol. 1, pp. 274–281, December 2012

69. Li, X., Guo, L., Zhao, Y.E.: Tag-based social interest discovery. In: Proceedings of the 17th International Conference on World Wide Web, WWW 2008, pp. 675–684. ACM, New York (2008)

70. Lin, Y., Lin, H., Jin, S., Ye, Z.: Social annotation in query expansion: a machine learning approach. In: Proceedings of the 34th International ACM SIGIR Conference on Research and Development in Information Retrieval, SIGIR 2011, pp. 405–414. ACM, New York (2011)

71. Lipczak, M., Milios, E.: The impact of resource title on tags in collaborative tagging systems. In: Proceedings of the 21st ACM Conference on Hypertext and Hypermedia, HT 2010, pp. 179–188. ACM, New York (2010)

72. Macdonald, C., Ounis, I.: Usefulness of quality click-through data for training. In: Proceedings of the 2009 Workshop on Web Search Click Data, WSCD 2009, pp. 75–79. ACM, New York (2009)

73. Manning, C.D., Raghavan, P., Schütze, H.: Introduction to Information Retrieval. Cambridge University Press, New York (2008)

74. Marinho, L.B., Nanopoulos, A., Schmidt-Thieme, L., Jäschke, R., Hotho, A., Stumme, G., Symeonidis, P.: Social tagging recommender systems. In: Ricci, F., Rokach, L., Shapira, B., Kantor, P.B. (eds.) Recommender Systems Handbook, pp. 615–644. Springer, Boston, MA (2011). https://doi.org/10.1007/978-0-387-85820-3_19

75. Markines, B., Cattuto, C., Menczer, F.: Social spam detection. In: Fetterly, D., Gyöngyi, Z. (eds.) Proceedings of the 5th International Workshop on Adversarial Information Retrieval on the Web AIRWeb, ACM International Conference Proceeding Series, pp. 41–48 (2009)

76. Marlow, C., Naaman, M., Boyd, D., Davis, M.: HT06, tagging paper, taxonomy, Flickr, academic article, to read. In: Proceedings of the Seventeenth Conference on Hypertext and Hypermedia, pp. 31–40. ACM (2006)

77. Mathes, A.: Folksonomies-Cooperative Classification and Communication Through Shared Metadata, Computer Mediated Communication, LIS590CMC (Doctoral Seminar). Graduate School of Library and Information Science, University of Illinois Urbana-Champaign, December 2004

78. Mika, P.: Ontologies are us: a unified model of social networks and semantics. In: Gil, Y., Motta, E., Benjamins, V.R., Musen, M.A. (eds.) ISWC 2005. LNCS, vol. 3729, pp. 522–536. Springer, Heidelberg (2005). https://doi.org/10.1007/11574620_38

79. Morrison, P.J.: Tagging and searching: search retrieval effectiveness of folksonomies on the world wide web. Inf. Process. Manag. **44**, 1562–1579 (2008)

80. Navarro Bullock, B., Jäschke, R., Hotho, A.: Tagging data as implicit feedback for learning-to-rank. In: Proceedings of the ACM WebSci 2011, June 2011

81. Neubauer, N., Obermayer, K.: Hyperincident connected components of tagging networks. SIGWEB Newsl. 4:1–4:10 (2009)

82. Neubauer, N., Wetzker, R., Obermayer, K.: Tag spam creates large non-giant connected components. In: Proceedings of the 5th International Workshop on Adversarial Information Retrieval on the Web, AIRWeb 2009, pp. 49–52. ACM, New York (2009)

83. Newman, M.: Power laws, Pareto distributions and Zipf's law. Contemp. Phys. **46**(5), 323–351 (2005)

84. Niebler, T., Becker, M., Zoller, D., Doerfel, S., Hotho, A.: FolkTrails: interpreting navigation behavior in a social tagging system. In Proceedings of the 25th ACM International on Conference on Information and Knowledge Management, CIKM 2016. ACM, New York (2016)

85. Noll, M.G.: Understanding and leveraging the social web for information retrieval. Ph.D. thesis, Universität Potsdam, April 2010

86. O'Reilly, T.: What is web 2.0. design patterns and business models for the next generation of software, September 2005. http://www.oreillynet.com/pub/a/oreilly/tim/news/2005/09/30/what-is-web-20.html,. Stand 12.5.2009

87. Overell, S., Sigurbjörnsson, B., van Zwol, R.: Classifying tags using open content resources. In: Proceedings of the Second ACM International Conference on Web Search and Data Mining, WSDM 2009, pp. 64–73. ACM, New York (2009)

88. Page, L., Brin, S., Motwani, R., Winograd, T.: The pagerank citation ranking: bringing order to the web. Technical report 1999-66, Stanford InfoLab, November 1999

89. Papadopoulos, S., Kompatsiaris, Y., Vakali, A.: A graph-based clustering scheme for identifying related tags in folksonomies. In: Bach Pedersen, T., Mohania, M.K., Tjoa, A.M. (eds.) DaWaK 2010. LNCS, vol. 6263, pp. 65–76. Springer, Heidelberg (2010). https://doi.org/10.1007/978-3-642-15105-7_6

90. Papadopoulos, S., Vakali, A., Kompatsiaris, Y.: Community detection in collaborative tagging systems. In: Pardede, E. (ed.) Community-Built Databases, pp. 107–131. Springer, Heidelberg (2011). https://doi.org/10.1007/978-3-642-19047-6_5

91. Peng, J., Zeng, D.D., Zhao, H., Wang, F.-Y.: Collaborative filtering in social tagging systems based on joint item-tag recommendations. In: Proceedings of the 19th ACM International Conference on Information and Knowledge Management, CIKM 2010, pp. 809–818. ACM, New York (2010)

92. Petersen, C., Simonsen, J.G., Lioma, C.: Power law distributions in information retrieval. ACM Trans. Inf. Syst. **34**(2), 8:1–8:37 (2016)
93. Quintarelli, E.: Folksonomies: power to the people, June 2005
94. Rendle, S., Schmidt-Thieme, L.: Pairwise interaction tensor factorization for personalized tag recommendation. In: Davison, B.D., Suel, T., Craswell, N., Liu, B. (eds.) WSDM, pp. 81–90. ACM (2010)
95. Sen, S., Lam, S.K., Rashid, A.M., Cosley, D., Frankowski, D., Osterhouse, J., Harper, F.M., Riedl, J.: Tagging, communities, vocabulary, evolution. In: Proceedings of the 2006 20th Anniversary Conference on Computer Supported Cooperative Work, CSCW 2006, pp. 181–190. ACM, New York (2006)
96. Spiteri, L.: Structure and form of folksonomy tags: the road to the public library catalogue. Webology **4**(2) (2007)
97. Stumme, G.: A finite state model for on-line analytical processing in triadic contexts. In: Ganter, B., Godin, R. (eds.) ICFCA 2005. LNCS (LNAI), vol. 3403, pp. 315–328. Springer, Heidelberg (2005). https://doi.org/10.1007/978-3-540-32262-7_22
98. Teevan, J., Ramage, D., Morris, M.R.: # TwitterSearch: a comparison of microblog search and web search. In: Proceedings of the Fourth ACM International Conference on Web Search and Data Mining, pp. 35–44. ACM (2011)
99. Vallet, D., Cantador, I., Jose, J.M.: Personalizing web search with folksonomy-based user and document profiles. In: Gurrin, C., He, Y., Kazai, G., Kruschwitz, U., Little, S., Roelleke, T., Rüger, S., van Rijsbergen, K. (eds.) ECIR 2010. LNCS, vol. 5993, pp. 420–431. Springer, Heidelberg (2010). https://doi.org/10.1007/978-3-642-12275-0_37
100. von Ahn, L., Dabbish, L.: Labeling images with a computer game. In: Proceedings of the SIGCHI Conference on Human Factors in Computing Systems, CHI 2004, pp. 319–326. ACM, New York (2004)
101. Wagner, C., Strohmaier, M.: The wisdom in tweetonomies: acquiring latent conceptual structures from social awareness streams. In: Proceedings of the Semantic Search 2010 Workshop (SemSearch 2010), April 2010
102. Wal, T.V.: Explaining and showing broad and narrow folksonomies. Blog post, February 2005
103. Wartena, C.: Automatic classification of social tags. In: Lalmas, M., Jose, J., Rauber, A., Sebastiani, F., Frommholz, I. (eds.) ECDL 2010. LNCS, vol. 6273, pp. 176–183. Springer, Heidelberg (2010). https://doi.org/10.1007/978-3-642-15464-5_19
104. Watts, D.J., Strogatz, S.: Collective dynamics of 'small-world' networks. Nature **393**, 440–442 (1998)
105. Wetzker, R., Zimmermann, C., Bauckhage, C.: Analyzing social bookmarking systems: a del.icio.us cookbook. In: Mining Social Data (MSoDa) Workshop Proceedings, ECAI 2008, pp. 26–30, July 2008
106. Wetzker, R., Zimmermann, C., Bauckhage, C.: Detecting trends in social bookmarking systems: a del. icio. us endeavor. In: Exploring Advances in Interdisciplinary Data Mining and Analytics: New Trends, pp. 34–51 (2011)
107. Wille, R.: Restructuring lattice theory: an approach based on hierarchies of concepts. In: Rival, I. (ed.) Ordered sets. NATO ASIC, vol. 83, pp. 445–470. Springer, Dordrecht (1982). https://doi.org/10.1007/978-94-009-7798-3_15
108. Willinge, W., Alderson, D., Doyle, J.C.: Mathematics and the internet: a source of enormous confusion and great potential. Technical report 5, May 2009

109. Wu, H., Zubair, M., Maly, K.: Harvesting social knowledge from folksonomies. In: Proceedings of the Seventeenth Conference on Hypertext and Hypermedia, HYPERTEXT 2006, pp. 111–114. ACM, New York (2006)
110. Wu, X., Zhang, L., Yu, Y.: Exploring social annotations for the semantic web. In: Proceedings of the 15th International Conference on World Wide Web, WWW 2006, pp. 417–426. ACM Press, New York (2006)
111. Yazdani, S., Ivanov, I., AnaLoui, M., Berangi, R., Ebrahimi, T.: Spam fighting in social tagging systems. In: Aberer, K., Flache, A., Jager, W., Liu, L., Tang, J., Guéret, C. (eds.) SocInfo 2012. LNCS, vol. 7710, pp. 448–461. Springer, Heidelberg (2012). https://doi.org/10.1007/978-3-642-35386-4_33
112. Yeung, C.M.A.: From user behaviours to collective semantics. Ph.D. thesis, University of Southampton (2009)
113. Yin, D., Hong, L., Davison, B.D.: Exploiting session-like behaviors in tag prediction. In: Proceedings of the 20th International Conference Companion on World Wide Web, WWW 2011, pp. 167–168. ACM, New York (2011)
114. Yue, Z., He, D.: Collaborative information search. In: Brusilovsky, P., He, D. (eds.) Social Information Access. LNCS, vol. 10100, pp. 108–141. Springer, Cham (2018)
115. Zhou, D., Lawless, S., Wade, V.: Improving search via personalized query expansion using social media. Inf. Retr. 15(3–4), 218–242 (2012)
116. Zoller, D., Doerfel, S., Jäschke, R., Stumme, G., Hotho, A.: On publication usage in a social bookmarking system. In: Proceedings of the 2015 ACM Conference on Web Science, WebSci 2015, pp. 67:1–67:2. ACM, New York (2015)

10
Rating-Based Collaborative Filtering: Algorithms and Evaluation

Daniel Kluver[1]([✉])(iD), Michael D. Ekstrand[2], and Joseph A. Konstan[1](iD)

[1] GroupLens Research, Department of Computer Science and Engineering,
University of Minnesota, Minneapolis, USA
{kluver,konstan}@cs.umn.edu
[2] People and Information Research Team (PIReT),
Department of Computer Science, Boise State University, Boise, USA
michaelekstrand@boisestate.edu

Abstract. Recommender systems help users find information by recommending content that a user might not know about, but will hopefully like. Rating-based collaborative filtering recommender systems do this by finding patterns that are consistent across the ratings of other users. These patterns can be used on their own, or in conjunction with other forms of social information access to identify and recommend content that a user might like. This chapter reviews the concepts, algorithms, and means of evaluation that are at the core of collaborative filtering research and practice. While there are many recommendation algorithms, the ones we cover serve as the basis for much of past and present algorithm development. After presenting these algorithms we present examples of two more recent directions in recommendation algorithms: learning-to-rank and ensemble recommendation algorithms. We finish by describing how collaborative filtering algorithms can be evaluated, and listing available resources and datasets to support further experimentation. The goal of this chapter is to provide the basis of knowledge needed for readers to explore more advanced topics in recommendation.

1 Introduction

One problem with online collections is *information overload* - when presented with too much information people have trouble making informed decisions. While the tools for searching, visualizing, and navigating these large collections introduced in previous chapters of this book can help users find content, even these tools can be insufficient if an online collection is big enough, or if the user is unsure of exactly what content they are interested in. Ideally, a system should know what kind of items each user is interested in without ever being told. Then the system can focus on presenting each user only those items that they are most likely to be interested in.

This idea has led to a proliferation of strategies for helping users focus only on the items they will like. The most basic strategy is to focus on the most

© Springer International Publishing AG, part of Springer Nature 2018
P. Brusilovsky and D. He (Eds.): Social Information Access, LNCS 10100, pp. 344–390, 2018.
https://doi.org/10.1007/978-3-319-90092-6_10

popular items, or those that are reviewed most favorably by other users. While not personalized for a given user, these strategies can quickly guide users to the best content the system has to offer. It isn't even that hard to do basic personalization within these simple strategies. For example, if a system knows what genres of music a user tends to listen to, then the system can focus on presenting popular artists from that genre.

In the early 1990s, these strategies set the stage for *collaborative filtering recommendation systems*. The insight behind collaborative filtering recommender systems is that people have relative stable tastes. Therefore, if two people have agreed in the past they will likely continue to agree in the future. A key part of this insight – and the major difference between this and the personalization strategies that came before – is that it does not matter why two users agree. They could share tastes in books because they both like the same style of book binding, or they could share taste in movies due to a nuance of directing; a collaborative filtering recommender system does not care. So long as the two users continue agreeing, we can use the stated preferences of one user to predict the preferences of another. Since we need very little supplementary information, collaborative filtering algorithms are applicable in a wide range of possible circumstances.

Since the mid 1990s, collaborative filtering recommender systems have become very popular in industry. Companies like Amazon, Netflix, Google, Facebook and many others, have deployed collaborative filtering algorithms to help their users find things they would enjoy. The popularity of these deployments has pushed the field of recommender systems, leading to faster, more accurate recommender systems. These improvements have been coupled to changes in how we think about deploying collaborative filtering systems to support users. Collaborative filtering systems were originally seen as a filter which could separate the interesting items from the uninteresting ones, hence the term collaborative *filtering*. As the field has advanced, it become more common to think of these algorithms as recommending a short list of the best items for a user. Even if there are plenty of good items that go unrecommended, the a recommendation algorithm is doing its job if it's list contains the best of the best for a given user.

Like other forms of social navigation, collaborative filtering algorithms rely on the connections and patterns made possible by large bodies of users. Despite this, collaborative filtering algorithms are not typically social in the traditional sense: while one user's behavior does directly affect other users' experiences, this is rarely made clear to the users. Most recommender system users are blissfully unaware of how their actions benefit not only themselves, but other users with similar tastes.

This chapter describes the foundational collaborative filtering algorithms and methods for evaluating these algorithms for use in a given application. In particular we will focus on *rating based* algorithms, in which user preference is measured by numerical ratings. After presenting the most common algorithms for rating based collaborative filtering, we will present two more recent approaches, a learning to rank algorithm and ensemble methods. While these approaches are still popularly deployed today, many extensions and applications of the algorithms we describe go well beyond what we can present here. Later chapters in this book are dedicated to recommendation algorithms that leverage social connections (Chap. 11 [44]), social tags (Chap. 12 [8]), user reviews (Chap. 13 [51]), implicit (non-rating) preference feedback (Chap. 14 [35]), and ways to recommend new social connections (Chap. 15 [25]). The goal of this chapter is to describe the foundational algorithms that are built upon in these later chapters. We also cover topics such as algorithm evaluation that are relevant throughout the following chapters.

1.1   Examples of Recommender Systems

There are many different ways recommendation algorithms can be incorporated into an online service. The most simple is the "streaming" style service, which is oriented around a stream of recommended content. Two examples of this are streaming music services like Pandora[1] and the Jester joke recommender[2]. Screenshots of these services is shown in Figs. 1 and 2. Both services share the same design: the user is presented with content (music or jokes). After each item the user is given the opportunity to evaluate the item. These evaluations influence the algorithm which then picks the next song or joke. This process repeats until the user leaves. Jester is known to use a collaborative filtering algorithm [24]. Interested readers can find more information about jester and even download a rating dataset for experimentation from the Jester web page. As a commercial product, less is known about Pandora's algorithm. However, it is reasonable to assume that they are using a hybrid algorithm that combines collaborative filtering information with their catalog of song metadata.

A quite different way to use recommendation algorithms can be seen in catalog based websites like MovieLens[3]. MovieLens is a movie recommender developed by the GroupLens research lab. On the surface MovieLens is similar to other movie catalog websites such as the Internet Movie DataBase (IMDB) or The Movie DataBase (TMDB). All three have pages dedicated to each movie detailing information about that movie and search features to help users find information about a given movie. MovieLens goes further, however, by employing a collaborative filtering algorithm. MovieLens encourages users to rate any movie they have seen, MovieLens then users these ratings to provide personalized predicted ratings which it shows alongside a movie's cover art in both movie search and detail pages. These predictions can help users rapidly decide if it is worth learning more about a movie. Users can also ask MovieLens to produce a list of recommended movies, with the top 8 most recommended movies for a user being centrally positioned on the MovieLens home page, this can be seen in Fig. 3.

A third common way to use recommendation algorithms is in e-commerce systems, perhaps the most notable being Amazon[4]. Amazon is an online store

---

[1] https://www.pandora.com/.
[2] http://eigentaste.berkeley.edu/.
[3] https://movielens.org/.
[4] https://www.amazon.com/.

Fig. 1. Screenshot of the Pandora music streaming service

One day the first grade teacher was reading the story of the Three Little Pigs to her class. She came to the part of the story where the first pig was trying to accumulate the building materials for his home. She read, "...and so the pig went up to the man with the wheelbarrow full of straw and said, 'Pardon me sir, but may I have some of that straw to build my house?'"

The teacher paused then asked the class, "And what do you think that man said?"

One little boy raised his hand and said, "I know...he said, 'Holy Shit! A talking pig!'"

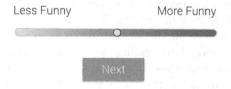

Fig. 2. Screenshot of the Jester joke recommender

Fig. 3. Screenshot of the MovieLens home page

which started as a bookstore, but has since diversified to a general purpose online storefront. While the average user may not notice the recommendations in Amazon (or at the very least may think little of them) much of the Amazon storefront is determined based on recommendation algorithms. A screenshot of the Amazon main page for one author is shown in Fig. 4 with recommendation features highlighted. Since only a small proportion of users use reviews on Amazon it is likely that Amazon uses data beyond ratings in their collaborative filtering algorithm. Unlike MovieLens, getting information and recommendation is not the primary motivation of Amazon users. Therefore, while the basic interfaces may be similar, the way recommendations are used, and the algorithm properties that a system designer might look for, will be different.

As these examples show, recommendation algorithms can be useful in a wide range of situations. That said, there are some commonalities: each service has some way of learning what users like. In MovieLens and the streaming services users can explicitly rate how much they like a movie, joke, or song. In Amazon purchase records and browsing history can be used to infer user interests. Each service also has some way of suggesting one or more item to the user based on their recommendation algorithm. It will be helpful to keep these examples in mind as they will help anchor the more abstract algorithm details covered in this chapter to a specific context of use.

Fig. 4. Screenshot of the Amazon home page

1.2   A Note on the Organization of Recommendation Algorithms

Every year dozens of new recommendation algorithms are introduced. It should be no surprise, therefore, that there have been various attempts organizing these algorithms into a taxonomy or classification scheme for collaborative filtering algorithms. The purpose of any such organization is to allow better communication about how an algorithm works, and what other algorithms it is similar to, by describing where that algorithm is in a taxonomy.

To some degree these classifications have been useful; chapters in this book, for example, are organized based on important distinctions between different types of algorithms. Other distinctions that have been made are less useful, either because algorithms have advanced to the point where a distinction has no meaning, or because the classification itself has been used inconsistently. In this chapter we will restrict ourselves to categorizations that we feel are useful for communication. That said, we note that other works on recommender systems that a reader might explore are still organized under some of these traditional taxonomies. Therefore we will introduce some of these distinctions now so the reader can be aware of them if they wish to read other resources on recommendation algorithms.

One important distinction that has been made between algorithms is between collaborative filtering algorithms (like those discussed in this chapter) and content-based algorithms. Collaborative filtering algorithms, as was described earlier, operate by finding patterns in user behavior that can be used to predict future behavior. The traditional example of this would be that two users tend to like the same things, therefore when one use likes something, we can predict the other user will as well. *Content based* algorithms, on the other hand, focus on relationships between users and the content they like. A traditional example of this would be an algorithm that learns which genres of music a user likes and recommends songs from that genre. While still meaningful, the line between collaborative and content based filtering has become somewhat blurry as modern algorithms have sought to combine the strengths of both algorithms. Readers can still expect to see this distinction made in new publications (including this one) as the algorithms that are both content based and collaborative filtering algorithms are still in the minority.

Within the specific range of collaborative filtering algorithms, the most common taxonomy separates so-called model-based algorithms and memory-based algorithms. The division was first made in a 1998 paper [9] where memory-based algorithms were defined as those that operate over the entire dataset, where model-based algorithms are those that use the dataset to estimate a model which can then be used for predictions. For recommender systems this split is problematic as many algorithms can be described sufficiently as a memory-based algorithm or a model-based algorithm depending on how the algorithm is optimized and deployed.

More recently this same distinction has been used more usefully to separate based on the basic design of an algorithm [66]. Model-based algorithms are those that use machine learning techniques to fit a parametrized model, while memory-based algorithms search through the training data to find similar examples (users or items). These examples are then aggregated to compute recommendations.

While still common we personally find this latter separation does not do a great job of communicating about the distinctions between algorithms. Therefore we will eschew this taxonomy and present algorithms grouped, and labeled, by their mathematical structure or motivation. In the next section we will cover the basic concepts and mathematical notation that will be used throughout this chapter. The section after that will describe *baseline algorithms*: simple algorithms which seek to capture broad trends in rating data. Section 4 will describe *nearest neighbor algorithms*: the group of algorithms that have historically been called memory based algorithms, which work by finding similar examples which are used in computing recommendations. Section 5 will describe *Matrix Factorization Algorithms*: a group of algorithms that share a common and powerful mathematical model inspired by matrix factorization. Section 6 will describe *Learning to Rank Algorithms*: algorithms that focus on ranking possible recommendations, instead of predicting what score a user will give a particular item. Section 7 will briefly mention other groups of algorithms which we do not explore in depth: *graph based algorithms*, *linear regression based algorithms*, and *probabilistic algorithms*. Section 8 will describe *ensemble methods*: ways to combine multiple recommenders. Finally, Sect. 9 will explore metrics and evaluation procedures for collaborative filtering algorithms.

2   Concepts and Notation

In this section we will discuss the core concepts and mathematical notations that will be used in our discussion of recommendation algorithms (summarized in Table 1). The two most central objects in a recommendation system are the *users* the system recommends to and the *items* the system might recommend. These terms are purposely domain neutral as different domains often have domain specific terms for these concepts.

One *user* represents one independently tracked account for recommendation. Typically this represents one system account, and is assumed to represent one person's tastes. We will denote the set of all users as U with $u, v, w \in U$ being individual users from the set.

One *item* represents one independently tracked thing that can be recommended. In most systems its obvious what services or products should map to an item in the recommendation algorithm; in an e-commerce system like Amazon or Ebay, each product should an item. In a movie recommender each movie should be an item. In other domains there might be more uncertainty; in a music recommender should each song be an item (and recommended individually) or should each an album be an item? We will denote the set of all items as I with $i, j, k \in I$ being individual items from the set.

Most traditional collaborative filtering recommender systems are based on *ratings*: numeric measures of a user's preference on an item. Ratings are collected from users on a given *rating scale* such as the 1-to-5 star scale used in MovieLens,

Table 1. Summary of mathematical notation

Users	The set of all profiles in the system	U
	A profile in the system, usually one person	$u, v, w \in U$
Items	The collection of things being recommended	I
	A member of the collection of items	$i, j, k \in I$
Rating	A measure of a user's preference for an item	$r_{ui} \in R$
User's ratings	The set of all items rated by one user	I_u
	The vector of all ratings made by one user	$\mathbf{r}_u$
Item's ratings	The set of all users who have rated one item	U_i
	The vector of all ratings made on one item	$\mathbf{r}_i$
Score/Prediction	An algorithm's score assigned to a user and item	$S(u, i)$

(a) Unary rating scales from Facebook (left) and Twitter (right)

(b) Binary rating scale from YouTube

(c) 5 point rating scale from MovieLens, and a larger, 10 point scale from IMDB

(d) A very large continuous rating scale used in the Jester joke recommender

Fig. 5. Examples of various rating scales used in the wild. (https://www.facebook.com/, https://twitter.com/, https://www.youtube.com)

Amazon, and many other websites or the ten star scale used by IMDB[5] (see Fig. 5c for examples). Rating scales are almost always designed so that larger numbers indicate more preference: a user should like a movie they rated 5/5 stars more than they like a movie they rated 4/5 stars. The rating user u assigns to item i will be denoted r_{ui}.

While there are many different rating scales that have been used, the choice of rating scale is often not relevant for a recommendation algorithm. Some interfaces, however, deserve special consideration. Small scales which can only take one or two values such as a binary (thumbs up, thumbs down) scale (see Fig. 5b) or unary "like" scales (see Fig. 5a) may require special adaptions when applying algorithms designed with a larger scale in mind. For example, when working with the unary "like" scale it can be important to explicitly treat non-response as a form of rating feedback.

While the algorithms in this chapter are focused on rating-based approaches, it is important to understand that they only need a numeric measure of preference. In this way these algorithms can be used with for many different forms of preference feedback. Chapter 14 in this book [35] covers recommendation based on implicit feedback (measurements of user preference that are not ratings). One of the strategies covered are ways to convert common types of data into ratings that can be used with the algorithms in this chapter.

Recommender systems often organize the set of all ratings as a sparse *rating matrix* R, which is a $|U| \times |I|$ matrix. R only has values at $r_{ui} \in R$ where user u has rated item i; for all other pairs of u and i, r_{ui} is blank. The set of all items rated by a user u is $I_u \subset I$, the collection of all ratings by one user can also be expressed as a sparse vector $\mathbf{r}_u$. Similarly, the set of all users who have rated one item i is $U_i \subset U$, and the collection of these ratings can be expressed as a sparse vector $\mathbf{r}_i$.

---

[5] http://www.imdb.com/.

As the rating matrix R is sparsely observed, one of the ways collaborative filtering algorithms can be viewed is *matrix completion*. Matrix completion is the task of filling in the missing values in a sparse matrix. In the recommendation domain this is also called the *prediction task* as filling in unobserved ratings is equivalent to predicting what a target user u would rate an item i. Rating predictions can be used by users to quickly evaluate an unknown item: if the item is predicted highly it might be worth further consideration.

Like we described in MovieLens, one use of predictions is to sort items by predicted user preference. This leads to the view of a recommender algorithm as generating a ranking score. The *ranking task* is to generate a personalized ranking of the item set for each user. Any prediction algorithm can be used to generate rankings, but not all ranking algorithms produce scores that can be thought of as a prediction. Algorithms that focus specifically on the ranking task are known as *learning-to-rank* recommenders and will be discussed towards the end of this chapter.

Both prediction and ranking oriented algorithms produce a score for each user and item. Therefore, we will use the syntax $S(u, i)$ to represent the output of both types of algorithm. Almost every algorithm we discuss in this chapter will produce output as a score for each user and item.

One of the most interesting applications of collaborative filtering recommendation technology is the *recommendation task*. The *recommendation task* is to generate a small list of items that a target user is likely to want. The simplest approach to this is *Top-N recommendation* which takes the N highest ranked items by a ranking or prediction algorithm. More advanced approaches involve combining ranking scores with other factors to change the properties of the list of recommendations.

For each task there are associated metrics which can be used to evaluate the algorithm. Prediction algorithms can be evaluated by the accuracy of their prediction, and ranking algorithms can be compared to how users rank items. Recommendation algorithms are a particular challenge to evaluate, however, as users are sensitive to properties such as the diversity of the recommendations, or how novel the recommended items are. Evaluation is an important concept in the study of recommender systems, especially as some algorithms are partially defined by specific evaluation metrics. We will discuss evaluation approaches in more detail in Sect. 9.

3 Baseline Predictors

Before describing true collaborative filtering approaches, we will first discuss *baseline predictors*. Baseline predictors are the most simple approaches for rating. While a baseline predictor is rarely the primary prediction algorithm for a recommender system, the baseline algorithms do have their uses. Due to their simplicity, baseline predictors are often the most reliable algorithms in extreme conditions such as new users [37]. Because of this, baseline predictions are often used as a fallback algorithm in cases where a more advanced algorithm might fail.

Baseline predictions can also be used to establish a minimum standard of performance when comparing new algorithms and domains. Finally, baseline prediction algorithms are often incorporated into more advanced algorithms, allowing the advanced algorithms to focus on modeling deviations from the basic, expected patterns that are already well captured by a baseline prediction.

The most basic baseline is the global baseline, in which one value is taken as the prediction for every user and item $S(u, i) = \mu$. While any value of μ is possible, taking μ as the average rating minimizes prediction error and is the standard choice. The global baseline can be trivially improved by using a different constant for every item or user leading to the item baseline $S(u, i) = \mu_i$ and the user baseline $S(u, i) = \mu_u$ respectively. In the item baseline μ_i is an estimate of the item i's average rating. This allows the item baseline to captures differences between different items. In particular, some items are widely considered to be good, while others are generally considered to be bad. In the user baseline μ_u is an estimate of the user u's average rating. This allows the user baseline to capture differences between how users tend to use the rating scale. Because most rating scales are not well anchored, two users might use different rating values to express the same preference for an item.

This discussion leads us to the generic form of the baseline algorithm, the user-item bias model, given in Eq. 1.

$$S(u, i) = \mu + b_u + b_i \tag{1}$$

Equation 1 has three variables: μ, the average rating in the system; b_i, the item bias representing if an item is, on average, rated better or worse than average; and b_u the user bias representing whether the user tends to rate high or low on average. By combining all three baseline models we are able to simultaneously account for differences between users and items, albeit in a very naive way. This equation is sometimes referred to as the personalized mean baseline as it is *technically* a personalized prediction algorithm, even though the personalization is very minimal.

This model can be learned many ways [40], but are most easily learned with a series of averages, with μ being the average rating, b_i being the item's average rating after subtracting out μ and b_u being the user's average rating after subtracting out μ and b_i [21]. The following equations can be used to compute μ, b_i and b_u:

$$\mu = \frac{\sum_{r_{ui} \in R} r_{ui}}{|R|} \tag{2}$$

$$b_i = \frac{\sum_{u \in U_i} (r_{ui} - \mu)}{|U_i|} \tag{3}$$

$$b_u = \frac{\sum_{i \in I_u} (r_{ui} - b_i - \mu)}{|I_u|} \tag{4}$$

One problem that can lead to poor performance from the user-item baseline is when a user or item has very few ratings. Predictions made on only a few ratings can be very unreliable, especially if the prediction is extremely high or low. One way to fix this is to introduce a damping term β to the numerator of the computation. Motivated by Bayesian statistics, this term will shrink the bias terms towards zero when the number of ratings for an item is small while having a negligible effect when the number of ratings is large.

$$b_i = \frac{\sum_{u \in U_i}(r_{ui} - \mu)}{|U_i| + \beta} \tag{5}$$

$$b_u = \frac{\sum_{i \in I_u}(r_{ui} - b_i - \mu)}{|I_u| + \beta} \tag{6}$$

Damping parameter values of 5 to 25 have been used in the past [22,37], but for best results β should be re-tuned for any given system.

4 Nearest Neighbor Algorithms

The first collaborative filtering algorithms were nearest neighbor algorithms. These algorithms work by finding similar items or users to the user or item we wish to make predictions for, and then uses ratings on these items, or by these users, to make a prediction. While newer algorithms have been designed, these algorithms are still in use in many live systems. The simplicity and flexibility of these basic approaches, combined with their competitive performance, makes them still important algorithms to understand.

Readers with a background in general machine learning approaches may be familiar with nearest-neighbor algorithms, as these algorithms are a standard technique in machine learning. That said, there are many important details in how recommender systems experts have deployed the nearest neighbor algorithm in the past. These details are the result of careful study in how to best predict ratings in the recommender system domain.

4.1 User-User

Historically, the first collaborative filtering algorithms were user-based nearest neighbors algorithm, sometimes called the user to user algorithm or user-user for short [58]. This is the most direct implementations of the idea behind collaborative filtering, simply find users who have agreed with the current user in the past and use their ratings to make predictions for the current user. User-based nearest neighbor algorithms were quite popular in early recommender systems, but they have fallen out of favor due to scalability concerns in systems with many users.

The first step of the user-user algorithm for a given user u and item i is to generate a set of users who are similar to u and have rated item i. The set of similar users is normally referred to as the user's neighborhood N_u, with the subset who have rated an item i being N_{ui}. Once we have the set of neighbors we can take a weighted average of their ratings on an item as the prediction. Therefore the most important detail in the user-based nearest neighbor algorithm is the similarity function $sim(u, v)$.

A natural and widely-used choice for $sim(u, v)$ is a measurement of the correlation between the ratings of the two users; usually, this takes the form of Pearson's r [27]:

$$sim(u, v) = \frac{\sum_i (r_{ui} - \mu_u)(r_{vi} - \mu_v)}{\sqrt{\sum_i (r_{ui} - \mu_u)^2} \sqrt{\sum_i (r_{vi} - \mu_v)^2}} \qquad (7)$$

Alternatively, the rank correlation in the form of Spearman's ρ (the Pearson correlation of ranks rather than values) or Kendall's τ (based on the number of concordant and discordant pairs) can be used. In addition to statistical measures, vector space measures such as the cosine of the angle between the users' rating vectors can be used:

$$sim(u, v) = \frac{\mathbf{r}_u \cdot \mathbf{r}_v}{\|\mathbf{r}_u\|_2 \|\mathbf{r}_v\|_2} \qquad (8)$$

In non-rating-based systems, the Jaccard coefficient between the two users' purchased items is a natural choice.

In most published work, Pearson correlation has produced better results than either rank-based or vector space similarity measures for rating-based systems [9,27]. However, Pearson correlation does have a significant weakness for rating data: it ignores items that only one the users has rated. In the extreme, if two users have only rated two items in common their correlation would be 1. Unfortunately, its unlikely that two users who do not watch the same things would truly be that similar. In general, correlations based on a small number of common ratings trend artificially towards extreme values. While these similar neighbors may have high similarity scores, they often do not perform well as neighbors whose similarity scores are based on a larger number of ratings.

Significance weighting [27] addresses this problem by introducing a multiplier to reduce the measured similarities between between users who have not rated many of the same items. The significance weighting strategy is to multiply the similarity by $\frac{\min(|I_u \cap I_v|, T)}{T}$, where T is a threshold of "enough" co-rated items. This causes the similarity to linearly decrease between users with fewer than T common rated items. Past work has found $T = 50$ to be reasonable, with larger values showing no improvement [27].

There is a natural, parameter-free way to dampen similarity scores for users with few co-occurring items. If items a user has not rated are treated as having the user's average rating, rather than discarded, the Pearson correlation can be computed over all items. When computing with sparse vectors, this can be realized by subtracting each user's mean rating from their rating vectors, then comparing users by taking the cosine between their centered rating vectors and assuming missing values to be 0. This results in the following formula:

$$sim(u, v) = \frac{\hat{r_u} \cdot \hat{r_v}}{\|\hat{r_u}\|_2 \|\hat{r_v}\|_2} \qquad (9)$$

$$= \frac{\sum_{i \in I_u \cap I_v} (r_{ui} - \mu_u)(r_{vi} - \mu_v)}{\sqrt{\sum_{i \in I_u} (r_{ui} - \mu_u)^2} \sqrt{\sum_{i \in I_v} (r_{vi} - \mu_v)^2}} \qquad (10)$$

This is equivalent to the Pearson correlation, except that *all* of each users' ratings contributes to their term in the denominator, while only the common ratings are counted in the numerator (due to the normalized value for missing ratings being 0). The result is similar to significance weighting. Similarity scores are damped based on the fraction of rated items that are in common; if the users have 50% overlap in their rated items, their resulting similarity will be greater than the similarity between users with the same common ratings but only 20% overlap due to additional ratings of other items. This method has the advantage of being parameter-free, and has been seen to perform at least as well [15, 16, 20].

After picking a similarity function, the next step in predicting for a user u and item i is to compute the similarity between user u and every other user. For systems with many users approximations such as randomly sampling users [29] can improve performance, possibly at a tradeoff of prediction accuracy. Once similarities have been computed the system must choose a set of similar users N_u. Various approaches can be taken here, from using all users to a limited number or only those that are sufficiently similar. Past evaluations have suggested that using the 20 to 60 most similar users performs well and avoids excessive computations [27]. Additionally, by filtering the users the algorithm can avoid the noise that would be introduced by the lower quality neighbors.

Once the algorithm has a set of neighbors N_u the prediction for item i is simply the weighted average of the neighboring users' ratings. Direct averages, without the weighting term, have been used in the past, but tend to perform worse. Let N_{ui} refer to the subset of N_u containing all users in N_u who have rated item i.

$$S(u, i) = \frac{\sum_{v \in N_{ui}} sim(u, v) * r_{vi}}{\sum_{v \in N_{ui}} |sim(u, v)|} \tag{11}$$

These predictions can often be improved by incorporating basic normalization into the algorithm. For example, since users have different average ratings we can take a weighted average of the item's offset from the user's average rating.

$$S(u, i) = \mu_u + \frac{\sum_{v \in N_{ui}} sim(u, v) * (r_{vi} - \mu_v)}{\sum_{v \in N_{ui}} |sim(u, v)|} \tag{12}$$

More advanced normalization is also possible by using z score normalization in which all ratings are first reduced by the user's average rating, and then divided by the standard deviation in the users rating.

The user-based nearest neighbors approach tends to produce good predictions, but is often outperformed by newer algorithms. In this regard, the algorithm is listed here mostly for reference value. Other than its accuracy, one core issue with the performance of the user-based recommender is its slow predict time performance. Most modern recommender systems have a very large number of users which makes finding neighborhoods of users expensive. For good results neighborhood finding should be done online and cannot be extensively cached for performance improvements. This computation makes user-based nearest neighbors very slow for large scale recommender systems, but it remains viable option for a recommender system with many items but relatively few users.

4.2 Item-Item

The item-based nearest neighbor algorithm (sometimes called the item to item or item-item algorithm for short) is closely related to user-user [61]. Where user-user works by finding users similar to the given user and recommending items they liked, item-item finds items similar to the items the given user has previously liked and uses those to make recommendations. Instead of computing similarities between users, item-item computes similarities between items, and uses an average rating over the item neighborhood to make predictions. Unlike user-user, item-item is well suited to modern systems which have many more users than items. This allows item-item some key performance optimizations over user-user which we will address shortly.

To make a prediction for user u and item i item-item first computers a neighborhood of similar items N_{ui}. In practice it is common to limit this neighborhood to only the k most similar items. $k = 30$ is a common value from the academic research, however, different systems may require different settings for optimal performance [61]. Item-item then takes the weighted average of user u's ratings on the items in this neighborhood, and uses that as a prediction.

$$S(u,i) = \frac{\sum_{j \in N_{ui}} sim(i,j) * r_{uj}}{\sum_{j \in N_{ui}} |sim(i,j)|} \tag{13}$$

This equation can be enhanced by subtracting a baseline predictor from the ratings r_{ui} so the algorithm is only predicting deviation from baseline. If this is done, the baseline should be added back in after the fact to make a prediction. Note this will have no effect if the global or per-user baselines are used. The following equation shows how a baseline predictor could be subtracted, using $B(u,i)$ to represent the baseline

$$P_{u,i} = B(u,i) + \frac{\sum_{j \in N_{ui}} sim(i,j) * (r_{uj} - B(u,j))}{\sum_{j \in N_{ui}} |sim(i,j)|} \tag{14}$$

Generally item-item uses the same similarity functions as user-user, simply replacing the user ratings with item ratings.

The cosine similarity metric is the most popular similarity metric for item-item recommendation. Past work has shown that cosine similarity performs better than other traditionally studied similarity functions [61].

The key to getting the best quality predictions using cosine similarity is normalizing the ratings [61]. Evaluations have shown that subtracting the user's average rating from the ratings before computing similarity leads to substantially better recommendations. In practice we have found that subtracting the item baseline or the user-item baseline leads to improvements in performance [20].

$$sim(i,j) = \frac{\sum_{u \in U} (r_{ui} - B(u,i)) * (r_{uj} - B(u,j))}{\sqrt{\sum_{u \in U} (r_{ui} - B(u,i))^2} * \sqrt{\sum_{u \in U} (r_{uj} - B(u,j))^2}} \tag{15}$$

Both Pearson and Spearman similarities have been tried for item-item prediction, but do not tend to perform better than cosine similarity [61]. Just like

Pearson similarity for the user-user algorithm, signficance weighting can improve prediction quality when using Pearson similarity with item-item.

One key advantage of the item-item algorithm is that item similarities and neighborhoods can be shared between users. Since no information about the given user is used in computing the list of similar items, there is no reason that the values cannot be cached and re-used with other users. Furthermore, in systems where the set of users is much larger than the set of items, we would expect the average item to have very many ratings. Many ratings per items leads to relatively stable item similarity scores, meaning that these can be cached for a much larger amount of time than user-user similarity scores.

This insight has led to the common practice of precomputing the item-item similarity matrix. With a precomputed list of similar items the specific item neighborhood used for prediction can be found with a quick linear scan, using only information about the given user's past ratings. This speeds up predict-time computation drastically, making item-item more suitable for modern interactive systems than the user-user algorithm.

The cost of this speed-up is a regularly-occurring "model build" in which the similarity model is recomputed. This frequently is done nightly to ensure that new items are included in the model and are available to recommend. Since this model build is not interactive, it can run on a separate bank of machines from the live system and be scheduled to avoid peak system use.

The improvements from precomputing similarities can be made even larger by truncating the stored model. For reasonably large systems storing the whole item-item similarity matrix can take a lot of space. Many items have low to no similarity with all but a small percent of the system. Since these dissimilar items will almost never be used in an actual item neighborhood, there is no point to store them. By keeping only the most useful potential neighbors, the model size on disk and in memory can be reduced and predict time performance can be increased. Therefore it is common to keep only the n most similar items for any item in the model as "potential neighbors". Past work has shown that larger models do perform slightly better than smaller models, but that the advantages disappear after some point. In the original work on item-item, the point at which a larger model has no benefit is around 100 to 200 items [61]. Work based on larger datasets have also found larger models (500 items or more) to more effective; suggesting that, like all other parameters, the best value for the model size will vary from system to system [17,37].

With the various tweaks and optimizations the research community has found since item-item was first published, item-item can be a strong algorithm for recommendation. While it is slightly outperformed by newer algorithms, it is still very competitive when well tuned [18,19]. Furthermore, item-item is easier to implement, modify, and explain to users than most other recommendation algorithms. For these reasons, item-item is still a competitive algorithm for large scale recommender systems, and still sees modern deployment despite more recent, slightly more accurate, algorithms being offered [59].

Variants. Nearest-neighbor algorithms are the best-known approaches for collaborative filtering recommendation. Because of this, they have been modified in many interesting ways. One variant is an inversion of the user-user algorithm: the K-furthest neighbors algorithm by Said et al. [60] The K-furthest neighbors algorithm makes neighborhoods based on the least similar users, instead of the most similar users. The idea behind this is to enhance the diversity of the recommendations made. User evaluations comparing nearest neighbor recommendation to furthest neighbor recommendation shows that the two are relatively close in user satisfaction, even if the predictions made by nearest neighbor recommendation are much more accurate.

Another interesting variant to nearest neighbor recommendation is Bell and Koren's Jointly Derived Neighborhood Interpolation Weights approach [3]. The key insight of this approach is that the quality of the similarity function directly determines the quality of the user recommendation in a neighborhood model. Therefore, these similarity scores should be directly optimized, instead of relying on ad-hoc similarity metrics. One key advantage of this is that the similarity scores can be jointly optimized, which makes the algorithm more robust to interactions involving multiple neighbors. A similar approach has also been taken by Ning and Karypis' SLIM algorithm [49].

Many variants of nearest-neighbor algorithms use some external source of information to inform the similarity function used. One example of this is the trust aware recommendation framework [46], which re-weights user similarity scores based on an estimated degree of trust between two users. In this way the algorithm bases predictions on more trusted users. This same approach could be used with other forms of information such as content based similarity information.

5   Matrix Factorization Algorithms

Nearest-neighbor algorithms are good at capturing pairwise relationships between users or items. They cannot, however, take advantage of broader structure in the data, such as the idea that five different items share a common topic, or that a user's ratings can be explained by their interest in a particular feature. To explicitly represent this type of relationship requires a fundamentally different approach to recommendation. One such approach is the use of *latent feature models* such as the popular family of *matrix factorization* algorithms.

Rather than modeling individual relationships between users or items, latent feature models represent each user's preference for items in terms of an underlying set of k features. Each user can then be described in terms of their preference for each latent feature, and each item can be described in terms of its relevance to each feature. These item and user feature scores can then be combined to predict the user's preference for future items.

All matrix factorization algorithms encode each user's preference numerically in k-dimensional vectors $\mathbf{p}_u$ and each item's relevance to features in k-dimensional vectors $\mathbf{q}_i$. We will use p_{uf} to indicate the value representing a user's preference for a given feature f and q_{if} to indicate the value representing an item's relevance to feature f. Once these vectors are computed, we can compute a user u's preference for a particular item i as the linear combination of the user feature vector and the item feature vector.

$$S(u,i) = \mathbf{p}_u \cdot \mathbf{q}_i = \sum_{f=1}^{k} p_{uf} q_{if} \tag{16}$$

Under this equation, $S(u,i)$ will be high if and only if those feature u prefers (with high scores is $\mathbf{p}_u$) are also those feature i is relevant to (with high scores in $\mathbf{q}_i$).

It is common to organize the vectors $\mathbf{p}_u$ into a $|U| \times k$ matrix named P and the vectors $\mathbf{q}_i$ into a $|I| \times f$ matrix named Q. This allows all scores for a given user to be computed in a single matrix operation $\mathbf{s}_u = \mathbf{p}_u \times Q^T$. Likewise, all scores for every user can be computed as $S = P \times Q^T$ These operations may be more efficient than repeatedly computing $S(u,i)$ in some linear algebra packages.

As with neighborhood based algorithms, this approach can easily be improved by directly accounting for a user's average rating and an item's average rating. This can be done as before, by normalizing ratings against a baseline predictor. However, it is much more common to introduce the bias terms directly into the model, and to learn these values simultaneously with learning matrices P and Q. This results in a biased matrix factorization model [40]:

$$S(u,i) = \mu + b_u + b_i + \mathbf{p}_u \cdot \mathbf{q}_i \tag{17}$$

The goal in matrix factorization algorithms is to find the vectors $\mathbf{p}_u$ and $\mathbf{q}_i$ (as well as extra terms like μ, b_u, b_i) that lead to the best scoring function for a given metric. One interesting difference between this and a nearest neighbor style algorithm is that the same core model and scoring equation algorithm can lead to many different algorithms depending on how $\mathbf{p}_u$ and $\mathbf{q}_i$ are learned. We will be presenting three approaches for learning $\mathbf{p}_u$ and $\mathbf{q}_i$, in this section we will see how to optimize $\mathbf{p}_u$ and $\mathbf{q}_i$ for prediction accuracy. In the following section we will address an algorithm that learns $\mathbf{p}_u$ and $\mathbf{q}_i$ to optimize how accurately the algorithm ranks pairs of items.

The algorithms we describe are a few of many possible matrix factorization algorithms. One of the interesting aspects of this model is that it has become a standard starting point for many novel algorithm modifications. SVD++ [40] and SVDFeature [12], for example, extend Eq. 17 by adding terms to incorporate implicit feedback and additional user or item feature information. By combining Eq. 17 with new terms to accommodate new data, and new ways of optimizing the model for different goals, many interesting algorithm variants are possible.

5.1   Training Matrix Decomposition Models With Singular Value Decomposition

One way to train a matrix factorization model for predictive accuracy, and the reason these are often called SVD algorithms, is with a truncated singular value decomposition (SVD) of the ratings matrix R.

$$R \approx P\Sigma Q^{\mathrm{T}} \tag{18}$$

Where P is an $|U| \times k$ matrix of user-feature preference scores, Q is an $|I| \times k$ matrix of item-feature relevance vectors, and Σ is a $k \times k$ diagonal matrix of global feature weights, called singular values. In a true algebraic SVD, P and Q are orthogonal, and this product is the best rank-k approximation for the original matrix R. This means that the matrices P and Q can be used to produce scores that are optimized to make accurate predictions of unknown ratings.

Singular value decompositions are not bounded to a particular k; the number of non-zero singular values will be equal to the rank of the matrix. However, we can truncate the decomposition by only retaining the k largest singular values and their corresponding columns of P and Q. This accomplishes two things: first, it greatly reduces the size of the model, and second, it reduces noise.

Ratings are known to contain both signal about user preferences and random noise [38]. If the ratings matrix is a combination of signal and noise, then consistent and useful signals will contribute primarily to the high-weight features while the random noise will primarily contribute to the lower wright features. For convenience, the columns of P and Q are often stored in a pre-weighted form so that Σ is not needed as a separate matrix. With this we see that the scoring function is simply $S(u,i) = \mathbf{p}_u \cdot \mathbf{q}_i$.

There are two important and related difficulties with the singular value decomposition for training a matrix factorization model. First, it is only defined over complete matrices, but most of R is unknown. In a rating-based system, this problem can be addressed by imputation, or assuming a default value (e.g. the item's mean rating) for unknown values [62]. If the ratings matrix is normalized by subtracting a baseline prediction before being decomposed, then the unknown values can be left as 0's and the normalized matrix can be directly decomposed with standard sparse matrix methods.

The second difficulty is that the process of computing a singular value decomposition is very computationally intensive and does not scale well to large matrices. Unlike with the first problem there is no natural solution to this. Because of this problem it is uncommon for matrix decomposition algorithms to operate based on a pure singular value decomposition.

Despite these limitations, using a singular value decomposition to compute $\mathbf{p}_u$ and $\mathbf{q}_i$ is still an easy way to build a basic collaborative filtering algorithm for experimentation. Optimized algorithms for computing matrix decompositions can be found in mathematical computing packages such as MATLAB. However, for the reasons mentioned above this is not a reasonable approach for production scale recommender systems.

5.2   Training Matrix Decomposition Models With Gradient Descent

The goal of matrix factorization is to produce an effective model of user preference. Therefore the algebraic structure (singular value decomposition) is more of a means to an end rather than the end itself. In practice it is common to sidestep the problems inherent in using a singular value decomposition and instead directly optimize P and Q against some metric over our training data. This way we can simply ignore missing data opening up a range of speed-ups over a singular value decomposition. Simon Funk pioneered this approach to great affect by using gradient descent to train P and Q to optimize the popular mean squared error accuracy metric [22]. Similar algorithms are now a common strategy for latent factor style recommendation algorithms [41].

Since our goal is a matrix decomposition with minimal mean squared error, we can learn a decomposition by treating the problem as an optimization problem: learn matrices P $(m \times k)$ and Q $(n \times k)$ such that predicting the known ratings in R with the multiplication PQ^T has minimal (squared) error. As mean squared error is easily differentiable, optimization is normally done via either stochastic gradient descent or alternating least squares.

Stochastic gradient descent is a general purpose optimization approach used in machine learning to optimizing a mathematical model for a given loss function or metric, so long as the metric is easy to take the derivative of. First the computer starts with an arbitrary initial value for the model parameters, in this case the matrices P and Q as well as the bias terms. Then, it iterates through each training point, in our case a user, item and rating: (u, i, r_{ui}). Based on this point it computes an update to the model parameters that will reduce the error made on that training point. The specific update rules are derived by taking the derivative of an error function with respect to the training point, in this case the error function is the squared error. These updates are repeated many times until the algorithm converges upon a local optimum.

The update rules to train a biased matrix factorization model to minimize squared error using gradient descent are:

$$\epsilon_{ui} = \mu + b_u + b_i + \mathbf{p}_u \cdot \mathbf{q}_i - r_{ui} \tag{19}$$

$$\mu \quad = \mu + \lambda(\epsilon_{ui} - \gamma\mu) \tag{20}$$

$$b_u \quad = b_u + \lambda(\epsilon_{ui} - \gamma b_u) \tag{21}$$

$$b_i \quad = b_i + \lambda(\epsilon_{ui} - \gamma b_i) \tag{22}$$

$$P_{uf} = P_{uf} + \lambda(\epsilon_{ui} * Q_{if} - \gamma P_{uf}) \tag{23}$$

$$Q_{if} = Q_{if} + \lambda(\epsilon_{ui} * P_{uf} - \gamma Q_{if}) \tag{24}$$

To apply these these update rules we first compute ϵ_{ui}, which represents the prediction error: $S(u, i) - r_{ui}$. This update rule should be applied to each of the k features. Then, the update for each variable should be computed before applying the updates.

The gradient descent process uses a learning rate λ that controls the rate of optimization (0.001 is a common value), and γ is a regularization term (0.02 is a common value). This regularization term penalizes excessively large user-feature and item-feature values to avoid overfitting. This update rule should be applied until some stopping condition is reached, the most common stopping condition being a specified number of iterations. To get the best performance, k, λ, and γ, and the stopping condition should be hand tuned using the evaluation methodologies discussed in Sect. 9.

Once learned, the set of variables μ, b_u, b_i, P and Q serves as the model for the algorithm. Given these values creating a prediction is as easy as applying Eq. 17. Like with item-item this model is normally computed offline. Traditionally the model is rebuilt daily or weekly (depending on how long it takes to rebuild a model and how actively new ratings, items, and users are added to the system). With this type of recommender model it is also possible to perform online updates [57] which allow a model to account for ratings, items, and users added after the model is built with a minimal loss of accuracy. In practice online and offline model updates can be combined to balance between complete optimization and interactive data use.

Matrix factorization approaches provide a memory-efficient and, at recommendation time, computationally efficient means of producing recommendations. Computing a score in a matrix factorization algorithm requires only $O(K)$ work (assuming that the factors are precomputed and stored for $O(1)$ lookup), this is true no matter how many items, users, or ratings are involved. Furthermore, by taking account of the underlying commonalities between users and items that are reflected in users' preferences and behaviors it makes very accurate predictions. Because of this matrix factorization algorithms are very popular and are one of the most common algorithms in the research literature.

6 Learning to Rank

As the original collaborative filtering algorithms focused on the prediction task, most of the research into the recommendation task has been designed around how we can use predictions to make good quality recommendations. The most common approach to doing this is also the most obvious: sort items by prediction. This approach is called the "Top-N recommendation", and is still used by systems like MovieLens and remains quite popular today. That said, other approaches have been developed for directly targeting the quality of a recommendation list.

Learning-to-rank algorithms are a recently popular class of algorithms from the broader field of machine learning. As their name suggests, these algorithms directly learn how to produce good rankings of items, instead of the indirect approach taken by the Top-N recommendation strategy. While there are several approaches, most learning-to-rank algorithms produce rankings by learning a ranking function. Like a prediction, a ranking function produces a score for each pairing of user and item. Unlike predictions, however, the ranking score has no deliberate relationship with rating values, and is only interesting for its ranked order.

Because learning-to-rank algorithms are designed around ranking and recommendation tasks, instead of prediction, they often outperform prediction algorithms at the recommendation task. However, because their output has no relation to the prediction task, they are incapable of making predictions. Many recommender systems do not display predicted rating to users; in such systems a learning-to-rank algorithm can lead to a much more useful recommender system.

The heart of most learning to rank algorithms is a specific way to define ranking or recommendation quality. Unlike prediction algorithms, where "accuracy" is easy to define, there are many ways to define ranking and recommendation quality. Furthermore, unlike prediction errors, ranking and recommendation errors are poorly suited for use in optimization. A small change to a model might lead to a small, but measurable prediction accuracy change but have no effect on the output ranking. Therefore the core work in many learning-to-rank algorithms is in designing easy-to-optimize measurements that approximate common ranking metrics. Once these new metrics are defined, standard optimization techniques can be applied to standard recommendation models such as the biased matrix factorization model from Eq. 17.

Learning-to-rank is an active area for research into recommender system algorithms with new algorithms being developed every year. To get a taste of this type of algorithm we will explain the *Bayesian Personalized Ranking* (BPR) algorithm [56]. First published in 2009, BPR is one of the earliest and most influential learning-to-rank algorithms for collaborative filtering recommendation. BPR will be discussed again in Chap. 14 in this book [35], as it was originally designed for implicit preference information. We discuss it here because it is trivial to modify for use with rating data, and the structure and development of the BPR algorithm serves as a good example of learning-to-rank algorithms in general.

6.1   BPR

BPR is a pairwise learning-to-rank algorithm, which means that it tries to predict which of two items a user will prefer. If BPR can accurately predict that a user will prefer one item over another we can use that prediction strategy to rank items and form recommendations. This approach has two advantages to prediction based training: first, BPR tends to produce better recommendations than prediction algorithms. Secondly, BPR can use a much wider range of training data. As long as we can deduce from user behavior that one item is preferred over another, we can use that as a training point.

BPR was originally designed for use with implicit unary forms of preference feedback, instead of ratings. For example, with unary data such as past purchases we can generate pairs by assuming that all purchased items are liked better than all other items. With a traditional ratings dataset we can generate training points by taking pairs of items that the user rated, but assigned different ratings to.

To predict that a user will prefer one item over another, BPR tries to learn a function $P(i >_u j)$ – the probability that user u prefers item i to item j. If $P(i >_u j) > 0.5$ then, according to the model, user u is more likely to prefer i over j than they are to prefer j over i. Therefore if $P(i >_u j) > 0.5$ we would want to rank item i above item j. There are many different functions that could be used for P, BPR uses the popular logistic function.

The logistic function allows us to shift focus from computing a probability to computing any number x_{uij} which represents a user's relative preference for i over j (or j over i if x_{uij} happens to be negative). The logistic transformation then defines P as $P(i <_u j) = 1/(1 + e^{-x_{uij}})$. While we could build algorithms to directly optimize x_{uij} its easier to change variables one more time by defining $x_{uij} = S(u,i) - S(u,j)$ for some scoring function S. The way the logistic function works we have $P(i >_u j) > 0.5$ if and only if $S(u,i) > S(u,j)$. Furthermore, the probability P will be more confident (closer to 0 or 1) if $S(u,i)$ is substantially greater or smaller than $S(u,j)$. Therefore to optimize $P(i >_u j)$ for predictive accuracy we need to optimize S so that it ranks items correctly. For the same reason, once we train S we can use S directly for ranking.

Based on this formalization we arrive at the BPR optimization criteria, which is the function BPR seeks to optimize. The optimization criteria depends on some scoring function $S(u,i)$, and a collection of training points (u,i,j) which represent that u has expressed a preference for item i over item j. Given these, the optimization criteria is the product of the probability BPR assigns to each observed preference $P(i >_u j) = 1/(1 + e^{-(x_{ui} - x_{uj})})$. A good ranking function should maximize these probabilities, therefore we seek to maximize performance against this criteria. The full derivation of this, as well as the complete optimization criteria can be found the original BPR paper by Rendle et al. [56].

Almost any model can be used for the scoring function S. All that is required is that the derivative of S with respect to its model parameters can be found. Therefore any algorithm that can be trained for accuracy using gradient descent can also be trained using BPR's optimization criteria to effectively rank items. We will cover BPR-MF, which uses a matrix factorization model. In particular, we will give update rules for the non-biased matrix factorization seen in Eq. 16. Only minor modifications would be needed to derive update rules for a biased matrix factorization model.

As with all matrix factorization models we have two matrices P and Q representing user and item factor values which need to be optimized so that $S(u,i) = \mathbf{p}_u \cdot \mathbf{q}_i$ provides a good ranking. P and Q can be optimized for the BPR optimization criteria using stochastic gradient descent by applying the following update rules:

For a given training sample (u,i,j) representing the knowledge that user u prefers item i over item j:

$$\epsilon_{uij} = \frac{e^{-(S(u,i)-S(u,j))}}{1 + e^{-(S(u,i)-S(u,j))}} \tag{25}$$

$$P_{uf} = P_{uf} + \lambda \left(\epsilon_{uij} * (Q_{if} - Q_{jf}) + \gamma * P_{uf} \right) \tag{26}$$

$$Q_{if} = Q_{if} + \lambda \left(\epsilon_{uij} * P_{uf} + \gamma * Q_{if} \right) \tag{27}$$

$$Q_{jf} = Q_{jf} + \lambda \left(-\epsilon_{uij} * P_{uf} + \gamma * Q_{jf} \right) \tag{28}$$

Where λ is the learning rate and γ is the regularization term. Like the equations for updating a traditional matrix factorization algorithm, ϵ_{uij} in this equation represents the degree to which S does or does not correctly rank items i and j.

When training BPR algorithms, the order in which training points are taken can have a drastic impact on the rate at which the algorithm converges. Rendle et al. [56] showed that the naive approach of taking training points grouped by user can be orders of magnitude slower than an approach that takes randomly chosen training points. Therefore, for simplicity, Rendle et al. recommend training the algorithm by selecting random training points (with replacement) and applying the update rules above. This process can be repeated until any preferred stopping condition, such as iteration count, has been reached.

Unsurprisingly, the BPR-MF algorithm is much better than classic algorithms at ordering pairs of items under the AUC metric. On other recommender metrics BPR-MF only shows modest improvements over traditional algorithms. The trade-off of this, however, is that BPR-MF, like most learning-to-rank algorithms, cannot make predictions. Theoretically, advanced techniques could be used to turn the ranking score into a prediction, however, in practice we find this does not lead to prediction improvements over prediction centered algorithms. Therefore for a website that uses both recommendations and predictions using separate algorithms for the two tasks might be essential.

7 Other Algorithms

The algorithms highlighted in this chapter provide an overview of the most influential and important recommender systems algorithms. While these few algorithms provide a basic grounding of most recommender algorithms, there are many more algorithms than covered in this chapter. Briefly, we want to mention a few other key approaches and techniques for generating personalized recommendations that have proved successful in the past.

7.1 Probabilistic Models

Probabilistic algorithms, such as those based on a Bayesian belief network, are a popular class of algorithms in the machine learning field. These algorithms have also seen increasing popularity in the recommender systems field recently. Many probabilistic algorithms are influenced by the PLSI (Probabilistic LSI) [30] and LDA (latent Dirichlet allocation) [7] algorithms. The basic structure of these models is to assume that there are k distinct clusters or profiles. Each profile has a distinct probability distribution over movies describing the movies that cluster tends to watch and, for each movie a probability distribution over ratings for that cluster. Instead of directly trying to cast a user into only one cluster, each user is a probabilistic mixture of all clusters [31]. This can be thought of as a type of latent feature model, each user has a value for each of k clusters (features) and each movie has a preference score associated with each of k clusters (features). One of the key advantages of these probabilistic models is that they are easier to update with new user or item information, due to the wealth of standard training approaches for probabilistic models [63].

7.2 Linear Regression Approaches

Many algorithms have incorporated linear regression techniques into their formulation. For example, the original work introducing the item-item algorithm experimented with using regression techniques in addition to the similarity computations. For each pair of items a linear regression is used to find the best linear transformation between the two items. This transformation would then be applied to get an adjusted rating to be used in the weighted average. While this showed some promise for very sparse systems the idea showed little promise for more traditional recommender systems.

A more recent implementation of this idea is the Slope One recommendation algorithm [45]. In slope one we compute an average offset between all pairs of items. We then predict an item i by applying the offset to every other rating by that user, and performing a weighted average. The slope one algorithm has some popularity, especially as a reference algorithm, due to how simple it is to implement and motivate. Outside of nearest neighbor approaches, linear regression approaches are also a common way to combine multiple scoring functions together, or combine collaborative filtering output with other factors to create ensemble recommenders.

7.3 Graph-Based Approaches

Graph-based recommender algorithms leverage graph theoretic techniques and algorithm to build better recommender systems. Although uncommon, these algorithms have been a part of recommender systems research community since the early days [1]. In traditional recommender system websites like MovieLens or Netflix there is rarely a natural graph to consider, therefore these algorithms tend to impose a graph by connecting users to items they have rated. By also connecting items using content information you can use graph-based algorithms to combine content and collaborative filtering approaches [32,52].

Some services, however, have both recommendation features and social network features. In these websites it is natural to assume that a person to person connection is an indicator of trust. This leads to the set of trust based recommendation algorithms in which a person to person trust network is used as part of the recommendation process. Many of these algorithms use graph based propagation of trust as a core part of a recommendation algorithm [33,46].

8 Combining Algorithms

Most recommender system deployments do not directly tie the scores output by one of the above algorithms to their user interface. While these algorithms perform well, there usually are further improvements that can be made by combining the output of these algorithms with other algorithms or scores. Generally there are three reasons to perform these modifications: business logic, algorithm accuracy/precision, and recommendation quality. The first of these reasons – business logic – is both the most simple, and ubiquitous. Many recommender systems modify the output of their algorithm to serve business purposes such as "do not recommend items that are out of stock" or "promote items that are on sale".

The second of these reasons – algorithm accuracy – leads researchers to develop and deploy ensemble algorithms. Ensemble algorithms are techniques for combining arbitrarily algorithms into one comprehensive final algorithm. The final algorithm normally performs better than any of its constituent algorithms independently. The design, development, and training of ensemble algorithms is a large topic in the broader machine learning field. As a comprehensive discussion of ensemble algorithms is out of scope for this chapter we will try to give a brief overview to the application of ensembles in the recommender systems field.

The final reason to combine algorithms – recommendation quality – is more complicated. Properties like novelty and diversity have a large impact on how well users like recommendations. These properties are very hard to deliberately induce in collaborative filtering algorithms as they can be in tension with recommending the best items to a user. Several algorithms have been developed to modify a recommendation algorithm's output specifically for these properties. While these algorithms are not ensembles in the traditional sense they are another way to moderate the behavior of a recommendation algorithm based on some other measure. These algorithms will be discussed after describing strategies for ensemble recommendation.

8.1   Ensemble Recommendation

The most basic approach to an ensemble algorithm is a simple weighted linear combination between two algorithms S_a and S_b

$$S(u, i) = \alpha + \beta_a * S_a(u, i) + \beta_b * S_b(u, i) \tag{30}$$

The simplest way to learn this is to have the developer directly specify the weightings. While this may sound naive there are several places where this can be appropriate, especially when the parameters are picked based on difficult to optimize metrics such as a user survey.

Linear Regression. A more attractive technique to train a linear model may be to use traditional linear regression to learn the best α and β parameters to optimize accuracy. You could imagine simple training S_a and S_b on all available training ratings and then using their predictions and the same training ratings to predict α and β. Unfortunately, this is not recommended – the core issue is that the same ratings should not be used when training the sub-algorithms and when training the ensemble as it leads to overfitting.

Ideally you want to train the ensemble on ratings that the sub-algorithms have not seen so that the ensemble is trained based on the out-of-sample error of each algorithm. The easiest way to do this would be to randomly hold out some small percent of training data (say 10%) and to use that withheld data to train the regression to minimize squared error. The problem with this approach to training a linear ensemble, however, is that it withholds a large amount of data, which might effect the overall algorithm performance.

Stacked Regression. An alternate approach, without this problem is Breiman's stacked regression algorithm [10]. Breiman's algorithm trains regression parameters by first producing a k subsets of the dataset by traditional (ratings based) crossfolding. Then each sub-algorithm is trained independently on each of the k folds of the algorithm. Due to the way crossfold validations are made this means that each of the original training ratings has an associated algorithm which has never seen that rating. When generating sub-algorithm predictions for any given training point (as needed to learn the linear regression) the algorithm which has never seen that training point is used. Once the linear regression has been trained the sub-algorithms should then be re-trained using the overall set of data. For a more comprehensive description of this procedure consult Breiman's original paper [10].

All ensemble algorithms work best when very different algorithms are being combined. So an ensemble between item-item (with cosine similarities) and item-item (with Pearson similarities) is unlikely to show the same improvement as an ensemble between item-item and a content-based algorithm [17].

In MovieLens, for example, we could imagine making a simple content-based algorithm by computing each user's average rating for each actor and director. While not necessarily the best algorithm, this content based algorithm would likely outperform a collaborative filtering algorithm for movies with very few ratings. Therefore, an ensemble of a content based algorithm and a collaborative filtering algorithm might show improved accuracy over a content based algorithm on its own.

This example does lead to one interesting observation: there are many cases where we would want to create an ensemble and we know the conditions when one algorithm might perform better than another. The actor-based recommender would be our best recommender when we have next to no ratings for a movie, as we get more ratings, however, we should trust the collaborative filtering algorithm more. A linear weighting scheme does not allow for this type of adjustment.

Feature Weighted Linear Stacking. The feature weighted linear stacking algorithm, introduced by Sill el al., is an extension to Breiman's linear stacked regression algorithm [64]. Feature weighted linear stacking allows the relative weights between algorithms to vary based on features like number of ratings on an item. This algorithm is most noteable for being very popular during the Netflix prize competition, a major collaborative filtering accuracy competition, where it was the key facet of the second best algorithm. In feature weighted linear stacking algorithms are linearly related as per Eq. 30, the difference is that the weights β are themselves a linear combination of the extra features. Sill et al. show that this model can be learned by solving a system of linear equations using any standard toolkit for solving systems of linear equations. Details of this solution, especially including information to assist in scaling are provided in the paper by Sill et al. [64].

While ensemble methods have provided much better predictive accuracy than single algorithm solutions, and could theoretically be applied to learning-to-rank problems as well, it should be noted that they can also become much more complex than a traditional recommender. While it is easy to overlook technical complexity when designing an algorithm, technical complexity can be a significant barrier to actually deploying large ensembles in the field. Notably, after receiving code for a 107-algorithm ensemble Netflix went on to only actually implement two of these sub-algorithms [2]. Ensemble methods can be much more complex and time invasive to keep up to date and can require much more processing when making predictions which can lead to slower responses to users. Therefore, when considering ensembles, especially very large ones, designers should consider if the improved algorithm accuracy is worth the increased system complexity.

8.2   Recommending for Novelty and Diversity

This brings us to the third reason that a recommender system might modify the scores output from a recommendation algorithm: to increase the quality of recommendations as reported by users. Research into recommendation systems has shown that selecting only good items is not enough to ensure that a user will find a recommendation useful [69,70]. Many other properties can effect how useful a recommendation is to a user. Two that have been shown to be important, and have been the focus of some research are novelty and diversity.

Novelty and diversity are properties of a recommendation that measure how the items relate to each other or the user. Novelty refers to how unexpected or unfamiliar the user is with their recommendations [53]. If a recommendation only contains obvious recommendations they are neither novel, nor useful at helping a user find new items. Diverse recommendations cover a large range of different items. One flaw with many recommender systems is that their recommendations are all very similar to each other, which limits how useful the recommendations are. For example, a top-8 recommendation consisting of only Harry Potter movies would neither be novel (as those movies are quite well known), nor would it be diverse (since the recommendation only represents a small niche of the user's presumably broader interests).

There have been several different approaches to modifying an algorithm's score to favor (or avoid) properties like novelty or diversity. We will focus on two broad strategies, the first is well suited to combining algorithms with item specific metrics such as how novel an item is, the second is well suited to metrics measured over the entire recommendation list. While these strategies have been pioneered for use with novelty and diversity, they can be applied with any metric. For example, when recommending library items, users may be disappointed by recommendations on item that have a waitlist and cannot be borrowed immediately. These strategies could be used to modify recommendations to favor items without a waitlist, increasing user satisfaction. Specific metrics for novelty and diversity will be discussed alongside algorithm evaluation in Sect. 9.4.

The most common way to combine algorithms with some metric to increase user satisfaction is to use a simple linear combination between the original algorithm score S_a and a some item level measurement of interest [6,67,69]. For example, the number of users who have rated an items $|U_i|$ (or its inverse $1/|U_i|$)

is normally measured to allow manipulating novelty. By blending the score from an algorithms with $1/|U_i|$ we can promote items that have fewer ratings and enhance the novelty of recommended items. These scores are normally used only when ranking, typically an unmodified rating prediction is still used even when the prediction is blended for recommendation.

The other major approach for modifying recommendations is an iterative re-ranking approach. In these approaches items are added to a recommendation set one at a time, with the ranking score recomputed after each step. This approach has two advantages. First, re-ranking allows hard constraints when selecting items. For example, the iterative function could reject any more than two movies by a given director. Secondly, the iterative approach allows measurements such as the average similarity of an item with the other items already chosen for recommending. This is often necessary when manipulating diversity, as diversity is a property of the recommendation, not one specific item. This was in fact the approach taken by the first paper to address diversity in recommendation [65]. The cost of this approach over re-scoring is slightly higher recommend time costs.

A primary example of the iterative re-ranking algorithm is the diversity adding algorithm introduced by Ziegler et al. [70]. To generate a top 10 recommendation list, this algorithm first picks the top 50 items for a user as candidate items. The size of the candidate set represents a trade off between run time cost and flexibility of the algorithm to find more diverse items. From the top 50 items, the best predicted item is immediately added to the recommendation. After that, for each remaining candidate the algorithm computes an sum of the similarity between that candidate and each item in the recommendation. The algorithm then sorts by inverse similarity sum to get a dissimilarity rank for each potential item. The overall item weight is then a weighted average of the prediction rank and the dissimilarity rank (with weights chosen beforehand). The item with the smallest score by this weight is added to the recommendation. This process repeats, with updated similarity scores, until the desired ten item recommendation list has been made.

Re-scoring and re-ranking algorithms are an easy way to promote certain properties in recommendations. Both algorithms can be relatively inexpensive to run, and can be added on top of an existing recommender. Additionally these approaches are very easy to re-use for a large variety of different recommendation metrics beyond just novelty and diversity.

9   Metrics and Evaluation

The last topic we will discuss in this chapter is how to evaluate a recommendation algorithm. This is an important concept in the recommender systems field as it gives us a way to compare and contrast multiple algorithms for a given problem. Given that there is no one "best" recommendation algorithm, it is important to have a way to compare algorithms and see which one will work best for a given purpose. This is true both when arguing that a new algorithm is better than previous algorithms, and when selecting algorithms for a recommender system.

One key application of evaluations in recommender systems is *parameter tuning*. Most recommendation algorithms have variables, called *parameters* which are not optimized as part of the algorithm and must be set by the system designer. Parameter tuning is the process of tweaking these parameters to get the best performing version of an algorithm. For example, when deploying a matrix factorization algorithm the system designer must choose the best number of features for their system. While we have tried to list reasonable starting points for each parameter of the algorithms we have discussed in this chapter, these are at best a guideline, and readers should carefully tune each parameter before relying on a recommendation algorithm.

In this section we will mostly focus on *Offline evaluation* methodologies. Offline evaluations can be done based only on a dataset and without direct intervention from users. This is opposed to *online evaluation* which is a term for evaluations done against actual users of the system, usually over the internet.

Offline evaluations are a common evaluation strategy from machine learning and information retrieval. In an offline evaluation we take the entire ratings dataset and split it into two pieces: a *training dataset* ($Train$), and a *test dataset* ($Test$); there are several ways of doing this, which we describe in more detail shortly. Algorithms are trained using only the ratings in the training set and asked to predict ratings, rank items, or make a recommendation for each user These outputs are then evaluated based on the ratings in the test dataset using a variety of metrics to assess the algorithm's performance.

Just as there are several different goals for recommendation (prediction, ranking items, recommending items) there are many different ways to evaluate recommendations. Furthermore, while evaluating prediction quality may be straightforward (how well does the prediction match the rating), there are many different ways to evaluate if a ranking or recommendation is correct. Therefore, there are a great number of different *evaluation metrics* which score different aspects of recommendation quality. These can be broadly grouped into *prediction metrics*, which evaluate how well the algorithm serves as a predictor, *ranking quality metrics*, which focus on the ranking of items produced by the recommender, *decision support metrics*, which evaluate how well the algorithm separates good items from bad, and metrics of *novelty and diversity* that evaluate how novel and diverse the recommendations might appear to users. Depending on how an algorithm might be used metrics from one or all of these sections might be used in an evaluation, and performance on several metrics might need to be balanced when deciding on a best algorithm.

9.1   Prediction Metrics

The most basic measurement we can take of an algorithm is the fraction of items it can score, known as the *prediction coverage metric* or simply the *coverage metric* for short. The coverage metric is simply the percent of user item pairs in the whole system that can be predicted. In some evaluations coverage is only computed over the test set. Beyond convenience of computation, this modification focuses more explicitly on how often the algorithms cannot produce a score for items that the user might be interested in (as evidenced by the user rating that item) [27]. This metric is of predominantly historic interest, as most modern algorithm deployments use a series of increasingly general baseline algorithms as a fallback strategy to ensure 100% coverage. That said, coverage can still be useful when comparing older algorithms, or looking at just one algorithm component in isolation.

Assuming that an algorithm is producing predictions, the next most obvious measurement question is how well its predictions match actual user ratings. To answer this we have two metrics *Mean Absolute Error* (MAE) and *Root Mean Squared Error* (RMSE). In the following two equations *Test* is a set containing the test ratings r_{ui} and the associated users and items.

$$MAE(Test) = \frac{\sum_{(u,i,r_{ui}) \in Test} |S(u,i) - r_{ui}|}{|Test|} \tag{31}$$

$$RMSE(Test) = \sqrt{\frac{\sum_{(u,i,r_{ui}) \in Test} (S(u,i) - r_{ui})^2}{|Test|}} \tag{32}$$

Both MAE and RMSE measure the amount of error made when predicting for a user, and are on the same scale as the ratings. The biggest difference between these two metrics is that RMSE assigns a larger penalty to large prediction errors when compared with MAE. Since large prediction errors are likely to be the most problematic, RMSE is generally preferred.

Both the RMSE and MAE metrics measure accuracy on the same scale as predictions. This can be normalized to a uniform scale by dividing the metric value by the size of the rating scale ($maxRating - minRating$), yielding the normalized mean average error (nMAE) and normalized root mean squared error (nRMSE) metrics. This is rarely done, however, as comparisons across different recommender systems are often hard to correctly interpret do to differences in how users use those systems.

Prediction metrics can be computed for an entire system or individually for each user. RMSE, for example, can be computed for the system by averaging over all test ratings, or computed per user by averaging over each user's test ratings. Due to the differences between people, most algorithms work better for some users than they do for others. Measuring per-user error scores lets us understand and measure how accurate the system is for each user. By averaging the per user metric values we can then get a second measure of the overall system.

While the difference between per-user error and system-wide (or by rating) error may seem trivial it can be very important. Most deployed systems have power users who have rated many more items than the average user. Because they have more ratings these users tend to be overrepresented in the test set, leading to these users being given more weight when estimating system accuracy. Averaging the per user accuracy scores avoids this issue and allows the system to be evaluated based on its performance for all users.

9.2 Ranking Quality

A more advanced way of evaluating an algorithm is to ask if the way it orders or ranks items is consistent with user preferences. There are several ways of approaching this, the most basic being to simply compute the Spearman ρ or Pearson r correlation coefficients between the predicted ratings and test set ratings for a user. As noted by Herlocker et al. [27] Spearman's ρ is imperfect because it works poorly when the user rates many items at the same level. Additionally, both metrics assign equal importance to accuracy at the beginning of the list (the best items) and the end (the worst items). Realistically, however, we want a ranking metric that is most sensitive at the beginning of the ranking, where users are likely to look, and less sensitive to errors towards the end of the list, where users are unlikely to look.

A more elegant metric for evaluating ranking quality is called the *discounted cumulative gain metric* DCG. DCG tries to estimate the value a user will receive from a list. It does this by assuming each item gives a value represented by its rating in the test set, or no value if unrated. To make DCG focus on the beginning of the list, these values are discounted logarithmically by their rank, so the maximum gain of items later in the list is smaller than the potential gain early in the list. The DCG metric is defined as follows:

$$dcg(u, Rec) = \sum_{i \in Rec} \frac{r_{ui}}{max(1, log_b(k_i))} \tag{33}$$

Where k_i refers to the rank order of i, Rec is an ordered list of items representing an algorithms recommendation for the user u, and b is the base of the logarithm. While different values are possible, DCG is traditionally computed with $b = 2$. Other values of b have not been shown to yield meaningfully different results [36].

The DCG metric is almost always reported normalized as the *normalized discounted cumulative gain metric* nDCG. nDCG is simply the DCG value normalized to the 0–1 range by dividing by the "optimal" discounted cumulative gain value which would be given by any optimal ranking

$$ndcg(u) = \frac{dcg(u, prediction)}{dcg(u, ratings)} \tag{34}$$

A similar metric to this is the *half life utility metric* [9]. Half life utility uses a faster exponential discounting function. The half life utility is as follows:

$$HalfLife(u) = \sum_{i \in Rec} \frac{max(r_{ui} - d, 0)}{2^{(k_i - 1)/(\alpha - 1)}} \tag{35}$$

Half life utility has two parameters. The first is d which is a score that should represent the neutral rating value. A recommendation for an item with score d should neither help nor hurt the user, while any item rated above d should be good recommendations. d should also be used as the "default" rating value for

r_{ui} where a user does not have a rating for that item, In this way unrated items are assigned a value of 0. The α variable controls the speed of exponential decay and should be set so that an item at rank α has a roughly 50% chance of actually being seen by the user.

One common modification on these metrics is to absolutely limit the recommendation list size. For example, the nDCG@n metric is taken by computing the nDCG over the top-n recommendations only. Any item in the test set but not recommended is ignored in the computation. This is reasonable if you know there is a hard limit to how many items users can view, or if you only care about the beginning n elements of the list.

9.3   Decision Support Metrics

Another common approach to evaluating recommender systems is to use metrics from the information retrieval domain such as precision and recall [14]. These metrics treat the recommender as a classifier with the goal of separating good items from other items. For example, a good algorithm should only recommend good items (precision) and should be able to find all good items (recall). Decision support metrics give us a way of understanding how well the recommendation could support a user in deciding which items to consume.

The basic workflow of all decision support metrics is to first perform a recommendation, then compare that recommendation against a previously selected "relevant item set". The relevant item set represents those items that we know to be good items to recommend to a user. You then count how many of the recommended items were relevant and how many were not.

For items in a larger scale system, a choice needs to be made about which items to consider relevant. The easiest choice would be to take all rated items in the test set as good items, which evaluates an algorithm on its ability to select items that the user will see. More useful, however, is the practice of choosing a cutoff such as four out of five stars at which we consider a recommendation 'good enough'. Best practices recommend testing with multiple similar cutoffs to ensure that results are robust across various choices for defining relevant items. Evaluation results that favor one algorithm with items rated 4 and above as "good", but another algorithm if 4.5 and above are "good", deserve more careful consideration.

Once this decision has been made there remains an issue of how to treat the remaining "non relevant" items. In traditional information retrieval work it is often reasonable to assume that every item that is not known to be relevant can be considered not relevant, and therefore bad to recommend. This assertion is much less reasonable in the recommender system domain, while some of these items are known to be rated poorly, many more have simply never been rated. There are likely many good items for each user that has not been rated and would therefore be considered not relevant. It has been argued that not having complete knowledge of which items a user would like may make these metrics inaccurate or suffer from a bias [4,13,28]. Ultimately, this problem has not been solved, and most evaluations settle for the assumption most non-rated items are not good to recommend, and the evaluation bias caused by this will be minor.

Related to the above issue of how to treat not relevant items is the question of how to compute recommendations. There have been various different approaches taken, and these have been shown to lead to different outcomes in the evaluation [5]. Commonly recommendations are done by taking the top-n predicted items, in which case these metrics would be labeled with that n such as $precision@20$ for precision computed over the top-20 list. n should be picked to match interface practices, so if only eight items are shown to a user, algorithms should be evaluated by their $precision@8$.

The other important consideration is which candidate set of items the recommendations should be over. Many different candidate set options have been used, but the most common options are either all items, or the relevant item set plus a random subsample of not relevant items. Some work has used the set of items that the user rated in the test set as a recommendation candidate set; while this does avoid the issue of how to treat unrated items, this evaluation methodology also provides different results which are believed to be less indicative to user satisfaction with a recommender [5].

Once the set of good items has been picked and recommendations have been generated, the next step is to compute a *confusion matrix* for each user. A confusion matrix is a two by two matrix counting how many of the relevant items were recommended, how many of the relevant items were not recommended, and so forth (Table 2).

Table 2. A confusion matrix

	Good items	Not good items
Recommended items	True positives (tp)	False positives (fp)
Other items	False negatives (fn)	True negatives (tn)

There are several metrics to compute based on this given confusion matrix.

- *precision* - $\frac{tp}{tp+fp}$ - The percent of recommended items that are good
- *recall* (also known as *sensitivity*) - $\frac{tp}{tp+fn}$ - The percent of good items that are recommended
- *false positive rate* - $\frac{fp}{fp+tn}$ - the percent of not good items that are recommended
- *specificity* - $\frac{tn}{fp+tn}$ - The percent of not good items that are not recommended

These metrics, especially precision and recall, are traditionally reported and analyzed together. This is because precision and recall tend to have an inverse relationship. An algorithm can optimize for precision by making very few recommendations, but doing that would lead to a low recall. Likewise, an algorithm might get high recall by making very many recommendations, but this would

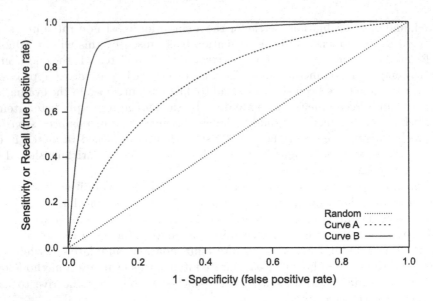

Fig. 6. An example ROC curve. Image used with permission from [21].

lead to low precision. An ideal algorithm will therefore want to balance these two properties finding a balance that recommends predominantly good items, and recommends almost all of the good items. To make finding this balance easier researchers often look at the *F-score*, which is the harmonic mean between precision and recall

$$F = \frac{2 * precision * recall}{precision + recall} \tag{36}$$

Another way to summarize this information is with the use of an ROC curve. An *ROC curve* is a plot of the recall on the y-axis against the false positive rate on the x-axis. An example ROC curve is given in Fig. 6. The ROC curve will have one point for every recommendation list length from recommending only one item to recommending all items. When the recommendation list is small, we expect a small false positive rate but also a small recall (hitting the point $(0, 0)$). Alternatively, when all items are recommended the false positive rate and recall will both be 1. Therefore the ROC curve normally connects the point $(0, 0)$ to $(1, 1)$. The ROC curve can be used to evaluate an algorithm broadly, with a good algorithm approaching the point $(0, 1)$ which means it has almost no false positives, while still recalling almost every good item. To make this property easier to compare numerically it is also common to compute the area under the ROC curve, referred to as the AUC metric. It has been pointed out [56] that the AUC metric also measures the fraction of pairs of items that are ranked correctly.

As mentioned earlier, it is often incorrect to assume that items that are not rated highly by a user are by definition bad items to recommend. That does not mean, however, that there are no bad recommendations, just like we can say highly rated (4 or more stars out of 5) items are clearly good, we can say poorly rated items (one or two stars out of five) are clearly bad. Using this insight, we can define the *fallout* metric. In a typical information retrieval evaluation fallout is the same as the false positive rate. In a recommender system evaluation, however, we can explicitly focus on how often bad items are recommended, and compute the percent of recommended items that are known to be bad. If one algorithm has a significantly higher fallout than another we can assume that it is making significant mistakes at a higher rate, and should be avoided.

One issue with the precision metric is that, while it rewards an algorithm for recommending good items, it does not care where those items are in a recommendation. Generally, we want good items recommended as early in the list as possible. To evaluate this we can use the *mean average precision metric* (MAP), which is the mean of the *average precision* over every user. The average precision metrics takes the average of the precision at each of the relevant items in the recommendation. If an item is not recommended then it contributes a precision of 0.

$$MAP@N = \frac{\sum_{u \in U} averagePrecision@N(u)}{|U|} \tag{37}$$

$$averagePrecision@N(u) = \frac{1}{|goodItems|} \sum_{i \in goodItems} precision@rank(i) \tag{38}$$

By taking the mean average precision we place more importance on the early items in the list than the later items, as the first item is used in all N precision computations, while the last one is only used in one.

Another approach for checking that good items are early in the recommendation is the *mean reciprocal rank metric* (MRR). Instead of looking at how many good items or bad items an algorithm returns, mean reciprocal rank looks at how many items the user has to consider before finding a good item. For any user, their rank ($rank_u$) is the position in the recommendation of the first relevant item. Based on this we can take reciprocal rank as $1/rank_u$ and mean reciprocal rank is the average reciprocal rank over all users. A larger mean reciprocal rank means that the average user should have to look at fewer items before finding an item they will enjoy.

$$MRR = \frac{\sum_{u \in U} 1/rank_u}{|U|} \tag{39}$$

9.4   Novelty and Diversity

There are several other properties of a recommendation that can be measured. The most commonly discussed are novelty and diversity. These properties are believed to be very important in determining whether a user will find a set of recommendations useful, even if they are unrelated to the pure quality of the recommendation. Understanding the effect these properties have on user satisfaction is still one of the ongoing directions in recommender systems research [18].

It is important to compare these metrics along with other metrics, such as accuracy or decision support metrics, as large values for these metrics are often seen along with large losses in quality. At the extreme a random recommender would have very high novelty and diversity, but would score bad on all other metrics. Generally speaking, good algorithms are those that increase novelty or diversity without meaningfully decreasing other measures of quality.

Novelty refers to how unexpected or unfamiliar the user is with their recommendations [53]. Recommendations that mostly contain familiar items are not considered novel recommendations. Since the goal of a recommender is to help its users find items they would not otherwise see, we expect that a good recommender should have higher novelty. The most obvious, and common, way to estimate novelty offline is to rely on some estimate of how well known an item is. The count of users who have rated an item, referred to as the item's popularity, is commonly used for this. More popular items are assumed to be better-known and less novel to recommend [11].

Diverse recommendations cover a large range of different items. One flaw with many recommender systems is that they focus too heavily on some small set of items for a given user [34]. So knowing that a user liked a movie from the Star Wars franchise, for example, might lead the algorithm to only recommend science fiction to a user, even if that user likes adventure films in general. The most common way of understanding how diverse a set of recommendations is, is to measure the total or average similarity between all pairs of items in the recommendation list [68, 70]. This inter-list similarity (ILS) measure can be seen as an inverse of diversity, the more similar recommended items are, the less diverse the recommendation is. Ideally a similarity function that is based on the item itself, or item meta-data is used, as it allows diversity to be measured independent of properties of the ratings and predictions. Where sim is the similarity function, the diversity of a recommendation list Rec can be defined as:

$$ILS(Rec) = \frac{2}{|Rec| * (|Rec| - 1)} \sum_{i \in Rec} \sum_{j \neq i \in Rec} sim(i, j) \qquad (40)$$

9.5 Structuring an Offline Evaluation

The heart of a good offline evaluation is how the train and test datasets are generated. Without a good process for splitting train and test datasets, the evaluation can fail to produce results, produce misleading results, or produce statistically insignificant results. To avoid this, simple standard approaches for generating train and test datasets have been developed. The standard approach to producing train and test datasets is user-based K-fold crossfolding.

In user-based K-fold crossfolding the users are split into K groups, where typical values of K are 5 or 10. For each of the K groups we generate a new train and test dataset split. For dataset split n, all users not in fold n are considered train users and all their ratings are allocated to the training dataset. The users in group n are then test users, and their ratings are split so that some can be in the train dataset (to inform the algorithm about that user's tastes) and the rest go to the test dataset. Typically either a constant number of ratings, such as ten per user, or a constant fraction, such as ten percent of ratings per user, are allocated for testing. Testing items can either be chosen randomly, or the most recent items can be chosen to emphasize the importance of the order in which a user makes ratings.

User-based K-fold crossfolding has several benefits. First, by performing a user based evaluation we know that each user will be evaluated once and only once. This ensures that our conclusions give equal weight to each user, with no user evaluated more or less than the others. Secondly, by ensuring that there are a large number of training users we know that we are evaluating the algorithm under a reasonably realistic condition, with a reasonable amount of training data. Finally, through replication we can measure statistical confidence around our metric values, as each train and test can be considered relatively independent. With user-based crossfolding it is common to treat each user as its own independent sample of the per-user error when computing statistical significance.

There are several other approaches for structuring an offline evaluation that have been used in the past. These approaches are typically designed to focus the evaluation on a single factor, or to support new and interesting metrics. For example, Kluver and Konstan used a modified crossfold strategy to look at how the algorithm changes as it learns more about users [37]. The key insight in this strategy is to initially perform a crossfold where the maximum number of training points is retained. New test sets with fewer training points per user can then be made by subsampling the training set, but leaving the test alone. By keeping the test set constant across different sample sizes, biases related to the size of the test set are avoided and a fair comparison can be drawn between algorithm properties at different number of training ratings for a user.

Another interesting approach is the temporal evaluation [42, 43]. Temporal evaluations have been used to look at properties of the recommender system over time as the collaborative filtering changes. In a temporal evaluation the dataset is split into N equal sized temporal windows. This allows $N - 1$ evaluations to be done for each window after the first where the train set is all windows before the given window, and the test set is the target window. Applying normal metrics this way can give you an understanding of how an algorithm might behave over time in a real deployment [42]. Temporal analysis also allows for interesting new metrics such as temporal diversity [43] that measure properties of how frequently recommendations change over time.

9.6 Online Evaluations

Not every comparison can be done without users. While offline evaluations are good for rapid development, online testing is needed to truly understand how a given system's users will respond to a given algorithm. Because the users are a central part of the recommender system, an algorithm that works well in an offline evaluation may have unexpected properties when it interacts with real users. Since it is these properties that determine which algorithm makes users most happy, we recommend that offline evaluation be used to choose a small set of algorithms and tunings that are then compared using a final online evaluation.

There are several ways to perform an online evaluation, each with its own benefits and drawbacks:

Lab Study - In a lab study users are brought into a computer lab and asked to go through a series of steps. These steps may involve using a real version of the website, an interactive survey, or an in depth interview. An example lab study might bring users into the lab so eye tracking can be performed to evaluate how well a recommender chooses eye catching content for the home page. Lab studies give the experimenter a large amount of control over what the user does. The drawback of this flexibility, however, is that lab studies often do not create a realistic environment for how a recommender system might be used. Additionally, lab studies are typically limited in how many users they can involve as they often require space and supervision from an experimenter.

Virtual Lab Study - Virtual lab studies are similar to lab studies but are performed entirely online and without the direct supervision of the experimenter. This deployment trades some of the control of a lab study for much larger scale, as virtual lab studies can involve many more users in the same amount of time as a lab study. These normally take the form of purpose built web services that interact with the recommender and guide the user through a series of actions and questions. While almost any form of data about user behavior and preference can be used with a virtual lab study, surveys are particularly popular. An example virtual lab study might guide users through rating on several different interfaces and then survey users to find out which they prefered. A well designed survey can be easy for a user to complete remotely, and very informative about how users evaluate the recommender system.

Online Field Study - Online field studies focus on studying how people use a deployed recommender system. Generally there are two approaches to online field studies. In the first, existing log data from a deployed system is used to try to understand how users have been behaving on the system. For example, rating data from a system could be analyzed to understand how often users enter a rating under a current system.

Alternatively, a change can be made to a deployed system to answer a specific question. New forms of logging could be introduced, or new experimental features deployed for a trail period. This can allow answering more specific questions about user behavior. For example, a book recommender system might set the goal that 80% or more of users find a book within 5 min of accessing the service. Logging could be added to the system to allow measuring how often users find books, and in how long so that performance can be directly compared with this goal.

Online field studies are ideal for understanding how a system is used or for measuring progress against some goal for the recommender system as a whole. Online field studies are not as suited for comparing different options. Likewise, online field studies are limited by their connection to a deployed live system, which might preclude studying possibly disruptive changes. Finally, online field studies often do not allow asking follow-up questions without resorting to a secondary evaluation, for example, while a researcher might know what users do in a situation, they will not be able to ask why.

A-B Test - A-B tests can be seen as an extension of an online field study. In an A-B test two or more versions of an algorithm or interface are deployed to a given website, with any user seeing only one of these versions. By tracking the behavior of these users, a researcher can identify differences in how users interact with the algorithm in a realistic setting. An example A-B test might deploy two algorithms to a recommender service for two months and then look at user retention rates. If one algorithm leads to fewer users having another visit, then we can say that algorithm is likely worse. Being an experimental extension to online field studies, A-B tests share the same weaknesses: it can be hard to understand what is causing the results that it finds.

Within these options virtual lab studies are most common when the goal is to understand why users perceive algorithms differently, and A-B tests are preferred when the goal is simply to pick the "best" algorithm by one or more user behavior metrics.

Online evaluations are a complicated subject, and the description here only scratches the surface. Several texts are available that go into depth on the various ways to design a user experiment to answer any number of questions, including those of algorithm performance. In particular, we recommend the book "Ways of knowing in HCI" [50], which contains in depth coverage of a broad range of research methods which are applicable in this scenario. With that said, there are some considerations that are specific to recommender systems which we will discuss.

Since all recommender systems require a history of user data to work with, ensuring that this information is available is important when considering the design of a study. This decision is closely tied to how participants for the study will be recruited. If recruiting from an existing recommender system a lab study or virtual lab study can simply request user account information and use that to access ratings. Better yet, in a virtual lab study, links to the study could be pre-generated with a user-specific code so that users do not need to log in.

If recruitment for a lab study does not come from an existing recommender system, than typically the first stage of the study is to collect enough ratings to make useful recommendations. As this is the same problem faced when onboarding users to a recommender system this is a well studied problem. The common solution is to present users with a list of items to rate [23,54,55]. There are many different ways to pick which order to show items in to optimize how useful the ratings are for recommending, the time that it takes to get a certain number of ratings, or both. For a good review of this field of work see the 2011 study by Golbandi et al. [23], which also includes an example of an adaptive solution to collecting useful user ratings. If time is not a major factor, however, a design where users search to pick items to represent their tastes may have some benefits [48].

Just as with offline evaluations, comparing reasonable algorithms is essential to producing useful results. Starting with offline evaluation methodologies can be a good way to make sure only the best algorithms are compared. When comparing novel algorithms, it can often help to include a baseline algorithm as a comparison point whose behavior is relatively well known.

When using a survey, often as part of a virtual lab study, it is important to choose well written questions. Some questions can be ambiguous making it hard to assign meaning to their answers. A survey question is useless to an experimenter if many users do not understand it. To support experimenters, several researchers have put together frameworks to support online, user centered evaluations of recommender systems [39,53]. These frameworks split the broader subject of how well the algorithm works into smaller factors tuned for specific properties of a recommender system. Each of these factors are also associated with well designed survey questions which are known to accurately capture a user's opinion. When possible we recommend using one of these frameworks as a resource when designing an online evaluation.

9.7   Resources for Algorithm Evaluation

There are plenty of resources available to help explore ratings-based collaborative filtering recommender systems. In particular, there are open source recommender algorithm implementations, and rating datasets available for non-commercial use. These resources have been developed in an effort to help reduce the cost of research and development in collaborative filtering.

In recent years there has been a push in the recommender systems community to support reproducible recommender systems research. One major result of this call has been several open source collections of recommender systems algorithms such as LensKit[6], Mahout[7], and MyMediaLight[8]. By publishing working code for an algorithm the researchers can ensure that every detail of an algorithm is public, and that two people comparing the same algorithm don't get different results due to implementation details. More importantly, however, these toolkits allow programmers who are not recommender systems experts to use and learn about recommender systems algorithms and benefit from the work of the research community.

The other noteworthy resource for exploring and evaluating recommender systems is the availability of public ratings datasets. These datasets are released by live recommender systems to allow people without direct access to a recommender service and its user base to participate in recommender system development. Some notable examples of datasets are the MovieLens movie rating datasets[9], described in detail in Harper and Konstan's 2015 paper [26], the Jester joke rating dataset[10] described in the 2001 paper by Goldberg et al. [24], the Book-Crossing book rating dataset[11] introduced in 2005 paper by Ziegler et al. [70] and the Amazon product rating and review dataset[12] first introduced in

---

[6] https://lenskit.org/.
[7] https://mahout.apache.org/.
[8] http://mymedialite.net/.
[9] available at http://grouplens.org/datasets/movielens/.
[10] available (with updates) at http://eigentaste.berkeley.edu/dataset/.
[11] available at http://www2.informatik.uni-freiburg.de/~cziegler/BX/.
[12] available at http://jmcauley.ucsd.edu/data/amazon/.

McAuley and Leskovec's 2013 paper [47]. These datasets, and many more available online, are available for anyone to download and use to learn about, and experiment with, recommender system algorithms. Most of the recommender systems toolkits have code for loading these datasets and performing offline evaluations already built.

While most directly applicable towards offline evaluations, these resources can also help with online evaluations. Open source libraries can be used to quickly prototype recommender systems either for incorporation in an existing live system or for a lab or virtual lab study. Likewise, research datasets can be used to power a collaborative filtering algorithm for use in a lab or virtual lab study design in which new ratings will be collected from research participants. This can allow anyone with access to research participants to perform online evaluations to deeply understand how users react to collaborative filtering technologies.

10   Conclusions

In this chapter we have presented the central concepts, algorithms, and means of evaluation in ratings based collaborative filtering. While recommendation systems construed broadly is still an active area of research, research on pure ratings-based collaborative filtering algorithms is becoming more rare. Indeed, it is increasingly rare to see a new pure ratings-based algorithm make a significant improvement in offline evaluations. Instead, research into collaborative filtering recommender systems has started focusing on new problems and new sources of information. Many of these more recent directions for collaborative filtering research have become the basis for the future chapters in this book.

The next five chapters of this book will explore more advanced techniques for recommendation based on different forms of information and recommendation tasks. Taken as a whole, these chapters serve as an excellent introduction to the state of the art in collaborative filtering recommender systems. Chapter 11 will explore recommendation based on online social networking [44], Chap. 12 will explore recommendation based user volunteered tags [8], Chap. 13 will explore recommendation based on publicly shared user opinions on sites like Amazon or Twitter [51], Chap. 14 will explore recommendation based on implicit feedback [35], and finally, Chap. 15 will explore how to use a recommender algorithms to recommend person-to-person connections in online social websites [25].

References

1. Aggarwal, C.C., Wolf, J.L., Wu, K.L., Yu, P.S.: Horting hatches an egg: a new graph-theoretic approach to collaborative filtering. In: Proceedings of the Fifth ACM SIGKDD International Conference on Knowledge Discovery and Data Mining, KDD 1999, pp. 201–212. ACM. (1999). https://doi.org/10.1145/312129.312230
2. Amatriain, X., Basilico, J.: Netflix recommendations: beyond the 5 stars (part 1). http://techblog.netflix.com/2012/04/netflix-recommendations-beyond-5-stars.html

3. Bell, R.M., Koren, Y.: Scalable collaborative filtering with jointly derived neighborhood interpolation weights. In: Proceedings of the 2007 Seventh IEEE International Conference on Data Mining, ICDM 2007, pp. 43–52. IEEE Computer Society (2007). https://doi.org/10.1109/ICDM.2007.90

4. Bellogin, A.: Performance prediction and evaluation in recommender systems: an information retrieval perspective. Ph.D. thesis. Universidad Autnoma de Madrid (2012)

5. Bellogin, A., Castells, P., Cantador, I.: Precision-oriented evaluation of recommender systems: an algorithmic comparison. In: Proceedings of the Fifth ACM Conference on Recommender Systems, RecSys 2011, pp. 333–336. ACM (2011). https://doi.org/10.1145/2043932.2043996

6. Bieganski, P., Konstan, J., Riedl, J.: System, method and article of manufacture for making serendipity-weighted recommendations to a user, 25 December 2001. US Patent 6,334,127

7. Blei, D.M., Ng, A.Y., Jordan, M.I.: Latent Dirichlet allocation. J. Mach. Learn. Res. 3, 993–1022. http://dl.acm.org/citation.cfm?id=944919.944937

8. Bogers, T.: Tag-based recommendation. In: Brusilovsky, P., He, D. (eds.) Social Information Access. LNCS, vol. 10100, pp. 441–479. Springer, Cham (2018)

9. Breese, J.S., Heckerman, D., Kadie, C.: Empirical analysis of predictive algorithms for collaborative filtering. Technical report MSR-TR-98-12, Microsoft Research, May 1998. http://research.microsoft.com/apps/pubs/default.aspx?id=69656

10. Breiman, L.: Stacked regressions. Mach. Learn. **24**(1), 49–64 (1996). https://doi.org/10.1023/A:1018046112532

11. Celma, Ò., Herrera, P.: A new approach to evaluating novel recommendations. In: Proceedings of the 2008 ACM Conference on Recommender Systems, RecSys 2008, pp. 179–186. ACM (2008). https://doi.org/10.1145/1454008.1454038

12. Chen, T., Zhang, W., Lu, Q., Chen, K., Zheng, Z., Yu, Y.: SVDFeature: a toolkit for feature-based collaborative filtering. J. Mach. Learn. Res. **13**(1), 3619–3622 (2012). http://dl.acm.org/citation.cfm?id=2503308.2503357

13. Cremonesi, P., Garzottto, F., Turrin, R.: User effort vs. accuracy in rating-based elicitation. In: Proceedings of the Sixth ACM Conference on Recommender Systems, RecSys 2012, pp. 27–34. ACM (2012). https://doi.org/10.1145/2365952.2365963

14. Cremonesi, P., Koren, Y., Turrin, R.: Performance of recommender algorithms on top-n recommendation tasks. In: Proceedings of the Fourth ACM Conference on Recommender Systems RecSys 2010, pp. 39–46. ACM (2010). https://doi.org/10.1145/1864708.1864721

15. Ekstrand, M.: Similarity functions for user-user collaborative filtering. http://grouplens.org/blog/similarity-functions-for-user-user-collaborative-filtering/

16. Ekstrand, M.: Similarity functions in item-item CF. https://md.ekstrandom.net/blog/2015/06/item-similarity/

17. Ekstrand, M., Riedl, J.: When recommenders fail: predicting recommender failure for algorithm selection and combination. In: Proceedings of the Sixth ACM Conference on Recommender Systems, RecSys 2012, pp. 233–236. ACM (2012). https://doi.org/10.1145/2365952.2366002

18. Ekstrand, M.D., Harper, F.M., Willemsen, M.C., Konstan, J.A.: User perception of differences in recommender algorithms. In: Proceedings of the 8th ACM Conference on Recommender Systems, RecSys 2014, pp. 161–168. ACM (2014). https://doi.org/10.1145/2645710.2645737

19. Ekstrand, M.D., Kluver, D., Harper, F.M., Konstan, J.A.: Letting users choose recommender algorithms: an experimental study. In: Proceedings of the 9th ACM Conference on Recommender Systems, RecSys 2015, pp. 11–18. ACM (2015). https://doi.org/10.1145/2792838.2800195

20. Ekstrand, M.D., Ludwig, M., Konstan, J.A., Riedl, J.T.: Rethinking the recommender research ecosystem: reproducibility, openness, and LensKit. In: Proceedings of the Fifth ACM Conference on Recommender Systems, RecSys 2011, pp. 133–140. ACM (2011). https://doi.org/10.1145/2043932.2043958

21. Ekstrand, M.D., Riedl, J.T., Konstan, J.A.: Collaborative filtering recommender systems. Found. Trends Hum.-Comput. Interact. 4(2), 81–173 (2011). https://doi.org/10.1561/1100000009

22. Funk, S.: Netflix update: try this at home. http://sifter.org/~simon/journal/20061211.html

23. Golbandi, N., Koren, Y., Lempel, R.: Adaptive bootstrapping of recommender systems using decision trees. In: Proceedings of the Fourth ACM International Conference on Web Search and Data Mining, WSDM 2011, pp. 595–604. ACM (2011). https://doi.org/10.1145/1935826.1935910

24. Goldberg, K., Roeder, T., Gupta, D., Perkins, C.: Eigentaste: a constant time collaborative filtering algorithm. Inf. Retrieval 4(2), 133–151 (2001). https://doi.org/10.1023/A:1011419012209

25. Guy, I.: People recommendation on social media. In: Brusilovsky, P., He, D. (eds.) Social Information Access. LNCS, vol. 10100, pp. 570–623. Springer, Cham (2018)

26. Harper, F.M., Konstan, J.A.: The MovieLens datasets: history and context. ACM Trans. Interact. Intell. Syst. 5(4), 19:1–19:19 (2015). https://doi.org/10.1145/2827872

27. Herlocker, J., Konstan, J.A., Riedl, J.: An empirical analysis of design choices in neighborhood-based collaborative filtering algorithms. Inf. Retrieval 5(4), 287–310 (2002). https://doi.org/10.1023/A:1020443909834

28. Herlocker, J.L., Konstan, J.A., Terveen, L.G., Riedl, J.T.: Evaluating collaborative filtering recommender systems. ACM Trans. Inf. Syst. 22(1), 5–53 (2004). https://doi.org/10.1145/963770.963772

29. Hill, W., Stead, L., Rosenstein, M., Furnas, G.: Recommending and evaluating choices in a virtual community of use. In: Proceedings of the SIGCHI Conference on Human Factors in Computing Systems, CHI 1995, pp. 194–201. ACM Press/Addison-Wesley Publishing Co. (1995). https://doi.org/10.1145/223904.223929

30. Hofmann, T.: Probabilistic latent semantic indexing. In: Proceedings of the 22nd Annual International ACM SIGIR Conference on Research and Development in Information Retrieval, SIGIR 1999, pp. 50–57. ACM (1999). https://doi.org/10.1145/312624.312649

31. Hofmann, T.: Latent semantic models for collaborative filtering. ACM Trans. Inf. Syst. 22(1), 89–115 (2004). https://doi.org/10.1145/963770.963774

32. Huang, Z., Chung, W., Chen, H.: A graph model for E-commerce recommender systems. J. Am. Soc. Inf. Sci. Technol. 55(3), 259–274 (2004). https://doi.org/10.1002/asi.10372

33. Jamali, M., Ester, M.: TrustWalker: a random walk model for combining trust-based and item-based recommendation. In: Proceedings of the 15th ACM SIGKDD International Conference on Knowledge Discovery and Data Mining, KDD 2009, pp. 397–406. ACM (2009). https://doi.org/10.1145/1557019.1557067

34. Jannach, D., Lerche, L., Gedikli, F., Bonnin, G.: What recommenders recommend – an analysis of accuracy, popularity, and sales diversity effects. In: Carberry, S., Weibelzahl, S., Micarelli, A., Semeraro, G. (eds.) UMAP 2013. LNCS, vol. 7899, pp. 25–37. Springer, Heidelberg (2013). https://doi.org/10.1007/978-3-642-38844-6_3

35. Jannach, D., Lerche, L., Zanker, M.: Recommending based on implicit feedback. In: Brusilovsky, P., He, D. (eds.) Social Information Access. LNCS, vol. 10100, pp. 510–569. Springer, Cham (2018)

36. Järvelin, K., Kekäläinen, J.: Cumulated gain-based evaluation of IR techniques. ACM Trans. Inf. Syst. **20**(4), 422–446 (2002). https://doi.org/10.1145/582415.582418

37. Kluver, D., Konstan, J.A.: Evaluating recommender behavior for new users. In: Proceedings of the 8th ACM Conference on Recommender Systems, RecSys 2014, pp. 121–128. ACM (2014). https://doi.org/10.1145/2645710.2645742

38. Kluver, D., Nguyen, T.T., Ekstrand, M., Sen, S., Riedl, J.: How many bits per rating? In: Proceedings of the Sixth ACM Conference on Recommender Systems, RecSys 2012, pp. 99–106. ACM (2012). https://doi.org/10.1145/2365952.2365974

39. Knijnenburg, B., Willemsen, M., Gantner, Z., Soncu, H., Newell, C.: Explaining the user experience of recommender systems. User Model. User-Adap. Interact. **22**(4), 441–504. https://doi.org/10.1007/s11257-011-9118-4

40. Koren, Y.: Factorization meets the neighborhood: a multifaceted collaborative filtering model. In: Proceedings of the 14th ACM SIGKDD International Conference on Knowledge Discovery and Data Mining, KDD 2008, pp. 426–434. ACM. https://doi.org/10.1145/1401890.1401944

41. Koren, Y., Bell, R., Volinsky, C.: Matrix factorization techniques for recommender systems. Computer **42**(8), 30–37 (2009). https://doi.org/10.1109/MC.2009.263

42. Lathia, N., Hailes, S., Capra, L.: Temporal collaborative filtering with adaptive neighbourhoods. In: Proceedings of the 32nd International ACM SIGIR Conference on Research and Development in Information Retrieval, SIGIR 2009, pp. 796–797. ACM (2009). https://doi.org/10.1145/1571941.1572133

43. Lathia, N., Hailes, S., Capra, L., Amatriain, X.: Temporal diversity in recommender systems. In: Proceedings of the 33rd International ACM SIGIR Conference on Research and Development in Information Retrieval, SIGIR 2010, pp. 210–217. ACM (2010). https://doi.org/10.1145/1835449.1835486

44. Lee, D., Brusilovsky, P.: Recommendations based on social links. In: Brusilovsky, P., He, D. (eds.) Social Information Access. LNCS, vol. 10100, pp. 391–440. Springer, Cham (2018)

45. Lemire, D., Maclachlan, A.: Slope one predictors for online rating-based collaborative filtering. In: Proceedings of SIAM Data Mining (SDM 2005) (2005). https://arxiv.org/abs/cs/0702144

46. Massa, P., Avesani, P.: Trust-aware recommender systems. In: Proceedings of the 2007 ACM Conference on Recommender Systems, RecSys 2007, pp. 17–24. ACM (2007). https://doi.org/10.1145/1297231.1297235

47. McAuley, J., Leskovec, J.: Hidden factors and hidden topics: Understanding rating dimensions with review text. In: Proceedings of the 7th ACM Conference on Recommender Systems, pp. 165–172. RecSys 2013. ACM (2013). https://doi.org/10.1145/2507157.2507163

48. McNee, S.M., Lam, S.K., Konstan, J.A., Riedl, J.: Interfaces for eliciting new user preferences in recommender systems. In: Brusilovsky, P., Corbett, A., de Rosis, F. (eds.) UM 2003. LNCS (LNAI), vol. 2702, pp. 178–187. Springer, Heidelberg (2003). https://doi.org/10.1007/3-540-44963-9_24

49. Ning, X., Karypis, G.: Slim: Sparse linear methods for top-n recommender systems. In: Proceedings of the IEEE 11th International Conference on Data Mining, ICDM 2011, pp. 497–506, December 2011. https://doi.org/10.1109/ICDM.2011.134
50. Olson, J.S., Kellogg, W.A. (eds.): Ways of Knowing in HCI. Springer, New York (2014). https://doi.org/10.1007/978-1-4939-0378-8
51. O'Mahoney, M., Smyth, B.: From opinions to recommendations. In: Brusilovsky, P., He, D. (eds.) Social Information Access. LNCS, vol. 10100, pp. 480–509. Springer, Cham (2018)
52. Phuong, N.D., Thang, L.Q., Phuong, T.M.: A graph-based method for combining collaborative and content-based filtering. In: Ho, T.-B., Zhou, Z.-H. (eds.) PRICAI 2008. LNCS (LNAI), vol. 5351, pp. 859–869. Springer, Heidelberg (2008). https://doi.org/10.1007/978-3-540-89197-0_80
53. Pu, P., Chen, L., Hu, R.: A user-centric evaluation framework for recommender systems. In: Proceedings of the Fifth ACM Conference on Recommender Systems, RecSys 2011, pp. 157–164. ACM (2011). https://doi.org/10.1145/2043932.2043962
54. Rashid, A.M., Albert, I., Cosley, D., Lam, S.K., McNee, S.M., Konstan, J.A., Riedl, J.: Getting to know you: learning new user preferences in recommender systems. In: Proceedings of the 7th International Conference on Intelligent User Interfaces, IUI 2002, pp. 127–134. ACM (2002). https://doi.org/10.1145/502716.502737
55. Rashid, A.M., Karypis, G., Riedl, J.: Learning preferences of new users in recommender systems: an information theoretic approach. ACM SIGKDD Explor. Newslett. 10(2), 90–100 (2008). https://doi.org/10.1145/1540276.1540302
56. Rendle, S., Freudenthaler, C., Gantner, Z., Schmidt-Thieme, L.: BPR: Bayesian personalized ranking from implicit feedback. In: Proceedings of the Twenty-Fifth Conference on Uncertainty in Artificial Intelligence, UAI 2009, pp. 452–461. AUAI Press. http://dl.acm.org/citation.cfm?id=1795114.1795167
57. Rendle, S., Schmidt-Thieme, L.: Online-updating regularized kernel matrix factorization models for large-scale recommender systems. In: Proceedings of the 2008 ACM Conference on Recommender Systems, RecSys 2008, pp. 251–258. ACM (2008). https://doi.org/10.1145/1454008.1454047
58. Resnick, P., Iacovou, N., Suchak, M., Bergstrom, P., Riedl, J.: GroupLens: an open architecture for collaborative filtering of netnews. In: Proceedings of the 1994 ACM Conference on Computer Supported Cooperative Work, CSCW 1994, pp. 175–186. ACM (1994). https://doi.org/10.1145/192844.192905
59. Rogers, S.K.: Item-to-item recommendations at Pinterest. In: Proceedings of the 10th ACM Conference on Recommender Systems, RecSys 2016, pp. 393–393. ACM (2016). https://doi.org/10.1145/2959100.2959130
60. Said, A., Fields, B., Jain, B.J., Albayrak, S.: User-centric evaluation of a k-furthest neighbor collaborative filtering recommender algorithm. In: Proceedings of the 2013 Conference on Computer Supported Cooperative Work, CSCW 2013, pp. 1399–1408. ACM (2013). https://doi.org/10.1145/2441776.2441933
61. Sarwar, B., Karypis, G., Konstan, J., Riedl, J.: Item-based collaborative filtering recommendation algorithms. In: Proceedings of the 10th International Conference on World Wide Web, WWW 2001, pp. 285–295. ACM. https://doi.org/10.1145/371920.372071
62. Sarwar, B.M., Karypis, G., Konstan, J.A., Riedl, J.T.: Application of dimensionality reduction in recommender system - a case study. In: WebKDD 2000. http://citeseerx.ist.psu.edu/viewdoc/summary?doi=10.1.1.29.8381
63. Shan, H., Banerjee, A.: Generalized probabilistic matrix factorizations for collaborative filtering. In: IEEE International Conference on Data Mining, pp. 1025–1030. IEEE Computer Society (2010). https://doi.org/10.1109/ICDM.2010.116

64. Sill, J., Takacs, G., Mackey, L., Lin, D.: Feature-weighted linear stacking. arXiv:0911.0460 [cs], http://arxiv.org/abs/0911.0460
65. Smyth, B., McClave, P.: Similarity vs. diversity. In: Aha, D.W., Watson, I. (eds.) ICCBR 2001. LNCS (LNAI), vol. 2080, pp. 347–361. Springer, Heidelberg (2001). https://doi.org/10.1007/3-540-44593-5_25
66. Su, X., Khoshgoftaar, T.M.: A survey of collaborative filtering techniques. Adv. Artif. Intell. (2009). http://www.hindawi.com/journals/aai/2009/421425/abs/
67. Weng, L.T., Xu, Y., Li, Y., Nayak, R.: Improving recommendation novelty based on topic taxonomy. In: 2007 IEEE/WIC/ACM International Conferences on Web Intelligence and Intelligent Agent Technology Workshops, pp. 115–118 (2007)
68. Zhang, M., Hurley, N.: Avoiding monotony: improving the diversity of recommendation lists. In: Proceedings of the 2008 ACM Conference on Recommender Systems, RecSys 2008, pp. 123–130. ACM (2008). https://doi.org/10.1145/1454008.1454030
69. Zhang, Y.C., Saghdha, D.Ò., Quercia, D., Jambor, T.: Auralist: introducing serendipity into music recommendation. In: Proceedings of the Fifth ACM International Conference on Web Search and Data Mining, WSDM 2012, pp. 13–22. ACM (2012). https://doi.org/10.1145/2124295.2124300
70. Ziegler, C.N., McNee, S.M., Konstan, J.A., Lausen, G.: Improving recommendation lists through topic diversification. In: Proceedings of the 14th International Conference on World Wide Web, WWW 2005, pp. 22–32. ACM (2005). https://doi.org/10.1145/1060745.1060754

11
Recommendations Based on Social Links

Danielle Lee[1]([✉]) [iD] and Peter Brusilovsky[2] [iD]

[1] Department of Software, Sangmyung University, 31, Sangmyungdae-gil,
Dongnam-gu, Cheonan-si, Chungcheongnam-do 31066, Korea
suleehs@gmail.com
[2] School of Computing and Information, University of Pittsburgh,
135 North Bellefield Avenue, Pittsburgh, PA 15260, USA
peterb@pitt.edu

Abstract. The goal of this chapter is to give an overview of recent works on the development of social link-based recommender systems and to offer insights on related issues, as well as future directions for research. Among several kinds of social recommendations, this chapter focuses on recommendations, which are based on users' *self-defined* (i.e., explicit) social links and suggest items, rather than people of interest. The chapter starts by reviewing the needs for social link-based recommendations and studies that explain the viability of social networks as useful information sources. Following that, the core part of the chapter dissects and examines modern research on social link-based recommendations along several dimensions. It concludes with a discussion of several important issues and future directions for social link-based recommendation research.

1 Introduction

The remarkable popularity of social media encourages Web users to participate in various online activities, and generates data on an unprecedented scale. Given the exponentially growing volumes of social media data, personalized recommendation technologies serve as a solution for the information overload problem, since they offer a positive user experience with more relevant content and give a competitive advantage to the social media business [122]. Paradoxically, while increasing the information overload, social media also offers a source of data to address it. Social media systems encourage users to contribute various kinds of information (such as tags, reviews, and social links) that could be efficiently used to improve the quality of recommendations. This chapter focuses on one kind of this socially contributed information: users' *self-defined* (i.e., explicit) social links.

Compared to the early Web era when users stayed in isolation, the new Web is increasingly social. A growing number of social media systems enabled Web users to explore socially-shared information, find people of interest, and establish various kinds of social connections with them. Web users' active participation in social media is motivated by both their information needs and their social desire for engagement and communication with likeminded people [98]. For many Web users, participation in social media becomes an essential part of their everyday lives. In particular, more and

P. Brusilovsky and D. He (Eds.): Social Information Access, LNCS 10100, pp. 391–440, 2018.
https://doi.org/10.1007/978-3-319-90092-6_11

more users share various resources with their social connections; for instance, movies to watch, books to read, academic papers to refer to, bookmarks to explore, music concerts to enjoy, and so on. Online social links provide a rich source of useful information and knowledge and, in turn, are used to propagate various kinds of information [141]. According to prominent social science theories (homophily and social influence, introduced in Sect. 2), we have a strong tendency to associate and bond with others who are similar to us, and are affected by our social links. They may cause changes in our attitudes, beliefs, and behavioral propensities [23]. In seeking useful sources to enrich users' information preference models and to acquire desirable information for personalized recommendations, researchers are more and more frequently considering users' online social links. This chapter reviews this area of research, which focuses on recommender technologies that leverage users' self-designed social links to recommend desirable *items*, not *people* of interest. It attempts to dissect and examine existing social link-based recommendation technologies, as well as to discuss related problems.

The type of recommendations reviewed in this chapter could be classified as *social recommendations,* an actively evolving field of research at the crossroads of recommender systems and social technologies. In addition to recommendation based on explicit social links reviewed in this chapter, this field include technologies that utilize *implicitly* inferred social links as foundations of recommendations; suggest person(s) to be connected (as introduced in Chap. 15 [51]); and recommend items to a *group of users* (for more detailed explanations of the various kinds of social recommendations, refer to Sect. 2).

The chapter is organized as following: Sect. 2 systematically reviews various ways to employ online social networks for personalized recommendations. Section 3 discusses the challenges of traditional collaborative filtering technology and provides a rationale for using online social links as an information source. Sections 4 and 5 provide an overview of multiple aspects of social link-based recommendations including algorithms, and Sect. 6 discusses other issues related to social link-based recommendations.

2   A Range of Definitions of Social Recommendations

Given the widely ranging definitions of social recommendation studies, before we embark on a discussion of social link-based recommendations, it is worth taking a look at the various streams of related research. It helps to more clearly limit the scope of social link-based recommendation technologies to be considered in this chapter. Because personalized recommendations on social media are a currently evolving matter [116], social information has been used for recommendation in different ways. For instance, one direction of research focuses on generating recommendations of information *items*, while another direction focuses on recommending *people* to connect with or *groups* to join. The kinds of social networks used in recommendations also vary; they range from explicit social connections among humans to artificial networks of intelligent machines. When considering the target information of recommendation and the types of social links used for recommendation, we could classify social recommendation research into the following five directions.

- Item recommendation using users' explicit (i.e. self-defined) social links
- Item recommendation using users' implicit social links
- Recommendations of trustworthy communication partners (i.e. machines) using artificial links
- People recommendation (based on all kinds of sources, including social links)
- Item recommendation in social media systems without social links

The studies in the first direction exploited social links that were explicitly created by target users as the foundations for their recommendations. This chapter mainly focuses on these kinds of social recommendations. The explicitly defined social connections are important, because the connections develop social phenomena through social interactions, as introduced in the previous section. That is, once users establish social connections explicitly, they tend to pay attention to their partners' online activities. Hence, users are easily affected by their social connections, and as a result, recommender systems are able to take advantage of social interactions and the resultant social phenomena in the personalization process. The typical collaborative filtering recommendations assume that the roles of all users in a recommender system are equal, hence the term 'role uniformity [124].' However, understanding recommendations and accepting suggestions is a decision-making process. Every user has different interests, levels of knowledge, and especially, different social context and social roles. For this reason, target users (i.e. recipients of recommendations) expect transparent explanations about where the recommended information comes from.

The second direction comprises studies of recommendations that are based on users' implicit social links. Implicit social connections between users are not established by the users themselves, but instead are inferred by analytical approaches (such as machine learning) by examining users' various online interactions (e.g. reviewing the same products, commenting on the same items, befriending similar group of people, etc.). Depending on the machine-learning method, different kinds of implicit social partners can be identified. Thus, the choice of machine learning methods defines users' implicit social networks, to some extent. Lumbreras and Gavaldà [79], Pitsilis and Knapskog [102] and Victor et al. [129] belong to this area of recommendations. A practical reason why researchers chose implicit social networks instead of explicit ones is the lack of available data sources, which includes explicit links (i.e., includes both users' favorite items and their social connections).

The studies that belong to the third direction focus on how to select reliable communication partners (i.e. computer agents) to exchange information between machines, not to take advantage of social links among human beings. Depending on the degree to which one machine is trustworthy to another machine, the researchers interpreted machine-to-machine connections as a trust-based network and computed how the trust values are inferred and propagated. Lam [71] and Shi et al. [111] provide examples of this in action.

The fourth direction aims to suggest people of interest, instead of information items or products of interest. The recommended people could be a male/female to date [2, 70, 103], a person to befriend [4, 37], or a colleague to work with [77, 138]. Chap. 15 [51] offers a thorough review of people recommendations.

Finally, there are some studies that do not use any social links directly, but are referred to as 'social recommendations,' since they use data sources provided by social media and use collective intelligent mechanisms. The technologies in these studies are traditional collaborative filtering or content-based recommendations without any social context. Bellogin and Parapar [9], Debnath et al. [31], Diaz-Aviles et al. [35], Messenger and Whittle [91], Pazos Arias et al. [100], Sanchez et al. [106], Yoon et al. [143], Zhou et al. [148] and Ziqi et al. [150] are all examples of this type of study. Some of these studies are reviewed in other chapters of this book, such as Chap. 12 [12], Chap. 13 [97], and Chap. 14 [57].

Finally, we should mention studies on 'group-based recommendation', which we have not included in the list above, since these studies are not referred to as social recommendations. Group-based recommendations aggregate the preferences of a group of people and suggest recommendations to a whole group, not an individual member [21]. For example, once users enter a room, Flytrap identifies each of them via RFID, aggregates all their preferences together, computes the probabilistic values of all songs for recommendations, and finally plays music for people in the same physical space [25]. Other existing group-based recommendations can suggest TV programs or movies to watch for a whole family [61, 66], venues and routes to travel together [27], restaurants to enjoy together [99], points of interest based on users' visited locations [132], recipes to cook for a family [10], or a community to join [3, 134]. Some of these studies account for links within the group and bear some similarities to the studies reviewed in this chapter.

In this broader context, we limit the main body of this chapter to the studies that satisfy four major criteria: (1) they should suggest items of interest; (2) the target recipients of their recommendations should be individual users; (3) the personalization should account for the opinions of users' social connections; and (4) users should explicitly define their social networks.

3   Background

3.1   Challenges of Traditional Collaborative Filtering Recommendations

Collaborative filtering is currently the most popular among several core recommendation technologies. Originally envisioned as *'word-of-mouth'* automation [110], it starts with finding a neighborhood of likeminded users (known as *peers*) who have similar interests to a target user (i.e., a recipient of recommendations) and then recommends items to the target user that are favored by the peers. An extensive review of rating-based collaborative filtering can be found in Chap. 10 of this book [63]. The inherent ability of collaborative filtering to harness ratings of multiple automatically identified peers (who are not even known to the target user) gives this technology an impressive power—while also serving as a source of several recognized problems [11].

The first problem is a striking vulnerability of collaborative filtering to *shilling attacks* and *copy-profile attacks*. In order to reinforce their own ratings and intentionally distort recommendation predictions toward selected directions, a malicious user can create multiple bogus user profiles and insert fake user-item ratings, which gives these users the capability to promote or defame a certain product. It is referred to as a 'shilling attack' [49, 146, 151]. A malicious user who wants to bias or damage recommendations given to *a specific user* can create very similar rating profiles by copying the specific user's ratings. Naturally, a collaborative recommender system will pick these bogus "users" as perfect peers, and any new items positively rated by these peers will be highly recommended to the specific user [89]. It is known as a 'copy-profile attack'.

The second problem is known as the '*cold-start/new-user problem*' [24, 89]. It refers to a situation when recently joined users have not yet rated enough items. In this situation, recommender systems can't comprehend users' preferences reliably and can't generate quality recommendations. A similar problem known as the 'gray sheep problem' affects users with eccentric taste (so-called 'gray sheep users'). Since these users have low taste similarity with other users, collaborative recommenders might not be able to find useful peers for a grey sheep user, even with a considerable number of ratings [117].

Many collaborative filtering systems are also affected by '*data-sparsity problem*' [11, 96]. When the number of items in a system is relatively high in respect to the number of users, user-item rating matrix is too sparse to find sufficient number of co-rated items among users. In particular, data sparsity makes it hard to use collaborative filtering in cases when items have a short life cycle (e.g., job openings, events, or news articles). These items might simply have too little time to accumulate enough ratings before their values expire, which makes the user-item rating distribution comparatively sparse. Finally, classic collaborative filtering incurs *high computational costs*, because it compares one target user's taste with the tastes of all other users [116].

The problems reviewed above stem from one of the two core principles of collaborative filtering: the automatic selection of anonymous peers based on rating similarity. Consequently, one way to resolve these problems could be to unlock the black box peer selection that is hidden from recipient users of recommendations and to enable users to participate in their recommendation process. Recommendations based on users' self-defined social links could be considered as a step in this direction.

3.2   Online Social Networks: A Useful Source of Information

The idea to use users' social links as an information source for their recommendations is based on two preeminent theories about sociality: homophily and social influence.

Homophily indicates that people tend to make social connections with other people who possess similar characteristics with them, for instance, age, sex, religion, ethnicity, educational and occupational class, social positions, and so forth, in a process of '*social selection*' [29]. Traditional homophily-related research has explored offline social networking and was focused on the similarity of people's personal characteristics and social status (i.e., *status homophily*) [23]. In contrast, recent research on homophily in online social networks context focused on homophily according to users' perceived values and internal knowledge states (i.e. *value homophily*). Friedkin [41] suggested that the desire to be connected to similar people stems from our tendency to use those who are similar to us as a reference group and compare ourselves with the references to get information or make a decision. Besides, due to the ease of communication, shared knowledge, and other factors that make interactions more comfortable, people are more likely to establish social connections with similar people than with dissimilar people [90, 137].

While homophily explains how and why people select their social partners, *social influence* explains how people's various aspects are affected by their social links. Social influence represents a situation where people' attitudes, beliefs, and behavioral propensities are affected by and adopted from their social ties [29]. Deutsch and Gerard [33] distinguished social influences into two distinct processes according to the expected results of the influences—*normative influence* and *informational influence*. *Normative influence* is a tendency to conform to the positive expectation of their friends or group members, with social desire to be a member of a social group or to seek social approval. This desire is intimately connected to a psychological burden and, time and again, the influence transforms into coercive compulsion and compliance [149]. Meanwhile, *informational influence* refers to "an influence to accept information from another as (trustworthy) evidence about objective reality [127, p. 35]." This type of influence is evoked by ambiguity and uncertainty of reality. When people are not sure about an accurate view of reality and if they are acting in a right way or not, they seek conformity from other people who are similar and have expertise and credibility [128]. In the context of online social networks, which are frequently used as a source of new information, information influence becomes critical, since it influences users' ability to collect accurate and useful information. As a result, the choice of social links is critical to social media users. In the modern connected world, users become more and more knowledgeable as to how to choose social partners as useful information sources. Furthermore, the information side of social influence is one of the reasons why social capital is so valuable and why viral marketing on social media works [28, 38].

The cumulative studies have demonstrated an interplay between homophily and social influence in social networks. For instance, when two socially associated people are alike, their reciprocal influence is stronger and lasts longer than for a pair who are less similar to each other [40]. Brzozowski et al. [20] empirically proved the homophily and social influence of three kinds of social connections: friends, allies and foes. The study was performed using data from an online forum where users share opinions and vote on various controversial political topics (resolves). In this context, both social ties and users' ideologies are important to consider. In this chapter, *friends* are based on personal familiarity (i.e. strong ties), while *allies* do not necessarily have a personal acquaintance, but share similar ideologies (i.e., weak ties). *Foes* are the users with whom a user doesn't ideologically agree (i.e., a negative tie). The authors examined which *resolves* users vote on and how they vote. They found that, when users voted on a certain resolve, their friends were more likely to vote on the same resolve than allies and foes. Interestingly, users tended to avoid resolves that their foes already voted on. However, users voted in more similar patterns with their allies than with their friends or foes. Of course, users agreed the least with their foes [20]. We interpreted these results to mean that all of the three ties played certain roles as social filters. Both friends and foes are important in the choice of items to consider, and allies are important when deciding how to vote.

Baartarjav et al. [7] presented an exemplary study demonstrating homophily in online space, in particular in online groups. In this study based on Facebook data, the authors focused on the personal traits of online group members, such as age, gender, religion, living area, political opinions, and others. They built clusters solely based on these traits and found the distinguishable characteristics of each group. Depending on the discovered characteristics, they recommended a group to join, and their recommendation accuracy was 73% on average. Therefore, similar people were socially associated as co-members of the same group [7].

Several studies explored social influences in social media applications. Singla and Richardson [112] tested the co-relationship between instant messenger logs and the similarities of search queries. Search interests of people who frequently exchanged instant messages were more similar than the interests of random pairs. Moreover, the longer they communicated, the more similar they were. Swamynathan et al. [120] compared the influence on users' satisfaction, depending on the different social identities of transaction partners. Specifically, they focused on Overstock auctions, an online auction site. In the system, users can have two kinds of relationships: their personal connections (friends) and business connections (once a user buys a product, the seller becomes a business connection). The authors found that users bought products mostly from total strangers; only 2% of transactions were made between two friends who were directly connected. However, user evaluations showed that users were much more likely to be satisfied with transactions made with personal connections than with their business connections or strangers. In addition, the degree of satisfaction was decreased along with the increase of social distance [120].

4 Multiple Dimensions of Social Link-Based Recommendations

With the purpose of systematically analyzing and classifying various social link-based recommendations, we identified six dimensions, as shown in Fig. 1.

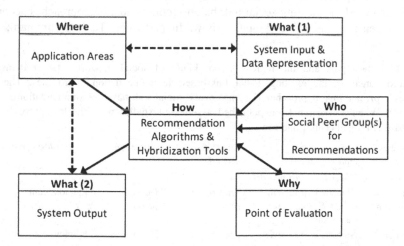

Fig. 1. Various dimensions of social link-based recommendations. The solid lines denote direct interactions with recommendation algorithms and the dotted lines represent indirect association with recommendation algorithms.

- The first dimension is *what kind of data is used as a source for personalization* and *in which format the input data was represented.* It identifies the kinds of *input* data the recommender system used and how the data was interpreted as user preference models.
- The kind of input data like above is directly related to target items and applications of the recommendation technologies, which is the second dimension. The second dimension is *what kind of item was recommended.* It classifies the *output* items that each study targeted.
- The third dimension is to explore *where* social network-based recommendations *are used.* The dimension identifies *applications* in which social link-based recommendation technologies were deployed or tested.
- The fourth dimension is *who the social peers were on which the recommendations were built.* The review identified the kinds of social networks used as the foundations of recommendations. We also discussed why each type of social network was used in recommendations.
- The fifth dimension is *how the recommendations are generated.* We explore this dimension from two perspectives: the kind of basic recommendation algorithms and whether and how multiple algorithms were hybridized.
- The last dimension is *why the recommendation technology was assessed.* Among the various evaluation criteria of recommendations, the review considers in which aspect each study assessed its proposed social link-based recommendations.

The arrow lines of Fig. 1 show the flow of data. Recommender systems receive user data input (i.e. what (1)) and social links used in recommendations (i.e. who) from applications (i.e. where). Once the recommendation computation is done (i.e. how), the systems produce target items (i.e. what (2)). Depending on the results assessed through evaluations (i.e. why), recommender systems iteratively update their recommendation algorithms and procedures to optimize their performance.

We analyzed the existing social link-based recommendation approaches according to the dimensions, as Tables 1 and 3 show. In particular, Table 1 summarizes the

Table 1. Data inputs and the representation, kinds of social networks, target items, and application areas of the existing social link-based recommendation approaches. The table includes *representative* approaches (i.e., not every social link-based recommendation study is included). When a research team published a series of studies with similar approaches, we grouped them together into one row.

Approaches	Input and data representation	Kinds of social networks	Target items	Systems/data sources
Recommendations based on trust networks				
Al-Sharawneh and Williams [5]	Numeric user-to-item ratings & users' trust statements	Trust-based network	General products	Epinions.com
Chen et al. [24]	Numeric user-to-review ratings & users' trust/distrust statements	Trust-based network	Reviews to refer to	Epinions.com

(*continued*)

Table 1. (*continued*)

Approaches	Input and data representation	Kinds of social networks	Target items	Systems/data sources
Chia and Pitsilis [26]	Numeric user-to-item ratings & users' trust statements	Trust-based network	General products	Epinions.com
Deng et al. [32]	Numeric user-to-item ratings & users' trust statements	Trust-based network	General products	Epinions.com
Golbeck et al. [43–45]	Numeric user-to-item ratings & users' numeric trust statements	Trust-based network	Movies	FilmTrust
Jamali and Ester [54, 55]	Numeric user-to-item ratings & users' trust statements	Trust-based network	General products	Epinions.com
Jamali and Ester [56]	Numeric user-to-item ratings, users' trust statements or the list of users' friends	Trust-based networks & friendship	(1) General products (2) Movies	(1) Epinions.com (2) Flixster
Ma et al. [80]	Numeric user-to-item ratings & users' trust statements	Trust-based networks	General products	Epinions.com
Ma et al. [82]	Numeric user-to-item ratings & users' trust statements	Trust-based networks	General products	Epinions.com
Ma et al. [83]	Numeric user-to-item ratings, the users' trust statements or the list of users' friends	Trust-based networks & friendship	(1) General products (2) Movies	(1) Epinions.com (2) Douban
Massa and Avesani [88]	Numeric user-to-item ratings & users' trust statements	Trust-based network	General products	Epinions.com
Moradi et al. [95]	Numeric user-to-item ratings & users' trust statements	Trust-based networks & friendship	(1) General products (2) Movies	(1) Epinions.com (2) FilmTrust
Moradi and Ahmadian [94]	Numeric user-to-item ratings & users' trust statements	Trust-based networks & friendship	(1) General products (2) Movies	(1) Epinions.com (2) FilmTrust
Symeonidis et al. [121]	Numeric user-to-item ratings, the users' trust statements or the list of users' friends	Trust-based network & friendship	(1) General products (2) Movies	(1) Epinions.com (2) Flixster
Victor et al. [130]	Numeric user-to-item ratings & users' trust statements	Trust-based network	General products	Epinions.com
Wang et al. [135]	Numeric user-to-item ratings, the users' trust statements or the list of users' friends	Trust-based network & friendship	(1) General products (2) Movies (3) Movies, books, and music	(1) Epinions.com (2) Flixster (3) Douban

(*continued*)

Table 1. (*continued*)

Approaches	Input and data representation	Kinds of social networks	Target items	Systems/data sources
Yang et al. [140]	Numeric user-to-item ratings & users' trust statements	Trust-based network	General products	Epinions.com
Yuan et al. [145]	Numeric user-to-item ratings & users' trust statements	Trust-based network	General products	Epinions.com
Recommendations based on friendship networks				
Bellogín et al. [8]	Numeric user-to-item ratings & the list of users' friends	Friendship	Movies	Filmtipset
Bonhard et al. [14]	Users' demographic profile, numeric user-to-item ratings & the list of users' friends	Friendship	Movies	MovieMatch
Carrer-Neto et al. [22]	Binary user-to-item ratings (like/dislike), various types of metadata of movies & the list of users' friends	Friendship	Movies	Experimented with 10 student participants and Movies Metadata from IMDB
De Meo et al. [30]	Numeric user-to-item ratings & list of users' friends	Friendship (college students)	Movies	Experimented with 37 college student participants
Groh and Ehmig [47]	Numeric user-to-item ratings and a list of users' friends	Friendship	Local night clubs	Lokalisten
Gürsel and Sen [50]	Users' social tags and comments on items and the list of users' friends	Friendship	Photos	Flickr
Jiang et al. [58]	Content of various information items (e.g. blogs/microblogs, photos, videos) and the sharing records within users' social networks and the list of users' reciprocal or unidirectional social links	Friendship and unidirectional relations	(1) Blogs, photos and video links (2) Microblogs	(1) Renren (2) Tencent Weibo
Knijnenburg et al. [65]	Facebook users' "likes" records on music artists and the list of users' friends	Friendship	Music artists	Experimented with 267 Facebook users, and the test bed system was a Facebook music recommender system
Konstas et al. [67]	Users' music play-counts, users' tags and the list of their friends	Friendship	Music	Last.fm
Liu and Lee [78]	Numeric user-to-item ratings and list of users' friends	Friendship	Digital items for personal websites	Experimented with 27 users of a Korean SNS (Cyworld) and their online friends.

(*continued*)

Table 1. (*continued*)

Approaches	Input and data representation	Kinds of social networks	Target items	Systems/data sources
Sinha and Swearingen [114]	Numeric user-to-item ratings and the list of users' friends	Friendship	(1) Books (2) Movies	Experimented with 19 students; recommendations came from (1) Amazon, Sleeper & RatingZones for books, and (2) Amazon, Reel.com, and MovieCritics for movies
Wang et al. [133]	Users' check-in records of visited locations & the list of users' friends	Friendship	Locations of interests	Brightkite and Gowalla

Recommendations based on other types of social networks

Guy et al. [52]	Users' tags, bookmarks, comments, organizational charts, the list of users' online friends and various working activities	Professional colleagues (in various working contexts) and online friends	Internet or intranet pages of interests, blogs, and communities	Experimented with 290 subjects; The testbed system was a social application suite for a company
Lee and Brusilovsky [73, 76]	Users' bookmarks and their group membership	Watching links (directed) and group membership	Academic articles	Citeulike
Macedo et al. [84]	Users' event attendance records, context information of events, and the list of users' groups	Group membership	Regional events	Meetup.com
Sun et al. [119]	Users' bookmarks, the social tags & list of users' groups	Group membership	Internet pages of interests	Delicious
Yuan et al. [144]	List of users' favorite artists, and the list of users' friends and groups	Group membership & friendship	Music artists	Last.fm
Zhang et al. [147]	Users' friends and followers, and activities of users' social links on social media	Friendship and following network	Social links' online activities	Social links and the online activity information from Facebook and Twitter. Experimented with 10 participants

distribution of the existing approaches from the first to the fourth dimensions, whereas Table 3 is intended to provide the overview of the existing approaches in respect to recommendation algorithms (fifth dimension) and the evaluation methods (sixth dimension).

4.1   Input Data Types of Social Link-Based Recommendations

In a typical collaborative filtering scenario, there are users $U \in \{u_1, u_2, ..., u_n\}$ and items $I \in = \{i_1, i_2, ..., i_l\}$. Each user u has a set of items I_u on which he some form of opinion (known also as feedback). r_{ui} denotes the user u's opinion on an item i_l. The r_{ui} can be explicit, implicit, or descriptive, as Table 2 shows.

Table 2. Various indications of users' opinions

Input type		Description	Examples
Explicit opinion	Unary	Like	Bookmarks and 'Like' on Facebook
	Binary	Like or dislike	Thumb up or thumb down on YouTube
	Numeric	5, 7 or 10 Likert scales	Movie ratings
Implicit opinion		No predetermined value ranges. Ranges from zero to maximum number of instances	Purchases, play-counts of videos and songs, click-through and check-in records of a location
Descriptive opinion		No predetermined value ranges or text values	Tags, movie descriptions, music descriptions

Explicit opinion is usually directly expressed by the user within a scale that is defined by the system, and can range from coarse to fine-grained. Unary explicit feedback such as Facebook-style 'like' buttons is on the coarse-grained side, while another example of coarse-grained unary feedback is a bookmark. A bookmark expresses a user's interest in an item; however, the absence of the bookmark does not necessarily indicate that she is not interested. A binary rating allows users to clearly separate likes and dislikes from unrated items. Finer-grained ratings can be expressed using a numeric rating scale. Most typical are scales that range from 1–5 or 1–10 (i.e. 1 usually indicates 'dislike a lot' and 5 or 10 means 'like a lot') [108]. Systems also collect user activity data from various sources to implicitly infer users' interests, but implicit indicators do not have clear value ranges. We cannot predict how many times a user has purchased a product or how many times a user has listened to a song. Hence, the implicit indications often require normalization of the values. For more detailed explanations of explicit and implicit data types, see Chap. 14 [57].

Most of the 36 approaches in Table 1 (24 studies, 66.6%) used numeric ratings, while six approaches used unary ratings based on bookmarks or binary ratings. The remaining six approaches used implicit indicators. For instance, to recommend locations of interest, Wang et al. [133] used users' check-in records of visited locations and the timestamps, while Konstas et al. [67] used users' music play-counts to recommend a song to enjoy.

Explicit and implicit preferences show whether and how much a user is interested in an item. However, relatively simple coarse-grained ratings or implicit preferences do not carry enough useful information. To add more direct information about users' preferences, some studies used extra textual indications to show the reasons of a user's interests or how the user understands an item. Carrer-Neto et al. [22] borrowed movie-related metadata from IMDB and combined it with users' binary ratings. In this way, they expanded the users' binary preferences into multi-dimensional preferences. Macedo et al. [84] considered not only users' event attendance records, but also the various contextual information of events, such as topics of events, their locations, and temporal information. Sun et al. [119] and Lee and Brusilovsky [73, 76] used social tags to improve the quality of bookmark-based recommendations. Bonhard et al. [14] took advantage of users' demographic information to recommend movies to watch.

4.2 Applications and Target Items of Social Link-Based Recommendations

The type of social links used by personalized recommenders to a considerable extent depends on the applications (i.e., a specific social system) and the target items that this application recommends. Figures 2 and 3 review the item types that social link-based recommendations have targeted, as well as applications that engage in social link-based recommendations.

Figure 2 shows that social link-based recommendations most often focus on recommending *general products*. The popularity of the "general products" category is defined by the position of Epinions.com[1] system as the most heavily used source of data for social link-based recommendations (see Fig. 3): among 18 studies based on trust-based networks, 17 studies used Epinions.com data sets (see Table 1 and Fig. 3). The popularity is caused by the fact that this is the first publicly available data source that contains not only user-to-item ratings but also user-to-user explicit social relations. Several versions of the Epinions.com data set have been used in social link-based recommendation studies and, as of 2015, three versions of this dataset are available online[2,3] The former two data sets were collected and shared by Massa et al. [87]. The first version contains about 49K users, 139K items, 664K ratings, and 487K trust-based relationships, while the second version contains a larger volume of data (132k users, 1. 6M items, 13.7M reviews, and 841K relationships). The difference in the data volume aside, the users' ratings of the second version are about other users' reviews, not about items. In addition, the positive trust relations constitute the trust network in the first data set, but the second data includes both positive and negative trust relations. Among 18 studies based on trust-based networks shown in Table 1, Al-sharawneh and Williams [5], Chia and Pitsilis [26], Deng et al. [32], Jamali and Ester [54, 55], Moradi et al. [94, 95], Symeonidis et al. [121], Victor et al. [130], Wang et al. [135] and Yuan et al. [144] used the first Epinions.com dataset, while one study [24] used the second dataset. The third Epinions.com dataset was collected and shared by Domingos et al. [105]. It has a large number of users and items (71K users, 104K items, 571K

---

[1] Epinions.com aims to review a wide range of products from digital gadgets, appliances, sports gears, toys, movies, books, songs and more. None of the studies using Epinions.com datasets clearly stated the product category of the target items; hence we classified the target items as general products.

[2] http://www.trustlet.org/wiki/Epinions_dataset.

[3] http://alchemy.cs.washington.edu/data/epinions/.

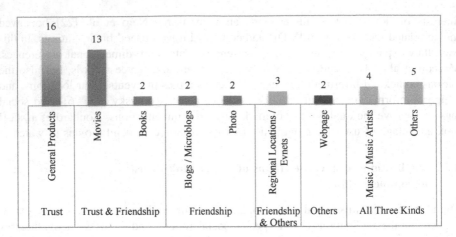

Fig. 2. Types of targeted items in social link-based recommendations. Some application data provides several types of items.

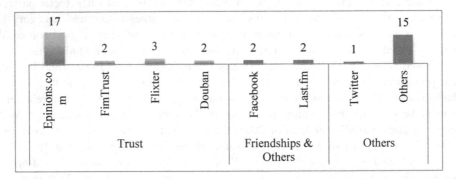

Fig. 3. Data sources of social link-based recommendations. Some studies used several data sources.

user-to-item ratings, and more than 500K trust-based relationships). Jamali and Ester [56] and Yang et al. [140] also used this dataset.

Movies, the second-most popular type of target items, have the longest history as a target of recommendations. The early pioneer recommender system, MovieLens, publicly shared the data set that has been used in a number of recommendations-related studies (and is stll used). The popularity of movies as the target items of personalized recommendations has been further increased by the Netflix Prize[4]. Researchers took advantage of various data sources for social link-based movie recommendations: FilmTrust [43–45, 94, 95][5], Flixster [56, 121, 135], Douban[6] [83, 135], Filmtipset [8],

---

[4] http://www.netflixprize.com.
[5] http://www.librec.net/datasets.html.
[6] http://dl.dropbox.com/u/17517913/Douban.zip.

MovieMatch [14], IMDB [22], Amazon.com, Reel.com and MovieCritics [114]. Since some of these systems use trust links and some use friendship links, both of these link types have been explored as a source of data to recommend movies.

Music or music artists recommendations explored the value of several types of social links: trust, friendship, and group membership. Douban [135], Last.FM [67, 144] and Facebook [65] were used as a context for music recommendations. Last.FM is a social music website. In this system, users are able to listen to music, add tags, and make friendships with other users as well. Several Last.FM datasets are available with and without the social network information[7,8,9]. Music is a challenging type of items to generate precise recommendations because user feedback is highly subjective. However, our music taste is influenced by our social connection, and we often share our favorite songs with our friends. In this context, recommendation approaches based on social links can provide valuable insights for music or music artist recommendations.

Another two related types of target items that could benefit from using social links for recommendation are *places of interest* (POI) and *events*. Locations and events to visit are inherently social. When we want to enjoy a Friday night at a local restaurant or go to a music concert, we invite friends or family to go along. In one of the pioneering works in this field, Wang et al. [133] has explored how to recommend places of interests via users' check-in locations and their online social networks. The authors started their study by proving the positive correlation between users' online social connections and their visiting records, and found that a considerable number of users visited locations that their friends or their friends of friends have visited before. In their study, based on the *Random Walk with Restart* algorithm, they demonstrated that when users make a decision to visit a place, information about the whereabouts of their online friends could reinforce the quality of personalized POI recommendations. Groh and Ehmig [47] used friendship information to recommend local night clubs. Their results showed that recommendations based on friends' data are better than suggestions that are based on traditional anonymous peers. A few years ago, Yelp, a social system for finding and rating local businesses (such as restaurants, bars, coffee shops, etc.), announced a recommendation challenge[10]. A number of researchers have participated in the challenge, and several rounds of evaluations have been completed. However, work focused on recommending POIs, such as restaurants or coffee shops, using online social connections are still rare. Some of this work is reviewed in Chap. 16 of this book [17].

Overall, the review of target item types and application for which social links were explored as a source of information for recommendation demonstrates that the work in this area is distributed quite unevenly. While there is a large concentration of work with some types of items where research data are officially released as a dataset or can be crawled (such as research based on Epinions.com datasets), other systems and item types received too little attention. It is interesting to note that the amount of research in the context of a specific social system is not quite correlated with its popularity.

---

[7] http://www.dtic.upf.edu/~ocelma/MusicRecommendationDataset/lastfm-1K.html.

[8] http://labrosa.ee.columbia.edu/millionsong/lastfm.

[9] http://socialcomputing.asu.edu/datasets/Last.fm.

[10] http://www.yelp.com/dataset_challenge.

For example, some highly popular social media systems such as Facebook and Twitter have rarely been used to explore the role of social links for recommendations. This may be due to the difficulty to crawl large volume of data on the applications and the wide diversity of items. In Table 1, two studies using Facebook or Twitter [65, 147] evaluated the quality of link-based recommendations with relatively small groups of human subjects.

4.3   Types of Social Connections Employed in Link-Based Recommendations

Figure 4 groups the work on social link-based recommendations reviewed in Table 1 by the type of social links that are used to generate recommendations. It shows that trust networks and friendship connections are the most popular social links to provide personalized access to information.

Trust is an asymmetric relationship. When a user u trusts another user v, the trusted user v does not necessarily trust user u. In this aspect, trust is different from friendships, which require mutual agreement on the relationship; user v does not need to approve user u's trust or reciprocally trust the user u. Besides, the trusting party (i.e., the user u) can state how much their trusted parties (i.e., user v) are trustworthy. The trust relationship also enables users to define negative relations (distrust) by using negative trustworthiness values. Golbeck suggested that, while the traditional definition of 'trust' is lexically related to security and reliability, 'social trust' in the Web 2.0 era is the broader definition of trust and is related to 'a matter of opinion and perspective'. Hence, information can be aggregated, sorted and filtered through social trust [42]. Based on this suggestion, in the course of her studies, Golbeck demonstrated that users prefer recommendations from trusted parties to traditional collaborative filtering recommendations [43–45].

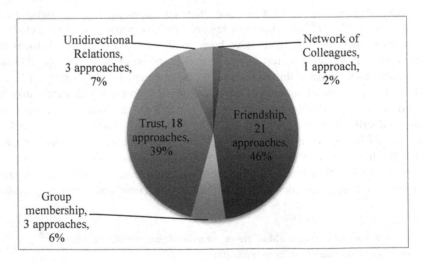

Fig. 4. Types of social links employed in personalized recommendations. Some approaches were based on more than one kind of social network.

Friendship is a reciprocal relationship. Compared with other types of online social connections introduced above, friendship relies on personal familiarity, which is a foundation upon which many friendships are rooted, in particular friendships existing offline. According to homophily and social influence theories introduced in Sect. 3.2, it is well-known that "strong ties", i.e., those people who engage in frequent interactions and have many overlapping connections, tend to be similar in various ways. Friendship-based recommendations actively use this social pattern.

The other types of online social connections are various object-centered unidirectional relations and network of colleagues. Wellman suggested that, in the Web 2.0 era, various new 'less bounded' online relationships would emerge [136]. *Following relationship* is a typical example of newly emerging online social networks. The typical examples of this relationship include "following" on Twitter, "watching" on CiteU-Like, "network" on Delicious, or "following" on Google + . Users of the Web 2.0 have found it easier to know who knows what through social media. However, it is a burden to contact people who possess the desired knowledge via their personal ties [62]. Many social bookmarking systems, which help users to manage and share interesting information, as well as blogging systems, which aim to post online journals to express their opinions, offer users this special kind of online sociability without any need to ask for a user's consent to be followed. Once users find other users whose information collections are useful, they are allowed to follow or watch the users' activities continuously. The relationships do not require any offline interactions or any mutual agreement for being connected. Some studies [72, 74, 75] have provided positive evidence that the relations met the similarity attraction hypothesis [92] and held transitivity power [93]. A high degree of similarity was embedded in *following* relations and the similarity decreased with the increase of distance, which means that users in unidirectional relations built their connections focusing on their objects of interest. As a result, these kinds of social networks could be classified as *object-centered social networks*. Breslin and Decker said that the social links connecting users via items of interests may be more long-lived than the relationships that did not share any items of interest [19].

Group membership is an online social connection established between members of the same online group or community. It is typically a highly object-centered social association, since group activities are usually centered on a certain topic. An online group is usually a community of interest or practice; for instance, a fan club of a musician, a community of Hadoop programmers, an online forum for students taking the same course, or an online space for members of the same project. Users engaged in group-based networks distribute topic-relevant information or contribute topic-relevant activities. The theory of communal sharing explains the social dynamics of group membership, as group members think they share common substances. Before online social networks emerged and proliferated, communal sharing relationships represented close relationships, such as kinship ties [39]. In the current Web 2.0 era where relationships are becoming flattened and less bounded, however, the sense of communal sharing can be applied to online group activities. Group members treat information objects as their shared assets and are willing to share what they need or contribute what they can. They usually do not expect to receive something back in return for their contributions. In addition, members do not pay attention to the portion of contributions that are made by each individual member. Being a member of a group is sufficient, since they are able to use the resources that the group is sharing [39]. Therefore, membership of an online group is informative in understanding users' information needs and personalizing their information space.

The analysis of social link types used in social link-based recommendations shows that the work is distributed quite unevenly. The dominant majority of work is focused on just two types of connections. Among the studies listed in Table 1, more than 85% are based on either trust networks or friendship. Moreover, as indicated above, almost all of the works on trust-based networks reviewed in Table 1 used the Epinions.com data set, with the exception of Golbeck et al. [43–45]. When considering a wide variety of online sociability, we believe that more work should be focused on proving the expandability of social link-based recommendations to more diverse and less explored social networks, as well as to new applications and domains.

Finally, our review shows that early works of social link-based recommendations were focused on an assumption that our social links have equal influence over us. However, recent works demonstrated that the reality is more sophisticated: different types of social links have different degrees of influence. As a way to account for this inequality, researchers distilled various social properties in social link-based recommendations. For example, Arazy et al. [6] suggested distinguishing four kinds of social properties—the degree of homophily, tie strength (the frequency of interactions), trust, and reputation—and integrated these properties into the nearest neighbor-based social link-based approach. Although the proposed approach is a conceptual framework without any empirical evaluation of the performance, it provided meaningful insights about how to incorporate various social network properties with personalized recommendations. Another type of social property, social influence, has been used in social link-based recommendations. In order to include the property as a part of a social link-based recommendation algorithm, for instance, Jiang et al. [58] computed the degree of social influence by considering social association (i.e., whether a given pair of two users are friends or one user followed another user) and the distribution of items shared between two given users. In particular, because the data sources of this study do not contain users' numeric/binary ratings for items, the authors relied on the content of items and estimated the social influence at the level of topical distributions of each item. Gürsel and Sen [50] considered the unequal influence of our friends on different topics of interest in producing recommendations. Chia et al. [26] took advantage of the 'experience level' of social connections as an extra social property in personalized recommendations, along with an intuition that we are more likely to seek advice from others who more experienced than us. The authors defined the experience level by combining explicit trust statements and the numbers of rated items.

5 Algorithms for Recommendations Based on Social Links

A unique feature of social recommendation algorithms is the engagement of social connections to generate or enhance recommendations. Understanding different approaches that could be used to leverage social links for recommendation is most critical for the developers of social recommendation approaches. This section attempts to combine three goals: explain most important types of social link-based recommendation algorithms in sufficient detail; classify existing research on social link-based recommendation from the prospect of employed algorithms; and provide representative examples of using each major algorithm type. We classified the algorithms used for social link-based recommendations according to Tang and Liu's classification of traditional approaches [123] with one specific addition—direct friend-to-friend recommendations. The classification includes the nearest-neighbor approaches, graph structure-based approaches, matrix factorization, and hybrid recommendations. Table 3 reviews the existing research on social link-based recommendations from the perspective of employed algorithms and the evaluation criteria. Figure 5 offers a visual summary of the algorithms and their use in the reviewed works.

As mentioned above, a major motivation for social link-based recommendations was to solve various problems of collaborative filtering by substituting or complementing anonymous like-minded peers used in traditional collaborative filtering with the explicitly defined social connections of the target user. Therefore, most of the social link-based recommendations use one of the traditional collaborative filtering algorithms as a basis and modify it to improve the quality of recommendation by infusing social links. As Fig. 5 depicts, the most popular algorithms are the nearest neighbor-based approaches and matrix factorization, which are also widely used in collaborative filtering. Due to this tight connection between traditional collaborative filtering and social link-based recommendations, Chap. 10 of this book [63], which reviews classic rating-based collaborative filtering, could provide useful background reading for this chapter; yet the presentation below is designed to be self-containing.

Table 3. Descriptions of recommendation algorithms and the evaluation criteria. The table includes representative approaches (i.e., not every social link-based recommendation study is included). When a research team published a series of studies with similar approaches, we grouped them together into one row.

Approaches	Description of recommendation algorithms	Evaluation criteria
Trust network-based recommendations		
Al-Sharawneh and Williams [5]	**Nearest Neighbor-Based Recommendations:** Authors defined global credibility values of users by combining their trustworthiness and expertise. The credibility is intended to identify global opinion leaders, and, at the final stage to compute the rating predictions, credibility values of leaders were multiplied to the probability values, which were calculated by the nearest-neighbor algorithm of conventional collaborative filtering	MAE

(continued)

Table 3. (*continued*)

Approaches	Description of recommendation algorithms	Evaluation criteria
Chen et al. [24]	**Clustering-Based Recommendations:** Based on users' ratings, the authors constructed clusters and within each cluster, they identify experts using reputation scores. The reputation scores were computed by the PageRank algorithm on trust-based networks. Among the identified experts, distrusted ones were excluded. Then by using a cluster, which is highly related to a candidate item and a target user's ratings, recommendation probability was computed	Coverage rate, MAE, computational time, precision, recall and F-measure
Chia and Pitsilis [26]	**Nearest Neighbor-Based Recommendations:** Authors used various sorting tactics—information similarity, experience level and trustworthiness—in selecting target users' social peers. Once they chose different groups of social peers, conventional Pearson correlation-based rating predictions were applied (refer to Eq. 3)	MAE, precision, recall, F-measure and coverage
Deng et al. [32]	**Graph Structure-Based Recommendations:** Modified Random Walk (i.e. RelevantTrustWalker) with trust relevancy. Instead of a random selection of social peers, each walk is selected according to trust relevancy, which combines users' trust statements and information similarities	RMSE, coverage, F-measure and computational time and cold-start user problem
Golbeck et al. [43–45]	**Nearest Neighbor-Based Recommendations:** For a given candidate item, the authors aggregated raters of the candidate item, through the Breadth First Search of users' trust-based networks. In order to propagate unknown trust values from target users to their indirectly trusted party, the TidalTrust metric was used (refer to Eqs. 5–7). Explicit and inferred trust values of social peers were multiplied to determine peers' ratings on the candidate item	Rating prediction accuracy (Absolute difference between actual ratings and predicted ratings)
Jamali and Ester [54, 55]	**Graph Structure-Based Recommendations:** The Random Walk of a trusted network and item-based Collaborative filtering with weighted hybridization, so-called 'TrustWalker'	RMSE, coverage, F-measure, and cold-start user problem

(*continued*)

Table 3. (*continued*)

Approaches	Description of recommendation algorithms	Evaluation criteria
Jamali and Ester [56]	**Matrix Factorization:** Matrix Factorization combined with users' trust-based networks—SocialMF. In particular, latent feature vectors of users were weighted by average ratings of users' direct trusted social links	RMSE, computational time, and cold-start user problem
Ma et al. [80]	**Matrix Factorization:** In matrix factorization, the authors considered target users' latent factors, user-to-user trust-based influences	MAE and RMSE
Ma et al. [82]	**Matrix Factorization:** The authors applied matrix factorization technique not only to a user-to-item matrix, but also to a user-to-user social network. Then, social factor matrix and the confidence values of trust statements were incorporated into the training function to minimize the sum of squared errors	MAE
Ma et al. [83]	**Matrix Factorization:** Matrix Factorization with two types of social regularization incorporates the tastes of target users' social links	MAE and RMSE
Massa and Avesani [88]	**Nearest Neighbor-Based Recommendations:** User-to-user similarities of collaborative filtering were replaced with trust values between users and their direct and indirect trust-based links. In order to propagate unknown trust values from target users to their indirectly trusted party, the MoleTrust metric was used	MAE, MAUE (Mean Absolute User Error), rating coverage, user coverage, and cold-start user problem
Moradi et al. [95]	**Clustering of Social Graphs:** As the first step, authors found sparse sub-graphs consisting of dissimilar nodes (i.e., other users) and, as the second step, the nodes were used as the initial centers of the clustering algorithm. Once the system found a fine set of clusters, as the last step to generate recommendations, the authors computed the recommendation probabilities of candidate items using a similar function between a target user and other users in the target user's cluster	MAE, RMSE, precision, recall, F-measure, and coverage

(*continued*)

Table 3. (*continued*)

Approaches	Description of recommendation algorithms	Evaluation criteria
Moradi and Ahmadian [94]	**Nearest Neighbor-Based Recommendations:** By combining trust-based social links and the anonymous top N nearest neighbors, authors built an initial trust network and generated a set of recommendations using the Pearson correlation-based nearest-neighbor approach. The authors applied a reliability measure to the generated recommendations, and when the reliability value did not exceed a certain threshold, they concluded that the initial trust network is not a reliable reference for making good recommendations. Therefore, the authors restructured the trust network, and iterated this network reconstruction until the reliability value of recommendations exceeded the predetermined threshold.	MAE, MAUE, item coverage, user coverage, cold-start user problem, opinionated users, black-sheep users, controversial items, and niche items
Symeonidis et al. [121]	**Hybrid Recommendation:** Weighted hybrid recommendations to combine item rating-based similarity and social structure-based similarity	RMSE, precision and recall
Victor et al. [130]	**Nearest Neighbor-based Recommendations:** Used both collaborative filtering-based anonymous peers and trust peers. EnsembleTrustCF (refer to Eq. 8)	MAE, RMSE, coverage, and controversial items
Wang et al. [135]	**Matrix Factorization:** Matrix Factorization where latent factors of a target's friends were combined with the given target user's latent factors by inner product	MAE and RMSE
Yang et al. [140]	**Matrix Factorization:** According to the categories of users' rated items, the authors subdivided users' social links into smaller social network matrix and trust values within the smaller social matrix were employed in matrix factorization	MAE and RMSE
Yuan et al. [145]	**Graph Structure-Based Recommendation:** Target users' direct and indirect social links were chosen as social peers using trust propagation distances (i.e. average path length property of the Epinions.com trust network), and the graph distance-based weight was employed to determine a traditional collaboration filtering-based rating prediction	MAE

(*continued*)

Table 3. (*continued*)

Approaches	Description of recommendation algorithms	Evaluation criteria
Friends-based recommendations		
Bellogín et al. [8]	**Graph Structure-Based Recommendation and Hybrid Recommendation:** (1) recommendations based on purely users' social networks using a breadth-first search algorithm; (2) hybrid recommendations based on anonymous peers and direct social connections using feature combination hybridization; and (3) a hybrid approach based on Random Walk with Restarts	Precision, recall and NDCG, user coverage and utility (the user ratio who received at least one correct suggestion)
Bonhard et al. [13, 14]	**Direct Recommendation:** Suggested movies rated by target users' friends with explanations of recommending friends' identity	Uptake rates of recommended items and recommendation explanations
Carrer-Neto et al. [22]	**Hybrid Recommendation:** Authors generated recommendations using knowledge base (i.e. movie-relevant ontology). Then by using the weighted hybrid recommendations, the target users' preferences and preferences of users' friends were combined	Precision, recall, and F-measure
De Meo et al. [30]	**Matrix Factorization:** Matrix Factorization combined with social distances between target users and their social links	RMSE
Groh and Ehmig [47]	**Nearest Neighbor-Based Recommendations:** Anonymous peers of collaborative filtering were replaced with users' direct friends. The remaining processes to calculate user-to-user similarity and compute the prediction probabilities of candidate items with cosine similarity are the same as in collaborative filtering	MAE and F-measure
Gürsel and Sen [50]	**Hybrid Recommendation:** They inferred categories of items via users' social tags. Using Bayes's theorem, the recommender chose items to suggest from the set of photos posted by target users' friends and that belonged to categories of users' interests	Precision and recall
Jiang et al. [58]	**Matrix Factorization:** MF combined with users' reciprocal or unidirectional social networks and latent topics of items—namely, ContextMF. Item-to-item and user-to-user similarities were calculated by topic distributions of items. User-to-user social influence and the degree of interactions were taken into account as well	MAE, RMSE, Kendall's ranking coefficient, Spearman's rho, acceptance ratio of recommended items

(*continued*)

Table 3. (*continued*)

Approaches	Description of recommendation algorithms	Evaluation criteria
Knijnenburg et al. [65]	**Nearest Neighbor-Based Recommendation:** Recommendations based on the similarity between target users and their friends. Pearson correlations between every pair of a target user and his friends and the similarities were aggregated as the weight for each candidate items	Inspectability (i.e. transparency), the feeling of control, and user satisfaction
Konstas et al. [67]	**Graph Structure-Based Recommendation:** The authors built a graph consisting of users, music tracks, and tags as nodes. Social relationships among users, users' music play-counts, and the frequencies of users' tags were also considered to add edge weights. Then, they applied the Random Walk with Restart algorithm to calculate the preferences of target users for a candidate item	Precision
Liu and Lee [78]	**Nearest Neighbor-Based Recommendations:** As the nearest neighbor groups, anonymous peers with collaborative filtering were simply combined with target users' online friends without any additional weight. Then the traditional nearest neighbor-based recommendation algorithm was applied withing the groups	MAE
Sinha and Swearingen [114]	**Direct Recommendations:** Suggested items favored by target users' friends.	Usefulness, satisfaction with recommended items, trustworthiness of the system, and various issues of recommendation explanations
Wang et al. [133]	**Graph Structure-Based Recommendations:** Random walk with restart algorithm was applied to one graph consisting of only target users' direct friends or another graph consisting of not only the target users' direct friends, but also their nearest neighbors whose interests are similar to the target users but are not socially associated	Precision, recall, and utility
Recommendations based on other types of social networks		
Guy et al. [52]	**Nearest Neighbor-Based Recommendation:** The authors specified ad-hoc user-to-item weights for users' various online activities (e.g., authorship of a blog, community membership, commenting, and bookmarking). Then, they multiplied user-to-user relationship strength to the ad-hoc user-to-item strength with time decay factor	The degree of interest and usefulness of recommendation explanations

(continued)

Table 3. (*continued*)

Approaches	Description of recommendation algorithms	Evaluation criteria
Lee and Brusilovsky [73, 76]	**Matrix Factorization:** The authors built a matrix factorization that consisted only of users' social links	Precision and recall
Macedo et al. [84]	**Hybrid Recommendation:** In order to cope with the data sparsity in the target domain (i.e. events), the authors consider various kinds of contextual information, such as group-based social context, event content, locations of events, and temporal information of events	NDCG ranking Evaluation Metric (@10)
Sun et al. [119]	**Matrix Factorization:** The authors clustered users' friends into subgroups, and also clustered items into subgroups, according to user-item bookmark/tag similarities. Then, a subgroup of friends, whose tastes are similar to target users, and a subgroup of items, which are similar to users' favorite items, were integrated into matrix factorization, using an individual-based regularization approach.	Precision and recall
Yuan et al. [144]	**Matrix Factorization:** User-item matrix factorization combined with users' friendship and membership in online groups. This friendship information was fused with user-item latent vectors via regularization and the group membership information was fused with the same latent user-item vectors via a user-group matrix factorization	Precision and recall
Zhang et al. [147]	**Content-Based Recommendations:** Based on content terms of users' online activities and their social link activities, the authors computed the term frequencies and found activities of which the content is similar to users' favorite activities	Precision

For the rest of this chapter, we will use the following notations. R is the user-item rating matrix, $R \in \{R_{ui}\}_{l \times n}$ where n and l denote the number of users and items, respectively. r_{ui} is the rating of an item i given by a user u. $\hat{r}_{ui}$ denotes the predicted rating of user u for a candidate item i, which is picked by recommendation algorithms as a presumably favorable item. The range of ratings varies, according to the recommender systems: from numeric to unary ratings.

5.1 Direct Friend-to-Friend Recommendations

As a part of daily interactions with our friends, we share a lot of opinions about everything in our lives; for instance, we may recommend a good book to read, a reliable mechanic to fix our cars, or ask for various opinions that could range from a movie to enjoy, a good restaurant to visit, an e-commerce site to explore and an apparel to buy. Early-generation recommender systems implemented this advice-seeking process and 'word-of-mouth' phenomenon among offline social connections systematically. They enable users to directly send an interesting item to online friends as a recommendation without any systematic computation.

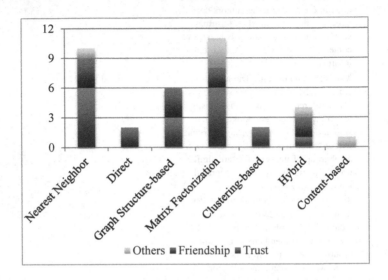

Fig. 5. Summary of social link-based recommendation algorithms

As one of the first related projects, Tapestry allowed users to send items of interest with annotations to their friends and colleagues [46]. Another early project let users send interesting research papers directly to other colleagues using a so-called 'active push approach' [86]. Dugan and colleagues also invented an interesting 'Dogear Game'—by borrowing the concept of direct friend-to-friend recommendations. Internet bookmarks were the target item of this game, and given a list of bookmarks, users were asked to guess to whom each bookmark belongs among their colleagues. When users make a correct guess, they score a point. Otherwise, while losing game points, they were asked to suggest the corresponding bookmark to the colleague as a recommendation. Even though users frequently incorrectly identified the creator of bookmarks, the authors suggested that there might be a good reason why users believed that the Internet page would interest one of their colleagues, which would make it a good recommendation [36].

In spite of the early efforts, the systems relying on the direct exchange of information within a "small world" found it difficult to retain users and to keep them actively contributing to the recommendation process. With these reasons, the conventional collaborative filtering recommendations have employed the "word of mouth" in more extensive and systematic ways. However, as in Sect. 3.1, this method has introduced a variety of problems, which were caused by a reliance on anonymous peers and their tastes. The success of social media sites and the abundance of online social networks are the contributing factors in making social link-based recommendations regain attention. Several studies have also demonstrated the advantage of social link-based recommendation over collaborative filtering.

In one of the approaches reviewed in Table 3, Sinha and Swearingen [113] compared the users' perceptional difference between machine-generated recommendations and friend-to-friend recommendations. Rather than producing system recommendations using their own algorithms, they relied on recommendations provided by third-party applications, such as Amazon, RatingZones, Reel.com, and MovieCritics. In order to acquire friends' recommendations, the authors asked their participants to provide the names of three friends who could provide reasonable recommendations to them and suggested items favored by participants' friends. The study revealed that participants found friends' recommendations to be more useful, satisfying, and trustworthy than system-generated recommendations [113]. Bonhard et al. [13] examined a similar question: whether the perceived familiarity with sources who sent recommendations would affect the target recipients' acceptance of the recommendations or not. The authors randomly chose recommendations from either target users' friends or strangers, along with information about the senders of the recommendations. Specifically, the information about each sender included their name, demographic profile (occupations, age, favorite movie genres, and hobbies) and overlapped ratings with target user. Through these detailed explanations, the authors tested three factors: (1) personal familiarity to the senders; (2) similarity of demographic profiles; and (3) rating similarity, on the choice of recommendations. The result showed that users overwhelmingly chose the recommendations from their friends and felt most trustworthy and confident about their choice [13]. Guy and the co-authors [52] obtained similar results in their experiments with recommendations based on professional colleague networks.

5.2 Nearest Neighbor-Based Recommendation Approach

The nearest neighbor-based approach consists of three steps: (1) calculating the similarity of a target user to other users in a recommender system; (2) selecting a small set of like-minded "peers"; and (3) computing the prediction probability of candidate items. For the first step, several similarity measures, such as Pearson correlation, cosine similarity, Spearman's rank correlation, Jaccard similarity, and log-likelihood similarity can be used. Pearson correlation (Eq. 1) and cosine similarity (Eq. 2) are the most popular types of recommendation algorithms. In these equations, $sim_{u,v}$ is the similarity between user u and user v; $I_{u,v}$ is the set of items co-rated by both user u and user v; r_u and r_v are the ratings sets of user u and user v, respectively; and $\overline{r_u}$ and $\overline{r_v}$ are the mean values of the users' rating set.

$$sim_{u,v} = \frac{\sum_{j \in I_{u,v}} \left(r_{uj} - \overline{r_u} \right) \left(r_{vj} - \overline{r_v} \right)}{\sqrt{\sum_{j \in I_{u,v}} \left(r_{uj} - \overline{r_u} \right)^2} \sqrt{\sum_{j \in I_{u,v}} \left(r_{vj} - \overline{r_v} \right)^2}} \qquad (1)$$

$$sim_{u,v} = \frac{\sum_{j \in I_{u,v}} r_{uj} \cdot r_{vj}}{\sqrt{\sum_{j \in I_{u,v}} r_{uj}^2} \sqrt{\sum_{j \in I_{u,v}} r_{vj}^2}} \qquad (2)$$

In the second step, according to the computed similarities and a threshold value θ, when u's similarity with v is larger than θ (i.e. $sim_{u,v} > \theta$), the user v will become one of u's anonymous peers ($v \in N_u$).

The last step is to predict any missing ratings of candidate items for the target user u. Candidate items are those favorite items of user u's peers that are not yet rated by the target user. Equation 3 shows the equation that is used to predict the user u's missing rating on a candidate item j (i.e., $\widehat{r_{uj}}$), N_{uj} denotes u's peers who rated item j.

$$\widehat{r_{uj}} = \overline{r_u} + \frac{\sum_{v \in N_{uj}} sim_{u,v}\left(r_{vj} - \overline{r_v}\right)}{\sum_{v \in N_{uj}} sim_{u,v}} \qquad (3)$$

A natural approach to adapting this nearest-neighbor approach to the social recommendation context is to change its second step by using the target user's social connections instead of automatically selected similar "peers". Another approach is to combine these peers with a user's explicitly selected social connections. Among papers reviewed in Table 3 and Fig. 5, ten projects use variants of this idea.

For example, Groh and Ehmig [47] produced friendship-based recommendations by replacing target users' anonymous peers with their friends. In this study to recommend enjoyable Munich-area clubs, collaborative filtering recommenders can obtain a relatively large number of anonymous peers by changing the threshold θ, but the social link-based recommendations have to rely on a relatively lower number of friends. Hence, this study included not only users' direct friends but also their friends of friends as social connections for generating recommendations. The results shows that the social link-based recommendations performed better, especially when the ratings were very sparse, and produced more novel suggestions than conventional recommendations. The highly social context of this study (i.e., local clubs) was likely a contributing factor of this success, because when people go to a club, they usually do so in a company of friends [47].

In the studies reviewed in Table 3, authors rarely employed weighting to model links between a target user and her social connections. However, depending on the type of social networks, it could be useful to use weighed connections that are based on social dynamics or graph theory-based measurements (e.g., the frequency of interactions, clustering coefficient, the degree of betweenness, etc.). In particular, when incorporating trust-based social networks into personalized recommendations, trust values could be used as effective weights to identify the properties of the social connections (i.e. trust vs. distrust or strong ties vs. weak ties).

For example, Massa and Avesani [88] proposed the use of trust-based networks, instead of anonymous peers, to improve recommendations. In order to secure a sufficient number of trust-based connections, they expanded the scope of trust-based social links beyond directly trusted parties. In the studies based on Epinions.com, the authors explained that the number of user's directly trusted connections (9.88 on average) is much smaller than the number of possible "like-minded" peers (all other users who share common items with target users; 160.73 users on average). Therefore, they included distantly connected users in the trust network (i.e., those in d hop distances). However, users of the recommender system assigned trust values only to directly connected parties, and the system does not know how much a target user would trust users in more distant relationships. In order to estimate the propagated trust values of a target user to indirectly connected users, the authors calculated a trust metric, which they call *MoleTrust*. With the assumption that trust is a binary scale (1 means 'trust' and 0 means 'absence of trust') and when trust-based links were expanded up to a predetermined maximum distance d, the estimated trust value of a social link in x distance ($d \geq x$) is $(d - x + 1)/d$ is modeled as a linear decay operation in propagating trust by distance. Once a group of trust-based users was chosen and the propagated trust values were calculated, in the last step to predict the missing ratings of candidate items, the user-to-user similarity—$sim_{u,v}$—was substituted with trust values between target users and their directly or indirectly trusted connections—$trust_{u,v}$ as noted in Eq. 4, where T_{uj} denotes u's trust-based social links who rated the item j [88].

$$\widehat{r_{uj}} = \overline{r_u} + \frac{\sum_{v \in T_{uj}} trust_{u,v}\left(r_{vj} - \overline{r_v}\right)}{\sum_{v \in T_{uj}} trust_{u,v}} \tag{4}$$

The study showed that trust-based recommendations solved cold-start user problem, improved predictions and attenuated the computational complexity [88]. Al-sharawneh and Williams used a similar approach while generalizing the trust weights as users' credibility by fusing them with users' expertise [5].

Golbeck and the colleagues [43–45] introduced another trust-based recommendation approach based on a new trust metric: *TidalTrust*. While *MoleTrust* included all trust-based connections that have rated a candidate item and are reachable within a predetermined maximum distance, *TidalTrust* focused on trust-based connections that have the *shortest path* from a target user. Even within a shortest path a candidate user was not considered to be a trust-based connection when trust estimates were below a certain threshold. To be precise, in order to calculate the prediction probability of a candidate item j, the recommender system first performed the Breadth First Search to aggregate the list of raters of item j. If none of a target user's direct social links rated item j, the search continued to trust-based links within x hop distance until raters were found. Once raters of item j were found, the recommender system inferred the trust values of raters in x hop distance by aggregating all trust values from a target user's direct links to the raters until x distance and calculating the propagated trust values, as in Eq. 5.

$$trust_{uw} = \frac{\sum_{v \in T_u} trust_{u,v} trust_{v,w}}{\sum_{v \in T_u} trust_{u,v}} \tag{5}$$

$$T_u^{+x} = \{w | trust_{uw} \geq \tau\} \tag{6}$$

Here, trust$_{uw}$ denotes the propagated trust value from a target user u to an indirect trust-based connection w. Once the trust values were computed, the connections whose estimated trust values were larger than a threshold τ were selected as the trust-based peer of the user u within x distance, represented as T_u^{+x} (refer to Eq. 6). Once a trust-based peer of the user is chosen and the trust values are calculated, the rating prediction is computed, as shown in Eq. 7. The experiment with FilmTrust data demonstrated that the use of trust enhanced the quality of recommendations [43–45, 123].

$$\widehat{r_{uj}} = \frac{\sum_{w \in T_{uj}^{+x}} trust_{u,w} r_{wj}}{\sum_{w \in T_{uj}^{+x}} trust_{u,w}} \tag{7}$$

The authors of the studies used propagated trust values emphasized the merit of social link-based recommendations in terms of accuracy. However, because of the relatively smaller number of socially connected users, social link-based approaches might also have lower coverage (the ratio of users who received at least one recommendation) than collaborative filtering. In order to create the synergy effect between two approaches, Victor et al. [130] introduced *EnsembleTrustCF* (Eq. 8).

$$\widehat{r_{uj}} = \overline{r_u} + \frac{\sum_{v \in T_{uj}} trust_{u,v} (r_{vj} - \overline{r_v}) + \sum_{v \in N_{uj}} sim_{u,v} (r_{vj} - \overline{r_v})}{\sum_{v \in T_{uj}} trust_{u,v} + \sum_{v \in N_{uj}} sim_{u,v}} \tag{8}$$

This approach aimed to use both types of candidates—anonymous "like-minded" peers (N_{uj}) and trust-based social connections T_{uj}—that rated a candidate item j. In cases when a user is connected with a target user through a direct or indirect trust-based link *and* also belongs to the target user's anonymous peer group, the system only took into account the trust value and ignored the peer similarity weight. According to the experiment using Epinions.com data, the approach of Victor et al. [130] produced results with a higher accuracy than other trust-based recommendation approaches and better coverage than collaborative filtering. Moradi and Ahmadian [94] also combined trust-based social connections with anonymous peers at the final stage to choose a list of recommended items.

The idea to modify traditional collaborating filtering by combining user-explicit social connections with traditional anonymous peers has been also explored for other types of social links, such friendship and professional collaboration networks [52, 65, 78]. For example, in a study based on a Korean online social networking site, Cyworld, Liu, and Lee [78] compared recommendations produced by a typical CF approach (based on the nearest neighbors' preferences), a social link-based approach (based on friend's preferences), and a combined approach (based on the combination of both the nearest neighbors and users' friends). The naïve hybridization of anonymous peers and social connections performed the best [78].

5.3 Recommendation Algorithms Based on Matrix Factorization

Despite its popularity in commercial systems, the nearest neighbor-based algorithms suffered from sparsity problems that are typical in systems with a large number of items and a limited number of ratings given by each user. With a sparse user-to-items ratings set, it is frequently hard to find users with a sufficient rating overlap who could be considered to be users with similar taste. The sparsity problem can degrade the recommendation performance by reducing the number of users who can receive any recommendations (i.e., reduced user coverage). Matrix factorization algorithms have been proposed as a solution for these sparsity problems. While modern collaborative filtering algorithms use a range of approaches, including Bayesian belief networks, clustering (e.g., k-means, density-based methods, hierarchical clustering), regression-based approaches, Markov decision processes, latent semantic model, etc. [117], advanced social link-based recommendations predominantly use matrix factorization technologies. Among the projects reviewed in Table 3, eleven projects (more than 30%) are based on matrix factorization. The general idea of the matrix factorization technologies is to compress a large and noisy user-to-item rating matrix into a more dense latent space model. The reduced model is based on the *latent features* of users and items, even though these features are usually hard for a human to interpret. The model is trained and optimized using the existing user-to-item data, and later, the predicted ratings of users for new items are computed by using latent space [68].

$$\widehat{R} = PQ^T \tag{9}$$

where $\widehat{R}$ is the matrix of predicted ratings. If f is the number of latent features and there are sets of users $U \in \{u_1, u_2, ..., u_n\}$ and items $I \in = \{i_1, i_2, ..., i_l\}$, matrix $P \in \mathbb{R}^{f \times n}$ represents the connections between users, and latent features and matrix $Q \in \mathbb{R}^{f \times l}$ represents the connections between items and latent features. In other words, vector p_u indicates how much the corresponding user u is interested in each of the f features, whereas a vector q_i shows how much the corresponding item i is associated with each of the f latent features. To learn the matrices P and Q and optimize the model, recommender systems minimize the sum of squared errors between the existing ratings R and $\widehat{R}$ like the following.

$$\min_{P,Q} \sum_u \sum_i W_{ui} \cdot \left(R_{ui} - \widehat{R}_{ui}\right)^2 + \lambda\left(\|P\|_F^2 + \|Q\|_F^2\right) \tag{10}$$

$\|\cdot\|_F^2$ is the Frobenius norm and $\lambda > 0$ is the regularization parameter. W_{ui} is the weight indicating that, if user u rated item i, the value equals 1; otherwise, the value equals 0. This objective function can find the minimum values by using gradient descent methods [81, 139].

To systematically analyze recommendation approaches based on matrix factorization, we followed the classifications of Yang et al. [139] and Tang et al. [122] that distinguish co-factorization methods, ensemble methods, and regularization methods.

Co-factorization
The projects using a co-factorization approach collectively factorize the user-to-item rating matrix and the user-to-user social link matrix. Therefore, in this collective factorization, there are matrices P and Q, and an additional matrix S—$n \times n$ matrix of

user-to-user social links. The authors specifically assume that users' latent feature representation is based on their social links, and a user u is represented by a vector in both P and S. SoRec [82] proposed by Ma and the colleagues belongs to this group. In building matrix S, the authors substitute the trust values with confidence values of the trust relations, by borrowing the local authority and local hub concepts from PageRank. They increased the confidence of trusted relations when a user is trusted by many other users and decreased confidence when a user trusts a lot of other users. In this study, users' social information can be captured, as in the following.

$$\widehat{S} = PZ^T \tag{11}$$

where $\widehat{S}$ is the predicted social relationship and $Z \in \mathbb{R}^{f \times n}$ is the factor feature matrix. With the assumption that the users' preferences can be learned from both rating and social information (i.e., user latent-feature matrix P is used to predict the user-to-item matrix $\widehat{R}$ *and* the social relation matrix $\tilde{S}$), the authors minimized the sum of the squared errors using the following objective function.

$$\min_{P,Q,Z} \sum_u \sum_i W_{ui} \cdot \left(R_{ui} - \widehat{R}_{ui} \right)^2 + \alpha \sum_u \sum_{v \in T_u} W_{uv}^s \cdot \left(S_{uv} - \widehat{S}_{uv} \right)^2 \\ + \lambda \left(\|P\|_F^2 + \|Q\|_F^2 + \|Z\|_F^2 \right) \tag{12}$$

In this study, based on the Epinions.com data set and the trust-based social links, W_{uv}^s is the social weighting indicating that, if a user u trusts another user v, it equals 1; otherwise, it equals 0, and T_U is the set of users whom a user u trusts.

Ensemble Methods

Ensemble methods aim to predict users' missing ratings using a linear combination of ratings from the users and their social links. *Social Trust Ensemble* [80] proposed by Ma et al. belongs to this group. In the study, the authors suggested that it would be possible to predict users' ratings using the following equation by including both the target user u's predicted ratings on the candidate item i and the weighted sum of predicted ratings for the item i from all of user u's social links. The power of social links' ratings on the final prediction for the item i can be controlled by the parameter α.

$$\widehat{R}_{ui}^* = \alpha P_u Q_i^T + (1 - \alpha) \sum_{v \in T_u} S_{uv} P_v Q_i^T \tag{13}$$

In this situation, the training objective function to minimize the sum of the square error can be expressed by the following equation:

$$\min_{P,Q} \sum_u \sum_i W_{ui} \cdot \left(R_{ui} - \widehat{R}_{ui}^* \right)^2 + \lambda \left(\|P\|_F^2 + \|Q\|_F^2 \right) \tag{14}$$

Regularization Methods

Regularization methods attempt to guide the matrix factorization process by keeping users' preferences as close as possible to their friends' preferences. The majority of social link-based matrix factorization approaches reviewed in Table 3 belong to this group: *SocialMF* [56], *Social Regularization* [83], *CircleCon* [140], *PWS* [135], De Meo's [30], and *RSboSN* [119]. Social Regularization proposed by Ma et al. [83] is a regularization method to consider the tastes of target users' friends differently, depending on their similarity to the target users.

$$
\min_{P,Q} \sum_u \sum_i W_{ui} \cdot \left(R_{ui} - \widehat{R}_{ui} \right)^2 + \beta \sum_u \sum_{v \in T_u} sim_{u,v} \cdot (P_u - P_v)^2 \\
+ \lambda_1 \|P\|_F^2 + \lambda_2 \|Q\|_F^2
\tag{15}
$$

where $sim_{u,v}$ is the information similarity between user u and v and the parameter β is to control the impact of social information [83]. The study tested two versions of the algorithm by calculating $sim_{u,v}$ as either vector space similarity or Pearson correlation coefficient (see Eq. 1). The results showed that the Pearson correlation coefficient was a better choice.

SocialMF [56] incorporated the mechanism of trust propagation in matrix factorization. Based on the social influence theory, the authors suggested that a latent feature vector of a user should be dependent on all latent feature vectors of all of that user's direct neighbors. We can make a target user's latent feature vector dependent on all direct and indirect social links, with decay weights for distances between the corresponding target user and their social links, by minimizing the following expression after normalizing each row of the social matrix S to 1 [56].

$$
\min_{P,Q,S} \sum_u \sum_i W_{ui} \cdot \left(R_{ui} - \widehat{R}_{ui} \right)^2 + \lambda \left(\|P\|_F^2 + \|Q\|_F^2 \right) \\
+ \beta \sum_{v \in T_u} \left\| P_u - \sum_{v \in T_u} S_{uv} P_v \right\|_F^2
\tag{16}
$$

CircleCon modified the *SocialMF* model to account for the social influence according to item category. In this project, which was based on Epinions.com data, the authors divided users' trust-based social links into sub-networks according to the category of their rated items. Then, they trained a separate matrix factorization model for each category as the following objective function [140].

$$
\min_{P^C,Q^C,S^C} \sum_u \sum_i \left(R_{ui}^c - \widehat{R}_{ui}^c \right)^2 + \beta \sum_{v \in T_u} \left\| P_u^C - \sum_{v \in T_u} S_{uv}^C P_v^C \right\|_F^2 \\
+ \lambda \left(\|P^C\|_F^2 + \|Q^C\|_F^2 \right)
\tag{17}
$$

5.4   Graph-Based Recommendation Approaches

While graph-based recommendation approaches were originally explored for traditional recommendations, they have become especially popular in the area of social
link-based recommendations, because social links could be most naturally represented
as a social graph. Among the projects reviewed in Table 3, seven studies used various
graph-based recommendation algorithms (Bellogín et al. [8], Deng et al. [32], Jamali
and Ester [54, 55], Konstas et al. [67], Wang et al. [133], and Yuan et al. [145]).

Bellogín et al. [8] suggested a social recommendation approach based on users'
friendship networks and using a Breadth-First Search algorithm. As a comparison, the
authors also fused users' reviewed item lists with users' friendship networks to compute
random walk (RW) and random walk with restart (RWR)[11]. According to the empirical
evaluation, the proposed social recommendations produced more accurate suggestions
than conventional CF recommendations with the RW and RWR algorithms [8].

As a way to integrate two different types of recommendations with one another—
trust-based recommendations and item-based collaborative filtering—Jamali and Ester
used an RW model called *TrustWalker* [54, 55] in the context of the Epinions.com data
set. Their proposed algorithm starts with RW on the trust network. Among a target user's
directly trusted connections, the algorithm finds raters of a given candidate item. If there
are any raters of the item, the rating value is returned; otherwise, the algorithm expands
the search to trusted users of the directly trusted links. This process continues recursively
until rating values of the candidate item are found among a target user's directly and
indirectly trusted links. However, in order to prevent walking too far in the trust-based
network, if a directly or indirectly trusted user rated an item which is quite similar to the
candidate item and the similarity weighted by the distance is above a certain threshold, the
algorithm stops the walking and returns the ratings of similar items [54, 55].

In contrast to other RW-based approaches that randomly select the steps of each
walk, Deng et al. [32] proposed a different RW-based recommendation approach,
called *RelevantTrustWalker*, which chooses the next movement according to trust
statement and information similarity (i.e., trust relevancy). The information similarity
was specifically calculated based on latent vectors computed by matrix factorization. In
Konstas et al. [67] and Wang et al. [133], the authors expanded the trust network
graphs by including additional types of nodes and links. Konstas et al. [67] used a
graph that included not only users' online friends, but also their favorite music and the
associated social tags. The edge value of user-to-music relations was the play-count
numbers of songs and the edge values of users-to-tags relations was the frequency of
the users' tag usage [67]. Wang et al. used a graph with both users' direct friends and
their anonymous peers [133]. Both studies [67, 133] applied the RWR algorithm to the
constructed graphs.

5.5   Advanced Hybrid Recommendation Approaches

Despite the various problems of the conventional collaborative filtering reviewed in
Sect. 3.1, its simplicity and good performance still garners significant attention. As one

---

[11] For more detailed information about random walk and random walk with restart, refer to [126].

effort to leverage the power of collaborative filtering recommendations while also using the benefits of social link-based recommendations, several researchers have explored hybrid recommendation approaches by fusing users' online social links with collaborative filtering. Simple hybrid recommendations are common, while sometimes people might prefer to explore more advanced hybrid recommendation approaches.

In developing their social link-based recommendation algorithms, Gürsel and Sen [50] emphasized that users have different preferences for their friends within different topics of interests. In their study using the Flickr dataset, Gürsel and Sen classified Flickr photos into ten categories, using users' social tags. Specifically, the authors composed a topical dictionary of social tags for each category and calculated the probability of the number of social tags for an item being associated with each topical dictionary. Then, the authors counted how many times a target user commented on a friend's photo that belonged to one of the ten categories, and calculated the probability of the target user liking an item posted by one of their friends using a Bayes theorem. This study interpreted users' comments on photos to mean their interests on the photos; hence, they are positive feedback. In the empirical evaluations, the performance of the proposed social network-based recommendation algorithm was significantly better than content-based recommendations that used users' social tags and random sampling-based recommendations with respect to both precision and recall [50]. This study contributed to a better understanding of social link-based recommendations by substantiating how to incorporate users' topical preferences on social peers into information personalization.

Several hybrid recommendation studies have fused users' social context with content-focused metadata, such as social tags or text descriptors of items. In particular, social media-related studies paid a significant amount of attention to social tags. Social tags are usually considered to be users' cognitive descriptors of the tagged items (an important form of collective intelligence) and have a lot of implications for various information access-related tasks.

Guy et al. [52, 53] used both online social networks and social tags as an information source to generate recommendations. In these studies, based on a company's social application suite, users were able to manage and share various social media items, such as Web sites of interest, wiki pages, blogs, files, and communities of interest. Users add social tags not only on those social media items, but also in relation to other users. To compile the list of a user's social connections, the system used the organization's HR chart, bookmarking and social tagging activities, and other systems to produce various types of links (e.g. friends, colleagues, other users who assigned tags on the same target users, commenters of users' bookmarked items, and so on). Social tags that a user assigned to items or were assigned to that user by other users were aggregated as text descriptors about that user's preference. The results demonstrated that both sources are valuable to increase quality, reinforce diversity, and to offer explanations of recommendations.

Pera and Ng [101] introduced a hybrid approach that fused the metadata properties of books with users' social context. First, in order to choose candidate books to recommend, social tag-based content similarities between the candidate books and target users' favorite books were counted, rather than the similarities of content derived directly from the books (such as from the titles, abstracts, or the authors' names). In the subsequent process, they aggregated the ratings of the candidate items given by the target users' friends. They also computed the similarity between friends' tastes as a reflection of the tastes of the target users. In the experiment using LibraryThing, social link-based recommendations were contrasted with the collaborative filtering recommendations provided by Amazon and content-based recommendations provided by LibraryThing were used as a baseline. The results show that the quality of the hybrid recommendations combining metadata information and friend relationships was better than other two baseline approaches in terms of precisions and ranks.

Carrer-Neto et al. [22] proposed a different hybrid recommendation approach rooted in a domain knowledge base and users' social networks. In their study, which was performed in a movie recommendation context, various kinds of metadata, such as genre, producer, actor, director, location, and award, were considered to build user preference models. When predicting favorable items, the recommender system accounted for not only a target user's preference model, but also models of their social connections. In contrast to other studies, the use of social connections' preferences deteriorated the overall quality of recommendations. This study hinted that users' sociality is not a one-size-fits-all solution for improving all kinds of recommendations, and it is critical to choose a right way to fuse the sociality with personalized recommendations.

6 Problems and Prospects of Social Recommendations

6.1 Evaluation of Social Recommendation

Evaluation is an important aspect of research on recommender systems. Serious attention to evaluation enables the field to prosper and mature through the development of gradually better and more powerful approaches. In this book, a detailed coverage of various approaches used to evaluate recommender systems is offered in Chap. 10 [63]. Instead of duplicating this information, this subsection attempts to provide a brief summary of evaluation approaches that have been used for social link-based recommendation and to attract attention to a relative lack of user-centered approaches. Figure 6 summarizes the evaluation criteria used in the studies that were reviewed in Table 3. As the figure shows, evaluation of existing social link-based recommendations is predominantly focused on objective quantifiable measures collected through automatic data-driven evaluation. The assessed categories include predictive accuracy (mean absolute error (MAE), root mean squared error (RMSE)), classification accuracy (precision, recall, F-measure, etc.), ranking (Spearman's rho, Kendall's coefficient, normalized discounted cumulative gain (NDCG)), coverage, and efficiency [109]. There are comparatively few efforts to understand users' subjective opinions about social link-based recommendations. In other words, user prospects in relation to social link-based recommendations are rarely considered. Among the 36 studies reviewed in Table 3, only four studies engaged human subjects with the evaluation process. In particular, none of the trust-based recommendations was assessed through human-subject studies.

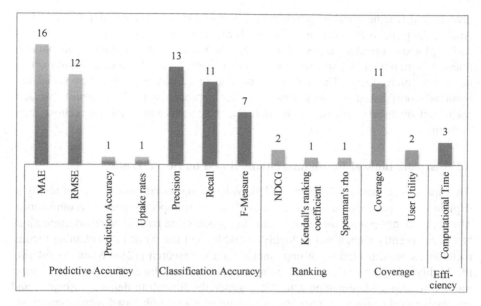

Fig. 6. Objective evaluation criteria used in social link-based recommendations.

As one of the early efforts, Sinha and Swearingen [114] explored both the subjective and qualitative user perceptions of recommendations. To compare the quality of direct friend-to-friend recommendations and machine-generated recommendations, the authors measured various qualitative evaluation criteria, including users' *psychological burden* or efforts to initiate their recommendations (the amount of data users have to input to receive recommendations), the time to receive recommendations, their perceived usefulness and trustworthiness, and other interface-related issues of recommendations and explanations, such as ease of use, navigability, and color schemes, among others. In the field of recommender systems, there are a number of commonly recognized subjective quality measures, such as novelty, serendipity, confidence [108], perceived usefulness, and trustworthiness [104]. Unfortunately, there are too few user studies of social link-based recommendations to reliably determine the user-perceived value of these approaches. For instance, one study suggested that social link-based recommendations deliver more novel recommendations than collaborative filtering [47], but another study suggested the opposite [113]. We believe that the next generation of research on social link-based recommendations should pay more attention to user-centered quality evaluation.

There is also an insufficient volume of evaluations that examines to what extent social link-based recommenders address the known problems in the field of recommender systems. As reviewed in Sect. 3.1, previous work on social recommender systems was motivated to a considerable extent by the various weaknesses of traditional collaborative filtering approaches. However, among these weaknesses, only the cold-start and data sparsity problems have been sufficiently addressed when evaluating social recommenders. For example, Al-sharawneh and Williams [5], Deng et al. [32], Jamali and Ester [54–56], Massa and Avesani [88], and Moradi and Ahmadian [94] investigated whether their proposed approaches can solve the cold-start user problem. Among the projects reviewed in Table 3, Moradi and Ahmadian [94] is the only study that explored a broader variety of problems associated with traditional collaborative filtering including cold-start users, opinionated users, black-sheep users, controversial items, and niche items. The authors were able to demonstrate that their social recommendations, based on the nearest-neighbor approach with the reliability weights calculated by users' trust-based social links, were sufficiently able to address these problems.

6.2   Explanations of Recommendations Based on Social Links

An important advantage of recommendations based on social links is a better ability to explain how recommendations are generated and why a specific item is recommended. The ability to understand the recommendation process and individual recommendations has been recently recognized as highly valuable, and the work on explaining recommendations has emerged as an important stream of research [125]. When considering the highly complex mathematical computations and the black-box process of most modern recommendation approaches, it is generally difficult to deliver reasonable and persuasive explanations. In contrast, the nature of social link-based recommendations based on users' self-defined social links makes social recommendation relatively easy to explain and comprehend. While at the moment, only a fraction of research on social recommendation has explored explanation approaches and the value of explanations, it is certainly a promising direction to pursue.

The most natural and explored explanation approach in the area of social link-based recommendation is connecting each recommended item with the target user's social connections that were used to select it. Bourke and the colleagues [18] executed an experiment to compare users' acceptance of recommendations, either with or without the explanations about the source of their recommendations. In a study with Facebook users, the authors suggested favorable movies and TV shows, according to the tastes of the participants' online friends. The results showed that participants gave significantly higher ratings when they could see the sources of their recommendations [18]. The series of work by Bonhard et al. [13, 15] and another work by Guy et al. [52] also positively proved that recommendation explanation increases user satisfaction and overall acceptance of recommendations.

Knijnenburg et al. [65] explored a more extensive explanation approach that was based on an interactive visualization. The main goal of this work was to develop a visualization that makes the *whole process* of social recommendation transparent, explainable, and controllable. It also investigates whether users' perceived inspectability and controllability increases the positive impressions of recommendations. In a visual interface that was originally applied for music artist recommendation in a Facebook context, the authors used an interactive graph (Fig. 7) to transparently display the sources of their recommendations to the recipients. The left side of the graph shows the target user's favorite artists, and by clicking on entities on the graph, the system shows how the recommendations in the right side list are generated. The transparency of the recommendation process was designed to increase its *inspectability*. The authors also enabled *controllability* by allowing the users to interactively adjust the weight of both

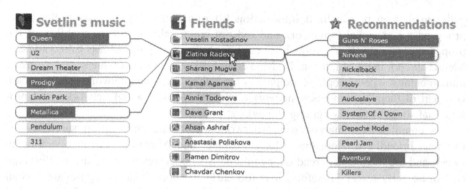

Fig. 7. An interactive graph explaining recommendations using users' favorite information, along with that of their friends [65].

their favorite artists and their friends. Questionnaire-based evaluations with 267 Facebook users showed that both inspectability and controllability have a positive effect on the user experience: they result in increased understandability and improved perception about recommendation quality. Later, the authors attempted to generalize this visual recommendation approach and apply it to other recommendation areas, such as job recommendation [16].

6.3 Cross-System and Multidimensional Online Social Networks

Due to the explosive popularity of online social networking systems (SNSs), users have been enjoying a plethora of online social networks. Even though online users have already been participating in miscellaneous SNSs, the unique features and functionalities of newly emerging SNSs may entice users into joining another. However, the current SNS market has raised several problems. Along with addiction to SNSs, violations of privacy, cyber bullying, and the spread of malevolent information, there is 'walled garden' problem [142]. The problem indicates that users' profiles and online social network information exclusively exist on a single SNS and are not shared with other systems. Hence, users' online profiles, shared information, and social connections may be scattered across many different SNSs. In order to stay socially active, users have to interact with different subsets of online friends on different SNSs. Even though studies like Subrahmanyam et al. [118] insisted that online users take advantage of various SNSs to strengthen different aspects of social networks, doing so evidently makes online users feel overloaded.

The scattering of online profiles and social networks across several SNSs also cause cold-start user problems. In one system, a user might have diligently established connections and shared interesting information for a considerable time. However, when that user moves to another system, all this information collection is ignored and the user needs to start again with an empty profile. To cope with this problem, several efforts have been proposed. Vu et al. [131] also introduced an exemplary system to aggregate a variety of social data—such as users' textual comments, their friends and groups information, and their online profiles—into a single framework, and use the gathered data in personalized information filtering or sharing. However, the paper proposed a conceptual framework and did not empirically assess the viability of the proposed personalization approach. Even though the main purpose was to recommend friends and not items of interest, De Meo et al. [30] also focused on the social nature of human beings, where people participate in more than one social network (e.g., a person is a part of their kinship-based networks, friendship-based social networks, and professional social networks, among others).

As commercial solutions, social network/social data aggregation services—for instance, FriendFeed, Hootsuite, Flock, Postano, and Alternion—are in operation to sweep and organize data spread over multiple SNSs. However, all of these efforts are still in an early stage. Therefore, it is too early to expect that the aggregated social networks across multiple SNSs have been used in personalizing users' information access. Even so, personalized recommendation based on multi-SNS social networks is a promising and necessary direction to pursue.

On the other hand, existing social link-based recommendation approaches are not quite ready to operate with a variety of links imported from different social networking applications. As shown in Table 1, the majority of existing approaches were developed to work with exactly one type of social link, although similar approaches are sometimes independently explored with different link types. To a large extent, this issue is related to the lack of datasets and systems that include multiple types of links; with no truly multi-dimensional data, it is hard to evaluate approaches that use more than one link type. However, in several cases, different types of social connections within one system are available, yet they are frequently ignored in the current body of work. For instance, a number of social systems (including Facebook) enable their users to socially associate with other users not only via direct connections (i.e., friendship) but also via membership in the same group. Social media systems also provide users the functionality to separate online social links into multiple sets. Google+ users, for instance, are able to sort their connections into several separate groups, such as friends, family, acquaintances, colleagues and more. On some social media systems, we can even freely define and name the different kinds of online social links.

As the number of different social links that connect users within and across social systems increases, it is becoming more and more important to develop social recommendation approaches that can work with many kinds of social links in parallel. The main problem in this context is the integration of different link types. Let's imagine a situation where our target user A is a friend of another user B and is in a co-authorship relation with user C. If a social system can use both of these connection types, is it okay to simply assign equal weights to both User B and C in order to generate recommendations for user A? Otherwise, how can we put different weights on multiple types of social links within a system so as to accurately gauge target user A's preferences? Kazienko et al. [60] presented an early study on this topic. The authors built multidimensional social networks based on users' various activities on the Flickr photo sharing system and used the multidimensional network in personalized recommendation. They established and used five kinds of social connections: one direct social network derived from users' contact list and four indirect object-centered social connections inferred by their behaviors on Flickr (two users added tags on the same items, joined the same group(s), marked one another's photo(s) as their favorite, and commented on the same items.). The value representing the strength of each link was calculated and added separately. This study provided insights on generating personalized recommendations, based on multidimensional social networks.

6.4   Privacy in Online Social Networks

Due to the prevalence of cheap and easy communication tools, social media users have enjoyed opportunities for meeting new friends, expanding their online social networks, gaining new and relevant information, and propagating their opinions, among other activities [48]. In order for users to leverage these values of online social media, they have to provide and gradually expand their personal information in the form of 'user profiles'. Most social media systems ask users to share personally identified or identifiable data, including personal histories, such as academic or professional affiliations, personal traits and tastes, or information preferences with other users. The fun, useful, and innovative nature of social media frequently causes users to ignore various risks related to revealing their personal and social information online. However, several studies [1, 48, 107] have reported that online users are becoming more and more conscious of and protective about their privacy. A study conducted in 2005 [48] used a sample of 4,540 users who shared identifiable names, phone numbers, personal images, and characteristics (e.g. their current residence, dating preferences and relationship status, or political views and various interests). Among them, only three users changed their profile visibility and only 1.2% of them changed their privacy settings. However, in another study conducted in 2011 [34], 33% of 1.5 million users had changed their profiles to private. Despite the remarkably increased recognition of privacy among SNS users, there are reportedly some technical leakages that are out of users' control, like so-called 'silent listeners'. Third-party applications and online advertisers can take advantage of users' profile information without receiving explicit consent to do so [115]. Even in cases when a system enables a reliable level of privacy protection and a user enables those options, it might not be sufficient, since a user still remains the weakest spot in the system. A striking case reportedly showed how easily online profiles could be compromised. A college student wrote a computer program to systematically send friend request messages to 250,000 American Facebook users, and one-third of them accepted those friend requests [59]. It is reasonable to assume that some malicious users could do the same, and access our private information by pretending to be our online friends. The invaded user profiles can, of course, put users at a variety of risks and attacks [69].

In this book, an extensive treatment of privacy issues in the context of social information access is provided in Chap. 2 [64]. The main goal of this section is to focus on the issues that specifically connect social recommendations and privacy. Indeed, because social link-based recommendations strongly rely on users' social links to recommend items, it has been recognized that the availability of social link-based recommendations might add additional challenges to the problem of privacy in SNSs. For instance, a user's shopping history, which is generally hidden even from friends, could be leaked to her friends through social link-based recommendations. Worse, our distrust in some friends (which we might not be eager to reveal) could be leaked out implicitly as a part of the social recommendation process. The issues of privacy in social recommendation are gradually becoming more critical, and are causing some researches to focus on privacy-sensitive recommendation approaches. In some cases, social recommendation approaches might have to be modified to ensure a desired level of privacy; in other cases, new approaches should be developed to address privacy concerns. A pioneer work of Machanavajjhala et al. [85] investigated the correlation between recommendation accuracy and the degree of privacy preservation, and substantiated that good and private recommendations are difficult to implement via Wikipedia and Twitter datasets. However, their study aimed to recommend social links to connect, not information items or products.

7 Conclusion

This chapter has focused on a specific information personalization technology in the context of social information access: recommending relevant information items using explicit user-defined social links. Social links are an important type of socially contributed information, and their use for generating recommendations is currently the principal approach to leverage the power of this information for better information access. The goal of this chapter is to provide an extensive overview of current research on social link-based recommendations while specifically emphasizing "how to" issue, i.e., recommendation algorithms. We started with a brief overview of problems associated with traditional collaborative filtering, as well as arguments in favor of using social links for recommendations. We also presented the background rationale for online social dynamics and various branches of social recommendations. We classified and reviewed existing social link-based recommendations according to the kinds of social networks used in recommendations, target applications/data sources, recommendation algorithms, and evaluation criteria. We also reviewed and explained the main classes of recommendation algorithms used in social recommendations: direct friend-to-friend recommendations, nearest neighbor recommendations, graph structure-based recommendations, and matrix factorization techniques. We also separately reviewed hybrid recommendations that attempted to fuse traditional recommendation approaches with social link-based recommendation. Finally, we discussed several emergent issues or social recommender systems, and connected them with possible areas for future research and possible improvements of social recommendations.

References

1. Acquisti, A., Taylor, C., Wagman, L.: The economics of privacy. J. Econ. Lit. **54**(2), 442–492 (2016)
2. Akehurst, J., et al.: CCR: a content-collaborative reciprocal recommender for online dating. In: Proceedings of the Twenty-Second International Joint Conference on Artificial Intelligence, vol. 3. AAAI Press (2011)
3. Akther, A., Kim, H.-N., Rawashdeh, M., El Saddik, A.: Applying latent semantic analysis to tag-based community recommendations. In: Kosseim, L., Inkpen, D. (eds.) AI 2012. LNCS (LNAI), vol. 7310, pp. 1–12. Springer, Heidelberg (2012). https://doi.org/10.1007/978-3-642-30353-1_1

4. Al-Oufi, S., Kim, H.-N., Saddik, A.E.: A group trust metric for identifying people of trust in online social networks. Expert Syst. Appl. **39**(18), 13173–13181 (2012)
5. Al-sharawneh, J., Williams, M.: Credibility-aware web-based social network recommender: follow the leader. In: Proceedings of the 2nd ACM RecSys Workshop on Recommender Systems and the Social Web, Barcelona, Spain (2010)
6. Arazy, O., Kumar, N., Shapira, B.: Improving social recommender systems. IT Prof. **11**(4), 38–44 (2009)
7. Baatarjav, E.-A., Phithakkitnukoon, S., Dantu, R.: Group recommendation system for Facebook. In: Meersman, R., Tari, Z., Herrero, P. (eds.) OTM 2008. LNCS, vol. 5333, pp. 211–219. Springer, Heidelberg (2008). https://doi.org/10.1007/978-3-540-88875-8_41
8. Bellogín, A., et al.: An empirical comparison of social, collaborative filtering, and hybrid recommenders. ACM Trans. Intell. Syst. Technol. **4**(1), 1–29 (2013)
9. Bellogin, A., Parapar, J.: Using graph partitioning techniques for neighbour selection in user-based collaborative filtering. In: Proceedings of the Sixth ACM Conference on Recommender Systems, pp. 213–216. ACM, Dublin (2012)
10. Berkovsky, S., Freyne, J.: Group-based recipe recommendations: analysis of data aggregation strategies. In: Proceedings of the Fourth ACM Conference on Recommender Systems, pp. 111–118. ACM, Barcelona (2010)
11. Bobadilla, J., et al.: Recommender systems survey. Knowl.-Based Syst. **46**, 109–132 (2013)
12. Bogers, T.: Tag-based recommendation. In: Brusilovsky, P., He, D. (eds.) Social Information Access. LNCS, vol. 10100, pp. 441–479. Springer, Cham (2018)
13. Bonhard, P., Sasse, M.: 'Knowing me, knowing you'—using profiles and social networking to improve recommender systems. BT Technol. J. **24**(3), 84–98 (2006)
14. Bonhard, P., Sasse, M.A., Harries, C.: "The devil you know knows best": how online recommendations can benefit from social networking. In: Proceedings of the 21st British CHI Group Annual Conference on HCI 2007: People and Computers XXI: HCI... But Not as We Know It, vol. 1, pp. 77–86. British Computer Society: University of Lancaster, United Kingdom (2007)
15. Bonhard, P., Sasse, M.A., Harries, C.: "The devil you know knows best": how online recommendations can benefit from social networking. In: Proceedings of the 21st British HCI Group Annual Conference on People and Computers: HCI... But Not as We Know It, vol. 1, pp. 77–86. British Computer Society: University of Lancaster, United Kingdom (2007)
16. Bostandjiev, S., et al.: LinkedVis: exploring social and semantic career recommendations. In: Proceedings of the 2013 International Conference on Intelligent User Interfaces, pp. 107–116. ACM, Santa Monica (2013)
17. Bothorel, C.C., et al.: Location recommendation with social media data. In: Brusilovsky, P., He, D. (eds.) Social Information Access. LNCS, vol. 10100, pp. 624–653. Springer, Cham (2018)
18. Bourke, S., McCarthy, K., Smyth, B.: Power to the people: exploring neighbourhood formations in social recommender system. In: Proceedings of the Fifth ACM Conference on Recommender Systems, pp. 337–340. ACM, Chicago (2011)
19. Breslin, J., Decker, S.: The future of social networks on the internet: the need for semantics. IEEE Internet Comput. **11**(6), 86–90 (2007)
20. Brzozowski, M.J., Hogg, T., Szabo, G.: Friends and foes: ideological social networking. In: Proceeding of the Twenty-Sixth Annual SIGCHI Conference on Human Factors in Computing Systems, pp. 817–820. ACM, Florence (2008)
21. Cantador, I., Castells, P.: Group recommender systems: new perspectives in the social web. In: Pazos Arias, J.J., Vilas, A.F., Díaz Redondo, R.P. (eds.) Recommender Systems for the Social Web. Intelligent Systems Reference Library, vol. 32, pp. 139–157. Springer, Heidelberg (2012). https://doi.org/10.1007/978-3-642-25694-3_7

22. Carrer-Neto, W., et al.: Social knowledge-based recommender system. Application to the movies domain. Expert Syst. Appl. **39**(12), 10990–11000 (2012)
23. Centola, D., et al.: Homophily, cultural drift, and the co-evolution of cultural groups. J. Conflict Resolut. **51**(6), 905–929 (2007)
24. Chen, C.C., et al.: An effective recommendation method for cold start new users using trust and distrust networks. Inf. Sci. **224**, 19–36 (2013)
25. Chen, J., Liu, Y., Li, D.: Dynamic group recommendation with modified collaborative filtering and temporal factor. Int. Arab J. Inf. Technol. (IAJIT) **13**(2), 294–301 (2016)
26. Chia, P.H., Pitsilis, G.: Exploring the use of explicit trust links for filtering recommenders: a study on Epinions.com. Inf. Media Technol. **6**(3), 871–883 (2011)
27. Christensen, I., Schiaffino, S., Armentano, M.: Social group recommendation in the tourism domain. J. Intell. Inf. Syst. **47**(2), 209–231 (2016)
28. Chu, S.-C., Kim, Y.: Determinants of consumer engagement in electronic word-of-mouth (eWOM) in social networking sites. Int. J. Advert. **30**(1), 47–75 (2011)
29. Crandall, D., et al.: Feedback effects between similarity and social influence in online communities. In: Proceeding of the 14th ACM SIGKDD International Conference on Knowledge Discovery and Data Mining, pp. 160–168. ACM, Las Vegas (2008)
30. De Meo, P., et al.: Improving recommendation quality by merging collaborative filtering and social relationships. In: 2011 11th International Conference on Intelligent Systems Design and Applications (ISDA) (2011)
31. Debnath, S., Ganguly, N., Mitra, P.: Feature weighting in content based recommendation system using social network analysis. In: Proceeding of the 17th International Conference on World Wide Web, pp. 1041–1042. ACM, Beijing (2008)
32. Deng, S., Huang, L., Xu, G.: Social network-based service recommendation with trust enhancement. Expert Syst. Appl. **41**(18), 8075–8084 (2014)
33. Deutsch, M., Gerard, H.: A study of normative and informational social influences upon individual judgement. J. Abnorm. Soc. Psychol. **51**(3), 629–636 (1955)
34. Dey, R., Jelveh, Z., Ross, K.: Facebook users have become much more private: a large-scale study. In: 2012 IEEE International Conference on Pervasive Computing and Communications Workshops (PERCOM Workshops) (2012)
35. Diaz-Aviles, E., et al.: Real-time top-n recommendation in social streams. In: Proceedings of the Sixth ACM Conference on Recommender Systems, pp. 59–66. ACM, Dublin (2012)
36. Dugan, C., et al.: The dogear game: a social bookmark recommender system. In: Proceedings of the 2007 International ACM Conference on Supporting Group Work, pp. 387–390. ACM, Sanibel Island (2007)
37. Eirinaki, M., Louta, M.D., Varlamis, I.: A trust-aware system for personalized user recommendations in social networks. IEEE Trans. Syst. Man Cybern.: Syst. **44**(4), 409–421 (2014)
38. Ellison, N.B., et al.: Cultivating social resources on social network sites: Facebook relationship maintenance behaviors and their role in social capital processes. J. Comput.-Mediat. Commun. **19**(4), 855–870 (2014)
39. Fiske, A.P.: The Four elementary forms of sociality: framework for a unified theory of social relations. Psychol. Rev. **99**(4), 689–723 (1992)
40. Fond, T.L., Neville, J.: Randomization tests for distinguishing social influence and homophily effects. In: Proceedings of the 19th International Conference on World Wide Web, pp. 601–610. ACM, Raleigh (2010)
41. Friedkin, N.E.: Structural bases of interpersonal influence in groups: a longitudinal case study. Am. Sociol. Rev. **58**(6), 861–872 (1993)

42. Golbeck, J.: Introduction to computing with social trust. In: Golbeck, J. (ed.) Computing with Social Trust. Human-Computer Interaction Series, pp. 1–5. Springer, London (2009). https://doi.org/10.1007/978-1-84800-356-9_1
43. Golbeck, J.: Trust and nuanced profile similarity in online social networks. ACM Trans. Web 3(4), 1–33 (2009)
44. Golbeck, J., Hendler, J.: FilmTrust: movie recommendations using trust in web-based social networks. In: 3rd IEEE Consumer Communications and Networking Conference, CCNC 2006 (2006)
45. Golbeck, J., Hendler, J.: Inferring binary trust relationships in Web-based social networks. ACM Trans. Internet Technol. 6(4), 497–529 (2006)
46. Goldberg, D., et al.: Using collaborative filtering to weave an information tapestry. Commun. ACM 35(12), 61–70 (1992)
47. Groh, G., Ehmig, C.: Recommendations in taste related domains: collaborative filtering vs. social filtering. In: Proceedings of the 2007 International ACM Conference on Supporting Group Work, pp. 127–136. ACM, Sanibel Island (2007)
48. Gross, R., Acquisti, A.: Information revelation and privacy in online social networks. In: Proceedings of the 2005 ACM Workshop on Privacy in the Electronic Society, pp. 71–80. ACM, Alexandria (2005)
49. Gunes, I., et al.: Shilling attacks against recommender systems: a comprehensive survey. Artif. Intell. Rev. 42(4), 767–799 (2014)
50. Gürsel, A., Sen, S.: Producing timely recommendations from social networks through targeted search. In: Proceedings of the 8th International Conference on Autonomous Agents and Multiagent Systems, vol. 2, pp. 805–812. International Foundation for Autonomous Agents and Multiagent Systems, Budapest (2009)
51. Guy, I.: People recommendation on social media. In: Brusilovsky, P., He, D. (eds.) Social Information Access. LNCS, vol. 10100, pp. 570–623. Springer, Cham (2018)
52. Guy, I., et al.: Personalized recommendation of social software items based on social relations. In: The 2009 ACM Conference on Recommender Systems, RecSys 2009. ACM, New York (2009)
53. Guy, I., et al.: Social media recommendation based on people and tags. In: Proceedings of the 33rd International ACM SIGIR Conference on Research and Development in Information Retrieval, pp. 194–201. ACM, Geneva (2010)
54. Jamali, M., Ester, M.: TrustWalker: a random walk model for combining trust-based and item-based recommendation. In: Proceedings of the 15th ACM SIGKDD International Conference on Knowledge Discovery and Data Mining, KDD 2009. ACM, Paris (2009)
55. Jamali, M., Ester, M.: Using a trust network to improve top-N recommendation. In: Proceedings of the Third ACM Conference on Recommender Systems, pp. 181–188. ACM, New York (2009)
56. Jamali, M., Ester, M.: A matrix factorization technique with trust propagation for recommendation in social networks. In: Proceedings of the Fourth ACM Conference on Recommender Systems, pp. 135–142. ACM, Barcelona (2010)
57. Jannach, D., Lerche, L., Zanker, M.: Recommending based on implicit feedback. In: Brusilovsky, P., He, D. (eds.) Social Information Access. LNCS, vol. 10100, pp. 510–569. Springer, Cham (2018)
58. Jiang, M., et al.: Social contextual recommendation. In: Proceedings of the 21st ACM International Conference on Information and Knowledge Management, pp. 45–54. ACM, Maui (2012)
59. Jump, K.: A new kind of fame. In: Missourian, 21 July 2008

60. Kazienko, P., Musial, K., Kajdanowicz, T.: Multidimensional social network in the social recommender system. IEEE Trans. Syst. Man Cybern. Part A Syst. Hum. **41**(4), 746–759 (2011)
61. Kim, H.-N., Rawashdeh, M., Saddik, A.E.: Tailoring recommendations to groups of users: a graph walk-based approach. In: Proceedings of the 2013 International Conference on Intelligent user Interfaces, pp. 15–24. ACM, Santa Monica (2013)
62. Klein, A., Ahlf, H., Sharma, V.: Social activity and structural centrality in online social networks. Telematics Inform. **32**(2), 321–332 (2015)
63. Kluver, D., Ekstrand, M., Konstan, J.: Rating-based collaborative filtering: algorithms and evaluation. In: Brusilovsky, P., He, D. (eds.) Social Information Access. LNCS, vol. 10100, pp. 344–390. Springer, Cham (2018)
64. Knijnenburg, B.P.: Privacy in social information access. In: Brusilovsky, P., He, D. (eds.) Social Information Access. LNCS, vol. 10100, pp. 19–74. Springer, Cham (2018)
65. Knijnenburg, B.P., et al.: Inspectability and control in social recommenders. In: Proceedings of the Sixth ACM Conference on Recommender Systems, pp. 43–50. ACM, Dublin (2012)
66. Kompan, M., Bielikova, M.: Personalized recommendation for individual users based on the group recommendation principles. Stud. Inform. Control **22**(3), 331–342 (2013)
67. Konstas, I., Stathopoulos, V., Jose, J.M.: On social networks and collaborative recommendation. In: Proceedings of the 32nd International ACM SIGIR Conference on Research and Development in Information Retrieval, pp. 195–202. ACM, Boston (2009)
68. Koren, Y., Bell, R.: Advances in collaborative filtering. In: Ricci, F., Rokach, L., Shapira, B. (eds.) Recommender Systems Handbook, pp. 77–118. Springer, Boston (2015). https://doi.org/10.1007/978-1-4899-7637-6_3
69. Krishnamurthy, B., Wills, C.E.: On the leakage of personally identifiable information via online social networks. In: Proceedings of the 2nd ACM Workshop on Online Social Networks, pp. 7–12. ACM, Barcelona (2009)
70. Kutty, S., Chen, L., Nayak, R.: A people-to-people recommendation system using tensor space models. In: Proceedings of the 27th Annual ACM Symposium on Applied Computing, pp. 187–192. ACM, Trento (2012)
71. Lam, C.: SNACK: incorporating social network information in automated collaborative filtering. In: Proceedings of the 5th ACM Conference on Electronic Commerce, pp. 254–255. ACM, New York (2004)
72. Lee, D.: How to measure the information similarity in unilateral relations: the case study of *Delicious*. In: Proceedings of the International Workshop on Modeling Social Media, pp. 1–4. ACM, Toronto (2010)
73. Lee, D., Brusilovsky, P.: Improving recommendations using watching networks in a social tagging system. In: The Proceedings of iConference 2011, Seattle, WA, USA (2011)
74. Lee, D.H., Brusilovsky, P.: Does trust influence information similarity? In: Proceedings of Workshop on Recommender Systems & the Social Web, the 3rd ACM International Conference on Recommender Systems. ACM, New York (2009)
75. Lee, D.H., Brusilovsky, P.: Social networks and interest similarity: the case of CiteULike. In: Proceedings of the 21th ACM Conference on Hypertext and Hypermedia. ACM, Toronto (2010)
76. Lee, D.H., Brusilovsky, P.: Using self-defined group activities for improving recommendations in collaborative tagging systems. In: Proceedings of the Fourth ACM Conference on Recommender Systems, pp. 221–224. ACM, Barcelona (2010)
77. Lee, D.H., Brusilovsky, P., Schleyer, T.: Recommending collaborators using social features and MeSH terms. In: ASIST 2011 Annual Meeting, New Orleans, LA, USA (2011)

78. Liu, F., Lee, H.J.: Use of social network information to enhance collaborative filtering performance. Expert Syst. Appl. **37**(7), 4772–4778 (2010)
79. Lumbreras, A., Gavaldà, R.: Applying trust metrics based on user interactions to recommendation in social networks. In: Proceedings of the 2012 International Conference on Advances in Social Networks Analysis and Mining (ASONAM 2012), pp. 1159–1164. IEEE Computer Society (2012)
80. Ma, H., King, I., Lyu, M.R.: Learning to recommend with social trust ensemble. In: Proceedings of the 32nd International ACM SIGIR Conference on Research and Development in Information Retrieval, pp. 203–210. ACM, Boston (2009)
81. Ma, H., King, I., Lyu, M.R.: Learning to recommend with explicit and implicit social relations. ACM Trans. Intell. Syst. Technol. **2**(3), 1–19 (2011)
82. Ma, H., et al.: SoRec: social recommendation using probabilistic matrix factorization. In: The 17th ACM Conference on Information and Knowledge Management: CIKM 2008. ACM Press, Napa Valley (2008)
83. Ma, H., et al.: Recommender systems with social regularization. In: Proceedings of the Fourth ACM International Conference on Web Search and Data Mining, pp. 287–296. ACM, Hong Kong (2011)
84. Macedo, A.Q., Marinho, L.B., Santos, R.L.T.: Context-aware event recommendation in event-based social networks. In: Proceedings of the 9th ACM Conference on Recommender Systems, pp. 123–130. ACM, Vienna (2015)
85. Machanavajjhala, A., Korolova, A., Sarma, A.D.: Personalized social recommendations: accurate or private. Proc. VLDB Endow. **4**(7), 440–450 (2011)
86. Maltz, D., Ehrlich, K.: Pointing the way: active collaborative filtering. In: Proceedings of the SIGCHI Conference on Human Factors in Computing Systems, pp. 202–209. ACM Press/Addison-Wesley Publishing Co., Denver/Colorado (1995)
87. Massa, P., Avesani, P.: Trust-aware bootstrapping of recommender systems. In: Proceedings of ECAI 2006 Workshop on Recommender Systems (2006)
88. Massa, P., Avesani, P.: Trust-aware recommender systems. In: Proceedings of the 2007 ACM Conference on Recommender Systems, pp. 17–24. ACM, Minneapolis (2007)
89. Massa, P., Avesani, P.: Trust metrics in recommender systems. In: Golbeck, J. (ed.) Computing with Social Trust. Human-Computer Interaction Series, pp. 259–285. Springer, London (2009). https://doi.org/10.1007/978-1-84800-356-9_10
90. McPherson, M., Lovin, S., Cook, J.: Birds of a feather: homophily in social networks. Ann. Rev. Sociol. **27**, 415–445 (2001)
91. Messenger, A., Whittle, J.: Recommendations based on user-generated comments in social media. In: 2011 IEEE Third International Conference on Social Computing (SocialCom), Privacy, Security, Risk and Trust (PASSAT) (2011)
92. Monge, P.E., Contractor, N.S.: Homophily, proximity, and social support theories. In: Theories of Communication Networks, pp. 223–239. Oxford, New York (2003)
93. Monge, P.E., Contractor, N.S.: Network concepts, measures, and the multitheoretical, multilevel analytic framework. In: Theories of Communication Networks, pp. 29–77. Oxford, New York (2003)
94. Moradi, P., Ahmadian, S.: A reliability-based recommendation method to improve trust-aware recommender systems. Expert Syst. Appl. **42**(21), 7386–7398 (2015)
95. Moradi, P., Ahmadian, S., Akhlaghian, F.: An effective trust-based recommendation method using a novel graph clustering algorithm. Phys. A **436**, 462–481 (2015)
96. Moshfeghi, Y., Piwowarski, B., Jose, J.M.: Handling data sparsity in collaborative filtering using emotion and semantic based features. In: Proceedings of the 34th International ACM SIGIR Conference on Research and Development in Information Retrieval, pp. 625–634. ACM, Beijing (2011)

97. O'Mahoney, M., Smyth, B.: From opinions to recommendations. In: Brusilovsky, P., He, D. (eds.) Social Information Access. LNCS, vol. 10100, pp. 480–509. Springer, Cham (2018)

98. Oh, S., Syn, S.Y.: Motivations for sharing information and social support in social media: a comparative analysis of Facebook, Twitter, Delicious, YouTube, and Flickr. J. Assoc. Inf. Sci. Technol. 66(10), 2045–2060 (2015)

99. Park, M.-H., Park, H.-S., Cho, S.-B.: Restaurant recommendation for group of people in mobile environments using probabilistic multi-criteria decision making. In: Lee, S., Choo, H., Ha, S., Shin, I.C. (eds.) APCHI 2008. LNCS, vol. 5068, pp. 114–122. Springer, Heidelberg (2008). https://doi.org/10.1007/978-3-540-70585-7_13

100. Groh, G., Birnkammerer, S., Köllhofer, V.: Social recommender systems. In: Pazos Arias, J.J., et al. (eds.) Recommender Systems for the Social Web. Intelligent Systems Reference Library, vol. 32, pp. 3–42. Springer, Heidelberg (2012). https://doi.org/10.1007/978-3-642-25694-3_1

101. Pera, M.S., Ng, Y.-K.: With a little help from my friends: generating personalized book recommendations using data extracted from a social website. In: Proceedings of the 2011 IEEE/WIC/ACM Joint Conference on Web Intelligent (WI 2011), Lyon, France, pp. 96–99 (2011)

102. Pitsilis, G., Knapskog, S.J.: Social trust as a solution to address sparsity-inherent problems of recommender systems. In: Proceedings of ACM RecSys 2009 Workshop on Recommender Systems and the Social Web, New York (2009)

103. Pizzato, L., et al.: Recommending people to people: the nature of reciprocal recommenders with a case study in online dating. User Model. User-Adap. Inter. 23, 1–42 (2012)

104. Pu, P., Chen, L., Hu, R.: Evaluating recommender systems from the user's perspective: survey of the state of the art. User Model. User-Adap. Inter. 22(4–5), 317–355 (2012)

105. Richardson, M., Domingos, P.: Mining knowledge-sharing sites for viral marketing. In: Proceedings of the Eighth ACM SIGKDD International Conference on Knowledge Discovery and Data Mining. ACM (2002)

106. Sanchez, F., et al.: Social and content hybrid image recommender system for mobile social networks. Mob. Netw. Appl. 17, 1–14 (2012)

107. Saridakis, G., et al.: Individual information security, user behaviour and cyber victimisation: an empirical study of social networking users. Technol. Forecast. Soc. Chang. 102, 320–330 (2016)

108. Schafer, J.B., Frankowski, D., Herlocker, J., Sen, S.: Collaborative filtering recommender systems. In: Brusilovsky, P., Kobsa, A., Nejdl, W. (eds.) The Adaptive Web. LNCS, vol. 4321, pp. 291–324. Springer, Heidelberg (2007). https://doi.org/10.1007/978-3-540-72079-9_9

109. Shani, G., Gunawardana, A.: Evaluating recommendation systems. In: Ricci, F., Rokach, L., Shapira, B., Kantor, P.B. (eds.) Recommender Systems Handbook, pp. 257–297. Springer, Boston (2011). https://doi.org/10.1007/978-0-387-85820-3_8

110. Shardanand, U., Maes, P.: Social information filtering: algorithms for automating "word of mouth". In: Proceedings of the SIGCHI Conference on Human Factors in Computing Systems, pp. 210–217. ACM Press/Addison-Wesley Publishing Co., Denver (1995)

111. Shi, Y., Larson, M., Hanjalic, A.: Towards understanding the challenges facing effective trust-aware recommendation. In: Proceedings of the 2nd ACM RecSys Workshop on Recommender Systems and the Social Web, Barcelona, Spain (2010)

112. Singla, P., Richardson, M.: Yes, there is a correlation: - from social networks to personal behavior on the web. In: The 17th International Conference on World Wide Web, WWW 2008. ACM, Beijing (2008)

113. Sinha, R., Swearingen, K.: Comparing recommendations made by online systems and friends. In: DELOS Workshop on Personalisation and Recommender Systems in Digital Libraries. Dublin City University, Ireland (2001)
114. Sinha, R., Swearingen, K.: Comparing recommendations made by online systems and friends. In: Proceedings of the DELOS-NSF Workshop on Personalization and Recommender Systems in Digital Libraries (2001)
115. Stutzman, F., Gross, R., Acquisti, A.: Silent listeners: the evolution of privacy and disclosure on Facebook. J. Priv. Confid. **4**(2), 2 (2013)
116. Su, X.: Collaborative filtering: a survey. In: 2015 4th International Conference on Reliability, Infocom Technologies and Optimization (ICRITO) (Trends and Future Directions) (2015)
117. Su, X., Khoshgoftaar, T.M.: A survey of collaborative filtering techniques. Adv. Artif. Intell. **2009**, 4 (2009)
118. Subrahmanyam, K., et al.: Online and offline social networks: Use of social networking sites by emerging adults. J. Appl. Dev. Psychol. **29**(6), 420–433 (2008)
119. Sun, Z., et al.: Recommender systems based on social networks. J. Syst. Softw. **99**, 109–119 (2015)
120. Swamynathan, G., et al.: Do social networks improve e-commerce? A study on social marketplaces. In: Proceedings of the First Workshop on Online Social Networks, pp. 1–6. ACM, Seattle (2008)
121. Symeonidis, P., Tiakas, E., Manolopoulos, Y.: Product recommendation and rating prediction based on multi-modal social networks. In: Proceedings of the Fifth ACM Conference on Recommender Systems, pp. 61–68. ACM, Chicago (2011)
122. Tang, J., Chang, Y., Liu, H.: Mining social media with social theories: a survey. SIGKDD Explor. Newsl. **15**(2), 20–29 (2014)
123. Tang, J., Hu, X., Liu, H.: Social recommendation: a review. Soc. Netw. Anal. Min. **3**(4), 1113–1133 (2013)
124. Terveen, L., McDonald, D.W.: Social matching: a framework and research agenda. ACM Trans. Comput.-Hum. Interact. **12**(3), 401–434 (2005)
125. Tintarev, N., Masthoff, J.: Effective explanations of recommendations: user-centered design. In: Proceedings of the 2007 ACM Conference on Recommender Systems, pp. 153–156. ACM, Minneapolis (2007)
126. Tong, H., Faloutsos, C., Pan, J.-Y.: Random walk with restart: fast solutions and applications. Knowl. Inf. Syst. **14**(3), 327–346 (2008)
127. Turner, J.C.: Social Influence. Brooks/Cole, Pacific Grove (1991)
128. Turner, J.C., Reynolds, K.J.: Self-categorization theory. In: Fiske, S.T., Gilbert, D.T., Lindzey, G. (eds.) Handbook of Social Psychology. Wiley, Hoboken (2010)
129. Victor, P., et al.: Trust Networks for Recommender Systems, pp. 91–107. Atlantis Press (2011)
130. Victor, P., et al.: A comparative analysis of trust-enhanced recommenders for controversial items (2009)
131. Vu, X.T., Abel, M.-H., Morizet-Mahoudeaux, P.: A user-centered and group-based approach for social data filtering and sharing. Comput. Hum. Behav. **51, Part B**, 1012–1023 (2015)
132. Wang, H., Li, G., Feng, J.: Group-based personalized location recommendation on social networks. In: Chen, L., Jia, Y., Sellis, T., Liu, G. (eds.) APWeb 2014. LNCS, vol. 8709, pp. 68–80. Springer, Cham (2014). https://doi.org/10.1007/978-3-319-11116-2_7
133. Wang, H., Terrovitis, M., Mamoulis, N.: Location recommendation in location-based social networks using user check-in data. In: Proceedings of the 21st ACM SIGSPATIAL International Conference on Advances in Geographic Information Systems. pp. 374–383. ACM, Orlando (2013)

134. Wang, J., et al.: Recommending Flickr groups with social topic model. Inf. Retrieval **15**(3–4), 278–295 (2012)

135. Wang, Z., Yang, Y., Hu, Q., He, L.: An empirical study of personal factors and social effects on rating prediction. In: Cao, T., Lim, E.-P., Zhou, Z.-H., Ho, T.-B., Cheung, D., Motoda, H. (eds.) PAKDD 2015. LNCS (LNAI), vol. 9077, pp. 747–758. Springer, Cham (2015). https://doi.org/10.1007/978-3-319-18038-0_58

136. Wellman, B.: Computer networks as social networks. Science **293**(5537), 2031–2034 (2001)

137. Weng, J., et al.: TwitterRank: finding topic-sensitive influential Twitterers. In: Proceedings of the Third ACM International Conference on Web Search and Data Mining, pp. 261–270. ACM, New York (2010)

138. Xu, Y., et al.: Combining social network and semantic concept analysis for personalized academic researcher recommendation. Decis. Support Syst. **54**(1), 564–573 (2012)

139. Yang, X., et al.: A survey of collaborative filtering based social recommender systems. Comput. Commun. **41**, 1–10 (2014)

140. Yang, X., Steck, H., Liu, Y.: Circle-based recommendation in online social networks. In: Proceedings of the 18th ACM SIGKDD International Conference on Knowledge Discovery and Data Mining, pp. 1267–1275. ACM, Beijing (2012)

141. Ye, M., Liu, X., Lee, W.-C.: Exploring social influence for recommendation: a generative model approach. In: Proceedings of the 35th International ACM SIGIR Conference on Research and Development in Information Retrieval, pp. 671–680. ACM, Portland (2012)

142. Yeung, C.A., et al.: Decentralization: the future of online social networking. In: W3C Workshop on the Future of Social Networking Position Papers. Citeseer (2009)

143. Yoon, H., et al.: Social itinerary recommendation from user-generated digital trails. Pers. Ubiquit. Comput. **16**(5), 469–484 (2012)

144. Yuan, Q., Chen, L., Zhao, S.: Factorization vs. regularization: fusing heterogeneous social relationships in top-n recommendation. In: Proceedings of the Fifth ACM Conference on Recommender Systems, pp. 245–252. ACM, Chicago (2011)

145. Yuan, W., et al.: Improved trust-aware recommender system using small-worldness of trust networks. Knowl.-Based Syst. **23**(3), 232–238 (2010)

146. Zhang, F.-G.: Preventing recommendation attack in trust-based recommender systems. J. Comput. Sci. Technol. **26**(5), 823–828 (2011)

147. Zhang, J., Wang, Y., Vassileva, J.: SocConnect: a personalized social network aggregator and recommender. Inf. Process. Manag. **49**(3), 721–737 (2013)

148. Zhou, X., et al.: The state-of-the-art in personalized recommender systems for social networking. Artif. Intell. Rev. **37**(2), 119–132 (2012)

149. Zhu, S., Chen, J.: E-commerce use in urbanising China: the role of normative social influence. Behav. Inf. Technol. **35**(5), 357–367 (2016)

150. Ziqi, W., et al.: Recommendation algorithm based on graph-model considering user background information. In: 2011 Ninth International Conference on Creating, Connecting and Collaborating through Computing (C5) (2011)

151. Zou, J., Fekri, F.: A belief propagation approach for detecting shilling attacks in collaborative filtering. In: Proceedings of the 22nd ACM International Conference on Information and Knowledge Management, pp. 1837–1840. ACM, San Francisco (2013)

12
Tag-Based Recommendation

Toine Bogers(⊠) (iD)

Department of Communication and Psychology,
Aalborg University Copenhagen, Copenhagen, Denmark
toine@hum.aau.dk

1 Introduction

Recommender systems are a type of personalized information filtering technology that aim to identify which items in a catalog might be of interest to a user. Historically, recommender systems have been applied to many different domains, such as movies, music, books, news, images, and websites. Recommendations can be generated using a variety of information sources related to both the user and the items: past user preferences and ratings, purchase history, demographic information, the user's social environment, the metadata characteristics of the products, or any combination of these sources.

In the early days of the field, recommender systems focused mostly on using user ratings [48] or item metadata [72] to generate recommendations. However, in recent years researchers have focused more and more on going beyond these data sources. Extensions to recommender systems include using information about relationships between users (such as friendship [93] or trust [103]) and information about the context of the recommendation process [1].

Social Tagging. Another source of information that has been used to generate more relevant recommendations are so-called *tags*. Tags are a product of *social tagging*[1], a information classification paradigm where the users themselves are given the power to describe and categorize content for their own purposes using tags. Tags are keywords that typically describe characteristics of the object they are applied to, and can be made up of one or more words. In addition to topical keywords, users are also free to tag objects for self-reference, task organization, and with subjective keywords. Users are free to apply any type and any number of tags to an object, resulting in true bottom-up classification. This is in stark contrast to the top-down classification that indexing schemes such as controlled vocabularies provide [71]. Although there is no inherent grouping or hierarchy in the tags assigned by users, some researchers have attempted to classify tags into

---

[1] This chapter focuses on how tags can be applied to improve the recommendation process. For more information about the paradigm of social tagging, we refer the reader to Mathes [71] and Hammond et al. [46]. For other applications of tags to information access, we refer the reader to Chapters 1 [15], 6 [27], and 9 [75] in this book.

© Springer International Publishing AG, part of Springer Nature 2018
P. Brusilovsky and D. He (Eds.): Social Information Access, LNCS 10100, pp. 441–479, 2018.
https://doi.org/10.1007/978-3-319-90092-6_12

different categories. Bischoff et al. [8], Golder and Huberman [43], Sen et al. [89], and Xu et al. [115] all provide examples of such tag taxonomies. Like recommender systems, social tagging has been applied to many domains such as music (e.g., Last.FM[2] [60,101]), movies (e.g., MovieLens[3], [31,33]), books (e.g., LibraryThing[4], [22,31]), websites (e.g., Delicious[5], [13,110]), and scientific articles (e.g., CiteULike[6], [13,98]).

Tag-based Recommendation. The popularity of social tagging, and the ease with which tags can be generated, assigned, and collected, has sparked significant research interest in tags and their possible applications. One such application is *tag-based recommendation*—generating better recommendations by incorporating tags into the recommendation process. Many different recommendation tasks can be supported using tags, depending on which objects are used to find relevant objects of another type. Figure 1 shows an overview of at least nine different possible recommendation tasks. For example, the classic application of recommendation algorithms is to find relevant items given a particular user. Tags could then be used as an additional resource to generate these recommendations. Another example could be locating item experts, which could be bolstered by using the tags describing those items.

Organization of this chapter. Other chapters in this volume focus on several of these recommendation tasks shown in Fig. 1. For example, Chapters 6 [27] focuses on how tags can be related to each other to provide efficient browsing of tag networks and hierarchies, whereas Chapters 9 [75] explains in detail how tags can be used to support searching for information.

The main focus of this chapter is on tag-based *item recommendation*—generating better recommendations for content to consume by incorporating tags into the recommendation process. First, in Sect. 2, we cover the preliminaries and notation needed for the common ground to compare the different algorithms and approaches. We then provide a detailed overview of the most important and influential approaches to item recommendation in Sects. 3–7. We group the approaches by the class of recommendation algorithm used, similar to the classification used in Chapters 10 [59] in this book. Section 8 briefly addresses the related task of recommending tag to apply to content on social bookmarking websites and provide a brief overview of popular algorithms. We conclude in Sect. 9 with a discussion of the future challenges for tag-based recommendation.

2   Preliminaries

In order to properly discuss and compare the different approaches to tag-based recommendation, we introduce and define some common concepts and notation in this section, based in part on notation by Bogers [10] and Clements et al. [22].

---

[2] http://www.last.fm.
[3] http://www.movielens.org.
[4] http://www.librarything.com.
[5] http://www.delicious.com.
[6] http://www.citeulike.org.

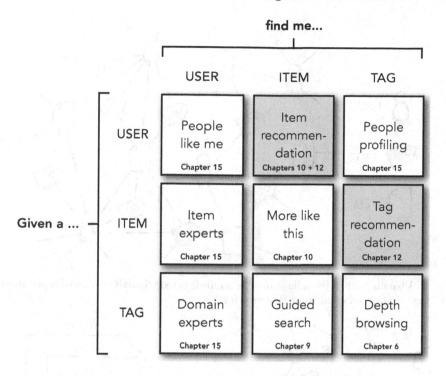

Fig. 1. Overview of different social recommendation tasks for users, items, and tags. In addition to the tasks that directly focus on tags, the other tasks could also be strengthened by using the information encoded in tasks, such as the tag-based item recommendation covered in this chapter. Figure adapted from Bogers [10] and Clements et al. [24].

When social tagging is used to aid in the categorization and description of a website's content, users typically first post items to their personal profiles—either implicitly through consumption or explicitly through bookmarking—and can then choose to label them with one or more tags. The aggregation of all tags assigned by all the users of a system is typically referred to as a *folksonomy* and serves as an extra annotation layer that connects users and items. We define a folksonomy to be the undirected tripartite graph that emerges from this collaborative annotation of items. Figure 2 visualizes this undirected tripartite network: it contains three different types of nodes—users, items and tags—with the ternary relations between them represented by edges between the nodes.

The resulting ternary relations that make up the tripartite graph can be represented as a 3D matrix of users, items, and tags, as shown in the top right of Fig. 3. We refer to this 3D matrix (or tensor) as $\mathcal{D}(u_k, i_l, t_m)$. Each matrix element $d(k, l, m)$ indicates if user u_k (with $k = \{1, \ldots, K\}$) tagged item i_l (with $l = \{1, \ldots, L\}$) with tag t_m (with $m = \{1, \ldots, M\}$), where a value of 1 indicates that the ternary relation is present in the folksonomy.

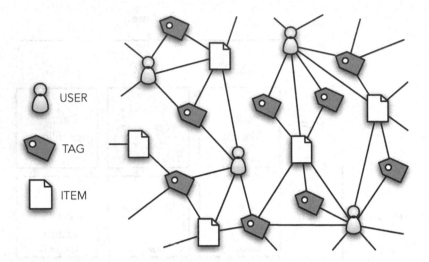

Fig. 2. Visualization of the folksonomy as an undirected tripartite graph of users, items, and tags. Figure adapted from Bogers [10].

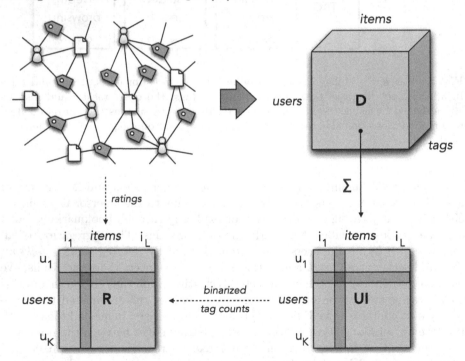

Fig. 3. Representing the folksonomy graph as a 3D matrix. Aggregation over the tag dimension of $\mathcal{D}$ gives us matrix **UI**, containing the tag counts for each user-item pair. The ratings matrix **R** is typically derived either (1) directly from the tripartite graph itself (as the weights on user-item edges), or (2) by binarizing the tag count values in UI. Figure adapted from Bogers [10].

Conventional recommender systems, and in particular collaborative filtering approaches, operate only on the user-item matrix to generate recommendations. This user-item matrix $\mathbf{R}(u_k, i_l)$ (shown in the bottom left of Fig. 3) contains numerical user preference information: to what degree did users consume and/or prefer which item(s). User preferences can come in the form of *explicit* ratings, when they are entered directly by the user, or *implicit* preference scores, when they are inferred from user behavior. The numerical representation of these user preferences can take many forms: from star ratings on a five-point scale to unbounded play counts or binary consumption patterns. These preference scores typically form the edge weights of the user-item edges in the tripartite graph. We can extract the ratings matrix $\mathbf{R}(u_k, i_l)$ for all user-item pairs directly from the tripartite graph. The individual elements of $\mathbf{R}$ are denoted by $r_{k,l}$ and each user is represented in this matrix by its user profile row vector $\overrightarrow{u_k}$, which lists the items that user u_k added to their profile. Items are represented by the column vectors of $\mathbf{R}$ which represent the item profile vectors $\overrightarrow{i_l}$ that contain all users that have added item i_l.

As shown in the bottom right of Fig. 3, we can also extract a user-item matrix from $\mathcal{D}$ by aggregating over the tag dimension. We then obtain the $K \times L$ user-item matrix $\mathbf{UI}(u_k, i_l) = \sum_{m=1}^{M} \mathcal{D}(u_k, i_l, t_m)$, specifying how many tags each user assigned to each item. Individual elements of $\mathbf{UI}$ are denoted by $x_{k,l}$. In case ratings information is not available, we can create a makeshift ratings matrix $\mathbf{R}$ by binarizing $\mathbf{UI}$. An example vector of tag frequency counts such as $[0, 14, 3, 1, 0, 5]$ would then be converted to $[0, 1, 1, 1, 0, 1]$. The actual construction of $\mathbf{R}$ depends both on the recommendation algorithm and the data set used.

Similar to the way we defined $\mathbf{UI}$, we can also aggregate the content of $\mathcal{D}$ over the user and the item dimensions. These aggregations are commonly used in tag-based recommendation algorithms. Figure 4 visualizes these 2D projections in the users' and items' tag spaces.

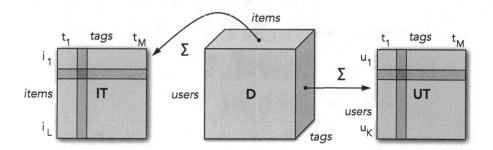

Fig. 4. Deriving tagging information at the user level as **UT**, and the item level as **IT**, by aggregating over the item and user dimensions respectively. Figure adapted from Bogers [10].

We define the $K \times M$ user-tag matrix $\mathbf{UT}(u_k, t_m) = \sum_{l=1}^{L} \mathcal{D}(u_k, i_l, t_m)$, specifying how often each user used a certain tag to annotate their items. Individual elements of **UT** are denoted by $y_{k,m}$. We define the $L \times M$ item-tag matrix $\mathbf{IT}(i_l, t_m) = \sum_{k=1}^{K} \mathcal{D}(u_k, i_l, t_m)$, indicating how many users assigned a certain tag to an item. Individual elements of **IT** are denoted by $z_{l,m}$. Binary versions of **UT** and **IT** are denoted as $\mathbf{UT}_{\text{binary}}$ and $\mathbf{IT}_{\text{binary}}$. The row vectors of the **UT** and **IT** matrices represent the user tag profiles $\overrightarrow{d_k}$ and item tag profiles $\overrightarrow{f_l}$ respectively. They list which tags have been assigned by a user, or to an item by all of its users.

3   Item Recommendation

Item recommendation is the task of recommending interesting items to a user based on a record of their past preferences. It is arguably the most popular recommendation task and many different classes of algorithms have been proposed in the past. Formally, the goal of each of the item recommendation algorithms discussed in this section is top-N recommendation: producing a ranking of all items i_l that are not yet in the profile of the *active*[7] user u_k (i.e., $x_{k,l} = \emptyset$). To this end, we predict a score $\widehat{x}_{k,l}$ for each item that represents the likelihood of that item being relevant for the active user. The final recommendations for a user are generated by ranking all items i_l by their predicted score $\widehat{x}_{k,l}$.

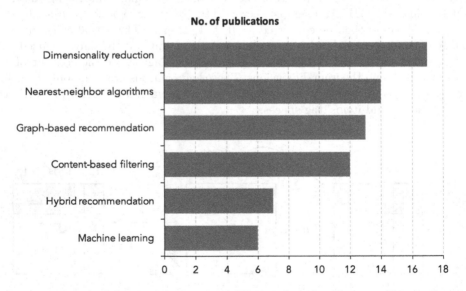

Fig. 5. Bar chart of the usage frequency of different algorithm classes used in tag-based item recommendation.

The next four sections present an overview f the different ways that tags have been integrated into algorithms for item recommendation. We group these

---

[7] The active user is the user the system is currently generating recommendations for.

approaches by the class of recommendation algorithm used, inspired by the classification scheme used in Chapters 10 [59] in this book. In addition, we add two classes of algorithms that are grouped together as miscellaneous approaches in Chapters 10 [59]: graph-based approaches and machine learning approaches. These approaches are commonly used for tag-based item recommendation and deserve their own separate treatment. Figure 5 shows the popularity of the different classes of tag-based item recommendation algorithms among the publications reviewed in this chapter.

Section 4 discusses the different tag-based variants of collaborative filtering algorithms: nearest-neighbor algorithms, algorithms that employ dimensionality reduction techniques, and graph-based recommendation approaches. Section 5 discusses tag-enhanced content-based filtering approaches, whereas Sect. 6 covers tag-based approaches that employ machine learning to generate recommendations. Section 7 describes the different hybrid approaches to tag-based recommendation that have been proposed in recent years.

4   Collaborative Filtering

Collaborative Filtering (CF) algorithms attempt to mimic the process of word-of-mouth recommendations by using the preferences of a group of users to provide recommendations for an individual user. These preferences can be explicit or implicit and they are recorded in the **R** and **UI** matrices, which traditionally serve as input for CF algorithms. CF algorithms can be subdivided into different classes: nearest-neighbor algorithms (Sect. 4.1), algorithms that employ dimensionality reduction techniques (Sect. 4.2), and graph-based recommendation approaches (Sect. 4.3). For each of these classes of CF algorithms, we first describe the standard algorithm, followed by their tag-enhanced variants.

4.1   Nearest-Neighbor Algorithms

Nearest-neighbor (NN) algorithms form a popular subclass of recommendation algorithms due to their simplicity, easy extensibility, and long history in a relatively young research field. They are also known as *memory-based* or *lazy* recommendation algorithms, because they defer the actual computational effort of predicting a user's interest in an item to the moment a user requests a set of recommendations. The training phase of an NN algorithm consists of simply storing all user preferences into memory. There are two variants of NN-based recommendation and both are based on the k-Nearest Neighbor algorithm from the field of machine learning [3]: *user-based* and *item-based* NN recommendation.

A user-based NN algorithm matches the active user's profile vector $\vec{u_k}$ in **R** against the other user profile vectors to find those neighboring users that the active user is most similar to in terms of consumption or rating behavior. This is typically done using similarity metrics such as the cosine similarity or Pearson's correlation coefficient. Once this neighborhood has been identified, all items in the neighboring user profiles that are unknown to the active user are considered as possible recommendations and sorted by their frequency in that neighborhood. A weighted aggregate of these frequencies is then used to generate the final list of recommendations [48]. The item-based variant of NN recommendation focuses on finding the most similar items instead of the most similar users. For each of an active user's item profile vectors $\vec{i_l}$ from $\mathbf{R}$, the neighborhood of most similar items is calculated. Items are considered to be similar when the same set of users has purchased them or rated them highly. Each of the top k neighbors is placed on a candidate list along with its similarity to the active user's item. A weighted aggregate of these neighboring items' similarity scores is then used to generate the list of recommendations [87]. For more details we refer the reader to Sect. 2.2 of Chapters 10 [59] in this volume or Desrosiers and Karypis [26], who provide a thorough overview of NN algorithms.

The general idea behind tag-based NN algorithms is to use the tags to aid in the calculation of the user and/or item similarities, or in some cases even replace the use of the $\mathbf{R}$ or $\mathbf{UI}$ matrices entirely. Instead, the $\mathbf{UT}$ and $\mathbf{IT}$ matrices are used to calculate the similarities. The main argument for using these projections of $\mathcal{D}$ is that they tend to be less sparse, because for every item that is liked or rated by a user more than one tag tends to be used to describe that item. Bogers [10] report the average number of tags assigned to an item to be in the range of 3.1 to 8.4, depending on the website and domain. This suggests a potential reduction in sparsity by at least a factor three.

Several researchers have proposed using the $\mathbf{UT}$ matrix to calculate better user similarities, albeit with mixed results: the effect of using the $\mathbf{UT}$ matrix appears to depend on the specific algorithm used as well as the domain. Firan et al. [34] were among the first to propose using the $\mathbf{UT}$ matrix when they examined the problem of music recommendation using Last.FM data. They compare a traditional user-based NN algorithm with a variant that uses the tags assigned to a user's tracks to calculate user similarity. They find that tag-based user profiles perform worse than the traditional baseline algorithm. In comparison, Nakamoto et al. [73, 74] find a positive effect of tag-based user similarities, albeit compared to a weaker baseline than the one used by Firan et al. [34]. They propose a user-based NN algorithm with tag-based similarities calculated on a $\mathbf{UT}$ matrix containing tag frequency counts. Zhao et al. [127] also report a tag-based variant of a user-based NN algorithm outperforming a traditional user-based NN algorithm on a data set collected from the social bookmarking system Dogear. They calculate user similarities on the $\mathbf{UT}$ matrix by calculating the semantic similarity between stemmed tags using Wordnet to calculate user similarities. Parra and Brusilovsky [78] use the BM25 term weighting formula [83] to calculate user similarities based on tag-based profiles. In a small-scale pilot study, they report performance comparable to more traditional NN algorithms. Amer-Yahia et al. [4] propose using both the binarized $\mathbf{UI}$ matrix and the $\mathbf{UT}$ matrix to calculate shared overlap between users of Delicious on which items they have consumed and how they have tagged them. The neighborhood of similar users is then used to generate recommendation without any weighting, as is common in regular user-based NN approaches. However, they perform no evaluation of their algorithm. Kim et al. [57] proposed a variant of the traditional user-based NN algorithm where user similarity is calculated based on semantic overlap on tags. They employ the **UT** matrix, but in their similarity calculations they take into account tag ambiguity and synonymity.

Many approaches have attempted to compare and/or combine both approaches by using both the **UT** and the **IT** matrix. Bogers and Van den Bosch [13], for instance, propose a tag-based version of the traditional user-based and item-based NN algorithms. For each of these two variants, they compare five different similarity metrics: the Dice coefficient, the Jaccard overlap, and the cosine similarity, the latter both with and without applying a tf·idf weighting scheme. An extensive evaluation on data sets collected from BibSonomy[8], CiteULike, and Delicious showed that, while a user-based NN algorithm did not benefit from using user similarities calculated on the **UT** matrix, an item-based NN algorithm that used item similarities from the **IT** matrix outperformed the traditional baseline algorithms and provided the best performance overall. Zeng and Li [119] also compare user-based and item-based NN algorithms with similarities calculated using the **UT** and **IT** matrices respectively. They evaluate their methods on three small subsets of a Delicious data set and find the opposite of Bogers and Van den Bosch [13]: user-based NN based on tags outperforms all other approaches.

Another popular approach to using information from both the **UT** and **IT** matrices is the *tag-aware fusion* algorithm of Tso-Sutter et al. [101]. Tag-aware fusion is an adaptation of the standard NN algorithm that fuses information from all three matrices together: **R**, **UT**, and **IT**. While conceptually similar to the approach by Bogers and Van den Bosch [13], it calculates the user and item similarities in a different manner. Their approach consists of two steps: (1) *similarity calculation* and (2) *similarity fusion*. In the first step, they calculate the user and item similarities using the **R** matrix, but extend this user-item matrix by including user tags as items and item tags as users. Figure 6 illustrates this process. Effectively, this means they concatenate a user's profile vector $\vec{u_k}$ with that user's tag vector $\vec{d_k}$, which is taken from **UT**$_{\text{binary}}$. For item-based filtering, the item profile vector $\vec{i_l}$ is extended with the tag vector for that item $\vec{f_l}$, also taken from **IT**$_{\text{binary}}$.

By extending the user and item profile vectors with tags, sparsity is reduced when calculating the user and item similarities, when compared to using only preference data from **R** to calculate these similarities. Adding the tags also reinforces the transaction information that is already present in $\vec{u_k}$ and $\vec{i_l}$. At the end of this phase, they use both the user-based and item-based NN algorithms with cosine similarity to generate items recommendations. In the second phase of their approach, similarity fusion, Tso-Sutter et al. fuse the predictions of the user-based and item-based NN algorithms together as a linearly weighted sum of the separate predictions. Tso-Sutter et al. test their approach on a self-crawled data set from Last.FM against a baseline NN algorithm based on usage similarity.

---

[8] http://www.bibsonomy.org.

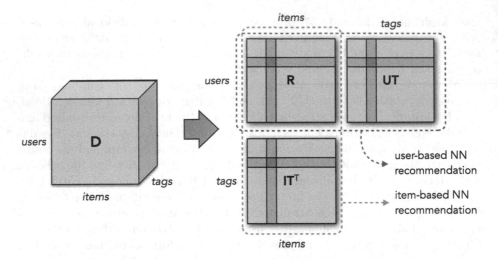

Fig. 6. Extending the user-item matrix for tag-aware fusion. For user-based filtering, the **UT** matrix is appended to the normal **R** matrix so that the tags serve as extra items during the user similarity calculations. It does so by including user tags as items and item tags as users. For item-based filtering, the transposed **IT** matrix is appended to the normal **R** matrix so that the tags serve as extra users in the item similarity calculations. Figure adapted from Bogers [10] and Tso-Sutter et al. [101].

They report that in their experiments they find no performance improvements using these extended similarities in their separate user-based and item-based variants. They do report significant improvements of their fused approach over their baseline runs, showing that their fusion method is able to capture the tripartite relationship between users, items, and tags effectively.

Several other approaches exist that were inspired by traditional NN algorithms, despite not being true NN algorithms themselves. For instance, Guy et al. [45] propose a tag-based recommender system for recommending bookmarks, blogs, communities, files, and wikis in the Lotus Connections platform. They propose a linearly weighted combination of a traditional user-based NN algorithm and a variant that uses the **UT** and **IT** matrices to calculate the association strength between user-tag and tag-item pairs. The weighted combination allows them to compare the two individual algorithms as well as three hybrid combinations. They find that, in their corporate setting, using tagging activity generates better recommendations than using social relations between users. Liang et al. [66,68] proposed an approach similar in spirit to that of Tso-Sutter et al. [101] by using the **UI**, **UT** and **IT** matrices to calculate user and item similarities. Using an unconventional evaluation paradigm on self-crawled Amazon and CiteULike data sets, they show that their method outperforms the tag-aware fusion method, although later comparisons do not appear to corroborate these findings [10]. Liang et al. [67] later combined their tag-based approach with one that calculates item similarities based on the distance in Amazon's item taxonomy. They report that this hybrid outperforms its individual components.

4.2   Dimensionality Reduction

Dimensionality reduction (DR) algorithms are a subclass of CF algorithms that have rapidly grown in popularity in recent years. They are often considered to provide state-of-the-art performance and, in addition to NN algorithms, are commonly used as a baseline algorithm against which to measure the performance of new algorithms. Different types of DR algorithms exist, but they all have in common that they attempt to reduce the complexity of the $\mathbf{R}$ matrix by transforming both the users and items to the same lower-dimensional latent factor space [62]. Sarwar et al. [88] were among the first to suggest the use of a DR algorithm called Latent Semantic Indexing for the purpose of recommendation. However, it was the Netflix Prize[9] that served as the catalyst for research on CF algorithms that employ DR techniques. We refer the reader to Sect. 2.3 of Chapters 10 [59] in this volume for a more detailed explanation of dimensionality reduction algorithms. Additionally, we refer to Koren and Bell [62] for a comprehensive overview of the most influential DR algorithms for item recommendation.

One of the more popular tag-based CF algorithms that uses DR techniques is the *tensor reduction* algorithm by Symeonidis et al. [98]. As we recall from Sect. 2, a tensor is a multidimensional matrix and the tripartite folksonomy graph can be represented as the 3D matrix (or tensor) $\mathcal{D}$. A popular method for producing lower-rank approximations of two-dimensional matrices is the Singular Value Decomposition (SVD), which has a long history in the field of IR [7] and has also been used successfully for recommendation [62]. The SVD of a two-dimensional matrix such as the ratings matrix $\mathbf{R}$ (visualized in Fig. 7) can be written as the product of three matrices:

$$\mathbf{R} = \mathbf{U} \times \mathbf{S} \times \mathbf{V}^{\top}, \tag{1}$$

where $\mathbf{U}$ is the user matrix containing the left singular vectors of $\mathbf{R}$, $\mathbf{V}^{\top}$ is the transpose of the item matrix containing the right singular vectors of $\mathbf{R}$, and $\mathbf{S}$ is the diagonal matrix of ordered singular values of $\mathbf{R}$. By keeping only the largest n singular values in $\mathbf{S}$, calculating the same SVD product results in $\widehat{\mathbf{R}}$, which is a lower-rank approximation of $\mathbf{R}$, but with the same dimensions. $\widehat{\mathbf{R}}$ then contains the predicted ratings by user u_k for item i_l in this lower-dimensional latent factor space.

Symeonidis et al. use a method called Higher-Order Singular Value Decomposition (HOSVD), which generalizes SVD to multi-dimensional matrices, such as the 3-order tensor $\mathcal{D}$. HOSVD 'matricizes' a tensor by building a 2D matrix representation in which all column or vectors are stacked after each other, depending on which mode is unfolded (or which dimensions are flattened) [98]. Tensor $\mathcal{D}$ is unfolded in all three modes, after which regular SVD can then be applied on these three matrix unfoldings. Figure 8 illustrates this unfolding process.

---

[9] The Netflix Prize was an open competition organized by Netflix which ran from 2006 to 2009. The aim was to develop the best recommendation algorithm to predict users' ratings for movies offered by Netflix. See http://www.netflixprize.com for more information.

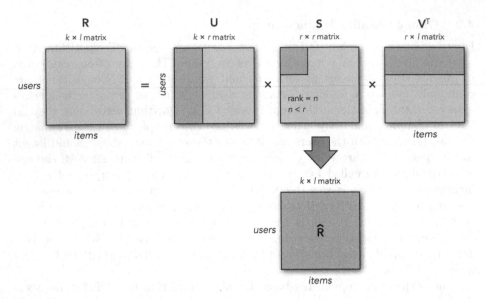

Fig. 7. Applying the Singular Value Decomposition to a ratings matrix **R**. The lower-rank approximation $\widehat{\mathbf{R}}$ is produced by multiplying the rank-reduced versions of **U**, **S**, and $\mathbf{V}^{\top}$.

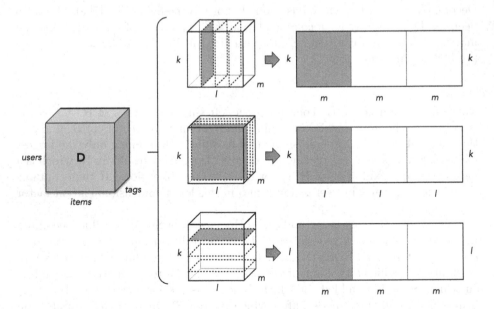

Fig. 8. Unfolding the tensor $\mathcal{D}$ in all three modes. Regular SVD can then be applied on the three resulting matrix unfoldings. The reduced tensor $\widehat{\mathcal{D}}$ is then reconstructed by refolding the reduced matrix unfoldings.

Symeonidis et al. reduce the dimensionality of the three resulting singular value matrices by 50%. They then reconstruct the reduced tensor $\widehat{\mathcal{D}}$ by refolding the reduced matrix unfoldings. Symeonidis et al. apply their approach to the task of music recommendation. In this case, the elements of the reconstructed tensor $\widehat{\mathcal{D}}$ measure represent a quadruplet $\{u, t, i, p\}$, where p is the probability that user u will tag track i with tag t. For item recommendation, the sum of the weights associated with each $\{u, t\}$ pair represents the interest of the user in that item. However, tensor reduction can also be used to support other recommendation tasks, such as suggesting relevant tags [96] or recommending related users. Symeonidis et al. [98] evaluate their tensor reduction for item recommendation on data sets from Last.FM [98] and later BibSonomy [97] and find that it outperformed the FolkRank algorithm by Hotho et al. [51] (see also Sect. 4.3). Rafailidis and Daras [80] propose an adaptation of the tensor reduction algorithm. They expand the set of tags assigned by users and assigned to items using Rocchio's relevance feedback mechanism, which decreases the sparsity of $\mathcal{D}$. After tag expansion they use k-Means clustering to generate tag clusters in order to reduce the tag dimension of $\mathcal{D}$. Finally, they apply the tensor reduction algorithm to this reduced version of $\mathcal{D}$. They evaluate their approach on an unnamed image labeling data set and show it to be more efficient and effective than the original tensor reduction algorithm. Peng et al. [79] propose a tensor reduction algorithm that uses the Tucker decomposition method to reduce the dimensionality of the latent factor space instead of HOSVD like Symeonidis et al. [98]. Evaluations using data sets from BibSonomy, CiteULike, and Delicious show that it significantly outperforms the tag-aware fusion method of Tso-Sutter et al. [101].

Wetzker et al. [109], later extended in Said et al. [86], take a Probabilistic Latent Semantic Analysis (PLSA) approach, which assumes a latent lower dimensional topic model. They extend the original PLSA algorithm [50] to integrate tags by estimating the topic model from both user-item occurrences as well as item-tag occurrences, and then linearly combine the output of the two models. They test their approach on data sets based on CiteULike and Delicious, and find that it significantly outperforms a popularity-based algorithm [86, 109]. They also show that model fusion yields superior results independent of the number of latent factors.

Latent Dirichlet Analysis (LDA) is topic modeling algorithm and a variant of PLSA where the topic distributions are assumed to have a Dirichlet prior [9]. Wang et al. [106] propose a tag-enhanced approach that applies LDA to both the **UT** and **IT** matrices. The probabilities that a user and an item belong to the detected topics are combined to predict the final rating of user u_k for item i_l. They test their approach on the MovieLens data set and find that it outperforms state-of-the-art algorithms that do not use tagging information. A tag-enhanced version of LDA is also proposed by Zhang et al. [122], who apply it to entire tripartite graph instead of the two-dimensional projections **UT** and **IT**. They show that their algorithm outperforms both the FolkRank [51] and the tensor reduction [98] algorithms on a CiteULike data set.

Many adaptations of matrix factorization (MF) models have been proposed that integrate tags into the approximation of the latent factor space, so that both user-item interactions as well as item-tag and user-tag interactions are modeled in that space. Luo et al. [70], for instance, propose a tag-augmented version of matrix factorization, which integrates the latent factors of the item tags and ratings to provide a better approximation of the lower-rank $\mathbf{R}$ matrix. They report that their approach outperforms tag-aware fusion [101] on the MovieLens data set. Xin et al. [113] propose a similar extension of probabilistic matrix factorization (PMF) that integrate the latent factors of the item tags and ratings. Experiments on the MovieLens data set also show it to outperform non-tag-aware PMF. Zhen et al. [128] propose an approach called *TagiCoFi*, which integrates tags into the PMF model. They use tagging information to calculate a user-user similarity matrix. These user similarities are then used to regularize the MF procedure for the purpose of making the user-specific latent feature vectors as similar as possible if two users have the same tagging behavior. Experimental results on the MovieLens data set demonstrate that TagiCoFi outperforms traditional PMF approach. Another tag-enhanced improvement upon the PMF model is proposed by Yin et al. [117], who develop a generalized latent factor model based on a Bayesian approach. It allows them to model the relations between different data types, such as tags and comments. Their Bayesian PMF model outperforms non-Bayesian and non-tag-enhanced versions of PMF on data sets from BibSonomy and Flickr.

In addition to tag-based recommendation algorithms that operate in a single domain, several algorithms have also been proposed for cross-domain recommendation: recommending items in one domain (partly) by using information from another domain. Tags are a commonly used data source in cross-domain recommendation, as they link different domains together. One example could be generating movie recommendations for an active user based on how that user and other users like him or her have rated books. Tags could then be used to link together books and movies about the same topics. Many such tag-based variants of cross-domain recommendation algorithms have already been proposed. Enrich et al. [31], for instance, propose three MF algorithms that utilize tags as a form of implicit feedback to enhance the item factors. One variant only exploits the tags the active user has assigned to the target item, another exploits all tags assigned by all users to that item, and a third variant utilizes only the most discriminative tags. Through an evaluation on MovieLens and LibraryThing data sets, they show all variants to outperform a single-domain MF algorithm. Fernández-Tobías et al. [33] further extend this work by proposing an MF algorithm that enriches both the user and item factors using tags to generate better cross-domain recommendations. Using the same MovieLens and LibraryThing data sets, they demonstrate that their approach outperforms those of Enrich et al. [31].

4.3   Graph-Based Recommendation

Graph-based recommendation algorithms view the item preferences of a large group of users as bipartite network of user and item nodes. This network structure is then used to generate recommendations for which unseen items may also be interesting to user. Examples of successful graph-based recommendation algorithms include the approaches by Aggarwal et al. [2], Baluja et al. [5], and Yildirim and Krishnamoorthy [116]. Tag-enhanced graph-based algorithms include tags as an additional node type and generate recommendations based on the resulting tripartite network (as visualized earlier in Fig. 2). Some proposed algorithms even add additional node types by drawing in specific categories of item metadata, such as actors in the case of movie recommendation.

The first graph-based recommendation algorithm to operate on the tripartite graph is the *FolkRank* algorithm by Hotho et al. [51]. Like the PageRank algorithm originally proposed by Page et al. [77], it follows the random surfer model and calculates the importance weights of the network nodes as the probability that an idealized random surfer can be found at those nodes after a walk of infinite length. This surfer normally follows the edges between the nodes, but occasionally teleports to another random node. Figure 9 visualizes this random walk, weighted by the transition probabilities between users, items and tags.

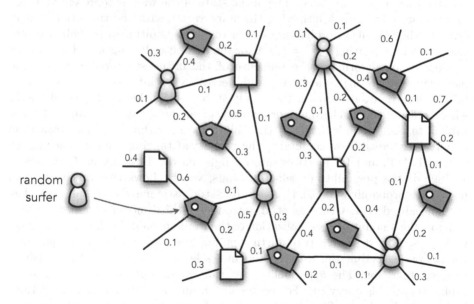

Fig. 9. Visualization of the FolkRank algorithm where a random surfer selects a specific object type—in this case a tag—as a starting point for their random walk across the folksonomy graph, governed by the transition probabilities.

These probabilities can be found by calculating the steady-state distribution of the network. In addition to adding an extra node type, Hotho et al. include preference vectors to initialize the starting point(s) of the random walk. This allows for personalized recommendations of items, tags, and even other users by specifying which node(s) the random walk should depart from. By calculating the difference between the rankings with and without such preference vectors, they can generate personalized recommendations. Hotho et al. compare their FolkRank algorithm to a version of PageRank adapted to the tripartite folksonomy network and find that FolkRank outperforms it on their BibSonomy data set. In addition to the original reference [51], FolkRank is discussed in more detail in the context of tag-based social search in Chapters 9 [75] in this book.

Another random-walk-based approach is the personalized Markov random walk algorithm by Clements et al. [22]. It is easily extensible and, like FolkRank, it allows for the execution of many different recommendation tasks, such as recommending related users, interesting tags, or similar items using the same elegant model. Clements et al. represent the tripartite graph of user, items, and tags, created by all transactions and tagging actions, as a transition matrix $\mathbf{A}$. A random walk is a stochastic process where the initial condition is known and the next state is given by a certain probability distribution. $\mathbf{A}$ contains the state transition probabilities from each state to the other. Figure 10 illustrates how $\mathbf{A}$ is constructed.

A random walk over this social graph is then used to generate a ranked list of the items to be recommended. The initial state of the walk is represented in the *initial state vector* $\vec{v_0}$. Multiplying the state vector with the transition matrix gives us the transition probabilities after one step; multi-step probabilities are calculated by repeating $\overrightarrow{v_{n+1}} = \vec{v_n} \cdot \mathbf{A}$ for the desired walk length n. The number of steps taken determines the influence of the initial state vector versus the background distribution: a shorter walk decreases the influence of $\mathbf{A}$. A walk of infinite length $(\overrightarrow{v_\infty})$ results in the steady state distribution of the social graph, which reflects the background probability of all nodes in the graph. This is similar to (but not identical to) the FolkRank algorithm [51]. The transition matrix $\mathbf{A}$ is created by combining the usage and tagging information present in the $\mathbf{R}$, $\mathbf{UT}$, and $\mathbf{IT}$ matrices into a single matrix. In addition, Clements et al. include the possibility of self-transitions, which allows the walk to stay in place with probability $\alpha \in [0, 1]$. $\mathbf{A}$ is a row-stochastic matrix, i.e., all rows of $\mathbf{A}$ are normalized to 1. Clements et al. introduce a third model parameter θ that controls the amount of personalization of the random walk. In their experiments with personalized search, two starting points are assigned in the initial state vector: one selecting the user u_k and one selecting the tag t_m they wish to retrieve items for. The θ parameter determines the influence of the personal profile versus this query tag. For personalized, unguided item recommendations it would suffice to set θ to 0, whereas a θ of 1 would result in pure tag-based search. After n steps, the item ranking is produced by taking the item transition probabilities from $\vec{v_n}$ for the active user $(\vec{v_n}(K + 1, \ldots K + L)$, highlighted in green in Fig. 10) and rank-ordering them by probability after removal of the items already owned by the active user. Clements et al. test their approach on data sets based on LibraryThing and BibSonomy [21–24] and find it to outperform

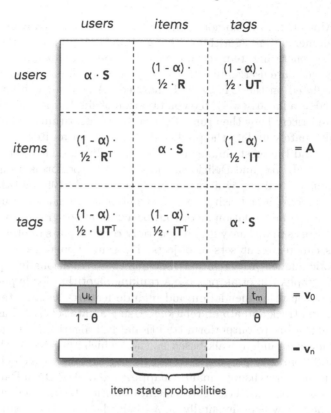

Fig. 10. Clements et al. [22] construct the transition matrix **A** by combining the **R**, **UT**, and **IT** matrices and their transposed versions. Self-transitions are incorporated by super-imposing a diagonal matrix of ones **S** on the transition matrix, multiplied by the self-transition parameter α. In the initial state vector the θ parameter controls the amount of personalization for the active user. Figure adapted from Clements et al. [22].

traditional NN algorithms, although others have been unable to confirm this under different evaluation conditions and on other data sets [10].

Bogers [11] proposes an extension of the personalized Markov random walk algorithm by Clements et al. [22]. Focusing on the movie domain, he includes two additional node types, actors and genres, and proposes to use random walks to generate recommendations on this multipartite network. He does not experimentally evaluate the proposed approach, but it would not be limited to only the movie domain. Many different types of metadata and contextual information could be included for any type of domain. Zhang et al. [125,126] propose a graph-based algorithm based on integrated diffusion—a method similar to spreading activation, where the influence of a node is spread iteratively over its neighboring nodes. They propose a linearly weighted combination of integrated diffusion on the bipartite networks represented by the **UI** and **IT** matrices respectively. Zhang et al. evaluate their approach on data sets based on BibSonomy, Delicious, and MovieLens, but do not compare their approach to any other state-of-the-art recommendation algorithms. Zhang et al. [124] propose a PageRank-like algorithm for generating item recommendations on the tripartite user-item-tag graph. They calculate user similarities as a linearly weighted combination of similarities calculated on the **UI** and **UT** matrices. Analogously, item similarities are calculated as a linearly weighted combination of similarities calculated on the **UI** and **IT** matrices. They then run a Pagerank-like algorithm on the original **UI** matrix to alternately update the predicted score $\hat{x}_{k,l}$ for an item using the most similar items and the most similar users. They evaluate their on data sets from BibSonomy, CiteULike, and Delicious, and show it outperforms both traditional NN algorithms as well as tag-aware fusion [101], PLSA [109] and the approach by Zhen et al. [128], albeit using an unconventional evaluation setup. Bu et al. [16] propose a recommendation algorithm based on hyper-graphs instead of regular graphs. Edges in ordinary graphs can only connect pairs of objects, whereas hyper-edges can represent sets of objects. By using hyper-graphs, it becomes easier to model higher-order relations, such as those traditionally represented in multipartite graphs. Bu et al. propose a ranking algorithm for hyper-graphs for the task of music recommendation and include features such as tags, acoustic similarities, and track and album relations. Using a data set from Last.FM, they show their algorithm to outperform traditional NN algorithms. Cantador et al. [19] also propose a random-walk-based method which operates on the tripartite graph. In addition, they propose the use of a tag classifier, which classifies tags into different categories using semantic mappings to Linked Open Data resources. Some of the commonly used categories include entities, tasks, opinions, locations, and time. They show experimentally on a Flickr data set, that reducing the tag space by mapping tags to higher-level categories can improve recommendation performance for specific tag categories, such as those related to item content and the context of use. Feng and Wang [32] also propose a random-walk algorithm where the edge weights are optimized using a learning-to-rank approach, i.e., a machine learning classifier is used to learn the optimal edge weights. They show their algorithm to outperform the unoptimized version as well as FolkRank [51] on datasets from Delicious and Last.FM.

5   Content-Based Filtering

Content-based filtering (CBF) algorithms, also known as content-based recommendation, typically focus on building a representation of the content in a system and then learning a profile of the user's interests. The content representations are then matched against the user's profile to find the items that are most relevant. Usually, content-based filtering for recommendation is approached as either (1) an IR problem, where document representations have to be matched to user representations on textual similarity; or (2) as a machine learning problem, where the textual content of the representations is incorporated as feature vectors, which are used to train a prediction algorithm. Examples of the IR approach include Whitman and Lawrence [111] and Bogers and Van den Bosch [12]; examples of the machine learning point of view include Lang [65] and Mooney and

Roy [72]. CBF has traditionally been most successful in text-heavy domains, such as recommending books, scientific articles, and websites. Automatically annotating multimedia content remains challenging, which is why tags present an opportunity for applying CBF to such domains as well. Tags can be seen as condensed textual descriptions of the objects they annotate, which makes them very suitable for use in CBF approaches.

In addition to proposing such tag-based variants of user-based and item-based NN algorithms, Bogers and Van den Bosch [13] also propose two different CBF algorithms: *profile-centric* and *post-centric matching*. In the profile-centric matching approach—visualized in Fig. 11—all of the metadata and tags assigned to the active user's past items are aggregated into a single textual representation of that user's interests. This means that tags assigned by other users are also included in the active user's profile. Item representations are similarly constructed by aggregating all metadata and tags assigned to that item in its lifetime. They then match the user's profile with all item representations using an IR engine based on language modeling. After removing the items already in the active user's profile, the final relevance-ordered list of item recommendations remains.

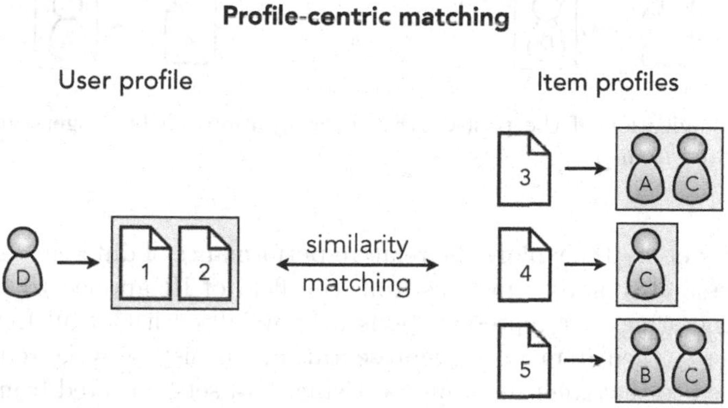

Fig. 11. Visualization of the profile-centric filtering approach by Bogers and Van den Bosch, adapted from [13].

Post-centric matching takes place at the level of individual posts and is visualized in Fig. 12. Here, a user's profile consists of a set of individual posts with general metadata, but user-specific tags. Each of these posts forms a separate textual representation of the object to be indexed by the IR engine. In the matching phase, each of these user's posts serves as a separate query with the most relevant matching posts being retrieved. This leads to a list of matching posts in order of similarity for each of the active user's posts. They then calculate a rank-corrected sum of normalized similarity scores for each item, thereby producing the final list of recommended items for the active user. Bogers and Van den Bosch evaluate their approach on data sets from BibSonomy, CiteULike, and

Delicious. They find that, while the best approach seemed to be dependent on the data set, post-centric matching seems to outperform profile-centric matching, most likely because the item representations are denser and therefore easier to match.

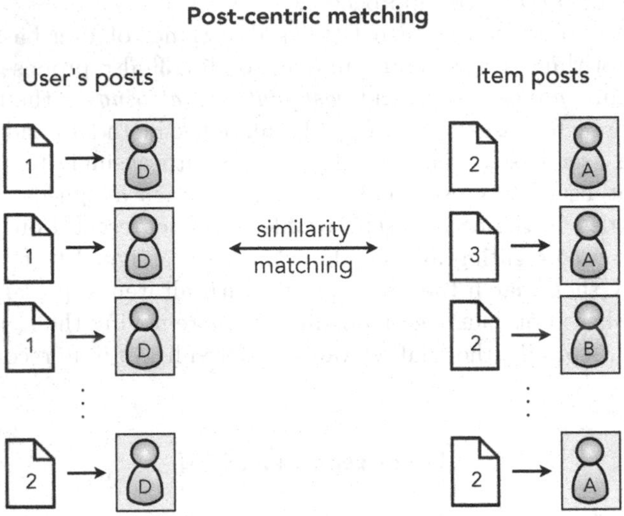

Post-centric matching

User's posts   Item posts

similarity
matching

Fig. 12. Visualization of the post-centric filtering approach by Bogers and Van den Bosch, adapted from [13].

Cantador et al. [18] explore the relative performance of different term weighting schemes and similarity metrics from the field of IR applied to CBF. They propose generating user representations as a weighted list of all tags assigned by that user to their items. Item representations consists of weighted lists of all tags assigned to that item by all users. Using data sets collected from Delicious and Last.FM, they find that the best performance is achieved using the BM25 term weighting scheme to calculate the tag weights and using cosine similarity to match user and item profiles. A similar algorithm is proposed by Durao and Dolog [28], who also compare different term weighting schemes and overlap metrics in a small-scale user-based pilot evaluation using data from Delicious. Szomszor et al. [99] propose a CBF approach that ranks the unseen items for an active user by the overlap between the tags assigned to those items and the active user's tag cloud. They reported that their CBF approach outperformed a popularity-based algorithm on the task of movie recommendation based on the Netflix data set, which they augmented by harvesting the tags belonging to each movie from IMDB[10]. Jomsri et al. [54] propose a similar approach for the task of paper recommendation, albeit without any experimental evaluation. Kim et al. [58] also propose calculating tag overlap between items to provide

---

[10] http://www.imdb.com.

item-to-item recommendations for a tourism recommender that suggests which sightseeing locations to visit next. Wartena et al. [107] propose a topic-aware, tag-based CBF algorithm that generates recommendations for each of the topics detected in a user's profile. They propose calculating item similarities or similarity between user profiles and items as the Jensen-Shannon divergence between their tag distributions. In a proposed of item-based NN variant, the similarity is used to find the most similar items to the active user's past interests. Their CBF approach uses the tag-based divergence to calculate the match between the user's tag-based profile and the item representations. Wartena et al. then propose topic-aware variants of these algorithms by clustering the tags in a user's profiles into a set of interests. For each of these interests, the aforementioned similarities are calculated to provide a more diversified set of recommendations. They show through experiments on a data set from LibraryThing that the topic-aware algorithms provide better performance in terms of accuracy as well as diversity [107,108].

CBF approaches, by virtue of relying on textual matches, tend to suffer from the so-called *vocabulary problem* [35]: people use different terms to describe the same objects. Without knowledge of synonyms and polysemous words, any system that performs word-level matching will have a much higher failure rate. Common strategies for overcoming this problem include using controlled vocabularies or resources such as WordNet that introduce knowledge about semantic relationships between words. Tags suffer from the same problem, as argued in other places in this book. Several semantic CBF approaches have been proposed to address this problem. Hung et al. [52], for instance, compute similarity between tags by constructing a tag-to-tag matrix, which is derived from the **UI** matrix. Individual cells in the tag-to-tag matrix specify how often two tags have been assigned together by a user. Recommendations are generated by calculating the overlap between a user's tags and the tags assigned to an item, taking the tag similarities into account. They evaluate their approach on a data set collected from Delicious, but do not compare it to other state-of-the-art approaches. De Gemmis et al. [25] describe a tag-based CBF approach for the cultural heritage domain. They use Wordnet, a lexical database that links related words together, to identify relevant concepts in the content of the items to be recommended, which results in a repository of disambiguated document representations. They then use a Naive Bayes classifier to learn a probabilistic model of the user's disambiguated interests. This semantic user profile is then matched against the semantic item representations to locate the most relevant items for the active user. Finally, Tatlı and Birtürk [100] propose a content-based music recommender that uses DBPedia, a Linked Open Data version of the Wikipedia, to match tags to other related musical genres. They generate user profiles by aggregating all of their tags; item representations consist of all tags assigned to a particular track. After applying the tf·idf weighting scheme to all terms, similarity between profiles and representations is calculated using the cosine similarity. They evaluate their approach on a Last.FM data set augmented with profile information from Facebook and IMDB, and find it outperforms an SVD approach.

6   Machine Learning

Machine learning is the field of computer science concerned with teaching computers how to perform a specific task by detecting salient patterns in the data, followed by associating those patterns to specific outcomes or predictions. While not as popular as CF and CBF approaches, there have been a handful of researchers that have proposed using machine learning for tag-based recommendation. Kim et al. [56], for instance, use the **UT** matrix to generate tag profiles for each user. These tags then serve as input to a Naive Bayes classifier, which predicts conditional probabilities for which items the user might like given the user's profile tags and the item-tag co-occurrence counts in the **IT** matrix. Unfortunately, their approach is not compared to other methods, which makes it hard to judge its effectiveness. Like Kim et al. [56], Vatturi et al. [102] also propose a tag-based recommendation algorithm based on a Naive Bayes classifier for recommending Web pages in the Dogear system. The authors describe three variants, which take as feature input a bag-of-words representation of all the tags assigned by a user. Implicit ratings then serve as class labels, with more recently bookmarked items receiving a higher rating. The difference between the first two variants—a general-interest classifier and a current-interest classifier—is the length of the time interval that results in positive ratings. The third variant was a disjunctive combination of the other two algorithms' predictions and provided the best performance. In their 2011 follow-up article, Kim et al. propose a variant of their 2010 algorithm that incorporates ratings information. The valence of a user's item rating is propagated to the tags assigned to that item. These tags, along with their propagated positive and negative weights, then again serve as features for a Naive Bayes classifier. On an artificially collected ratings data set based on IMDB, they show that their approach outperforms traditional user-based and item-based NN algorithms. Guan et al. [44] propose an machine learning approach that attempts to learn a two-dimensional representation of the tripartite graph, after which they recommend other items that are closer to the user in that compressed 2D space. They evaluate their approach on data sets from CiteULike and Delicious and find that it outperforms traditional NN and SVD algorithms.

Sen et al. [90] propose a type of tag-based recommendation algorithms called *tagommenders*. In the first stage of *tagommendation*, they propose and explore different methods for inferring user preferences for specific tags based on a variety of explicit and implicit signals of interest in items and tags, such as searches for and clicks on specific tags, ratings for movies tagged with specific tags, as well as a Bayesian generative model for predicting how users rate movies with specific tags. In the second stage these inferred tag preferences are normalized and used in five proposed recommendation algorithm. Two of these algorithms are content-based algorithms for producing item rankings only, not ratings prediction. They calculate the cosine similarity between user-preferred tags and the most representative tags for a movie, with or without a popularity prior. The other three algorithms use both implicit and explicitly inferred tag preferences and combine them with cosine similarity, general linear combination, and

Regression Support Vector Machines. The latter algorithm trains a linear regression function for each user-item pair using Regression Support Vector Machines (SVM) and then attempts to predict the item's rating by that user based on the assigned tags and their ratings. They evaluate their algorithms on a MovieLens data set and find that the Regression SVMs algorithm outperforms all other algorithms, including an standard SVD approach. Their work is followed up by Gedikli and Jannach [38], who test a similar approach using the ratings that tags have received in the MovieLens data set to improve movie recommendation. They also find that training a linear regression function for each user-item pair using Regression SVMs outperforms item-based NN and a standard SVD approach.

Finally, Chatti et al. [20] compare 16 different tag-based recommendation algorithms for use in a personal learning environment: two tag-augmented NN algorithms that use the **UT** and **IT** matrices to calculate similarities (similar to, e.g., Bogers and Van den Bosch [13]); four hybrid NN algorithms that operate on low-rank approximations of the **UT** and **IT** matrices after application of LSA; the two algorithms propose by Kim et al. [56]; and eight different clustering algorithms that cluster similar users or items together and recommend the unseen items from those clusters; and two approaches based on association rule-mining. They compare the 16 algorithms on a small dataset collected from the personal learning environment website and find that item-based k-Means clustering produced the best results.

7   Hybrid Recommendation

Hybrid recommenders combine aspects of different (types of) recommendation algorithms with the aim of creating a recommendation algorithm that can leverage the strengths of its component algorithms while alleviating their weaknesses. Burke [17] provides a taxonomy of seven different methods for creating hybrid recommendation algorithms, which we reproduce here in Table 1. We will classify the hybrid algorithms discussed in this section according to this taxonomy; for more information about the different hybridization methods, we refer the reader to Burke [17].

Several hybrid recommendation algorithms which incorporate tags in one or more ways have been proposed. Some of these approaches we were discussed in earlier sections, such as the tag-aware fusion approach by Tso-Sutter et al. [101], which is a weighted hybrid approach. Chatti et al. [20] also propose two hybrid NN algorithms, which that operate on low-rank approximations of the **UT** and **IT** matrices after application of LSA. These are examples of a feature-augmented hybrid algorithms.

In addition to proposing tag-based variants of user-based and item-based NN algorithms and two CBF approaches, Bogers and Van den Bosch [13] also propose two different hybrids of CBF and NN algorithms: *user-centric hybrid filtering* and *item-centric hybrid filtering*. In the former method, user similarities are calculated as the textual overlap in metadata belonging to the items in

Table 1. A taxonomy of recommender system combination methods, as given by Burke [17].

Hybridization method	Description
Mixed	Recommendations from several different recommenders are presented at the same time
Switching	The system switches between recommendation techniques depending on the current situation
Feature combination	Features from different recommendation data sources are thrown together into a single recommendation algorithm
Cascade	One recommender refines the recommendations given by another
Feature augmentation	The output from one technique is used as an input feature to another technique
Meta-level	The model learned by one recommender is used as input to another
Weighted	The scores of several recommendation techniques are combined together to produce a single recommendation

those users' profiles. These user similarities are then plugged into the user-based NN algorithm. For the item-centric variant, item similarities are calculated and applied in a similar manner, making them both examples of a weighted hybrid. Bogers and Van den Bosch [13] evaluate their hybrid algorithms on data sets from BibSonomy, CiteULike, and Delicious, and show that they are competitive with the tag-augmented NN and the pure CBF algorithms.

Bogers and Van den Bosch [14] later expand upon their earlier work [13] by comparing the eight different recommendation approaches directly as well as experimenting with six different methods for producing weighted combinations of the eight different algorithms. They use three different methods for performing *results fusion* from the field of IR, where the results of different retrieval algorithms on the same collection are combined. Analogously, here the results of different recommendation algorithms on the same data set are combined. The first of these three methods, CombSUM, simply takes the sum of the normalized scores of the individual runs for an item i_l. CombMNZ takes the sum of the normalized scores, multiplied by the number of hits of an item, i.e., how often it appears in the recommendation lists. The third method, CombANZ, takes the sum of the normalized scores of an item and divides it by the number of hits of that item. The CombSUM, CombMNZ, and CombANZ methods are all unweighted, which means that the preference weights for each run are equal. In addition to these unweighted methods, Bogers and Van den Bosch [14] also examine the benefits of weighted variants of CombSUM, CombMNZ, and CombANZ, where the optimal weights for each contributing algorithm are determined using a random-restart hill climbing algorithm. Through an experimental evaluation on data sets from BibSonomy, CiteULike, and Delicious, the authors show that it is often better to combine approaches that use different data representations, such as tags and metadata, than it is to combine approaches that only vary in the algorithms they use. They argue that the best results are obtained when both of these aspects of the recommendation task are varied in the combination process.

Gemmell et al. [39] propose a linearly weighted hybrid of four different NN algorithms: (1) user-based and (2) item-based NN recommendation on the **R** matrix; (3) user-based NN where user similarities are calculated on the **UT** matrix; and (4) item-based NN with item similarities calculated on the **IT** matrix. The recommendation scenario they target is one where the user provides the system with a single query tag, which then is the anchor for the recommendations. While this approach could also be categorized as tag-based search, we discuss it here because the proposed algorithms could also be used for 'pure' item recommendation. In addition to the weighted hybrid, Gemmell et al. [39] also adapt *pair-wise interaction tensor factorization* (PITF), an algorithm by Rendle et al. [82] originally designed for tag recommendation. Adapting PITF to the task of tag-based item recommendation task of Gemmell et al. [39] involves flipping the roles of items and tags: instead of recommending tags for a particular item, they recommend items for a specific tag. PITF uses a special case of Tucker decomposition of the tensor $\mathcal{D}$, which does not induce the tensor itself but instead provides a ranking of items for each user-tag pair. In their experiments on six data sets—Amazon, BibSonomy, CiteULike, Delicious, Last.FM, and MovieLens—they find that the linearly weighted hybrid outperforms all individual algorithms consistently, included the state-of-the-art PITF approach [39,40].

Xu et al. [114] propose a layered hybrid approach to tag-based recommendation. At the top level, their approach is a feature-augmented hybrid of a user-based NN algorithm that takes as input a linearly weighting combination of two sources of user similarities. The first source of user similarities is LDA applied to the **IT** matrix. The topic distributions generated by LDA are then used to reduce the dimensionality of the user profile vectors from the **UT** matrix, with user similarities calculated as the cosine similarity between two reduced user profile vectors. The second algorithm is a another feature-augmented hybrid itself: hierarchical agglomerative clustering is applied to the **IT** matrix to generate tag clusters. Again, the dimensionality of the user profile vectors from the **UT** matrix is reduced using these tag clusters. Xu et al. [114] show their hybrid algorithm to outperform to other tag-aware algorithms on a filtered version of the MovieLens data set and a data set collected from MedWorm, a medical website containing a collection of blog posts organized by subject.

Finally, Zhang et al. [120] propose a tiered, feature-augmented hybrid algorithm for tag-based recommendation of scientific articles. First, they construct a tag-tag matrix to cluster synonymous tags together based on their co-occurrence in user profiles and item profiles. Based on the user profiles from the clustered **UT** matrix, user similarities are then calculated using a standard user-based NN approach on the **UT** matrix. These user similarities then serve as input features for a CBF algorithm that operates on the tag-clustered **IT** matrix. Zhang et al. investigate the effects of neighborhood size on their approach on a filtered data set from CiteULike, but do not compare their approach with any other methods.

8   Tag Recommendation

While tag recommendation is not the focus of this chapter, it is important to mention that several of the approaches described are not limited to only provide recommendations for which items to consume; several of them can also be used to perform the other recommendation tasks shown in Fig. 1. For instance, another popular recommendation task for social tagging systems is suggesting which tags to apply to a newly bookmarked or consumed item. By virtue of their design, the tensor reduction approach of [98], FolkRank [51][11], and the personalized Markov random walk algorithm by [22] can all provide tag recommendations without any necessary adjustment or re-training. Like with tag-based item recommendation, there is an equally rich variety in algorithms for tag recommendation proposed in recent years. These algorithms can be categorized into the same algorithm classes as the ones used in this chapter. Table 2 contains a brief list of some of the more influential algorithms for tag recommendation, categorized by algorithm type. These are meant as a starting point only; we refer the reader to the individual papers for more details.

Table 2. An overview of tag recommendation algorithms, organized by algorithm type.

Algorithm class	Example approaches
Collaborative filtering	
- Nearest-neighbor	Garg and Weber [36,37], Givon and Lavrenko [41], Hamouda and Wanas [47], Lu et al. [69]
- Dimensionality reduction	Krestel and Fankhouse [64], Krestel et al. [63], Lu et al. [69], Rendle and Schmidt-Thieme [81,82], Si and Sun [92], Symeonidis et al. [96], Wetzker et al. [110], Yin et al. [118]
- Graph-based recommendation	Jäschke et al. [53], Song et al. [94,95], Zhang et al. [121,123]
Content-based filtering	Godoy and Amandi [42], Zhang et al. [121]
Machine learning	Eck et al. [29,30], Garg and Weber [36], Kataria [55], Ness et al. [76], Song et al. [94], Vojnovic et al. [104], Wang et al. [105], Xia et al. [112]
Hybrid recommendation	Godoy and Amandi [42], Zhang et al. [123]

---

[11] See also Chap. 9 [75] in this book for a more detailed description of FolkRank.

9   Conclusions

In the previous sections, we have provided a comprehensive overview of the most popular algorithms for item recommendation that incorporate tags. Many tag-aware algorithms have been proposed for each of the classes of recommendation algorithms. While not all algorithms have been thoroughly evaluated in equal measure, there does seem to plenty of evidence for the beneficial effect of including tags in the recommendation process. Whenever authors compare their tag-augmented algorithms to state-of-the-art algorithms that only operate on the $\mathbf{R}$ matrix, the tag-based algorithms tend to provide significantly better performance.

This does not mean that every possible way of incorporating tags is equally fruitful—indeed, Bogers [10] report that a small number of the proposed algorithms do not outperform baseline algorithms that do not make use of tagging information. Reporting negative results is also much less common, so unsuccessful tag-augmented algorithms are rarely published. As a result, there is no clear overview of which techniques do *not* work. However, for the majority of the proposed approaches, generating recommendations by using the tripartite user-item-tag graph does produce more accurate recommendations.

Unfortunately, while it is clear that tags do offers benefits, it remains unclear what the best-performing algorithms are. Two papers may even contradict each other with regard to the relative performance of two algorithms—the PMF approach that handily outperformed a NN algorithm in one experiment might be outclassed in another. There are three main issues that prevent us from being able to recommend which tag-algorithms to use in which situation: (1) a lack of comparisons with state-of-the-art approaches; (2) variation in data sets used for evaluation; and (3) a lack of standardized evaluation setup for tag-based item recommendation.

Lack of comparisons with the state-of-the-art. One obvious obstacle to gaining a clear idea of which algorithm works best is the lack of proper comparisons with state-of-the-art approaches. A handful of the approaches discussed in this chapter do not include any evaluation or do not compare their algorithm with any other methods. However, many of the ones that do, use weak or otherwise inappropriate baselines, such as popularity-based recommendation or traditional NN algorithms.

Another problem is that the proposed tag-aware algorithms are often only compared with their tag-unaware siblings. While this is a necessary comparison to show the value of tagging information, it does not provide any information about where the algorithm stands in comparison to other tag-based algorithms. While this does not hold for the first tag-based algorithms that were proposed—as there were no other tag-based algorithms to compare them with—future research on tag-based algorithms should aim to compare against state-of-the-art tag-based algorithms, such as Symeonidis et al. [98] or Tso-Sutter et al. [101].

Variation in data sets. Another problem for comparing approaches is the variety in data sets used in the experimental evaluations. Different papers will use data sets collected from different websites at different times, filtered in different ways. While a handful of publicly available data sets exist, these are rarely used, which makes a fair comparison between algorithms effectively impossible.

To examine the severity of this problem, we took stock of all the data sets used in evaluating all item recommendation algorithms discussed in Sects. 3–7. For each of the data sets, we recorded where they were collected and whether the data set was publicly available or not. If the same data set was used to evaluate an extended version of a particular algorithm, then that data set was only counted once. Figure 13 shows a bar chart of the usage frequencies of the different data sets for evaluating tag-based recommendation algorithms, differentiated by their availability.

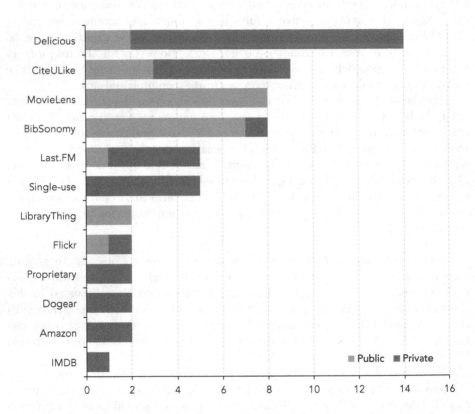

Fig. 13. Bar chart of the usage frequency of different data sets used in evaluating tag-based recommendation algorithms. Light blue denotes the number of times a publicly available version of each data set was used; dark blue denotes the frequency of use of private version. Data sets used only once are listed under 'Single-use' and unspecified data sets are listed under 'Proprietary'. (Color figure online)

Figure 13 shows that social bookmarking websites, such as Delicious, CiteU-Like and Bibsonomy, are the most popular sources for evaluation data sets together with the MovieLens movie recommendation data set[12]. The public availability of the BibSonomy and MovieLens data sets has resulted in them seeing heavy use in evaluation, which increases the likelihood of meaningful comparisons between algorithms. While Wetzker et al. [109] have made their Delicious data set publicly available, most researchers resort to crawling their own Delicious data set. CiteULike make anonymized dumps of their database available, but most articles do not mention the exact version of the dump that was used, making direct comparisons impractical. Figure 13 shows that only 24 out of 60 reported evaluations make use of publicly available data sets, which reinforces the problematic nature of this issue. To help address this issue, we have included a list of publicly available data sets for tag-based recommendation (and other tasks shown in Fig. 1) in Appendix A.

Lack of standardized evaluation. Finally, the last main obstacle to conducting a fair comparison between different algorithms is the lack of a standardized evaluation setup. Comprehensive overviews on how to evaluate a recommender system exist [49,91]. However, these do not prescribe a specific evaluation setup. As a result, many elements of the evaluation setup show great variation across approaches, such as the recommendation task (top-N recommendation vs. ratings prediction, the number of folds in n-fold cross-validation, and the metrics used to evaluate the recommendation list (precision, recall, NDCG, MAE, or RMSE). This results in a lack of comparability, which has been flagged as a serious problem before by, for instance, Said and Bellogín [84]. A possible solution to this problem could be the use of toolkits designed to increase the reproducibility of recommendation experiments, such as the RIVAL toolkit proposed by Said and Bellogín [85].

A lot of research effort has been dedicated to solving the problem of tag-based recommendation in recent years. We believe that what is needed now is not more algorithms, but instead a comprehensive, structured comparison of the existing algorithms on a variety of different publicly available data sets.

Appendix A: Data Sets for Tag-Based Recommendation

Below is an overview of the most commonly used publicly available recommendation data sets that include tagging information, organized by source.

---

[12] While there are multiple versions of the MovieLens data set available for download at http://grouplens.org/datasets/movielens/, only the MovieLens 10M data set contains tagging information. This is the data set we refer to as MovieLens in this book chapter.

BibSonomy

Benz et al. [6] make several dumps of the BibSonomy system available at http://www.kde.cs.uni-kassel.de/bibsonomy/dumps/.

CiteULike

CiteULike makes dumps of their folksonomy (user-item-tag relations with timestamps) available on their website at http://www.citeulike.org/faq/data.adp.

Delicious

Wetzker et al. [109] have made their Delicious data sets publicly available at http://www.dai-labor.de/en/irml/datasets/delicious/.

Flickr

Cantador et al. [19] have made their Flickr data set publicly available at http://mir.dcs.gla.ac.uk/resources/.

Last.FM

Konstas et al. [60] have made a Last.FM data set available at http://mir.dcs.gla.ac.uk/resources/.

LibraryThing

Two different tagging data sets based on LibraryThing have been made available. Clements et al. [22] made their LibraryThing data set available at http://www.macle.nl/tud/LT/. A recommendation data set including tagging information from LibraryThing has been made available as part of the Social Book Search track at CLEF [61]. See http://social-book-search.humanities.uva.nl/ for more information.

MovieLens

The GroupLens research group have a long history of making data sets from their movie recommender system available. The latest two versions, MovieLens 10M and MovieLens 20M, also contain tagging information, although only the former has been used for evaluation of tag-based recommender systems so far. More information on how to obtain these data sets can be found at http://grouplens.org/datasets/movielens/.

References

1. Adomavicius, G., Tuzhilin, A.: Context-aware recommender systems. In: Ricci, F., Rokach, L., Shapira, B., Kantor, P.B. (eds.) Recommender Systems Handbook, pp. 217–253. Springer, Boston, MA (2011). https://doi.org/10.1007/978-0-387-85820-3_7
2. Aggarwal, C.C., Wolf, J.L., Wu, K.L., Yu, P.S.: Horting hatches an egg: a new graph-theoretic approach to collaborative filtering. In: Proceeding of the 5th ACM SIGKDD International Conference on Knowledge Discovery and Data Mining, KDD 1999, pp. 201–212. ACM, New York (1999)
3. Aha, D.W., Kibler, D., Albert, M.K.: Instance-based learning algorithms. Mach. Learn. **6**(1), 37–66 (1991)

4. Amer-Yahia, S., Galland, A., Stoyanovich, J., Yu, C.: From del.icio.us to x.qui.site: recommendations in social tagging sites. In: Proceedings of the 2008 ACM SIG-MOD International Conference on Management of Data, SIGMOD 2008, pp. 1323–1326. ACM, New York (2008)

5. Baluja, S., Seth, R., Sivakumar, D., Jing, Y., Yagnik, J., Kumar, S., Ravichandran, D., Aly, M.: Video suggestion and discovery for YouTube: taking random walks through the view graph. In: Proceedings of the 17th International Conference on World Wide Web, WWW 2008, pp. 895–904. ACM, New York (2008)

6. Benz, D., Hotho, A., Jäschke, R., Krause, B., Mitzlaff, F., Schmitz, C., Stumme, G.: The social bookmark and publication management system bibsonomy - a platform for evaluating and demonstrating Web 2.0 research. VLDB 19(6), 849–875 (2010)

7. Berry, M.W., Dumais, S.T., O'Brien, G.W.: Using linear algebra for intelligent information retrieval. SIAM Rev. 37(4), 573–595 (1995)

8. Bischoff, K., Firan, C.S., Nejdl, W., Paiu, R.: Can all tags be used for search? In: Proceedings of the seventeenth international conference on information and knowledge management, CIKM 2008, pp. 193–202. ACM, New York (2008)

9. Blei, D.M., Ng, A.Y., Jordan, M.I.: Latent Dirichlet allocation. J. Mach. Learn. Res. 3(4–5), 993–1022 (2003)

10. Bogers, T.: Recommender systems for social bookmarking. Ph.D. thesis, Tilburg University, December 2009

11. Bogers, T.: Movie recommendation using random walks over the contextual graph. In: Adomavicius, G., Tuzhilin, A. (eds.) Proceedings of the RecSys CARS 2010 workshop (2010)

12. Bogers, T., Van den Bosch, A.: Comparing and evaluating information retrieval algorithms for news recommendation. In: Proceedings of the 2007 ACM Conference on Recommender Systems, RecSys 2007, pp. 141–144. ACM, October 2007

13. Bogers, T., Van den Bosch, A.: Collaborative and content-based filtering for item recommendation on social bookmarking websites. In: Jannach, D., Geyer, W., Freyne, J., Anand, S.S., Dugan, C., Mobasher, B., Kobsa, A. (eds.) Proceedings of the ACM RecSys 2009 workshop on Recommender Systems and the Social Web, pp. 9–16, October 2009

14. Bogers, T., Van den Bosch, A.: Fusing recommendations for social bookmarking websites. Int. J. Electron. Commer. 15(3), 33–75 (2011)

15. Brusilovsky, P., Smyth, B., Shapira, B.: Social search. In: Brusilovsky, P., He, D. (eds.) Social Information Access. LNCS, vol. 10100, pp. 213–276. Springer, Cham (2018)

16. Bu, J., Tan, S., Chen, C., Wang, C., Wu, H., Zhang, L., He, X.: Music recommendation by unified hypergraph: combining social media information and music content. In: Proceedings of the International Conference on Multimedia, MM 2010, pp. 391–400. ACM, New York, October 2010

17. Burke, R.: Hybrid recommender systems: survey and experiments. User Model. User-Adap. Inter. 12(4), 331–370 (2002)

18. Cantador, I., Bellogín, A., Vallet, D.: Content-based recommendation in social tagging systems. In: Proceedings of the Fourth ACM Conference on Recommender Systems, RecSys 2010, pp. 237–240. ACM (2010)

19. Cantador, I., Konstas, I., Jose, J.M.: Categorising social tags to improve Folksonomy-based recommendations. Web Semant. Sci. Serv. Agents World Wide Web 9(1), 1–15 (2011)

20. Chatti, M.A., Dakova, S., Thus, H., Schroeder, U.: Tag-based collaborative filtering recommendation in personal learning environments. IEEE Trans. Learn. Technol. **6**(4), 337–349 (2013)
21. Clements, M.: Personalized access to social media. Ph.D. thesis, Delft University of Technology, December 2010
22. Clements, M., de Vries, A.P., Reinders, M.J.T.: Optimizing single term queries using a personalized Markov random walk over the social graph. In: Alonso, O., Zaragoza, H. (eds.) Proceedings of the ECIR Workshop on Exploiting Semantic Annotations in Information Retrieval (ESAIR 2008), pp. 18–24 (2008)
23. Clements, M., de Vries, A.P., Reinders, M.J.T.: The influence of personalization on tag query length in social media search. Inf. Process. Manag. **46**(4), 403–412 (2010)
24. Clements, M., de Vries, A.P., Reinders, M.J.T.: The task dependent effect of tags and ratings on social media access. ACM Trans. Inf. Syst. **28**, 21 (2010)
25. De Gemmis, M., Lops, P., Semeraro, G., Basile, P.: Integrating tags in a semantic content-based recommender. In: Proceedings of the 2008 ACM Conference on Recommender Systems, RecSys 2008, pp. 163–170. ACM, New York (2008)
26. Desrosiers, C., Karypis, G.: A comprehensive survey of neighborhood-based recommendation methods. In: Ricci, F., Rokach, L., Shapira, B., Kantor, P.B. (eds.) Recommender Systems Handbook, pp. 107–144. Springer, Boston, MA (2011). https://doi.org/10.1007/978-0-387-85820-3_4
27. Dimitrov, D., Helic, D., Strohmaier, M.: Tag-based navigation and visualization. In: Brusilovsky, P., He, D. (eds.) Social Information Access. LNCS, vol. 10100, pp. 181–212. Springer, Cham (2017)
28. Durao, F., Dolog, P.: A personalized tag-based recommendation in social web systems. In: CEUR Workshop Proceedings, vol. 485, pp. 40–49 (2009)
29. Eck, D., Bertin-Mahieux, T., Lamere, P.: Autotagging music using supervised machine learning. In: Proceedings of the 8th International Conference on Music Information Retrieval, ISMIR 2007 (2007)
30. Eck, D., Lamere, P., Bertin-Mahieux, T., Green, S.: Automatic generation of social tags for music recommendation. Adv. Neural Inf. Process. Syst. **20**, 385–392 (2008)
31. Enrich, M., Braunhofer, M., Ricci, F.: Cold-start management with cross-domain collaborative filtering and tags. In: Huemer, C., Lops, P. (eds.) EC-Web 2013. LNBIP, vol. 152, pp. 101–112. Springer, Heidelberg (2013). https://doi.org/10.1007/978-3-642-39878-0_10
32. Feng, W., Wang, J.: Incorporating heterogeneous information for personalized tag recommendation in social tagging systems. In: Proceedings of the 18th ACM SIGKDD International Conference on Knowledge Discovery and Data Mining, KDD 2012, pp. 1276–1284. ACM, New York, August 2012
33. Fernández-Tobías, I., Cantador, I.: Exploiting social tags in matrix factorization models for cross-domain collaborative filtering. In: Bogers, T., Koolen, M., Cantador, I. (eds.) Proceedings of the 1st Workshop on New Trends in Content-based Recommender Systems, pp. 34–41. CEUR-WS.org (2014)
34. Firan, C.S., Nejdl, W., Paiu, R.: The benefit of using tag-based profiles. In: Proceedings of the 2007 Latin American Web Conference, pp. 32–41, October 2007
35. Furnas, G.W., Landauer, T.K., Gomez, L.M., Dumais, S.T.: The vocabulary problem in human-system communication. Commun. ACM **30**(11), 964–971 (1987)
36. Garg, N., Weber, I.: Personalized, interactive tag recommendation for Flickr. In: Proceedings of the 2008 ACM Conference on Recommender Systems, RecSys 2008, pp. 67–74. ACM, New York, October 2008

37. Garg, N., Weber, I.: Personalized tag suggestion for Flickr. In: Proceedings of the 17th International Conference on World Wide Web, WWW 2008, pp. 1063–1064. ACM, New York (2008)

38. Gedikli, F., Jannach, D.: Improving recommendation accuracy based on item-specific tag preferences. ACM Trans. Intell. Syst. Technol. (TIST) 4(1), 11–19 (2013)

39. Gemmell, J., Schimoler, T., Mobasher, B., Burke, R.: Tag-based resource recommendation in social annotation applications. In: Konstan, J.A., Conejo, R., Marzo, J.L., Oliver, N. (eds.) UMAP 2011. LNCS, vol. 6787, pp. 111–122. Springer, Heidelberg (2011). https://doi.org/10.1007/978-3-642-22362-4_10

40. Gemmell, J., Schimoler, T., Mobasher, B., Burke, R.: Resource recommendation in social annotation systems: a linear-weighted hybrid approach. J. Comput. Syst. Sci. 78(4), 1160–1174 (2012)

41. Givon, S., Lavrenko, V.: Predicting social-tags for cold start book recommendations. In: Proceedings of the Third ACM Conference on Recommender Systems, RecSys 2009, pp. 333–336. ACM, New York, October 2009

42. Godoy, D., Amandi, A.: Hybrid content and tag-based profiles for recommendation in collaborative tagging systems. In: Proceedings of the 2008 Latin American Web Conference, LA-WEB 2008, pp. 58–65. IEEE (2008)

43. Golder, S.A., Huberman, B.A.: Usage patterns of collaborative tagging systems. J. Inf. Sci. 32(2), 198–208 (2006)

44. Guan, Z., Wang, C., Bu, J., Chen, C., Yang, K., Cai, D., He, X.: Document recommendation in social tagging services. In: Proceedings of the 10th International Conference on World Wide Web, WWW 2010, pp. 391–400. ACM, New York, April 2010

45. Guy, I., Zwerdling, N., Ronen, I., Carmel, D., Uziel, E.: Social media recommendation based on people and tags. In: Proceedings of the 33rd International ACM SIGIR Conference on Research and Development in Information Retrieval, pp. 194–201. ACM (2010)

46. Hammond, T., Hannay, T., Lund, B., Scott, J.: Social bookmarking tools (I) - a general review. D-Lib Mag. 11(4), 1082–9873 (2005)

47. Hamouda, S., Wanas, N.: PUT-tag: personalized user-centric tag recommendation for social bookmarking systems. Soc. Netw. Anal. Min. 1(4), 377–385 (2011)

48. Herlocker, J.L., Konstan, J.A., Borchers, A., Riedl, J.: An algorithmic framework for performing collaborative filtering. In: Proceedings of the 22nd Annual International ACM SIGIR Conference on Research and Development in Information Retrieval, SIGIR 1999, pp. 230–237. ACM, New York (1999)

49. Herlocker, J.L., Konstan, J.A., Terveen, L.G., Riedl, J.T.: Evaluating collaborative filtering recommender systems. ACM Trans. Inf. Syst. 22(1), 5–53 (2004)

50. Hofmann, T.: Latent semantic models for collaborative filtering. ACM Trans. Inf. Syst. 22(1), 89–115 (2004)

51. Hotho, A., Jäschke, R., Schmitz, C., Stumme, G.: Information retrieval in folksonomies: search and ranking. In: Sure, Y., Domingue, J. (eds.) ESWC 2006. LNCS, vol. 4011, pp. 411–426. Springer, Heidelberg (2006). https://doi.org/10.1007/11762256_31

52. Hung, C.C., Huang, Y.C., Hsu, J.Y., Wu, D.K.C.: Tag-based user profiling for social media recommendation. In: Proceedings of the 2008 AAAI Workshop on Intelligent Techniques for Web Personalization & Recommender Systems. AAAI Press, July 2008

53. Jäschke, R., Marinho, L., Hotho, A., Schmidt-Thieme, L., Stumme, G.: Tag recommendations in Folksonomies. In: Kok, J.N., Koronacki, J., Lopez de Mantaras, R., Matwin, S., Mladenič, D., Skowron, A. (eds.) PKDD 2007. LNCS (LNAI), vol. 4702, pp. 506–514. Springer, Heidelberg (2007). https://doi.org/10.1007/978-3-540-74976-9_52

54. Jomsri, P., Sanguansintukul, S., Choochaiwattana, W.: A framework for tag-based research paper recommender system: an IR approach. In: 2010 IEEE 24th International Conference on Advanced Information Networking and Applications Workshops, pp. 103–108. IEEE (2010)

55. Kataria, S.: Recursive neural language architecture for tag prediction, March 2016. arXiv:1603.07646v1, cs.IR

56. Kim, H.N., Ji, A.T., Ha, I., Jo, G.S.: Collaborative filtering based on collaborative tagging for enhancing the quality of recommendation. Electron. Commer. Res. Appl. 9(1), 73–83 (2010)

57. Kim, H.N., Roczniak, A., Lévy, P., El Saddik, A.: Social media filtering based on collaborative tagging in semantic space. Multimedia Tools Appl. 56(1), 63–89 (2012)

58. Kim, J., Kim, H., Ryu, J.: TripTip: a trip planning service with tag-based recommendation. In: CHI 2009 Extended Abstracts on Human Factors in Computing Systems, pp. 3467–3472. ACM (2009)

59. Kluver, D., Ekstrand, M.D., Konstan, J.A.: Rating-based collaborative filtering: algorithms and evaluation. In: Brusilovsky, P., He, D. (eds.) Social Information Access. LNCS, vol. 10100, pp. 344–390. Springer, Cham (2017)

60. Konstas, I., Stathopoulos, V., Jose, J.M.: On social networks and collaborative recommendation. In: Proceedings of the 32nd Annual International ACM SIGIR Conference on Research and Development in Information Retrieval, SIGIR 2009, pp. 195–202. ACM (2009)

61. Koolen, M., Bogers, T., Kamps, J., Kazai, G., Preminger, M.: Overview of the INEX 2014 social book search track. vol. 1180, pp. 462–479. CEUR-WS.org (2014)

62. Koren, Y., Bell, R.: Advances in collaborative filtering. In: Ricci, F., Rokach, L., Shapira, B., Kantor, P.B. (eds.) Recommender Systems Handbook, pp. 145–186. Springer, Boston, MA (2011). https://doi.org/10.1007/978-0-387-85820-3_5

63. Krestel, R., Fankhause, P.: Language models and topic models for personalizing tag recommendation. In: Proceedings of the 2010 IEEE/ACM International Conference on Web Intelligence-Intelligent Agent Technology, WI-IAT 2010, pp. 82–89 IEEE (2010)

64. Krestel, R., Fankhauser, P., Nejdl, W.: Latent Dirichlet allocation for tag recommendation. In: Proceedings of the Third ACM Conference on Recommender Systems, RecSys 2009, pp. 61–68. ACM, New York, October 2009

65. Lang, K.: NewsWeeder: learning to filter netnews. In: Proceedings of the 12th International Conference on Machine Learning, ICML 1995, pp. 331–339. Morgan Kaufmann, San Mateo (1995)

66. Liang, H., Xu, Y., Li, Y., Nayak, R.: Tag based collaborative filtering for recommender systems. In: Wen, P., Li, Y., Polkowski, L., Yao, Y., Tsumoto, S., Wang, G. (eds.) RSKT 2009. LNCS (LNAI), vol. 5589, pp. 666–673. Springer, Heidelberg (2009). https://doi.org/10.1007/978-3-642-02962-2_84

67. Liang, H., Xu, Y., Li, Y., Nayak, R.: Personalized recommender system based on item taxonomy and folksonomy. In: Proceedings of the Nineteenth International Conference on Information and Knowledge Management, CIKM 2010, pp. 1641–1644. ACM, New York, October 2010

68. Liang, H., Xu, Y., Li, Y., Nayak, R., Tao, X.: Connecting users and items with weighted tags for personalized item recommendations. In: Proceedings of the 21st ACM Conference on Hypertext and Hypermedia, HT 2010, pp. 51–60. ACM, New York, June 2010

69. Lu, C., Hu, X., Park, J., Huang, J.: Post-based collaborative filtering for personalized tag recommendation. In: Proceedings of the 2011 iConference, iConference 2011, pp. 561–568. ACM, New York, February (2011)

70. Luo, X., Ouyang, Y., Xiong, Z.: Improving neighborhood based collaborative filtering via integrated folksonomy information. Pattern Recogn. Lett. **33**(3), 263–270 (2012)

71. Mathes, A.: Folksonomies - Cooperative Classification and Communication through Shared Metadata, Technical report (2004)

72. Mooney, R.J., Roy, L.: Content-based book recommending using learning for text categorization. In: Proceedings of the Fifth ACM Conference on Digital Libraries, DL 2000, pp. 195–204. ACM, New York (2000)

73. Nakamoto, R., Nakajima, S., Miyazaki, J., Uemura, S.: Tag-based contextual collaborative filtering. In: Proceedings of the 18th IEICE Data Engineering Workshop (2007)

74. Nakamoto, R., Nakajima, S., Miyazaki, J., Uemura, S., Kato, H., Inagaki, Y.: Reasonable tag-based collaborative filtering for social tagging systems. In: Proceeding of the 2nd ACM Workshop on Information Credibility on the Web, WICOW 2008, pp. 11–18. ACM, New York (2008)

75. Navarro Bullock, B., Hotho, A., Stumme, G.: Accessing information with tags: search and ranking. In: Brusilovsky, P., He, D. (eds.) Social Information Access. LNCS, vol. 10100, pp. 310–343. Springer, Cham (2017)

76. Ness, S.R., Theocharis, A., Tzanetakis, G., Martins, L.G.: Improving automatic music tag annotation using stacked generalization of probabilistic SVM outputs. In: Proceedings of the 17th ACM International Conference on Multimedia, MM 2009, pp. 705–708. ACM, New York, October 2009

77. Page, L., Brin, S., Motwani, R., Winograd, T.: The PageRank citation ranking: bringing order to the web, Technical report (1998)

78. Parra, D., Brusilovsky, P.: Collaborative filtering for social tagging systems: an experiment with CiteULike. In: Proceedings of the Third ACM Conference on Recommender Systems, RecSys 2009, pp. 237–240. ACM, New York (2009)

79. Peng, J., Zeng, D.D., Zhao, H., Wang, F.: Collaborative filtering in social tagging systems based on joint item-tag recommendations. In: Proceedings of the Nineteenth International Conference on Information and Knowledge Management, CIKM 2010, pp. 809–818. ACM, New York, October 2010

80. Rafailidis, D., Daras, P.: The TFC Model: tensor factorization and tag clustering for item recommendation in social tagging systems. IEEE Trans. Syst. Man Cybern. Syst. **43**(3), 673–688 (2013)

81. Rendle, S., Schmidt-Thieme, L.: Factor models for tag recommendation in Bibsonomy. In: Proceedings of 2009 ECML/PKDD Discovery Challenge Workshop, pp. 235–243 (2009)

82. Rendle, S., Schmidt-Thieme, L.: Pairwise interaction tensor factorization for personalized tag recommendation. In: Proceedings of the Third ACM International Conference on Web Search and Data Mining, WSDM 2010, pp. 81–90. ACM, New York, February 2010

83. Robertson, S.: The Probabilistic Relevance Framework: BM25 and Beyond. Foundations and Trends in Information Retrieval, vol. 3. Now Publishers Inc. (2009)

84. Said, A., Bellogín, A.: Comparative recommender system evaluation: benchmarking recommendation frameworks. In: Proceedings of the 2015 ACM Conference on Recommender Systems, RecSys 2015, pp. 129–136. ACM Press, New York (2014)

85. Said, A., Bellogín, A.: RIVAL: a toolkit to foster reproducibility in recommender system evaluation. In: Proceedings of the 2015 ACM Conference on Recommender Systems, RecSys 2015, pp. 371–372. ACM Press, New York (2014)

86. Said, A., Wetzker, R., Umbrath, W., Hennig, L.: A hybrid PLSA approach for warmer cold start in folksonomy recommendation. In: Proceedings of the ACM RecSys 2009 Workshop on Recommender Systems & The Social Web (2009)

87. Sarwar, B., Karypis, G., Konstan, J.A., Riedl, J.T.: Item-based collaborative filtering recommendation algorithms. In: Proceedings of the 10th International Conference on World Wide Web, WWW 2001, pp. 285–295. ACM, New York (2001)

88. Sarwar, B., Karypis, G., Konstan, J.A., Riedl, J.T.: Incremental SVD-based algorithms for highly scaleable recommender systems. In: Proceedings of the Fifth International Conference on Computer and Information Technology (ICCIT 2002) (2002)

89. Sen, S., Lam, S.K., Rashid, A.M., Cosley, D., Frankowski, D., Osterhouse, J., Harper, F.M., Riedl, J.T.: Tagging, communities, vocabulary, evolution. In: Proceedings of the 2006 20th Anniversary Conference on Computer Supported Cooperative Work, CSCW 2006, pp. 181–190. ACM, New York (2006)

90. Sen, S., Vig, J., Riedl, J.: Tagommenders: connecting users to items through tags. In: Proceedings of the 18th International Conference on World Wide Web, WWW 2009, pp. 671–680. ACM, New York April 2009

91. Shani, G., Gunawardana, A.: Evaluating recommendation systems. In: Ricci, F., Rokach, L., Shapira, B., Kantor, P.B. (eds.) Recommender Systems Handbook, pp. 257–297. Springer, Boston, MA (2011). https://doi.org/10.1007/978-0-387-85820-3_8

92. Si, X., Sun, M.: Tag-LDA for scalable real-time tag recommendation. J. Comput. Inf. Syst. **6**, 23–31 (2009)

93. Smyth, B., Coyle, M., Briggs, P.: Communities, collaboration, and recommender systems in personalized web search. In: Ricci, F., Rokach, L., Shapira, B., Kantor, P.B. (eds.) Recommender Systems Handbook, pp. 579–614. Springer, Boston, MA (2011). https://doi.org/10.1007/978-0-387-85820-3_18

94. Song, Y., Zhang, L., Giles, C.L.: Automatic tag recommendation algorithms for social recommender systems. ACM Trans. Web **5**(1), 4–31 (2011)

95. Song, Y., Zhuang, Z., Li, H., Zhao, Q., Li, J., Lee, W.C., Giles, C.L.: Real-time automatic tag recommendation. In: Proceedings of the 31st Annual International ACM SIGIR Conference on Research and Development in Information Retrieval, SIGIR 2008, pp. 515–522. ACM, New York (2008)

96. Symeonidis, P., Nanopoulos, A., Manolopoulos, Y.: Tag recommendations based on tensor dimensionality reduction. In: Proceedings of the 2008 ACM Conference on Recommender Systems, RecSys 2008, pp. 43–50. ACM, New York (2008)

97. Symeonidis, P., Nanopoulos, A., Manolopoulos, Y.: A unified framework for providing recommendations in social tagging systems based on ternary semantic analysis. IEEE Trans. Knowl. Data Eng. **22**(2), 179–192 (2010)

98. Symeonidis, P., Ruxanda, M., Nanopoulos, A., Manolopoulos, Y.: Ternary semantic analysis of social tags for personalized music recommendation. In: Proceedings of the 9th International Conference on Music Information Retrieval, ISMIR 2008, pp. 219–224 (2008)

99. Szomszor, M., Cattuto, C., Alani, H., O'Hara, K., Baldassarri, A., Loreto, V., Servedio, V.D.P.: Folksonomies, the semantic web, and movie recommendation. In: Proceedings of the ESWC Workshop on Bridging the Gap between Semantic Web and Web 2.0 (2007)

100. Tatlı, İ., Birtürk, A.: A tag-based hybrid music recommendation system using semantic relations and multi-domain information. In: Proceedings of the 2011 IEEE 11th International Conference on Data Mining Workshops, ICDMW 2011, pp. 548–554. IEEE (2011)

101. Tso-Sutter, K.H., Marinho, L.B., Schmidt-Thieme, L.: Tag-aware recommender systems by fusion of collaborative filtering algorithms. In: Proceedings of the 2008 ACM symposium on Applied computing, SAC 2008, pp. 1995–1999. ACM, New York (2008)

102. Vatturi, P.K., Geyer, W., Dugan, C., Muller, M., Brownholtz, B.: Tag-based filtering for personalized bookmark recommendations. In: Proceedings of the 17th ACM Conference on Information and Knowledge Management, pp. 1395–1396. ACM (2008)

103. Victor, P., De Cock, M., Cornelis, C.: Trust and recommendations. In: Ricci, F., Rokach, L., Shapira, B., Kantor, P.B. (eds.) Recommender Systems Handbook, pp. 645–675. Springer, Boston, MA (2011). https://doi.org/10.1007/978-0-387-85820-3_20

104. Vojnovic, M., Cruise, J., Gunawardena, D., Marbach, P.: Ranking and suggesting tags in collaborative tagging applications, Technical report MSR-TR-2007-06, Microsoft Research, February 2007

105. Wang, H., Shi, X., Yeung, D.Y.: Relational stacked denoising autoencoder for tag recommendation. In: Proceedings of the Twenty-Ninth AAAI Conference on Artificial Intelligence, AAAI 2015, pp. 3052–3058. AAAI Press, January 2015

106. Wang, Z., Wang, Y., Wu, H.: Tags meet ratings: improving collaborative filtering with tag-based neighborhood method. In: Proceedings of the 2010 IUI Workshop on Social Recommender Systems (2010)

107. Wartena, C., Brussee, R., Wibbels, M.: Using tag co-occurrence for recommendation. In: Proceedings of the 2009 Ninth International Conference on Intelligent Systems Design and Applications, pp. 273–278. IEEE (2009)

108. Wartena, C., Wibbels, M.: Improving tag-based recommendation by topic diversification. In: Clough, P., Foley, C., Gurrin, C., Jones, G.J.F., Kraaij, W., Lee, H., Mudoch, V. (eds.) ECIR 2011. LNCS, vol. 6611, pp. 43–54. Springer, Heidelberg (2011). https://doi.org/10.1007/978-3-642-20161-5_7

109. Wetzker, R., Umbrath, W., Said, A.: A hybrid approach to item recommendation in folksonomies. In: Proceedings of the WSDM 2009 Workshop on Exploiting Semantic Annotations in Information Retrieval, ESAIR 2009, pp. 25–29. ACM, New York (2009)

110. Wetzker, R., Zimmermann, C., Bauckhage, C., Albayrak, S.: I Tag, You Tag: translating tags for advanced user models. In: Proceedings of the 2010 International Conference on Web Search and Data Mining (WSDM) (2010)

111. Whitman, B., Lawrence, S.: Inferring descriptions and similarity for music from community metadata. In: Proceedings of the 2002 International Computer Music Conference, pp. 591–598 (2002)

112. Xia, X., Lo, D., Wang, X., Zhou, B.: Tag recommendation in software information sites. In: Proceedings of the 10th Working Conference on Mining Software Repositories, MSR 2013, pp. 287–296. IEEE Press, May 2013

113. Xin, L., Ouyang, Y., Zhang, X.: Improving latent factor model-based collaborative filtering via integrated folksonomy factors. Int. J. Uncertainty Fuzziness Knowl.-Based Syst. **19**(02), 307–327 (2011)
114. Xu, G., Gu, Y., Dolog, P., Zhang, Y., Kitsuregawa, M.: SemRec: a semantic enhancement framework for tag based recommendation. In: Burgard, W., Roth, D. (eds.) Proceedings of the Twenty-Fifth AAAI Conference on Artificial Intelligence, AAAI 2011, pp. 1267–1272. AAAI Press (2011)
115. Xu, Z., Fu, Y., Mao, J., Su, D.: Towards the semantic web: collaborative tag suggestions. In: Proceedings of the WWW 2006 Collaborative Web Tagging Workshop (2006)
116. Yildirim, H., Krishnamoorthy, M.S.: A random walk method for alleviating the sparsity problem in collaborative filtering. In: Proceedings of the 2008 ACM Conference on Recommender Systems, RecSys 2008, pp. 131–138. ACM, New York (2008)
117. Yin, D., Guo, S., Chidlovskii, B., Davison, B.D., Archambeau, C., Bouchard, G.: Connecting comments and tags: improved modeling of social tagging systems. In: Proceedings of the Sixth ACM International Conference on Web Search and Data Mining, WSDM 2013, pp. 547–556. ACM, New York, February 2013
118. Yin, D., Xue, Z., Hong, L., Davison, B.D.: A probabilistic model for personalized tag prediction. In: Proceedings of the 16th ACM SIGKDD International Conference on Knowledge Discovery and Data Mining, KDD 2010, pp. 959–968. ACM, New York, July 2010
119. Zeng, D., Li, H.: How useful are tags? — an empirical analysis of collaborative tagging for web page recommendation. In: Yang, C.C., et al. (eds.) ISI 2008. LNCS, vol. 5075, pp. 320–330. Springer, Heidelberg (2008). https://doi.org/10.1007/978-3-540-69304-8_32
120. Zhang, M., Wang, W., Li, X.: A paper recommender for scientific literatures based on semantic concept similarity. In: Buchanan, G., Masoodian, M., Cunningham, S.J. (eds.) ICADL 2008. LNCS, vol. 5362, pp. 359–362. Springer, Heidelberg (2008). https://doi.org/10.1007/978-3-540-89533-6_44
121. Zhang, N., Zhang, Y., Tang, J.: A tag recommendation system for folksonomy. In: Proceedings of the 2nd ACM Workshop on Social Web Search and Mining, SWSM 2009, pp. 9–16. ACM, New York, November 2009
122. Zhang, Y., Zhang, B., Gao, K., Guo, P., Sun, D.: Combining content and relation analysis for recommendation in social tagging systems. Phys. A **391**(22), 5759–5768 (2012)
123. Zhang, Y., Zhang, N., Tang, J.: A collaborative filtering tag recommendation system based on graph. In: Proceedings of 2009 ECML/PKDD Discovery Challenge Workshop (2009)
124. Zhang, Z., Zeng, D.D., Abbasi, A., Peng, J., Zheng, X.: A random walk model for item recommendation in social tagging systems. ACM Trans. Manag. Inf. Syst. **4**(2), 1–24 (2013)
125. Zhang, Z.K., Liu, C., Zhang, Y.C., Zhou, T.: Solving the cold-start problem in recommender systems with social tags. EPL (Europhys. Lett.) **92**(2), 28002 (2010)
126. Zhang, Z.K., Zhou, T., Zhang, Y.C.: Personalized recommendation via integrated diffusion on user–item–tag tripartite graphs. Phys. A **389**(1), 179–186 (2010)
127. Zhao, S., Du, N., Nauerz, A., Zhang, X., Yuan, Q., Fu, R.: Improved recommendation based on collaborative tagging behaviors. In: Proceedings of the 13th International Conference on Intelligent User Interfaces, IUI 2008, pp. 413–416. ACM (2008)

128. Zhen, Y., Li, W.J., Yeung, D.Y.: TagiCoFi: tag informed collaborative filtering. In: Proceedings of the Third ACM Conference on Recommender Systems, RecSys 2009, pp. 69–76. ACM, New York, October 2009

13
From Opinions to Recommendations

Michael P. O'Mahony[(⊠)] [iD] and Barry Smyth

Insight Centre for Data Analytics, School of Computer Science,
University College Dublin, Dublin, Ireland
michael.omahony@ucd.ie

Abstract. Traditionally, recommender systems have relied on user preference data (such as ratings) and product descriptions (such as metadata) as primary sources of recommendation knowledge. More recently, new sources of recommendation knowledge in the form of social media information and other kinds of user-generated content have emerged as viable alternatives. For example, services such as Twitter, Facebook, Amazon and TripAdvisor provide a rich source of user opinions, positive and negative, about a multitude of products and services. They have the potential to provide recommender systems with access to the fine-grained opinions of real users based on real experiences. This chapter will explore how product opinions can be mined from such sources and can be used as the basis for recommendation tasks. We will draw on a number of concrete case-studies to provide different examples of how opinions can be extracted and used in practice.

Keywords: Recommender systems · Opinion mining
Sentiment analysis

1 Introduction

Traditionally, recommender systems have relied on user preference data and product descriptions as the primary sources of recommendation knowledge. For example, collaborative recommendation approaches [8,29,60,64,66] rely on the former to identify a neighbourhood of like-minded users to a target user to act as a source of product recommendations (see also Chap. 10 of this book [38]). Alternatively, content-based recommendation approaches [46,57,67] select products for recommendation because they are similar to those that the target user has liked in the past (see also Chap. 12 of this book [7]). These approaches have worked well when suitable sources of recommendation knowledge is available, such as user-item ratings or item meta-data, but there are many circumstances where these approaches are less successful. For example, collaborative filtering systems work well when there are large communities of active users leading to rich user profiles to drive the recommendation process. But they are less successful when dealing with new users or where there is a sparsity of preference or

© Springer International Publishing AG, part of Springer Nature 2018
P. Brusilovsky and D. He (Eds.): Social Information Access, LNCS 10100, pp. 480–509, 2018.
https://doi.org/10.1007/978-3-319-90092-6_13

ratings data. Content-based techniques are effective when rich product descriptions are available but are less successful when more limited product information can be gathered.

One approach to dealing with the shortcomings of these conventional approaches has been to develop hybrid recommender systems that attempt to combine collaborative and content-based ideas. Such hybrid approaches [10] (see also Chap. 12 [7] of this book) are able to compensate for the short comings of any individual approach in isolation have proven to be successful in practice. This hybridization approach is of course just one strategy for improving recommender system competence. In this paper we consider an alternative by harnessing new types of recommendation knowledge that is increasingly available online.

Recently novel, alternative sources of recommendation knowledge in the form of social media information (see Chap. 11 [40] of this book) and other kinds of user-generated content have emerged. For example, services such as Twitter, Facebook, Amazon and TripAdvisor provide a rich source of user opinions, positive and negative, about a multitude of products and services. This chapter will explore how product opinions can be mined from such sources and can be used as the basis for recommendation tasks. With this in mind we describe three related case-studies to describe different ways to extract and use this type of information in a recommendation context.

2   Sources of Recommendation Knowledge

Recommender systems have traditionally leveraged two sources of data—ratings or meta-data—in order to generate make suggestions to a target user[1]. Different algorithms have been developed to take advantage of these different types of data, offering different advantages, disadvantages, tradeoffs and compromises; see [8, 29,46,57,60,64,66,67]. Indeed, some systems combine these data sources to offer hybrid approaches [10]. In this section we will briefly outline these conventional approaches to recommendation before exploring new sources of recommendation knowledge in the form of user-generated content.

2.1   Collaborative Filtering

The well-known collaborative filtering style of recommender system [8,29,60, 64,66] relies on ratings data provided by users. Each item is associated with a set of user ratings and each user is profiled in terms of their item ratings. Effectively a collaborative filtering system starts with a *user-item ratings matrix* in which each user-item combination can be associated with a rating; although in practice these ratings matrices tend to be extremely sparsely populated because most users only rate a tiny fraction of available items. An extensive discussion of collaborative filtering recommender systems can be found in Chap. 10 of this book [38].

---

[1] See Chaps. 11 and 12 [7,40] of this book for other examples of recommendation knowledge.

As described in Chap. 14 of this book [34], ratings can be explicit (directly provided by users) or implicit (inferred from user behaviour). For example, Netflix explicitly encourages users to rate movies on a 5-star scale. On the other hand, ratings can be also inferred by interpreting various types of user behaviours from purchasing a product (a highly positive 'rating') or selecting a link for more product detail (a moderately positive 'rating') to eliminating a product from a list (a negative 'rating'). In each case the power of collaborative filtering stems from its ability to translate these item ratings into user recommendations by identifying users (or items) with similar ratings histories. In one form of collaborative filtering, *user-based collaborative filtering*, items are suggested for the target user because they have been liked by *other users* with similar rating histories [8,29]. Alternatively, *item-based collaborative filtering* adopts a more item-centred perspective by suggesting items for the target user that have similar ratings histories to *other items* that the target user has liked [63].

Both user-based and item-based collaborative filtering approaches generate recommendations directly from the ratings matrix. Other approaches attempt to uncover latent factors that exist within the ratings space and use these as the basis for recommendation. For example, matrix factorisation approaches seek to identify latent features that are shared between items and users. They do this by factoring the user-item ratings matrix into separate user and item matrices which map users and items to a set of k latent features respectively [39]. Then a rating for item i by user u can be predicted by computing the dot product of the uth column of the user matrix and the ith row of the item matrix.

2.2   Content-Based and Hybrid Recommendation

In contrast to the ratings-based techniques of collaborative filtering, content-based recommenders leverage item meta-data in order to make recommendations. The meta-data for an item is typically composed of a set of descriptive features, keywords, or tags (see Chap. 12 of this book [7] for an extensive discussion on tag-based recommendation). For example, in a movie recommender a movie might be represented in terms of its genre, the lead actors, the director etc. Recommendations are generated by selecting items that are similar (based on meta-data) to those that the user has liked in the past; see [46,57]. A number of variations of content-based techniques have been proposed including case-based recommendation [67], which relies on structured feature-based item descriptions, and textual recommenders, which use less structured item descriptions.

On their own collaborative filtering and content-based techniques have a number of pros and cons. The former work well in the absence of rich item meta-data, for example, but require large, mature communities of users with extended ratings histories. The latter are less dependent on mature user communities to get started but do require detailed meta-data, which may be difficult or expensive to acquire. Collaborative filtering approaches have trouble recommending new items until such time as they have been rated by a minimum number of users, whereas content-based techniques can recommend new items from the outset. Other challenges exist when it comes to dealing with users with unusual tastes or generating diverse and novel recommendations. In response to these pros and cons, researchers have considered various ways to combine collaborative filtering and content-based approaches [7, 10].

Ratings-data, meta-data, and other forms of item content are widely used sources of recommendation data. However, the rise of the social web and the proliferation of user-generated content in the form of user reviews provides new opportunities for recommender systems research, and in this chapter we explore how this form of content can be used as a new type of recommendation data.

2.3   User-Generated Content for Recommendation

The rise of the so-called social web has seen an explosion in user-generated content, from short-form status updates to long-form reviews and blog posts. This content is typically noisy and unstructured but it has the potential to act as a rich source of user opinions about products and services. If we can mine these opinions then we may be able to harness them for a new form of recommender system (see also Chap. 5 of this book [21] for a discussion on social navigation). To do this researchers have been turning their attention to developing techniques for mining user-generated content to, for example, identify opinion sentiment, identify product features, and even combining sentiment and features to generate richly opinionated product descriptions and user profiles that can be used in recommendation.

One important focus for research has been the application of sentiment analysis techniques to user-generated content [69]. Sentiment analysis encompasses different areas such as sentiment classification [9, 48, 56], which seeks to determine whether the semantic orientation of a piece of text is positive or negative (and sometimes, neutral). In [59] it was demonstrated that these sentiment classification models can be topic-dependant, domain-dependant and temporally-dependant and suggested that training with data which contains emoticons can make these models more independent. Another area within sentiment analysis is subjectivity classification [73], which classifies text as subjective (i.e. it contains author opinions) or objective (i.e. it contains factual information). Ultimately the ability to understand the polarity and perspective of an opinion (positive or negative, subjective or objective) is a key enabling technology for opinion mining.

Increasingly, product reviews provide a rich source of user opinions and it is now common practice to research our purchases by reading reviews prior to making a final buying decision. Using natural language processing, opinion mining, and sentiment analysis techniques it is now possible to mine reviews to identify features that are being discussed and the precise nature of the discussion. Accordingly we can generate a much richer picture of a product or service by understanding how users feel about certain features or by identifying entirely new features that are unlikely to appear in a regular description of the product. So, for example, by mining a hotel review we might learn that the hotel has an excellent business centre and also realise that its restaurant serves a delicious eggs benedict. Much of the initial work in this area has focused on extracting features from electronic products such as cameras or MP3 players, where the set of product features is typically more restricted, hence representing a more tractable problem, compared to other domains such as movies or books. In more recent work [33], where feature extraction and opinion mining is performed on more complex (from a feature perspective) movie reviews, the authors first attempt to identify the set of key features that authors discuss by applying clustering techniques; a Latent Dirichlet Allocation approach was found to provide the best results. While some research [31,58] applies feature extraction techniques, such as point-wise mutual information or feature-based summarisation in a domain-independent context, it is argued in [11] that a domain-dependent approach is preferable, leading to a more precise feature set, and describe an approach based on a taxonomy of the domain product features.

A methodology for building a recommender system by leveraging user-generated content is described in [74]. In this work, the authors propose a hybrid of a collaborative filtering and a content-based approach to recommend hotels and attractions, where the collaborative filtering component utilises the review text to compute user similarities in place of traditional preference-based similarity computations. Another early attempt to build a recommender system based on user-generated review data is described in [1]. In that work an ontology is used to extract concepts from camera reviews based on users' requests about a product; for example, "I would like to know if Sony361 is a good camera, specifically its interface and battery consumption". In this case, the features *interface* and *battery* are identified, and for each of them a score is computed according to the opinions (i.e. polarities) of other users and presented to the user. Similar ideas are presented in [2], which look at using user-generated movie reviews from IMDb in combination with movie meta-data (e.g. keywords, genres, plot outlines and synopses) as input for a movie recommender system. Their results show that user reviews provide the best source of information for movie recommendations, followed by movie genre data. Further, the authors in [71] leverage opinions mined from online reviews to enhance user preference models for use in collaborative recommender systems. Experiments indicate the approach outperforms baselines algorithms with respect to accuracy and recall.

While the research and techniques described above have focused primarily on long-form review text, recent work has also considered the analysis of short-form reviews, such as micro-blog messages. For instance, Twitter messages are classified as positive, negative or neutral in [55] by creating two classifiers: a neutral-sentiment classifier and a polarity (negative or positive) classifier. Moreover, the effect of different attribute sets on sentiment classification for short-form and long-form reviews is compared in [6]. Results show that while classification accuracy for long-form reviews can benefit from using more complex attribute sets (for example, bigrams and POS tagging), this is not the case for short-form reviews where simpler attributes based on unigrams alone were sufficient from a performance perspective. Further, mining users' interests and hot topics from micro-blog posts have also been investigated in recent research [3,5].

2.4   Review Filtering, Quality and Spam

While product reviews are undoubtedly useful from a recommendation and user profiling perspective, reviews can however vary greatly in their quality and helpfulness. For example, reviews can be biased or poorly authored, while others can be very balanced and insightful. For this reason, the ability to accurately identify helpful reviews would be a useful, albeit challenging, feature to automate. While some services are addressing this by allowing users to rate the helpfulness of each review, this type of feedback can be sparse and varied, with many reviews, particularly the more recent ones, failing to attract any feedback. Hence the need exists to develop automated approaches to classify review helpfulness.

In this regard, a significant body of work has been carried out on the classification of product review helpfulness. For example, one approach to review classification has been proposed in [37], which considered feature sets relating to the structural, lexical, syntactic, semantic and some meta-data properties of reviews. Of these features, score, review length and unigram (term distribution) were among the most discriminating. Reviewer expertise was found to be a useful predictor of review helpfulness in [44], capturing the intuition that people interested in a particular genre of movies are likely to author high quality reviews for movies within the same or related genres. Timeliness of reviews was also important, and it was shown that (movie) review helpfulness declined as time went by. The use of credibility indicators was proposed in [72] in relation to topical blog post retrieval. Some of the indicators considered were text length, the appropriate use of capitalisation and emoticons in the text, spelling errors, timeliness of posts and the regularity at which bloggers post; such indicators were found to significantly improve retrieval performance in this work. Research in relation to sentiment and opinion analysis [69] is also of interest in this regard. For example, the classification of reviews for sentiment using content-based feature sets was considered in [4], where a study based on TripAvisor reviews demonstrated the effectiveness of this approach. Additional related work can be found in [30,51,52].

The need to identify *malicious* or *biased* reviews has also been considered in recent times. Such reviews can be well written and informative and so appear to be helpful. However these reviews often adopt a biased perspective that is designed to help or hinder sales of a target product [43]. Thus, a number of approaches have been proposed in the literature to identify such biased reviews. For example, a machine learning approach to spam detection is described in [42] that is enhanced by information about the spammer's identify as part of a two-tier co-learning approach. On a related topic, network analysis techniques are used in [50] to identify recurring spam in user generated comments associated with YouTube videos; in this work discriminating comment *motifs* are identified that are indicative of spambots. For other work in this area, see for example [35,36,47,54,70].

In this chapter, we begin with two case-studies which focus on leveraging product reviews for recommendation. The first case-study (Sect. 3) presents a recommendation approach which is inspired by ideas from the area of information retrieval. In this approach, users and products are modelled using a bag of words approach, and user profiles act as queries against product indices to generate recommendation lists. The second case-study (Sect. 4) presents a more sophisticated approach in which user opinions expressed in reviews are mined to construct an experiential case representation for products. With this representation, products which are not only similar to, but are *better* than (from a sentiment perspective) previous liked items can be recommended to users. Finally, in the third case-study (Sect. 5), the problem of review helpfulness classification is discussed, and one approach from the literature to address this problem is described in detail.

3   Case Study 1 – Mining Recommendation Knowledge from Product Reviews

As mentioned above, a key issue with conventional collaborative and content-based recommenders is that oftentimes neither user ratings nor item meta-data are available in sufficient quantity to effectively drive either approach. In this case study, a third source of recommendation data—namely, user-generated content relating to products—is explored as the basis for an alternative content-based approach to recommendation. In particular, user and item profiles are constructed from product reviews and recommendations are made using traditional item representation, term weighing and similarity techniques from the area of information retrieval.

A significant challenge associated with this approach is the inherently noisy nature of product reviews. For example, while some reviews can be comprehensive and informative, others are overly brief, off-topic or biased. Nonetheless, product reviews are plentiful, and range from the long-form reviews found on sites such as TripAdvisor and Amazon to opinions expressed by users in short-form on micro-blogging sites such as Twitter. In this case study, reviews from a Twitter-like service called Blippr are considered, where reviews are in the form of 160-character text posts. Figure 1 shows an example of a typical review posted on Blippr. In what follows, the review-based recommender proposed in [26] is described (see also [22, 25]), and an evaluation of the approach is presented which shows that comparable performance to more conventional recommendation approaches is achieved when applied to a range of product domains.

3.1   Review-Based Recommendation Approach

In this section, the main steps of the approach, based on ideas from information retrieval, are described: (1) how users and products are represented and (2) how this representation is used for the purposes of recommendation. In addition, a benchmark approach, inspired by the collaborative filtering approach to recommendation, is described.

Fig. 1. A review of the movie *The Dark Night* from Blippr.

3.1.1   Index Creation

Two indices, representing users and products, are created as the basis for the approach as follows.

Product Index. Consider a product P_i which is associated with a set of reviews, $Reviews(P_i) = \{r_1, ..., r_j\}$. In turn, each review r_u is made up of a set of terms, $Terms(r_u) = \{t_1, ..., t_v\}$. Thus, each product can be represented as a set of terms using a bag-of-words style approach [62] consisting of all the terms in the reviews associated with it as per Eq. 1.

$$P_i = \{t \in r : r \in Reviews(P_i)\}. \tag{1}$$

In this way individual products can be viewed as documents made up of the set of terms (words) contained in their associated reviews. An index of these documents can be created such that documents (that is products) can be retrieved based on the terms that are present in their reviews. Moreover, terms that are associated with a given product can be *weighted* based on how representative or informative these terms are with respect to the product in question; here, the *term frequency–inverse document frequency* (TF-IDF) [62] and the BM25 (also referred to as *Okapi* weighting) [61] term-weighting schemes are considered. Briefly, in the case of the TD-IDF scheme (see Eq. 2), the weight of a term t_j in a product P_i, with respect to some collection of products $\mathbf{P}$, is proportional to the frequency of occurrence of t_j in P_i (denoted by $tf(t_j, P_i)$), but inversely proportional to the frequency of occurrence of t_j in $\mathbf{P}$ overall, thereby giving preference to terms that help to discriminate P_i from the other products in the collection. For details regarding the BM25 scheme, see [26].

$$\text{TF-IDF}(P_i, t_j, \mathbf{P}) = \frac{tf(t_j, P_i)}{\sum_{t_k \in P_i} tf(t_k, P_i)} \times \text{idf}(t_j, \mathbf{P}), \tag{2}$$

$$\text{idf}(t_j, \mathbf{P}) = \log\left(\frac{|\mathbf{P}|}{|\{P_k \in \mathbf{P} : t_j \in P_k\}|}\right). \tag{3}$$

Thus a term-based index of products $\mathbf{P}$ can be created, such that each entry $\mathbf{P_{ij}}$ encodes the importance of term t_j in product P_i, where term weights are calculated according to TF-IDF (Eq. 4) or BM25 (Eq. 5).

$$\mathbf{P_{ij}} = \text{TF-IDF}(P_i, t_j, \mathbf{P}). \tag{4}$$

$$\mathbf{P_{ij}} = \text{BM25}(P_i, t_j, \mathbf{P}). \tag{5}$$

User Index. A similar approach to that above is used to create the user index. Specifically, each user U_i is represented as a document made up of the terms in their posted reviews as per Eq. 6, where $Reviews(U_i)$ denotes the reviews posted by user U_i. As before, a user index, $\mathbf{U}$, consisting of all users is created, such that each entry $\mathbf{U_{ij}}$ encodes the importance of term t_j for user U_i, once again using the TF-IDF or BM25 weighting schemes as per Eqs. 7 and 8, respectively.

$$U_i = \{t \in r : r \in Reviews(U_i)\}. \tag{6}$$

$$\mathbf{U_{ij}} = \text{TF-IDF}(U_i, t_j, \mathbf{U}). \tag{7}$$

$$\mathbf{U_{ij}} = \text{BM25}(U_i, t_j, \mathbf{U}). \tag{8}$$

3.1.2   Product Recommendation

In the above, two types of index for use in recommendation are described: an index of users and an index of products, based on the terms in their associated reviews. This suggests the following recommendation strategies. First, a *user-based* approach can be implemented in which the *target user's* profile from the user index acts as a query against the product index to produce a ranked-list of similar products (the target user's reviews are first removed from the product index to ensure that no bias is introduced into the process); see Fig. 2. Different variations of this approach can be considered by using different weighting schemes (TF-IDF and BM25) to index and query the index[2]. Further, term stemming can be applied to the data to improve the match between query and index terms.

In addition, to provide a benchmark for the above index-based approaches, a *community-based* approach based on collaborative filtering ideas [66] can be implemented. A set of similar users (or *neighbours*) is first identified, by using the target user profile as a query on the user index, and then the preferred products of these neighbours are ranked based on their frequency of occurrence in neighbour profiles; see Fig. 3.

3.2   Evaluation

The recommendation performance provided by the review-based and benchmark algorithms described above is presented. The datasets used in the evaluation are first described, followed by the metrics used to measure performance.

---

[2] Lucene (http://lucene.apache.org) is used in the subsequently described experiments to provide the term-weighting and querying functionality.

Input: Target user U_T, user index **U**, product index **P**, number of products to retrieve n
Output: Top n product recommendations

```
1.    UserBasedRecommendation (U_T, U, P, n)
2.    Begin
3.          query ← U.get(U_T)           // Return term vector for U_T in U
4.          recs ← P.retrieve(query)     // Retrieve ranked list of
                                         // products from P based on query
5.          return recs.first(n)         // Return top n recommendations
6.    End
```

Fig. 2. User-based recommendation algorithm.

Input: Target user U_T, user index **U**, product index **P**, number of products to retrieve n, neighbourhood size k
Output: Top n product recommendations

```
1.    CommunityBasedRecommendation (U_T, U, M, n, k)
2.    Begin
3.          query ← U.get(U_T)           // Return term vector for U_T in U
4.          users ← U.retrieve(query)    // Get ranked list of similar users
5.          neighs ← users.first(k)      // Get the top k most similar
                                         // users as neighbours
6     recs ← {}                          // Get all neighbours' products
7.    for each n ∈ neighs
8.          recs ← recs ∪ n.products()   // Add products from current
                                         // neighbour to recommendation set
9.    end
10.   return recs.sort(score(.,.), n)    // Return top n most frequently
11.  end                                 // occurring products
                                         // score(P_i, neighs) = ∑ occurs(P_i, n),
                                         //                      n ∈ neighs
                                         // where occurs(P_i, n) = 1 if P_i is
                                         // present in n and 0 otherwise
```

Fig. 3. Community-based recommendation algorithm.

3.2.1 Datasets

The evaluation is based on reviews extracted from the Blippr service, which allows users to review products from a number of different domains. Reviews (or blips) are in the form of 160-character text messages, and users must also supply an accompanying rating on a 4-point rating scale: *love it, like it, dislike it* or *hate it*. Data was collected using the Blippr API in April 2010, capturing reviews written in the English language before that date. Preprocessing of reviews is performed, such as removing stopwords, special symbols (?, *, & etc.), digits and multiple repetitions of characters in words (e.g. *goooood* is reduced to *good*). Further, only reviews which have *love it* ratings are considered (i.e. where users have expressed the highest sentiment toward products) since we wish to recommend products which are actually liked by users. Note that reviews which express negative sentiment could also be considered to identify products which are disliked by users and which should not be recommended; however, such an approach is not examined here.

The experiments use Blippr data relating to four product types: *movies, books, applications (apps)* and *games*. Products with at least three reviews and users that have authored at least five reviews are selected. See Table 1 for dataset statistics.

Table 1. Evaluation dataset statistics.

	Movies	*Apps*	*Books*	*Games*
# Products	1,080	268	313	277
# Users	542	373	120	164
# Reviews	15,121	10,910	3,003	3,472

3.2.2 Metrics

Precision and *recall*, which have been widely used in the field of information retrieval, are used to evaluate recommendation accuracy. These metrics have been adapted to evaluate the accuracy of a set of recommended products [64] and are defined as follows:

$$\text{Precision} = \frac{|T \cap R|}{|R|}, \tag{9}$$

$$\text{Recall} = \frac{|T \cap R|}{|T|}, \tag{10}$$

where T and R are the test and recommended sets for each user, respectively.

We also evaluate recommendation *coverage*, which measures the number of products that a recommender is capable of making recommendations for (as a percentage of the total number of products in the system) [27]. In general, the ability of an algorithm to make recommendations for large numbers of (relevant) products is a desirable system property, so as to avoid situations in which only a limited number of items (e.g. popular items) are ever capable of being recommended.

3.2.3 Results

To evaluate the recommendation algorithms, separate product and user indices are first created for each of the four datasets according to the approach described in Sect. 3.1. The main objective is to compare the performance of the user-based approach with that of the community-based benchmark. In the case of the user-based approach, the performance of two term-weighting schemes is considered: TF-IDF and BM25. Further, to determine if term stemming has any effect on the performance of the user-based approach, versions of TF-IDF weighting with (TF-IDF+) and without stemming (TF-IDF) are compared.

For each dataset, a leave-one-out approach is used where each user in turn acts as the target user (as per Sect. 3.1) and precision and recall scores are computed for different recommendation-list sizes ranging from 5 to 30 items. Results are presented in Fig. 4 for the *movies* and *books* datasets. The results show that there is a clear benefit for the user-based recommendation strategies compared to the community-based approaches. For example, in the case of the *books* dataset using recommendation lists of size 5, the best user-based approach enjoys a precision of 0.44. In contrast, the best performing community-based approach (CB-10), where 10 similar users are selected as the basis for recommendation, achieves a precision of 0.32.

For all datasets, TF-IDF with and without stemming provide similar results; with stemming applied, TF-IDF performs marginally better for most datasets. For the larger datasets (*movies* and *apps*), the performance provided by BM25 is very close to that of TF-IDF, but is seen to fall off for the smaller datasets (*books* and *games*); see [26] for more details.

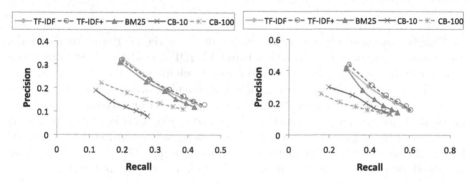

Fig. 4. Precision–recall for the *movies* (left) and *books* (right) datasets for user-based (TF-IDF vs. TF-IDF+ vs. BM25) and community-based (CB-10 vs. CB-100) recommendation.

In Fig. 5(left), the precision and recall provided by the user-based approach using TF-IDF are compared across the four datasets. It can be seen that the best performance is achieved for the *apps* dataset where, for example, precision and recall values of 0.54 and 0.37 are achieved, respectively, compared to values of 0.42 and 0.29 for the *books* dataset (these values correspond to recommendation lists of size 5). Also shown in this figure is the mean number of reviews (blips) per product for each dataset; it can be seen that these values correlate well with the precision ($r = 0.84$) and recall ($r = 0.83$) performance achieved for the datasets. This seems a reasonable finding, since it indicates that richer product indices (i.e. products are described by a greater number of reviews) lead to better recommendation performance. However, since the datasets used in the evaluation contain short-form reviews and relatively small numbers of users and products, further analysis is required to draw general conclusions in this regard.

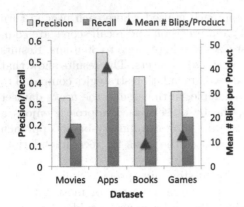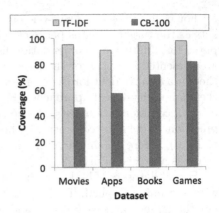

Fig. 5. Precision–recall (recommendation list sizes of 5) provided by user-based recommendation using TF-IDF and mean number of reviews (blips) per product vs. dataset (left) and coverage provided by recommendation approaches vs. dataset (right).

Finally, coverage performance is shown in Fig. 5(right). Here, trends for the user-based recommendation strategy using TF-IDF and for the best performing community-based approach using 100 nearest neighbours (CB-100) are shown. It can be seen that the user-based approach provides almost complete coverage for all datasets, well in excess of that given by the community-based approach, particularly for the larger datasets (*movies* and *apps*). This is a positive finding in respect of the utility of reviews as a source of recommendation data. It should be noted that other forms of coverage (see, for example, [27,65]) have also been proposed; however, an analysis of such criteria is not considered here.

3.2.4   Discussion

This case-study investigates how user-generated content can be used as a new source of recommendation knowledge. An approach is proposed to represent users and products based on the terms in their associated reviews using techniques from information retrieval. An evaluation performed on short-form, and inherently noisy, reviews from a number of product domains shows promising results. The work described here is related to a growing body of research on the potential for user-generated content to provide product recommendations; for example, enriching user and item profiles by using sentiment analysis and feature extraction techniques, classification of reviews by product category to facilitate personalisation and search [23,24], and the potential for cross-domain recommendation, where indices created using reviews from one domain are used to recommend products from other domains. For other work in this area, see [1,2,33].

4   Case Study 2 – Opinionated Recommendation

The previous case-study described an approach to leveraging the text of short-form user-generated product reviews directly for recommendation. Indeed, user-generated reviews have previously been used in a number of recommendation contexts: as part of collaborative filtering approaches to provide virtual ratings [75]; for user profiling [49]; and in content-based recommendation [20].

   In this case-study we focus on the type of long-form product reviews typically found on sites like Amazon and TripAdvisor and we describe how these reviews can be used to generate complete item descriptions, in the absence of meta-data or as a complement to meta-data. Crucially, we make the point that these review-based item descriptions are *experiential* in nature—they describe the real experience of users—rather than capturing the type of technical/catalog features that are more common in conventional meta-data representations. We describe how item descriptions are created and how review content can be used to infer opinion sentiment which can be used in a novel way during recommendation. Accordingly items can be selected and ranked not only on the basis that they *have* a given feature (e.g. *Free Wifi* in a hotel), but also based on whether the *opinion* of reviewers about these features is positive or negative (see Fig. 6).

Fig. 6. A hotel page on TripAdvisor showing ratings and catalogue meta-data (e.g. *Free Wifi, Breakfast Buffet, Air Conditioning* etc.) for the property. Features mentioned in a sample review are also highlighted (where green and red denote positive and negative sentiment, respectively.) (Color figure online)

4.1   From Reviews to Recommendation

An overview of the approach is presented in Fig. 7, which highlights the core opinion mining and recommendation components involved.

Briefly, we start with a set of reviews for some product/item P and any available meta-data. The reviews are mined to identify and extract product features using some straightforward NLP techniques. Next we analyse the sentiment of these features based on the text of the reviews. The combination of features and sentiment for each product, plus its meta-data (if available), is combined to produce a product/item description. Given a new user query (i.e. the current item the user is looking at), the recommendation component retrieves and ranks a set of matching items based on a combination of feature similarity and sentiment. In what follows we will describe each of these steps in more detail and provide some concluding evidence in support of the efficacy of this approach for recommendation.

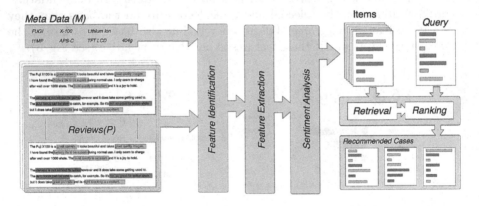

Fig. 7. An overview of the experiential product recommendation architecture.

4.2   Identifying Review Features

One straightforward way to identify candidate features is to use simple NLP methods to look for certain patterns of words. For example, bi-grams in reviews which conform to one of two basic part-of-speech co-location patterns can be considered as features—an adjective followed by a noun (AN) or a noun followed by a noun (NN). In the former case, bi-grams whose adjective is a sentiment word (e.g. *excellent, terrible* etc.) in the sentiment lexicon used in our approach [32] are excluded. Separately, single-nouns can also be considered as features after eliminating nouns that are rarely associated with sentiment words in reviews as per [32].

4.3 Evaluating Feature Sentiment

For each feature we evaluate its sentiment based on the sentence containing the feature within a given review. We use a modified version of the *opinion pattern mining* technique proposed by Moghaddam and Ester [45] for extracting opinions from unstructured product reviews. Once again we use the sentiment lexicon from [32] as the basis for this analysis. For a given feature F_i and corresponding review sentence S_j from review R_k, we determine whether there are any sentiment words in S_j. If there are not then this feature is marked as *neutral* from a sentiment perspective. If there are sentiment words then we identify the word w_{min} which has the minimum word-distance to F_i.

Next we determine the part-of-speech (POS) tags for w_{min}, F_i and any words that occur between w_{min} and F_i. The POS sequence corresponds to an *opinion pattern*. For example, in the case of the bi-gram feature *noise reduction* and the review sentence, *"...this camera has great noise reduction..."* then w_{min} is the word *"great"* which corresponds to an opinion pattern of *JJ-FEATURE* as per Moghaddam and Ester [45]. After a complete pass of all features over all reviews, we can compute the frequency of all recorded opinion patterns. To filter spurious opinion patterns that rarely occur, a pattern is deemed to be valid if it occurs more than the average number of occurrences over all patterns. For valid patterns we assign sentiment to F_i based on the sentiment of w_{min}, subject to whether S_j contains any negation terms within a 4-word-distance of w_{min} [31]. If there are no such negation terms then the sentiment assigned to F_i in S_j is that of the sentiment word in the sentiment lexicon; otherwise this sentiment is reversed. If an opinion pattern is deemed not to be valid (based on its frequency), then we assign a *neutral* sentiment to each of its occurrences within the review set.

4.4 From Review Features to Item Descriptions

For each product P we have a set of features $F(P) = \{F_1, ..., F_m\}$ that have been either identified from the meta-data associated with P or that have been discussed in the various reviews of P, *Reviews(P)*. For each feature F_i we compute its *popularity*, which is given by the fraction of reviews it appears in (see Eq. 11). Also, we compute the *sentiment* associated with each feature; i.e. how often it is mentioned in reviews in a positive, neutral, or negative manner (see Eq. 12, where $Pos(F_i, P)$, $Neg(F_i, P)$, and $Neut(F_i, P)$ denote the number of times that feature F_i has positive, negative and neutral sentiment in the reviews for product P, respectively). In this way, each item/product can be represented as the aggregate of its features and their popularity and sentiment data as in Eq. 13.

$$Pop(F_i, P) = \frac{|\{R_k \in Reviews(P) : F_i \in R_k\}|}{|Reviews(P)|}. \tag{11}$$

$$Sent(F_i, P) = \frac{Pos(F_i, P) - Neg(F_i, P)}{Pos(F_i, P) + Neg(F_i, P) + Neut(F_i, P)}. \tag{12}$$

$$Item(P) = \{[F_i, Sent(F_i, P), Pop(F_i, P)] : F_i \in F(P)\}. \tag{13}$$

4.5   Recommending Products

Unlike traditional content-based recommenders—which tend to rely exclusively on similarity in order to rank products with respect to some user profile or query—the above approach accommodates the use of feature sentiment, as well as feature similarity, during recommendation; see [13,16]. Briefly, a candidate product C can be evaluated against a query product Q (i.e. the current product the user is looking at) according to a weighted combination of similarity and sentiment as per Eq. 14. $Sim(Q, C)$ is a traditional similarity metric such as cosine similarity, producing a value between 0 and 1, while $Sent(Q, C)$ is a sentiment metric producing a value between -1 (negative sentiment) and +1 (positive sentiment).

$$Score(Q, C) = (1 - w) \times Sim(Q, C) + w \times \left(\frac{Sent(Q, C) + 1}{2} \right). \qquad (14)$$

4.5.1   Similarity Assessment

For the purpose of similarity assessment a standard cosine similarity metric based on feature popularity scores can be used, as per Eq. 15; see also, for example [57].

$$Sim(Q, C) = \frac{\sum\limits_{F_i \epsilon F(Q) \cup F(C)} Pop(F_i, Q) \times Pop(F_i, C)}{\sqrt{\sum\limits_{F_i \epsilon F(Q)} Pop(F_i, Q)^2} \times \sqrt{\sum\limits_{F_i \epsilon F(C)} Pop(F_i, C)^2}}. \qquad (15)$$

4.5.2   Sentiment Assessment

Sentiment is somewhat unusual in a recommendation context but its availability offers an additional way to compare products, based on a feature-by-feature sentiment comparison as per Eq. 16. We can say that feature F_i is *better* in C than in Q if F_i in C has a higher sentiment score than it does in Q.

$$better(F_i, Q, C) = \frac{Sent(F_i, C) - Sent(F_i, Q)}{2}. \qquad (16)$$

Accordingly we can calculate an overall better score at the product level by aggregating the individual better scores for the product features. We can do this by computing the average better scores across the *union* of features of Q and C, assigning non-shared features a neutral sentiment score of 0. This is captured in Eq. 17; see also the work of [14] for a second variation on this scoring metric. This approach gives due consideration to the *residual* features in the query and candidate products, that is, those features that are unique to the query or candidate products.

$$Sent(Q, C) = \frac{\sum_{F_i \in F(Q) \cup F(C)} better(F_i, Q, C)}{|F(Q) \cup F(C)|}. \qquad (17)$$

4.6   Evaluation

Finally in this case-study we provide some evaluation results taken from [15] to demonstrate the utility of this approach to opinion mining in recommendation.

The data for this experiment was sourced from TripAdvisor during September 2013. We focused on hotel reviews from six different cities across Europe, Asia, and the US; here, for reasons of space, we consider just two cites, London and Chicago. The data is summarised in Table 2, where we show the total number of reviews per city (*#Reviews*), the number of hotels per city (*#Hotels*), as well as including statistics (mean and standard deviation) on the number of features extracted from the reviews per hotel (*RF*). We can see that this approach to opinion mining produces product descriptions that are rich in features; on average London and Chicago hotels are represented by more than 31 and 28 features per hotel, respectively.

Table 2. Dataset statistics.

City	#Reviews	#Hotels	$\mu(\sigma)_{RF}$
London	62,632	717	31.8 (5.5)
Chicago	11,091	125	28.6 (5.0)

4.6.1   Methodology

To evaluate our approach to recommendation we adopt a standard *leave-one-out* methodology. For each city dataset, we treat each hotel in turn as a query case Q and generate a set of top-5 recommendations according to Eq. 14 using different values of w (0 to 1 in increments of 0.1) in order to test the impact of different combinations of similarity and sentiment; we refer to this approach as RF. Then we compare our recommendations to those produced natively by TripAdvisor (TA) using two comparison metrics. First, we calculate the average *query similarity* between each set of recommendations (RF and TA) and Q using a Jaccard similarity metric. Second, we compare the two sets of recommendations based on the TripAdvisor user ratings to calculate a *ratings benefit* as per Eq. 18; for example, a ratings benefit of 0.1 means that our RF recommendation list enjoys an average rating score that is 10% higher that those produced by the default TripAdvisor approach (TA).

$$Ratings\, Benefit(RF, TA) = \frac{\overline{Rating(RF)} - \overline{Rating(TA)}}{\overline{Rating(TA)}}. \tag{18}$$

4.6.2   Results

Figure 8 show the results for the London and Chicago hotels, graphing the average ratings benefit (RB) and average query similarity (QS) against different levels of w. Each graph also shows the average query similarity for the TA recommendations (the upper black horizontal solid line), and the region between the upper and lower horizontal lines corresponds to the region of 90% similarity; that is, query similarity scores that fall within this region are 90% as similar to the target query as the default recommendations produced by TA. The intuition here is that query similarity scores which fall below this region run the risk of compromising too much query similarity to be useful as *more-like-this* recommendations.

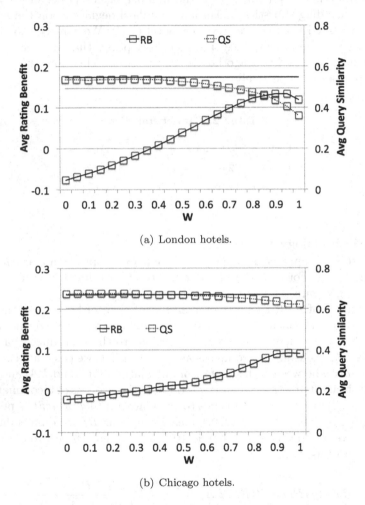

(a) London hotels.

(b) Chicago hotels.

Fig. 8. Ratings benefit (RB) and query similarity (QS) versus w, which controls the relative influence of similarity and sentiment on recommendation ranking scores (see Eq. 14)

4.7   Results Discussion

There are a number of observations that can be made about these results. First, as w increases we can see that there is a steady increase in the average ratings benefit. In other words, as we increase the influence of sentiment in the scoring function (Eq. 14), we tend to produce recommendations that offer better overall ratings than those produced by TA. Therefore combining similarity and sentiment in recommendation delivers a positive effect overall.

We can also see that as w increases there is a gradual drop in query similarity. In other words, as we increase the influence of sentiment (and therefore decrease the influence of similarity) in the scoring function (Eq. 14), we tend to produce recommendations that are less similar to the target query. On the one hand, this is a way to introduce more diversity [68] into the recommendation process with the added benefit, as above, that the resulting recommendations tend to enjoy a higher ratings benefit compared to the default TripAdvisor recommendations (TA). But on the other hand, there is the risk that too great a drop in query similarity may lead to products that are no longer deemed to be relevant by the end-user. For this reason, we have (somewhat arbitrarily) chosen to prefer query similarities that remain within 90% of those produced by TA.

We can usefully compare the recommendation approaches by noting the average ratings benefit available at the value of w for which the query similarity of a given approach crosses this 90% (TA) query similarity threshold. For example, in Fig. 8, for London hotels we can see that the query similarity for the RF approach falls below the 90% threshold at about $w = 0.7$ and this corresponds to a ratings benefit of 0.1. In the case of Chicago query similarity never dips below this 90% threshold and a maximum ratings benefit of just under 0.1 at $w = 0.9$. Thus, we can conclude that our approach is capable of providing recommendations which enjoy higher ratings compared to those provided by TA, which maintaining a high degree of similarity to the user query.

5   Case Study 3 – Review Helpfulness Classification

In the above case-studies, two approaches are described which leverage product reviews for recommendation purposes. However, not all reviews are equally informative and comprehensive, and hence the need to rank reviews for products and to filter less helpful reviews—both to validate the data used as input to recommender systems and to facilitate users to navigate though the thousands of reviews that are often available for popular products. In order to address the issue of information overload in the review space, sites such as TripAdvisor and Amazon allow users to provide manual feedback on review helpfulness; for example, by allowing other consumers to post comments about reviews, to report abuse in cases where review content is considered inappropriate and to indicate whether reviews are found to be helpful or not. While such approaches are of benefit to consumers in highlighting the most helpful reviews, they depend on the willingness of the community at large to contribute feedback and there is no guarantee that all reviews will receive feedback in sufficient quantities to provide a robust signal to consumers. Thus, the community has sought to address this problem by proposing *automated* approaches to classify review helpfulness and a significant body of work has been carried out in this area in recent times; see, for example, [17,28,30,44,53]. In this case-study, one approach from the literature [37] to automatically classify the helpfulness of reviews is described.

5.1 Classifying Review Helpfulness

In the approach presented in [37], the problem of classifying the helpfulness of review is formulated as a supervised classification task. Each review instance is represented by a number of feature sets and the ground truth is given by the review helpfulness as per Eq. 19:

$$h(r \in R) = \frac{rating_+(r)}{rating_+(r) + rating_-(r)}, \tag{19}$$

where $rating_+(r)$ and $rating_-(r)$ are the number of helpful and unhelpful (manually provided) votes for review r, respectively. Thus, once each review in the training set is translated into a feature-based instance representation, a model is learned which is then applied to classify the helpfulness, $h(r_t)$, of an unseen review instance, r_t. The feature sets and classification approach used are described in the following sections.

5.1.1 Feature Sets

Review instances consist of feature sets derived from distinct categories which are mined from individual reviews and from the wider community reviewing activity. The following feature sets, which are hypothesised to be predictive of review helpfulness, are considered in [37]:

– *Structural features* capture aspects of the review structure and formatting and include features such as review length, the number of sentences in the review, the mean sentence length, the percentage of sentences with questions, the number of exclamation marks contained in the review, and the number of HTML bold tags <b> and line breaks <br> in the review body.
– *Lexical features* concern the occurrence of words in reviews; in this case, the TF-IDF statistic of each unigram and bigram occurring in a review are calculated.
– *Syntactic features* capture the linguistic properties of a review by calculating the percentages of tokens that are open-class, nouns, verbs, verbs conjugated in the first person, adjectives or adverbs.
– *Semantic features* capture the intuition that helpful reviews are likely to contain critiques of particular product features (e.g. *capacity* and *zoom* in the case of MP3 players and digital camera products); thus, the number of lexical matches that occur for each product feature and the number of sentiment words in a review are calculated.

– *Meta-data features*, in contrast to the above feature sets, are based on knowledge that is independent of the review text; in this regard, two features that are related to the rating scores that often accompany the review text are considered – namely, the rating score assigned by the reviewer and the absolute difference between this score and the mean rating score assigned by all reviewers.

5.1.2 Ranking Reviews

Given the availability of training instances and once a classifier is trained, a set of reviews R for a given product can then be ranked in descending order of $h(r)$, $r \in R$. SVM regression [18] using a radial basis function (RBF) kernel is used in [37] as this combination was found to provide optimal performance.

5.2 Evaluation

In this section, the datasets used in the evaluation are first described, followed by a description of the evaluation methodology and metrics used. A summary of the key findings of the classification approach is then presented.

5.2.1 Datasets

Evaluation datasets consisting of reviews for all products from two product categories, *MP3 Players* and *Digital Cameras*, were sourced from Amazon. Following pre-processing (which included the removal of duplicate reviews, reviews for duplicate products, and reviews for which less than five helpful and unhelpful votes were available), two evaluation datasets were created; see Table 3 for statistics relating to these datasets.

Table 3. Evaluation dataset statistics (source [37]).

	MP3 players	*Digital cameras*
Total products	736	1,066
Total reviews	11,374	14,467
Average reviews/product	15.4	13.6
Min/max reviews/product	1/375	1/168

5.2.2 Methodology and Metrics

For each dataset, 10% of products were withheld in order to determine the optimal SVM kernel (RBF) and to tune kernel parameters. Thereafter, the remaining 90% of products were randomly divided into 10 sets, and a 10-fold cross-validation approach was applied to rank (as per Sect. 5.1) the reviews for each product in the test folds. Thus, a ranking for each product's review set is learned, which is compared to a ground truth ranking based on actual helpfulness votes extracted from Amazon.com. Spearman rank correlation is used to compare the learned and actual rankings for each product. Moreover, since in the course of ranking, the absolute helpfulness scores for reviews are learned by the classifier, Pearson correlation is also used to compare these absolute scores to ground truth scores obtained from Amazon.com.

5.2.3   Results

Results are shown in Table 4 for different combinations of features drawn from the subset of features which provides best performance; these features are review length (LEN), unigrams (UGR) and rating score (STR1). When used in isolation, these features provide similar performance. For both datasets, the best performing pair of features is the combination of review length and rating score. As can be seen, the combination of review length, unigrams and rating score features is optimal, achieving Spearman rank corrections of 0.656 and 0.595 for the *MP3 Players* and *Digital Cameras* datasets, respectively.

It is interesting to note the differences between the Spearman rank and Pearson correlation results; in all instances, the quality of the review rankings produced by the classifier (given by Spearman rank correlation) exceeded the performance of the classifier when learning absolute helpfulness scores (given by Pearson correlation). For example, in the case of the *MP3 Players* dataset, Spearman rank and Pearson correlations of 0.656 and 0.476 are seen using a combination of all three features, respectively. Given that learning the absolute helpfulness scores of reviews is a more difficult task, this finding is not surprising; moreover, the results also indicate that accurate rankings can be learned without learning the absolute helpfulness scores of reviews perfectly.

For further details on the evaluation and a discussion on the performance of various other feature combinations, see [37].

Table 4. Evaluation results (source [37]).

Feature combinations	MP3 players		Digital cameras	
	Spearman[†]	*Pearson*[†]	*Spearman*[†]	*Pearson*[†]
LEN	0.575 ± 0.037	0.391 ± 0.038	0.521 ± 0.029	0.357 ± 0.029
UGR	0.593 ± 0.036	0.398 ± 0.038	0.499 ± 0.025	0.328 ± 0.029
STR1	0.589 ± 0.034	0.326 ± 0.038	0.507 ± 0.029	0.266 ± 0.030
UGR+STR1	0.644 ± 0.033	0.436 ± 0.038	0.490 ± 0.032	0.324 ± 0.032
LEN+UGR	0.582 ± 0.036	0.401 ± 0.038	0.553 ± 0.028	0.394 ± 0.029
LEN+STR1	0.652 ± 0.033	0.470 ± 0.038	0.577 ± 0.029	0.423 ± 0.031
LEN+UGR+STR1	$\mathbf{0.656 \pm 0.033}$	$\mathbf{0.476 \pm 0.038}$	$\mathbf{0.595 \pm 0.028}$	$\mathbf{0.442 \pm 0.031}$

LEN = *Length*; UGR = *Unigram*; STR = *Stars*
[†]95% confidence bounds are calculated using 10-fold cross-validation

5.2.4 Discussion

User-generated reviews have become an important source of knowledge for consumers and are known to play an active role in decision making in many domains. However, given the thousands of reviews which can often accrue for popular products on sites such as Amazon and TripAdvisor, a new challenge arises—namely, how best to facilitate users to rapidly and effectively identify the most useful reviews. Hence the need for automatic approaches to identify review helpfulness to assist users by, for example, filtering less informative or comprehensive reviews from the user's view. The case study presented in this chapter highlights one approach in a significant body of work carried out in this area; further work on this problem can be found in [17, 30, 44, 51, 53].

6 Conclusions

Today, product reviews have become an important part of our online experience, assisting consumers to make informed choices and providing key insights to retailers about their product offerings. For example, Lee et al. report that 84 percent of Americans are influenced by online reviews when they are making purchase decisions [41]; see also [12, 76]. Further, many companies have now recognised that consumer reviews represent a new and important communication channel with their consumers, and they have begun monitoring online consumer reviews as a crucial source of product feedback [19]. Moreover, companies can predict their performance or sales according to this online feedback; for example, Duan et al. used Yahoo movie reviews and box office returns to examine the persuasive and awareness effects of online user reviews on the daily box office performance [19].

Increasingly, researchers are also leveraging product reviews for the purposes of user profiling and recommendation. In particular, reviews often capture detailed and nuanced user opinions for different kinds of products and services, and thus represent a plentiful, albeit noisy and unstructured, alternative source of recommendation knowledge to replace or complement the more conventional data sources such as product ratings and meta-data. In this chapter, we have presented two case-studies which describe particular approaches in which review data can be successfully leveraged for recommendation. Moreover, we have described an approach to estimate review quality in order to help users cope with the volume and variability of review content. Given the prevalence of user-generated content online and the valuable insights it provides to both consumers and retailers alike, this area of research presents many exciting opportunities for the future.

Acknowledgments. This work is supported by Science Foundation Ireland through the CLARITY Centre for Sensor Web Technologies under grant number 07/CE/I1147 and through the Insight Centre for Data Analytics under grant number SFI/12/RC/2289.

References

1. Aciar, S., Zhang, D., Simoff, S., Debenham, J.: Recommender system based on consumer product reviews. In: Proceedings of the 2006 IEEE/WIC/ACM International Conference on Web Intelligence (WI-IATW 2006), pp. 719–723. IEEE Computer Society, Washington, D.C. (2006)
2. Ahn, S., Shi, C.-K.: Exploring movie recommendation system using cultural metadata. In: Pan, Z., Cheok, A.D., Müller, W., Rhalibi, A.E. (eds.) Transactions on Edutainment II. LNCS, vol. 5660, pp. 119–134. Springer, Heidelberg (2009). https://doi.org/10.1007/978-3-642-03270-7_9
3. Angel, A., Koudas, N., Sarkas, N., Srivastava, D.: What's on the grapevine? In: Proceedings of the 35th SIGMOD International Conference on Management of Data (SIGMOD 2009), pp. 1047–1050. ACM, New York (2009)
4. Baccianella, S., Esuli, A., Sebastiani, F.: Multi-facet rating of product reviews. In: Boughanem, M., Berrut, C., Mothe, J., Soule-Dupuy, C. (eds.) ECIR 2009. LNCS, vol. 5478, pp. 461–472. Springer, Heidelberg (2009). https://doi.org/10.1007/978-3-642-00958-7_41
5. Banerjee, N., Chakraborty, D., Dasgupta, K., Mittal, S., Joshi, A., Nagar, S., Rai, A., Madan, S.: User interests in social media sites: an exploration with microblogs. In: Proceeding of the 18th ACM Conference on Information and Knowledge Management (CIKM 2009), pp. 1823–1826. ACM, New York (2009)
6. Bermingham, A., Smeaton, A.F.: Classifying sentiment in microblogs: is brevity an advantage? In: Proceedings of the 19th ACM International Conference on Information and Knowledge Management (CIKM 2010), pp. 1833–1836. ACM, New York (2010). https://doi.org/10.1145/1871437.1871741
7. Bogers, T.: Tag-based recommendation. In: Brusilovsky, P., He, D. (eds.) Social Information Access. LNCS, vol. 10100, pp. 441–479. Springer, Cham (2018)
8. Breese, J.S., Heckerman, D., Kadie, C.M.: Empirical analysis of predictive algorithms for collaborative filtering. In: Cooper, G.F., Moral, S. (eds.) Proceedings of the Fourteenth Conference on Uncertainty in Artificial Intelligence (UAI 1998), pp. 43–52. Morgan Kaufmann, Burlington (1998)
9. Brew, A., Greene, D., Cunningham, P.: Using crowdsourcing and active learning to track sentiment in online media. In: Proceedings of the 19th European Conference on Artificial Intelligence (ECAI 2010), pp. 145–150. IOS Press, Amsterdam, The Netherlands (2010). http://portal.acm.org/citation.cfm?id=1860967.1860997
10. Burke, R.: Hybrid recommender systems: survey and experiments. User Model. User-Adap. Inter. **12**(4), 331–370 (2002)
11. Cruz, F.L., Troyano, J.A., Enríquez, F., Ortega, F.J., Vallejo, C.G.: A knowledge-rich approach to feature-based opinion extraction from product reviews. In: Proceedings of the 2nd International Workshop on Search and Mining User-Generated Contents (SMUC 2010), pp. 13–20. ACM, New York (2010). https://doi.org/10.1145/1871985.1871990
12. Dhar, V., Chang, E.A.: Does chatter matter? The impact of user-generated content on music sales. J. Interact. Mark. **23**(4), 300–307 (2009). http://www.sciencedirect.com/science/article/pii/S1094996809000723
13. Dong, R., O'Mahony, M.P., Schaal, M., McCarthy, K., Smyth, B.: Sentimental product recommendation. In: Proceedings of the 7th ACM Conference on Recommender Systems, RecSys 2013, pp. 411–414. ACM, New York (2013). https://doi.org/10.1145/2507157.2507199

14. Dong, R., O'Mahony, M.P., Schaal, M., McCarthy, K., Smyth, B.: Combining similarity and sentiment in opinion mining for product recommendation. J. Intell. Inf. Syst. **46**(2), 285–312 (2016). https://doi.org/10.1007/s10844-015-0379-y

15. Dong, R., O'Mahony, M.P., Smyth, B.: Further experiments in opinionated product recommendation. In: Lamontagne, L., Plaza, E. (eds.) ICCBR 2014. LNCS (LNAI), vol. 8765, pp. 110–124. Springer, Cham (2014). https://doi.org/10.1007/978-3-319-11209-1_9

16. Dong, R., Schaal, M., O'Mahony, M.P., McCarthy, K., Smyth, B.: Opinionated product recommendation. In: Delany, S.J., Ontañón, S. (eds.) ICCBR 2013. LNCS (LNAI), vol. 7969, pp. 44–58. Springer, Heidelberg (2013). https://doi.org/10.1007/978-3-642-39056-2_4

17. Dong, R., Schaal, M., O'Mahony, M.P., Smyth, B.: Topic extraction from online reviews for classification and recommendation. In: Proceedings of the 23rd International Joint Conference on Artificial Intelligence (IJCAI 2013), pp. 1310–1316 (2013)

18. Drucker, H., Burges, C.J.C., Kaufman, L., Smola, A.J., Vapnik, V.N.: Support vector regression machines. In: Advances in Neural Information Processing Systems 9 (NIPS 1996). pp. 155–161. MIT Press (1996)

19. Duan, W., Gu, B., Whinston, A.B.: Do online reviews matter? – An empirical investigation of panel data. Decis. Support Syst. **45**(4), 1007–1016 (2008). https://doi.org/10.1016/j.dss.2008.04.001

20. Esparza, S.G., O'Mahony, M.P., Smyth, B.: Effective product recommendation using the real-time web. In: Bramer, M., Petridis, M., Hopgood, A. (eds.) Research and Development in Intelligent Systems XXVII, pp. 5–18. Springer, London (2011). https://doi.org/10.1007/978-0-85729-130-1_1

21. Farzan, R., Brusilovsky, P.: Social navigation. In: Brusilovsky, P., He, D. (eds.) Social Information Access. LNCS, vol. 10100, pp. 142–180. Springer, Cham (2018)

22. Garcia Esparza, S., O'Mahony, M.P., Smyth, B.: On the real-time web as a source of recommendation knowledge. In: Proceedings of the 4th ACM Conference on Recommender Systems (RecSys 2010), pp. 305–308. ACM, New York (2010)

23. Garcia Esparza, S., O'Mahony, M.P., Smyth, B.: Towards tagging and categorization for micro-blogs. In: Proceedings of the 21st Irish Conference on Artificial Intelligence and Cognitive Science (AICS 2010), pp. 122–131 (2010)

24. Garcia Esparza, S., O'Mahony, M.P., Smyth, B.: Further experiments in micro-blog categorization. In: Proceedings of the 22nd Irish Conference on Artificial Intelligence and Cognitive Science (AICS 2011), pp. 156–165 (2011)

25. Garcia Esparza, S., O'Mahony, M.P., Smyth, B.: A multi-criteria evaluation of a user-generated content based recommender system. In: Proceedings of the 3rd Workshop on Recommender Systems and the Social Web, 5th ACM Conference on Recommender Systems (RSWEB 2011) (2011)

26. Garcia Esparza, S., O'Mahony, M.P., Smyth, B.: Mining the real-time web: a novel approach to product recommendation. Knowl. Based Syst. **29**, 3–11 (2012)

27. Ge, M., Delgado-Battenfeld, C., Jannach, D.: Beyond accuracy: evaluating recommender systems by coverage and serendipity. In: Proceedings of the Fourth ACM Conference on Recommender Systems, RecSys 2010, pp. 257–260. ACM, New York (2010). https://doi.org/10.1145/1864708.1864761

28. Ghose, A., Ipeirotis, P.G.: Designing novel review ranking systems: predicting the usefulness and impact of reviews. In: Proceedings of the Ninth International Conference on Electronic Commerce, ICEC 2007, pp. 303–310. ACM, New York (2007). https://doi.org/10.1145/1282100.1282158

29. Herlocker, J.L., Konstan, J.A., Borchers, A., Riedl, J.: An algorithmic framework for performing collaborative filtering. In: Proceedings of the 22nd Annual International ACM SIGIR Conference on Research and Development in Information Retrieval (SIGIR 1999), pp. 230–237. ACM, New York (1999)

30. Hsu, C.F., Khabiri, E., Caverlee, J.: Ranking comments on the social web. In: Proceedings of the International Conference on Computational Science and Engineering (CSE 2009), vol. 4, pp. 90–97. IEEE (2009)

31. Hu, M., Liu, B.: Mining and summarizing customer reviews. In: Proceedings of the Tenth ACM SIGKDD International Conference on Knowledge Discovery and Data Mining (KDD 2004), pp. 168–177. ACM, New York (2004)

32. Hu, M., Liu, B.: Mining opinion features in customer reviews. In: Proceedings of the 19th National Conference on Artificial Intelligence, AAAI 2004, pp. 755–760. AAAI Press (2004). http://dl.acm.org/citation.cfm?id=1597148.1597269

33. Jakob, N., Weber, S.H., Müller, M.C., Gurevych, I.: Beyond the stars: exploiting free-text user reviews to improve the accuracy of movie recommendations. In: Proceeding of the 1st International CIKM Workshop on Topic-Sentiment Analysis for Mass Opinion (TSA 2009), pp. 57–64. ACM, New York (2009). https://doi.org/10.1145/1651461.1651473

34. Jannach, D., Lerche, L., Zanker, M.: Recommending based on implicit feedback. In: Brusilovsky, P., He, D. (eds.) Social Information Access. LNCS, vol. 10100, pp. 510–569. Springer, Cham (2018)

35. Jindal, N., Liu, B.: Opinion spam and analysis. In: Proceedings of the 2008 International Conference on Web Search and Data Mining, WSDM 2008, pp. 219–230. ACM, New York (2008). https://doi.org/10.1145/1341531.1341560

36. Jindal, N., Liu, B., Lim, E.P.: Finding unusual review patterns using unexpected rules. In: Proceedings of the 19th ACM International Conference on Information and Knowledge Management, CIKM 2010, pp. 1549–1552. ACM, New York (2010). https://doi.org/10.1145/1871437.1871669

37. Kim, S.M., Pantel, P., Chklovski, T., Pennacchiotti, M.: Automatically assessing review helpfulness. In: Proceedings of the 2006 Conference on Empirical Methods in Natural Language Processing (EMNLP 2006), pp. 423–430. Association for Computational Linguistics, Stroudsburg (2006). http://portal.acm.org/citation.cfm?id=1610075.1610135

38. Kluver, D., Ekstrand, M., Konstan, J.: Rating-based collaborative filtering: algorithms and evaluation. In: Brusilovsky, P., He, D. (eds.) Social Information Access. LNCS, vol. 10100, pp. 344–390. Springer, Cham (2018)

39. Koren, Y., Bell, R., Volinsky, C.: Matrix factorization techniques for recommender systems. Computer $42(8)$, 30–37 (2009). https://doi.org/10.1109/MC.2009.263

40. Lee, D., Brusilovsky, P.: Recommendations based on social links. In: Brusilovsky, P., He, D. (eds.) Social Information Access. LNCS, vol. 10100, pp. 391–440. Springer, Cham (2018)

41. Lee, J., Park, D.H., Han, I.: The different effects of online consumer reviews on consumers' purchase intentions depending on trust in online shopping mall: an advertising perspective. Internet Res. $21(2)$, 187–206 (2011). http://dblp.uni-trier.de/db/journals/intr/intr21.html#LeePH11

42. Li, F., Huang, M., Yang, Y., Zhu, X.: Learning to identify review spam. In: Proceedings of the Twenty-Second International Joint Conference on Artificial Intelligence, IJCAI 2011, pp. 2488–2493. AAAI Press (2011). https://doi.org/10.5591/978-1-57735-516-8/IJCAI11-414

43. Lim, E.P., Nguyen, V.A., Jindal, N., Liu, B., Lauw, H.W.: Detecting product review spammers using rating behaviors. In: Proceedings of the 19th ACM International Conference on Information and Knowledge Management, CIKM 2010, pp. 939–948. ACM, New York (2010). https://doi.org/10.1145/1871437.1871557

44. Liu, Y., Huang, X., An, A., Yu, X.: Modeling and predicting the helpfulness of online reviews. In: Proceedings of the 2008 Eighth IEEE International Conference on Data Mining (ICDM 2008), pp. 443–452. IEEE Computer Society, Pisa (2008)

45. Moghaddam, S., Ester, M.: Opinion digger: an unsupervised opinion miner from unstructured product reviews. In: Proceedings of the 19th ACM International Conference on Information and Knowledge Management, CIKM 2010, pp. 1825–1828. ACM, New York (2010). https://doi.org/10.1145/1871437.1871739

46. Mooney, R.J., Roy, L.: Content-based book recommending using learning for text categorization. In: Proceedings of the Fifth ACM Conference on Digital Libraries (DL 2000), pp. 195–204. ACM, New York (2000)

47. Mukherjee, A., Liu, B., Glance, N.: Spotting fake reviewer groups in consumer reviews. In: Proceedings of the 21st International Conference on World Wide Web, WWW 2012, pp. 191–200. ACM, New York (2012). https://doi.org/10.1145/2187836.2187863

48. Mullen, T., Collier, N.: Sentiment analysis using support vector machines with diverse information sources. In: Proceedings of the Conference on Empirical Methods in Natural Language Processing (EMNLP 2004), pp. 412–418 (2004)

49. Musat, C.C., Liang, Y., Faltings, B.: Recommendation using textual opinions. In: Proceedings of the 23rd International Joint Conference on Artificial Intelligence (IJCAI 2013), pp. 2684–2690. AAAI Press, Menlo Park (2013)

50. O'Callaghan, D., Harrigan, M., Carthy, J., Cunningham, P.: Network analysis of recurring YouTube spam campaigns. In: Proceedings of the Sixth International Conference on Weblogs and Social Media (ICWSM 2012), pp. 531–534 (2012)

51. O'Mahony, M.P., Cunningham, P., Smyth, B.: An assessment of machine learning techniques for review recommendation. In: Coyle, L., Freyne, J. (eds.) AICS 2009. LNCS (LNAI), vol. 6206, pp. 241–250. Springer, Heidelberg (2010). https://doi.org/10.1007/978-3-642-17080-5_26

52. O'Mahony, M.P., Smyth, B.: Learning to recommend helpful hotel reviews. In: Proceedings of the Third ACM Conference on Recommender Systems (RecSys 2009), pp. 305–308. ACM, New York (2009). https://doi.org/10.1145/1639714.1639774

53. O'Mahony, M.P., Smyth, B.: A classification-based review recommender. Knowl. Based Syst. 23(4), 323–329 (2010)

54. Ott, M., Choi, Y., Cardie, C., Hancock, J.T.: Finding deceptive opinion spam by any stretch of the imagination. In: Proceedings of the 49th Annual Meeting of the Association for Computational Linguistics: Human Language Technologies, HLT 2011, vol. 1, pp. 309–319. Association for Computational Linguistics, Stroudsburg (2011). http://dl.acm.org/citation.cfm?id=2002472.2002512

55. Pandey, V., Iyer, C.K.: Sentiment analysis of microblogs. Technical report, Stanford University (2009). http://www.stanford.edu/class/cs229/proj2009/PandeyIyer. pdf. Accessed Nov 2010

56. Pang, B., Lee, L., Vaithyanathan, S.: Thumbs up?: sentiment classification using machine learning techniques. In: Proceedings of the 2002 Conference on Empirical Methods in Natural Language Processing (EMNLP 2002), pp. 79–86. Association for Computational Linguistics, Morristown (2002)

57. Pazzani, M.J., Billsus, D.: Content-based recommendation systems. In: Brusilovsky, P., Kobsa, A., Nejdl, W. (eds.) The Adaptive Web. LNCS, vol. 4321, pp. 325–341. Springer, Heidelberg (2007). https://doi.org/10.1007/978-3-540-72079-9_10

58. Popescu, A.M., Etzioni, O.: Extracting product features and opinions from reviews. In: Proceedings of the Conference on Human Language Technology and Empirical Methods in Natural Language Processing (HLT 2005), pp. 339–346. Association for Computational Linguistics, Morristown (2005)

59. Read, J.: Using emoticons to reduce dependency in machine learning techniques for sentiment classification. In: Proceedings of the ACL Student Research Workshop (ACL 2005), pp. 43–48. Association for Computational Linguistics, Morristown (2005). http://portal.acm.org/citation.cfm?id=1628960.1628969

60. Resnick, P., Iacovou, N., Suchak, M., Bergstrom, P., Riedl, J.: GroupLens: an open architecture for collaborative filtering of netnews. In: Proceedings of the ACM Conference on Computer-Supported Cooperative Work (CSCW 1994), Chapel Hill, North Carolina, USA, pp. 175–186, August 1994

61. Robertson, S., Walker, S., Jones, S., Hancock-Beaulieu, M., Gatford, M.: Okapi at TREC-3. In: Text REtrieval Conference (TREC), pp. 109–126 (1996)

62. Salton, G., McGill, M.J.: Introduction to Modern Information Retrieval. McGraw-Hill Inc., New York (1986)

63. Sarwar, B.M., Karypis, G., Konstan, J.A., Riedl, J.: Item-based collaborative filtering recommendation algorithms. In: Proceedings of the 10th International World Wide Web Conference (WWW 2001), Hong Kong, pp. 285–295, May 2001

64. Sarwar, B., Karypis, G., Konstan, J., Riedl, J.: Analysis of recommendation algorithms for e-commerce. In: Proceedings of the 2nd ACM Conference on Electronic Commerce (EC 2000), pp. 158–167. ACM, Minneapolis, 17–20 October 2000

65. Shani, G., Gunawardana, A.: Evaluating recommendation systems. In: Ricci, F., Rokach, L., Shapira, B., Kantor, P.B. (eds.) Recommender Systems Handbook, pp. 257–297. Springer, Boston (2011). https://doi.org/10.1007/978-0-387-85820-3_8

66. Shardanand, U., Maes, P.: Social information filtering: algorithms for automating "word of mouth". In: Proceedings of the SIGCHI Conference on Human Factors in Computing Systems (CHI 1995), pp. 210–217. ACM Press/Addison-Wesley Publishing Co. (1995)

67. Smyth, B.: Case-based recommendation. In: Brusilovsky, P., Kobsa, A., Nejdl, W. (eds.) The Adaptive Web. LNCS, vol. 4321, pp. 342–376. Springer, Heidelberg (2007). https://doi.org/10.1007/978-3-540-72079-9_11

68. Smyth, B., McClave, P.: Similarity vs. diversity. In: Aha, D.W., Watson, I. (eds.) ICCBR 2001. LNCS (LNAI), vol. 2080, pp. 347–361. Springer, Heidelberg (2001). https://doi.org/10.1007/3-540-44593-5_25

69. Tang, H., Tan, S., Cheng, X.: A survey on sentiment detection of reviews. Expert Syst. Appl. **36**(7), 10760–10773 (2009)

70. Wang, G., Xie, S., Liu, B., Yu, P.S.: Review graph based online store review spammer detection. In: Proceedings of the 2011 IEEE 11th International Conference on Data Mining, ICDM 2011, pp. 1242–1247. IEEE Computer Society, Washington, D.C. (2011). https://doi.org/10.1109/ICDM.2011.124

71. Wang, W., Wang, H.: Opinion-enhanced collaborative filtering for recommender systems through sentiment analysis. New Rev. Hypermedia Multimed. **21**(3–4), 278–300 (2015)

72. Weerkamp, W., de Rijke, M.: Credibility improves topical blog post retrieval. In: Proceedings of the Association for Computational Linguistics with the Human Language Technology Conference (ACL-08:HLT), Columbus, Ohio, USA, pp. 923–931 (2008)

73. Wiebe, J., Riloff, E.: Creating subjective and objective sentence classifiers from unannotated texts. In: Gelbukh, A. (ed.) CICLing 2005. LNCS, vol. 3406, pp. 486–497. Springer, Heidelberg (2005). https://doi.org/10.1007/978-3-540-30586-6_53

74. Wietsma, R.T.A., Ricci, F.: Product reviews in mobile decision aid systems. In: Proceedings of Pervasive Mobile Interaction Devices (PERMID 2005) - Mobile Devices as Pervasive User Interfaces and Interaction Devices - Workshop in Conjunction with: The 3rd International Conference on Pervasive Computing (PERVASIVE 2005), Munich, Germany, pp. 15–18 (2005)

75. Zhang, W., Ding, G., Chen, L., Li, C., Zhang, C.: Generating virtual ratings from Chinese reviews to augment online recommendations. ACM Trans. Intell. Syst. Technol. 4(1), 9:1–9:17 (2013). https://doi.org/10.1145/2414425.2414434

76. Zhu, F., Zhang, X.M.: Impact of online consumer reviews on sales: the moderating role of product and consumer characteristics. J. Mark. 74(2), 133–148 (2010)

14
Recommending Based on Implicit Feedback

Dietmar Jannach[1]([✉])(iD), Lukas Lerche[2](iD), and Markus Zanker[3](iD)

[1] AAU Klagenfurt, 9020 Klagenfurt, Austria
dietmar.jannach@aau.at
[2] Department of Computer Science, TU Dortmund University,
44221 Dortmund, Germany
lukas.lerche@tu-dortmund.de
[3] Free University of Bozen-Bolzano, 39100 Bozen, Italy
mzanker@unibz.it

Abstract. Recommender systems have shown to be valuable tools for filtering, ranking, and discovery in a variety of application domains such as e-commerce, media repositories or document-based information in general that includes the various scenarios of Social Information Access discussed in this book. One key to the success of such systems lies in the precise acquisition or estimation of the user's preferences. While general recommender systems research often relies on the existence of *explicit* preference statements for personalization, such information is often very sparse or unavailable in real-world applications. Information that allows us to assess the relevance of certain items indirectly through a user's actions and behavior (*implicit feedback*) is in contrast often available in abundance. In this chapter we categorize different types of implicit feedback and review their use in the context of recommender systems and Social Information Access applications. We then extend the categorization scheme to be suitable to recent application domains. Finally, we present state-of-the-art algorithmic approaches, discuss challenges when using implicit feedback signals in particular with respect to popularity biases, and discuss selected recent works from the literature.

Keywords: Implicit user feedback · Recommender systems
Collaborative filtering

1 Introduction and Motivation

Recommendation is a key functionality on many modern websites and mobile applications. The task of recommendation components within applications is typically to point users to additional items of interest by ranking or filtering them according to the past preferences and the current contextual situation of these users.

© Springer International Publishing AG, part of Springer Nature 2018
P. Brusilovsky and D. He (Eds.): Social Information Access, LNCS 10100, pp. 510–569, 2018.
https://doi.org/10.1007/978-3-319-90092-6_14

Mainstream research in the field of Recommender Systems (RS) – as discussed in detail in Chap. 10 of this book [55] – was historically fueled by applications scenarios in which preference statements of users in the form of *explicit item ratings* are available [49]. This led to the development of sophisticated algorithms that are able to very accurately predict which rating a user would probably give to a certain item. Much of the power of these algorithms is based on the existence of large datasets of historical ratings in which for each user dozens of explicit ratings exist. Since the evaluation is often only done on this historical offline data, the algorithms are optimized to accurately "post-dict" recommendations rather than to predict (which may or may not overlap with real-world performance). While there exist a number of dedicated Social Web platforms on which users can rate movies, books, restaurants or other businesses, there are also many real-world application domains in which rating matrices are very sparse or even non-existent today [42,43]. For example, while some popular items on Amazon.com receive many ratings, most of the items in the catalog do not have any ratings. Also, in domains like friend discovery for social networks, people usually can not be rated explicitly.

When building personalized recommenders in such application domains we have to rely on *indirect ways* of assessing the interests and preferences of users by monitoring and interpreting their actions and behavior. In the research literature, these observations of a user's actions that are interpreted as *statements on the relevance of a particular item* are called *implicit feedback*. Sometimes also the term "nonintrusive" feedback is used because users are not explicitly stating their preferences, but these are derived from their observed actions. In a classic e-commerce setting, an example of a user action that might indicate a preference for an item is when the user views the detailed product description of an item or puts the item on a wish list. On media streaming platforms, the repeated consumption of a track or music video can be interpreted as an interest or preference of the user toward the track itself, the track's artist or the genre. On a social network, sharing a certain news story in a post might express the user's general interest in the topic, as further discussed in Chap. 11 of this book [63].

Implicit and explicit feedback are however not a set of boolean categories, but rather a continuum. Consider the case of a user playing a music track or sharing a news article. These actions can be interpreted as implicit feedback, i.e., the user might have a preference towards the track or the article contents. We might also infer from the user actions that he is interested in the track's artists or the topic of the news story. However, if the user (explicitly) gave the track a rating or "liked" the news article, we could also (implicitly) infer that he might be interested in the artist or topic. Therefore, when we speak of *implicit feedback*, we mean all kinds of interactions with the systems from which we can indirectly *infer* user preferences.

The amount of available implicit preference signals in reality can be huge. Today, every mouse move of a user can in theory be tracked in an online application. In the future, with the continuing development of the *Internet of Things* and users being "always-on" through mobile or wearable devices, even larger amounts of information about the users' behavior and about the objects with whom they interact with will be available.

Besides the technical challenge of efficiently processing such a constant stream of possibly large amounts of data, a number of further questions has to be addressed. These questions include, for example, which of the many types of signals should be used to build a preference profile and how to combine these signals with possibly existing explicit rating information. Furthermore, different signals might indicate a different "strength" of a preference, i.e., a purchase may count more than an item view action in an online store. Finally, implicit feedback signals are often positive-only and in addition we cannot be always sure that we interpret the signals correctly as, e.g., an online shopper can be disappointed later on with a purchase or was purchasing something for a friend.

Overall, recommendation based on implicit feedback in real-world applications is probably much more common than relying (solely) on explicit ratings, e.g., because the acquisition of ratings requires certain efforts from the user's side. A general problem of explicit ratings is that many users use ratings as a means to assess the *quality* of an item, e.g., a movie, than to express their *enjoyment*, which is probably more relevant in a recommendation scenario.[1] Recommendations based on the true user behavior might therefore in fact be more reliable than predictions that are based on explicit ratings in reality.

In this chapter we will review existing approaches and challenges of creating user profiles for recommendation based on implicit feedback. These types of pref-

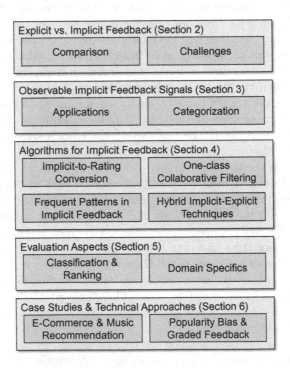

Fig. 1. Structural outline of the chapter

---

[1] http://finance.yahoo.com/news/netflix-wants-ditch-5-star-202428660.html.

erence signals are particularly common in Social Information Access scenarios discussed in this book. Think, e.g., of users who share, comment on, or tag resources in Social Web applications (Chap. 8 [35], Chap. 11 [63] and Chap. 12 of this book [8]); people who regularly check in at certain locations on location-based social networks (Chap. 16 of this book [12]); or music lovers who post their playlists and connect with other users on music streaming platforms (Chap. 11 of this book [63]). These and various other types of user actions can be used to create additional recommendations on the corresponding platforms.

Figure 1 shows the structural outline of this chapter. In Sect. 2, we characterize explicit and implicit feedback signals and discuss challenges when detecting and interpreting such feedback. We then give examples for implicit-feedback application domains of recommender systems in Sect. 3 and propose an extension to Kim and Oard's [85] classification scheme for implicit feedback. In Sect. 4, we review typical algorithmic approaches to deal with implicit feedback signals. Section 5 discusses evaluation aspects. In Sect. 6 we finally present a number of case studies and selected technical approaches from the literature.

2   Explicit vs. Implicit Feedback

A number of different types of input data can be used in the recommendation process and the research literature typically differentiates between explicit and implicit feedback. However, as mentioned in the introduction, implicit and explicit signals are not boolean categories, but a continuum, because explicit feedback can to some degree infer implicit preferences. In the following, we will characterize both implicit and explicit feedback, as well as differentiate them from other input that is neither. Lastly, we will address challenges when implicit feedback is used in recommender systems.

Explicit feedback in general corresponds to a deliberate, unambiguous, and intentional quality assessment by a user on the performance of a system. These assessments are obviously dependent on the application domain. For instance, in the context of a recommender system this feedback is typically related to the relevance of a specific item in a given situational context.

In contrast, implicit feedback lacks this user intention to provide an opinion to the system, but it subsumes all sorts of user actions or behavior which can be exploited by a system to infer the relevance of its propositions to users, i.e., estimate the positive or negative bias towards a specific item *or* towards items with similar characteristics. Obviously, the exploitation of implicit feedback comes at the cost of the uncertainty when interpreting it, i.e. first, the implicit signal may not always unanimously represent a positive or negative bias. For instance, viewing time can be an indicator of interest, of having problems to understand the content or can be the result of being distracted by other uncontrolled events. Second, even if the direction of an implicit feedback category is unanimous its quantification in order to aggregate it with other implicit signals and explicit preference statements adds uncertainty.

2.1 Explicit Feedback

The most prominent form of such feedback in the literature are user-provided *ratings*, e.g., on a 1-to-5 scale often displayed as "stars". In most settings, only one overall rating per item is available. In multi-criteria recommendation approaches, more fine-grained rating feedback regarding different quality dimensions of the items is used. The topic of rating-based collaborative filtering techniques is further discussed in Chap. 10 of this book [55].

Apart from star ratings, other common forms of explicit feedback are unary "like" or "recommend to a friend" statements as well as binary "thumbs up", "thumbs down" selections. In certain applications, we also find explicit negative user actions such as "banning" a track on a music streaming platform or blocking or hiding certain messages on a Social Web platform. Though the latter signals are unary or binary, they are not *implicit* feedback. Sometimes these aspects are confused as most implicit feedback algorithms only rely on unary or binary signals and can therefore be applied for these feedback types as well, as mentioned in the discussion of implicit feedback algorithms later on in Sect. 4.2.

Besides these directly processable preference expressions, there are other forms of explicit feedback which however require further analysis or which are application-specific. On the Social Web, users can for example express their opinions through reviews in natural language or by annotating items with tags that have a (known) positive or negative connotation, see also Chap. 13 of this book [86]. An example for an application-specific explicit feedback would be that a user of an online bookstore puts a book on a "recommended reading" list. Also adding a browser bookmark for a website *can* be an explicit statement in case the bookmark is put into a folder with a clear positive or negative connotation, e.g., "My Favorites".

The distinction between the different feedback types for these latter cases can however be a continuum and explicit statements might infer further implicit preferences. For example, any bookmarking action – independent of the fact that we potentially can unambiguously derive the users' quality assessment for the item – is never an explicit feedback signal, because the users' intention is *not* to inform the system about their preferences in the first place. Such an argument could also be raised for explicit star ratings, where the users' main intention might be to use the rating as a personal reminder for themselves or to share their experiences with other users and not state their opinion in the first place.

In addition, however, explicit rating information may be sparse as such ratings require extra work by the users who might not immediately see the benefit of specifying the preferences. Furthermore, providing an explicit rating requires a considerable amount of cognitive effort by the users and some might be challenged in expressing their preferences using a single rating on a pre-defined and often coarse scale, as reported, e.g., in the study in [124]. This study explored how different factors influence the utility of implicit relevance feedback in search systems and identified that the task complexity had a considerable impact on the users' preference to provide implicit or explicit feedback. In complex search tasks where users rarely identified fully relevant objects, implicit feedback was preferred, because the users' focus centered around the search task. Based on these findings one could hypothesize that users would prefer implicit feedback in domains where the primary task requires their full attention, such as in online shopping, while in media and entertainment domains they might be more willing to provide explicit feedback.

2.2 Implicit Feedback

As mentioned before, implicit feedback subsumes all sorts of user actions or behavior that were not intentionally executed in order to provide feedback on specific items or the system performance in general. However, these implicit signals can be observed and are worthwhile to exploit in order to infer a positive or negative user bias towards a specific item, towards items with specific characteristics, or towards a specific action taken by the system. Usually, one of the tasks when using implicit feedback is to find a suitable way of interpreting the feedback, for example, by mapping it onto a rating scale or by learning relative (pair-wise) preference models.

2.2.1 Observable User Actions

The typical types of such interpretable signals are observable user actions, e.g., when users view or purchase some items on an online store, when they select news articles of certain topics to be displayed, when they listen to a track on a music streaming portal, when they tag or bookmark a resource, or join a group on a social network. The user's navigation behavior – from category browsing to mouse and eye movements – represents another typical category of implicit feedback.

An early categorization of possible types of observable implicit feedback signals – focusing on information filtering and recommendation – can be found in [83]. This classification was later extended by Oard and Kim in [85], who identified three types of observable behavior: **Examination**, **Retention**, and **Reference**. Later on, in [84], a fourth category – **Annotation** – was added, which in some sense unifies implicit and explicit feedback based on the types of observable behavior [49]. In the bibliographical review presented in [52], the

Table 1. Summary of the five types of observable behavior, adapted from Oard and Kim in [84], [85], and [52].

Category	Examples of observable behavior
Examination	Duration of viewing time, repeated consumption, selection of text parts, dwell time at specific locations in a document, purchase or subscription
Retention	Preparation for future use by bookmarking or saving a named or annotated reference, printing, deleting
Reference	Establishing a link between objects. Forwarding a document and replying, creating hyperlinks between documents, referencing documents
Annotation	Mark up, rate or publish an object (includes explicit feedback)
Create	Write or edit a document, e.g. [13] or [39]

authors introduce a fifth dimension called *Create*, which relates, e.g., the user activity of writing or editing an original piece of information. The five categories of implicit feedback are summarized in Table 1.

We will take another look at these five types of observable behavior later on in Sect. 3.2 after we have reviewed recent research on implicit feedback in recommender systems and related fields. We will see that the development of recommendation technology over the last two decades suggests that this classification should be extended and present a suitable extension later on in Table 2.

2.2.2 User-Action-Related Indirect Preference Signals

Implicit feedback for an item can also be inferred from indirect preference signals that are based on explicit feedback (ratings) on related objects or from other user actions that are not directly related to a specific item. We use the term "preference signals" here as the user's actions usually can not be directly considered as feedback on a specific item. An explicit "like" expression for an artist on a social music platform can, for instance, be used as a positive signal for the artist's musical pieces in a music recommender system. Such types of information are usually exploited by content-based filtering recommender systems, which often rely on these forms of "indirect" preference signals.

2.2.3 User-Feature-Related Indirect Preference Signals

User demographics, the user's *current* location, or the user's item-independent embedding in a social network are usually *not* considered as implicit feedback. Depending on the application scenario, some of these user features *can* however represent indirect preference indicators, i.e., a form of implicit feedback, if the characteristics are the results of user actions that are at least indirectly related with the recommendation targets.

For example, in a restaurant recommender, information about the user's *past* geographic location and movement profile *can* be considered as a form of implicit preference signals in case the movement profile allows us to infer a restaurant preference of a specific user without having the user explicitly "checked in" to the restaurant. An in-depth discussion of location-based recommenders can be found in Chap. 16 of this book [12]. Also, the user's connections in a social network *can* be considered as implicit preference signals in particular when the goal is to recommend people or groups (see also Chap. 15 of this book [31]).

2.2.4 Discussion

As we have seen, apart from observable implicit feedback, a variety of additional preference signals can be used in the user profiling and recommendation process including in particular the users' demographics or other user characteristics that are independent of an individually recommended item. In the categorization of different feedback types, these signals are usually not considered as implicit feedback. Furthermore, user-independent, additional information about items – including information about item features or to which other items they are connected – is also not considered to be implicit feedback per se, but it might be useful in correctly interpreting implicit feedback signals such as listening or viewing actions. Similarly, contextual information about the users like the location or time when a specific explicit rating was issued, do not fall into this category of implicit feedback, but help to contextualize the collected feedback.

Overall, the distinction between explicit and implicit feedback and other types of information is not always consistent in the research literature and, as discussed, cannot be seen as a boolean categorization. However, one common aspect of all kinds of feedback, that is not explicitly meant to provide an opinion or a relevance assessment, is a set of specific challenges that will be discussed next.

2.3   Challenges of Using Implicit Feedback

When relying on implicit feedback, a number of challenges has to be addressed. This list is by far from being complete and we recommend to see also the discussion in [49].

2.3.1   Interpretation of Signal Strength

In many situations, several types of user actions have to be considered in parallel and the question on how to aggregate them turns up. Usually a uniform weighting strategy might not be appropriate. For example, in an e-commerce scenario a purchase action might be a stronger preference indicator than a repeated item visit. In addition, the different strengths of implicit feedback signals could be determined by additional post-processing steps, e.g., for identifying different degrees of friendship between users based on their observed communication patterns as done in [105].

2.3.2   Interpretation in Relation to Explicit Signals

Sometimes, both explicit and implicit feedback signals are available, but with different degrees of coverage of the item space. Therefore suitable ways of combining them are needed. Simple approaches in which all implicit actions are, e.g., interpreted as a "four star" rating on a five-item scale and this way transformed into explicit rating signals are popular but inappropriate as the rating database becomes "dominated" by the large amounts of implicit signals. Often the implicit feedback "scales", e.g., visit duration, track playcounts etc., are also incompatible with the five-point scales used for explicit feedback.

2.3.3   Transparency

When explicit feedback is available, it might be easier for the user to understand the rationale of the provided recommendations as they, e.g., can be used in system-generated explanations. Recommendations that result from implicit feedback signals might not be that obvious or plausible for the user. For example, showing a recommendation to a user with the explanation "because you rated [movie A] with 5 stars" might be more plausible than the explanation "because you watched [movie A]", as in the latter case the user might not have liked movie A after all.

2.3.4   Lack of Negative Signals

Implicit feedback is often "positive-only", i.e., we only can learn positive biases from a user's interaction with an item. This lack of negative signals often means that special types of algorithms (one-class collaborative filtering) have to be applied. This also leads to challenges when applying standard evaluation measures as no ground-truth about non-relevant items is available.

2.3.5   Data Not Missing at Random

In most domains, implicit feedback signals for the few very popular items are prevalent while feedback for niche items can be very sparse [77]. Therefore, the distribution of feedback is skewed in a long-tail shape. Building recommendation models based on such data can easily lead to a strong popularity bias ("blockbuster effect") and a "starvation" of the niche items.

2.3.6   Abundance of Data

The computation of sophisticated machine learning models can be challenging on large platforms even when only explicit ratings are considered. The amount of data points to be processed, if for example every single navigation action of a user is logged, makes this problem even worse. Furthermore, given the variety of available types of data points, it is not always clear which of the many signals are the most promising ones to retain and consider in the recommendation process.

On the other hand, while implicit feedback signals have some disadvantages when compared to explicit ratings, one advantage of implicit signals is that they can be collected from all users, while (sufficient amounts of) explicit rating information might in many domains only be available from a few "heavy" users. As a result, the models that are learned solely from explicit ratings might overrepresent some user groups.

3   Categories of Observable Implicit Feedback Signals

In this section we will review typical examples of applications from the research literature that use implicit feedback. We will then come back to the previously discussed categorization scheme for implicit feedback by Oard and Kim and propose an extension with additional types of user actions which have become observable due to technological advancements during the last years.

3.1   Types of Observed Behavior in Applications

Historically, one of the various roots of today's recommender systems lies in the field of Information Filtering, an area that dates back to the 1960s under the term "Selective Dissemination of Information" [37]. The main tasks of information filtering systems typically are to identify and rank documents within larger collections based on their presumed degree of relevance given the user's profile information. Recommender systems nowadays are used in various applications domains, e.g., e-commerce, media consumption and social networks.

In the following, we will give examples of research works from the recommender systems literature to illustrate the various (new) ways of how user actions and observable behavior can be interpreted and used in different application scenarios. The review of existing works will serve as a basis of our proposal to extend the categorization scheme of [85] in Sect. 3.2.

3.1.1 Navigation and Browsing Behavior

Monitoring how users navigate a website or how they use a (web-based) application is a very general type of observable user actions. Several early works that focused on implicit feedback aimed at the *dynamic content adaptation* by, e.g., generating links to possibly additionally relevant content or filter the available content according to the user's preferences.

Analyzing Dwelling Times for Information Filtering. As mentioned in Sect. 2, interpreting dwelling time as implicit feedback is a challenging task. One of the earlier works in the area of personalized information filtering that tries to rely on the observation of the users' behavior, e.g., dwelling times, to infer their interests is found in [82]. The authors' specific assumption was that users of their NetNews system will spend more time on interesting items than on noninteresting ones. To verify their hypothesis, they designed a study in which users had to read news articles during a period of several weeks and provide explicit ratings for the articles. The collected data then indeed showed that reading times are good indicators for the relevance of an article and that both the length and the readability of an article only had a limited impact on reading time. The news filtering systems discussed later in [57] or [107] had similar goals and the studies confirm that relying on reading times alone can help to generate accurate recommendations in many situations. Furthermore, as mentioned in [56], too short viewing times can also be interpreted as *negative* implicit feedback and not only as *missing positive feedback*. The complexity of the interpretation of dwelling time as positive or negative feedback will be further discussed in Sect. 4.1.3.

Monitoring Navigation Actions. Before the large success of WWW search engines, a number of proposals were made to help users with finding relevant websites based on the observation of their browsing behavior. An early approach of that type relying on the user's browsing behavior to infer the user's interest is the system "Letizia" [70]. The client-side looks at the links that are followed by a user, at initiated searches, or at bookmarking activities and applies content-based heuristics to find additional relevant web pages. Other early tools that are similar to the basic idea to customize recommendations based on the users joint navigation behavior (e.g., link selection) and document content similarities are described in [6] or, with a focus on personalized recommendations, [79].

Browsing Actions. In [56], a number of additional browsing-based interest indicators besides the following of hyperlinks are mentioned, including micro-level actions like scrolling, highlighting or the visual enlargements of objects. Depending on the installed equipment on the client side, one can also try to capture the eye gazes of the users [15] or approximate them by tracking the user's mouse movements [104]. From a technical perspective, server-side logging of client-side actions can nowadays be implemented very efficiently using AJAX-based micro requests. Further user interface level actions include requesting help or explanations for an object.

Web Usage Mining. In contrast to approaches that only rely on navigation or browsing logs of individual users, web usage mining systems aim to detect usage patterns in the logs of a larger user community using, e.g., clustering or association rule mining techniques. Personalization systems like WebPersonalizer presented in [80] for example try to match the current user's most recent navigation activities with "aggregated profiles" to generate personalized website recommendations.

Discussion. The "Social Information Access" aspect is most obvious in the last category (Web Usage Mining) where the behavior of other users in the community is directly exploited to make suggestions for the current user. Nonetheless, also the other presented techniques which were partially designed for individual-user settings can in principle be extended to consider the behavior of the community, e.g., by adding collaborative features within the server-side components.

3.1.2   Shopping Behavior

Implicit feedback signals in e-commerce applications – and also others as mentioned below – could in principle be considered as a subclass of the *navigation and browsing behavior*. However, in a commercial context specific *semantic meanings* can be attached to some navigation actions such as viewing an item or adding it to a wishlist or to the shopping basket, while usually not all navigation actions are considered to be relevant for exploitation.

Shopping Basket Analysis. Amazon.com's *Users who bought ... also bought ...* denotion of one of their recommendation lists characterizes the main idea of such approaches quite well. The general underlying concept is to find patterns in the shopping baskets of users [74]. Often, these patterns are identified using more general techniques like classic Association Rule Mining [2] or variations thereof, which can then be applied to make recommendations for the current user [73].

Shop Visitor Navigation Logs. Another category of recommendations on Amazon.com's site is named "Users who viewed ...", which expresses that also other types of user actions can be used for building user profiles on shopping sites. One difference to the above-mentioned general approaches based on navigation logs is, as said, that a purchase is a very distinctive action and one of the main business metrics to be optimized. Recent examples of works which aim to exploit the user's recent navigation behavior to predict the next shopping action include [44,102,111,119] and are often based on approaches that model sequential decision processes.

Discussion. In the past, academic researchers often converted explicit rating datasets into "purchase transactions", e.g., by considering five-star ratings as purchases, because not many public datasets were available. In recent years, we see an increased rate of works which are based on real-world shop navigation logs. Academic competitions like at the 2015 ACM RecSys Challenge[2] help to fuel these types of research as they are based on publicly available real-world datasets. With the emergence of the Social Web, more and more shopping platforms allow their users to comment, review, and share their experiences on the site, and a variety of other user-related data becomes available for specific tasks like next-basket predictions.

3.1.3 Media Consumption Behavior

Reading news online is, as described above, a classic information filtering scenario in which implicit feedback was explored. Other types of electronic media consumption in which implicit feedback recommendation systems were employed include the recommendation of (IP) TV programs based on viewing times, video recommendations using the watching behavior or music recommendation based on listening logs.

Implicit feedback signals related to media consumption often face additional challenges. Both for music and TV shows it is not always clear who – if anyone at all – in the household is currently watching or listening. In addition, user actions like a "skip" to the next track can be context dependent and interpreting it as a general negative assessment of the previous track might be misleading.

TV-Related Recommendations. Recommending based on implicit feedback in the context of TV programs was for instance explored in [26], where the viewing duration as in [41] was considered as an indicator for the signal strength and methods were proposed to deal with the uncertainty of the signal. The case of *linear* programs in contrast to video-on-demand services was, e.g., discussed in [133] where they also consider various information signals related to noise in the data and the new-item problem. In the deployed TiVo system [4], the fact that someone recorded a show is treated as an implicit feedback signal and combined with explicit binary feedback. According to the recent literature review in [120], implicit profiling is therefore the most common approach in this domain.

Music Recommendation. The use of implicit feedback signals for music recommendation and playlist generation will be discussed in more depth in Sect. 6.2. As an example, consider the work presented in [87] where the authors develop a multi-criteria music recommendation approach which utilizes both explicit as

---

[2] http://recsys.acm.org/recsys15/challenge/.

well as implicit feedback. Implicit feedback signals are inferred both for the overall rating of the track as well as for the criteria preferences (i.e., on music, lyrics and voice). As feedback signals the authors use the total time spent by users hearing a track, the number of accesses to an item and the actual play duration per listening event.

Another music-related approach is presented in [64], where the authors as in [18] rely on listening logs of users obtained from the Last.fm music platform as a basis for music recommendation. A specific aspect of their work is that their algorithms exploit additional (time-related) context information which they automatically derive from logs.

3.1.4   Social Behavior

With the development of the "participatory Web", social networks, and Web 2.0 technologies, users transformed from being pure information consumers to becoming also active content contributors. They now can explore the information space of the Web not only by accessing the structures provided by (classic) information providers, but also by using the behavior or content from other peers in their social networks as guidance. Typical interactions of this "social navigation" are, for example, commenting or posting on a social network or microblogging platform, tagging or bookmarking content on the Web or establishing social connections with other people [40].

Given these novel types of interactions, a number of additional preference signals can be used in recommendation processes. Some of these types of signals were anticipated in the *Annotation* and *Create* categories of observable behavior in [52,84]. Since the observable user actions on the Social Web are not necessarily directly related to a target object (such as "annotate" or "publish") but can signify also indirect preference indications, we introduce *"Social & Public Action"* as an additional category.

Tags and Bookmarks. Bookmarking or tagging items with keywords for own later use is a classic implicit feedback signal in Information Filtering. In the Social Web sphere, tags and bookmarks are now shared with others and can serve as a basis, e.g., to build tag-based recommender systems [28,33,110,131], see also Chap. 12 of this book [8].

Posts and Comments. Publishing information on social media in terms of a post or comment about an opinion or the own current activity is another type of implicit preference signal on the Social Web, as further discussed in Chap. 10 of this book [55]. Such often very short posts can be analyzed to build user profiles that reflect the user's interests [1]. The contents of posts was for example analyzed in [96] through a topic modeling technique with the goal to recommend other users to follow on the social network (see also Chap. 15 of this book [31]). Finally, the problem of filtering interesting items in a social "feed" corresponds to a classic collaborative information filtering problem with some additional challenges, e.g., that the content to be analyzed can be very short [117].

Structuring Objects. The organization of objects for later use is another observable user action mentioned in [84]. A typical example in the recommendation domain is when users share music playlists, which can serve as a basis for next-track music recommendation [11].

Connecting with Others. A final category of implicit feedback signals can be the user's embedding within a social network. One can analyze the user's social neighborhood, explicit or implicit trust signals, or the network topology [5] to recommend additional friends or followees, or inspect existing group memberships or channel subscriptions and their topics to recommend further groups or other items [32][3], as discussed in more detail in Chap. 15 of this book [31]. In [71], for instance, the followers of Twitter accounts are used to generate interest profiles in the context of the cold-start problem for app recommendation. Another application domain, personalized social search, exploits an individual's relations in a social network to compute more relevant query results [14].

3.1.5 Ubiquitous User Modeling

With the availability of modern smartphone devices and their various sensors as well as the emerging trend of the "Internet of Things", more and more information about the user's current location and environment becomes available. We propose to summarize these types of observable user actions under the umbrella term *"Physical Action"* in the extended classification scheme.

Location and Movement Profiles. The user's past and current movement profile can be a valuable indicator of the user's interests, as also discussed in Chap. 16 of this book [12] in the context of social media data. In [9], for example, the movement profiles and dwelling times of users in a museum are used as indicators for the user's interest in the individual exhibit objects. Other application domains in which the past locations of the users can be used for user profiling include in particular the tourism domain – think, e.g., of past visited places or GPS trajectories [132] as interest indicators – or leisure activities. In the mobile context a proactive approach has been advocated to enrich and ease the mobile experience by *"providing the right information, at the right time, and in the right form for the current context"* [115]. However, such a proactive system behavior exclusively relies on implicit user feedback and accurate observations of user actions in order to avoid an obtrusive system behavior [29]. The work in [67] analyzed, for instance, a user's activity (movement) from GPS logs in order to develop a proactivity model and determine when it is appropriate to interrupt the user and to provide an unrequested recommendation.

Note that in contrast to context-aware recommendations (CARS) we are not necessarily interested in the user's *current* location to make suitable recommendations, but rather rely on the observed user behavior and relationships between past user actions to determine the appropriate next steps.

---

[3] As indicated in Sect. 2, we consider such information only as implicit feedback if the signal is related to some target recommendation object.

Smart Homes. In the Internet of Things, all sorts of electronic devices, e.g., in a smart home, will be connected with the network and can represent additional sources of information about the environment of a user or with which devices the user has interacted with. One typical task in such a context is called "activity recognition", i.e., to estimate based on the available sensor data, e.g., from a mobile phone [23,95], which activity the user currently pursues and where he or she is located. Quite an amount of research on knowledge-based or learning-based activity recognition has been done in the area of *smart homes*, see e.g. [118] for an early work. While these types of sensor information have been largely ignored in the mainstream recommender systems literature, these preference-based adaptations of, for instance, light or music actors [53] in smart homes represent adaptive and personalized systems in their purest form that heavily rely on implicit user feedback. Some more recent examples include the automatic identification of users while watching TV in order to learn their interests [72] or the use of gaze tracking in combination with explicit ratings to derive content-based interest profiles [54].

3.2 An Extended Categorization of Observable User Behavior

Oard and Kim's early categorization scheme – *Examination, Retention, Reference* – was mainly focusing on document-centric applications and is in particular suitable when the goal is to recommend news messages, text documents, or web pages, see Table 2. This also holds for the additions in [49,84], *Annotation* and *Create*. With the wide-spread application of recommendation technology in all sorts of domains that we have observed in the last two decades, a variety of other types of implicit feedback signals have been successfully exploited since Oard and Kim's early work.

Table 2. Extension of the five types of observable behavior (see Table 1 and [52,84,85]) by two now categories: *Social & Public Action* and *Physical Action*.

Category	Examples of observable behavior
Examination	Duration of viewing time, repeated consumption, selection of text parts, dwell time at specific locations in a document, purchase or subscription
Retention	Preparation for future use by bookmarking or saving a named or annotated reference, printing, deleting
Reference	Establishing a link between objects. Forwarding a document and replying, creating hyperlinks between documents, referencing documents
Annotation	Mark up, rate or publish an object (includes explicit feedback)
Create	Write or edit a document, e.g. [39] or [13]
Social & Public Action	Public posting, commenting and communicating, activity posts, following and connecting with people, joining groups, expressing trust
Physical Action	Observed user actions that can be interpreted as feedback towards objects of the physical world. Being at a location, roaming profiles and dwelling time, other recognizable activities in the physical world (e.g., smart homes, Internet of Things)

Based on our review of application scenarios for implicit feedback in recommender systems, we suggest to extend the existing classification scheme with additional observable user actions. They are related to (a) the user's social behavior and (b) the increased availability of data for "ubiquitous" user modeling. The new items are shown in Table 2 with a detailed description. They meet the requirements of new behavioral patterns that emerged with the widespread availability of connected mobile devices and social functionalities on the Web. Keep in mind that in practice the types of observable behavior can overlap. For example, the *Social & Public Actions* "posting" and "rating" articles on a social network can also be seen as *Create* and *Annotation* actions.

4   Algorithms for Implicit Feedback Situations

In this section, we will discuss algorithmic approaches to generate recommendations based on implicit feedback. As mentioned earlier, interpreting implicit feedback can be difficult and we will first discuss techniques to transform and encode preference signals as explicit feedback to be able to use standardized recommender system algorithms for rating prediction. After that, we will briefly present selected examples of collaborative filtering algorithms that are especially designed to deal with "one-class" only feedback signals. Then, we will examine methods to find frequent patters in implicit feedback in more detail and finally show examples of hybrid algorithms that try to combine explicit ratings with implicit feedback.

4.1   Converting Implicit Signals to Ratings

As discussed in Sect. 2.3, implicit feedback is often "positive-only", i.e., no or only minimal information is given about items that were disliked by the users. Also, there are often multiple signals and different kinds of feedback, e.g., when a user visits an item detail page in an online store multiple times, bought some items and placed other items to a "wishlist". In addition, the "rating matrix" of implicit feedback is most of the time very sparse. Therefore, the available data consists of few positive signals that are sometimes hard to interpret and numerous unlabeled examples.

To deal with such situations, many so-called "One-Class Collaborative Filtering" techniques were proposed in the literature, some of which we discuss later in Sect. 4.2. In this section, we show an alternative to this approach, which is the transformation of the given data into two-class or multilevel numerical "ratings" and the creation of a user-item rating matrix. Such a transformation then allows us to apply standard recommendation techniques which were originally designed for explicit ratings. In the following, we will discuss ways of transforming implicit feedback signals into numerical rating values. Further discussion on rating-based collaborative filtering approaches can be found in Chap. 10 of this book [55].

4.1.1 Problems of Basic Transformation Strategies

The first step of a basic implicit-to-numerical transformation is to add a virtual rating of "1" to the user-item rating matrix for each observed user-item interaction. Different options exist to deal with the unknown data points and the missing negative feedback, each of them having certain drawbacks as discussed in [88].

All Missing as Negative (AMAN). In this approach, all non-observed items are treated as a "0" rating and thus as negative feedback. The resulting user-item-matrix contains only ones and zeros, see Fig. 2. When a machine learning model is fitted to this data, the distribution between the two classes – 0 and 1 – is strongly biased toward the negative feedback, since there are only few positive entries in the matrix. Any rating prediction technique for explicit feedback might tend to always predict 0. Usually, regularization methods are used to prevent this kind of overfitting but the ratio of negative to positive feedback in the data is however still problematic [101].

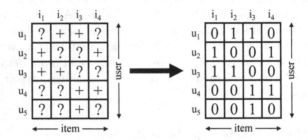

Fig. 2. AMAN: transformation of implicit feedback to explicit "0" and "1" ratings [88,101].

All Missing as Unknown (AMAU). Alternatively, the missing data points could be treated as unknowns, see Fig. 3. A rating prediction algorithm therefore only operates on the positive ratings, i.e., ignores all missing data. Since the whole dataset only consists of "1" ratings, the distribution is biased toward the positive feedback. Without proper regularization, this would result in a trivial solution and typical explicit feedback algorithms would tend to always predict 1 [116].

4.1.2 Discerning Negative from Unknown Signals

To avoid the drawbacks of the *extreme* ways to deal with unknown examples – labeling them as negative or ignoring them – more advanced approaches assume that there might be *some* negative examples in the unknown data. If these could be labeled properly, existing explicit feedback approaches could be employed.

Several ways to guess which of the unknown entries could be negative feedback have been introduced in the past, some of them as part of a one-class collaborative filtering techniques which we will discuss in Sect. 4.2.

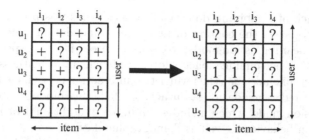

Fig. 3. AMAU: transformation of implicit feedback to explicit unknown and "1" ratings [88,101].

A simple approach is to randomly sample negative examples from the unknowns, as done, e.g., in [101]. To learn the ranking of items, their approach uses positive-negative item-pairs for each user. However, since there is no negative feedback in the data, they select a random unknown item for each (positive) feedback to create the pairs. More elaborate schemes use statistical [92] or weighting-based [88] approaches to choose negative samples in a way that the distribution of the resulting set of negative ratings resembles the set of the positive ratings.

As an alternative to inferring negative ratings, users could be asked to give some negative (and positive) feedback, e.g., in an initial interaction phase with the system. However, this would be considered explicit feedback and might be perceived as a burden by users [88,94].

4.1.3 Converting Graded Implicit Feedback to Ratings

Instead of converting implicit feedback signals into explicit 0/1 ratings, some proposals in the literature adopt more fine-grained strategies. Since in many application settings different types of user behavior can be observed, the idea is to assign a different "strength" to each type of signal, i.e., to encode the different levels of *graded* relevance feedback as ratings.

In a study on recommendations in an online mobile games store [42], for example, the authors used explicit ratings – which were only sparsely available – and in addition considered view and purchase actions, which were transformed into explicit rating values. On a scale from -2 to $+2$, view events were interpreted as 0 and purchase events as $+1$. Explicit positive (negative) ratings where considered as a $+2$ (-2) rating. The choice of this encoding was done somewhat arbitrarily and led to a very skewed distribution of the rating values, as there are many more view events than purchase events.

A time-based approach of assigning numerical values to implicit feedback signals was proposed by Lee et al. [65] in the context of a recommender system for wallpaper images. Purchase information was used as an implicit signal and the strength of the signal was determined based on the release date of an item and the point in time when the user made the purchase. The authors then used a time-based decay function to promote more recent events received higher scores.

An approach of combining different feedback types was presented by Parra and Amatriain [93] in the context of music recommendation. The authors propose to use a linear regression model to combine three different aspects of implicit feedback signals – personal feedback, global feedback and recentness – into a rating score. They conclude that the former two interaction types have the strongest impact on the recommendation accuracy. In [94], this model is extended to a logistic regression model which includes a number of additional variables related to consumption behavior as well as demographic data.

In [59], another work in the field of music recommendation, items that have both explicit ratings and observed user actions are exploited to learn which types of implicit feedback can be mapped onto which ratings. The user actions that were interpreted as implicit feedback consist of play counts and play percentages, listening date and time, number of skips and next-track statistics. Subsequently, the system rates items that did not receive any explicit feedback with a naive Bayesian classification based on the implicit signals that the item received.

The examples above show that transforming signals of different types of user behavior into one single (rating) score largely depends on the respective domain and cannot be generalized easily. Sometimes, it may not be possible to map different kinds of feedback, e.g., viewing and buying an item, to a linear rating scale. Other signals may be difficult to interpret, for example, a short dwelling time for an item detail page could be interpreted as negative feedback because the user seems to be not interested in the item. However, it could also mean that the user already knows the item and does not need to look at the page again without an indication of positive or negative feedback. On the other hand, a long dwelling time does not necessarily correspond to positive feedback. A user might have lost interest and abandoned the page without actively leaving it because the item was not relevant anymore. In the following section, we therefore discuss the correlation between implicit signals and explicit ratings in more detail.

4.1.4   Correlating Implicit Feedback with Explicit Ratings

How to encode different types of feedback into numerical scores can, as discussed, be challenging and is sometimes done in an arbitrary manner. Several researchers have therefore investigated the relationship between explicit ratings and implicit feedback actions, including [19,93,94,99,112]. Depending on the domain and experimental setup, the obtained results are however not always consistent.

In [19], the results of a laboratory study are reported in which users were first asked to freely browse the Web for 30 min and subsequently had to rate each visited page with respect to how interesting its contents were. The recorded user actions, such as mouse movements and dwelling times, were then compared with the collected explicit ratings. The analysis revealed that the time spent and the scrolling activity on a web page correlates with explicit ratings. Other indicators, however, such as mouse movement and clicks, had no clear relation to the participant's interest.

Zhang and Callan [129] report the results of a user study on a web-based news filtering system. The participants had to read personalized news for one hour per day over a period of 4 weeks and assess the articles according to multiple dimensions, such as relevance, novelty and readability. After the study, each participant also completed a questionnaire about the topics of the articles that they read. In addition, the same user actions as in [19] were recorded and the authors similarly concluded that dwelling time and scrolling activity correlate the most with the explicit ratings. However, they also state that the answers of the questionnaire about topic interests are much more correlated with the explicit feedback and therefore advise that in real-world settings the users should initially be asked about their topics of interest.

Building on the insights and log data from this study and the work from [19], the authors of [134] propose a Bayesian modeling technique to combine implicit and explicit feedback signals. Their results, however, indicate that the implicit feedback was unstable and possessed only limited predictive value. Thus, the combination of both feedback types was only marginally better than when using explicit feedback alone.

The results of a similar study on the relationship between various types of browsing actions and explicit interest statements are reported in [112]. The strongest correlations with the explicit ratings were found for the indicator "time of mouse movement relative to reading time" and the number of visited links on the page. Note that mouse movements were not considered to be a good indicator according to the study [19] discussed above that however did not put the mouse movements in relation to the dwelling time.

More recently, Parra et al. [93, 94] report on their attempt to correlate implicit and explicit preferences in the music domain. The authors first carried out a survey in which the participants rated tracks from Last.fm. This information was used to derive preference patterns and biases, e.g., whether users generally prefer recent or popular tracks. The insights of the survey were then used to design a linear regression model to predict ratings from *what they call* implicit feedback signals. In fact, the authors rather adopt an approach based on metadata to learn which features of the liked items are particularly relevant to the users.

Finally, in some domains implicit feedback seems to be more meaningful than explicit preference information. In [99], Pizzato et al. use the data of 21.000 users of an online dating platform and compare the predictive accuracy of different input types. In contrast to most of the other works reviewed so far, their results show that explicit preference statements are often incomplete or imprecise and recommending based on implicit feedback can be more accurate. This emphasizes once more that the interpretation of implicit feedback can be highly dependent on the respective domain.

4.2   One-Class Collaborative Filtering Techniques

The naive conversion strategies to generate (binary) numerical scores from implicit feedback have their drawbacks, e.g., converting all unobserved data points into zeros (*All Missing as Negative*, AMAN) or leaving them as unknowns (*All Missing as Unknown*, AMAU) both result in a class imbalance problems and standard rating prediction techniques tend to always predict 0 or 1, respectively. Therefore, more sophisticated techniques were proposed to deal with positive-only feedback in the literature.

These so-called *one-class collaborative filtering techniques (OCCF)* [88] are algorithms that only need one single type of signals. They usually interpolate which of the missing data points could be negative feedback or try to guess if a user prefers one item over a different, unknown one. Typically, OCCF techniques are used in domains where only unary implicit feedback is available. As discussed earlier in Sect. 2.1, some of them are also applicable when dealing with unary explicit feedback, such as "likes" on a social network, as well. In the following, we will present examples of selected OCCF techniques in more detail – wALS, Random Graphs, BPR, CLiMF – and briefly review other related approaches.

4.2.1 wALS and sALS-ENS

In [88], the authors introduce two strategies to handle the missing feedback in a way that is somewhere in between the two extremes of AMAU and AMAN. In the first strategy, a low-rank approximation X of the "rating" matrix R is calculated and in the objective function, a confidence weight is used to express the probability that a signal is (correctly) interpreted as positive or negative. A weight of 1 is assigned to the positive data points, since they are known beforehand. The unknown, missing values, on the other hand, have a confidence value lower than 1, because some of them have the chance to be negative samples. The following equation shows the objective function.

$$\mathcal{L}(U, V) = \sum_{ij} W_{ij}(R_{ij} - U_i V_j^T)^2 + \lambda(||U||_F^2 + ||V||_F^2) \tag{1}$$

The low-rank approximation X of R is decomposed to $X = UV^T$ and can be used for prediction. To prevent overfitting, the objective function has a regularization term. The matrix W is the non-negative weight matrix that assigns confidence values to the observations and the optimization problem is solved by an Alternating Least Squares algorithm (ALS), hence the name wALS (weighted ALS).

The characteristics of this algorithm are influenced by the choice of W. For $W = 1$, the confidence for all data points would be 1 and therefore the strategy would be equivalent to AMAN, where all unknown are treated as negatives. The authors propose three different weighting schemes W for the unknown entries W_{ij} in the user-item interaction matrix which are described in Table 3.

Calculating a large approximative low-rank matrix is however computationally intensive and, in addition, the class imbalance problem is still present, because there are still many more negative than positive samples.

Therefore, with sALS-ENS the authors propose a more advanced way to consider all (known) positive examples from the data and add a subsample of negative feedback based on a sampling probability matrix. They propose three sampling strategies that behave similar to the ones used for the weighting matrix of wALS. As a result, a smaller rating matrix is generated that can be used as a

Table 3. Weighting schemes for OCCF.

Weighting scheme	Confidence matrix	Description
Uniform	$W_{ij} = \delta$	All missing entries are assigned a fixed confidence weight δ between 0 and 1
User-oriented	$W_{ij} \propto \sum_j R_{ij}$	Higher confidence is set for "heavy" users as they are assumed to know the item catalog better and therefore have discarded unknown items with a higher probability
Item-oriented	$W_{ij} \propto m - \sum_i R_{ij}$	Higher confidence is set for items which received more interactions, i.e., unpopular items have a higher probability to be discarded as negative. Here, m is the number of users

basis for calculating the low-rank approximation of R via ALS. The experiments show that the wALS approach is slightly superior in terms of accuracy but considerably slower than sALS-ENS. An approach similar to wALS has been proposed in [41] that also uses a weighting term for the implicit observations.

4.2.2   Random Graphs

In a similar spirit, Paquet and Koenigstein [92] model the unknown negative feedback using a random graph. The approach is based on a bipartite graph G that contains edges $g_{mn} = 1$ between users m and items n when observed implicit feedback is available.

The additional assumption is however that, although a user m has interacted with an item n, there should be some other items that the user considered but discarded as not relevant. The authors therefore model a second *hidden* graph H that is also bipartite and contains edges $h_{mn} = 1$ whenever a user m considered an item n. In addition, $g_{mn} = 1 \Rightarrow h_{mn} = 1$ holds, i.e., if a user m accepted an item n, then it was considered before.

Therefore, G is a subgraph of H and all the other edges of H are the considered items that were discarded as not relevant, i.e., the negative feedback. Since the negative feedback is unobserved, the authors use the following popularity-based sampling strategy to generate the edges in H that represent the negative feedback.

For each user m with d_m observations of positive feedback, additional d_m edges of negative feedback are randomly sampled from a distribution $M(\pi)$ based on the popularity of all items. Instead of using a popularity distribution with $\pi_n = d_n$, where d_n is the number of times that there was positive feedback for an item n, the authors assume that popular items are generally more liked, i.e., have less negative feedback. Therefore, for the popular items, less negative examples should be sampled for H and the distribution is modified in the following way: $\pi_n = d_n^{\gamma}$ with $\gamma = 1 - \log d_{max} / \log r$, where $d_{max} = max\{d_n\}$.

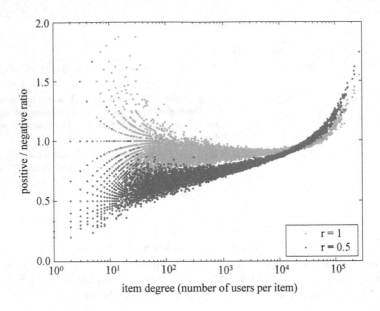

Fig. 4. Ratio of positive and negative edges in H [92].

The parameter r controls the ratio between sampled negative and known positive samples in H, as can be seen in Fig. 4. In addition, the sampling procedure is done "without replacement", i.e., if the sampling draws an already known positive example $h_{mn} = 1 \wedge g_{mn} = 1$, no negative sample $h_{mn} = 0$ is added. Therefore, the ratio is skewed for the most popular items, as there is a higher chance to draw a positive example for them, which results in a lower amount of generated negative feedback signals.

An advantage of this approach is that it is easily extensible with richer feedback signals. For example, the hidden graph H could also be populated with implicit negative examples gathered from other sources, e.g., a visited detail pages of an item without a subsequent purchase or a purchase of an equivalent item could indicate $h_{mn} = 1 \wedge g_{mn} = 0$. Similarly, information about items that could have never been considered $h_{mn} = 0$ could be included in the graph, e.g., because the item was not listed in the shop at the time the user was active.

To generate recommendations with the two graphs G and H the authors propose a bilinear collaborative filtering model with matrix factorization which can estimate the probability of accepting an item as relevant after considering it under $p(g_{mn} = 1 | h_{mn} = 1)$. The model is designed to be largely agnostic of the popularity of the items. Therefore, the popularity bias of the recommendations can be reduced. More details on the model have been discussed in [92]. The proposed approach was developed and deployed in the context of the Microsoft Xbox Live environment, in which one particular challenge lies in the large amounts of data that have to be processed.

4.2.3 Bayesian Personalized Ranking (BPR)

BPR [101] – as already mentioned in Chap. 10 of this book [55] – deals with the one-class CF problem by turning it into a ranking task and implicitly assuming that users prefer items they have interacted with over other unknown items.

In some sense, BPR therefore creates artificial negative feedback in a similar spirit as the approaches discussed so far. However, instead of applying rating-prediction techniques using the implicit feedback data BPR ranks the candidate items for a user without calculating a "virtual" rating.

The overall goal of the algorithm is to find a personalized total ranking $>_u \subset I^2$ for all users $u \in U$ and pairs of items $(i,j) \in I^2$ that has to satisfy the properties of a total order (totality, antisymmetry, transitivity).

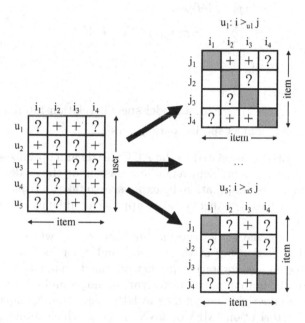

Fig. 5. Transformation of implicit feedback to pairwise preferences for each user [101].

To model the negative feedback, Rendle et al. [101] use a pair-wise interpretation of the positive-only feedback. The general idea is that a user's positive feedback for an item is interpreted as the user's preference of this item over all other items that the user did not give feedback for. As shown in Fig. 5, the positive-only feedback is thus transformed to positive and negative feedback in pairs of items (i,j) where the user preferred i over j (positive), or j over i (negative). If the user interacted with both items or none of them, no additional information can be deduced for the pair (i,j). The different pairs of items form the training data D_S for the BPR algorithm and can be formalized as triples of a user and an item pair:

$$D_S := \{(u, i, j) | i \in I_u^+ \wedge j \in I \setminus I_u^+\}$$
$$\text{with} \tag{2}$$
$$I_u^+ : \text{items with implicit feedback from } u$$

To create a personalized ranking of items, the authors introduce a general optimization criterion called BPR-OPT, which is derived through a Bayesian analysis of the problem and which aims to maximize the posterior probability $p(\Theta| >_u) \propto p(>_u |\Theta)p(\Theta)$ where Θ is the parameter vector of the underlying algorithmic model. The optimization criterion, including substitutions for smoothing, is formulated as follows:

$$\begin{aligned}
\text{BPR-OPT} &:= \ln p(\Theta| >_u) \\
&= \ln p(>_u |\Theta)p(\Theta) \\
&= \sum_{(u,i,j)\in D_S} \ln \sigma(\hat{x}_{uij}) - \lambda_\Theta ||\Theta||^2 \\
&\text{with}
\end{aligned} \tag{3}$$

$$\sigma(x) = 1/(1 + e^{-x})$$
$$\hat{x}_{uij} := \hat{x}_{ui} - \hat{x}_{uj} : \text{a model-specific relationship function}$$
$$\lambda_\Theta : \text{model specific parameters}$$

The BPR-OPT criterion is related to the AUC metric and optimizes it indirectly. To solve the optimization problem, a gradient descent on the model parameters Θ can be used. Since it is computationally expensive to take all triples $(u, i, j) \in D_S$ into account, a stochastic gradient descent approach randomly chooses the triples uniformly from D_S.

By decomposing the model specific function $\hat{x}_{uij}$ – which is a real-valued function for the relationship of the items i and j for user u – into $\hat{x}_{ui}$ and $\hat{x}_{uj}$, existing techniques for rating prediction can be applied to calculate the two terms. In [101], both a matrix factorization model and a kNN approach are presented as the underlying model for the BPR algorithm. Compared to standalone Matrix Factorization (MF) or kNN models, which minimize the rating prediction error, the BPR-OPT criterion instead ensures that the item ranking is optimized.

4.2.4   CLiMF

Collaborative Less is More Filtering is another approach for ranking optimization in one-class CF settings [114]. CLiMF aims to directly optimize a smoothed version of the Mean Reciprocal Rank (MRR) metric to achieve an optimal ranking of the top-n items. In comparison to BPR, both algorithm optimize a smoothed version of a ranking metric. BPR, however, implicitly assumes negative feedback in the data, while CLiMF only uses the positive feedback signals.

The optimization target of CLiMF, the reciprocal rank RR_i of a recommendation list for a user i, represents the position of the earliest occurrence of a relevant item for the user. For N items it can be defined as:

$$RR_i = \sum_{j=1}^{N} \frac{Y_{ij}}{R_{ij}} \prod_{k=1}^{N} (1 - Y_{ik}\mathbb{I}(R_{ik} < R_{ij}))$$

with

$Y_{ij} = 1$ if i interacted with j, else 0

R_{ij} = rank of item j in list of user i

$\mathbb{I}(x) = 1$ if $x = true$, else 0

(4)

In essence, the formula only calculates $1/R_{ij}$ for the first relevant item j for user i. However, directly optimizing the reciprocal rank with standard optimization functions – like gradient descent – is not possible, since it is a non-smooth function. Therefore, the authors introduce a smoothed approximation of RR_i which can be optimized. To that end, the indicator function $\mathbb{I}(x)$ and the rank $1/R_{ij}$ are substituted by the following approximations:

$$\mathbb{I}(R_{ik} < R_{ij}) \approx g(f_{ik} - f_{ij})$$
$$1/R_{ij} \approx g(f_{ij})$$

with

$$g(x) = 1/(1 + e^{-x})$$
$$f_{ij} = \langle U_i, V_j \rangle$$

(5)

Here, the predictor function for the relevance score f_{ij} is based on a factor model of the latent user and item factor vectors U_i and V_j. Although inserting the substitutions of Eq. 5 in Eq. 4 creates a smooth approximation of the reciprocal rank and could in theory be optimized, the optimization task has a complexity of $O(N^2)$, i.e., is quadratic with the number of items, which is not practically feasible in most domains. It is, however, possible to derive a lower bound of the reciprocal rank which can be optimized with a lower complexity [114]:

$$L(U_i, V) = \sum_{j=1}^{N} Y_{ij}[\ln g(f_{ij}) + \sum_{k=1}^{N} \ln(1 - Y_{ik}g(f_{ik} - f_{ij}))]$$

(6)

The optimization function of CLiMF (Eq. 6) has two terms that are maximized: (1) Y_{ij} and (2) the rest of the equation in square brackets. Maximizing the first term promotes the relevant items. Maximizing the second term optimizes the ranking by learning latent factors. As discussed in [121], this can also lead to a diversification of the recommendation results. Equation 7 shows the final regularized optimization function for all users. It can be optimized with a stochastic gradient descent approach and a complexity of $O(dS)$ with S being the number of observed positive feedback examples.

$$F(U, V) = \sum_{i=1}^{M} \sum_{j=1}^{N} Y_{ij}[\ln g(U_i^T V_j) + \sum_{k=1}^{N} \ln(1 - Y_{ik}g(U_i^T V_k - U_i^T V_j))]$$
$$- \frac{\lambda}{2}(||U||^2 + ||V||^2)$$

(7)

Later on, the authors proposed a generalized version of CLiMF called xCLiMF which is able to deal with situations where a relevance level for the feedback is available [113]. A similar generalization to deal with graded relevance feedback was also proposed for BPR in [66], which will be discussed later in Sect. 6.3.2.

4.2.5 Other One-Class Classification Approaches

The problem of positive-only data can also be found in the field of classification when we only have positively labeled training data. Support Vector Machines (SVM) are a typical method that was originally designed for two-class classification tasks and requires labeled input data. In [109], Schölkopf et al. develop a theoretical foundation to apply support vector machines (SVM) to unlabeled, one-class data.

These one-class SVM (1-SVM) are able to identify a region in the item-space where most of the "positive" examples are located. Likewise, the other regions of the item-space can be labeled as "negative". A practical implementation of 1-SVM for recommender systems and a benchmark on the MovieLens data is presented in [125]. Similar examples for classification tasks without the need for negative training samples are [51,126], where the goals are to classify websites and text based on positive and unlabeled data only.

A density estimation that is similar to the one of Schölkopf et al. [109] is presented in [7]. The authors introduce a model to estimate high-density areas of the data points in the item-space. Additionally, the model assumes that not all the data points are positive feedback but could also be negative examples. If there is positive feedback as well as unlabeled examples in the data, it is possible to solve the one-class classification problem by applying the expectation-maximization (EM) algorithm, see, e.g., [24,123]. In [69], finally, unlabeled data points are treated as negative examples. This transforms the problem into a problem of "learning with noise" and is solved with a regression approach to model a linear function as a classifier.

4.3 Frequent Patterns in Implicit Feedback

One of the most prominent examples of a recommendation system is Amazon.com's list of shopping proposals labeled "Customers who bought ... also bought ...". The label suggests that the contents of the non-personalized but item-dependent list are based on the analysis of the buying behavior of Amazon's customer base and the detection of item co-occurrence patterns.

In these classic *Shopping Basket Analysis* settings, the goal is to find sets of items that are frequently purchased together. The input to the analysis is a set of purchase *transactions* where each transaction contains a set of items that were bought together, e.g., in one shopping session.

4.3.1 Association Rule Mining

Technically, the identification of such patterns can be accomplished with the help of *Association Rule Mining* (ARM) techniques [2]. An association rule has the form $A \Rightarrow B$, where A and B are sets of items and the arrow can express something like "whenever A was purchased, also B was purchased"; typically, the strength of a rule is expressed in terms of the measures *support* and *confidence*.

Following the description of [108], *Association Rule Mining* can be formally defined as follows. Let $T = \{t_1, \dots, t_m\}$ be the set of all transactions and $I = \{i_1, \dots, i_n\}$ the set of all available items. Each transaction t consists of a subset of items $t \subseteq I$. A transaction t could therefore represent a shopping basket of items that was bought by a customer. Let $A, B \subseteq I$ and $A \cap B = \emptyset$, i.e., A and B are also subsets of I but have no items in common. An *association rule* is defined as the implication $A \Rightarrow B$. It expresses that whenever the items contained in A are included in the transaction t, then items contained in B will also be included in t. The left side of a rule $A \Rightarrow B$ is often called the *rule body* or *antecedent* while the right side is the *rule head* or *consequent*.

As can be seen from the definition above, each co-occurrence of two or more items in a transaction can be expressed as an association rule. However, not all association rules are helpful, e.g., two items could only have occurred together once in a single transaction, and the goal of *Association Rule Mining* is to find only those rules in a set of transactions T that are meaningful. To quantify the significance of association rules, various measures have been introduced in the past[4] but the most widely-used measures are *support* and *confidence*.

The *support* of a set of items A is the proportion of transactions $t \in T$ that contains A, i.e., the transactions where $A \subseteq t$. It can also interpreted as the probability of the co-occurrence of all items in A in a transaction.

$$supp(A) = \frac{|\{t \in T; A \subseteq t\}|}{|T|} \tag{8}$$

The *confidence* of an association rule is then defined as the ratio between the number of transactions that contain $A \cup B$ in relation to the number of transactions that only contain A. Therefore, the *confidence* is the conditional probability of B given A. It can therefore be expressed with the support of A and $A \cup B$.

$$conf(A \Rightarrow B) = \frac{supp(A \cup B)}{supp(A)} \tag{9}$$

Usually a minimum support and a minimum confidence has to be satisfied by an association rule to be considered significantly meaningful for the transactions. Therefore, when generating the association rules, first a threshold for the support is applied to find the most frequent sets of items in the transactions. However, when there are n items in the set of all items I, the number of possible subsets that have to be considered is $2^n - 1$, excluding the empty set. For a large number of items n, considering all combinations individually is not feasible. Efficient calculation of the support is however possible by exploiting the *downward-closure property* [2]: If an itemset A is frequent according to some support threshold, then all of its subsets $A' \subseteq A$ are also frequent for that threshold. Likewise, if

---

[4] http://michael.hahsler.net/research/association_rules/measures.html.

an itemset A is not frequent according to some support threshold, then all of its supersets $A' \supseteq A$ are also not frequent for that threshold. After the most frequent itemsets have been found, a confidence threshold is used to determine the most important association rules.

To detect these rules automatically and for large amounts of data, a variety of algorithms was proposed over the last decades to find frequent patterns and derive association rules, starting with the Apriori algorithm [2] that uses the *downward-closure property*, over more efficient schemes like FP-Growth [34] to techniques that find patterns in parallel [68] or are able to identify rules for niche items [73]. For cases in which the sequence of the item interactions is relevant, *Sequential Pattern Mining* [3] can be applied.

4.3.2   Recommending with Association Rules

Once the rules are determined, recommending based on association rules can be done, e.g., in the e-commerce domain, as follows. First, we determine the set of the current user's (recently) purchased or viewed items and then look for rules in which these items appear in the antecedent (the left hand side, A). The elements appearing in the right hand sides (B) of the corresponding rules then form the set of possible recommendations. Items for recommendation can then be ranked with a $score_{ui}$ based on different heuristics, e.g., by using the confidence of the rules that are applicable to the subsets A of the items I_u that a user u purchased and that lead to the inclusion of an individual item i.

$$score_{ui} = \sum_{A \subseteq I_u} conf(A \Rightarrow i) \qquad (10)$$

A specific aspect to consider in the recommendation domain is that we are not necessarily interested in the strongest rules, as they might lead to obvious recommendations, but rather in rules for unexpected patterns or niche items. Also, depending on the domain, different kinds of association rules can be mined. In [73], for example, the score used to rank the items was calculated using both user associations ("user u_1 likes an item" $\Rightarrow$ "user u_2 likes an item") and items associations ("item i_1 is liked" $\Rightarrow$ "item i_2 is liked"). When recommending, their approach ranks the items by a $score_{ui}^{User}$ for users that are above a fixed support threshold, i.e., users that already have left some feedback in the system. The score is calculated both from the support and confidence of the user associations. However, A is now a subset of the users U_i that liked the item i.

$$score_{ui}^{User} = \sum_{A \subseteq U_i} supp(A \cup u) \cdot conf(A \Rightarrow u) \qquad (11)$$

For users below the support threshold, i.e., cold-start users that only left little feedback, item association rules are used to calculate the item ranking similar to Eq. 10. To find niche items, the item association rules are however mined for each item as a fixed *consequent* and only the subset of transactions T' that contains the *consequent* is used to calculate the support of each item. Since T' is usually small compared to T, the resulting support for the items is higher. Otherwise, rules for new or niche items would be filtered out, since their support would often be below the general support threshold over all transactions T.

Amazon.com's "Customers who bought" recommendations can be generated in a similar way with Association Rule Mining. The particularity of such an approach is that only frequent itemsets of size two are required – which means that simple co-occurrence counts can be sufficient – and that these recommendations can already be provided in the context of the customer view of one particular item.

Association Rule Mining techniques have been applied in different recommendation scenarios in the literature. Examples include the identification of navigation patterns in the context of Web Usage Mining [80], the identification of rules exploiting item characteristics in e-commerce [98], the recommendation of next tracks in music playlist generation [11, 36], or in the context of e-learning [27]. Association Rules and "co-visitation counts" also serve as a basis for the YouTube video recommendation system [21].

4.4   Hybrid Implicit-Explicit Techniques

In some domains both explicit and implicit feedback signals are available. For example, in most online stores users can rate products and at the same time their navigation behavior is logged by the system. In the following sections, we will discuss some approaches that combine explicit and implicit feedback or use the explicit rating of an item as an additional implicit input signal. Besides the discussed methods, many ways to hybridize explicit and implicit feedback have been proposed in the literature. Some focus on the specifics of certain domains, e.g., the music domain [48, 59], TV programs [4, 127] or web pages [129]. Others propose new techniques to combine the different types of feedback, for example, when using matrix factorization [75, 97].

4.4.1   Hybrid Neighborhood and MF Models

In application domains where explicit ratings are available, matrix factorization (MF) techniques can nowadays be seen as the state-of-the art for efficient and accurate rating prediction. In [60] Koren proposes to combine classic neighborhood models and MF for explicit ratings with implicit feedback. To that end, four hybridization strategies are introduced that build on each other: (1) a neighborhood model, (2) Asymmetric-SVD, (3) SVD++ and (4) an integrated model.

The first model is based on the classic way of predicting a rating for a user, e.g., by aggregating the ratings of similar items weighted by their similarity $\hat{r}_{ui} = \sum r_{uj} \cdot sim_{ij}$. The complete neighborhood model is defined as follows:

$$\begin{aligned}
\hat{r}_{ui} = {}& \mu + b_u + b_i \\
& + |R^k(i; u)|^{-\frac{1}{2}} \sum_{j \in R^k(i;u)} (r_{uj} - b_u - b_j) w_{ij} \\
& + |N^k(i; u)|^{-\frac{1}{2}} \sum_{j \in N^k(i;u)} c_{ij}
\end{aligned} \tag{12}$$

Besides the overall average rating μ and the user and item biases b_u and b_i, the *neighborhood model* includes all explicit $R^k(i; u)$ and implicit ratings $N^k(i; u)$ of user u for the k nearest neighbors of item i, see Formula 12. For each item-item combination of i with its neighbors j, the sum is weighted with the factors w_{ij} and c_{ij} which model the strength of the relationship between i and j and are not given by a similarity function but learned in an alternating least squares learning phase discussed in [60].

The second model is *Asymmetric-SVD* in which the computationally expensive neighborhood calculation of Formula 12 is substituted by an MF approach and the rating prediction is therefore changed to:

$$\hat{r}_{ui} = \mu + b_u + b_i$$
$$+ q_i^T \left(|R(u)|^{-\frac{1}{2}} \sum_{j \in R(u)} (r_{uj} - b_u - b_j) x_j \right. \tag{13}$$
$$\left. + |N(u)|^{-\frac{1}{2}} \sum_{j \in N(u)} y_j \right)$$

Instead of directly looking at all neighbors of item i to calculate a prediction, an "SVD-like" lower rank decomposition of the rating matrix is introduced. Compared to traditional SVD, e.g., $\hat{r}_{ui} = \mu + b_u + b_i + q_i^T p_u$, there are no user-wise latent factors p_u in this model. Instead, p_u is approximated and replaced with a term (between parenthesis in Formula 13) over all explicit $R(u)$ and implicit $N(u)$ ratings of user u. The parameters x_j and y_j are now latent item weights that are learned in the optimization process. As a side note, compared to p_u in the classic SVD approach, the three model parameters, q_i, x_j, y_j, are not user-dependent. Therefore, the model can directly predict ratings for new users without being re-trained.

The third model, *SVD++*, simplifies the *Asymmetric-SVD* model by reintroducing the latent factors p_u for each user u, but only for the explicit feedback. *SVD++* is defined as follows:

$$\hat{r}_{ui} = \mu + b_u + b_i + q_i^T \left(p_u + |N(u)|^{-\frac{1}{2}} \sum_{j \in N(u)} y_j \right) \tag{14}$$

The final model combines both the *SVD++* and the *neighborhood model* into an *integrated model*. The underlying reason is that neighborhood models perform well when detecting localized relationships between few specific items but fail to capture the overall structure in a large set of ratings [60]. MF techniques, on the other hand, behave complementary. The hybrid approach is defined as:

$$\hat{r}_{ui} = \mu + b_u + b_i + q_i^T \left(p_u + |N(u)|^{-\frac{1}{2}} \sum_{j \in N(u)} y_j \right)$$

$$+ |R^k(i;u)|^{-\frac{1}{2}} \sum_{j \in R^k(i;u)} (r_{uj} - b_{uj}) w_{ij} \tag{15}$$

$$+ |N^k(i;u)|^{-\frac{1}{2}} \sum_{j \in N^k(i;u)} c_{ij}$$

In their evaluation the authors used the Netflix dataset and generate implicit feedback by transforming the explicit ratings. They compare their methods against the classic neighborhood model $\hat{r}_{ui} = \sum r_{uj} \cdot sim_{ij}$ and SVD, and conclude that by adding implicit feedback, the recommendation accuracy can be significantly improved compared to the baselines. Also, $SVD++$ performed better than $Asymmetric\text{-}SVD$ when the implicit feedback was generated from explicit feedback. However, the authors state that for domains where implicit feedback is available, $Asymmetric\text{-}SVD$ should in theory be more accurate.

4.4.2 Collaborative Feature-Combination

In [128] an approach to combine multiple (explicit and implicit) aspects of the user model was proposed. Classic CF approaches only take one type of rating data (explicit ratings) into account and are consequently challenged in cold-start situations. The proposed *collaborative feature-combination* recommender can help to deal with these challenges by considering existing implicit feedback – e.g. the navigation history of a user – if explicit feedback is not available. The general idea is to extend the single-category neighborhood calculation to multiple relevance-ordered feature dimensions. Then, the ranking score for the recommendation can be calculated as follows.

$$rec_{fch^*}(i, u, d_t, d_{rec}) = \frac{\sum_{v \in N_u} score_{i,v}}{|N_u|}$$

with

$$score_{i,v} = sim_{fch^*}(u, v, d_t) \text{ if } i \in R_{d_{rec},v} \preceq R_{d_t} \text{ else } 0$$

$$sim_{fch^*}(u, v, d_t) = \sum_{d \preceq d_t} w_d \times cos(\overrightarrow{R_{d,u}}, \overrightarrow{R_{d,v}}) \tag{16}$$

$cos(\overrightarrow{a}, \overrightarrow{b})$: cosine similarity

$R_{d,u}$: rating vector for feature dimension d and user u

R_{d_t} : threshold feature dimension

w_d : feature dimension weight

In this equation, the recommendation score is the average of the weighted cosine similarity $score_{i,v}$ over all users N_u. The similarity $sim_{fch^*}(u, v, d_t)$ is calculated as a weighed combination over the feature dimensions, e.g., the implicit feedback of observed *buy*, *context*, *view* or *navigation* actions. In addition, the feature dimensions are ordered, for example by their predictive performance, i.e., *buy* $\prec$ *context* $\prec$ *view* $\prec$ *navigation*. For the recommendation, a *threshold feature dimension* R_{d_t} has to be specified and the algorithm only uses feature dimensions that have a higher predictive accuracy than the threshold dimension. For example, by using the dimension *context* as the threshold, only implicit feedback of *buy* and *context* actions is included and the other (less meaningful) feature dimensions *view* and *navigation* are excluded in the calculation. The approach is therefore capable to gradually include different types of implicit feedback signals in the prediction model.

4.4.3   Bayesian Adaptive User Profiling

Similar to the collaborative feature combination approach, the authors of [134] propose a method to avoid the cold-start problem by simultaneously taking explicit and implicit feedback into account to model the user profile in a hierarchical Bayesian approach. Initially, there is only little explicit feedback available. Therefore, for new users, the model automatically focuses on the "cheap" implicit feedback and the collaborative information gathered from other users. The authors use a general Bayesian model to formalize this as follows:

$$
\begin{aligned}
& f^u \sim P(f|\theta) \\
& y = f^u(x) \\
& \quad \text{with} \\
& f^u : \text{model of user } u \\
& x : \text{item} \\
& y : \text{rating}
\end{aligned}
\tag{17}
$$

From a general perspective, the user model is a function f^u for each user u that estimates a rating y for each item x and is modeled as a prior distribution on some parameters θ. In addition, the user model is personalized by learning from a sample dataset D_u for each user that consists of item/rating-pairs. With Bayes' Rule, the general model can be extended to:

$$
\begin{aligned}
P(f^u|\theta, D_u) &= \frac{P(D_u|f^u, \theta)P(f^u|\theta)}{P(D_u|\theta)} \\
&= P(f^u|\theta) \prod_{i=1}^{N_u} \frac{P(f^u(x_i^u) = y_i^u|f^u)}{P(f^u(x_i^u) = y_i^u|\theta)}
\end{aligned}
\tag{18}
$$

$$
\begin{aligned}
& \quad \text{with} \\
& D_u = \{(x_i^u, y_i^u)|i = 1 \dots N_u\} \\
& N_u : \text{number of training samples for } u
\end{aligned}
$$

For each user u, the belief about the user model is also based on the training data D_u. Equation 18 shows that the user model depends both on the model's prior probability $P(f^u|\theta)$ and the data likelihood given the user model f^u. If the number of training samples N_u for a user is small, i.e., there is little explicit feedback available, the prior probability based on the observed behavior and other users is the major contributor of the final model. For the prior, the authors use a hierarchical Gaussian network which is further discussed in [134].

4.4.4 Reciprocal Compatibility

In [99], explicit and implicit feedback is used in the domain of online dating as a two-step approach. The explicit feedback, which consists of features like age and body type that the user prefers, is used to filter the possible recommendation results. The ranking of user profiles, on the other hand, is based on the implicit feedback – viewing user profiles, sending and replying to messages – by calculating a "reciprocal compatibility score". This similarity measure is calculated as follows:

$$recip\_compat(u, v) = \frac{2}{compat(u, v)^{-1} + compat(v, u)^{-1}}$$

with

$$compat(u, v) = \sum_{i=1}^{n} \sum_{j=1}^{k_i} \frac{f_{u,i,j}}{k_i} \times P(v, i, j)$$

(19)

Here, $P(v, i, j)$ indicates that some feature A_i (e.g., *body type*) has a certain value a_{ij} (e.g., *slim*) in the profile of user v. The factor $f_{u,i,j}$ is the implicit preference of user u for that features value a_{ij}, e.g., the number of times the user viewed the profile of a *slim* user. Therefore, this approach uses the observed user behavior to weight the preference of explicitly given features.

5   Evaluation Aspects

In this section, we will discuss aspects related to the accuracy evaluation for implicit feedback situations. We will limit our discussion to settings in which we only have "positive" user-item interactions (unary feedback) available for learning and evaluation, i.e., there are no negative signals and no graded positive feedback signals. See also Chap. 10 of this book [55] for a more detailed discussion of evaluation metrics for ranking tasks. Furthermore, we assume that we apply the usual procedure of splitting the available data into training and test data and perform cross-validation, a sort of repeated subsampling.

As in this chapter previously discussed, the problem is that we usually cannot know if entries in the user-item interaction matrix are missing because the user did not like the item (and therefore, e.g., did not purchase it) or the user was simply not aware of the item.

5.1   Recommendation as a Classification and Ranking Task

One basic functionality of a recommendation algorithm is to classify each item into one out of two categories, i.e., predict if a user will like it or not. If we treat the empty cells as containing zeros and the others as ones, we can use any classification technique and compare their performance using standard measures from the literature. Given the outcome of the classification process, we can compare the results with the "ground truth" in the test set and categorize each test outcome in one of four groups: True Positives (TP), True Negatives (TN), False Positives (FP), and False Negatives (FN) as shown in Table 4.

Table 4. Contingency table of classification outcomes.

	Relevant	Non-relevant
Recommended	True Positives	False Positives
Not recommended	False Negatives	True Negatives

Assessing Unranked Recommendations. Based on this categorization, we can compute the True Positive Rate ("How many of the *ones* in the test set have been correctly predicted as such", a.k.a. Recall), the True Negative Rate and so on. In recommendation settings, the measurement is typically done with a predefined list length and the Information Retrieval measures Precision and Recall. Precision measures how many of the elements placed into a top-k recommendation by an algorithm were truly "relevant". The Recall indicates how many of *all existing relevant items* made it into the top-k list. The values for Precision and Recall depend on the length of the top-k list. The Recall value will typically increase with longer list lengths while Precision might decrease at the same time. The F-measure can therefore be used as weighted combination of Precision and Recall using, e.g., the harmonic mean of the values.

Assessing Ranked Recommendation Lists. In most recommendation and information retrieval scenarios, the first few elements of the top-k lists receive the most attention from users. Precision and Recall do not take the position of "hits" (i.e. True Positives) into account, but we know that changing the list length will influence their values. A good classifier should therefore be able to place the hits at the top of the ranked list and will lead to high Recall values already for short lists. A poorer classifier might have more False Positives at the top and achieves similar Recall values only at longer list lengths (at the cost of lower Precision).

To assess how different classifiers perform using different thresholds (list lengths), Precision-Recall curves can be used, where Precision is plotted as a function of Recall. Receiver Operating Characteristic (ROC) curves are a related visual approach, which can be derived based on the position of the True Positive Rate and the False Positive Rate in the ROC space under varying threshold levels.

In order to make the comparison of such curves easier, researchers use the measures Mean Average Precision as the average Precision at each threshold level or the Area Under the Curve (AUC) for ROC curves[5]. A discussion of the

---

[5] The BPR-OPT criterion used in the previously described BPR method has a close correspondence to the AUC measure.

similarities and differences between ROC curves and Precision-Recall curves can be found in [22]. A particular aspect according to this work is that ROC curves tend to overestimate the performance of a classifier when compared to the usage of Precision-Recall curves.

Single-Item Rank Position Measurements. The *Mean Reciprocal Rank* (MRR) measure considers the position of the first hit in a ranked list of items. Its value is computed as the multiplicative inverse of the position (rank) where a hit occurred, e.g., if the hit occurred at the third position, the MRR will be 1/3. The MRR was used as an optimization goal for example in the CLiMF method [114].

Another evaluation variant for the top-k recommendation was used in [20,60]. The idea of the protocol is to evaluate each relevant item of each user in the test set individually by combining it with n (e.g., $n = 100$) other items for which the ground truth is not known, i.e., for which no rating exists from the given user. The task of the recommender is to rank these $n+1$ items. Recall is then determined at a certain list length k and can be zero or one, depending on if the test set item was in the top-k list or not. Precision is defined as *Recall/k* in this setting. To make the measure independent of the list length, sometimes the *percentile* rank of the hit is reported (*Mean Percentile Rank*), see [17] or [41].

5.2   Domain-Dependent Evaluation Approaches

The most popular measures in the Recommender Systems literature (Precision, Recall, RMSE/MAE) [47] have the advantage that they are domain-independent ways of comparing different classifiers or recommendation algorithms. Unfortunately, the choice of the specific evaluation measure often seems arbitrary in the literature even though it clearly depends on the domain which measure should be optimized, e.g., if the problem is to find "one good item" or "find all good items" etc. [38]. The RMSE, for example, has been heavily used since the Netflix prize even though it is not fully clear if high prediction accuracy comes with high user satisfaction or increased sales.

Problem-specific measures in implicit feedback settings have, e.g., been proposed in the domains of e-commerce recommendations and music playlist recommendation. The ACM RecSys 2015 Challenge, for instance, was based on time-stamped and sessionized user activity logs from a online store and the participants had to accomplish two tasks. For each session in the test set, the task was to predict if a visitor will purchase an item and if so, which item will be bought. The used evaluation measure was then a combined score which (a) takes into account how often a session with a purchase action was properly predicted and if the purchased item was correctly predicted and (b) penalizes all wrong predictions of sessions with a purchase action. The winning strategy in the contest used a two-phase classification approach that in the first step identifies sessions that will probably include at least one purchase and in the second step detects individual items that will likely be bought in these sessions. A high accuracy of the task was achieved by considering time-based aspects, i.e., the time and day when users tend to do a purchase.

Another evaluation protocol when using time-stamped activity logs was proposed in [44]. The specific research question was to estimate how fast different recommendation approaches adapted their buying proposals to the short-term interests of customers. The protocol therefore includes a mechanism to incrementally "reveal" the most recent clicks to a recommender. Different strategies were evaluated on an e-commerce dataset and the results indicated that considering short-term interests and the most recent user actions are crucial to achieve high accuracy, but also that having long-term preference models is important. In both discussed e-commerce settings, user sessions and activity time-stamps are crucial for being able to compare different recommendation strategies in a more realistic way.

In the music domain, the Recall was, for instance, used to compare different strategies of generating playlist continuations (next-track recommendations), see [11] or [36]. The input to the recommendation algorithms were hand-crafted playlists shared by users. The specific setup is to predict the very last track of each playlist in the test set. Differently to standard approaches, very long list lengths were used in the evaluation process (e.g., 300 items). In particular for the problem of playlist generation, several other measures were proposed in the literature, including the *Average Log Likelihood* that an algorithm will produce a "natural" playlist as well as other list-based measures that aim to determine quality factors like homogeneity or the smoothness of transitions, see [11].

5.3   Discussion

Offline evaluation designs as discussed above in general have a number of limitations independent of the type of input data, i.e., implicit or explicit feedback, such that we cannot be sure if optimizing for high accuracy leads to the desired effects from an application perspective or that accuracy measures do not capture other possible quality factors like diversity or novelty. When only implicit feedback signals are available, some further aspects should however be considered.

The Lack of the Ground Truth for Negative Feedback. The classic IR measures Precision and Recall were designed with the idea that we can organize the prediction outcome in the contingency table (Table 4), i.e., we know the expected outcome for the ranked items, and research was often based on manually labeled documents. In RS applications the ground truth for most items is however unknown, which means that in particular the Recall measure has to be interpreted with care [38] and – depending on the specific way of measurement – can only be considered as a lower bound of the accuracy of a recommender in reality [11, 114].

To evaluate their implicit-feedback algorithms, some researchers use common rating-based datasets (e.g., from MovieLens) and convert them to binary ratings using an arbitrary rating threshold[6]. Every rating below the threshold is considered a negative feedback. Although such an evaluation can give us some insight about algorithm performance for a two-class explicit-feedback setting, it might not be truly representative for the performance of algorithms when only positive feedback is available. Furthermore, depending on how the items with missing ground truth are counted when determining the values for Precision and Recall, completely different algorithm rankings can result from an comparative evaluation [46].

Finally, when converting explicit feedback to binary feedback, also the standard prediction accuracy measures (RMSE, MAE) can be applied in particular for cases where the recommendation algorithms do not only output binary predictions. Again, since the underlying data is not truly a one-class collaborative filtering problem, the results of such an offline comparison might not be representative for the true performance of the different techniques.

In the other direction of implicit-explicit mappings, one could try to transform different forms of implicit feedback to different levels of explicit feedback. A "purchase" could be a five-star rating, an item view correspond to three stars etc. Then all forms of rating-based algorithms can be applied and evaluated in terms of error measures such as the RMSE. Again, the outcome of the evaluation process might depend on the particular encoding of the implicit signals. Furthermore, the number of available ratings at each level might be very different, i.e., there might be orders of magnitude more view events than purchases, such that constantly predicting a "three" might be a competitive strategy when error measures are used.

Popularity Biases. Even when explicit rating information is available, we often see that the distribution of the ratings is quite skewed and the more positive ratings dominate. Even more, research suggests that ratings are not missing at random [77], which can lead to undesired biases in the recommendation models. At the same time, we often see a very skewed distribution regarding how many ratings an individual item received, i.e., there is typically a long-tail distribution in which a small set of all items obtains a large fraction of all existing ratings.

One particular problem in that context is that in implicit feedback settings there is a huge number of "negatives" when we try to compute Precision and Recall. If we rank ten thousands of items, the chances of placing one of the few dozen "true positives" in a top-10 list are very low[7]. In some works, researchers therefore rely on very long top-k lists to avoid that they have to compare very tiny Recall values (like 0.003 vs. 0.004) [11,36], or report (cumulative) probability percentages and use subsets of items to be ranked [61].

Generally, as a result, recommending the most popular items to everyone can be a baseline that is hard to beat [20]. Being better than such a simple approach

---

[6] This was for example done for the evaluation of the implicit-only algorithm BPR, see http://www.mymedialite.net/examples/item_recommendation_datasets.html.

[7] Many more of the top-ranked elements might be relevant for the user, but no explicit information is given.

in terms of Precision and Recall can often only be achieved by using algorithms which themselves can exhibit a strong popularity bias like BPR [46]. In reality, recommending only popular items can certainly be of limited value for the user. In some works like [114], researchers consider the n most popular items as being irrelevant for the user and do not "count" them in their evaluation. Although this might be reasonable from a practical perspective, the choice of the top n elements to be ignored appears somewhat arbitrary in particular if it does not consider the individual user's knowledge about certain items.

6   Case Studies and Selected Technical Approaches

In this section, we will discuss case studies and selected technical approaches from different domains in more detail to obtain a deeper understanding of approaches based on implicit feedback and how researchers deal with the specific challenges. The case studies are selected from the e-commerce and music application domains as in these domains implicit feedback is often prevalent. In addition, we present technical approaches for implicit feedback algorithms that address the popularity bias of implicit feedback and the support of graded feedback.

6.1   Recommending Based on Activity Logs in E-Commerce

We will focus on two recent approaches that use the activity logs of Zalando, a larger European online retailer for fashion products, as described in [44,119]. We chose these two works as they are based on a real-world dataset[8] that contains information that is (a) typically available for many online shops and (b) corresponds to what companies might share with researchers as no sensitive information is contained in the logs. Furthermore, the log contains *all* user interactions for a given time period[9] and is not limited to a particular user group, e.g., *heavy* users. The social aspect when generating recommendations in this setting is the collective behavior of the website users which is analyzed to identify patterns in the navigation and buying behavior.

6.1.1   Data Aspects

The user activity log contains more than 24 million recorded user actions of different types (item views, purchases, cart actions). Most of the actions – about 20 million – are item views and about 1 million actions correspond to purchases. The user actions are related to more than 1.5 million *sessions*, which comprise sequential actions within a certain time frame. Each log entry comprises a limited amount of information about the item itself like the category, price range or color. The actions were performed by about 800.000 anonymous users. The catalog of

---

[8] The data is not publicly available.
[9] The data was sampled in a way that no conclusions about visitor or sales numbers can be drawn.

products (including product variants) appearing in the log is huge and consists of around 150.000 items.

The dataset exemplifies several of the challenges mentioned in Sect. 2.3, including, e.g., the abundance of data[10], the general sparseness with respect to the available purchase data as the majority of users has never made any purchase, or the problem of the interpretation of the strength of the different signals.

On the other hand, such datasets allow us to perform analyses and design algorithms which are much closer to the demands of real-world recommendation systems than the non-contextualized ex-post prediction of missing entries in a user-item rating matrix, which is the most common evaluation setup in research [47].

6.1.2 Topic Detection for User Sessions

In [119], the authors present an approach to automatically infer the "topic" or short-term shopping goal for the individual user sessions. The proposed approach for topic detection is based on Markov Decision Processes (MDP) and can be easily transformed to serve as a topic-driven recommendation technique or MDP-based recommender system [111].

The basic idea of their approach is to view each session as a sequence of item attributes as shown in Fig. 6. The example for instance shows that the user has only viewed items from the category "shirt". The shirts had however different colors and different price ranges. The general topic (shopping goal) to be inferred is shown on the right hand side of the figure, i.e., the user looked for dark-colored shirts for women in any price range.

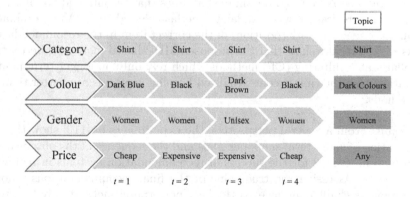

Fig. 6. Viewing a session as a sequence of attributes [119].

Approach. Technically, the idea is to model the topic detection problem as a reinforcement-learning problem based on Markov Decision Processes. The

---

[10] The data sample was taken within a very limited period of time.

observed sequences of actions – in this case sequences of item features – are therefore used to train a model to predict the most likely next observation (state). The distribution of the item attribute values in the user session are considered the *topic* of the session which can then be leveraged in the recommendation process. The learned models are strictly session-dependent, i.e., no long-term profile of the individual user is built in this approach.

Relying on MDPs for the recommendation task was done, although in a different form, for example in [111]. The particular challenge however lies in the computational complexity of such an approach given the huge amounts of items and possible states. In [119], this scalability problem is addressed by using *factorized* MDPs. They rather model sequences of item attribute values than sequences of observed interactions with items and learn such fMDPs independently for each attribute. Furthermore, an approximation technique is used in the optimization phase to avoid scalability problems in terms of memory requirements.

As a result of the (approximate) optimization process we obtain probabilities which express the most likely next observed attribute values. This information can be used to extract the topic of the session as well as to rank items based on their particular item features.

Results. In their empirical evaluation, the authors first compare different strategies for topic detection. The results show that their method is highly accurate in predicting the topic (around 90%, depending on the length of the observed history) and much better than the compared baselines, among them a simple Markov process based on the frequencies of item clicks.

Second, a comparison was made for the recommendation tasks where the baselines include both models that rely on the long-term user profile and latent factor techniques as well as more simple baselines that recommend popular items or items that are feature-wise similar to the last viewed item. As an evaluation measure, the *average rank* (position) of the correct item in the recommendations was used. The results show that the proposed MDP-based method is better than the collaborative filtering (CF) methods which rely only on longer-term models. In addition, also the simple contextualized baseline methods are better than the CF methods.

Discussion. From a general perspective, the experiments in [119] show that the consideration of short-term shopping goals and the sequence of the observed user actions can be crucial for the success of recommendation systems in real-world environments. Assessing the true value of the final recommendations unfortunately remains challenging as even the best performing method only lead to an average rank of 15,000 (due to the large item assortment in the shop).

The results also indicate that optimizing for long-term goals alone as done in the state-of-the-art baseline methods can be insufficient. Overall, at least in the e-commerce domain, using implicit feedback data with time information might help us to develop models which are much closer to real-world requirements than models that generate time- and situation-agnostic predictions for missing items in the rating matrix. Furthermore, the work highlights scalability limitations of existing approaches when it comes to deal with real-world datasets. For the first set of experiments, the authors merely used a few percent of the available data to be able to perform the optimization. For the larger dataset, unfortunately no information is provided on computation times for model building and generating recommendations.

A related case study can be found in Chap. 16 of this book [12] for context-aware recommendations of places. There, a post-filtering approach according to the user's short-term goals is employed to create an intention-based ranking of nearby relevant locations.

6.1.3   Evaluating the Combination of Short-Term and Long-Term Models

The discussion in the previous section indicated the importance of generating recommendations that consider the recent short-term user intent while solely exploiting long-term preference models might be insufficient. In fact, many of the recommendations of popular e-commerce sites like Amazon.com are either simply reminders of recently viewed items or recommendations that are connected to the currently viewed item ("Users who viewed ... also viewed ...").

In [44], the authors aim to quantify the effectiveness of such comparably simple recommendation strategies and furthermore analyze the possible benefits of combining them with optimized long-term models. One further goal is to assess how quickly the different strategies adapt their recommendations after the most recent user action in a session.

Evaluation Protocol. Since standard evaluation setups in the research literature do not cover situations in which time-ordered session logs are available, the authors propose a parameterizable and domain-independent evaluation protocol as shown in Fig. 7.

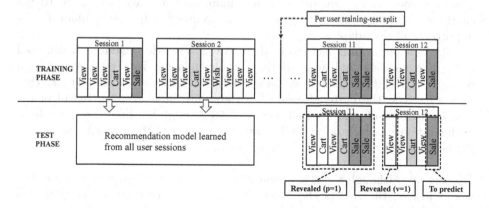

Fig. 7. Proposed evaluation protocol [44].

The general idea is to split the data as usual into training and test data while maintaining the order of the log entries. The task in the test phase is then to predict which item will be purchased in a session of the training set. In contrast to similar protocols, e.g., the one used in the ACM RecSys 2015 Challenge, the idea is now to vary the amount of information that a recommender is allowed to see from the *current* and *previous* session. In one setup, we could for example reveal the first 2 item views of the current session and all user actions of the preceding session. Using this extra information, it is for example possible to assess the effectiveness of a strategy in which the most recently viewed items are displayed. As a success measure, the Recall can be used which indicates if the purchased item was in the top-k list.

Algorithms and Results. A number of different algorithms were used, including the one-class CF method BPR described in Sect. 4.2.3 as well as the more recent Factorization Machines approach of [100]. These long-term preference modeling approaches were then combined with a number of short-term adaptation strategies, including approaches which recommend (a) the most recently viewed items, (b) items that are content-wise similar to those viewed in the current context, (c) generally popular items, or (d) items that co-occurred with the currently viewed ones in past transactions. Combinations of the different short-term strategies were tested as well.

Similar to the findings reported in [119], the results show that standard CF methods like Factorization Machines do not perform well at all in this evaluation setup and only the BPR method, which has a comparably strong popularity bias, outperforms the popularity-based baseline when no context information is available.

All short-term adaptation strategies on the other hand however immediately led to a strong increase in terms of the Recall even when a weak baseline strategy was used and only the first two item views in a session were revealed. Although the comparison of context-agnostic long-term models and the short-term strategies is in some sense "unfair" as a few more user actions are known to the short-term strategy the strong increase in accuracy helps us to quantify the importance of the adaptation process.

The best-performing method in the end was a hybrid technique which used BPR as a baseline and adapted the recommendation lists by favoring both recently viewed items as well as items whose features are similar to those that were viewed in the current session[11]. In absolute numbers, the Recall of the baseline method of 0.40 was increased to 0.66 through the hybrid method for a configuration in which only the first two item views of the current session and the last two preceding sessions were revealed.

Discussion. Although the short-term adaptations in the experimental analysis were effective, the results also show that the choice of a strong baseline and the capability of understanding the user's long-term preferences are important. On the other hand, while the results of the log-based analysis emphasize the importance of considering short-term interests, it is not fully clear whether the

---

[11] The importance of feature-based similarities was also the basis in [119].

"winning" models fulfill the business goals of the shop owner in the best possible way. The BPR method, for example, can exhibit a comparably strong tendency of recommending popular items and is probably not very helpful when the goal of the recommendation component in a shop is to guide the customers to long-tail items or to help them discover additional or new items in the catalog.

Reminding users of recently viewed items shows to be very effective, e.g., because users might have a tendency to postpone their buying decisions for at least another day in order to sleep on them. However, while the strategy leads to good results in terms of the Recall, it is unclear if the recommendations generate any additional revenue for the shop owner.

In the work in [44], user actions like "add-to-wishlist" or "put-in-cart" were not considered and more work is required to understand (a) how to weight these user actions in comparison to, e.g., view actions and (b) whether or not it is reasonable from an application perspective to remind users on the items in their carts or wishlists.

6.2 Next-Track Music Recommendations from Shared Playlists

Many implicit feedback techniques were developed in the context of media services where the media consumption behavior of users was monitored and analyzed to understand their preferences. A special form of a social action for the users on several media platforms like Last.fm or YouTube is that they can share so-called *playlists* (ordered sequences of music tracks or videos) with others.

In this section, we will highlight some aspects of music playlist generation and in particular approaches that rely on playlists shared by other users. The particularity of this recommendation problem are that

1. shared playlists represent a form of feedback which is not related to one single item but to the whole recommendation list,
2. and that the recommendation outcome is usually not a list of items where the user should find at least one relevant element but rather a list of items which should be sequentially consumed by a user.

A shared playlist has characteristics of explicit and implicit feedback. Regarding the individual tracks, one can usually safely assume that the user who shares the list likes at least most of the individual tracks in the playlist. In addition, the organization of the tracks in a playlist is also an indicator that the collection of tracks leads to a certain listening experience for the user because playlist (mix) creators typically have a certain underlying theme or goal in mind when selecting and ordering the tracks [11].

In this section, we will address the problem of "next-track music recommendation" with a particular focus on approaches that use shared playlists as a social-based input. The general problem setting is that we are given a sequence of recently listened tracks or a "playlist beginning" and the task is to create a continuation. This problem corresponds to generating a virtually endless jukebox or radio station for a given seed track or artist.

The problem can be considered as consisting of two subtasks. The first one is to identify tracks that generally match the playlist beginning, e.g., in terms of the musical genre. The second task is to bring the tracks into a certain order, e.g., to achieve smooth transitions between the tracks or to avoid that multiple tracks of the same artist appear in the playlist.

6.2.1   Identifying the Right Tracks

Various techniques to select good tracks for a playlist continuation have been proposed in the past. They can be distinguished, for example, by their underlying objective when selecting the tracks.

Popularity-Based Approaches. In [78], McFee et al. propose a very simple baseline method in the context of the Million Song Dataset (MSD) challenge[12] called "same artists – greatest hits (SAGH)". The simple idea is to take the set of artists appearing in the listening history of the user and predict that users will listen to the greatest hits (most popular tracks) of the same artists in the future.

A comparison with a baseline based on recommending the generally most popular tracks and the BPR algorithm shows that the simple strategy outperforms the other techniques on different accuracy measures like Mean Average Precision, Precision or the Mean Reciprocal Rank. This suggests that popularity and concentration biases have to be considered. In their dataset, half of the user-track interactions were related to only 3% of all the tracks.

A straightforward extension to the SAGH scheme was proposed for playlist continuations in [10] called "collocated artists – greatest hits" (CAGH). Instead of only playing the greatest hits of the artists in the playlists, the top hits of similar artists are recommended, where the similarity between two artists can be computed based on the co-occurrence of the tracks of the artists on the existing playlists.

Co-listening Events and Nearest-Neighbors. In the Million Song Dataset challenge the task was to continue the listening histories of thousands of users in the test set for which the first half of their history was revealed. The challenge indicates that relying on co-listening events is a particularly well-suited strategy, i.e., quite simple collaborative filtering techniques that look for users who (frequently) listened the same tracks in the past, see e.g., [30], or search for track co-listening patterns that occurred often in the training data worked quite well while matrix factorization approaches did not.

While the MSD challenge focused on full listening histories, playlist generation techniques are often designed to continue an individual given playlist beginning without knowledge about the creator's listening history. The focus in contrast to, e.g., the MSD challenge is to predict the immediate next tracks. For such cases, different forms of considering track co-occurrences were proposed and evaluated in the literature, including for example Sequential Patterns [10], latent Markov Embeddings [81] and *k-nearest-neighbor* (kNN) methods [36].

---

[12] http://labrosa.ee.columbia.edu/millionsong/challenge.

Using the kNN method with large neighborhood sizes is particularly effective. The basic idea when applied to the playlist generation problem is to encode all playlists as binary vectors, in which the vector elements are the available tracks and a vector contains a 1 in case a track appeared in the playlist beginning. Based on this encoding, similar playlists can be found by computing the cosine similarity between the playlist beginning and all training playlists. The *score* of the recommendable track can then be computed based on how often it appears in similar playlists, weighted by the similarity score.

Using Additional Information. The construction of a playlist can be governed by different guiding "themes", which can be related to the artists or genres of the tracks, musical features like tempo or loudness, as well social aspects like the popularity of the tracks.

In [36], the idea is to try to identify the "topic" of a playlist based on the social tags attached to the tracks. Such tagging information can for example be obtained from the Last.fm music platform. The particular idea in [36] is to determine a set of latent topics for each track and then mine the training dataset of playlists for topic transitions. These patterns are then combined with a kNN method and slight increases on the hit rate were observed when small neighborhood sizes for the baseline were used.

A multifaceted track scoring approach was proposed in [45], where again the kNN method was used as a baseline which can be combined with a number of additional scores. Each additional score expresses the match of each track according to a certain criterion, e.g., if the track matches the tempo of the playlist beginning, if it has the same popularity, or if the track has similar tags attached as tracks that the user frequently puts into the playlists. This last aspect can therefore be used to introduce a personalization aspect into the playlist continuation process.

Discussion. Most of the presented approaches were (at least) evaluated in terms of Recall. A common setup is that the last track of each playlist in the test set is hidden. Recall is then measured by determining if the hidden track appears in the top-n list returned by the "playlister". The results for three different sets of playlists shared by users presented in [45] indicate that using a combination of features (co-occurrence, "content" based on social tags, etc.) can be a promising strategy to obtain high accuracy.

6.2.2 Generating Coherent Continuations

Finding suitable tracks to continue the recent listening history or playlist is often only one part of the problem. Achieving a high hit rate (Recall) might therefore not be sufficient as an optimization goal because other criteria like track transitions or homogeneity can be important, too [130]. Furthermore, already a few "wrong" tracks placed in the playlist continuation can have measurable impact on the quality experience of the listener [16].

In the literature, for example, a number of optimization-based approaches have been proposed – see [11] for an overview – which try to create playlists that obey certain user-defined constraints regarding, e.g., the start and end track, the number of different artists etc. A problem of some of these approaches lies in their limited scalability as the optimization problem becomes very complex given that there are usually millions of possible tracks that can be included.

In [130], a method was proposed that combines multiple algorithms with the goal of obtaining a high accuracy while at the same time observing other desired quality factors like diversity, novelty, and in particular serendipity.

Similar goals were in the focus of the work presented in [45], where the authors present a generic optimization-based re-ranking procedure that can be used to balance different optimization goals. The technique takes an accuracy-optimized playlist continuation as input and systematically exchanges top-ranked elements with lower-ranked elements in case the lower-ranked element represents a better "match" for the most recent playlist, e.g., in terms of the tempo or the general topic. An empirical evaluation on different datasets showed that the re-ranking technique not only leads to more homogeneous playlist continuations that better match the playlist beginnings, but also can help to further increase the accuracy of the track selection process.

6.3   Considering Application-Specific Requirements in BPR

BPR is a state-of-the-art ranking algorithm for one-class collaborative filtering situations (Sect. 4.2). Since its original presentation in [101], several variations and extensions were proposed in the literature, e.g., to make the algorithm better suited for certain application requirements. In this section, we will look at some of these proposals in more detail. Among others, we will discuss an approach to counteract the popularity bias of the algorithm and present algorithm extensions to deal with graded relevance feedback.

Besides the discussed methods, multiple other enhancements to BPR were proposed in the literature. The improvements are for example related to including the temporal information, the social connection or the item taxonomy [25,50,102]. Some approaches also extend the two-dimensional user-item perspective of BPR towards additional dimensions [62,76,103]. In terms of the pairwise item-item relations, there are some approaches that introduce the concept of group-wise relations [89,90].

6.3.1   Dealing with the Popularity Bias

Some one-class collaborative filtering algorithms discussed in Sect. 4.2 use different strategies to create artificial negative feedback signals. In most cases, some kind of weighting or sampling scheme is used to derive negative feedback from the structure of the observed interactions. In the wALS, sALS-ENS, Random Graph and BPR approaches, the created negative examples were chosen in a way that was inversely proportional to the popularity of the items, i.e., the algorithms assume that popular items are more acceptable.

While in general this assumption seems plausible, it can however lead to a popularity bias in the recommendations, i.e., the algorithms tend to recommend popular items to everyone. In [46], the authors compared a number of recommendation techniques regarding popularity and concentration criteria and showed that BPR strongly tends to focus on the most popular items. Although popularity-biased recommendations can lead to high values in terms of Precision and Recall [46], the bias might be undesired in specific application settings.

In BPR, the popularity bias emerges from the specific way the algorithm takes samples to learn the preference relations. As discussed in 4.2, the BPR algorithm optimizes the set of model parameters Θ with a stochastic gradient descent procedure by randomly sampling triples (u, i, j) from D_S. The distribution of the observed (positive feedback) signals is typically non-uniform, i.e., the popularity of the items has a long-tail shape. As a result, sampling the triples randomly from all observed interactions leads to many sampled triples (u, i, j) where the item i belongs to the more popular items. The item j, however, is randomly sampled over all items and thus more likely to be from the long-tail of unpopular items. As a result, the gradient descent algorithm updates the model parameters with many triples (u, i, j) that contain a (popular, unpopular) item pair and therefore BPR favors popular items in the recommendation step. In [46], an adapted sampling strategy was introduced that counters the popularity bias of BPR.

Approach. Instead of applying random uniform sampling, a modified distribution function ϕ is used to sample the triples (u, i, j). The non-uniform sampling with a distribution ϕ biases the sampling probability of the items i in a way that more unpopular items are sampled. As a result, the model is updated more often with tuples (u, i, j) where i is less popular. Figure 8 shows the shape of a sampling function ϕ that can be used to sample the positive feedback signals of items i (dashed line) in comparison to the popularity distribution of the item space (solid line). Function ϕ is a monotonously decreasing distribution function that samples more popular items with a lower probability.

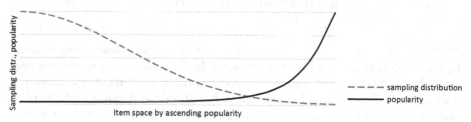

Fig. 8. Sampling distribution for items i with positive feedback compared to their popularity. The x-axis represents the items space by ascending popularity. The y-axis contains the distribution of the item popularity (solid line) and sampling probability of function ϕ (dashed line). Popular items are sampled less frequently.

Different distribution functions for ϕ are possible and in [46] a normal distribution $\phi(\omega)$ with a mean of $\mu = 0$ and a standard deviation $\sigma = \frac{|L_u|}{\omega}$ is used. Here, $|L_u|$ is the number of rated items of user u.

$$\phi(\omega) = \mathcal{N}\left(0, \left(\frac{|L_u|}{\omega}\right)^2\right) \tag{20}$$

The strength of the counter-bias when sampling can be chosen by varying the breadth of the function with the parameter ω, e.g., increasing ω leads to a narrower distribution $\phi(\omega)$ and less popular items are sampled for i. On the other hand, setting $\omega < 1$ leads to a more uniform selection of items.

Results and Discussion. The proposed sampling method [46] was compared with the original implementation by Rendle et al. [101] on the MovieLens and Yahoo movie datasets. The results show that there is an (expected) trade-off between recommendation accuracy and the popularity bias. When increasing the width ω of the sampling distribution, thus focusing on more unpopular items, the *average popularity* and the *Gini index* (to measure the concentration of recommendations) decrease, which means that the algorithm no longer recommends the same set of blockbusters to everyone. At the same time, however, also Precision and Recall decrease, but at a much lower rate. A 5% decrease in accuracy can for example be traded in for a 10% reduction of the overall popularity of the recommended items. The actual size of the desired effect can be determined based on the specific application requirements.

6.3.2   Supporting Graded Relevance Feedback

In many domains, implicit feedback occurs not only as a binary indicator but in a graded form. In the e-commerce domain, for example, different observable user actions like item views or purchases are interest indicators of different strength. The repeated consumption of items in an online media service is another indicator which should be interpreted as a stronger signal than a single consumption event.

In its original form, BPR only supports binary feedback but there are some extensions that allow different graded signal strengths to be considered when learning the preference relations.

BPRC.   To that end, Wang et al. [122] introduced a confidence weight in the objective function of the BPR-OPT criterion. In their *BPR with confidence* approach, the confidence score originates from the problem setting of recommending social network messages and is calculated based on the difference of reception times of two messages. Thus, the optimization criterion is extended as follows to BPRC-OPT:

$$\text{BPRC-OPT} = \sum_{(u,i,j) \in D_S} \ln \sigma(c_{uij} \hat{x}_{uij}) - \lambda_\Theta ||\Theta||^2$$

$$\text{with} \tag{21}$$

$$c_{uij} = \frac{1}{t_i - t_j} : \text{confidence weight}$$

The confidence weight c_{uij} is the inverse of the difference between the reception times t_i and t_j of two messages. If the time between two messages is long, the confidence weight c_{uij} lowers their impact $\hat{x}_{uij}$ in the training phase of the model.

Since the confidence values are given by the application setting, the optimization is analogous to BPR.

The approach was benchmarked against classic kNN, MF and BPR on a dataset from Sina Weibo[13], which is a similar microblogging platform as Twitter. The BPRC approach has the same recommendation characteristics as BPR but has a higher accuracy in terms of Precision and Recall. In addition, the authors report that the confidence-based method greatly outperforms many other algorithms in a cold-start scenario.

ABPR. In [91], a similar generalization of BPR for so-called "heterogeneous implicit feedback" has been proposed. This *Adaptive Bayesian Personalized Ranking* (ABPR) has the ability to model and reduce uncertainty for different types of observed feedback. In their work, the authors have a problem setting with two types of implicit feedback, transactions and examinations, i.e., item purchase events and item click events in an online store. A naive approach would be to assume that both types of user actions are equivalent positive implicit feedback. However, viewing a product page is not necessarily positive feedback, and using only the transactions would result in sparse training data. Therefore, like the confidence-extension for BPR [122], the ABPR approach assumes confidence weights for both types of feedback. The optimization criterion in thus extended to:

$$\text{ABPR-OPT} = \sum_{(u,i,j)\in D_S} f_{uij}^T(c_{ui}, \Theta) + \lambda_E f_{uij}^E(c_{ui}, \Theta) - \lambda_\Theta \|\Theta\|^2$$

with

$$f_{uij}^T, f_{uij}^E : \text{estimation function for transaction, examination}$$

$$c_{ui} : \text{individual confidence weight}$$

$$\lambda_E : \text{global confidence weight}$$

(22)

The optimization criterion now depends on both the estimation of the transactions f_{uij}^T and examinations f_{uij}^E. Furthermore, the impact of the examinations is controlled by a global confidence weight parameter λ_E and the individual confidence weights c_{ui} determine the impact for each triple transaction or examination triple (u, i, j). Compared to the BPRC approach where the messages had a time stamp that determined the confidence, in this setting the weights are not deduced directly from some meta-data and are instead determined in the optimization process. When there is a transaction for user u and item i in the training data, the confidence weight is assumed to be 1. Otherwise it is initially unknown and learned in the extended stochastic gradient descent algorithm discussed in [91]. This is in some sense similar to the graph-based approach for one-class implicit feedback discussed in Sect. 4.2.2.

ABPR was benchmarked on MovieLens and Netflix datasets. In terms of accuracy and ranking metrics, it performs significantly better than classic BPR that uses both types of implicit feedback in a naive way.

---

[13] http://www.weibo.com.

BPR^{++}. In [66], a graded preference relation scheme is introduced to extend the set of triples D_S used for training the model in the gradient descent phase. This new set D_S^{++} includes additional triples based on the graded observed feedback. Similar to the BPRC and ABPR approaches discussed before, the goal of the BPR^{++} technique is to adapt BPR to non-binary feedback, e.g., the confidence in the interaction, the number of times an interaction was observed, the recency of the interaction or the type of an interaction. The enlarged training set is defined as follows:

$$D_S^{++} := \{(u,i,j)|pweight(u,i) > pweight(u,j), i \in I, j \in I\} \qquad (23)$$

The function $pweight(u,i)$ is a real-valued preference weight function that models the strength of the interaction between user u and item i, e.g., confidence, time or rating. If there is no interaction between u and i, the preference weight is 0. Compared to the original set of training triples as shown before in Eq. 2, the extended set D_S^{++} contains all triples of D_S. In addition, triples that would have been ignored in BPR because both items had observed feedback can now appear in D_S^{++} if they have a different preference weight.

The number of these additional triples $D_S^{++} \setminus D_S$ is however comparably small which means that these triples will not be often considered when using the random sampling strategy of BPR. The authors therefore introduce a weighted sampling approach, similar to the one presented in [46], which increases the sampling probability for the new triples in $D_S^{++} \setminus D_S$.

The BPR^{++} approach was benchmarked against the original BPR technique on two e-commerce datasets with implicit feedback and a MovieLens dataset with ratings. Different preference weight functions where used in the experiments to model the strength of the relevance including the interaction time, the number of interactions and – for the MovieLens data – the ratings. The results show that on the implicit feedback datasets the use of the interaction time with BPR^{++} significantly increased the accuracy (Precision@10 and Recall@10) when compared to BPR. On the MovieLens dataset this is also true when the preference strength is modeled by ratings. The number of interactions, however, did not have an impact on the recommendation accuracy.

Besides being able to improve the prediction accuracy, the adapted sampling strategy helps reducing the time needed for the gradient descent procedure to converge.

7   Summary and Outlook

Historically, fueled by competitions and publicly available datasets, research on recommender systems was largely focused on applications that are based on explicit user feedback. Only in recent years approaches based on implicit feedback have received more attention due to their high practical relevance. This chapter provided an overview on the use of implicit feedback in recommender systems. We presented examples of typical application scenarios and extended an established categorization of observable behavior for new domains that emerged with the Social Web and ubiquitous systems. The chapter furthermore reviewed state-of-the-art algorithmic approaches for implicit feedback as well as typical challenges that arise when dealing with implicit preference signals.

As an outlook on future developments all three key challenges identified by [58] particularly apply to recommendation systems operating on implicit user feedback: *scalability, better exploitation of user-contributed content* and *research infrastructure*, meaning more efficient mechanisms to evaluate the suitability of research contributions for actual live systems.

In our view, in particular the third point of assessing the practical value contribution of theoretical concepts and algorithms for the interaction with real users creates the most diverse opportunities for future research. A recommender system based on explicit feedback requires the user to actively input feedback to a system in order to receive recommendations in return. In an implicit feedback recommender the user is monitored by a system that might therefore even proactively provide recommendations based on its observation of the user behavior and assumptions about the user's next goals. Thus, not only aspects of user experience but also the perceived intrusiveness receives particular importance in the context of exploiting implicit user feedback in real-world systems.

In general, the issue of the internal validity and reproducibility of research results is of particular relevance for research contributions on implicit user feedback. In [106] serious doubts about the internal validity of published results on explicit feedback datasets and their reproducibility with different algorithm libraries have been raised. Moreover, works on implicit feedback datasets often use proprietary data and commonly available benchmark datasets and algorithm libraries are still missing. Thus, we conclude this chapter with the hope that it will help to stimulate further progress on these lines.

References

1. Abel, F., Gao, Q., Houben, G.-J., Tao, K.: Semantic enrichment of Twitter posts for user profile construction on the social web. In: Antoniou, G., Grobelnik, M., Simperl, E., Parsia, B., Plexousakis, D., De Leenheer, P., Pan, J. (eds.) ESWC 2011, Part II. LNCS, vol. 6644, pp. 375–389. Springer, Heidelberg (2011). https://doi.org/10.1007/978-3-642-21064-8_26
2. Agrawal, R., Imieliński, T., Swami, A.: Mining association rules between sets of items in large databases. In: Proceedings of the 1993 ACM SIGMOD International Conference on Management of Data, ACM SIGMOD 1993, pp. 207–216 (1993)
3. Agrawal, R., Srikant, R.: Mining sequential patterns. In: Proceedings of the Eleventh International Conference on Data Engineering, ICDE 1995, pp. 3–14 (1995)
4. Ali, K., van Stam, W.: TiVo: making show recommendations using a distributed collaborative filtering architecture. In: Proceedings of the Tenth ACM SIGKDD International Conference on Knowledge Discovery and Data Mining, KDD 2004, pp. 394–401 (2004)
5. Armentano, M.G., Godoy, D., Amandi, A.: Topology-based recommendation of users in micro-blogging communities. J. Comput. Sci. Technol. **27**(3), 624–634 (2012)

6. Armstrong, R., Freitag, D., Joachims, T., Mitchell, T.: WebWatcher: a learning apprentice for the world wide web. In: AAAI Technical Report SS-95-08, pp. 6–12 (1995)

7. Ben-David, S., Lindenbaum, M.: Learning distributions by their density levels: a paradigm for learning without a teacher. J. Comput. Syst. Sci. **55**, 171–182 (1997)

8. Bogers, T.: Tag-based recommendation. In: Brusilovsky, P., He, D. (eds.) Social Information Access. LNCS, vol. 10100, pp. 441–479. Springer, Cham (2018)

9. Bohnert, F., Zukerman, I.: Personalised viewing-time prediction in museums. User Model. User-Adapt. Interact. **24**(4), 263–314 (2014)

10. Bonnin, G., Jannach, D.: Evaluating the quality of playlists based on hand-crafted samples. In: Proceedings of the 2013 International Society for Music Information Retrieval, ISMIR 2013, pp. 263–268 (2013)

11. Bonnin, G., Jannach, D.: Automated generation of music playlists: survey and experiments. ACM Comput. Surv. **47**(2), 26:1–26:35 (2014)

12. Bothorel, C., Lathia, N., Picot-Clemente, R., Noulas, A.: Location recommendation with social media data. In: Brusilovsky, P., He, D. (eds.) Social Information Access. LNCS, vol. 10100, pp. 624–653. Springer, Cham (2018)

13. Budzik, J., Hammond, K.: Watson: anticipating and contextualizing information needs. In: 62nd Annual Meeting of the American Society for Information Science, pp. 727–740 (1999)

14. Carmel, D., Zwerdling, N., Guy, I., Ofek-Koifman, S., Har'El, N., Ronen, I., Uziel, E., Yogev, S., Chernov, S.: Personalized social search based on the user's social network. In: Proceedings of the 18th ACM Conference on Information and Knowledge Management, CIKM 2009, pp. 1227–1236 (2009)

15. Castagnos, S., Jones, N., Pu, P.: Eye-tracking product recommenders' usage. In: Proceedings of the 2010 ACM Conference on Recommender Systems, RecSys 2010, pp. 29–36 (2010)

16. Chau, P.Y.K., Ho, S.Y., Ho, K.K.W., Yao, Y.: Examining the effects of malfunctioning personalized services on online users' distrust and behaviors. Decis. Support Syst. **56**, 180–191 (2013)

17. Chen, W.Y., Chu, J.C., Luan, J., Bai, H., Wang, Y., Chang, E.Y.: Collaborative filtering for Orkut communities: discovery of user latent behavior. In: Proceedings of the 18th International Conference on World Wide Web, WWW 2009, pp. 681–690 (2009)

18. Cheng, Z., Shen, J.: Just-for-me: an adaptive personalization system for location-aware social music recommendation. In: Proceedings of International Conference on Multimedia Retrieval, ICMR 2014, pp. 185:185–185:192 (2014)

19. Claypool, M., Le, P., Wased, M., Brown, D.: Implicit interest indicators. In: Proceedings of the 6th ACM International Conference on Intelligent User Interfaces, IUI 2001, pp. 33–40 (2001)

20. Cremonesi, P., Koren, Y., Turrin, R.: Performance of recommender algorithms on top-N recommendation tasks. In: Proceedings of the 2010 ACM Conference on Recommender Systems, RecSys 2010, pp. 39–46 (2010)

21. Davidson, J., Liebald, B., Liu, J., Nandy, P., Van Vleet, T., Gargi, U., Gupta, S., He, Y., Lambert, M., Livingston, B., Sampath, D.: The YouTube video recommendation system. In: Proceedings of the 2010 ACM Conference on Recommender Systems, RecSys 2010, pp. 293–296 (2010)

22. Davis, J., Goadrich, M.: The relationship between precision-recall and ROC curves. In: Proceedings of the 23rd International Conference on Machine Learning, ICML 2006, pp. 233–240 (2006)

23. De Pessemier, T., Dooms, S., Martens, L.: Context-aware recommendations through context and activity recognition in a mobile environment. Multimed. Tools Appl. **72**(3), 2925–2948 (2014)

24. Denis, F.: PAC learning from positive statistical queries. In: Richter, M.M., Smith, C.H., Wiehagen, R., Zeugmann, T. (eds.) ALT 1998. LNCS (LNAI), vol. 1501, pp. 112–126. Springer, Heidelberg (1998). https://doi.org/10.1007/3-540-49730-7_9

25. Du, L., Li, X., Shen, Y.-D.: User graph regularized pairwise matrix factorization for item recommendation. In: Tang, J., King, I., Chen, L., Wang, J. (eds.) ADMA 2011, Part II. LNCS (LNAI), vol. 7121, pp. 372–385. Springer, Heidelberg (2011). https://doi.org/10.1007/978-3-642-25856-5_28

26. Gadanho, S.C., Lhuillier, N.: Addressing uncertainty in implicit preferences. In: Proceedings of the 2007 ACM Conference on Recommender Systems, RecSys 2007, pp. 97–104 (2007)

27. Garcia, E., Romero, C., Ventura, S., Castro, C.: An architecture for making recommendations to courseware authors using association rule mining and collaborative filtering. User Model. User-Adapt. Interact. **19**(1–2), 99–132 (2009)

28. Gedikli, F., Jannach, D.: Improving recommendation accuracy based on item-specific tag preferences. ACM Trans. Intell. Syst. Technol. **4**(1), 11:1–11:19 (2013)

29. Gil, M., Pelechano, V.: Exploiting user feedback for adapting mobile interaction obtrusiveness. In: Bravo, J., López-de-Ipiña, D., Moya, F. (eds.) UCAmI 2012. LNCS, vol. 7656, pp. 274–281. Springer, Heidelberg (2012). https://doi.org/10.1007/978-3-642-35377-2_38

30. Glazyrin, N.: Music recommendation system for million song dataset challenge. CoRR abs/1209.3286 (2012)

31. Guy, I.: People recommendation on social media. In: Brusilovsky, P., He, D. (eds.) Social Information Access. LNCS, vol. 10100, pp. 570–623. Springer, Cham (2018)

32. Guy, I., Zwerdling, N., Carmel, D., Ronen, I., Uziel, E., Yogev, S., Ofek-Koifman, S.: Personalized recommendation of social software items based on social relations. In: Proceedings of the 2009 ACM Conference on Recommender Systems, RecSys 2009, pp. 53–60 (2009)

33. Guy, I., Zwerdling, N., Ronen, I., Carmel, D., Uziel, E.: Social media recommendation based on people and tags. In: Proceedings of the 33rd International ACM SIGIR Conference on Research and Development in Information Retrieval, SIGIR 2010, pp. 194–201 (2010)

34. Han, J., Pei, J., Yin, Y.: Mining frequent patterns without candidate generation. In: Proceedings of the 2000 ACM SIGMOD International Conference on Management of Data, SIGMOD 2000, pp. 1–12 (2000)

35. Han, S., He, D.: Network-based social search. In: Brusilovsky, P., He, D. (eds.) Social Information Access. LNCS, vol. 10100, pp. 277–309. Springer, Cham (2018)

36. Hariri, N., Mobasher, B., Burke, R.: Context-aware music recommendation based on latent topic sequential patterns. In: Proceedings of the 2012 ACM Conference on Recommender Systems, RecSys 2012, pp. 131–138 (2012)

37. Hensley, C.B.: Selective dissemination of information (SDI). In: Proceedings of the 1963 Spring Joint Computer Conference, AFIPS 1963, pp. 257–262 (1963)

38. Herlocker, J.L., Konstan, J.A., Terveen, L.G., Riedl, J.T.: Evaluating collaborative filtering recommender systems. ACM Trans. Inf. Syst. **22**(1), 5–53 (2004)

39. Hill, W.C., Hollan, J.D., Wroblewski, D., McCandless, T.: Edit wear and read wear. In: Proceedings of the ACM SIGCHI Conference on Human Factors in Computing Systems, CHI 1992, pp. 3–9 (1992)

40. Höök, K., Benyon, D., Munro, A.J.: Designing Information Spaces: The Social Navigation Approach. Springer Science & Business Media, London (2012). https://doi.org/10.1007/978-1-4471-0035-5
41. Hu, Y., Koren, Y., Volinsky, C.: Collaborative filtering for implicit feedback datasets. In: Proceedings of the 2008 Eighth IEEE International Conference on Data Mining, ICDM 2008, pp. 263–272 (2008)
42. Jannach, D., Hegelich, K.: A case study on the effectiveness of recommendations in the mobile internet. In: Proceedings of the 2009 ACM Conference on Recommender Systems, RecSys 2009, pp. 205–208 (2009)
43. Jannach, D., Karakaya, Z., Gedikli, F.: Accuracy improvements for multi-criteria recommender systems. In: Proceedings of the 13th ACM Conference on Electronic Commerce, EC 2012, pp. 674–689 (2012)
44. Jannach, D., Lerche, L., Jugovac, M.: Adaptation and evaluation of recommendations for short-term shopping goals. In: Proceedings of the 2015 ACM Conference on Recommender Systems, RecSys 2015, pp. 211–218 (2015)
45. Jannach, D., Lerche, L., Kamehkhosh, I.: Beyond "hitting the hits": generating coherent music playlist continuations with the right tracks. In: Proceedings of the 2015 ACM Conference on Recommender Systems, RecSys 2015, pp. 187–194 (2015)
46. Jannach, D., Lerche, L., Kamehkhosh, I., Jugovac, M.: What recommenders recommend: an analysis of recommendation biases and possible countermeasures. User Model. User-Adapt. Interact. **25**(5), 427–491 (2015)
47. Jannach, D., Zanker, M., Ge, M., Gröning, M.: Recommender systems in computer science and information systems – a landscape of research. In: Huemer, C., Lops, P. (eds.) EC-Web 2012. LNBIP, vol. 123, pp. 76–87. Springer, Heidelberg (2012). https://doi.org/10.1007/978-3-642-32273-0_7
48. Jawaheer, G., Szomszor, M., Kostkova, P.: Comparison of implicit and explicit feedback from an online music recommendation service. In: Proceedings of the 1st International Workshop on Information Heterogeneity and Fusion in Recommender Systems, HetRec 2010, pp. 47–51 (2010)
49. Jawaheer, G., Weller, P., Kostkova, P.: Modeling user preferences in recommender systems: A classification framework for explicit and implicit user feedback. ACM Trans. Interact. Intell. Syst. **4**(2), 8:1–8:26 (2014)
50. Kanagal, B., Ahmed, A., Pandey, S., Josifovski, V., Yuan, J., Garcia-Pueyo, L.: Supercharging recommender systems using taxonomies for learning user purchase behavior. Proc. VLDB Endow. **5**(10), 956–967 (2012)
51. Ke, T., Yang, B., Zhen, L., Tan, J., Li, Y., Jing, L.: Building high-performance classifiers using positive and unlabeled examples for text classification. In: Wang, J., Yen, G.G., Polycarpou, M.M. (eds.) ISNN 2012, Part II. LNCS, vol. 7368, pp. 187–195. Springer, Heidelberg (2012). https://doi.org/10.1007/978-3-642-31362-2_21
52. Kelly, D., Teevan, J.: Implicit feedback for inferring user preference: a bibliography. SIGIR Forum **37**(2), 18–28 (2003)
53. Khalili, A., Wu, C., Aghajan, H.: Autonomous learning of user's preference of music and light services in smart home applications. In: Behavior Monitoring and Interpretation Workshop at German AI Conference (2009)
54. Kliegr, T., Kuchar, J.: Orwellian eye: video recommendation with Microsoft Kinect. In: Proceedings 21st European Conference on Artificial Intelligence, ECAI/PAIS 2014, pp. 1227–1228 (2014)

55. Kluver, D., Ekstrand, M., Konstan, J.: Rating-based collaborative filtering: algorithms and evaluation. In: Brusilovsky, P., He, D. (eds.) Social Information Access. LNCS, vol. 10100, pp. 344–390. Springer, Cham (2018)
56. Kobsa, A., Koenemann, J., Pohl, W.: Personalised hypermedia presentation techniques for improving online customer relationships. Knowl. Eng. Rev. **16**(2), 111–155 (2001)
57. Konstan, J.A., Miller, B.N., Maltz, D., Herlocker, J.L., Gordon, L.R., Riedl, J.: GroupLens: applying collaborative filtering to usenet news. Commun. ACM **40**(3), 77–87 (1997)
58. Konstan, J.A., Riedl, J.: Recommender systems: from algorithms to user experience. User Model. User-Adapt. Interact. **22**(1–2), 101–123 (2012)
59. Kordumova, S., Kostadinovska, I., Barbieri, M., Pronk, V., Korst, J.: Personalized implicit learning in a music recommender system. In: De Bra, P., Kobsa, A., Chin, D. (eds.) UMAP 2010. LNCS, vol. 6075, pp. 351–362. Springer, Heidelberg (2010). https://doi.org/10.1007/978-3-642-13470-8_32
60. Koren, Y.: Factorization meets the neighborhood: a multifaceted collaborative filtering model. In: Proceedings of the 14th ACM SIGKDD International Conference on Knowledge Discovery and Data Mining, KDD 2008, pp. 426–434 (2008)
61. Koren, Y.: Factor in the neighbors: scalable and accurate collaborative filtering. ACM Trans. Knowl. Discov. Data (TKDD) **4**(1), 1:1–1:24 (2010)
62. Krohn-Grimberghe, A., Drumond, L., Freudenthaler, C., Schmidt-Thieme, L.: Multi-relational matrix factorization using Bayesian personalized ranking for social network data. In: Proceedings of the Fifth ACM International Conference on Web Search and Data Mining, WSDM 2012, pp. 173–182 (2012)
63. Lee, D., Brusilovsky, P.: Recommendations based on social links. In: Brusilovsky, P., He, D. (eds.) Social Information Access. LNCS, vol. 10100, pp. 391–440. Springer, Cham (2018)
64. Lee, D., Park, S., Kahng, M., Lee, S., Lee, S.: Exploiting contextual information from event logs for personalized recommendation. In: Lee, R. (ed.) Computer and Information Science 2010. SCI, vol. 317, pp. 121–139. Springer, Heidelberg (2010). https://doi.org/10.1007/978-3-642-15405-8_11
65. Lee, T.Q., Park, Y., Park, Y.T.: A time-based approach to effective recommender systems using implicit feedback. Expert Syst. Appl. **34**(4), 3055–3062 (2008)
66. Lerche, L., Jannach, D.: Using graded implicit feedback for Bayesian personalized ranking. In: Proceedings of the 2014 ACM Conference on Recommender Systems, RecSys 2014, pp. 353–356 (2014)
67. Lerchenmüller, B., Wörndl, W.: Inference of user context from GPS logs for proactive recommender systems. In: AAAI Workshop Activity Context Representation: Techniques and Languages. AAAI Technical Report WS-12-05 (2012)
68. Li, H., Wang, Y., Zhang, D., Zhang, M., Chang, E.Y.: PFP: parallel FP-growth for query recommendation. In: Proceedings of the 2008 ACM Conference on Recommender Systems, RecSys 2008, pp. 107–114 (2008)
69. Li, X.-L., Liu, B.: Learning from positive and unlabeled examples with different data distributions. In: Gama, J., Camacho, R., Brazdil, P.B., Jorge, A.M., Torgo, L. (eds.) ECML 2005. LNCS (LNAI), vol. 3720, pp. 218–229. Springer, Heidelberg (2005). https://doi.org/10.1007/11564096_24
70. Lieberman, H.: Letizia: an agent that assists web browsing. In: Proceedings of the 14th International Joint Conference on Artificial Intelligence, IJCAI 1995, vol. 1, pp. 924–929 (1995)

71. Lin, J., Sugiyama, K., Kan, M.Y., Chua, T.S.: Addressing cold-start in app recommendation: latent user models constructed from Twitter followers. In: Proceedings of the 36th International ACM SIGIR Conference on Research and Development in Information Retrieval, SIGIR 2013, pp. 283–292 (2013)

72. Lin, K.H., Chung, K.H., Lin, K.S., Chen, J.S.: Face recognition-aided IPTV group recommender with consideration of serendipity. Int. J. Future Comput. Commun. 3(2), 141–147 (2014)

73. Lin, W., Alvarez, S.A., Ruiz, C.: Efficient adaptive-support association rule mining for recommender systems. Data Min. Knowl. Disc. 6(1), 83–105 (2002)

74. Linden, G., Smith, B., York, J.: Amazon.com recommendations: item-to-item collaborative filtering. IEEE Internet Comput. 7(1), 76–80 (2003)

75. Liu, N.N., Xiang, E.W., Zhao, M., Yang, Q.: Unifying explicit and implicit feedback for collaborative filtering. In: Proceedings of the 19th ACM International Conference on Information and Knowledge Management, CIKM 2010, pp. 1445–1448 (2010)

76. Liu, Y., Zhao, P., Sun, A., Miao, C.: AdaBPR: a boosting algorithm for item recommendation with implicit feedback. In: Proceedings of the 24th International Joint Conference on Artificial Intelligence, IJCAI 2015 (2015)

77. Marlin, B.M., Zemel, R.S., Roweis, S., Slaney, M.: Collaborative filtering and the missing at random assumption. In: Proceedings of the 23rd Conference on Uncertainty in Artificial Intelligence, UAI 2007, pp. 267–275 (2007)

78. McFee, B., Bertin-Mahieux, T., Ellis, D.P., Lanckriet, G.R.: The million song dataset challenge. In: Proceedings of the 21st International Conference Companion on World Wide Web, WWW 2012 Companion, pp. 909–916 (2012)

79. Mladenic, D.: Personal WebWatcher: design and implementation. Technical report IJS-DP-7472. J. Stefan Institute, Slovenia (1996)

80. Mobasher, B., Cooley, R., Srivastava, J.: Automatic personalization based on web usage mining. Commun. ACM 43(8), 142–151 (2000)

81. Moore, J.L., Chen, S., Joachims, T., Turnbull, D.: Learning to embed songs and tags for playlist prediction. In: Proceedings of the 2012 International Society for Music Information Retrieval Conference, ISMIR 2012, pp. 349–354 (2012)

82. Morita, M., Shinoda, Y.: Information filtering based on user behavior analysis and best match text retrieval. In: Proceedings of the 17th Annual International ACM SIGIR Conference on Research and Development in Information Retrieval, SIGIR 1994, pp. 272–281 (1994)

83. Nichols, D.M.: Implicit rating and filtering. In: Proceedings of 5th DELOS Workshop on Filtering and Collaborative Filtering (1997)

84. Oard, D.W.: Modeling information content using observable behavior. In: Proceedings of the 64th Annual Conference of the American Society for Information Science and Technology (2001)

85. Oard, D.W., Kim, J.: Implicit feedback for recommender systems. In: Proceedings of the 1998 AAAI Workshop on Recommender Systems (1998)

86. O'Mahoney, M., Smyth, B.: From opinions to recommendations. In: Brusilovsky, P., He, D. (eds.) Social Information Access. LNCS, vol. 10100, pp. 480–509. Springer, Cham (2018)

87. Palanivel, K., Sivakumar, R.: A study on implicit feedback in multicriteria e-commerce recommender system. J. Electron. Commer. Res. 11(2), 140–156 (2010)

88. Pan, R., Zhou, Y., Cao, B., Liu, N.N., Lukose, R., Scholz, M., Yang, Q.: One-class collaborative filtering. In: Proceedings of the 2008 Eighth IEEE International Conference on Data Mining, ICDM 2008, pp. 502–511 (2008)

89. Pan, W., Chen, L.: CoFiSet: collaborative filtering via learning pairwise preferences over item-sets. In: Proceedings of the 2013 SIAM International Conference on Data Mining, pp. 180–188 (2013)

90. Pan, W., Chen, L.: GBPR: group preference based Bayesian personalized ranking for one-class collaborative filtering. In: Proceedings of the 23rd International Joint Conference on Artificial Intelligence, IJCAI 2013, pp. 2691–2697 (2013)

91. Pan, W., Zhong, H., Xu, C., Ming, Z.: Adaptive Bayesian personalized ranking for heterogeneous implicit feedbacks. Knowl.-Based Syst. **73**, 173–180 (2015)

92. Paquet, U., Koenigstein, N.: One-class collaborative filtering with random graphs. In: Proceedings of the 22nd International Conference on World Wide Web, WWW 2013, pp. 999–1008 (2013)

93. Parra, D., Amatriain, X.: Walk the talk: analyzing the relation between implicit and explicit feedback for preference elicitation. In: Proceedings of the 19th ACM International Conference on User Modeling, Adaptation, and Personalization, UMAP 2011, pp. 255–268 (2011)

94. Parra, D., Karatzoglou, A., Yavuz, I., Amatriain, X.: Implicit feedback recommendation via implicit-to-explicit ordinal logistic regression mapping. In: Proceedings of the CARS 2011 (2011)

95. Partridge, K., Price, B.: Enhancing mobile recommender systems with activity inference. In: Houben, G.-J., McCalla, G., Pianesi, F., Zancanaro, M. (eds.) UMAP 2009. LNCS, vol. 5535, pp. 307–318. Springer, Heidelberg (2009). https://doi.org/10.1007/978-3-642-02247-0_29

96. Pennacchiotti, M., Gurumurthy, S.: Investigating topic models for social media user recommendation. In: Proceedings of the 20th International Conference Companion on World Wide Web, WWW 2011, pp. 101–102 (2011)

97. Pilászy, I., Zibriczky, D., Tikk, D.: Fast ALS-based matrix factorization for explicit and implicit feedback datasets. In: Proceedings of the 2010 ACM Conference on Recommender Systems, RecSys 2010, pp. 71–78 (2010)

98. Pitman, A., Zanker, M.: An empirical study of extracting multidimensional sequential rules for personalization and recommendation in online commerce. In: 10. Internationale Tagung Wirtschaftinformatik, pp. 180–189 (2011)

99. Pizzato, L., Chung, T., Rej, T., Koprinska, I., Yecef, K., Kay, J.: Learning user preferences in online dating. In: European Conference on Machine Learning and Principles and Practice of Knowledge Discovery in Databases, ECML-PKDD, Preference Learning Workshop (2010)

100. Rendle, S.: Factorization machines with libFM. ACM Trans. Intell. Syst. Technol. **3**(3), 57:1–57:22 (2012)

101. Rendle, S., Freudenthaler, C., Gantner, Z., Schmidt-Thieme, L.: BPR: Bayesian personalized ranking from implicit feedback. In: Proceedings of the Twenty-Fifth Conference on Uncertainty in Artificial Intelligence, UAI 2009, pp. 452–461 (2009)

102. Rendle, S., Freudenthaler, C., Schmidt-Thieme, L.: Factorizing personalized Markov chains for next-basket recommendation. In: Proceedings of the 19th International Conference on World Wide Web, WWW 2010, pp. 811–820. ACM (2010)

103. Rendle, S., Schmidt-Thieme, L.: Pairwise interaction tensor factorization for personalized tag recommendation. In: Proceedings of the Third ACM International Conference on Web Search and Data Mining, WSDM 2010, pp. 81–90 (2010)

104. Rodden, K., Fu, X., Aula, A., Spiro, I.: Eye-mouse coordination patterns on web search results pages. In: Extended Abstracts on Human Factors in Computing Systems, CHI EA 2008, pp. 2997–3002 (2008)

105. Roth, M., Ben-David, A., Deutscher, D., Flysher, G., Horn, I., Leichtberg, A., Leiser, N., Matias, Y., Merom, R.: Suggesting friends using the implicit social graph. In: Proceedings of the 16th ACM SIGKDD International Conference on Knowledge Discovery and Data Mining, KDD 2010, pp. 233–242 (2010)

106. Said, A., Bellogín, A.: Comparative recommender system evaluation: benchmarking recommendation frameworks. In: Proceedings of the 2014 ACM Conference on Recommender Systems, RecSys 2014, pp. 129–136 (2014)

107. Sakagami, H., Kamba, T.: Learning personal preferences on online newspaper articles from user behaviors. Comput. Netw. ISDN Syst. **29**(8–13), 1447–1455 (1997)

108. Sarwar, B., Karypis, G., Konstan, J., Riedl, J.: Analysis of recommendation algorithms for e-commerce. In: Proceedings of the 2nd ACM Conference on Electronic Commerce, EC 2000, pp. 158–167 (2000)

109. Schölkopf, B., Platt, J.C., Shawe-Taylor, J.C., Smola, A.J., Williamson, R.C.: Estimating the support of a high-dimensional distribution. Neural Comput. **13**(7), 1443–1471 (2001)

110. Sen, S., Vig, J., Riedl, J.: Tagommenders: connecting users to items through tags. In: Proceedings of the 18th International World Wide Web Conference, WWW 2009, pp. 671–680 (2009)

111. Shani, G., Heckerman, D., Brafman, R.I.: An MDP-based recommender system. J. Mach. Learn. Res. **6**, 1265–1295 (2005)

112. Shapira, B., Taieb-Maimon, M., Moskowitz, A.: Study of the usefulness of known and new implicit indicators and their optimal combination for accurate inference of users interests. In: Proceedings of the 2006 ACM Symposium on Applied Computing, SAC 2006, pp. 1118–1119 (2006)

113. Shi, Y., Karatzoglou, A., Baltrunas, L., Larson, M., Hanjalic, A.: xCLiMF: optimizing expected reciprocal rank for data with multiple levels of relevance. In: Proceedings of the 2013 ACM Conference on Recommender Systems, RecSys 2013, pp. 431–434 (2013)

114. Shi, Y., Karatzoglou, A., Baltrunas, L., Larson, M., Oliver, N., Hanjalic, A.: CLiMF: learning to maximize reciprocal rank with collaborative less-is-more filtering. In: Proceedings of the 2012 ACM Conference on Recommender Systems, RecSys 2012, pp. 139–146 (2012)

115. Sohn, T., Li, K.A., Griswold, W.G., Hollan, J.D.: A diary study of mobile information needs. In: Proceedings of the ACM SIGCHI Conference on Human Factors in Computing Systems, CHI 2008, pp. 433–442 (2008)

116. Srebro, N., Jaakkola, T., et al.: Weighted low-rank approximations. In: Proceedings of the Twentieth International Conference on Machine Learning, ICML 2003, pp. 720–727 (2003)

117. Sriram, B., Fuhry, D., Demir, E., Ferhatosmanoglu, H., Demirbas, M.: Short text classification in Twitter to improve information filtering. In: Proceedings of the 33rd International ACM SIGIR Conference on Research and Development in Information Retrieval, SIGIR 2010, pp. 841–842 (2010)

118. Tapia, E.M., Intille, S.S., Larson, K.: Activity recognition in the home using simple and ubiquitous sensors. In: Ferscha, A., Mattern, F. (eds.) Pervasive 2004. LNCS, vol. 3001, pp. 158–175. Springer, Heidelberg (2004). https://doi.org/10.1007/978-3-540-24646-6_10

119. Tavakol, M., Brefeld, U.: Factored MDPs for detecting topics of user sessions. In: Proceedings of the 2014 ACM Conference on Recommender Systems, RecSys 2014, pp. 33–40 (2014)

120. Véras, D., Prota, T., Bispo, A., Prudêncio, R., Ferraz, C.: A literature review of recommender systems in the television domain. Expert Syst. Appl. **42**(22), 9046–9076 (2015)
121. Wang, J., Zhu, J.: On statistical analysis and optimization of information retrieval effectiveness metrics. In: Proceedings of the 33rd International ACM SIGIR Conference on Research and Development in Information Retrieval, SIGIR 2010, pp. 226–233 (2010)
122. Wang, S., Zhou, X., Wang, Z., Zhang, M.: Please spread: recommending tweets for retweeting with implicit feedback. In: Proceedings of the 2012 Workshop on Data-Driven User Behavioral Modelling and Mining from Social Media, DUBMMSM 2012, pp. 19–22 (2012)
123. Ward, G., Hastie, T., Barry, S., Elith, J., Leathwick, J.R.: Presence-only data and the EM algorithm. Biometrics **65**(2), 554–563 (2009)
124. White, R.W., Ruthven, I., Jose, J.M.: A study of factors affecting the utility of implicit relevance feedback. In: Proceedings of the 28th Annual International ACM SIGIR Conference on Research and Development in Information Retrieval, pp. 35–42 (2005)
125. Yajima, Y.: One-class support vector machines for recommendation tasks. In: Ng, W.-K., Kitsuregawa, M., Li, J., Chang, K. (eds.) PAKDD 2006. LNCS (LNAI), vol. 3918, pp. 230–239. Springer, Heidelberg (2006). https://doi.org/10.1007/11731139_28
126. Yu, H., Han, J., Chang, K.C.C.: PEBL: positive example based learning for web page classification using SVM. In: Proceedings of the Eighth ACM SIGKDD International Conference on Knowledge Discovery and Data Mining, KDD 2002, pp. 239–248 (2002)
127. Yu, Z., Zhou, X.: TV3P: an adaptive assistant for personalized TV. IEEE Trans. Consum. Electron. **50**(1), 393–399 (2004)
128. Zanker, M., Jessenitschnig, M.: Collaborative feature-combination recommender exploiting explicit and implicit user feedback. In: Proceedings of the IEEE Conference on Commerce and Enterprise Computing, CEC 2009, pp. 49–56 (2009)
129. Zhang, Y., Callan, J.: Combining multiple forms of evidence while filtering. In: Proceedings of the Conference on Human Language Technology and Empirical Methods in Natural Language Processing, HLT 2005, pp. 587–595 (2005)
130. Zhang, Y.C., Séaghdha, D.O., Quercia, D., Jambor, T.: Auralist: introducing serendipity into music recommendation. In: Proceedings of the Fifth ACM International Conference on Web Search and Data Mining, WSDM 2012, pp. 13–22 (2012)
131. Zhao, S., Du, N., Nauerz, A., Zhang, X., Yuan, Q., Fu, R.: Improved recommendation based on collaborative tagging behaviors. In: Proceedings of the 13th ACM International Conference on Intelligent User Interfaces, IUI 2008, pp. 413–416 (2008)
132. Zheng, V.W., Zheng, Y., Xie, X., Yang, Q.: Towards mobile intelligence: learning from GPS history data for collaborative recommendation. Artif. Intell. **184/185**, 17–37 (2012)
133. Zibriczky, D., Hidasi, B., Petres, Z., Tikk, D.: Personalized recommendation of linear content on interactive TV platforms: beating the cold start and noisy implicit user feedback. In: ACM Workshop on TV and Multimedia Personalization, UMAP 2012 (2012)
134. Zigoris, P., Zhang, Y.: Bayesian adaptive user profiling with explicit & implicit feedback. In: Proceedings of the 15th ACM International Conference on Information and Knowledge Management, CIKM 2006, pp. 397–404 (2006)

15
People Recommendation on Social Media

Ido Guy[1,2]([✉]) [iD]

[1] Ben-Gurion University of the Negev, Beer Sheva, Israel
idoguy@acm.org
[2] eBay Research, Netanya, Israel

Abstract. The social web has brought about many new types of recommender systems. One of the most important is recommendation of
people, which bears many unique characteristics and challenges. In this
chapter, we will review much of the research that has studied people
recommendation in social media. The three main types of people recommendation are based on the presumed level of relationship of the user
with the recommended individuals and thereby the goal of the recommendation: from recommending familiar people the user may invite to
their network or meet at a place, through recommending interesting people the user may subscribe to or follow, to recommending similar people
the user may want to get familiarize with. We will demonstrate each of
these recommendation types and the techniques used to address them
through different case studies. We will also discuss related research areas,
summarize key aspects, and suggest future directions.

Keywords: Followee recommendation · Friend recommendation
People recommendation · People recommender systems
Profile matching · Recommending people · Social matching
Stranger recommendation

1 Introduction

People recommendation is one of the most important and fascinating types of recommendation within the broader domain of social recommender systems (SRSs),
i.e., recommender systems that target the social media domain [39, 42, 43]. There
are many reasons as for why recommendation of people to themselves deserves
special attention, holds distinctive characteristics, and poses unique challenges
compared to other domain of recommender systems. Terveen and Mcdonald
[110] were the first to discuss in depth why recommendation of people to people, which they termed "social matching" are of special interest and earn their
own "domain". They distinguished people recommendation from systems that
recommend items to people as they require to reveal some amount of personal
information about the recommended individuals, which raises issues of privacy,
trust, reputation, and interpersonal attraction.

© Springer International Publishing AG, part of Springer Nature 2018
P. Brusilovsky and D. He (Eds.): Social Information Access, LNCS 10100, pp. 570–623, 2018.
https://doi.org/10.1007/978-3-319-90092-6_15

The emergence of social media introduced websites in which people play a central role [69]. As such, the relationship among people, i.e., the underlying social network, be it an explicit network of friends/followers, or an implicit network of people with common interests or shared goals, is a key part of social media websites. In social network sites (SNSs) [27], the network serves as the spinal cord, driving the diffusion and virality of the site, and also its key features and functionality. The user's set of connections is exposed as part of their profile page and serves as evidence for their social capital and often their social status [6,65]. In addition, the news feed or social stream a user gets on sites such as Facebook, LinkedIn, and Twitter, which is the principal source of information in these sites, is based on the user's set of connections [2,49,54,89,105]. On the other hand, users typically share their own information, such as photos, posts, and links, with their set of network connections. It is therefore not surprising that social media sites are putting many efforts to promote connections within their networks and that the number of overall connections is one of the most common measurement for the success of SNSs.

As the network is such a key component in social media websites, people recommendation has become the most effective mechanism for encouraging connections and growing the network. Widgets such as "people you may know" [52] on Facebook and LinkedIn and "people you may follow" [59] on Twitter have become an organic part of these sites. While not published, it is believed that a substantial portion of the connections in these sites are driven by the people recommendation widgets. As other types of SRSs, people recommender and social media have a symbiotic relationship: on the one hand, as mentioned, social media sites depend on people recommendations to flourish and succeed; on the other hand, people recommenders rely on new types of data that social media introduces and often makes public, such as co-authorship of shared documents (wikis, shared files, etc.), tags, comments, 'likes', and others.

The area of people recommendation has substantially evolved in the past few years, with growing number of sites using this technology, growing number of use cases driving it, and growing number of techniques employed to address the new challenges. Other than enriching the network, sites are employing people recommendation techniques for matchmaking users, take advantage of mobile devices to recommend people in a specific location or event, suggest collaborators or teammates, and more. In this chapter, we will review the different types of people recommendation, the techniques used to address them, and their effects on the sites and on users.

2   People Recommendation on Social Media - A Framework

2.1   Fundamental Techniques

Various methods are used to produce people recommendation. At a high level, three approaches are normally applied, separately or in a mix, to provide people recommendations: (1) **graph-based** techniques consider the graph representation of the network (with different possible semantics of the relationships represented by the edges of the graph). Different algorithms may be applied taking advantage of the graph representation and applying techniques such as social network analysis or link analysis; (2) **interaction-based** techniques take into account the interaction of users with content. Social media enables many different types of interactions, each with its own unique characteristics, such as commenting, tagging, joining, voting, or 'liking' [41,53,85]. These interactions form different types of implicit user relationships that can be leveraged for recommendation; (3) **content-based** techniques use the actual content (usually textual) associated with users, typically by an authorship semantics, to derive potential relationships between users and make recommendations.

2.2   Network Types

As mentioned, the target for people recommendation in social media websites is the underlying social network. This network may have several key characteristics that determine its nature and influence the appropriate technique to be used:

- **Explicit vs. implicit.** At the core of many social media websites, and especially SNSs, is an explicit (articulated) network of users. In this case, users need to explicitly connect to each other, for example by sending and accepting invitations or by following other users. In these sites, the network plays a more central role in the overall site's functionality. The user's connections (or a subset) are typically presented as part of the user's profile page and also determine the information s/he will receive on the site (tweets, posts, newsfeed updates, etc.). In other sites, users are not explicitly required to connect to other users. For example, in YouTube, users can subscribe to channels, but no explicit social network is formed among users. Still, sharing is a big part of YouTube's functionality and implicit networks exist among users based on the channels they subscribe to, the videos they watch, or the searches they perform. These implicit networks can be utilized for video recommendation [22] and other features.

  A substantial part of the work on people recommendation in social media focuses on recommendation for explicit networks. As we already discussed, the recommendation for explicit networks is usually aimed for enriching and expanding the network with more connections. In the implicit case, recommendations usually aim to increase sharing and interaction. In both cases, the indirect desired effects of people recommendation also include growing user engagement on the site, for example by increasing content consumption and production, time spent on the site, and returned rates.

- **Symmetric vs. asymmetric.** In some sites, such as Facebook and LinkedIn, a relationship between two users is reciprocated, which makes the underlying graph undirected. In such a case, one user typically sends an invitation to connect to another user, who needs to accept the invitation. Once the other user accepts, the two are reciprocally connected on the site. On the other hand, asymmetric (directed) relationships, such as on Twitter, Tumblr, or Pinterest, allow one user to "subscribe to" or "follow" another user. The other user does not need to follow the first user back and thus many asymmetric relationships are formed. The degree distribution in asymmetric networks tends to be particularly skewed, with popular individuals (e.g., celebrities on the web or executives in the enterprise [56,74]) often having a particularly high number of subscribers or followers. Therefore, different signals are required for providing recommendations in asymmetric networks, e.g., recommendation of familiar people may not be the best strategy in such a case.

- **Confirmed vs. non-confirmed.** Some of the sites require the other side's agreement for connecting or following, while others do not. Typically, symmetric networks require such confirmation and as long as it has not been received, no connection is formed. Asymmetric networks do not usually require a confirmation and any user can choose to follow any other user, however there are exceptions to these norms. Typically, confirmed connections are harder to cancel, since there is more concern the other side would be aware of this removal. Softer mechanism, such as "muting" or "ignoring" are typically enabled. In the asymmetric case, it is more common to see unfollow or unsubscribe actions [73]. Yet, in both cases, the majority of connections usually last for a long period of time.
- **Ad-hoc vs. regular.** Some of the sites encourage connection for an ad-hoc purpose, such as for people to meet at an event or a place or for individuals to partner for a joint task, while others aim for a "regular" relationship, which is meant to last over months and years. Recommendations for ad-hoc networks are typically a one-time "tip" (e.g., meeting a friend at a given place and/or time), which is more similar to the case of traditional item recommendation (movies, music, etc.). Longer term effects may still take place due to a successful recommendation, but are less inherent to the recommendation process. On the other hand, recommendation for regular connections, such as on leading SNSs (Facebook, LinkedIn, Twitter) have a clear long term effect, especially if both users remain regularly active on the site. Once a recommendation is accepted, the user will constantly get updates from the recommended individual as part of their stream or feed. Even without any interaction (comments, likes, re-tweets, etc.), this is already a substantial effect; as mentioned in the previous paragraph, removing the connection is rather uncommon and therefore it is likely to persist for months and years.
- **Signed vs. unsigned.** Most social networks include just one type of "positive" connection, such as "friending" or "following". Yet, in recent years, a few social media websites, such as Epinions and Slashdot, have started to include types of negative connections [14,72,78]. Negative edges may have different semantics, such as disagreeing opinions [77], distrust [77], or regarding another person as having a lower status [85]. Recommender systems can use negative edges to refine their recommendation rankings (especially for graph-based techniques) and overrule certain recommendation options. In a more general sense, networks can assign a weight to relationships, which can also be negative in some cases. These weights can play an important role in the recommendation techniques and influence the ultimate goal of the recommendation.

2.3 Relationship Types

The different characteristics of people relationships in the different sites require different recommendation techniques. For example, a recommender for people to connect with on Facebook may seek to recommend familiar people, while a recommender for people to follow on Twitter may recommend people the user is interested in, even if they are not familiar. Recommending "celebrities" or popular people is probably a better strategy for a follower-followee network than for a friendship network. In this chapter, we review three main types of people recommendation for social media, defined based on the intended relationship type between the user and the recommended individuals. The relationship type reflects the nature of the individuals the system tries to predict and also the desired action as a result of the recommendation, which can be quite different between one type and another. For example, recommendation of familiar people may suggest sending an invitation to connect within an SNS, while a recommendation of a stranger may merely suggest viewing their profile page or blog post. Specifically, we distinguish among the following three relationship types:

- **Familiar people.** Recommendation of known individuals is often used for suggesting connections in confirmed symmetric networks, but also to recommend friends in locations or other specific contexts. Since the two involved individuals are already presumed to be familiar with each other, the goal is to lead to an action that would be beneficial for both of them, for example connecting formally in an SNS, or meeting each other at a given event.
- **Interesting people.** This type of recommendation is usually suitable for asymmetric networks, since the nature of an interest relationship is unidirectional: one person can be interested in another person, while the other person might not be interested (as opposed to familiarity and similarity relationships, which are symmetric in nature). It is normally required that the user would *know of* the recommended individual, but not necessarily *know* them. The goal of the recommendation is to provide the user with a possible source of information, for example for blog posts, recipes, or news, and the desired result is usually a "follow" or "subscribe" action in a directional network.
- **Similar people.** Recommendation of similar people usually aims to introduce a stranger, with whom the user shares a common interest or goal. This type of recommendation aims to create an initial form of "matchmaking" that would get at least one side (the user) interested in the other and ultimately increase their social awareness and social capital. The immediate action may be "softer" in this case, for example viewing created content, recent activity, resume, or a set of attributes related to the recommended person.

We note that a relationship between two individuals can evolve along these three relationship types (in reverse order): in the beginning two random individuals are likely to be strangers; when similar interests are revealed, one may become interested in the other; if the other is also interested, then they are likely to become familiar and ultimately friends.

As we already mentioned in this section, social media has reach data, with many types of content and users' relationships to content. Particularly, social media data can be used in different ways for mining relationships between people [34,47]. All three relationship types mentioned here – familiarity, interest, and similarity – can be mined from different signals in social media. For example, the SONAR system, built for enterprise relationship mining, maps different signals for each of these three relationship types. For instance, co-authoring a wiki page or sharing a file with another individual imply a familiarity relationship [45,47, 52]; commenting on a blog, following, person tagging, or file reading indicate asymmetric interest [66]; and using the same tags, bookmarking the same pages, commenting on the same posts, being members of the same (large) communities, or tagging the same people, indicate shared interests [46,55].

The remainder of this chapter is organized as follows. In the next three chapters, we review in detail the three types of people recommendation based on the intended relationship type, and discuss relevant use cases, many of them along a multi-year set of studies conducted within a large global enterprise. The following section discusses in brevity related research areas, including people search end expertise location, and link prediction in social networks. The chapter is concluded by discussing key aspects across the people recommendation domain and suggesting directions or future research.

3   Recommending Familiar People

Recommendation of familiar people aims at discovering people the user is likely to know and suggest an action that the user may want to perform. Since familiarity is a symmetric relationship in nature (both users know each other), the recommended action usually involves both users and requires the consent or mutual will of both, even if aimed to be initiated by the user who receives the recommendation. Recommending familiar people (often also referred to as "friend recommendation") is very different when performed for ad-hoc versus for regular networks. In ad-hoc networks, the recommendation usually involves some context, such as the location and time in which both users occur. Combining the fact that they are likely to know each other with the contextual features generates an opportunity, such as meeting at a conference or a shopping center, or collaborating on a joint task.

3.1   "Regular" Networks

For "regular" networks, the most fundamental and common scenario of recommending familiar people is recommendation of people to connect with in social network sites. The network of people in these sites serves as the basis for both the site's growth and functionality. For example, in Facebook, the largest SNS with over a billion of users, almost every piece of information, such as status updates, photos, or links, is shared with network connections, also known as "friends". SNSs therefore invest a lot of effort to promote and encourage more connections between their users.

The early approach for recommending people to connect to withing SNSs was based on import wizards, which allowed fetching contacts from email or instant messaging (IM) clients or from other SNSs. These wizards have quite a few weaknesses: first, they require the user's password for the applications from which contacts are imported, which might discourage using them; second, as email, IM, and SNS applications are becoming more and more numerous, one's contacts are likely to be scattered around many applications, which might make the whole import process tedious; and third, these wizards are mostly designed for one time use and do not notify users once new contacts to whom they are not connected yet show up.

Towards the end of the previous decade, people recommendation widgets started to emerge alongside the import wizard. Facebook and LinkedIn introduced "people you may know" widgets, which serve as integral part of their homepage and functionality ever since. These recommendations are also "pushed" to users after a new connection has been made. The suggested action for this type of recommendation is sending an invitation to connect within the SNS to the recommended individual. Thus, accepting the recommendation means clicking on a button or link that would trigger such an invitation. While very little has been published about these particular widgets, it is believed that they drive a substantial portion of Facebook and LinkedIn's new connections. Such people recommendation widgets have become a "must have" ingredient in almost every SNS since then.

Familiar people recommendation in regular networks has two unique characteristics:

(1) Reciprocity – the underlying network is symmetric and confirmed, which means that the other side (the individual who was recommended) needs to accept the invitation and only after both an invitation is sent and the other side confirmed, a connection is formed. That is, acceptance of a recommendation by the user for whom recommendation was provided is not enough for the recommendation to be successful. The user who receives the recommendation knows this, which makes the recommendation more challenging, as they might be a concern the other side will not approve; additional factors are coming into play, such as the interpersonal relationships, social status and reputation, trust, and personality.
(2) Lifecycle – the underlying network is a "regular" symmetric network. As mentioned in the introduction, it is quite rare to see connections unformed in such networks. The recommendations in such networks are therefore very likely to have a regular and ongoing influence for a long time. For example, connecting to a friend on Facebook means that not only do they appear on your list of friends when others view your profile, but also that you will regularly get updates in your newsfeed from that person and therefore remain updated with what is going on in their lives: posts, photos, personal events, etc. Similarly, they will keep track and stay updated with your news from this point onward.

3.2   Enterprise Case Studies

3.2.1   The "Do You Know?" Widget

The first study that focused on people recommendation in an SNS introduced the "do you know?" (DYK) widget [52]. The widget recommended people to connect to within an enterprise SNS. The recommendations were interaction-based and considered a wide variety of familiarity signals that could be mined within the studied enterprise: organizational chart relationships (peers, manager-employee, etc.), paper and patent co-authorship, project co-membership, blog commenting, person tagging, mutual connections, connection on another SNS, wiki co-editing, and file sharing. Since many of these signals were external to the SNS in which the widget appeared, they allowed providing recommendations for brand new users, even if their profile and network were still empty. For each such signal, a relationship score between two individuals was calculated. For example, for paper co-authorship, the score was based on the number of co-authored papers, their dates, the number of other co-authors, the length of the paper, its popularity, and similar factors (see [45, 47] for more details). The overall relationship score between two individuals was based on normalized summation of the different signals across all applications and served as the basis for ranking the recommended individuals. Ultimately, recommended individuals were presented by their relationship score to the user. Intuitively, the more relationship signals with another individual and the stronger their score were, the higher this individual appeared in the recommendation list for the user, provided they were not yet connected.

Figure 1 illustrates the widget, which allowed the user to scroll through a list of recommendations, showing the target person in the middle and the previous and next people on the list as smaller thumbnails, serving as teasers for scrolling. As opposed to typical "people you may know" widgets, the "do you know?" included elaborated explanations for each recommendation. The explanations indicated the counts per each of the signals mentioned above and further hovering over an evidence line allowed seeing the specific details (e.g., the wiki pages co-edited) and getting to the actual page of the evidence pieces.

The evaluation of the widget included a combination of quantitative and qualitative analysis. The quantitative part was based on a field study, where the use of the widget withing an enterprise SNS called Fringe was analyzed along a period of four months. The qualitative part was based on a user study that included a survey, interviews, and responses in the corporate blogging system.

The quantitative evaluation indicated that the effect on the site was dramatic. Before the deployment of the widget, the mechanism for inviting people to connect with was by viewing their profiles and, when relevant, clicking a button to invite them to connect. This was the usual mechanism for invitation in other SNSs, before the emergence of recommendation widgets. Figure 2 shows the comparison between the usage of the DYK widget and the usage of the existing mechanism during the four month trial. It can be seen that the number of invitations sent through the DYK widget was almost seven times higher than the number of invitations sent through user profiles. Also, the number of users sending invitations through the DYK widget was three times higher than the

Fig. 1. The "Do You know?" (DYK) Widget [52].

number of users inviting through the existing mechanism, indicating the growth is not only in the number of invitations, but also in the number of users who initiate connections. Out of the users who used the DYK widget, 77.4% did not use the traditional profile invitation method at all – the DYK widget was just enough for them as a mechanism for inviting people.

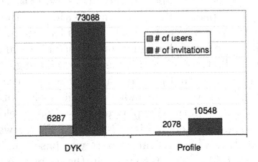

Fig. 2. DYK vs. Profile usage throughout the inspected period [52].

Figure 3 illustrates the substantial change in the average number of connections per user after the DYK was introduced: it grew almost by a factor of four for frequent users and a factor of six for all users. The qualitative analysis revealed very enthusiastic responses: one user explained *"I must say I am a lazy social networker, but Fringe was the first application motivating me to go ahead and send out some invitations to others to connect."* Another blogger wrote: *"I've NEVER seen such an easy way to invite someone. I mean, that rollover thingie*

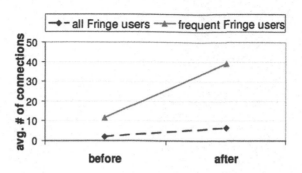

Fig. 3. Average number of invitations per user before and after the inspected period [52].

to invite people to connect with you is addictive. In a matter of seconds, I had sent invitations to 28 people. Me! The oh, I'm choosy, I don't send to anyone and everyone social networker."

The qualitative study also indicated that the rich and detailed explanations provided increased user trust in the system and made users feel more comfortable sending invitations. One user noted: *"If I see more direct connections I'm more likely to add them [...] I know they are not recommended by accident."* and another described: *"last week and even yesterday I looked through a couple [of recommended people] and it jumped out that there was unexpected evidence that allowed me to have grounding for the connection."* Another user explained: *"It's important to say why – helps me understand how its mind works. Then even if I don't want to connect I understand the reason for it and don't lose trust".*

Inspecting the usage of the DYK widget along the entire four-month period, indicated a decay of its use over time. Figure 4 shows the use of the Fringe site as well as the use of the DYK widget during the period. A day of DYK usage is defined as a day in which a user used it to invite at least one other person. It can be seen that the usage of both Fringe and the DYK widget decreases during the period. The effect of the enthusiasm in the first few weeks is especially prominent. The percentage of users who used the DYK out of those who logged into Fringe also decreases along the period: from about 25% during the first month to around 20% along the second and third months and around 17% during the fourth month. It appears that after users build their initial network, the recommendations may exhaust themselves. While new connections are always formed, and new employees may join the company, the recommendations remain useful on a less frequent basis. This is where other types of recommendations, which we will discuss in the following sections, may become valuable.

3.2.2 Algorithm Comparison

In a follow-up study [16], conducted within a different enterprise SNS, nicknamed Beehive, the interaction-based algorithm used by the DYK widget (called 'SONAR' for the underlying social aggregation system) was compared with three

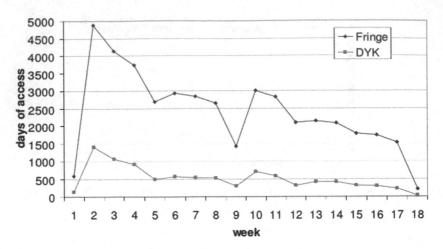

Fig. 4. Fringe and DYK accesses by week during the period [52].

other algorithms for people recommendation: (1) Content Matching (CM) – a pure content-based algorithm using the cosine similarity of the content created by both users: profile entries, status messages, photos' text, shared lists, job title, location, description, and tags. Word vectors were created by a simple TF-IDF procedure. Latent semantic analysis (LSA) was not shown to produce better results and was not applied since it does not yield intuitive explanations; (2) Content plus Link (CplusL) – combined content-based and graph-based techniques. In particular, the CM algorithm was hybridized with social links. A social link was defined as a sequence of 3 or 4 users, where for each pair of users in the sequence u1 and u2, either u1 is connected to u2, u2 is connected to u1, or u1 commented on u2's content; (3) Friend of Friends (FoF) – a graph-based algorithm that relies on the number of mutual friends, as done in many of the popular SNSs. The FoF algorithm was able to produce recommendations for only 57.2% of the users (compared to 87.7% for SONAR). Figure 5 shows the recommendation widget, which included explanations for each recommendation, such as common keywords or a social path.

Evaluation was based on a user survey and a controlled field study. Figure 6 shows the main survey results. The percentages of unknown people recommended by each algorithm are shown above the horizontal center line and the percentages of known people below it. The chart also shows the percentages of good versus not good recommendations in two different colors, broken down by known and unknown recommendations. It can be seen that the CM and CplusL algorithms produced mostly unknown people, while SONAR and FoF produced mostly known individuals. As could be expected, a higher portion of the recommended people who were familiar to the user were rated as good recommendations. Yet, the unknown recommended individuals were also marked as good in many cases. In such cases, these recommendations may help discover new potential friends, which may be a more valuable outcome than a good recommendation of a known individual.

Fig. 5. People recommender widget showing a person recommended using the CplusL algorithm [16].

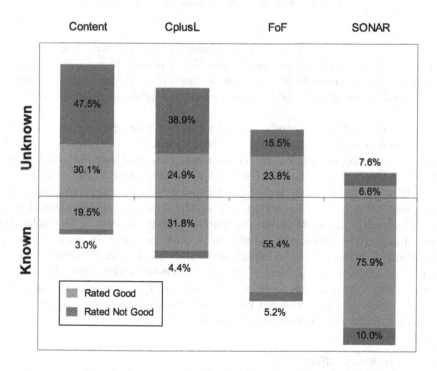

Fig. 6. Survey results for the four algorithms [16].

Table 1. Recommendations results in connect actions.

SONAR	FoF	CplusL	Content
59.7%	47.7%	40.0%	30.5%

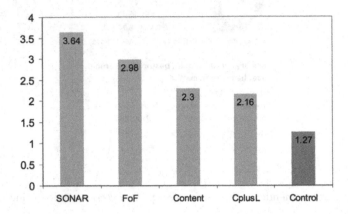

Fig. 7. Increase in the number of friends [16].

Survey comments confirmed that explanations are not only helpful but necessary for all recommendation algorithms. For example, *"I connect to people for a wide variety of contexts but not just because..."*, *"Always state why you are recommending someone"*, and *"I have to have a legitimate reason to connect to someone."*

The results of the field study, in which real recommendations were presented on the Beehive website, confirmed the trends inspected in the survey's results. Table 1 shows the percentage of recommendations that resulted in a connect action for the different algorithms. Post-hoc comparison (LSD) showed that SONAR had a significantly higher connection action rate than the other three algorithms. The connection action rate of the friend-of-friends algorithm was also significantly higher than the content matching algorithm. The field study also compared the increase in number of friends per user for each of the four algorithms as a results of the experiment. SONAR was most effective with an increase of 3.64 friends on average per user, followed by the other algorithms as shown in Fig. 7. A small increase in the number of friends was also observed for the control group (that received no recommendations), which can possibly be attributed to the advertisement of friend-related features.

3.2.3   Network Effects

A later study examined the recommendation impact on the network structure [21]. Since recommendations play such a key role in building the network during its early stages, they also substantially influence the structure of the generated network, its characteristics, and metrics. For example, Fig. 8 shows the average degree of recommended connections for each of the four algorithms, both for all recommended individuals and only for the accepted recommendations. FoF is the most biased towards high-degree connections, while CM does not have such bias: it often recommends users with few connections or even none at all. The high-degrees of FoF recommendations lead to a network with fewer nodes and highly skewed degree distribution compared to the network created by CM recommendations.

Another effect is on betweenness centrality, which measures the importance of nodes in the graph [11]: CM and SONAR were shown to generate higher delta in betweenness centrality for connecting users, compared to CplusL and FoF. This means that CM and SONAR produced more links between weakly connected or completely disconnected communities, while FoF and CplusL more often linked to people who were already part of the user's social circle. Demographic characteristics were also examined: CM was found to be most biased towards the same country, but least biased towards the same organizational unit, while SONAR substantially increased cross-country and intra-unit connections.

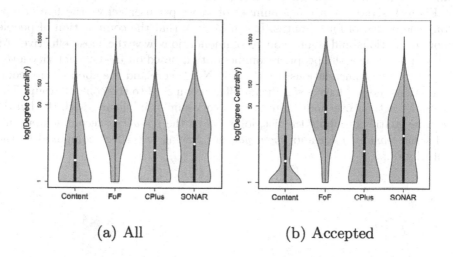

(a) All   (b) Accepted

Fig. 8. Degree of recommended connections across the four algorithms [21].

3.2.4 New Users

Another important use case for people recommendation in the enterprise was explored in a study by Freyne et al. [31]. The study explored recommendation tools for increasing the engagement of new users of the Beehive enterprise SNS. Specifically, two types of recommendations were examined. The first was recommendation of profile entries to produce, so that new users would contribute short content that tells more about themselves ('about you' entries). The second was people recommendation, in order to connect new users with existing users of the Beehive network. For people recommendation, the SONAR algorithm described before was used. SONAR's ability to take into account data external to the site, such as the organizational chart (available for almost every employee), project databases, and other social media applications, enabled to overcome the user cold start problem [3] and provide recommendations for brand new users of the Beehive site. The experiments examined live traffic of news users of Beehive and considered five groups according to the recommendations they received during the sign-up process: the *ctrl* group received no recommendations; the *about-you* group received recommendations of profile entries to produce and no people recommendations; the *ppl-familiarity* group received people recommendations only, ranked and selected by their SONAR score, which reflects the familiarity level with the user; the *ppl-active* group received people recommendations only, calculated by SONAR (top 100), but ranked by their activity level on Beehive; and the *ppl-familiarity+about-you* received both profile entry recommendations and people recommendations (calculated and ranked by SONAR). Engagement effects were examined during a period of four months after new users have signed up for Beehive. They included the number of page views during the period, the number of actions (contributions) during the period, and return rates to the site.

Figure 9 shows the average number of views per user across the five groups along the period of four months. It can be seen that the combination of people recommendations and about-you recommendations was the most effective. At second place comes the people recommendation based on Beehive activity level. The people recommendations based on SONAR score and the about-you recommendations did not have a significant effect compared to the control group. The results for actions over four months were very similar in nature, with the hybrid recommendations topping the list, followed by the *ppl-active* recommendations, and with *people-familar* and *about-you* not having any effect compared to the control group.

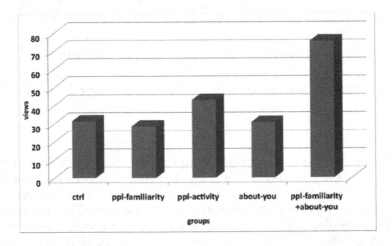

Fig. 9. Average number of views per user over four months [31].

The study also measured the return rates of users in each of the five groups to the Beehive site. For an estimation of the return rate, the study considered, for each week, the percentage of users who viewed or contributed to Beehive during that week or any following week along the four-month period. Figure 10 shows the results for the first 12 weeks of the study. Here the most effective group was found to be the *ppl-active* group, i.e., the group that received people recommendations only, ranked based on their level of activity in Beehive. For example, we can see a clear difference in the retention rates at the end of week 1, with the *about-you* and *ctrl* groups losing between 35% and 42% of users in comparison with the *ppl-active* group losing only 24%.

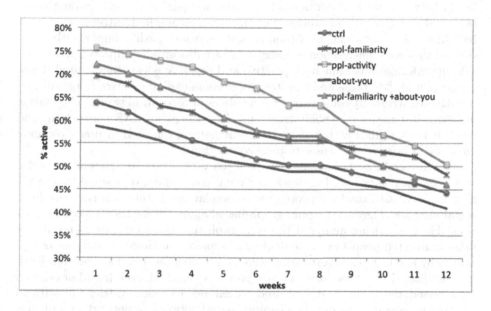

Fig. 10. Return rates over 12 weeks [31].

Overall, the results of the study proved that providing recommendations, and people recommendations in particular, are an effective way to influence new members to make more contributions and views and to return to the site more often. An interesting outcome of this study is that for new users, sorting people recommendations based on users' activity on the site, rather than based on their strength of relationship to the user, is substantially more effective. This indicates that it is more beneficial to recommend a weaker tie, but with higher engagement on the site, than a strong tie who is less active. This does not mean that this strategy is always the desired one for people recommendation. Yet, for new users it appears that the introduction of highly active users, who are also somewhat familiar, is the productive approach.

3.3   Different Friend Recommendation Approaches

Further studies of familiar people recommendation in SNSs have been published in recent years, typically addressing specific challenges that characterize special types of SNSs. One example is a study by Symeonidis et al. [107], which focused on friend recommendation in signed networks. They applied a graph-based approach that used a node similarity measure, which effectively captured the proximity between neighbor graph nodes. They also exploited global graph features by introducing transitive node similarity. Based on this, two individuals connected with a path had a high probability to know each other, proportionally to: (i) the length of the path they are connected with, and (ii) the degree of similarity between the neighbor nodes that form that pathway. Using evaluation on both real and synthetic networks, they showed their method outperformed basic baselines, such as "friend of a friend" and "shortest path". Importantly, they showed that recommendation accuracy can be substantially improved when considering information about both positive and negative edges. In a related study, Eirinaki et al. [26] presented a system for recommending positive (trustful) and negative (distrustful) connections to members of a social network. The system distinguished between explicit trust signals, such as an SNS connection, and implicit trust signals with a more transient nature, reflected in users' common interests as inferred from user-to-item connections.

Friend recommendation was also explored within smaller types of networks, such as a university social network, a virtual community, or a small local SNS. Silva et al. [102] examined people recommendations within a small-scale local social network. They used a pure graph-based approach for friend recommendation. Their algorithm analyzed the sub-graph composed by a user and all the other connected people by three degrees of separation. However, only users separated by two degrees of separation were candidates for friend suggestion. Evaluation showed the superiority of the proposed method over friend-of-a-friend recommendations. Du et al. [25] used an extended version of friend-of-a-friend to provide recommendations in a campus social network (a network of a university or a college). The extended version considered multiple relationship types, including common friends, common followed users, common followers, and common groups. To make recommendations more scalable, incremental relationship data, rather than the whole network, was used to create the freshest recommendation list. Recommendations were also explainable because complete records of common relationships were kept during data processing. As another example, the IntRank model [121] considered interaction attributes that may imply trust and friendship between members of virtual online communities. Specifically, its interaction-based algorithm tried to refine friendship indicators by considering individuals with whom the user has intensively interacted, through signals such as reply frequency, comment length, and response time. Evaluation was based on logistic regression analysis on interaction data from Slashdot and showed that IntRank is able to predict top friends with high accuracy. In a similar scope, the STrust set of algorithms [90] incorporated social trust for friend recommendation within an online well-being community, based on engagement, activity, and popularity in the community. Experiments showed that social trust based algorithms outperformed social graph based algorithms.

3.4   Ad-Hoc Networks

The literature on recommendation of familiar people on ad-hoc networks has also been growing in the recent years, with the emergence of mobile devices and the prevalence of social media and social networks in particular. On the one hand, contextual and particularly location data can enrich the data for people recommendation with important signals for familiarity relationships. On the other hand, people recommendation in a specific context, particularly a location, can make social media applications on mobile devices valuable to their users in many new ways.

One of the early studies introduced FriendSensing [94], a framework that used BlueTooth and other short-range technologies to "sense" and keep track of other devices within close proximity. Both the frequency and duration of two individuals being in close proximity were taken into account. A weighted graph was built accordingly and graph-based techniques, such as shortest path, PageRank, HITS, and markov chains, were applied to generate recommendations. Simulation-based evaluation indicated both frequency and duration perform similarly well and way beyond a random baseline.

A later study referred to privacy issues of collecting location information for friend recommendation, assuming that due to the sensitivity of location data, private social networks will become more common [98]. A method for computing the recommendation scores of all users within a certain radius of a target user in a privacy-preserving manner was developed. The friend recommendation method combined graph-based and interaction-based techniques, taking into account message exchange among users. Another approach for using location data for friend recommendation was introduced in Friendbook [116]. Mobile sensor data was used to infer user's daily routines and identify their lifestyles. Recommendations were then made based on similar lifestyles.

As mentioned, mobile devices not only contribute to data enrichment for friend recommendation, but also provide new scenarios for recommending familiar people. Perhaps the most common example is recommendation of friends in location-based social networks, such as Foursquare, where users share their location with friends [99]. With the real-time location of users, an individual can discover friends around their physical location to enable social activities in the physical world, e.g., inviting people to have dinner or go shopping [8], or meeting with familiar people who are visiting the same place, such as a museum or a stadium. More broadly, users' context on mobile devices, such as their presence (online, offline, etc.) or calendar status, can be used to provide dynamic friend recommendation for social networks on mobile devices [93].

An interesting scenario for familiar people recommendation on mobile networks was proposed by Grob et al. [36]. They presented a recommender system that suggests contacts in order to address them as a group, e.g., 'university colleagues', 'coworkers', 'family', or 'friends'. This grouping may be useful since communication may occur among the members of a certain community simultaneously (e.g., as nowadays enabled by WhatsApp groups [19]). A graph-based approach using clustering techniques was applied, and the authors proposed other potentially useful signals, such as tagging of contacts and manual grouping, and analysis of communication content and patterns. Experiments with a prototype application showed that a user's ego-graph contains a significant amount of community information, which can be extracted using clustering techniques. The evaluation also showed that the contact recommendations can save a considerable amount of time in the group initialization process on a mobile device. Furthermore, since recommendation accuracy only mildly decreased when data got sparse, this type of recommendation could be valuable for application that do not yet own a large user base.

4   Recommending Interesting People

In the Recommender Systems domain, interest relationships have been most commonly studied between people and items, where they serve as a key indicator for the potential success of an item's recommendation to a user [20], as detailed in Chapter "Recommending based on Implicit Feedback" of this book [67]. When discussing user-to-user relationships, interest can also indicate a successful recommendation. By far, the most commonly studied scenario for recommending interesting people is recommendation of people to follow in asymmetric networks. The "follow" action is most recognized with microblogging sites, where it forms the underlying network structure. The following user is referred to as the "follower", while the followed users is referred to as the "followee". The asymmetry of the network, and the fact that the other side does not need to confirm, leads to different uses of the network and satisfies different users needs than symmetric networks, in which confirmation is required. SNSs such as Facebook, have added a "follow" functionality of their own alongside their traditional "friending", to support these distinct user needs (e.g., receiving news and updates from a person without them receiving updates or without a confirmed status of being connected). While the main network allows keeping in touch with friends, colleagues, and acquaintances, the follow network allows to subscribe to users of interest, for example "ordinary" people can keep track with important individuals or celebrities.

The first study to differentiate person-to-person interest relationships from familiarity and similarity was conducted within the enterprise and focused on mining users' interest in other users from social media data sources [66]. The study defined "interest" as reflecting curiosity or care about another individual and observed that as opposed to familiarity and similarity it reflects a directional type of link. Since both familiarity and similarity do not necessarily reflect interest, and vice versa, the paper argued there is value in mining and analyzing this type of relationship. The study compared interest relationships, mined from blog commenting, file reading, following, and person tagging data, with the more traditional familiarity relationships, mined from co-authorship, membership, friendship, file sharing, and other signals, as described in the previous section. It found that the list produced by the interest data sources was highly dissimilar to the list of the user's most familiar people. Furthermore, the interest list indeed contained individuals who were more interesting to the target user than the most familiar ones, as indicated in a direct user survey.

4.1   Followee Recommendation

4.1.1   Early Studies

The first study that focused on recommending people to follow within a microblogging site introduced "Twittomender", a recommender of people to follow on Twitter [59]. The approach used Twitter data itself to generate the recommendations and supported both a search scenario (triggered by a user query) and a recommendation scenario (no query involved). Both content-based and graph-based strategies were explored: the content-based strategies examined user modeling by their own tweets (S1), the tweets of their followees (S2), the tweets of their followers (S3), and hybridization of the three (S4). The graph-style strategies examined representation of a user as a set of Twitter IDs: the user's followees (S5), the user's followers (S6), and hybridization of both (S7). Two more strategies examined hybridization of the content-based and graph-based strategies: a score-based hybridization of strategies S1 and S6 (S8) and a rank-based hybridization of all seven strategies S1 S7 (S9). The open source search engine Lucene was used to index users by their profile and TF-IDF was calculated in order to boost distinctive terms or users within the profile.

Evaluation was based on an offline dataset and a live trial. The offline dataset included 20,000 Twitter users, with 19,000 used as a training set and the remaining 1,000 as a test set. The different strategies were compared based on their ability to predict the user's actual followees. In other words, a successful recommendation was considered as a recommendation of a person the target user was known to follow. They examined both the precision of the entire recommendation list produced by each strategy and the position of the successful recommendations. Thus, the size of the recommendation list produced by each of the strategies played a factor in these metrics. Figure 11 shows the mean average precision and the mean average position for each of the nine recommendation strategies. As can be seen, there is a some trade-off between the precision-based and the position-based measures. One clear outcome was that graph-based strategies were more effective than content-based strategies. As we have already seen in the case of familiar people recommendation, pure content-based approaches tend to be noisy and thus often suffer from low accuracy. A slight advantage can be observed to profiles that were based on followers and followers' followers tweets, a somewhat surprising finding. Hybridization further improved the precision, but not necessarily the position.

A small-scale online experiment was also conducted using the Twittomender system. The chosen strategy was S9, which combined all different sources for profiling information. Each participant was presented with 30 recommended Twitter users and was asked to indicate whom they were likely to follow. Users already followed by the participant were filtered out. On average, participants indicated

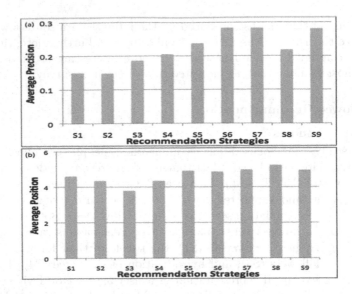

Fig. 11. Precision (a) and position (b) results of the nine strategies [59].

willingness to follow 6.9 users per recommendation list. 50.8% of the relevant recommendations were among the top 10 recommended individuals. The authors
have found these results encouraging for a first implementation of the followee
recommendation feature.

Another early study was performed within the workplace by Brzozowski and
Romero [15], who experimented with the WaterCooler enterprise SNS. During a
24-day live-trial period, they observed patterns of 110 users who followed 774 new
individuals. The strongest pattern found was of the form $A \leftarrow X \rightarrow B$, meaning
that sharing an audience (follower) with another person is a strong reason to
follow that person. In addition to network patterns, other signals were examined
between a user A (receives the recommendation) and a user B (recommended
individual), including the number of posts written by B that A has clicked on, the
number of replies made from A to B, and the similarity of profile tags between
A and B. Of these, only the number of replies was found to be a strong signal
for predicting that A would follow B.

The recommendations also included explanations that conveyed the specific
network pattern that led to the recommendation and statistics about the other
signals (number of replies, posts read). In their design recommendations at the
conclusion of the paper, the authors state that explanations were appreciated
by the vast majority of the users and in some cases were indicated to have made
the difference between accepting and ignoring the suggestion. They concluded
that while it was difficult for users to visualize the entire social graph, short
simple descriptions of the connections they share with a potential contact were
compelling in explaining why they may want to follow them.

Table 2. Reasons users follow people in the workplace.

I follow people who...	
Post interesting things	85%
I work with	77%
I've met	57%
I have similar interests	53%
Follow me	34%
Are high up in the org	26%

The experiment was followed by a user survey that received 56 responses. Table 2 lists respondents' reasons for following people on WaterCooler. For each reason, the percentage of confirming respondents is given. Interest tops the list and beats both familiarity-related signals (work with or having met) and similarity (shared interests). Employees may follow an interesting individual even when they are strangers, while on the other hand they are likely to avoid following users whose posts are mostly not interesting, regardless of their level of familiarity or similarity to them. For example, some participants indicated they will not follow a colleague if they feel they are too personal or social in their postings, e.g.: *"Some folks are a bit 'chatty' on Chatter or need to install the 'personal filter'. My favorite is the person who announced they are bored (for [many] employees to read). Second favorite example of filtering is some of [the] [location] folks who talk about non-work stuff fairly often to each other (lots of social niceties). Great that folks are social (World Cup chats are great examples), but we are at work..."* At the bottom of the list, with small percentages, are reciprocity and being high up in the organization. The former is expected in an asymmetric setting, but the latter should be regarded with care, since it is known that employees do have strong bias towards senior individuals in the company [49,53,56], even if they are not always aware of it or willing to admit it.

4.1.2 Twitter's WTF

A more recent paper revealed some details about the implementation of the followee recommender system in use by Twitter, called "Who to Follow" (WTF) [38]. From an architectural perspective, the authors noted the decision to process the entire Twitter follower-followee graph in memory using a single server, which contributed to the performance of the feature. They developed an open-source in-memory graph processing engine to traverse the Twitter graph and generate recommendations fast. The graph-based random walk algorithm examined two approaches: the first gave equal influence to every follower-followee edge, while the second gave each user the same importance, split among their associated edges, regardless of their number.

The algorithm itself was designed to work on directed graphs. The authors note that after much experimentation, their selected approach is based on the

SALSA algorithm [76], which was originally developed for web search result ranking. The algorithm constructs a bipartite graph between "hubs" on the one side and "authorities" on the other. Each step in the SALSA algorithm always traverses two links – one forward and one backward (or vice-versa) – so it remains on the same side of the bipartite graph. For the case of people recommendations, the "hubs" were initialized as the user's circle of trust, computed based on egocentric random walk, while the "authorities" were initialized with users that the "hubs" follow. Running multiple iterations of SALSA on this graph produced scores for each of its sides, which were used to rank them separately. The ranking of the "authorities" was interpreted as standard "interested in" user recommendations, while the ranking of the hubs was interpreted as user similarity and were used as a source of candidates for the "similar to you" feature on Twitter. This distinction is another evidence for the difference between interest and similarity relationships.

The authors speculate that the SALSA algorithm has been proven useful in production since it mimics the recursive nature of the actual problem: a user u is likely to follow those who are followed by users that are similar to u. These users are in turn similar to u if they follow the same (or similar) users. The SALSA iterations seem to operationalize this idea – providing similar users on one step and their followers on the next step. The random walk ensures equitable distribution of scores out of the nodes in both directions. Furthermore, the initialization of the bipartite graph ensures that similar users are selected from the circle of trust of the user, which is itself the product of a reasonably good user recommendation algorithm (personalized PageRank).

The evaluation of the WTF system was based on both offline experiments on retrospective data and online A/B ("bucket") testing on live traffic. The simplest metric used was the "follow-through rate", which measures the accuracy of the recommendations (number of generated follows divided by the number of impressions of the recommendations attributed to a particular condition). However, as the authors state, this measure does not capture recall, does not relate to user's lifecycle (new users are likely to be more receptive to following new individuals), and does not measure the quality of the recommendations in terms of creating a high-quality stream for the user that increases the overall engagement. The latter notion is approximated by a metric they call "engagement per impression", which quantifies the amount of engagement by the user on that recommendation in a specified time interval. Based on this metric, the impact of a recommendation can be estimated. The metric's downside is that it is available only after the specified time interval, which slows the speed at which deployed algorithms can be assessed. Among other things, these evaluation processes and metrics were used to compare different algorithms (personal page rank, SALSA, others), and, as already mentioned, indicated that SALSA is the preferable one.

4.1.3 Refined Approaches

In the recent few years, several studies have tried to further refine and advance the use cases and methods for recommending people to follow. A recent study argued that classifying users into target categories, depending, e.g., on their political affiliation, preferred football team, favorite coffee shop, etc., is valuable for this recommendation task, as the categories allow fine tuning, enable better efficiency, and support intuitive explanations [28]. Another study proposed design guidelines and an architecture for user recommendation on social bookmarking sites [85]. Social bookmarking systems allow users to store, share, and tag bookmarked resources. A few examples are Delicious, where the bookmarked resources are web pages; CiteULike, where users bookmark academic papers; and Flickr, where pictures can be referred to as the bookmarked resource. The proposed interaction-based recommendation framework was based on shared interests and considered common tags and common resources. As we have previously seen, however, relying solely on such signals may produce noisy recommendations. In another study, the use of Twitter lists for people recommendation was examined [60]. Lists allow the user to group other users based on user-defined topics or themes. Other users can subscribe to the list and benefit from its members' tweets. Twitter lists have become an important means for users to curate content of interest and for marketing departments to organize content and connect with communities. The study compared a content-based approach in which users were represented by the content of their tweets with an interaction-based approach, which represented users by the set of tags associated with the lists they were member of. The advantages of each approach and the benefits of combining them were demonstrated in an accuracy-focused evaluation.

4.1.4 Celebrity Recommendation

The skewed degree distribution of asymmetric networks gives special importance to "celebrities" or "stars" – individuals known to a large public, such as politicians, actors, singers, models, or athletes. In addition, as social platforms like micrblogging services become a medium for receiving news, opinions, and ideas, the interest in celebrities and famous individuals continues to grow [74]. Also, as celebrities are often very active in microblogging and similar media, they serve as the source of a large portion of an average user's feed, in spite of being a very small portion of the entire user population. Selecting the right celebrities to follow therefore becomes a key task in making the feed interesting and engaging for its consuming user and coping with the social overload problem [39]. In recent years, researchers have started to explore the domain of celebrity recommendation. The number of candidates for recommendations is much smaller in this case, since the total number of celebrities is small (the exact number depends on the definition of celebrities; the most basic option is by applying a threshold over the number of followers). This opens the door for new specialized methods for this kind of recommendation task.

The most elaborated approach for celebrity recommendation has been proposed by Ding et al. [24]. They found that almost 40% of Twitter's users and 90% of the users of Tencent Weibo (a Chinese microblogging site) follow more celebrities than "ordinary" users. They argued that users may not only be interested in the celebrities themselves, but also in the interests behind them. They therefore used the intra-relationships of celebrities to build a network of interests among them. This network is compact and efficient and generally denser than the all-user social network, since relationships among celebrities tend to be more frequent (e.g., singers to themselves or even singers to sports stars). In addition, as textual information for celebrities is more widely available, a rich celebrity profile can be built using information from Wikipedia, personal homepages, and other resources. The resulted model combined social network analysis with semantic analysis to learn user interests in celebrities.

Experiments were conducted over both Twitter and Tencent Weibo and considered both a warm start and a cold start (a new celebrity starting to use a site) situations. In both, the proposed algorithm (marked CSTR) was shown to outperform, in terms of both precision and recall, the following baselines: recommendation of the celebrities who have the most connections with the ones followed by the target user (SN); recommendation of celebrities with the largest number of followers (MP); and a version of the proposed algorithm that does not include the network of celebrities (CTR). For example, Table 3 shows the average precision at 20 (AP@20) across the four algorithms in both warm-start and cold-start situations. The authors conjectured that the key advantage of the algorithm is the use of the celebrity social network.

Table 3. AP@20 for warm-start and cold-start for CSTR and three baselines.

	Warm start		Cold start	
	Twitter	Tencent Weibo	Twitter	Tencent Weibo
CSTR	11.056	21.135	7.899	14.095
CTR	10.4	20.9	0.234	1.078
SN	5.117	6.191	3.347	6.371
MP	8.699	6.646	—	—

In another study of celebrity recommendation, Liu et al. [82] used an enhanced collaborative filtering model with adapted Jaccard similarity and integrated social status features to provide recommendations. Their model helped address the cold start problem and ease data sparsity issues.

4.2   Ad-Hoc Networks

While the majority of the literature on interesting people recommendation focuses on "regular" or permanent scenarios, several studies also discussed recommendation of interesting people for an ad-hoc connection. One interesting example is recommendation of people to mention on a tweet or a post. Particularly, researchers explored recommendation of people to mention on tweets. On Twitter (and other microblogging systems) users can mention other users on their tweet using the '@' sign. These mentions take an important role in user conversations and serve as a means for information sharing. The mentioned user receives a notification and their re-tweet may promote the diffusion and reach of the tweet. Thus, the "mention" feature provides an opportunity for ordinary users to improve the visibility of their tweets and go beyond their close audience of followers. Since a tweet is strictly limited in characters, the number of users one can mention is small and needs to be carefully considered.

The first in-depth study of the topic [114] introduced the "whom to mention" problem and put a special focus on recommending users to mention who will help disseminate the tweet. The goal was to favor influencers, who are not only likely to re-tweet, but whose re-tweet may also have a big impact on the reach of the original tweet. Two cases were distinguished: the first is mentioning a follower, in which case the mention serves as a useful notification, especially when the follower follows a large number of users (a typical case for influencers); the second is mentioning a non-follower, who may spread the tweet to a brand new audience, which often leads to further cascade diffusion.

The study used a machine learning approach to train a ranking model for mention recommendations that used graph-based, interaction-based, and content-based features. These features included the match between the given tweet and the interest profile of the recommended user, the relationship between the recommended user and the author of the tweet, and the influence of the recommended user. The user relationship features were based on re-tweet inter-actions and considered the content of the tweets one user has re-tweeted from another. Experiments over the large Chinese microblogging site Sina Weibo, showed that the best performance is achieved when all three types of features are used. The content-dependent features in user relationships were confirmed as having high effectiveness in the recommendation model. Ultimately, the "whom to mention" approach was shown to significantly improve the diffusion of a tweet based on various metrics.

A later study focused on using the mention feature for choosing the right audience for marketing and publishing purposes on Twitter [108]. More empha-sis was put on the topic of the tweets (e.g., the same publisher may mention different users for different promotion tweets) and on locating an audience with high response rate. The problem was referred to as a top-k ranking problem. SVM-based learning-to-rank model with social, content, location, and time-based features was applied and shown to achieve good performance. Another recent study applied text analysis techniques to provide mention recommenda-tions [35]. Specifically, translation models were used to expand the language and extract topics from both the tweet in question and previous tweets by candi-date "mentioners". Experiments were performed over Sina Weibo and showed that the new approach can outperform baseline models, including the "whom to mention" method [114].

The area of recommending interesting people is in its infancy and poses many new opportunities for future directions. The follow model used in most social media sites is still rather simplistic. The user can either follow an individual and receive all their content, or not follow them and not receive any of their updates (unless re-tweeted by others they follow). A finer grain version that allows to combine people-based and content-based selections can improve user experience (e.g., received updates from an individual only when they post about a certain topic; or better yet, when the system thinks it is an interesting post for the specific user). The pace at which a certain user can consume information can also come into play with such highly-personalized approaches [51]. These approaches and similar require deeper understanding of users, to avoid putting a lot of extra burden on them when building their optimal stream. This is where new recommendation techniques can come into play, for example recommending both people and associated topics to follow; recommending people and topics to unfollow; suggesting more dynamic settings based on context (time, location, etc.); and recommending based on the rate of items' appearance and the user's ability to consume them.

5   Recommending Similar People

The final type of relationship targeted by people recommendation is similarity. Similarity between people is at the core of one of the most popular recommendation techniques – user-based collaborative filtering – where it serves as the basis for item recommendation [12]; also see Chapter "Collaborative Filtering" of this book [71]. In our case, however, similar people are the target of the recommendation, rather than the means to produce it. In this recommendation task, the set of potential candidates can be much larger than in the other people recommendation tasks, since it is not limited to people the user knows. In fact, in many use cases it is desirable to avoid recommendation of familiar people and focus on strangers, as we will demonstrate later in this section. The underlying assumption under this type of recommendation follows the line of homophily (love of the same), i.e., the tendency of individuals to associate and bond with similar others [88]. Similarity can be derived from different signals, such as demographic characteristics (often available as part of the "profile" in social media sites), past activity, personality traits, or network characteristics. Due to the high number of potential candidates and the fact that they typically include strangers, this type of recommendation is often more exploratory in nature and might have substantially lower success rates than other people recommendation tasks. Arguably, however, the potential value of such a recommendation is substantially higher, since it exposes the target user to new individuals, which may increase their social awareness and ultimately their social capital and influence. Therefore, noisier recommendations may be more tolerable in this recommendation task; sometimes just one successful recommendation, be it, e.g., a potential new colleague or a date partner, can make a big difference.

5.1   Enterprise Case Studies

One of the key use cases focuses on recommending similar people within the enterprise. A first study on the topic mapped the similarity relationships that could be mined from enterprise social media [46]. Similarity indicators were categorized into three types: (1) common places: being member of the same community, commenting on the same blog post, or corresponding on the same forum thread; (2) common things: using the same tag, bookmarking the same page, or being tagged with the same tag; and (3) common people: having the same friend on an SNS; tagging the same person; or being tagged by the same person. Figure 12 depicts the overlap, as measured within the top-100 lists, among the similarity indicators and also indicators' overlap with the familiarity network, mined from familiarity indicators (org-chart, co-authorship, project co-membership, SNS friendship), as described in previous sections. It can be seen that the overlap within the three groups (people, things, places) is higher, justifying the proposed partitioning. The overlap with the familiarity list is generally low, but higher for the people group, as could be expected. It is especially high (over 25%) for common SNS friends – indeed we have seen that having common friends can be a productive signal for familiarity and is commonly used for friend recommendation.

	tagged_by	friending	tagged_with	tag_person	tag_usage	bookmarks	communities	blogs	forums
familiarity	9.43	26.21	12.84	10.16	4.43	4.12	6.01	5.22	2.62
tagged_by	100	14.97	10.17	4.95	3.12	2.61	3.38	3.04	1.33
friending	14.97	100	15.31	10.52	6.21	5.10	7.50	6.25	3.05
tagged_with	10.17	15.31	100	8.28	11.06	6.56	6.54	5.86	3.18
tag_person	4.95	10.52	8.28	100	4.87	3.59	2.65	3.97	1.54
tag_usage	3.12	6.21	11.06	4.87	100	14.29	4.34	3.46	1.61
bookmarks	2.61	5.10	6.56	3.59	14.29	100	3.44	3.01	1.41
communities	3.38	7.50	6.54	2.65	4.34	3.44	100	2.52	1.53
blogs	3.04	6.25	5.86	3.97	3.46	3.01	2.52	100	2.26
forums	1.33	3.05	3.18	1.54	1.61	1.41	1.53	2.26	100
average	*5.45*	*8.61*	*8.37*	*5.05*	*6.12*	*5.00*	*3.99*	*3.80*	*1.99*

Fig. 12. Mean *match@100* among the nine similarity sources and familiarity [46].

That study also examined the similarity groups in a pseudo-people recommendation setting, in which anonymized individuals were recommended. As the actual recommended person was anonymized, the focus was on the presented evidence, which included up to nine items of the same group of indicators (people, things, or places). Based on this evidence, participants were asked to rate four statements, corresponding to four scenarios, listed on the leftmost column of Table 4. Notice that the first two scenarios try to establish initial interest, as in reading a person's blog or looking at their bookmarks, based on similarity evidence. The last scenario, on the other hand, aims higher and tries to suggest SNS connection based on similarity. The participants in the survey were asked to rate the answer to each of the four statements on a 5-point Likert scale, ranging from "strongly disagree" to "strongly agree". The average ratings for the places, things, and people groups are presented in Table 4. Overall, it can be seen that for the first three scenarios, the average rating for the things group was highest, while for people it was lowest. For the fourth scenario, referring to SNS connection, the people group received higher ratings than the places group, and its highest across all scenarios (on the contrary, the things and places groups received their highest ratings for the blog reading scenario).

Table 4. Average rating for similarity groups in different scenarios.

	Things	Places	People
I am interested in reading this person's blog	3.9	3.7	3.6
I am interested in looking at this person's bookmarks	3.8	3.5	3.4
This person reflects a subset of my expertise	3.8	3.6	3.4
I would like to connect to this person on a SNS	3.8	3.6	3.7

The results of this study were used in a follow-up experiment that focused on recommending strangers within an organization [55]. The goal of the study was to recommend people the user does not know, but may want to get familiar with. This type of recommendations can be useful in many potential manners, such as, for getting help or advice, reach new opportunities, discover new routes for career development, learn about new assets that can be leveraged, connect with subject-matter experts and influencers, cultivate one's organizational social capital, and grow own reputation and influence within the organization. As mentioned before, recommendation of people to connect to within an SNS is mostly effective for the network-building phase. Afterwards, one's recommendations become staler, as the network becomes more stable and connection to others becomes less frequent. This is where stranger recommendation can become more relevant and complement the recommendation of familiar individuals, by suggesting people the user does not know, but may want to start getting acquainted with.

Figure 13 shows the user interface of the proposed recommender, called StrangerRS. Since it aimed at recommending strangers, more information about each person was presented, in the form of their full profile page (part A). Evidence for why this person may be interesting was also presented (part B).

It included common things and common places, and in some cases also common people, with that individual, e.g., common tags, common communities, common files, and common bookmarks. The action suggested by the recommender was not a connection within the SNS, since it is likely to be too soon to connect to a stranger, but rather it was suggested to view the person's profile, read their blog, or follow them (part C).

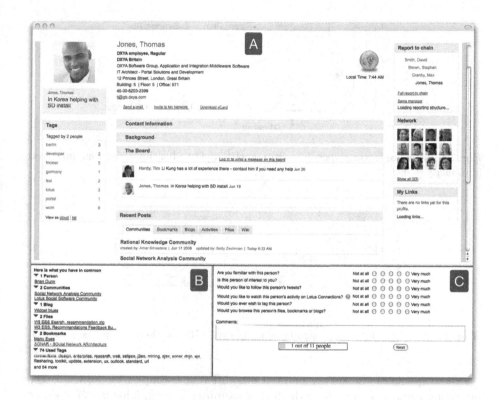

Fig. 13. User interface of the stranger recommender system [55].

A successful recommendation by StrangerRS was considered a recommendation of a stranger who might be interesting to the user. These two, almost contradicting, goals were not easy to satisfy and led to a much lower accuracy level than other people recommendations. Yet, supposedly, the value of a successful recommendation in this case is much higher, since this is no longer just about facilitating a connection to a known person, but rather about exposing the user to a new interesting person s/he was not even aware of. The method used for producing the recommendations was based on network composition – an arithmetic set operation between two types of networks: the extracted familiarity network was subtracted from the extracted similarity network to produce the recommendations. Jaccard index was the main measure used for similarity between two

individuals. Results of a survey that examined participants' response to part C
in Fig. 13 are depicted in Figs. 14 and 15. They indicated that two thirds of the
recommended individuals were indeed strangers, yet strangers who were significantly more interesting than a random stranger. In all, out of 9 recommendations
presented to each user, 67% included at least one stranger rated 3 or above in
terms of the user's interest, on a 5-point Likert scale.

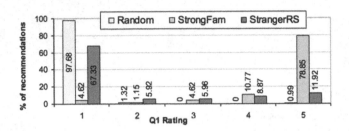

Fig. 14. Rating of "strangerness" for StrangerRS and two baselines: random and strong
familiarity [55].

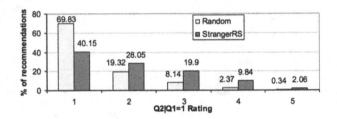

Fig. 15. Rating of interest in strangers for StrangerRS vs. the random baseline [55].

Participants of the survey pointed out different examples for how this type
of recommendations may help them. For example, one explained: *"This experiment is interesting, because I'm sure that in [our organization] there are people
with similar roles (and pains) that see similar customer expectation"*. Another
participant noted, with regards to a specific recommendation: *"She works for a
marquee customer in the Telecom sector I cover. Any lessons or best practices
she shares I would be very much interested in"* and another wrote with regards
to a recommended stranger: *"Works with implementations of products I work
with. Other key contacts are known by this person. Useful tags. Looks useful"*.
Other participants, however, described some of the recommendations as weird
and indicated they expected a higher level of accuracy of the recommendations,
e.g. *"only a few people that would be of interest at this time* and *"Many recommendations are way off. I'm used to get better recommendations."*

As found in the previous study [46], the people group had the highest overlap
with the familiarity list. StrangerRS, therefore, experimented with three options:
in the first, the people group of indicators was not considered neither as part of

the familiarity network nor as part of the similarity network; the second considered the people group as part of the similarity network; and the third considered the people group as part of the familiarity network. It was found that it is worth including the people group in either sides of the composition equation: when included as part of familiarity, StrangerRS produced more interesting strangers, and when included in the similarity list, StrangerRS yielded fewer strangers and more weak ties. These differences can help fine tune the recommender based on the specific requirements.

Recommendation of similar people in a work context can be extended to support team building [109]. In a study over SourceForge.Net, the largest database of open source software, researchers argued that compatible members tend to share similar programming styles, naming standards, design patterns, and so forth [106]. In their study, they built a collaboration network among developers, projects, and project properties, used as an infrastructure for recommending a list of top developers that are most compatible with a target user, based on their programming language skills and categories of past projects. A graph-based algorithm applying random walk with a restart procedure was used to generate the recommendations. Due to the task's difficulty, the evaluation measured the accuracy of the whole list of recommendations, rather than the accuracy of each recommendation separately (as we have seen, a similar approach was used as part of the evaluation of StrangerRS). Specifically, a recommendation list of k collaborators was considered successful for a target user and a given project if it was able to predict at least one collaborator with whom the user worked on the project. Experiments with different values of k showed accuracy ranging from 78.9% for $k = 5$ to 83.33% for $k = 20$. The approach was sensitive to the user's cold start problem and required at least 5 past projects per user in order to produce reasonable recommendations.

5.2   Refined Approaches

Similar people recommendation has also been proposed for academic collaboration. Lopes et al. [83] studied the recommendation of collaborators on an academic social network by combining graph-based similarity, called "global cooperation" with content-based similarity, called "global correlation". The "global cooperation" was based on a directed graph, in which an edge weight was determined by the number of joint publications between two authors, normalized by the target author's total number of publications. The "global correlation" considered the correlation between the weighted vectors of research areas in which two authors were involved. The weight of a research area for an author was determined by the author's number of publications in the area, divided by the author's total number of publications. A case study, performed over InWeb – Brazilian National Institute of Science and Technology for the Web – demonstrated how the approach may work in a real-world setting.

Content-based approaches are commonly used for the task of similar people recommendation. As opposed to recommendation of familiar and interesting people, where stronger evidence than matching keywords or topics is required for high accuracy, in this type of recommendation content-based techniques may go farther compromising accuracy for serendipity. Van Le et al. [113] proposed a profile modeling and matching approach based on the content of users' posts and comments on social media. Latent Dirichlet Allocation (LDA) was used for extracting latent topics from the content. Recommendations included the user's nearest neighbors based on topic similarity. Rastogi et al. [95] enhanced traditional content-based approaches for user similarity by considering not only the content of the posts, but also their sentiment. They observed that users may share similar interests, but have different opinions on them. Sentiment analysis was therefore used to understand the opinion of a user on a set of topics they were interested in. Based on both topics and opinion similarity, users on Facebook were recommended.

5.3   Ad-Hoc Networks

As already mentioned in the section about familiar people recommendation, the emergence of mobile devices, and the combination of context (particularly, location) data with social networks, open the door to new types of ad-hoc people recommendations. Recommendation of similar people who are not necessarily familiar to the user is especially common in ad-hoc networks, serving to match-make individuals for a variety of purposes. One of the classic scenarios is recommendation of people with similar interests at a conference. A lot of money and efforts are invested in organizing and traveling to conferences, where networking is one of the key goals, if not the primary one, yet still occurs on a rather random basis. Connecting to strangers at a conference may have different values, for example for an academic conference they may include getting research help or advice; acquiring new research opportunities; discovering new research directions; and learning about new research projects.

Find & Connect [18] used physical proximity information (via RFID badges) to recommend new contacts to a user at a conference. A field trial was conducted at UbiComp 2011 and indicated users consider historical physical encounter information to be most important when they want to meet a new contact. It was also found that homophily, reflected in shared research interests, attendance of the same activities, and sharing of the same contacts (along the lines of the "things, places, and people" paradigm described before), worked as a factor in users' decision to add new contacts at the conference. Another study of the topic described a system for recommending networking opportunities at a conference [63], combining analysis of past activity on social media (Facebook, Twitter, LinkedIn) with matchmaking of user profiles. People with whom a connection has already been established on an SNS were excluded from recommendation. The system was tested at a conference with over 1000 attendees and was shown of high potential to enhance networking experience. As another example, the SPARP algorithm hybridized similarity of interpersonal relationships (weak ties) with similarity of personality traits to suggest conference participants to each other [119]. The authors hypothesized that personality traits, such as openness, extroversion, agreeableness, conscientiousness, and neuroticism (emotional stability), should be highly considered in the establishment of an interactive scenario between participants at a conference. Evaluation against data from the ICWL 2012 conference showed the superiority of SPARP for predicting new ties, compared to two baselines that did not consider personality traits.

As in the case of familiarity, location data can also be used to enhance similarity inference between two users. The HGSM (hierarchical graph-based similarity measurement) framework [80] was among the first to propose the mining of similarity between users based on their historical spatio-temporal data. The framework took into account both the sequence of people's movement behaviors and the hierarchy of geographic spaces. It was evaluated by using GPS data collected by 65 volunteers over a period of 6 months and was shown to outperform related similarity measures, such as the cosine similarity and Pearson similarity. Zhen et al. [123] extended this framework to also include the visit popularity of a location. Evaluation showed that injecting this similarity measure into a collaborative filtering algorithm enhanced its general performance and ability to cope with the cold start problem.

There are other good examples for similar people recommendation for ad-hoc purposes. Saez-Trumper et al. [97] proposed techniques for guest recommendation to venues or events. They pointed out that SNSs such as Facebook have started to offer recommendation of guests that users might want to invite to events, such as law firm parties, birthday parties, or PR's club invitations. The approach taken was reducing the problem to the more traditional item recommendation task of recommending events or venues to users. They pointed at a data sparsity problem for this type of recommendation, and proposed to handle it using two means: first, explicitly differentiating "power users", who visit many places, and second, by considering the fact that people visit a venue not only because they like it, but also because they are close-by. Their new model was evaluated using Foursquare data for the city of London and showed that simple models, such as linear regression and a Bayesian model, produced accurate recommendations for all types of venues.

Another study proposed to accompany item recommendations, such as for movies or for dinner, with suitable activity partners [112]. A survey conducted found that activity partner recommendation can improve the success of item recommendation and increase users' excitement about the recommendation process as a whole. Recommendations were calculated based on social closeness, similar interests (derived from item preferences), and preference likelihood for the item in question (i.e., boosting users who are also likely to prefer the movie or restaurant). Evaluation using three location-based social networks compared different strategies for the activity partner recommendation task and demonstrated its value.

5.4   Dating Recommendations

Finally, it is worth mentioning that one of the most common uses of similar people recommendation is on dating sites. While not usually categorized under social media, dating sites also center around people, and their key task is to matchmake individuals for romantic purposes. Pizzato et al. [91] explored recommendations on dating sites and introduced the more general concept of "reciprocal recommenders" – people recommenders in which success can only occur when both sides like each other, or reciprocate. As a few examples for such recommenders, they give an employer-employee recommendation, mentor-mentee suggestions, business partner identification, and, the focus of their work, dating sites that aim to help a person meet a suitable partner. Similarly to the original social matching framework [110], the authors listed a few unique characteristics of reciprocal recommendations: as already mentioned, success is dependent on both ends; both sides need to provide their profiles so that matching can occur; typically, it is required that one individual will not be recommended to too many others; there is a strong need to avoid bad recommendations (as opposed to some other use cases of stranger recommendation mentioned before), since they might make users feel rejected; and, one successful recommendation is often all the user needs. Their most prominent algorithm for partner recommendation on dating sites is called RECON [92] and used content-based techniques over user profiles. Its uniqueness lied in the fact that it considered both users' content preference rather than just one side, as in traditional item recommendation. RECON was evaluated on a major Australian dating site, based on four weeks of training and two weeks of testing. Recommendations' success was determined based on their ability to predict user interaction. It was shown that accounting for reciprocity features improved accuracy and helped address the cold-start problem.

6   Related Research Areas

In this section, we will discuss two research areas that relate to people recommendation: link prediction and people search.

6.1   Link Prediction

Link prediction in social networks is a fertile research domain, which is closely related to people recommendation and has often been offered to enhance it. The seminal work by Liben-Nowell and Kleinberg [81] formalized it as a task to predict new interactions within a social network based on the existing set of interactions. Experimentation with paper co-authorship networks showed, using an unsupervised learning approach, that the network topology can be effectively used to predict future collaboration. Link prediction can be used for reconstruction of networks, evaluation of network evolving mechanism, and classification of networks [84]. In the social media domain, Leskovec et al. [77] developed models to determine the sign of links (positive or negative) in SNSs where interactions can be positive or negative (Epinions, Slashdot, Wikipedia). Fire et al. [30] experimented with five social media sites, including Facebook, YouTube, and Flickr, and proposed a set of graph-topology features for identifying missing links. This technique was shown to outperform common-friends and Jaccard's coefficient measures, implying it can also be useful for recommending new connections.

Scellato et al. [100] focused on location-based social networks and suggested a supervised learning framework to predict new links among users and places. In another study of mobile networks, Wang et al. [115] showed that combining network-based features with human mobility features (e.g., user movement across locations) can significantly improve link prediction performance using supervised learning. For more details, also see Chapter "Social Link-based Recommendations" of this book [75].

6.2   People Search and Expertise Location

The domain of people search focuses on a scenario triggered by a user query, where the returned results are people. For similar reasons to those discussed in this chapter, the case of searching for people bears unique characteristics compared to other search scenarios and therefore forms its own area of research. For a broader overview see Chater "Social Search" of this book [13].

As we have already seen, a few of the people recommendation studies also focused on people search scenarios [59]. The key difference between people search and people recommendation is that the former involves a user query. This indicates that the user initiated a request as they had some information need, rather than the system trying to "guess" or predict the user's needs [48]. It is thus usually the case that the people search scenario reflects an ad-hoc need, such as contacting a person, seeing how they look, finding how old they are, and similar. In addition to being initiated by the user, the query involved in the people search scenario reflects the user's intent or need and helps focus the response. For example, a search may include a name, an attribute, a topic, or a combination of these, which narrow the potential set of results to very specific individuals. The use cases for people search include searching for familiar people (e.g., looking for the phone number of a colleague or the latest news about a friend) and interesting people (e.g., searching for the martial status of a singer, the homepage of a politician, or the publication page of a researcher). The case of searching for similar people is more complex: people do not look for other individuals who are generally similar to them; rather, they may look for an individual based on a common interest, such as a joint attribute (e.g., elementary school) or a joint topic of interest. In some cases, the search does not fall under any of these three categories of relationship types and the searcher looks for someone with whom they have no relationship, based on a certain quality, role, or topic, expressed through the query.

Weerkmap et al. [117] were the first to introduce a comprehensive study of the people search vertical. They analyzed the logs of a Dutch people search engine over three dimensions: queries, sessions, and users. They distinguished between queries for high-profile individuals (celebrities or ones involved in trendy events) and low-profile individuals (friends, relatives, and complete strangers). Less than 4% of the queries involved a keyword, i.e., the vast majority were based on person names. Another study on the topic described Faces, a system that enabled efficient people search within a large organization [56]. Query log analysis of Faces indicated a strong bias towards searching for executives and other senior employees. The main goals of the searches were indicated to be the following: finding the contact details of an employee (phone number, email, etc.), exploring their organizational environment (managers, peers, reports), looking for their job description, looking for their full name, and looking for their photo. Huang et al. [64] studied people search usage on the LinkedIn SNS. They found that for name-based searches, users primarily click on one result and that closer network distance leads to higher click-through rates. In contrast, for non-name searches (e.g., by job title, skill, or company name), users are more likely to click on multiple results that are not in their existing connections, but with whom they have shared connections. People searching on Facebook has also been studied [104], revealing that females search for people proportionally more than males and that users submit more queries as they gain more friends. Also, 57.6% of the person name searches were for friends, 41.8% for non-friends, and 0.6% for self. It was also reveled that 32.3% of the top-1000 celebrity page queries were for musicians, 19.4% for public figures, 17.8% for actors or directors, 8.2% for entertainers, 7.4% for artists, and 7.3% for athletes. Additionally, it was found that the number of queries, their type, and portion of friends, substantially vary with different user demographics, such as age, gender, number of friends, and celebrity status. Hsieh et al. [62] focused on searching of a person of interest without knowing their name, but rather using attributes such as hometown, school, or work. They developed a method that considered, in addition to label match, the social proximity to the user, and the interaction with the query. Experiments using Facebook and Twitter data indicated their method outperformed a few baselines, especially when the number of labels was high.

A specific case of people search that received substantial attention in the literature over the years is expertise location (sometimes referred to as expert search, expert finding, or even expert recommendation, although it always involves a query by the user). Expertise location aims at searching for a person knowledgeable in a certain topic or domain. The user query in this case does not involve any hint about the person's name, but rather the corresponding topic or domain of expertise, sometimes accompanied with desired attributes of the expert, such as their company, country, or role. For relevance calculation, the two prevailing approaches have been the candidate-based approach, which builds profiles for candidate experts and ranks them based on their similarity with the query; and the document-based approach, which first finds documents relevant to the query and then locates the associated experts based on the documents in which each candidate expert is represented by [7]. Different studies have argued that experts should be ranked not only by their relevance to the query, but also based on their social proximity to the user. ReferralWeb [70] was one of the first systems to do so, allowing users to specify a search topic and a social criterion (e.g., people who are related by up to two degrees of separation). Expertise Recommender [86] filtered expert search results based on two elements of the user's network: organizational relationships and social relationships, gathered through ethnographic methods. Recent research has studied the use of social media for expertise location [41]. It suggested that the diversity of content types and user associations with content, their public and dynamic nature, and the fact that social media data also reflects social network information, make it a highly valuable source for expertise mining. The study distinguished between searching for people who are interested in a topic (e.g., for arranging a brainstorm or diffusing an idea) and searching for people who are experts in a topic. The Expertise Locator (EL) system demonstrated these ideas by presenting experts and social media-based evidence within an organization [120]. Experiments showed that the sources found most effective for mining expertise [41] were not necessarily the ones serving as best evidence. For example, microblogs were found to be a good source for expertise mining, but were not perceived as a strong evidence by users.

7 Summary and Future Directions

In this section, we discuss key topics related to people recommendation surfaced throughout the previous sections and suggest directions for future work.

7.1 Key Topics

Relationship Types. We categorized the people recommenders reviewed in this chapter according to the relationship type with the recommended users. As we discussed, a relationship between two strangers may evolve, through shared similarities, to interest, and if/when the interest is mutual, to familiarity. Both Familiarity and similarity are symmetric relationships, and the difference between them has been extensively discussed through the perspective of their use for collaborative filtering (e.g., [37,57,103]). Interest between two individuals, on the other hand, is an asymmetric relationship and is therefore typically suitable for recommendations that do not require the other side's consent. The reverse side of the asymmetric interest relationship focuses on identifying the people who are interested in a given individual and has been used to calculate reputation and influence [65]. However, it was not discussed in this chapter, as we are not aware of studies that recommended people who are interested in a target user. Anyhow, once a network of familiarity, similarity, or interest is formed, it can be used as a basis for further recommendations, such as for items or for groups. The semantics of each of the three network types is different and may lead to different types of recommendations, for example movies by friends might yield a different list than movies by followees.

The border between the three relationship types is not always precise and some signals may indicate multiple types. For example, commenting on a post may indicate both interest and familiarity; and being member of a community may indicate both familiarity and similarity. We have also seen that an algorithm may produce more than one relationship type: the followee recommendation algorithm developed for Twitter produced interesting people on one side of its bipartite graph (as the authorities) and similar people on the other side (as the hubs) [38]. Furthermore, some of the recommendation scenarios discussed in this chapter can make use of more than one relationship type: we have seen that recommendation of people on an SNS should start with suggesting familiar people, but later, when a user's network is established, may combine recommendation of strangers with shared similarity [16,52,55]. Familiarity and similarity may also both be applicable for other scenarios we discussed, such as team building [61,106] or activity partner finding [112]. When considering location for people recommendation, all relationship types are relevant. For example, at a conference, recommending friends, identifying interesting people, and suggesting strangers with similar interests, are all likely to be valuable for an attendees [18].

Network Composition. One way to combine different relationship types in one people recommender system is composition, i.e., performing an arithmetic operation between two (or more) sets of people who are related to the target user by different semantics. In this chapter, we saw the use of network composition for recommendation of strangers, either in the enterprise [55] or at a conference [63]. In both cases, the familiarity network was subtracted from the similarity network, in order to create a recommendation list that includes strangers with whom similar interests are shared. Due to factors such as homophily, the two lists are likely to have higher overlap than random, and thus their composition is essential to make sure the recommendations fulfill both the requirement of being strangers and the requirement of having shared interests. Another example can be considered in the case of friend recommendation on a "regular" SNS, where the network of friends (users to whom the target user is already connected) is subtracted from the broader familiarity network before producing the final list of recommendations [52]. Analogously, for followee recommendation, the list of individuals currently followed by the target user is subtracted from the broader list of interesting people [59]. Future work on people recommendation may explore other ways to compose recommendation lists considering multiple relationship types.

Recommendation Techniques. We pointed out three main techniques that are used in people recommender systems: graph-based, interaction-based, and content-based. We have seen the use of all three approaches across all three relationship type recommendations. We also observed that content based methods are usually more suitable for similar people recommendation, which is more speculative in nature and can benefit from the wide span yet noisy nature of content-based techniques. In many cases, researchers experimented with multiple techniques, compared them, and often combined them to produce optimal results. Underlying our high-level technique categorization are many different algorithms and methods, taken from diverse research areas including machine learning, statistics, information retrieval, natural language processing, and social network analysis.

Cold Start. The cold start problem is one of the key challenges in recommneder systems, and people recommendation is no exception. Due to the inherent reciprocity, the user cold start problem and the item cold start problem both occur when new users (who also serve as items in this case) join a social site. Quite a few studies evaluated their people recommender systems in cold start situations [24,36,82,92,106] and often found that a special treatment is required. We have seen different ways to cope with the cold start problem: aggregating information from external sites [31], using personality traits to cluster users [119], using location data for similarity inference [123], and considering reciprocity [92].

More broadly, social media sites that try to establish a network among users might run into a cold start problem, as an initial core network might be hard to form. One direction to explore in this case is the combination of people recommendation techniques with other methods. For example, gamification approaches were shown to have strong short-term effects on motivating and engaging users, but tend to lose their charm rather quickly [23,29,44,50]. Social systems may consider applying gamification at early stages in order to boost ties among users and establish a critical mass of connected users, which can then help provide recommendations to continue the network's growth.

Explanations. The value of explanations for recommender systems has been widely discussed [111]. For people recommendation, explanations play an important role, both in providing immediate reasoning for a suggestion to connect to another individual and in establishing long-term trust with the user. The use of explanations was demonstrated across all three relationship types: familiarity [16,52], interest [15,28], and similarity [55]. For example, when recommending strangers, explanations assist in showing the user why the person may be interesting and teasing for reading their blog or viewing their profile. When recommending friends to connect with in an SNS, we have seen that a person may feel more comfortable sending an invitation after an explanation list is presented and "proves" there is a common ground. It could be that due to the inherent reciprocity in such cases, explanations should also be shared with the recommended individual, in case a request to connect has been made by the other side.

Location. Recommendations that take advantage of user location, as available through state-of-the-art mobile devices, are growing in popularity. Location data can be used to enhance inference of relationships types, for example co-occurrence data has been used to derive both familiarity and similarity and was shown particularly useful in both cases [80,94,116,123]. Location data can also be useful to derive interest relationships, for example from attendance of music concerts or sports events. User location can also be used to recommend people in a specific context and has substantially contributed to the advancement of people recommendation for ad-hoc networks. Examples include recommending people at a conference [18,63,119] and suggesting people to invite to a dinner or to go shopping [8]. On the other hand, people relationships can be used to provide location recommendations, as described in Chapter "Location Recommendation with Social Media Data" of this book [10].

Location is just one example of contextual data that is made available through the use of mobile devices. Other contextual information, such as time, presence status, or device characteristics, has been used in the area of context-aware recommender systems [1] and can also be valuable for people recommendation [93]. Looking further into the future, wearable devices, such as glasses and watches, are likely to obtain access to even more personal information that on the one hand will provide more context for a recommender to work with, and on the other hand will require more advanced recommendation techniques, so these devices can work appropriately with minimum input from the user.

Trust and Reputation. Trust between individuals plays a key role in any recommendation process, but in people recommendation it may carry even higher importance, especially in the case of recommending familiar people. Trust is a personal and subjective quality that is hard to measure and assess. We have seen various attempts to estimate trust, from simply considering an SNS connection, to inspecting interaction signals such as reply rate or response time. When recommending familiar people, the decision whether to connect to them, regularly or for an ad-hoc purpose, may depend on trusting them [90, 121]. Particularly, in signed networks, trust and distrust serve for recommending friends and "foes" [26]. In addition, in some scenarios of similar people recommendation, such as dating or employment (a.k.a "reciprocal recommenders" [91]), trust can take an important part in the user's decision making process. For interest networks, the concept of reputation, which may be viewed as a form of global trust, typically plays a key role: more reputable individuals are likely to arouse more interest and thus be followed by more individuals, mentioned more frequently by others, and so forth [65]. We also saw that Twitter's "Who to Follow" algorithm starts with the user's "circle of trust", calculated based on a random walk over the user's egocentric network [38]. Generally, since trust and reputation carry some form of transitivity in a social network, graph-based random walk algorithms (e.g., PageRank, HIT, SALSA, folkrank) are involved in their calculation.

Network Effects. The vast majority of the work on people recommendation has focused on the value for the individual user, e.g., extending their list of friends or meeting similar individuals. As people recommendations aim at enriching and promoting a social network, their global effects on the network are also important to consider. This type of evaluation naturally requires more complete data along a longer period of time. We have seen one example of a study that explored the network effects in the case of familiar people recommendation within the enterprise [21]. It showed that different recommendation strategies can lead to a different network structure in terms of number of nodes, degree distribution, centrality, and cluster characteristics. When studying new people recommendation features, researchers and practitioners should also examine (and report) the effect on the network as whole, rather than just on individual users. While demonstrated for a familiarity network, such effects can also be relevant for other types of relationships, for example when recommending interesting people to build a "follow" network or when recommending similar people to build a team or a community. As we will soon discuss, this type of evaluation is part of a generally-required longer-term evaluation in the domain of people recommendation.

Recommendation Bias. In some of the studies about people recommenders, we have seen that bias towards active or popular individuals is desirable. For instance, in the case of recommendation to new enterprise SNS users [31], recommendation of highly active users can have a positive effect on the new user's engagement within the site. In the case of recommending people to mention on a tweet, recommending influential individuals can help the tweet's diffusion and the twitterer's reach [108,114]. On the other hand, we have also seen that such recommendations may lead, in the long run, to highly skewed degree distribution and other extreme network effects [21]. People recommenders should therefore evaluate their bias towards specific individuals and whether this bias is worthwhile, both from the perspective of the individual user and from the perspective of the system as whole.

7.2   Directions for Future Work

In this section, we point out several challenges that we believe are at the core of future people recommenders research.

Group Recommendation. The vast majority of people recommendation studies focus on recommending one person to another. The area of group recommendation studies recommender systems that are aimed for a set of individuals [4]. Various studies explored group recommendation for social media sites and content [32,96], but not for people. As stated by Terveen and Mcdonald in their work on social matching [110], people recommendation may also be useful for groups and communities and can facilitate participation. This can be relevant across all three relationship types discussed in this chapter. For example, group recommendation can help a group of school mates identify more individuals to invite to a reunion meeting; help a community of horror movie fans to identify individuals they want to follow; or help an online community of vegans in a city to identify new individuals who can be invited to join. Moreover, using people recommenders, communities can be created automatically on an ad-hoc basis, rather than being established by moderators who need to invite more moderators and members. The team building scenario discussed in this chapter [106] is one example that ties to group recommendation. However, since little work has been conducted to explore this area, it may serve as fertile ground for future research.

It is also worth mentioning that the other direction – recommendation of groups or communities to individuals users – has received its own attention in the literature, across different social media sites, such as Facebook [5], Orkut [17], and Flickr [122].

Privacy. Privacy considerations did not receive much attention in people recommender studies. In general, designers of people recommender systems should pay attention to the following aspects:

- Location data is particularly sensitive: when making recommendations, even if for friends, systems must carefully consider what information can be exposed and to whom. For example, the FindU system proposed profile matching schemes, where only minimal and necessary information was exposed regarding user's private information, such as location, in order to limit privacy risks. Formal security proofs were provided for the proposed protocols [79].

– While we discussed the value of explanations, they might also expose sensitive information about content or user interaction with content, and should be considered with care.

– For graph-based recommendation, when exposing graph patterns to the user, systems need to make sure that information not directly related to the user is public and allowed to be shared. The most obvious example is friend of a friend – this method relies on information the user might not be directly exposed to (who the friends of their friends are). Samanthula et al. [98] argued that social network information is likely to become more restricted since users will become more concerned about their privacy. They proposed a method for computing recommendation scores under the assumption that network information is restricted from the perspective of a given user.

– For stranger recommendations, when seeking for activity that identifies similar interests, there is a need to make sure this activity can be shared with the user from a privacy's perspective.

– People recommenders that make use of private data, such as the user's email or local files, should verify that sensitive information (e.g., a confidential email sent from another individual) is handled with care, even though the user already has access to it in another application. For example, a user might find it puzzling to find a sensitive email as part of a recommendation's explanation.

Interactivity. The people recommenders reviewed in this chapter support only basic interactivity, e.g., by providing explanations or learning from previous user behavior. Higher levels of interactivity would enable users to provide more instant and direct feedback about the type of recommendations they are interested in and the quality of recommendations they have already received. Interactivity can contribute to the transparency of the recommendation task and gives users more control over the recommendation process as a whole. A good example for an interactive system in a related domain is PeopleExplorer [58], which enabled exploratory people search. Users were allowed to specify their task objectives by selecting and adjusting key criteria, including the content relevance, the candidate authoritativeness, and the social similarity between the user and the candidates. A user-based evaluation showed that PeopleExplorer retrieved more relevant results, while requiring less effort from the users, compared to a non-interactive baseline. It should be noted that the emergence of intelligent personal assistants, such as Siri and Google Assistant [68], and the advancement in voice-enabled interfaces [40], make interactive dialogue systems more ubiquitous. It is therefore reasonable to assume that people recommender systems would be able to conduct a more natural type of dialogue with users, which should help them become more interactive.

Evaluation. People recommender studies typically use similar evaluation techniques and metrics to other recommender system studies. Evaluation techniques include offline analysis, in which the researchers work with a dataset and usually try to learn from part of it in order to predict the rest; surveys, which are especially common for people recommenders, in which participants are asked about the potential value of recommendations and can sometimes take real actions (from viewing a profile to inviting to connect); and online evaluations that usually use buckets ("A/B testing") to compare different strategies in a live system with real users. Each of these techniques has its own pros and cons. For example, offline evaluation allows experimenting with different settings at a low cost, and is therefore often used for initial experimentation and parameter tuning. However, it is based on retrospective analysis rather than real-time response. Surveys allow focusing on particular aspects, which sometimes cannot be measured in other settings, but are limited in scale and often suffer from participants' bias. A/B tests allow experimenting with real users in real-time at large scale, but entail high costs that are not always affordable and do not always allow speculative experimentation. We have seen many studies combining two or more of these techniques, as their pros and cons complement each other, and their combination produces a more reliable and robust evaluation.

Evaluation metrics generally range from accuracy, through error measurement (root mean squared error and mean average error), to information retrieval measures (from precision and recall to mean average precision and discounted cumulative gain). Many of the studies use the friend-of-a-friend (FoF) method as their baseline and show their superiority on top of it. Yet, FoF is no longer the state of the art and researchers should start using more advanced baselines to make their contribution compelling.

In the vast majority of the studies, as in other studies of recommender systems, the focus has been the accuracy of the recommendations, i.e., their ability to predict future connections (in offline evaluations) or produce recommendations that are accepted by users at high percentages (in surveys or online evaluations). However, it is agreed by the recommender system community that accuracy-focused metrics do not capture all desired aspects of a recommender and are therefore insufficient for a through evaluation [87]. Other metrics, such as diversity, novelty, and serendipity have been proposed to complement accuracy [33,49]. Typically, there is a trade-off between these measures and accuracy. In the people recommendation case, many of the scenarios for similar people recommendation produce relatively low accuracy and high serendipity. We have seen studies that therefore measure the hit rate of a list of recommendations (i.e., how many times it contains one good recommendation) rather than of each recommendation separately [55,106]. Underlying this method is the assumption that the value of a serendipitous recommendation of a stranger, be it a colleague, a person to meet at a conference, or a potential partner, is high, however this has not been scientifically proven.

Measuring the value of a recommendation involves analysis over time. We have seen studies measure network effects [21], engagement, and return rates [31] to evaluate the value of recommendations from a temporal perspective. We have also seen that industry-scale recommenders try to assess the value of a recommendation in terms of user engagement on the site as a result of the new connection [38]. Further methods are required, especially in the case of similar people recommendation, to asses their real value.

Another aspect to consider when evaluating the value of a recommendation is whether it has really led to an action the user would not have performed otherwise, and how much effort would the user have to put in order to reach the same outcome without the recommendation. Initial steps in these directions have started to appear within the broad recommender systems research community [101], and should be further developed for the specific case of people recommendations. As we discussed, recommendations can take two strangers all the way to becoming close friends. But currently there is no means to evaluate the value of each step in the process or the final outcome as a whole.

Emerging Domains. It is worth mentioning two additional domains that can serve as an interesting target for future work on people recommendation.

The healthcare domain has always been slow to adopt social technologies, among other things due to the special privacy concerns it entails. However, the value of recommendations in general and people recommendations in particular has especially high potential in this area. People recommenders in the healthcare domain can be used to suggest both doctors and patients. Since suggestions are likely to focus on strangers, recommendation of similar people is especially relevant in this case. Some initial steps in this direction have already been taken. For example, the PatinetsLikeMe [118] website aims to connect patients with one another for sharing, support, and research.

Recommender systems have been quite popular in the TV domain for many years. The Netflix prize advanced this domain even further [9]. However, as TVs continue to evolve into "smart TVs", they enable many more social elements, such as sharing and interaction between watchers, which make the new TVs a social medium on its own. This provides a highly interesting opportunity for people recommenders across all three relationship types. For example, connecting family members or friends when watching their favorite TV series, or connecting fans of a sports team while watching a game or a tournament.

References

1. Adomavicius, G., Tuzhilin, A.: Context-aware recommender systems. In: Ricci, F., Rokach, L., Shapira, B. (eds.) Recommender Systems Handbook, pp. 191–226. Springer, Boston, MA (2015). https://doi.org/10.1007/978-1-4899-7637-6_6
2. Agarwal, D., Chen, B.-C., Gupta, R., Hartman, J., He, Q., Iyer, A., Kolar, S., Ma, Y., Shivaswamy, P., Singh, A., Zhang, L.: Activity ranking in LinkedIn feed. In: Proceedings of the 20th ACM SIGKDD International Conference on Knowledge Discovery and Data Mining, KDD 2014, pp. 1603–1612. ACM, New York (2014)
3. Ahn, H.J.: A new similarity measure for collaborative filtering to alleviate the new user cold-starting problem. Inf. Sci. **178**(1), 37–51 (2008)

4. Amer-Yahia, S., Roy, S.B., Chawlat, A., Das, G., Yu, C.: Group recommendation: semantics and efficiency. Proc. VLDB Endow. **2**(1), 754–765 (2009)
5. Baatarjav, E.-A., Phithakkitnukoon, S., Dantu, R.: Group recommendation system for Facebook. In: Meersman, R., Tari, Z., Herrero, P. (eds.) OTM 2008. LNCS, vol. 5333, pp. 211–219. Springer, Heidelberg (2008). https://doi.org/10. 1007/978-3-540-88875-8_41
6. Bakshy, E., Hofman, J.M., Mason, W.A., Watts, D.J.: Everyone's an influencer: quantifying influence on Twitter. In: Proceedings of the Fourth ACM International Conference on Web Search and Data Mining, WSDM 2011, pp. 65–74. ACM, New York (2011)
7. Balog, K., Azzopardi, L., de Rijke, M.: Formal models for expert finding in enterprise corpora. In: Proceedings of the 29th Annual International ACM SIGIR Conference on Research and Development in Information Retrieval, SIGIR 2006, pp. 43–50. ACM, New York (2006)
8. Bao, J., Zheng, Y., Wilkie, D., Mokbel, M.F.: Recommendations in location-based social networks: a survey. Geoinformatica **19**(3), 525–565 (2015)
9. Bennett, J., Lanning, S.: The netflix prize. In: Proceedings of KDD Cup and Workshop, vol. 2007, p. 35 (2007)
10. Bothorel, C., Lathia, N., Picot-Clemente, R., Noulas, A.: Location recommendation with social media data. In: Brusilovsky, P., He, D. (eds.) Social Information Access. LNCS, vol. 10100, pp. 624–653. Springer, Cham (2018)
11. Brandes, U.: A faster algorithm for betweenness centrality. J. Math. Sociol. **25**(2), 163–177 (2001)
12. Breese, J.S., Heckerman, D., Kadie, C.: Empirical analysis of predictive algorithms for collaborative filtering. In: Proceedings of the Fourteenth Conference on Uncertainty in Artificial Intelligence, UAI 1998, pp. 43–52. Morgan Kaufmann Publishers Inc., San Francisco (1998)
13. Brusilovsky, P., Smyth, B., Shapira, B.: Social search. In: Brusilovsky, P., He, D. (eds.) Social Information Access. LNCS, vol. 10100, pp. 213–276. Springer, Cham (2018)
14. Brzozowski, M.J., Hogg, T., Szabo, G.: Friends and foes: ideological social networking. In: Proceedings of the SIGCHI Conference on Human Factors in Computing Systems, CHI 2008, pp. 817–820. ACM, New York (2008)
15. Brzozowski, M.J., Romero, D.M.: Who should i follow? Recommending people in directed social networks. In: Fifth International AAAI Conference on Weblogs and Social Media, ICWSM 2011 (2011)
16. Chen, J., Geyer, W., Dugan, C., Muller, M., Guy, I.: Make new friends, but keep the old: recommending people on social networking sites. In: Proceedings of the SIGCHI Conference on Human Factors in Computing Systems, CHI 2009, pp. 201–210. ACM, New York (2009)
17. Chen, W.-Y., Chu, J.-C., Luan, J., Bai, H., Wang, Y., Chang, E.Y.: Collaborative filtering for Orkut communities: discovery of user latent behavior. In: Proceedings of the 18th International Conference on World Wide Web, WWW 2009, pp. 681–690. ACM, New York (2009)
18. Chin, A., Xu, B., Yin, F., Wang, X., Wang, W., Fan, X., Hong, D., Wang, Y.: Using proximity and homophily to connect conference attendees in a mobile social network. In: 2012 32nd International Conference on Distributed Computing Systems Workshops (ICDCSW), pp. 79–87, June 2012

19. Church, K., de Oliveira, R.: What's up with WhatsApp? Comparing mobile instant messaging behaviors with traditional SMS. In: Proceedings of the 15th International Conference on Human-computer Interaction with Mobile Devices and Services, MobileHCI 2013, pp. 352–361. ACM, New York (2013)

20. Claypool, M., Le, P., Wased, M., Brown, D.: Implicit interest indicators. In: Proceedings of the 6th International Conference on Intelligent User Interfaces, IUI 2001, pp. 33–40. ACM, New York (2001)

21. Daly, E.M., Geyer, W., Millen, D.R.: The network effects of recommending social connections. In: Proceedings of the Fourth ACM Conference on Recommender Systems, RecSys 2010, pp. 301–304. ACM, New York (2010)

22. Davidson, J., Liebald, B., Liu, J., Nandy, P., Van Vleet, T., Gargi, U., Gupta, S., He, Y., Lambert, M., Livingston, B., Sampath, D.: The YouTube video recommendation system. In: Proceedings of the Fourth ACM Conference on Recommender Systems, RecSys 2010, pp. 293–296. ACM, New York (2010)

23. Deterding, S., Sicart, M., Nacke, L., O'Hara, K., Dixon, D.: Gamification: using game-design elements in non-gaming contexts. In: CHI 2011 Extended Abstracts on Human Factors in Computing Systems, CHI EA 2011, pp. 2425–2428. ACM, New York (2011)

24. Ding, X., Jin, X., Li, Y., Li, L.: Celebrity recommendation with collaborative social topic regression. In: Proceedings of the Twenty-Third International Joint Conference on Artificial Intelligence, IJCAI 2013, pp. 2612–2618. AAAI Press (2013)

25. Du, Z., Hu, L., Fu, X., Liu, Y.: Scalable and explainable friend recommendation in campus social network system. In: Li, S., Jin, Q., Jiang, X., Park, J.J.J.H. (eds.) Frontier and Future Development of Information Technology in Medicine and Education. LNEE, vol. 269, pp. 457–466. Springer, Dordrecht (2014). https://doi.org/10.1007/978-94-007-7618-0_45

26. Eirinaki, M., Louta, M., Varlamis, I.: A trust-aware system for personalized user recommendations in social networks. IEEE Trans. Syst. Man Cybern.: Syst. **44**(4), 409–421 (2014)

27. Ellison, N.B., et al.: Social network sites: definition, history, and scholarship. J. Comput.-Mediated Commun. **13**(1), 210–230 (2007)

28. Faralli, S., Stilo, G., Velardi, P.: Recommendation of microblog users based on hierarchical interest profiles. Soc. Netw. Anal. Mining **5**(1), 25 (2015)

29. Farzan, R., DiMicco, J.M., Millen, D.R., Dugan, C., Geyer, W., Brownholtz, E.A.: Results from deploying a participation incentive mechanism within the enterprise. In: Proceedings of the SIGCHI Conference on Human Factors in Computing Systems, CHI 2008, pp. 563–572. ACM, New York (2008)

30. Fire, M., Tenenboim, L., Lesser, O., Puzis, R., Rokach, L., Elovici, Y.: Link prediction in social networks using computationally efficient topological features. In: 2011 IEEE Third International Conference on and 2011 IEEE Third International Conference on Social Computing (SocialCom) Privacy, Security, Risk and Trust (PASSAT), pp. 73–80. IEEE (2011)

31. Freyne, J., Jacovi, M., Guy, I., Geyer, W.: Increasing engagement through early recommender intervention. In: Proceedings of the Third ACM Conference on Recommender Systems, RecSys 2009, pp. 85–92 (2009). ACM, New York

32. Gartrell, M., Xing, X., Lv, Q., Beach, A., Han, R., Mishra, S., Seada, K.: Enhancing group recommendation by incorporating social relationship interactions. In: Proceedings of the 16th ACM International Conference on Supporting Group Work, GROUP 2010, pp. 97–106. ACM, New York (2010)

33. Ge, M., Delgado-Battenfeld, C., Jannach, D.: Beyond accuracy: evaluating recommender systems by coverage and serendipity. In: Proceedings of the Fourth ACM Conference on Recommender Systems, RecSys 2010, pp. 257–260. ACM, New York (2010)

34. Gilbert, E., Karahalios, K.: Predicting tie strength with social media. In: Proceedings of the SIGCHI Conference on Human Factors in Computing Systems, CHI 2009, pp. 211–220. ACM, New York (2009)

35. Gong, Y., Zhang, Q., Sun, X., Huang, X.: Who will you "@"? In: Proceedings of the 24th ACM International on Conference on Information and Knowledge Management, CIKM 2015, pp. 533–542. ACM, New York (2015)

36. Grob, R., Kuhn, M., Wattenhofer, R., Wirz, M.: Cluestr: mobile social networking for enhanced group communication. In: Proceedings of the ACM 2009 International Conference on Supporting Group Work, GROUP 2009, pp. 81–90. ACM, New York (2009)

37. Groh, G., Ehmig, C.: Recommendations in taste related domains: collaborative filtering vs. social filtering. In: Proceedings of the 2007 International ACM Conference on Supporting Group Work, GROUP 2007, pp. 127–136. ACM, New York (2007)

38. Gupta, P., Goel, A., Lin, J., Sharma, A., Wang, D., Zadeh, R.: WTF: the who to follow service at Twitter. In: Proceedings of the 22nd International Conference on World Wide Web, WWW 2013, pp. 505–514. Republic and Canton of Geneva, Switzerland (2013). International World Wide Web Conferences Steering Committee

39. Guy, I.: Recommender Systems Handbook, chapter Social Recommender Systems, pp. 511–543. Springer, Boston (2015)

40. Guy, I.: Searching by talking: analysis of voice queries on mobile web search. In: Proceedings of the 39th International ACM SIGIR Conference on Research and Development in Information Retrieval, SIGIR 2016, pp. 35–44. ACM, New York (2016)

41. Guy, I., Avraham, U., Carmel, D., Ur, S., Jacovi, M., Ronen, I.: Mining expertise and interests from social media. In: Proceedings of the 22nd International Conference on World Wide Web, WWW 2013, pp. 515–526. Republic and Canton of Geneva, Switzerland (2013). International World Wide Web Conferences Steering Committee

42. Guy, I., Carmel, D.: Social recommender systems. In: Proceedings of the 20th International Conference Companion on World Wide Web, WWW 2011, pp. 283–284. ACM, New York (2011)

43. Guy, I., Chen, L., Zhou, M.X.: Introduction to the special section on social recommender systems. ACM Trans. Intell. Syst. Technol. 4(1), 7:1–7:2 (2013)

44. Guy, I., Hashavit, A., Corem, Y.: Games for crowds: a crowdsourcing game platform for the enterprise. In: Proceedings of the 18th ACM Conference on Computer Supported Cooperative Work and Social Computing, CSCW 2015, pp. 1860–1871. ACM, New York (2015)

45. Guy, I., Jacovi, M., Meshulam, N., Ronen, I., Shahar, E.: Public vs. private: comparing public social network information with email. In: Proceedings of the 2008 ACM Conference on Computer Supported Cooperative Work, CSCW 2008, pp. 393–402. ACM, New York (2008)

46. Guy, I., Jacovi, M., Perer, A., Ronen, I., Uziel, E.: Same places, same things, same people? Mining user similarity on social media. In: Proceedings of the 2010 ACM Conference on Computer Supported Cooperative Work, CSCW 2010, pp. 41–50. ACM, New York (2010)

47. Guy, I., Jacovi, M., Shahar, E., Meshulam, N., Soroka, V., Farrell, S.: Harvesting with sonar: the value of aggregating social network information. In: Proceedings of the SIGCHI Conference on Human Factors in Computing Systems, CHI 2008, pp. 1017–1026. ACM, New York (2008)

48. Guy, I., Jaimes, A., Agulló, P., Moore, P., Nandy, P., Nastar, C., Schinzel, H.: Will recommenders kill search? Recommender systems - an industry perspective. In: Proceedings of the Fourth ACM Conference on Recommender Systems, RecSys 2010, pp. 7–12. ACM, New York (2010)

49. Guy, I., Levin, R., Daniel, T., Bolshinsky, E.: Islands in the stream: a study of item recommendation within an enterprise social stream. In: Proceedings of the 38th International ACM SIGIR Conference on Research and Development in Information Retrieval, SIGIR 2015, pp. 665–674. ACM, New York (2015)

50. Guy, I., Perer, A., Daniel, T., Greenshpan, O., Turbahn, I.: Guess who? Enriching the social graph through a crowdsourcing game. In: Proceedings of the SIGCHI Conference on Human Factors in Computing Systems, CHI 2011, pp. 1373–1382. ACM, New York (2011)

51. Guy, I., Ronen, I., Raviv, A.: Personalized activity streams: sifting through the "river of news". In: Proceedings of the Fifth ACM Conference on Recommender Systems, RecSys 2011, pp. 181–188. ACM, New York (2011)

52. Guy, I., Ronen, I., Wilcox, E.: Do you know? Recommending people to invite into your social network. In: Proceedings of the 14th International Conference on Intelligent User Interfaces, IUI 2009, pp. 77–86. ACM, New York (2009)

53. Guy, I., Ronen, I., Zwerdling, N., Zuyev-Grabovitch, I., Jacovi, M.: What is your organization 'like'? A study of liking activity in the enterprise. In: Proceedings of the SIGCHI Conference on Human Factors in Computing Systems, CHI 2016. ACM, New York (2016)

54. Guy, I., Steier, T., Barnea, M., Ronen, I., Daniel, T.: Swimming against the streamz: search and analytics over the enterprise activity stream. In: Proceedings of the 21st ACM International Conference on Information and Knowledge Management, CIKM 2012, pp. 1587–1591. ACM, New York (2012)

55. Guy, I., Ur, S., Ronen, I., Perer, A., Jacovi, M.: Do you want to know? Recommending strangers in the enterprise. In: Proceedings of the ACM 2011 Conference on Computer Supported Cooperative Work, CSCW 2011, pp. 285–294. ACM, New York (2011)

56. Guy, I., Ur, S., Ronen, I., Weber, S., Oral, T.: Best faces forward: a large-scale study of people search in the enterprise. In: Proceedings of the SIGCHI Conference on Human Factors in Computing Systems, CHI 2012, pp. 1775–1784. ACM, New York (2012)

57. Guy, I., Zwerdling, N., Carmel, D., Ronen, I., Uziel, E., Yogev, S., Ofek-Koifman, S.: Personalized recommendation of social software items based on social relations. In: Proceedings of the Third ACM Conference on Recommender Systems, RecSys 2009, pp. 53–60. ACM, New York (2009)

58. Han, S., He, D., Jiang, J., Yue, Z.: Supporting exploratory people search: a study of factor transparency and user control. In: Proceedings of the 22nd ACM International Conference on Information & Knowledge Management, CIKM 2013, pp. 449–458. ACM, New York (2013)

59. Hannon, J., Bennett, M., Smyth, B.: Recommending Twitter users to follow using content and collaborative filtering approaches. In: Proceedings of the Fourth ACM Conference on Recommender Systems, RecSys 2010, pp. 199–206. ACM, New York (2010)

60. Hannon, J., McCarthy, K., Smyth, B.: Content vs. tags for friend recommendation. In: Bramer, M., Petridis, M. (eds.) Research and Development in Intelligent Systems XXIX, pp. 289–302. Springer, London (2012)
61. Hinds, P., Carley, K., Krackhardt, D., Wholey, D.: Choosing work group members: balancing similarity, competence, and familiarity. Organ. Behav. Hum. Decis. Process. **81**(2), 226–251 (2000)
62. Hsieh, H.-P., Li, C.-T., Yan, R.: I see you: person-of-interest search in social networks. In: Proceedings of the 38th International ACM SIGIR Conference on Research and Development in Information Retrieval, SIGIR 2015, pp. 839–842. ACM, New York (2015)
63. Huang, S., Anton, P.M.: Conference networking recommendations based on past online activity (2012)
64. Huang, S.-W., Tunkelang, D., Karahalios, K.: The role of network distance in LinkedIn people search. In: Proceedings of the 37th International ACM SIGIR Conference on Research and Development in Information Retrieval, SIGIR 2014, pp. 867–870. ACM, New York (2014)
65. Jacovi, M., Guy, I., Kremer-Davidson, S., Porat, S., Aizenbud-Reshef, N.: The perception of others: inferring reputation from social media in the enterprise. In: Proceedings of the 17th ACM Conference on Computer Supported Cooperative Work and Social Computing, CSCW 2014, pp. 756–766. ACM, New York (2014)
66. Jacovi, M., Guy, I., Ronen, I., Perer, A., Uziel, E., Maslenko, M.: Digital traces of interest: deriving interest relationships from social media interactions. In: Bødker, S., Bouvin, N., Wulf, V., Ciolfi, L., Lutters, W. (eds.) ECSCW 2011, pp. 21–40. Springer, London (2011). https://doi.org/10.1007/978-0-85729-913-0_2
67. Jannach, D., Lerche, L., Zanker, M.: Recommending based on implicit feedback. In: Brusilovsky, P., He, D. (eds.) Social Information Access. LNCS, vol. 10100, pp. 510–569. Springer, Cham (2018)
68. Jiang, J., Hassan Awadallah, A., Jones, R., Ozertem, U., Zitouni, I., Gurunath Kulkarni, R., Khan, O.Z.: Automatic online evaluation of intelligent assistants. In: Proceedings of the 24th International Conference on World Wide Web, WWW 2015, pp. 506–516. ACM, New York (2015)
69. Kaplan, A.M., Haenlein, M.: Users of the world, unite! the challenges and opportunities of social media. Bus. Horiz. **53**(1), 59–68 (2010)
70. Kautz, H., Selman, B., Shah, M.: Referral web: combining social networks and collaborative filtering. Commun. ACM **40**(3), 63–65 (1997)
71. Kluver, D., Ekstrand, M., Konstan, J.: Rating-based collaborative filtering: algorithms and evaluation. In: Brusilovsky, P., He, D. (eds.) Social Information Access. LNCS, vol. 10100, pp. 344–390. Springer, Cham (2018)
72. Kunegis, J., Lommatzsch, A., Bauckhage, C.: The Slashdot Zoo: mining a social network with negative edges. In: Proceedings of the 18th International Conference on World Wide Web, WWW 2009, pp. 741–750. ACM, New York (2009)
73. Kwak, H., Chun, H., Moon, S.: Fragile online relationship: a first look at unfollow dynamics in Twitter. In: Proceedings of the SIGCHI Conference on Human Factors in Computing Systems, CHI 2011, pp. 1091–1100. ACM, New York (2011)
74. Kwak, H., Lee, C., Park, H., Moon, S.: What is Twitter, a social network or a news media? In: Proceedings of the 19th International Conference on World Wide Web, WWW 2010, pp. 591–600. ACM, New York (2010)
75. Lee, D., Brusilovsky, P.: Recommendations based on social links. In: Brusilovsky, P., He, D. (eds.) Social Information Access. LNCS, vol. 10100, pp. 391–440. Springer, Cham (2018)

76. Lempel, R., Moran, S.: SALSA: the stochastic approach for link-structure analysis. ACM Trans. Inf. Syst. **19**(2), 131–160 (2001)
77. Leskovec, J., Huttenlocher, D., Kleinberg, J.: Predicting positive and negative links in online social networks. In: Proceedings of the 19th International Conference on World Wide Web, WWW 2010, pp. 641–650. ACM, New York (2010)
78. Leskovec, J., Huttenlocher, D., Kleinberg, J.: Signed networks in social media. In: Proceedings of the SIGCHI Conference on Human Factors in Computing Systems, CHI 2010, pp. 1361–1370. ACM, New York (2010)
79. Li, M., Yu, S., Cao, N., Lou, W.: Privacy-preserving distributed profile matching in proximity-based mobile social networks. IEEE Trans. Wirel. Commun. **12**(5), 2024–2033 (2013)
80. Li, Q., Zheng, Y., Xie, X., Chen, Y., Liu, W., Ma, W.-Y.: Mining user similarity based on location history. In: Proceedings of the 16th ACM SIGSPATIAL International Conference on Advances in Geographic Information Systems, GIS 2008, pp. 34:1–34:10. ACM, New York (2008)
81. Liben-Nowell, D., Kleinberg, J.: The link-prediction problem for social networks. J. Am. Soc. Inf. Sci. Technol. **58**(7), 1019–1031 (2007)
82. Liu, Q., Xiong, Y., Huang, W.: Integrating social information into collaborative filtering for celebrities recommendation. In: Selamat, A., Nguyen, N.T., Haron, H. (eds.) ACIIDS 2013. LNCS (LNAI), vol. 7803, pp. 109–118. Springer, Heidelberg (2013). https://doi.org/10.1007/978-3-642-36543-0_12
83. Lopes, G.R., Moro, M.M., Wives, L.K., de Oliveira, J.P.M.: Collaboration recommendation on academic social networks. In: Trujillo, J., Dobbie, G., Kangassalo, H., Hartmann, S., Kirchberg, M., Rossi, M., Reinhartz-Berger, I., Zimányi, E., Frasincar, F. (eds.) ER 2010. LNCS, vol. 6413, pp. 190–199. Springer, Heidelberg (2010). https://doi.org/10.1007/978-3-642-16385-2_24
84. Lü, L., Zhou, T.: Link prediction in complex networks: a survey. Phys. A: Stat. Mech. Appl. **390**(6), 1150–1170 (2011)
85. Manca, M., Boratto, L., Carta, S.: Friend recommendation in a social bookmarking system: design and architecture guidelines. In: Arai, K., Kapoor, S., Bhatia, R. (eds.) Intelligent Systems in Science and Information 2014. SCI, vol. 591, pp. 227–242. Springer, Cham (2015). https://doi.org/10.1007/978-3-319-14654-6_14
86. McDonald, D.W., Ackerman, M.S.: Expertise recommender: a flexible recommendation system and architecture. In: Proceedings of the 2000 ACM Conference on Computer Supported Cooperative Work, CSCW 2000, pp. 231–240. ACM, New York (2000)
87. McNee, S.M., Riedl, J., Konstan, J.A.: Being accurate is not enough: how accuracy metrics have hurt recommender systems. In: CHI 2006 Extended Abstracts on Human Factors in Computing Systems, CHI EA 2006, pp. 1097–1101. ACM, New York (2006)
88. McPherson, M., Smith-Lovin, L., Cook, J.M.: Birds of a feather: homophily in social networks. Ann. Rev. Sociol. **27**, 415–444 (2001)
89. Naaman, M., Boase, J., Lai, C.-H.: Is it really about me? Message content in social awareness streams. In: Proceedings of the 2010 ACM Conference on Computer Supported Cooperative Work, CSCW 2010, pp. 189–192. ACM, New York (2010)
90. Nepal, S., Paris, C., Pour, P., Bista, S., Freyne, J.: A social trust based friend recommender for online communities. In: 2013 9th International Conference Conference on Collaborative Computing: Networking, Applications and Worksharing (Collaboratecom), pp. 419–428, October 2013

91. Pizzato, L., Rej, T., Akehurst, J., Koprinska, I., Yacef, K., Kay, J.: Recommending people to people: the nature of reciprocal recommenders with a case study in online dating. User Model. User-Adap. Interact. **23**(5), 447–488 (2013)

92. Pizzato, L., Rej, T., Chung, T., Koprinska, I., Kay, J.: Recon: a reciprocal recommender for online dating. In: Proceedings of the Fourth ACM Conference on Recommender Systems, RecSys 2010, pp. 207–214. ACM, New York (2010)

93. Qiao, X., Li, X., Su, Z., Cao, D.: A context-awareness dynamic friend recommendation approach for mobile social network users. Int. J. Adv. Intell. **3**(2), 155–172 (2011)

94. Quercia, D., Capra, L.: FriendSensing: recommending friends using mobile phones. In: Proceedings of the Third ACM Conference on Recommender Systems, RecSys 2009, pp. 273–276. ACM, New York (2009)

95. Rastogi, S., Singhai, R., Kumar, R.: A sentiment analysis based approach to Facebook user recommendation. Int. J. Comput. Appl. **90**(16), 21–25 (2014)

96. Ronen, I., Guy, I., Kravi, E., Barnea, M.: Recommending social media content to community owners. In: Proceedings of the 37th International ACM SIGIR Conference on Research and Development in Information Retrieval, SIGIR 2014, pp. 243–252. ACM, New York (2014)

97. Saez-Trumper, D., Quercia, D., Crowcroft, J.: Ads and the city: considering geographic distance goes a long way. In: Proceedings of the Sixth ACM Conference on Recommender Systems, RecSys 2012, pp. 187–194. ACM, New York (2012)

98. Samanthula, B.K., Jiang, W.: A randomized approach for structural and message based private friend recommendation in online social networks. In: Can, F., Özyer, T., Polat, F. (eds.) State of the Art Applications of Social Network Analysis. LNSN, pp. 1–34. Springer, Cham (2014). https://doi.org/10.1007/978-3-319-05912-9_1

99. Scellato, S., Noulas, A., Lambiotte, R., Mascolo, C.: Socio-spatial properties of online location-based social networks. ICWSM **11**, 329–336 (2011)

100. Scellato, S., Noulas, A., Mascolo, C.: Exploiting place features in link prediction on location-based social networks. In: Proceedings of the 17th ACM SIGKDD International Conference on Knowledge Discovery and Data Mining, KDD 2011, pp. 1046–1054. ACM, New York (2011)

101. Sharma, A., Hofman, J.M., Watts, D.J.: Estimating the causal impact of recommendation systems from observational data. In: Proceedings of the Sixteenth ACM Conference on Economics and Computation, EC 2015, pp. 453–470. ACM, New York (2015)

102. Silva, N., Tsang, I.-R., Cavalcanti, G., Tsang, I.-J.: A graph-based friend recommendation system using genetic algorithm. In: 2010 IEEE Congress on Evolutionary Computation (CEC), pp. 1–7, July 2010

103. Sinha, R., Sinha, R., Swearingen, K.: Comparing recommendations made by online systems and friends. In: Proceedings of the DELOS-NSF Workshop on Personalization and Recommender Systems in Digital Libraries (2001)

104. Spirin, N.V., He, J., Develin, M., Karahalios, K.G., Boucher, M.: People search within an online social network: large scale analysis of Facebook graph search query logs. In: Proceedings of the 23rd ACM International Conference on Conference on Information and Knowledge Management, CIKM 2014, pp. 1009–1018. ACM, New York (2014)

105. Sun, E., Rosenn, I., Marlow, C., Lento, T.M.: Gesundheit! modeling contagion through Facebook news feed. In: Proceedings of the Third International Conference on Weblogs and Social Media, ICWSM 2009. AAAI Press, San Jose (2009)

106. Surian, D., Liu, N., Lo, D., Tong, H., Lim, E.-P., Faloutsos, C.: Recommending people in developers' collaboration network. In: 2011 18th Working Conference on Reverse Engineering (WCRE), pp. 379–388, October 2011
107. Symeonidis, P., Tiakas, E., Manolopoulos, Y.: Transitive node similarity for link prediction in social networks with positive and negative links. In: Proceedings of the Fourth ACM Conference on Recommender Systems, RecSys 2010, pp. 183–190. ACM, New York (2010)
108. Tang, L., Ni, Z., Xiong, H., Zhu, H.: Locating targets through mention in Twitter. World Wide Web 18(4), 1019–1049 (2015)
109. Tannenbaum, S.I., Beard, R.L., Salas, E.: Team building and its influence on team effectiveness: an examination of conceptual and empirical developments. In: Kelley, K. (ed.) Issues, Theory, and Research in Industrial/organizational Psychology. Advances in psychology, vol. 82, pp. 117–153. North-Holland (1992)
110. Terveen, L., McDonald, D.W.: Social matching: a framework and research agenda. ACM Trans. Comput. Hum. Interact. 12(3), 401–434 (2005)
111. Tintarev, N., Masthoff, J.: A survey of explanations in recommender systems. In: 2007 IEEE 23rd International Conference on Data Engineering Workshop, pp. 801–810, April 2007
112. Tu, W., Cheung, D.W., Mamoulis, N., Yang, M., Lu, Z.: Activity-partner recommendation. In: Cao, T., Lim, E.-P., Zhou, Z.-H., Ho, T.-B., Cheung, D., Motoda, H. (eds.) PAKDD 2015. LNCS (LNAI), vol. 9077, pp. 591–604. Springer, Cham (2015). https://doi.org/10.1007/978-3-319-18038-0_46
113. Van Le, T., Nghia Truong, T., Vu Pham, T.: A content-based approach for user profile modeling and matching on social networks. In: Murty, M.N., He, X., Chillarige, R.R., Weng, P. (eds.) MIWAI 2014. LNCS (LNAI), vol. 8875, pp. 232–243. Springer, Cham (2014). https://doi.org/10.1007/978-3-319-13365-2_21
114. Wang, B., Wang, C., Bu, J., Chen, C., Zhang, W.V., Cai, D., He, X.: Whom to mention: expand the diffusion of tweets by @ recommendation on micro-blogging systems. In: Proceedings of the 22nd International Conference on World Wide Web, WWW 2013, pp. 1331–1340. Republic and Canton of Geneva, Switzerland (2013). International World Wide Web Conferences Steering Committee
115. Wang, D., Pedreschi, D., Song, C., Giannotti, F., Barabasi, A.-L.: Human mobility, social ties, and link prediction. In: Proceedings of the 17th ACM SIGKDD International Conference on Knowledge Discovery and Data Mining, KDD 2011, pp. 1100–1108. ACM, New York (2011)
116. Wang, Z., Liao, J., Cao, Q., Qi, H., Wang, Z.: Friendbook: a semantic-based friend recommendation system for social networks. IEEE Trans. Mob. Comput. 14(3), 538–551 (2015)
117. Weerkamp, W., Berendsen, R., Kovachev, B., Meij, E., Balog, K., de Rijke, M.: People searching for people: analysis of a people search engine log. In: Proceedings of the 34th International ACM SIGIR Conference on Research and Development in Information Retrieval, SIGIR 2011, pp. 45–54. ACM, New York (2011)
118. Wicks, P., Massagli, M., Frost, J., Brownstein, C., Okun, S., Vaughan, T., Bradley, R., Heywood, J.: Sharing health data for better outcomes on PatientsLikeMe. J. Med. Internet Res. 12(2), e19 (2010)
119. Xia, F., Asabere, N., Liu, H., Chen, Z., Wang, W.: Socially aware conference participant recommendation with personality traits. IEEE Syst. J. PP(99), 1–12 (2014)

120. Yogev, A., Guy, I., Ronen, I., Zwerdling, N., Barnea, M.: Social media-based expertise evidence. In: Boulus-Rødje, N., Ellingsen, G., Bratteteig, T., Aanestad, M., Bjørn, P. (eds.) ECSCW 2015, pp. 63–82. Springer, Cham (2015). https://doi.org/10.1007/978-3-319-20499-4_4
121. Zhang, L., Fang, H., Ng, W.K., Zhang, J.: IntRank: interaction ranking-based trustworthy friend recommendation. In: 2011 IEEE 10th International Conference on Trust, Security and Privacy in Computing and Communications (TrustCom), pp. 266–273, November 2011
122. Zheng, N., Li, Q., Liao, S., Zhang, L.: Flickr group recommendation based on tensor decomposition. In: Proceedings of the 33rd International ACM SIGIR Conference on Research and Development in Information Retrieval, SIGIR 2010, pp. 737–738. ACM, New York (2010)
123. Zheng, Y., Zhang, L., Ma, Z., Xie, X., Ma, W.-Y.: Recommending friends and locations based on individual location history. ACM Trans. Web 5(1), 5:1–5:44 (2011)

16
Location Recommendation with Social Media Data

Cécile Bothorel[1], Neal Lathia[2], Romain Picot-Clemente[3(✉)] (iD),
and Anastasios Noulas[4]

[1] IMT Atlantique, Lab-STICC, Univ. Bretagne Loire, 29238 Brest, France
`cecile.bothorel@imt-atlantique.fr`
[2] Skyscanner, Edinburgh, UK
`neal.lathia@gmail.com`
[3] Saagie, Le Petit-Quevilly, France
`romain@saagie.com`
[4] School of Computing and Communications, University of Lancaster, Lancaster, UK
`a.noulas@lancaster.ac.uk`

Abstract. Smartphones with inbuilt location-sensing technologies are now creating a new realm for recommender systems research and pratice. In this chapter, we focus on recommender systems that use location data to help users navigate the physical world. We examine various recommendation problems: recommending new places, recommending the next place to visit, events to attend, and recommending neighbourhoods or large areas to explore further. Lastly, we discuss how (personalized) place search is analogous to web search. For each of these domains, we present relevant data, algorithms, and methods, and we illustrate how researchers are investigating them with examples from the literature. We close by summarizing key aspects and suggesting future directions.

Keywords: Location data · Venue data
Location-based social networks · Points-of-Interest (POIs)
Location recommendation · Recommendation using location data
New places recommendation · The next place to go recommendation
Events recommendation · Neighbourhoods recommendation
Place search

1 Introduction

As smartphone adoption continues to grow, interaction with online social networks and recommender systems while *on the go* is becoming increasingly commonplace. Services that were once exclusively web-based can now provide users with personalised information about their whereabouts: smartphones can guide users to events, neighbourhoods, and locations that may be of interest to them, allowing them to discover places that surround them. In doing so, location recommender systems capture the emerging trend of context-awareness in personalised

P. Brusilovsky and D. He (Eds.): Social Information Access, LNCS 10100, pp. 624–653, 2018.
https://doi.org/10.1007/978-3-319-90092-6_16

systems [3]; users can be given information that is relevant to where, when, and who they are.

The shift towards smartphone-centric, location-aware interaction is now widening the scope of recommender system research and practice. Social data is already widely used to improve user experiences in many application areas, e.g. helping users navigate electronic information spaces (See Chap. 5 of this book [19]). Location-based recommender systems allow researchers to apply similar techniques to help users navigate real, physical spaces. As previous chapters in this book have addressed, recommender systems have traditionally been designed to handle explicit preferences (e.g., ratings, tags in Chap. 10 [32]), implicit preferences captured via users' actions and behaviours (ref. Chaps. 13 [45] and 14 [28]), as well as social links [6]. Location-based recommender systems now introduce sensor-derived (i.e., GPS) location data, which adds a spatial layer to users' preferences, to this landscape. Recommender algorithms have usually centered around rating prediction for item ranking: adding in geography to preferences allows these systems to address different prediction questions, such as recommending the *next location* that a user may be interested in [11,48]. Finally, building on the growing literature around evaluating recommender systems, location-based recommenders introduce new challenges when assessing their quality.

In this chapter, we aim to introduce the broad area of location recommendation using social and smartphone data. We do not aim to provide a comprehensive review of the state-of-the-art; instead, our goal is to address different types of recommendations, and specifically focus, in each example, on the data, algorithms, techniques, and evaluation metrics that are applied.

1. **Data.** What sources of data can be used to build location recommenders? We describe a variety of datasets, and how they are processed into features that can be used when generating recommendations.
2. **Algorithms.** What algorithmic approaches have been applied to location recommendation? We describe the questions and methods that researchers have used to build location recommender systems, including collaborative filtering (deeply detailed in the dedicated Chap. 10 of this book [32]), graph-based approaches, and other supervised learning methods.
3. **Evaluation.** Metrics which have already been introduced in previous chapters (Chap. 10 [32]) are also used for location recommendation. We give an overview of the methods and metrics used in both online and offline evaluations of location recommendations at the end of the chapter. We also provide a special focus on different other methods when required in the scope of a specific problem or example.

We examine these three aspects—data, algorithms and evaluation—within four problem areas: recommending new places, recommending next place to go or how to use the context to provide recommendations in real-time, recommending events and neighbourhoods and finally place search. We close the chapter by enumerating some emergent themes, future applications and research questions.

2   Location Data

In this section, we review the kind of data that has been used when examining peoples' preferences and mobility. In particular, we focus on the types of data that can be sourced from location-based social networks and review sites (check-ins, tags, ratings), geo-located social streams, and mobility data from smartphones and urban sensors.

2.1   Location-Based Social Networks

The most prominent source of spatial user preferences are location-based social networks (LBSNs). Popular examples here include Foursquare, Gowalla, and Facebook Places (see Fig. 1). Many of these services inherit the same data-generating interaction methods that have been popularised by web-based recommendation and social sites. For example, users can add friends, and venues can be rated, tagged, categorised, and reviewed (sometimes referred to as 'comments' or 'tips'). However, the unique interaction method that has emerged in these services is the concept of 'checking in,' where users manually select a venue where they currently are. Users are encouraged to do so via gamification (e.g., gaining points by checking in, or obtaining 'mayorships' for having checked in the most frequently at a particular venue) or to simply broadcast their current location to their friends–research has shown that users opt to check in (or not) for a variety of reasons [39]. This data can be viewed as a unary flag denoting presence at (not necessarily preference for) a venue.

While location-based social networks share similarities with their web-based predecessors, social networks that are *not* LBSNs are now also adopting location-enhanced features. For example, Twitter geolocates tweets, and Flickr and Instagram geolocate photographs. All of these systems therefore collect data about users, their social connections, preferences for (some kind of) item, location, and time: the key ingredients required to build location recommendation services.

2.2   Venue Data

Location-based recommender systems rely on having a database of locations, or Points-of-Interest (POIs), that can be recommended. The recent literature has described a number of means of finding and inferring POIs from users' data. Foursquare, for example, maintains their POI database via crowd sourcing; the check-ins that users provide can then be used to uncover venues' spatio-temporal patterns [44]. Others, instead, infer them from implicit data. These include, for example, sourcing POIs by clustering geo-tagged photographs that users upload to services like Flickr [17]. These datasets can be used to automatically extract features of places and events [53]. Further information about the inferred items can be gathered by intersecting the location data with any available content and tags [31].

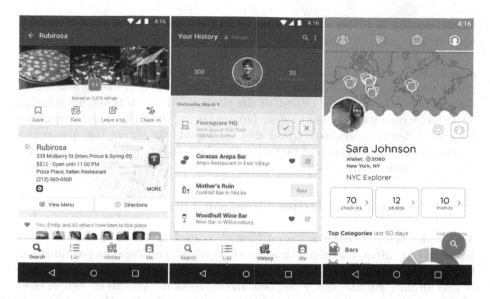

Fig. 1. LBSNs mobile applications: Foursquare City Guide and Foursquare Swarm. Retrieved from the Fourthsquare Press Kit on https://foursquare.com/about/.

2.3   Smartphones and Urban Sensors

Beyond the location-based social networks described above, researchers have leveraged smartphones to collect data via both sensors and participant's direct input. These include the phones' Global Position System (GPS) [68], proximity sensors [18], GSM traces [56], as well as Call Detail Records (CDRs) that are created when devices pair with cellular network communication towers [10]. These sources of data uncover a vast range of features about users' behaviours, including how far they tend to travel, their likely mode of travel, and the urban areas they frequent [54]. All of these data sources differ from one another in how easily and accurately they may be collected. Typically, sources such as GSM and CDRs are only available to mobile operators; GPS and similar on-board location services require a tailor made app-based data collector. While the former kind of data is typically coarse-grained, and GPS can provide much finer-grained samples (both spatially and temporally), fully efficient implementations are dependent on the needs of the underlying application. In particular, continuously querying a phone's GPS sensor will quickly degrade the device's battery: system designers need to trade-off between the sampling accuracy that they seek and the energy efficiency of their application [52].

There are a number of studies that indicate why these data sources are relevant to recommender systems. Froehlich *et al.* [22] found that mobility patterns correlate with users' preferences: people tend to frequent those places that they like; similarity between users can also be measured from location histories [37]. Moreover, GPS traces can be mined for 'interesting' locations [69] in order to recommend locations and activities [67]; further details of the algorithmic approaches appear in the following sections.

Beyond phones, however, urban areas are increasingly being digitised in a manner that allows for the collection of mobility data. For example, the London public transport system uses the Oyster card: a personal contact-less smart card that allows passengers to access all of the city's multi-modal transport systems. The Oyster card itself is used to store fares; in doing so, it uncovers the variance between different individuals' travel choices [34] and the extent that passengers overspend on public transport by failing to relate their travel behaviours to the fare most suited to them [33].

3   Recommending Using Location Data

The traditional recommendation problem is defined as one of predicting values for unrated content, and using those predictions to rank items as recommendations. In the location recommendation scenario, there are a number of variants of this formulation that can be applied. In this section, we focus on four of these problems: recommending *new* places to go to, recommending the *next* place to go, recommending *events* to attend and *neighbourhoods* to explore, and finally *place search*. For each problem we focus on the relevant data, current techniques and algorithms and provide illustrative examples of research or services to illustrate.

It is worth noting that traditional methods for recommending web items [2] remain relevant when recommending places, and that the LBSN-specific methods build on this foundation. We therefore briefly recap how traditional user-based collaborative filtering works; for more details, see Chap. 11. User-based collaborative filtering algorithms predict an item's score based on the similarity between users' ratings. Let u be the target user and U all users. The predicted score of u for an item i (from the set of all items I), denoted $\widehat{s}(u, i)$, is defined by the average rating of other users v on the item i, denoted $s(v, i)$, weighted by their usage similarity with u, denoted $sim(u, v)$:

$$\widehat{s}(u, i) = \frac{\sum_{v \in U} sim(u, v) \times s(v, i)}{\sum_{v \in U} sim(u, v)} \qquad (1)$$

We will use this notation in the subsections below.

Places are different from ordinary items due to their geographical attributes, which is a key feature to include when improving recommendation quality [67]. For example, researchers compared traditional recommender algorithms, from simple, non-personalized popular venue recommendation to make predictions based on a matrix factorization method, on two datasets (Gowalla and Foursquare), and found that the collaborative filtering approaches that have been successful in online recommendation scenarios, like matrix factorization, have not achieved a similar status with mobility data [43]. Geographical information can help to discriminate between otherwise similar places. For instance, it is likely that a very far place from a target user should not be relevant for her within a short time frame. Moreover, if the user has the choice between two identical shops at two different locations, the one closer to her usual shopping area is more likely to be preferred [66].

3.1 Recommending New Places

The main goal of recommending *new* places is to suggest interesting *previously unvisited* places to a target user. These places can be, for example, monuments, restaurants or any new places to discover–and apply both when a user is in their 'home' environment as well as in tourist contexts.

LBSNs have popularised the concept of new place recommendation; often, these are implemented using a 'show me only places that I have not been to before' feature. The recommendations in these kind of networks are not only based on preference and geography, but also use social links: assuming that friends tend to be like-minded (homophily), those unvisited venues that friends have checked in to may be useful recommendations. Hence, recommender systems of new places using social data try to improve traditional recommender systems by considering two additional dimensions beyond the usual preference (affinity) dimension: social and geographical dimensions.

In the following subsections, we present studies that have addressed the issue of new place recommendation. There are two main approaches that have been proposed in the literature: similarity-based and graph-based approaches. We present a detailed overview of both approaches.

3.1.1 Similarity-Based Approaches
One of the first studies that focused on place recommendation with social data is presented in [62]. The authors propose a memory-based approach that incorporates user preferences, social influence and geographical influence to generate recommendations. To do this, the score of a place for a given user is the combination of three scores: (1) a score proportional to the similarity of users in terms of check-ins, (2) a score calculated according to the similarity of friends in terms of check-ins and social network (common friends), and (3) the check-in probability of the user for this place according to based on that venue's proximity to previous check-ins and the overall distribution of distances between mutual check-ins. Each of the 3 predicted scores is computed based on an adaptation of Eq. (1). Since LBSNs do not always include ratings, similarity is measured between users with 'ratings' by users for places defined as, generally, being proportional to the number of check-ins (perhaps appropriately normalised).

More formally, the three scores in [62] are defined as follows:

1. **Usage-Based Score.** The first predicted score for a user u to place i, denoted $\widehat{s_{usage}}(u, i)$, is only based on the usage similarity (check-ins similarity). It is computed like in Eq. (1), with $sim(u, v)$ being the cosine similarity between the usage scores of users u and v:

$$sim_{usage}(u,v) = \frac{\sum_{j \in I} s(u,j) \times s(v,j)}{\sqrt{\sum_{j \in I}(s(u,j))^2} \times \sqrt{\sum_{j \in I}(s(v,j))^2}} \qquad (2)$$

2. **Social-Based Score.** A second predicted score, denoted $\widehat{s_{social}}(u,i)$, is related to the social influence and is computed by only considering friends (the set F_u) of the target user u for the similarity computation. The score is again computed as above; this time, the similarity between two friend users u and v, denoted $sim_{social}(u,v)$, is based both on usage similarity and social network similarity:

$$sim_{social}(u,v) = \eta \cdot \frac{|F_u \bigcap F_v|}{|F_u \bigcup F_v|} + (1 - \eta) \cdot \frac{|I_u \bigcap I_v|}{|I_u \bigcup I_v|} \qquad (3)$$

where $\eta \in [0,1]$ is a tunning parameter and I_u the place set of the user u_k. The first part of this weighted sum is the social network similarity and the second part is the usage similarity.

3. **Geographic-Based Score.** Finally, a third predicted score, denoted $\widehat{s_{geo}}(u,i)$, is calculated based on the distribution of distances between mutual check-ins. The score is inversely proportion to the distance between the target place and the typical places the user frequents. The first step consists in generating, from the history of visits, the distribution of distances between mutual check-ins on the whole set of users. The distribution is approximated by a function f depending on its shape. Some works [13,44] have shown that the overall distribution of distances globally looks like a power law function in location-based social networks as shown in Fig. 2. The computation detail of this approximation is provided in [62].

This function f produces a check-in probability in a place according to its distance with other check-ins, it is possible to calculate a global probability $\Pr(u,i)$ of check-in in a place i according to its distance from **all** check-ins of the target user u. This probability is the third score and is defined as follows:

$$\widehat{s_{geo}}(u,i) = \Pr(u,i) = \prod_{k \in I_u} f(distance(i,k)) \qquad (4)$$

where I_u is the set of places visited by the user u and $distance(i,k)$ is the distance in kilometer between the places i and k.

These three predicted scores are aggregated to generate one predicted score, denoted $\widehat{score}(u,i)$, using a weighted sum as follows:

$$\widehat{s}(u,i) = (1 - \alpha - \beta)\widehat{s_{usage}}(u,i) + \alpha\widehat{s_{social}}(u,i) + \beta\widehat{s_{geo}}(u,i) \qquad (5)$$

where α and β are two weighting parameters defined such that $0 \leq \alpha + \beta \leq 1$. For each user u, the unvisited places with the best predicted scores are selected as recommendations. These scores are combined using a weighted sum and calculated for each place in order to deduce recommendations. Their method is

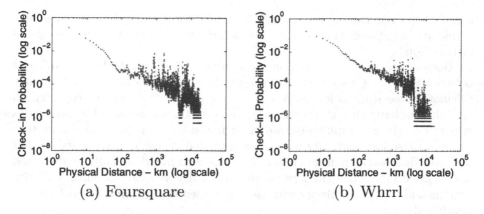

(a) Foursquare   (b) Whrrl

Fig. 2. Geographical influence probability distribution. Used with permission from [62].

tested on 2 datasets (Whrrl and Foursquare) and the results show that the recommendation of new places can be improved, in terms of recall and precision metrics, by taking social and geographical influence into account.

Recent research has proposed alternatives which may improve this method. For example, [66] have shown that the use of a personalized distribution according to the check-ins' behavior of the target user, instead of an overall distribution (for computing $\widehat{score}_{geo}$), improved the quality of recommendations. Moreover, [23,66] have shown that taking into account the distance between friends for computing the social similarity improves recommendation quality. A simple way for considering this distance influence between friends, could be the use of a coefficient to moderate the social similarity score sim_{social} into the social predicted score $score_{social}$, according to the distance between the target user and his/her friends.

Similarly, [23] proposes four 'social circles' for calculating the score of a given place. These circles respectively represent "close friends," "close non-friends," "distant friends" and "distant non-friends." These four scores are then combined in one score for ranking recommendations. Their tests on a Foursquare dataset show that there are social correlations in LBSNs, but that these correlations are more important between close non-friends users than friends regardless of their distance. In [66], the authors propose to compute scores of places considering geographic influence of places and considering geographic similarities between friends. The geographic similarities are based on places of residence of users. The closer places of residence of two users are, the more geographically similar they are. A first score is calculated as the average of the number of visits of friends, weighted by their geographic similarity with the target user. The final score of a place is an aggregation of the previous score (normalized) with the check-in probability in this place according to its distance with the previous check-ins of the user. This probability is based on a personalized distribution of distances between mutual check-ins, which is different from [62] that considers the global distribution of distances in the computation. They show that considering a personalized distribution gives significantly better recommendations, based on recall and precision metrics, than considering the global distribution.

Both of these methods include social influence when calculating similarity between the target user and other users. They differ, however, on how they calculate these similarities and on how they choose the users to consider. In [66], the similarity that is computed between the target user and his/her friends is based on the geographic position of their residence. In [62], two similarities are computed: one similarity with all users based on the mutual visits another with friends based on mutual visits and on common friends. Finally, as discussed above, [23] compute four similarity scores: similarity with close friends, similarity with distant friends, similarity with close non-friends and similarity with distant non-friends.

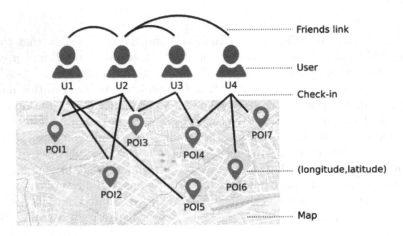

Fig. 3. Graph representation of user-user friendship and user-location check-in activity in a LBSN.

3.1.2 Graph-Based Approaches

A different approach, based on graph-mining, has been introduced in the recent literature. In [43], the authors propose to construct an undirected graph in which nodes are the users and places. Social relations between users are represented by edges between corresponding user nodes in the graph. Users' check-ins are represented by edges between corresponding users and places (The same graph is used by [62], see Fig. 3). For a target user, their recommendation method relies on a random walk algorithm that allows finding potentially interesting places that are indirectly connected to the user. Evaluations on Gowalla and Foursquare data demonstrated that the random walk method outperforms collaborative filtering approaches like matrix factorization.

In [47], the authors propose another graph-based method: the authors first construct a graph including social, check-ins and geographic information, in order to map relations between users and places.

The first step consists in creating this graph in which nodes represents users and places, and in which edges can link users with users, users with places, and places with places. To accomplish this, the authors initially create three distinct graphs: a usage (or frequency) graph denoted $\mathbf{F}$, a social graph denoted $\mathbf{S}$, and a geographic graph denoted $\mathbf{G}$. Then, these graphs are combined into a unique graph called $\mathbf{C}$. More formally:

1. The usage graph $\mathbf{F}$ is based on the check-ins of users in the different places. In this graph, nodes are either users or places. Edges connect users with the places they have visited (in which they have made one or several check-in(s)) and they are weighted according to the number of visits. The matrix F representing this graph is an $N \times M$ matrix (M is the number of places), in which F_{ui} is the number of time that user u has visited place i.

2. The social graph $\mathbf{S}$ is based on the friendship relations between users. In this graph, nodes represent users, and edges connect users with users if there are friendship relations between them in their social network. The matrix S representing this graph is a $N \times N$ symmetric matrix (N is the number of users), in which S_{uv} is 1 if there is a friendship relation between the user u and the user v, and is 0 otherwise.

3. The geographic graph $\mathbf{G}$ connects places together. The matrix G representing this graph is a $M \times M$ matrix; [47] proposes to construct this geographic graph by linking places together with weights that are the probabilities of check-ins to them, according to their mutual distance. The closer a place is from an other place, the higher the weight between these two places is. Similar to [62], the authors propose to first generate the distribution of distances between mutual check-ins on the whole set of users and approximating it with a power-law function f. Then the undirected transition probability $\Pr(i,j)$ between two places i and j is defined as follows:

$$\Pr(i,j) = f(distance(i,j)) \tag{6}$$

where $distance(i,j)$ is the distance in kilometers between the places i and j. Finally, the geographic matrix G is defined where G_{ij}, the weight on the edge between the places i and j, in the graph $\mathbf{G}$, is the transition probability between i and j. The authors merge the graphs $\mathbf{F}, \mathbf{S}, \mathbf{G}$ in one unique graph $\mathbf{C}$.

In this combined graph, nodes are either places or users, and edges can connect users together, places together or users with places:

$$C = \begin{pmatrix} \alpha S & \delta F \\ \delta F^{\mathsf{T}} & \gamma G \end{pmatrix} \tag{7}$$

where α, δ and γ are respectively the influence coefficients of the matrices S, F, and G in the matrix C, with $\alpha + \delta + \gamma = 1$. The Katz centrality method [30] is then used on the graph C for identifying non-obvious relations between users and places.

The authors propose to limit the computation of the Katz centrality to a maximum rank k. Finally, for each user u, the unvisited places with the best weights on the line of the user in the matrix $tKatz(C, k)_{12}$ are selected as recommendations to provide to the user. The authors compared their method with the memory-based method of [62] and show significant improvement of recommendations in terms of recall and precision metrics. Their results also show that the inclusion of social relationships can sometimes degrade recommendations. They conclude that the inclusion of social relationships in the recommendation process should be based on the propensity of the target user to be influenced by their social network.

Some modifications have been suggested for improving this graph-based approach. First, other methods of weight propagation such as the Random Walk method used in [43] can be tested and compared with the Katz centrality to select the most efficient one. As before, using a personalized distribution of distances between mutual check-ins for constructing the geographic matrix may improve the quality of recommendations. Moreover, weighting edges between friends in the social graph by taking into account the distances between them (as suggested in [23,66]) is an other simple way to possibly improve the recommendations.

3.2   Recommending the Next Place

The previous section describes approaches to discover new places. The implicit objective is to reduce the overall searching time and extract relevant places from exhaustive lists of potential locations, but independently of the current situation of users. An alternative is to consider that users may want a recommendation that takes into account their current location and/or context. For example: what is the best place to eat after visiting the Louvre in Paris? What is a good bar to go to after dining in Shoreditch in London? Much like the work focusing on users' current session within e-commerce platforms (Chap. 15), next-place prediction focuses on a setting with a short-term goal. LBSNs are now popularising the idea of the 'next place' by providing users with information about where others tend to check in *after* they have checked in to where the user has just checked in to.

The concept of next-place venue recommendation is closely tied to the idea of context-aware recommender systems. One the most cited definitions of *context* refers to *any information that can be used to characterize the situation of a user: location, identity, time and activity* [1]. Recently [3] provided an overview of the multifaceted notion of context in the field of recommender systems, and among many other applications, referred to the travel guides recommending restaurants or places of interest. Therefore, while next-place recommender systems use all of the same data described in the previous section, they may also source further data to describe a user's context. This could include, for example, weather, traffic, or public transit statuses. Furthermore, these systems may use data that has been inferred about the user's current activity/location; for example, that the user is currently in an unfamiliar city as a tourist. However, accurately modeling context-aware systems remains challenging: Which contextual factors need to be considered? How should they be encoded? Are these factors correlated? Which POIs are suitable to which context? In the following, we examine how some researchers have built context-aware recommenders.

In [42], the authors formalise next-place recommendation as a supervised learning problem. Given a user u whose current check-in is c (to venue l at time t), our aim is to rank the set of venues L so that the *next* venue to be visited by the user will be ranked at the highest possible position in the list. According to this setting, the next check-in problem is essentially a ranking task, where a ranking score r for all venues in L is computed. This approach takes a number of steps:

1. **Item pre-filtering.** The authors filter candidate venues by selecting those that are in the *same city* as l'. Pre- and post-filtering are methods used to select relevant items using contextual information [3]. In prefiltering approaches, contextual information is used *before* item ranks/scores are computed: only the relevant set of data is served to the recommender which then computes recommendations upon the selected input. In postfiltering approaches, no filtering is done on the input data: the recommender ranks items from the entire dataset, and the ones with few ratings in the target context are penalized.

2. **Feature extraction.** Given a set of user check-ins, a variety of descriptive features can be extracted. These include:
   - User features: the authors define features that count how frequently the user has visited nearby venues in the past, as well as how frequently the user has visited locations of a particular type (e.g., coffee shops) in the past. These features capture the assumption that users are likely to return to places that they visit often, as well as go to the kinds of places that they tend to go to.
   - Social features: the authors define a feature that is a count of how often that user's friends have visited nearby places. This feature captures the assumption that a user is likely to visit places where friends have been.
   - Global features: since users are likely to visit both popular and nearby places, the authors define features that consider how popular places are as well as how close they are to the current user's location. They consider both geographic as well as rank distance, which sorts venues based on the number of venues between the current and target place. More importantly, the authors also define a transition feature, which counts how frequently users transition from the current venue to the target one.
   - Temporal features: since some venues may be more appealing at certain hours than at others (e.g., a coffee shop in the morning, a nightclub in the evening), the authors further define some temporal features that enumerate how likely it is for a venue to be visited at a particular time of day.

3. **Supervised Learning.** Finally, given the features described above, the authors create a training and test set and evaluate how well various supervised learning approaches can predict where a user will check in next. Most notably, if the authors limited their training set to positive instances derived from the check-in data, the algorithms would be trained with highly biased data. They therefore produce a balanced dataset, including negative samples that are created by randomly selecting venues that the user has not visited.

In [11], the authors provide an example of another contextual system which also makes use of a prefiltering step. Polar[1] is a mobile application which suggests POIs adapted to the users preferences and their current needs. When users query the system, e.g., for a restaurant, the application encodes the current information such as time, location and speed, and sends it to the server along with the query.

Fig. 4. Polar: a mobile application with the top personalised POIs. Image retrieved from Polar's web site.

The recommendation engine receives the current context from the mobile device and enriches it with additional information such as weather and traffic reports from external public services. The engine then searches for, in this case, restaurants in its local database. Restaurants that match the user's profile and the current context are pre-filtered, then ranked and sent to the user. For example, the first result may be a French restaurant within 10 min walk from the user's location. Data extraction of POIs from the social web plays an important role in populating the local database. The aim of this backend process in Fig. 5 is to describe POIs by metadata such as addresses, zip code, hours, price range, phone number, accessibility information, take-out, outdoor seating, etc. An extensive data extraction process classifies the POIs into categories and qualify them by tags such as "French", "traditional", "cassoulet", "finest", "sophisticated", etc. The information is collected from specialized web sites, e.g. Yelp[2], TripAdvisor[3], from users' reviews posted and judged according to their usefulness, e.g. Yelp,

---

[1] http://www.dia.uniroma3.it/~ailab/?page_id=97.

[2] www.yelp.com.

[3] www.tripadvisor.com.

from tag-based bookmarking services, e.g. Delicious[4] and from location-based services like Google Maps.

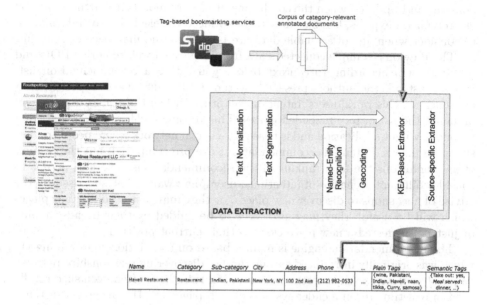

Fig. 5. Polar's backend process to describe POIs by metadata. Image retrieved from Polar's web site.

The *Polar* system is based on a standard feed-forward perceptron with one hidden layer. This network finds out how close the POIs are to the user's current context. Only the POIs close to the current location are retrieved from the local database (as per the prefiltering approach). This selection is presented to the input layer of the perceptron; the selected POIs are decomposed into twelve features such as "outdoor seating", "time before closing" or "categories". The contextual features also feed the input layer: the mode of transportation inferred from the mobile time and speed, the current activity among {working, traveling, other} inferred also from the distance to work and the speed, the weather {good, bad} deduced from pre-processing of external information and the time of the day {morning, lunch, afternoon, dinner, night} inferred from the time. The neural network outputs the classification of each POI into five classes, indicating how much they match the current context. The top personalised POIs are presented to the user with detailed information such as hours, price range, time to go and excerpts of users' reviews (see Fig. 4).

According to [11], research into these kinds of systems is non-trivial and highly time-consuming, since contextual variables need to be collected and encoded, and rules for their usage need to be defined. To assess the relevance of

---

[4] www.delicious.com.

various contextual variables, some systems rely on experts [60], even though these approaches ultimately rely on writing hand-coded rules. Machine learning-based approaches are preferable, because they can easily be extended into different domains and updated when the conditions of use are modified. Furthermore, by not relying on experts, less bias may be introduced to select the raw information to consider when any of available data are processed and may become relevant.

The previous example collected data from the social web to enrich POIs and prefilter them according to context before generating a recommendation list. In [48], instead, the authors present a system that makes use of a postfiltering approach. In this work, users' intentions, captured when they query the system, influence the list of recommended places. This recommender system is part of a mobile payment application and is based on a social network where people review shopping places. Potential recommendations are pre-computed, but the context of use conditions the final ranking of the recommendations. The idea here is to consider different scenarios of intentions: users may want to be surprised by the advise of new shops to discover new places, or they may just need relevant places which will be useful; they may also want to be guided by their friends' habits, or just be suggested a new place close to their current position.

The recommendation engine is mainly based on association rules mining [4], which was originally introduced as a method of discovering relationships between co-purchased items. More recently, association rules have often been successfully used in real-time recommender systems of web pages [21,41]. In typical systems, an offline component generates the association rules based on usage histories and an online component is in charge of providing real-time recommendations to users based on matching association rules with the current usages of users. The intention-based recommender system in [48] follows the same architecture, see Fig. 6. Recommending the next place to go can be decomposed into three steps:

1. **Offline Learning,** which is in charge of extracting a set of association rules r_i from the set S_k of all the past visiting sessions of all users. The association rules are extracted according to a *support* threshold to ensure that the extracted rules are verified a minimum number of times.
2. **Rule Selection.** Based on the association rules extracted in the learning part, the recommendation part is in charge of finding suitable places for a given user according to her profile, context and desired type of recommendation. This is done by selecting the rules that are related to the current session. The rules that generate places that have already been visited are eliminated. This extraction can be expensive, in terms of computational time, but does not need to be done online. Moreover the computation time can be limited by changing the support threshold value and/or by deploying using a scalable implementations [61].
3. **Rules Scoring,** where a score is generated in real-time (online) to the selected rules according to the type of recommendation expressed by the user. Frequently, association rules are scored and ranked using a confidence measure. When using this measure, top rules are those with the biggest probability

of happening. In our recommending context, top rules would generate popular, if not obvious, recommendations. Changing this measure allows to highlight different kinds of rules and thus different kinds of recommendations. For instance, the *surprise* measure [27] favors rules that are less obvious and more surprising. The authors propose to adapt the measure to the current user's intention: users have a choice between promoting surprise, combining surprise with friends' habits, prioritizing geographic criteria according to their willingness to travel, combining their geographic tolerance with friends' habits, etc. Four measures are proposed to express the geographic, social, surprise and popular criteria. The scoring function is implemented as a linear combination of these different measures so that the induced ranking of rules (and recommended places) maximizes the multi-criteria scores.

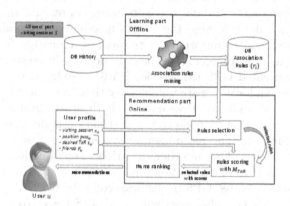

Fig. 6. A post-filtering approach with on-the-fly rules selection according to the user's choice of measure of interest, e.g. the surprise.

In this approach, the context of use is handled by S_{k_u}, the last past visited places during the session and the current position, but it could be extended by other factors, e.g., speed or whether the user is alone or not. We note that in this work the explicit interaction with the user about his intentions is of particular interest. Both implicit context and explicited intention influence the rules scoring and ranking. This contextual postfiltering step allows the system to adapt the recommendations to the current need. And offers the opportunity for users to try in real-time different kinds of recommendations. Of course, this is possible because the ranking computation itself is efficient and can be done online, and because the costly part (the learning step extracting the exhaustive list of potential rules) is done offline.

More extensive panorama of recommendation systems in location-based social networks can be found in [8]. This survey describes location recommendation systems, but also systems recommending LBSN users, activities, or social media. Additionally, they mention systems with more complex objectives, such as sequential location recommendations, an extension of the "next place" problems. These recommenders suggest *itineraries* of locations maximizing the number of interesting POI while minimizing the total travel time or energy consumption. Such systems are based on two types of data: the geo-tagged social media content like photos, or users' GPS trajectories. Sequences of locations can be extracted from geo-tagged photos shared on social media. Association rule mining and in particular sequence mining [5,15] can be used to infer popular sequential locations and recommend trips based on users' historical visiting patterns [57]. Given a location sequence and a time span, [58] goes further and constructs top-k itineraries routable on maps. Let us note an original work from Quercia et al. who introduce emotion scores collected from a crowd-sourcing platform where users vote on which streets scenes in London looks more beautiful, quiet, and happy [50]. Their graph-based recommender systems proposes itineraries that are not only short but also emotionally-pleasing. In addition to socially labelled locations, another source of data worth to mention are GPS trajectories. They are by nature rich data because they contain sequences of positions and time information, including the duration a user spent at a location. To model typical user's routes in an area, Yoon et al. propose a Location-Interest Graph built from multiple user-generated GPS trajectories [63,64]. Nodes are the locations and relationship between locations are weighted by time information. To recommend trips, they define a candidate itinerary as a path in this graph where the total time duration from the start node to the destination node does not exceed the user-specified duration. The recommender system computes a selection of trip candidates and their ranking in order to satisfy users' queries. They evaluated the efficiency of their recommendation method against baseline algorithms with a large set of user-generated GPS trajectories collected from 125 users in Beijing, China. [63] describes how they deal with raw GPS traces to get 119 locations from the detected 35,319 stay points. Those systems are fully automated, and even if they mine human behaviors to recommend trips, they provide *mathematically* optimal itineraries which are not always very *natural* and may not fit personal considerations. Yahi et al. propose Aurigo, a completely different approach where users build themselves their trips in a guided manner [59]. This interactive tour planner is based on a recommendation engine which selects an ordered list of POIs, uses Google API to trace the route between them, and invite the users to adapt the itinerary providing extra information such as social reviews about the places.

3.3   Recommending Events and Neighborhoods

The previous sections have largely focused on recommending *places*; in other words, the items that were being recommended were physical locations (e.g., a restaurant) that typically have a precise boundary and purpose. Urban areas, however, are highly heterogeneous: each of their neighbourhoods have a unique character and appeal, and every day there are thousands of one-off social events happening in different parts of town. In this section, we consider the problem of recommending events and neighbourhoods in urban settings. We focus on two particular case studies: recommending social events in Boston, USA [49], and recommending neighbourhoods in London (UK), San Francisco (USA), and New York (USA) [65].

3.3.1   Social Events

In [49], the authors describe experiments that explore the quality of social event recommendations based on coarse-grained mobile phone location data; in doing so, a number of challenges are discussed. The basic requirement for a social event recommender system is a dataset that captures which users have attended what social events; this assumes that attendance is in some way indicative of preference [22]. Mobile phone data does not readily provide such a mapping, yet inferences can be drawn from the location traces that they do capture by combining them with publicly-available event lists, such as those provided by popular tourist and ticket-sales web sites; the authors merged the mobile location traces with the events listed in the 'Boston Globe Calendar,' subject to a number of constraints. This step could be circumvented altogether by relying on other sources of data, such as ticket sales or location-based social network check-ins.

A second requirement for any personalised recommender service is a preference history for each user, that can be used to tailor the future recommendations. In [49], the authors recognise that creating such profiles is challenging; their user-event data is over 95% sparse. One way that they address this problem is by clustering users according to their home location, which was also inferred from their mobility traces. The resulting data, therefore, does not represent individuals' event attendance; rather, it represents where users who attend particular events tend to reside.

Given the dataset generated in the previous two steps, the final requirement is an algorithmic approach to recommending social events. The authors explore a number of approaches, ranging from non-personalised popularity-based events, ranking events by geographical distance, ranking events using a variant of TF-IDF to favour events that are locally popular but globally less so, and two k-NN approaches that are similar to the traditional user- and item-based approaches; in this case, they are referred to as the k-Nearest Locations and k-Nearest Events.

The most notable result from this work is that the 'geographically close' recommendations – which underlie many location-based services – produced the worst results when comparing the ranking percentile of recommendations. This result brings into question the assumption that many geographically-based systems take: that physical proximity will be proportional to preference; instead it is indicative that users may instead be willing to travel some distance to reach those events that they really like.

Going forward, there remain a number of open challenges. For example, many events are one-off or highly irregular (e.g., a particular band's concert in a city), and there are varying temporal constraints on events (e.g., a month-long museum exhibit vs. an evening outdoor cinema event). In this context, the problem of *item* cold-start, where a new event does not have sufficient preference data becomes more prominent. There may be opportunities to address this problem by merging it with online data. For example, events could be recommended based on online music preferences; however, these kind of cross-domain recommendations have not been largely addressed in the research literature [20]. Moreover, the results described here provide early evidence that there is a variety of factors that drive users to events (beyond location): social data could provide further insights into the social, temporal, and geospatial factors that influence users' attendance at events [24].

3.3.2 Neighbourhoods

While social events are largely characterised by their varying temporal nature, neighbourhoods differ from locations by varying in size and often having loosely defined borders. In [65], the authors explore a large, 5-month, set of Foursquare check-ins that were sourced via the public Twitter API. Their analysis uncovers a variety of inter-city characteristics of check-ins. In particular, there are areas that are more likely to attract tourists, while other areas tend to only attract locals, and the types and check-in times to venues vary between different urban areas.

Although neighbourhoods have traditionally been defined via political boundaries, the check-in data presents an opportunity to analyse neighbourhoods from the perspective of how urban residents use and navigate through their own city. The authors build a neighbourhood inference model by splitting each city into a grid, where each cell spans a 100×100 m area that can be described with a feature vector derived from the check-ins in that area (e.g., proportion of tourists, etc.). Each of these features was then used in conjunction with a variant of the OPTICS clustering algorithm [7] to create polygons that encapsulate areas of high relative density for the feature in question (e.g., high relative proportion of tourists).

Given a city that has been partitioned into neighbourhoods, that are each characterised by a variety of features, the next step is to build a recommender system. Such a system would, given a user profile, be able to suggest areas of the city that the given user would likely be interested in exploring further. The authors do so in a number of steps. First, they map user profiles to neighbourhoods, by creating a text summary of a neighbourhood based on the textual profile information provided by any user that has checked into that area. Then, given a user and his/her profile (note: this text could be replaced with a search query), the aim of the recommender is to provide a personalised ranking of urban neighbourhoods.

Evaluating the quality of neighbourhood recommendations remains challenging, in part due to the fact that there is (in a strict sense) no 'ground truth' with which to operate. In [65], the authors therefore compare two metrics of recommendation performance: first, the accuracy of recommendations (i.e., the ability for the recommender to highly rank those neighbourhoods that have a close match with the user's profile), and second, the 'area cost,' or the ability for the recommender to produce geographically specific recommendations. The authors find that there is a trade-off between these two metrics; however, when reverting back to a baseline that uses the U.S. census to define neighbourhood boundary data, accuracy is consistently lower than the area cost – providing evidence that defining neighbourhoods over social data can provide greater quality recommendations.

3.4   Personalised Place Search

Place search enables the discovery of places in the city in a manner that is analogous to web search (See the Social Search Chapter [14]). The two applications share the same modeling abstraction which entails the ranking of items according to relevance scores generated with respect to an input query generated on the user side.

As an example, consider a mobile user looking for a place to eat during the evening. He also desires that the place he eats is Italian so he opens up his favorite place discovery application (Foursquare, Google etc.) and queries for "Italian Food". The app's place search engine then returns the top-k most relevant places to the user's query. The user can then choose the places they are mostly interested in and retrieve further information about them (to-do tips, prices, menu, booking info etc.). What is more, they can apply filters to the place search list in order to exclude places that are not in vicinity (filtering by distance) or places that do not accept credit cards.

The example above assumes certain facts about the place search system. The use of input queries noted above implies the existence of a natural language processing system in operation to perform text matching between user queries and places. Textual representations of places in this context can be built using tips, comments or reviews left by other users at the place or simply by exploiting descriptive information about place types (Peruvian Food, Post Office, Swimming Pool etc.). Filters on the other hand can be applied by the user considering a potentially large number of different variables (distance, driving time, free wifi etc.) that are accessible through the application's interface. Yet the role of filters in place search, from a technical point of view, is superficial. Filters simply remove - from an existing ranked list of items - those that do not satisfy the selected constraints. The challenging task in the context of place search is in fact to build a relevant list of items in the first place by utilizing the multi-facet interaction of phone and user signals.

The ranking task in place search involves a set of challenges that need to be addressed to guarantee a graceful user experience. First and foremost, in large urban environments where location recommender systems are mostly useful, there may be thousands of places near the user, but they will eventually visit just one or a few in a given time. Place search is thus a prediction task characterized by a high class imbalance in the machine learning terminology. That is, there exists a single, or a few positive items out of many candidate ones. The hardness of the problem intensifies further due to data sparsity; typically only a few or no data points that describe user preferences for places are available. What are the techniques then that could enable an appropriate balance of geographic, social, historic place preference and temporal signals so as to offer a personalized place search experience to the end user. Moreover, how can this goal be achieved by also taking into account the constraints imposed for a quick, near real time, response?

Despite the advances in machine learning and statistical modeling techniques in recent years, standard recommendation, off-the-shelf solutions were not available to address the requirements of place search. As in the case of web search, a dedicated system that is designed in response to the requirements of the target application is required. From an algorithmic modeling perspective in a ranking task relevant scores for the items in the prediction list need to be obtained and consequently, the necessity emerges for the acquirement of two pieces of information: firstly, *labels* which describe how much a user likes the place they have visited and secondly, a set of information signals, or features, that can be exploited to represent a place. The features for place representation will be used as an input to the ranking algorithm which will process it to assign a relevance score to the place that will decide its position in the ranked list. The labels on the other hand are to be exploited as a means to ground truth, in particular to assess the goodness of the ranking method.

An algorithmic approach for place search is now demonstrated. Our demonstration here is not exhaustive, but serves instead as a didactic example for the reader. Furthermore, the algorithmic approaches applied in the context of place search are very much dependent on the type of label information that becomes available. Labels on place preference attitudes can be obtained in two ways. Either through explicit feedback where the users state through a feedback mechanism how much they like a place (e.g. star ratings) or through implicit feedback assuming that if a user has indicated their presence at a place (for instance through a check-in) then they must have liked it. While the first method allows for a more accurate monitoring of user to place preferences, the latter method is applicable in more real world scenarios and allows effectively for access to large amounts of feedback data.

In terms of the features used to represent a place, these can include information signals such as the geographic distance from current location, the popularity of the place or its category or type (Museum, Reggae Bar etc.). Those signals can be extended to account for the system's temporal dynamics (e.g. popularity during a day or hour of the week) or additionally, new ones can be devised that explicitly model the personal preferences of a user. The latter can include information about the user's social network or historical visits. Regardless of the specific information exploited however, the set of signals can be represented as a multi-dimensional vector $x \in R^d$, where d is the number of features or put otherwise, the dimensionality of the input vector.

Figure 7 the performance of various information signals (features) is demonstrated, in ranking the place visited next by a mobile user in Foursquare as described in [42]. One can observe that in terms of an APR (Average Percentile Rank) score, different features may offer better or worse performance with significant deviations at times, and perhaps more interestingly, their performance heavily depends on the hour of the week that the place search prediction task is

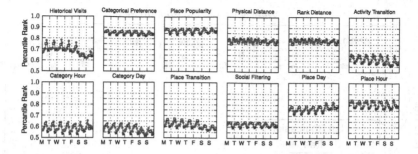

Fig. 7. Set of information signals evaluated for place search over different times of the week as described in [42].

carried out. As an example, take the *Historic Visits* features that ranks places according to the frequency a mobile user has visited them in the past. This feature performs worse in the weekends when users are more likely to deviate from their regular patterns and visit something new. Consequently, a different feature, such as the *Popularity* of nearby places may be more effective during this period. Ultimately, the goal for a place search engine would be to balance appropriately these features in a multivariate model so as to achieve an optimal performance. An example on how such method could be put forward in practice is provided in the next paragraph.

When users do not express explicit preferences to places, place search engineers may resort to implicit label collection method. A popular method corresponds to the extraction of binary labels that capture whether a user likes $(+1)$, or not (-1), a place. A common assumption in this setting is that when a user indicates voluntarily their presence at a place, through a check-in for example, they must like it. Following this paradigm one can then extract one positive label per user movement. A negative label can then be selected by picking a place in the city in a uniformly random manner. The latter is based on an additional assumption, that out of the thousands of places in a city, a user will only visit and like a few, and thus, a randomly chosen venue would hopefully be one that the user won't like. Practitioners are generally encouraged to explore their own label extraction techniques, but for the sake of simplicity, in the present chapter we will stick to the above method.

So with the goal being the exploitation of the union of individual features in order to perform place search, supervised models are trained assuming knowledge up to prediction time t'. For every check-in that took place before t', a training example $\mathbf{x}$ can be considered which encodes the values of the features of the visited venue (e.g., popularity, distance from previous venue, temporal activity scores) and whose label y is positive. Then, a negative labeled input can be retrieved by sampling at random across all other places in the city. Essentially, the aim is to teach the model what the crucial characteristics are that would allow to differentiate places that attract user check-ins from those which would not. This method of training a model by providing feedback in the form of user

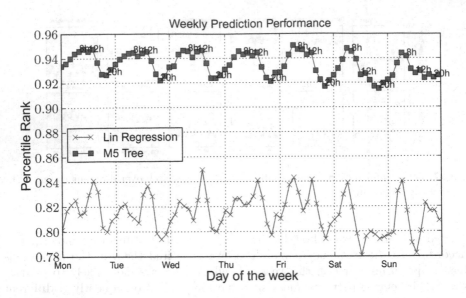

Fig. 8. Supervised learning models performance in place search over different hours of a week [42]

preference has been established in the past [16] and corresponds to an effective reduction of the *ranking problem* to a *binary classification task*.

As an example, consider two different supervised learning models to learn how feature vectors **x** correspond to positive and negative labels: linear ridge regression and M5 decision trees [51]. The linear model makes the assumption that the relationship between the feature vector **x** and the output label y is linear. The M5 model tree, instead, is an approach based on continuous decision-tree learning [51]. Figure 8 shows the performance of the two models in terms of *APR*. The M5 Decision Tree performs significantly better than the Linear Regressor, but also by a clear margin better than any of the individual features evaluated in the previous paragraph (see Fig. 7).

The method to rank places in a city presented above in the context of place search is not the only one. Foursquare's data science team in [55] demonstrates a learning-to-rank method for spatio-temporal place search. The authors exploit the rich set of spatial signals extracted from the mobility traces of Foursquare users, first to represent venues as a two-dimensional gaussian mixture model, and subsequently to probabilistically match a given data point to one of those venues. In [40] the authors present a matrix-factorisation based approach enhanced so place types are taken into account during place search. Collaborative filtering approaches has also been applied in [38], whereas in [9] the authors propose LearNext, a probabilistic framework in the context of place search targetted towards groups of users that are tourists.

4   Evaluating Location Recommendations

Most of the metrics used to evaluate recommendations of location recommender systems are the same as the ones used for traditional recommender systems. An extensive discussion of the metrics can be found in the Chap. 10 [32]. The same metrics, both offline and online, are also used for people recommendations (As introduced in the Chapter dedicated to this problem [26]) or other various kinds of social recommendations (For more detailed explanations, refer to the Chapter Social Link-based Recommendations [35]). We therefore do not intend here to show all the existing metrics for evaluating recommender systems of places, we only describe the ones used in the works previously presented:

- Precision: a measure of exactness, determines the fraction of recommended places that are really done by the user among all the proposed recommendations.

$$precision = \frac{|good\_recommendations|}{|all\_recommendations|} \qquad (8)$$

- Recall: a measure of completeness, determines the fraction of relevant recommendations out of all really visited places.

$$recall = \frac{|good\_recommendations|}{|all\_visited\_places|} \qquad (9)$$

- F1-measure: a metric to combine Precision and Recall into a single value for comparison purposes.

$$F_1 = 2 \times \frac{precision \cdot recall}{precision + recall} \qquad (10)$$

- Average Percentile Rank (APR): a measure used to assess the efficacy of an algorithm in a ranking task in terms of ranking the *wanted item* high in the list of candidate items. The following formula is used for the calculation

$$APR = \frac{\sum_{j-1}^{T} \frac{rank_i}{N}}{T} \qquad (11)$$

with T being the number of tests, $rank_i$ the position of the wanted item i in the prediction list and N the total number of items (places).

Beyond these general metrics, more specific ones exist for the domain of location prediction like the computation of the euclidean distance between the location predictions and the real positions collected by a mobile. Since all these metrics allow evaluating the predictive behavior of a recommender system in off-line condition, the relevance of such evaluation can be questioned. Is it useful to recommend a place the user will probably go without any recommendation? This is an open question in recommender systems in general.

When it is possible, on-line evaluations are preferred, such A/B testing (Again, see Chap. 10 [32]). In this kind of evaluation, two slightly different recommendation methods are competing. A set A of real users is provided recommendations from the first method and a set B receives recommendations from the second one. The recommendation method that statistically outperforms the other on a predefined target metric (number of visits, incomes, satisfaction, etc.) wins the competition.

5   Conclusion

Deploying social and personalised services on emerging sensor-enhanced technologies is set to transform recommender system research and practice. Traditional recommendation methods are not obsolete in location-based recommendations; the literature shows that methods that include geographic and social information can improve significantly the quality of recommendations.

In this chapter, we have focused on services that recommend places using social media data. Location data are different from ordinary items due to their geographic attributes. They bring key information about the user behavior and a new dimension to compute similarities between them. We have tackled four problems of location data recommendation: recommending new places, recommending next place to go or how to use the context to provide recommendations in real-time, recommending events and neighbourhoods and finally place search. For each of them, representative methods from the literature and the key concepts have been detailed in order to be easily reproduced and adapted to specific use cases. We also presented the main metrics used to evaluate these methods. It is worth noting that evaluating recommender systems offline evaluation should be taken carefully since experiments have shown that good offline recommendations can be bad in practice and inversely.

Beyond Places. As personalised systems continue to transition from online to offline domains, the opportunity to build recommender systems that go beyond places, events, and neighbourhoods increases. Many of these opportunities encompass revisiting what the recommender system's *user* or *item* may be. For example, social location data can be used to measure the impact of large-scale events on local businesses [25]; a recommender system could be built to recommend retail locations to prospective business owners [29]. In this case, the 'item' is a potential location. In another scenario, riders of public transit systems could be recommended the best fares [33] or routes to travel, based on their historical preferences and current service levels; in this case, the item may indeed be a route or ticket that has been ranked according to travel time or cost, rather than simply preference.

Geographic Information to Improve Non-spatial Recommendation. We can also find recommender systems of 'classical item', e.g. movies, which use location data to improve technical efficiency. Location-awareness is not dedicated to the location recommendation or place search. LARS [36] is an example of a system which considers spatial properties of 'users', 'ratings' and 'items'; LARS produces recommendations using spatial ratings for non-spatial items; location information is exploited in a pre-filtering manner; *user partitioning* and *travel penalty* techniques avoid exhaustively processing all recommendation candidates in order to maximize system scalability while not sacrificing recommendation quality.

Geographic Information and Recommendation Techniques to Improve User Experience in Other Domains. In a very different domain, Internet Providers and Content Delivery Networks constantly face the challenge of delivering more content whose size grows up to more users with increasingly needs for speed. One of the main idea of the research project VIPEER [12] is to use regional servers as caches with content replicas to decrease the load of the origin content server and decrease the downloading time to end-users; recommendation techniques (collaborative filtering, SVD or simple popularity based rankings) are applied along with spatial movies consumption datasets to predict the future downloads and manage the location of replicas in the different caches spread in the country.

Context-Awareness and Internet of Things. As seen in this chapter, and more generally in this book, taking into account more data about usages (friends, geographic coordinates) allows improving the relevance of recommendations. With the paradigm of the Internet of Things [46], a huge quantity of heterogeneous data from sensors all around the world is becoming more and more available about everything. These data are a wealth of information about people. They can inform about their mood, health, the weather, etc. No doubt that recommender systems will benefit directly from these new context data, since they bring a better understanding of usages, habits and allow finding more correlations between users.

References

1. Abowd, G.D., Dey, A.K., Brown, P.J., Davies, N., Smith, M., Steggles, P.: Towards a better understanding of context and context-awareness. In: Gellersen, H.-W. (ed.) HUC 1999. LNCS, vol. 1707, pp. 304–307. Springer, Heidelberg (1999). https://doi.org/10.1007/3-540-48157-5_29
2. Adomavicius, G., Tuzhilin, A.: Toward the next generation of recommender systems: a survey of the state-of-the-art and possible extensions. IEEE Trans. Knowl. Data Eng. **17**(6), 734–749 (2005)
3. Adomavicius, G., Tuzhilin, A.: Context-aware recommender systems. In: Ricci, F., Rokach, L., Shapira, B., Kantor, P.B. (eds.) Recommender Systems Handbook, pp. 217–253. Springer, Boston (2011). https://doi.org/10.1007/978-1-4899-7637-6_6
4. Agrawal, R., Imieliński, T., Swami, A.: Mining association rules between sets of items in large databases. In: ACM SIGMOD Record, vol. 22, pp. 207–216. ACM (1993)
5. Agrawal, R., Srikant, R.: Mining sequential patterns. In: Proceedings of the Eleventh International Conference on Data Engineering, pp. 3–14. IEEE (1995)
6. Amatriain, X., Pujol, J.: Data mining methods for recommender systems. In: Ricci, F., Rokach, L., Shapira, B., Kantor, P. (eds.) Recommender Systems Handbook. Springer, Boston (2011). https://doi.org/10.1007/978-0-387-85820-3_2

7. Ankerst, M., Breunig, M., Kriegel, H., Sander, J.: OPTICS: ordering points to identify the clustering structure. In: ACM SIGMOD, Philadelphia, USA (1999)

8. Bao, J., Zheng, Y., Wilkie, D., Mokbel, M.: Recommendations in location-based social networks: a survey. GeoInformatica **19**(3), 525–565 (2015)

9. Baraglia, R., Muntean, C.I., Nardini, F.M., Silvestri, F.: LearNext: learning to predict tourists movements. In: Proceedings of the 22nd ACM International Conference on Information and Knowledge Management, pp. 751–756. ACM (2013)

10. Becker, R., Caceres, R., Hanson, K., Isaacman, S., Loh, J., Martonosi, M., Rowland, J., Urbanek, S., Varshavsky, A., Volisky, C.: Human mobility characterization from cellular network data. Commun. ACM **56**(1), 74–82 (2013)

11. Biancalana, C., Gasparetti, F., Micarelli, A., Sansonetti, G.: An approach to social recommendation for context-aware mobile services. ACM Trans. Intell. Syst. Technol. **4**(1), 10:1–10:31 (2013)

12. Bothorel, C., Picot-Clemente, R., Simon, G., Li, Z., Michiardi, P., Hadjadj-Aoul, Y., Garnier, J.: Technical report: preliminary report on CDN/dCDN modeling and analysis. ANR Project Vipeer, Deliverable 44 (2012)

13. Brockmann, D., Hufnagel, L., Geisel, T.: The scaling laws of human travel. Nature **439**(7075), 462–465 (2006)

14. Brusilovsky, P., Smyth, B., Shapira, B.: Social search. In: Brusilovsky, P., He, D. (eds.) Social Information Access. LNCS, vol. 10100, pp. 213–276. Springer, Cham (2018)

15. Chand, C., Thakkar, A., Ganatra, A.: Sequential pattern mining: survey and current research challenges. Int. J. Soft Comput. Eng. **2**(1), 185–193 (2012)

16. Cohen, W.W., Schapire, R.E., Singer, Y.: Learning to order things. J. Artif. Intell. Res. **10**(1), 243–270 (1999)

17. Crandall, D., Backstrom, L., Huttenlocher, D., Kleinberg, J.: Mapping the world's photos. In: WWW, Madrid, Spain, April 2009

18. Eagle, N., Pentland, A.: Reality mining: sensing complex social systems. Pers. Ubiquit. Comput. **10**, 255–268 (2006)

19. Farzan, R., Brusilovsky, P.: Social navigation. In: Brusilovsky, P., He, D. (eds.) Social Information Access. LNCS, vol. 10100, pp. 142–180. Springer, Cham (2018)

20. Fernandez-Tobias, I., Cantador, I., Kaminskas, M., Ricci, F.: A generic semantic-based framework for cross-domain recommendation. In: Proceedings of the 2nd International Workshop on Information Heterogeneity and Fusion in Recommender Systems, Chicago, USA (2011)

21. Forsati, R., Meybodi, M., Neiat, A.G.: Web page personalization based on weighted association rules. In: 2009 International Conference on Electronic Computer Technology, pp. 130–135. IEEE (2009)

22. Froehlich, J., Chen, M.Y., Smith, I.E., Potter, F.: Voting with your feet: an investigative study of the relationship between place visit behavior and preference. In: Dourish, P., Friday, A. (eds.) UbiComp 2006. LNCS, vol. 4206, pp. 333–350. Springer, Heidelberg (2006). https://doi.org/10.1007/11853565_20

23. Gao, H., Tang, J., Liu, H.: gSCorr: modeling geo-social correlations for new check-ins on location-based social networks. In: Proceedings of the 21st ACM International Conference on Information and Knowledge Management, CIKM 2012, pp. 1582–1586. ACM, New York (2012)

24. Georgiev, P., Noulas, A., Mascolo, C.: The call of the crowd: event participation in location-based social services. In: Proceedings of the Eighth International AAAI Conference on Weblogs and Social Media, Ann Arbour, USA, June 2014

25. Georgiev, P., Noulas, A., Mascolo, C.: Where businesses thrive: predicting the impact of the olympic games on local retailers through location-based services data. In: Proceedings of the Eighth International AAAI Conference on Weblogs and Social Media, Ann Arbour, USA, June 2014

26. Guy, I.: People recommendation on social media. In: Brusilovsky, P., He, D. (eds.) Social Information Access. LNCS, vol. 10100, pp. 570–623. Springer, Cham (2018)

27. Hussain, F., Liu, H., Lu, H.: Relative measure for mining interesting rules. In: Proceedings of the Fourth European Conference on Principles and Practice of Knowledge Discovery in Databases, PKDD 2000, pp. 117–132. Citeseer (2000)

28. Jannach, D., Lerche, L., Zanker, M.: Recommending based on implicit feedback. In: Brusilovsky, P., He, D. (eds.) Social Information Access. LNCS, vol. 10100, pp. 510–569. Springer, Cham (2018)

29. Karamshuk, D., Noulas, A., Scellato, S., Nicosia, V., Mascolo, C.: Geo-spotting: mining online location-based services for optimal retail store placement. In: Proceedings of 19th ACM International Conference on Knowledge Discovery and Data Mining, Chicago, USA (2013)

30. Katz, L.: A new status index derived from sociometric analysis. Psychometrika 18(1), 39–43 (1953)

31. Kennedy, L., Naaman, M., Ahern, S., Nair, R., Rattenbury, T.: How flickr helps us make sense of the world: context and content in community-contributed media collections. In: ACM MM, Augsburg, Germany, September 2007

32. Kluver, D., Ekstrand, M., Konstan, J.: Rating-based collaborative filtering: algorithms and evaluation. In: Brusilovsky, P., He, D. (eds.) Social Information Access. LNCS, pp. 344–390. Springer, Cham (2018)

33. Lathia, N., Capra, L.: Mining mobility data to minimise travellers' spending on public transport. In: ACM KDD, San Diego, California, August 2011

34. Lathia, N., Froehlich, J., Capra, L.: Mining public transport usage for personalised intelligent transport systems. In: IEEE ICDM, Sydney, Australia, December 2010

35. Lee, D., Brusilovsky, P.: Recommendations based on social links. In: Brusilovsky, P., He, D. (eds.) Social Information Access. LNCS, pp. 391–440. Springer, Cham (2018).

36. Levandoski, J.J., Sarwat, M., Eldawy, A., Mokbel, M.F.: LARS: a location-aware recommender system. In: 2012 IEEE 28th International Conference on Data Engineering (ICDE), pp. 450–461. IEEE (2012)

37. Li, Q., Zheng, Y., Xie, X., Chen, Y., Liu, W., Ma, W.: Mining user similarity based on location history. In: International Conference on Advances in Geographic Information Systems, Santa Ana, USA (2008)

38. Lian, D., Zheng, V.W., Xie, X.: Collaborative filtering meets next check-in location prediction. In: Proceedings of the 22nd International Conference on World Wide Web Companion, pp. 231–232. International World Wide Web Conferences Steering Committee (2013)

39. Lindqvist, J., Cranshaw, J., Wiese, J., Jong, J., Zimmerman, J.: I'm the mayor of my house: examining why people use foursquare - a social-driven location sharing application. In: Proceedings of the SIGCHI Conference on Human Factors in Computing Systems, pp. 2409–2418. ACM (2011)

40. Liu, X., Liu, Y., Aberer, K., Miao, C.: Personalized point-of-interest recommendation by mining users' preference transition. In: Proceedings of the 22nd ACM International Conference on Conference on Information and Knowledge Management, pp. 733–738. ACM (2013)

41. Mobasher, B., Dai, H., Luo, T., Nakagawa, M.: Effective personalization based on association rule discovery from web usage data. In: Proceedings of the 3rd International Workshop on Web Information and Data Management, pp. 9–15. ACM (2001)
42. Noulas, A., Scellato, S., Lathia, N., Mascolo, C.: Mining user mobility features for next place prediction in location-based services. In: IEEE International Conference on Data Mining, ICDM 2012 (2012)
43. Noulas, A., Scellato, S., Lathia, N., Mascolo, C.: A random walk around the city: new venue recommendation in location-based social networks. In: Proceedings of the 2012 ASE/IEEE International Conference on Social Computing and 2012 ASE/IEEE International Conference on Privacy, Security, Risk and Trust, SOCIALCOM-PASSAT 2012, pp. 144–153. IEEE Computer Society, Washington, D.C. (2012)
44. Noulas, A., Scellato, S., Mascolo, C., Pontil, M.: An empirical study of geographic user activity patterns in foursquare. In: Adamic, L.A., Baeza-Yates, R.A., Counts, S. (eds.) ICWSM. The AAAI Press (2011)
45. O'Mahoney, M., Smyth, B.: From opinions to recommendations. In: Brusilovsky, P., He, D. (eds.) Social Information Access, LNCS. LNCS, vol. 10100, pp. 480–509. Springer, Cham (2018)
46. Perera, C., Zaslavsky, A., Christen, P., Georgakopoulos, D.: Context aware computing for the internet of things: a survey. IEEE Commun. Surv. Tutor. **16**(1), 414–454 (2014)
47. Picot-Clemente, R., Bothorel, C.: Recommendation of shopping places based on social and geographical influences. In: 5th ACM RecSys Workshop on Recommender Systems and the Social Web, RSWeb 2013, Hong Kong, Hong Kong SAR China, October 2013
48. Picot-Clemente, R., Bothorel, C., Lenca, P.: Contextual recommender system on a location-based social network for shopping places recommendation using association rules mining. In: The 6th Asian Conference on Intelligent Information and Database Systems, ACIIDS 2014, vol. 551, pp. 3–13. Springer, Cham (2014)
49. Quercia, D., Lathia, N., Calabrese, F., Lorenzo, G.D., Crowcroft, J.: Recommending social events from mobile phone location data. In: IEEE ICDM, Sydney, Australia, December 2010
50. Quercia, D., Schifanella, R., Aiello, L.M.: The shortest path to happiness: recommending beautiful, quiet, and happy routes in the city. In: Proceedings of the 25th ACM Conference on Hypertext and Social Media, HT 2014, pp. 116–125. ACM, New York (2014). http://doi.acm.org/10.1145/2631775.2631799
51. Quinlan, J.: Learning with continuous classes. In: AI 1992 (1992)
52. Rachuri, K., Mascolo, C., Musolesi, M.: Energy-accuracy trade-offs of sensor sampling in smart phone based sensing systems. In: Lovett, T., O'Neill, E. (eds.) Mobile Context Awareness: Capabilities Challenges and Applications Workshop. Springer, Copenhagen (2010). https://doi.org/10.1007/978-0-85729-625-2_3
53. Rattenbury, T., Good, N., Naaman, M.: Toward automatic extraction of event and place semantics from flickr tags. In: ACM SIGIR, pp. 103–110, July 2007
54. Ratti, C., Pulselli, R., Williams, S., Frenchman, D.: Mobile landscapes: using location data from cell phones for urban analysis. Environ. Plann. B **33**(5), 727–748 (2006)
55. Shaw, B., Shea, J., Sinha, S., Hogue, A.: Learning to rank for spatiotemporal search. In: Proceedings of the Sixth ACM International Conference on Web Search and Data Mining, pp. 717–726. ACM (2013)

56. Sohn, T., et al.: Mobility detection using everyday GSM traces. In: Dourish, P., Friday, A. (eds.) UbiComp 2006. LNCS, vol. 4206, pp. 212–224. Springer, Heidelberg (2006). https://doi.org/10.1007/11853565_13

57. Tai, C.H., Yang, D.N., Lin, L.T., Chen, M.S.: Recommending personalized scenic itinerarywith geo-tagged photos. In: 2008 IEEE International Conference on Multimedia and Expo, pp. 1209–1212. IEEE (2008)

58. Wei, L.Y., Zheng, Y., Peng, W.C.: Constructing popular routes from uncertain trajectories. In: Proceedings of the 18th ACM SIGKDD International Conference on Knowledge Discovery and Data Mining, pp. 195–203. ACM (2012)

59. Yahi, A., Chassang, A., Raynaud, L., Duthil, H., Chau, D.H.P.: Aurigo: an interactive tour planner for personalized itineraries. In: Proceedings of the 20th International Conference on Intelligent User Interfaces, IUI 2015, pp. 275–285. ACM, New York (2015). http://doi.acm.org/10.1145/2678025.2701366

60. Yang, S.J., Zhang, J., Chen, I.Y.: A JESS-enabled context elicitation system for providing context-aware web services. Expert Syst. Appl. **34**(4), 2254–2266 (2008)

61. Yang, X.Y., Liu, Z., Fu, Y.: Mapreduce as a programming model for association rules algorithm on Hadoop. In: 2010 3rd International Conference on Information Sciences and Interaction Sciences (ICIS), pp. 99–102. IEEE (2010)

62. Ye, M., Yin, P., Lee, W.C., Lee, D.L.: Exploiting geographical influence for collaborative point-of-interest recommendation. In: Proceedings of the 34th International ACM SIGIR Conference on Research and Development in Information Retrieval, SIGIR 2011, pp. 325–334. ACM, New York (2011)

63. Yoon, H., Zheng, Y., Xie, X., Woo, W.: Smart itinerary recommendation based on user-generated GPS trajectories. In: Yu, Z., Liscano, R., Chen, G., Zhang, D., Zhou, X. (eds.) UIC 2010. LNCS, vol. 6406, pp. 19–34. Springer, Heidelberg (2010). https://doi.org/10.1007/978-3-642-16355-5_5

64. Yoon, H., Zheng, Y., Xie, X., Woo, W.: Social itinerary recommendation from user-generated digital trails. Pers. Ubiquit. Comput. **16**(5), 469–484 (2012)

65. Zhang, A., Noulas, A., Scellato, S., Mascolo, C.: Hoodsquare: modeling and recommending neighbourhoods in location-based social networks. In: IEEE SocialCom, Washington D.C., September 2013

66. Zhang, J.D., Chow, C.Y.: iGSLR: personalized geo-social location recommendation: a kernel density estimation approach. In: Proceedings of the 21st ACM SIGSPATIAL International Conference on Advances in Geographic Information Systems, SIGSPATIAL 2013, pp. 334–343. ACM, New York (2013)

67. Zheng, V., Zheng, Y., Xie, X., Yang, Q.: Collaborative location and activity recommendations with GPS history data. In: ACM Proceedings of the 19th International Conference on World Wide Web, Raleigh, North Carolina, pp. 1029–1038, April 2010

68. Zheng, Y., Li, Q., Chen, Y., Xie, X., Ma, W.: Understanding mobility based on GPS data. In: ACM Ubicomp, Seoul, Korea (2008)

69. Zheng, Y., Zhang, L., Xie, X., Ma, W.: Mining interesting locations and travel sequences from GPS trajectories. In: WWW, Madrid, Spain, April 2008

Author Index